www.brookscole.com

www.brookscole.com is the World Wide Web site for Thomson Brooks/Cole and is your direct source to dozens of online resources.

At *www.brookscole.com* you can find out about supplements, demonstration software, and student resources. You can also send e-mail to many of our authors and preview new publications and exciting new technologies.

www.brookscole.com
Changing the way the world learns®

Chemistry in Focus

A MOLECULAR VIEW OF OUR WORLD

THIRD EDITION

Nivaldo J. Tro
Westmont College

Australia • Brazil • Canada • Mexico • Singapore • Spain
United Kingdom • United States

Chemistry in Focus: A Molecular View of Our World, Third Edition
Nivaldo J. Tro

Chemistry Acquisitions Editor: Lisa Lockwood
Development Editor: Sarah E. Lowe
Editorial Assistant: Lauren Oliviera
Technology Project Manager: Ericka Yeoman-Saler
Marketing Manager: Amee Mosley
Marketing Assistant: Michele Colella
Marketing Communications Manager: Bryan Vann
Project Manager, Editorial Production: Lisa Weber
Creative Director: Rob Hugel
Art Director: Lee Friedman
Print Buyer: Judy Inouye
Permissions Editors: Sarah Harkrader and Bob Kauser

Production Service: Lachina Publishing Services
Text Designer: The Davis Group, Inc.
Photo Researcher: Jane Sanders Miller
Copy Editor: Sara Black
Illustrators: Imagineering Media Services Inc., Lachina Publishing Services, The Clarinda Company, and Progressive Publishing Alternatives
Cover Designer: Raoul Ollman
Cover Image: Eastcott Momatiuk/Getty Images
Cover Printer: Seng Lee Press
Compositor: Lachina Publishing Services
Printer: Seng Lee Press

© 2007 Thomson Brooks/Cole, a part of The Thomson Corporation. Thomson, the Star logo, and Brooks/Cole are trademarks used herein under license.

ALL RIGHTS RESERVED. No part of this work covered by the copyright hereon may be reproduced or used in any form or by any means—graphic, electronic, or mechanical, including photocopying, recording, taping, web distribution, information storage and retrieval systems, or in any other manner—without the written permission of the publisher.

Printed in Singapore

3 4 5 6 7 09 08 07 06

Library of Congress Control Number: 2001012345

Student Edition: ISBN-13: 978-0-495-01956-5
ISBN-10: 0-495-01956-9

Thomson Higher Education
10 Davis Drive
Belmont, CA 94002-3098
USA

For more information about our products, contact us at:
Thomson Learning Academic Resource Center
1-800-423-0563

For permission to use material from this text or product, submit a request online at **http://www.thomsonrights.com**.
Any additional questions about permissions can be submitted by e-mail to **thomsonrights@thomson.com**.

To Annie

About the Author

Nivaldo J. Tro received his BA degree from Westmont College and his PhD degree from Stanford University. He went on to a post-doctoral research position at the University of California at Berkeley. In 1990, he joined the chemistry faculty at Westmont College in Santa Barbara, California. Professor Tro has been honored as Westmont's outstanding teacher of the year in 1994 and again in 2001. He was named Westmont's outstanding researcher of the year in 1996. Professor Tro lives in the foothills of Santa Barbara with his wife, Ann, and their four children, Michael, Alicia, Kyle, and Kaden. In his leisure time, Professor Tro likes to spend time with his family in the outdoors. He enjoys running, camping, surfing, and snowboarding.

Brief Contents

1. Molecular Reasons   2
2. The Chemist's Toolbox   28
3. Atoms and Elements   54
4. Molecules, Compounds, and Chemical Reactions   88
5. Chemical Bonding   116
6. Organic Chemistry   146
7. Light and Color   186
8. Nuclear Chemistry   212
9. Energy for Today   242
10. Energy for Tomorrow: Solar and Other Renewable Energy Sources   276
11. The Air Around Us   298
12. The Liquids and Solids Around Us: Especially Water   328
13. Acids and Bases: The Molecules Responsible for Sour and Bitter   360
14. Oxidation and Reduction   382
15. The Chemistry of Household Products   402
16. Biochemistry and Biotechnology   432
17. Drugs and Medicine: Healing, Helping, and Hurting   476
18. The Chemistry of Food   510
19. Nanotechnology   542

Appendix 1: Significant Figures   A-1
Appendix 2: Answers to Selected Exercises   A-5
Appendix 3: Answers to Your Turn Questions   A-31
Appendix 4: Scientific Notation   S-1

Glossary   G-1

Contents

CHAPTER 1 — Molecular Reasons 2

MOLECULAR THINKING
Why Should Nonscience Majors Study Science? 5

WHAT IF...
Observation and Reason 10

THE MOLECULAR REVOLUTION
Seeing Atoms 21

1.1 Firesticks 4
1.2 Molecular Reasons 5
1.3 The Scientist and the Artist 6
1.4 The First People to Wonder About Molecular Reasons 9
1.5 Immortality and Endless Riches 9
1.6 The Beginning of Understanding 10
1.7 The Classification of Matter 11
1.8 The Properties of Matter 15
1.9 The Development of the Atomic Theory 16
1.10 The Nuclear Atom 18
  CHAPTER SUMMARY 22
  CHEMISTRY ON THE WEB 23
  KEY TERMS 23
  EXERCISES 23
  FEATURE PROBLEMS AND PROJECTS 26

CHAPTER 2 — The Chemist's Toolbox 28

MOLECULAR THINKING
Feynman's Ants 31

THE MOLECULAR REVOLUTION
Measuring Average Global Temperatures 33

2.1 Curious About Oranges 30
2.2 Measurement 31
2.3 Scientific Notation 33
2.4 Units in Measurement 35
2.5 Converting Between Units 38
2.6 Reading Graphs 40
2.7 Problem Solving 44
2.8 Density: A Measure of Compactness 46
  CHAPTER SUMMARY 49
  CHEMISTRY ON THE WEB 49
  KEY TERMS 49
  EXERCISES 50
  FEATURE PROBLEMS AND PROJECTS 52

CHAPTER 3

Atoms and Elements — 54

WHAT IF. . .
Complexity Out of Simplicity 65

WHAT IF. . .
Philosophy, Determinism, and Quantum Mechanics 72

MOLECULAR THINKING
Is Breathing Helium Dangerous? 74

THE MOLECULAR REVOLUTION
The Reactivity of Chlorine and the Depletion of the Ozone Layer 74

3.1 A Walk on the Beach 56
3.2 Protons Determine the Element 58
3.3 Electrons 60
3.4 Neutrons 62
3.5 Specifying an Atom 63
3.6 Atomic Mass 64
3.7 Periodic Law 66
3.8 A Theory That Explains the Periodic Law: The Bohr Model 67
3.9 The Quantum Mechanical Model for the Atom 71
3.10 Families of Elements 73
3.11 A Dozen Nails and a Mole of Atoms 76
CHAPTER SUMMARY 80
CHEMISTRY ON THE WEB 81
KEY TERMS 81
EXERCISES 81
FEATURE PROBLEMS AND PROJECTS 85

CHAPTER 4

Molecules, Compounds, and Chemical Reactions — 88

WHAT IF. . .
Problem Molecules 95

MOLECULAR FOCUS
Calcium Carbonate 97

THE MOLECULAR REVOLUTION
Engineering Animals to Do Chemistry 107

MOLECULAR THINKING
Campfires 110

4.1 Molecules Cause the Behavior of Matter 90
4.2 Chemical Compounds and Chemical Formulas 90
4.3 Ionic and Molecular Compounds 92
4.4 Naming Compounds 96
4.5 Formula Mass and Molar Mass of Compounds 99
4.6 Composition of Compounds: Chemical Formulas as Conversion Factors 101
4.7 Forming and Transforming Compounds: Chemical Reactions 103
4.8 Reaction Stoichiometry: Chemical Equations as Conversion Factors 106
CHAPTER SUMMARY 111
CHEMISTRY ON THE WEB 112
KEY TERMS 112
EXERCISES 112
FEATURE PROBLEMS AND PROJECTS 115

CHAPTER 5

Chemical Bonding — 116

MOLECULAR THINKING
Fluoride 121

5.1 From Poison to Seasoning 118
5.2 Chemical Bonding and Professor G. N. Lewis 119
5.3 Ionic Lewis Structures 121
5.4 Covalent Lewis Structures 123

MOLECULAR FOCUS
Ammonia 129

THE MOLECULAR REVOLUTION
AIDS Drugs 136

5.5 Chemical Bonding in Ozone 129
5.6 The Shapes of Molecules 131
5.7 Water: Polar Bonds and Polar Molecules 135
    CHAPTER SUMMARY 141
    CHEMISTRY ON THE WEB 141
    KEY TERMS 142
    EXERCISES 142
    FEATURE PROBLEMS AND PROJECTS 145

CHAPTER 6

Organic Chemistry 146

THE MOLECULAR REVOLUTION
The Origin of Life 150

THE MOLECULAR REVOLUTION
Determining Organic Chemical Structures 166

WHAT IF. . .
Alcohol and Society 172

MOLECULAR FOCUS
Carvone 174

MOLECULAR THINKING
What Happens When We Smell Something 180

6.1 Carbon 148
6.2 A Vital Force 149
6.3 The Simplest Organic Compounds: Hydrocarbons 151
6.4 Isomers 160
6.5 Naming Hydrocarbons 162
6.6 Aromatic Hydrocarbons and Kekule's Dream 165
6.7 Functionalized Hydrocarbons 168
6.8 Chlorinated Hydrocarbons: Pesticides and Solvents 169
6.9 Alcohols: To Drink and to Disinfect 170
6.10 Aldehydes and Ketones: Smoke and Raspberries 172
6.11 Carboxylic Acids: Vinegar and Bee Stings 175
6.12 Esters and Ethers: Fruits and Anesthesia 176
6.13 Amines: The Smell of Rotten Fish 178
6.14 A Look at a Label 179
    CHAPTER SUMMARY 181
    CHEMISTRY ON THE WEB 181
    KEY TERMS 182
    EXERCISES 182
    FEATURE PROBLEMS AND PROJECTS 185

CHAPTER 7

Light and Color 186

MOLECULAR THINKING
Changing Colors 190

WHAT IF. . .
X-Rays—Dangerous or Helpful? 195

WHAT IF. . .
The Cost of Technology 201

WHAT IF. . .
The Mind–Body Problem 202

THE MOLECULAR REVOLUTION
Watching Molecules Dance 205

7.1 A New England Fall 188
7.2 Light 190
7.3 The Electromagnetic Spectrum 193
7.4 Excited Electrons 196
7.5 Identifying Molecules and Atoms with Light 198
7.6 Magnetic Resonance Imaging: Spectroscopy of the Human Body 199
7.7 Lasers 202
7.8 Lasers in Medicine 204
    CHAPTER SUMMARY 207
    CHEMISTRY ON THE WEB 208

MOLECULAR FOCUS
Retinal 206

KEY TERMS 208
EXERCISES 209
FEATURE PROBLEMS AND PROJECTS 211

CHAPTER 8 — Nuclear Chemistry 212

WHAT IF...
The Ethics of Science 226

THE MOLECULAR REVOLUTION
Fusion Research 230

MOLECULAR THINKING
Radiation and Smoke Detectors 233

WHAT IF...
Radiation—Killer or Healer? 237

8.1 A Tragedy 214
8.2 An Accidental Discovery 215
8.3 Radioactivity 216
8.4 Half-Life 219
8.5 Nuclear Fission 222
8.6 The Manhattan Project 224
8.7 Nuclear Power 226
8.8 Mass Defect and Nuclear Binding Energy 229
8.9 Fusion 230
8.10 The Effect of Radiation on Human Life 231
8.11 Carbon Dating and the Shroud of Turin 234
8.12 Uranium and the Age of the Earth 236
8.13 Nuclear Medicine 237
CHAPTER SUMMARY 238
CHEMISTRY ON THE WEB 239
KEY TERMS 239
EXERCISES 239
FEATURE PROBLEMS AND PROJECTS 241

CHAPTER 9 — Energy for Today 242

MOLECULAR THINKING
Campfire Smoke 259

MOLECULAR FOCUS
Sulfur Dioxide 263

MOLECULAR THINKING
Are Some Fossil Fuels Better Than Others? 267

THE MOLECULAR REVOLUTION
Taking Carbon Captive 268

9.1 Molecules in Motion 244
9.2 Our Absolute Reliance on Energy 245
9.3 Energy and Its Transformations: You Cannot Get Something for Nothing 247
9.4 Nature's Heat Tax: Energy Must Be Dispersed 249
9.5 Units of Energy 251
9.6 Temperature and Heat Capacity 254
9.7 Chemistry and Energy 256
9.8 Energy for Our Society 257
9.9 Electricity from Fossil Fuels 260
9.10 Smog 261
9.11 Acid Rain 263
9.12 Environmental Problems Associated with Fossil-Fuel Use: Global Warming 264
CHAPTER SUMMARY 270
CHEMISTRY ON THE WEB 270

KEY TERMS 271
EXERCISES 271
FEATURE PROBLEMS AND PROJECTS 274

CHAPTER 10 Energy for Tomorrow: Solar and Other Renewable Energy Sources 276

MOLECULAR FOCUS
Hydrogen 285

WHAT IF...
Politics and Energy 287

THE MOLECULAR REVOLUTION
Fuel Cell and Hybrid Electric Vehicles 291

WHAT IF...
Future Energy Scenarios 292

10.1 Earth's Ultimate Energy Source: The Sun 278
10.2 Hydroelectric Power: The World's Most Used Solar Energy Source 278
10.3 Wind Power 280
10.4 Solar Thermal Energy: Focusing and Storing the Sun 280
10.5 Photovoltaic Energy: From Light to Electricity with No Moving Parts 283
10.6 Energy Storage: The Plague of Solar Sources 285
10.7 Biomass: Energy from Plants 286
10.8 Geothermal Power 287
10.9 Nuclear Power 287
10.10 Efficiency and Conservation 288
10.11 2050 World: A Speculative Glimpse into the Future 290
CHAPTER SUMMARY 293
CHEMISTRY ON THE WEB 294
KEY TERMS 294
EXERCISES 295
FEATURE PROBLEMS AND PROJECTS 297

CHAPTER 11 The Air Around Us 298

MOLECULAR THINKING
Drinking from a Straw 304

MOLECULAR FOCUS
Ozone 318

THE MOLECULAR REVOLUTION
Measuring Ozone 318

11.1 Air Bags 300
11.2 A Gas Is a Swarm of Particles 301
11.3 Pressure 301
11.4 The Relationships Between Gas Properties 304
11.5 The Atmosphere: What Is in It? 308
11.6 The Atmosphere: A Layered Structure 310
11.7 Air Pollution: An Environmental Problem in the Troposphere 312
11.8 Cleaning Up Air Pollution: The Clean Air Act 313
11.9 Ozone Depletion: An Environmental Problem in the Stratosphere 316
11.10 The Montreal Protocol: The End of Chlorofluorocarbons 320
11.11 Myths Concerning Ozone Depletion 321
CHAPTER SUMMARY 323
CHEMISTRY ON THE WEB 323
KEY TERMS 324
EXERCISES 324
FEATURE PROBLEMS AND PROJECTS 326

CHAPTER 12

The Liquids and Solids Around Us: Especially Water   328

MOLECULAR THINKING
Making Ice Cream 332

MOLECULAR THINKING
Soap—A Molecular Liaison 338

MOLECULAR THINKING
Flat Gasoline 342

MOLECULAR FOCUS
Trichloroethylene (TCE) 350

WHAT IF. . .
Criticizing the EPA 353

12.1 No Gravity, No Spills 330
12.2 Liquids and Solids 331
12.3 Separating Molecules: Melting and Boiling 332
12.4 The Forces That Hold Us—and Everything Else—Together 334
12.5 Smelling Molecules: The Chemistry of Perfume 339
12.6 Chemists Have Solutions 341
12.7 Water: An Oddity Among Molecules 343
12.8 Water: Where Is It and How Did It Get There? 345
12.9 Water: Pure or Polluted? 345
12.10 Hard Water: Good for Our Health, Bad for Our Pipes 346
12.11 Biological Contaminants 348
12.12 Chemical Contaminants 348
12.13 Ensuring Good Water Quality: The Safe Drinking Water Act 351
12.14 Public Water Treatment 352
12.15 Home Water Treatment 354
  CHAPTER SUMMARY 356
  CHEMISTRY ON THE WEB 356
  KEY TERMS 357
  EXERCISES 357
  FEATURE PROBLEMS AND PROJECTS 359

CHAPTER 13

Acids and Bases: The Molecules Responsible for Sour and Bitter   360

MOLECULAR FOCUS
Cocaine 365

MOLECULAR THINKING
Bee Stings and Baking Soda 372

WHAT IF. . .
Practical Environmental Protection 376

THE MOLECULAR REVOLUTION
Neutralizing the Effects of Acid Rain 377

13.1 If It Is Sour, It Is Probably an Acid 362
13.2 The Properties of Acids: Tasting Sour and Dissolving Metals 362
13.3 The Properties of Bases: Tasting Bitter and Feeling Slippery 363
13.4 Acids and Bases: Molecular Definitions 365
13.5 Strong and Weak Acids and Bases 367
13.6 Specifying the Concentration of Acids and Bases: The pH Scale 368
13.7 Some Common Acids 369
13.8 Some Common Bases 372
13.9 Acid Rain: Extra Acidity from the Combustion of Fossil Fuels 374
13.10 Acid Rain: The Effects 375
13.11 Cleaning Up Acid Rain: The Clean Air Act Amendments of 1990 376
  CHAPTER SUMMARY 378
  CHEMISTRY ON THE WEB 378
  KEY TERMS 379
  EXERCISES 379
  FEATURE PROBLEMS AND PROJECTS 381

CHAPTER 14

Oxidation and Reduction 382

- 14.1 Rust   384
- 14.2 Oxidation and Reduction: Some Definitions   384
- 14.3 Some Common Oxidizing and Reducing Agents   388
- 14.4 Respiration and Photosynthesis   389
- 14.5 Batteries: Making Electricity with Chemistry   389
- 14.6 Fuel Cells   393
- 14.7 Corrosion: The Chemistry of Rust   395
- 14.8 Oxidation, Aging, and Antioxidants   397
  - CHAPTER SUMMARY   398
  - CHEMISTRY ON THE WEB   398
  - KEY TERMS   399
  - EXERCISES   399
  - FEATURE PROBLEMS AND PROJECTS   401

MOLECULAR THINKING
The Dulling of Automobile Paint   387

MOLECULAR FOCUS
Hydrogen Peroxide   388

THE MOLECULAR REVOLUTION
Fuel Cell Vehicles   395

WHAT IF...
The Economics of New Technologies and Corporate Handouts   396

CHAPTER 15

The Chemistry of Household Products 402

- 15.1 Cleaning Clothes with Molecules   404
- 15.2 Soap: A Surfactant   405
- 15.3 Synthetic Detergents: Surfactants for Hard Water   407
- 15.4 Laundry-Cleaning Formulations   409
- 15.5 Corrosive Cleaners   410
- 15.6 Hair Products   411
- 15.7 Skin Products   412
- 15.8 Facial Cosmetics   414
- 15.9 Perfumes and Deodorants: Producing Pleasant Odors and Eliminating Unpleasant Ones   415
- 15.10 Polymers and Plastics   419
- 15.11 Copolymers: Nylon, Polyethylene Terephthalate, and Polycarbonate   423
- 15.12 Rubber   424
  - CHAPTER SUMMARY   426
  - CHEMISTRY ON THE WEB   427
  - KEY TERMS   427
  - EXERCISES   428
  - FEATURE PROBLEMS AND PROJECTS   430

MOLECULAR FOCUS
Polyoxyethylene   409

MOLECULAR THINKING
Weather, Furnaces, and Dry Skin   413

WHAT IF...
Consumer Chemistry and Consumerism   418

THE MOLECULAR REVOLUTION
Conducting Polymers   423

CHAPTER 16

Biochemistry and Biotechnology 432

- 16.1 Brown Hair, Blue Eyes, and Big Mice   434
- 16.2 Lipids and Fats   434
- 16.3 Carbohydrates: Sugar, Starch, and Sawdust   440

MOLECULAR FOCUS
Raffinose   446

MOLECULAR THINKING
Wool 455

THE MOLECULAR REVOLUTION
Forensics 464

WHAT IF...
The Ethics of Therapeutic Cloning and Stem Cell Research 467

16.4 Proteins: More Than Muscle 446
16.5 Protein Structure 451
16.6 Some Common Proteins 454
16.7 Nucleic Acids: The Blueprint for Proteins 456
16.8 Recombinant DNA Technology 462
16.9 Cloning 465
CHAPTER SUMMARY 468
CHEMISTRY ON THE WEB 469
KEY TERMS 469
EXERCISES 469
FEATURE PROBLEMS AND PROJECTS 475

CHAPTER 17
Drugs and Medicine: Healing, Helping, and Hurting — 476

MOLECULAR THINKING
Generic or Name Brands? 482

MOLECULAR FOCUS
Azidothymidine (AZT) 485

WHAT IF...
The Controversy of Abortion 487

WHAT IF...
Alcoholism 491

WHAT IF...
The Danger of Street Drugs 496

WHAT IF...
Prescription Drug Abuse 502

THE MOLECULAR REVOLUTION
Consciousness 503

17.1 Love and Depression 478
17.2 Relieving Pain, Reducing Fever, and Lowering Inflammation 478
17.3 Killing Microscopic Bugs: Antibiotics 480
17.4 Antiviral Drugs and Acquired Immune Deficiency Syndrome 482
17.5 Sex Hormones and the Pill 486
17.6 Steroids 487
17.7 Chemicals to Fight Cancer 488
17.8 Depressants: Drugs That Dull the Mind 490
17.9 Narcotics: Drugs That Diminish Pain 493
17.10 Stimulants: Cocaine and Amphetamine 496
17.11 Legal Stimulants: Caffeine and Nicotine 498
17.12 Hallucinogenic Drugs: Mescaline and Lysergic Acid Diethylamide 499
17.13 Marijuana 501
17.14 Prozac and Zoloft: SSRIs 502
CHAPTER SUMMARY 505
CHEMISTRY ON THE WEB 506
KEY TERMS 506
EXERCISES 507
FEATURE PROBLEMS AND PROJECTS 508

CHAPTER 18
The Chemistry of Food — 510

MOLECULAR THINKING
Sugar Versus Honey 514

THE MOLECULAR REVOLUTION
Does Sugar Make Children Hyperactive? 516

18.1 You Are What You Eat, Literally 512
18.2 Carbohydrates: Sugars, Starches, and Fibers 513
18.3 Proteins 517
18.4 Fats, Oils, and Cholesterol 518
18.5 Caloric Intake and the First Law: Extra Calories Lead to Fat 521

WHAT IF...
The Second Law
and Food Energy 518

MOLECULAR FOCUS
Ammonium Nitrate 533

WHAT IF...
Pesticide Residues in Food—
A Cause for Concern? 536

18.6   Vitamins 524
18.7   Minerals 527
18.8   Food Additives 529
18.9   The Molecules Used to Grow Crops: Fertilizers and Nutrients 532
18.10 The Molecules Used to Protect Crops: Insecticides and Herbicides 534
       CHAPTER SUMMARY 538
       CHEMISTRY ON THE WEB 538
       KEY TERMS 539
       EXERCISES 539
       FEATURE PROBLEMS AND PROJECTS 541

CHAPTER 19

MOLECULAR FOCUS
Buckminsterfullerene 547

WHAT IF...
Value-Free Science 550

Nanotechnology   542

19.1 Extreme Miniaturization 544
19.2 Scanning Tunneling Microscope 545
19.3 Atomic Force Microscope 546
19.4 Buckyballs—A New Form of Carbon 547
19.5 Carbon Nanotubes 548
19.6 Nanomedicine 549
19.7 Nanoproblems 550
      CHAPTER SUMMARY 551
      CHEMISTRY ON THE WEB 551
      EXERCISES 552
      FEATURE PROBLEMS AND PROJECTS 552

Appendix 1: Significant Figures   A-1

Appendix 2: Answers to Selected Exercises   A-5

Appendix 3: Answers to Your Turn Questions   A-31

Appendix 4: Scientific Notation   S-1

Glossary   G-1

Index   I-1

Preface

To the Instructor

Chemistry in Focus is a text designed for a one-semester college chemistry course for students not majoring in the sciences. This book has two main goals: the first is to develop in students an appreciation for the molecular world and the fundamental role it plays in daily life; the second is to develop in students an understanding of the major scientific and technological issues affecting our society.

> *The two main goals of this book are for students to understand the molecular world and to understand the scientific issues that face society.*

A MOLECULAR FOCUS

The first goal is essential. Students should leave this course understanding that the world is composed of atoms and molecules and that everyday processes—water boiling, pencils writing, soap cleaning—are caused by atoms and molecules. After taking this course, a student should look at water droplets, salt crystals, and even the paper and ink of their texts in a different way. They should know, for example, that beneath the surface of a water droplet or a grain of salt lie profound reasons for each of their properties. From the opening example to the closing chapter, this text maintains this theme through a consistent focus on explaining the macroscopic world in terms of the molecular world.

The art program, a unique component of this text, emphasizes the connection between what we see—the macroscopic world—and what we cannot see—the molecular world. Throughout the text, photographs of everyday objects or processes are magnified to show the molecules and atoms responsible for them. The molecules within these magnifications are depicted using space-filling models to help students develop the most accurate picture of the molecular world. Similarly, many molecular formulas are portrayed not only with structural formulas but with space-filling drawings as well. Students are not meant to understand every detail of these formulas—since they are not scientists, they do not need to—rather, they should begin to appreciate the beauty and form of the molecular world. Such an appreciation will enrich their lives as it has enriched the lives of those of us who have chosen science and science education as our career paths.

CHEMISTRY IN A SOCIETAL AND ENVIRONMENTAL CONTEXT

The other primary goal of this text is to develop in students an understanding of the scientific, technological, and environmental issues facing them as citizens and consumers. They should leave this course

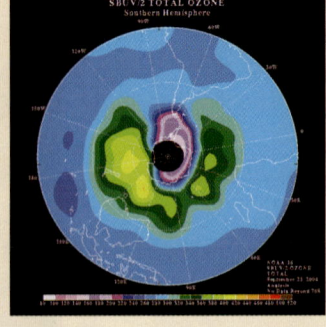

with an understanding of the impact of chemistry on society and on humankind's view of itself. Topics such as global warming, ozone depletion, acid rain, drugs, medical technology, and consumer products are covered in detail. In the early chapters, which focus primarily on chemical and molecular concepts, many of the box features introduce these applications and environmental concerns. The later chapters focus on these topics directly and in more detail.

MAKING CONNECTIONS

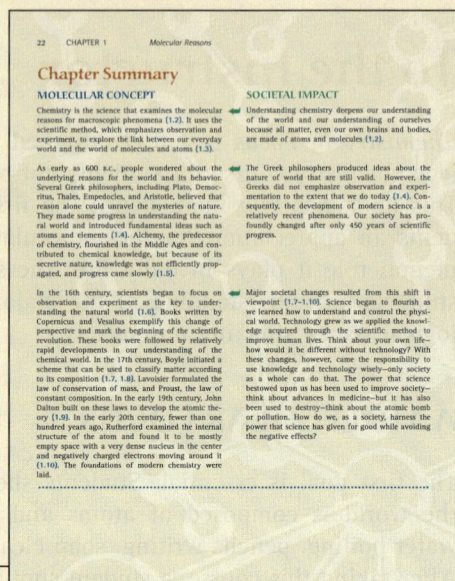

Throughout the text, I have made extensive efforts to help students make connections, both between the molecular and macroscopic world and between principles and applications. The chapter summaries are designed to reinforce those connections, particularly between chemical concepts and societal impact. The chapter summaries consist of two columns, one summarizing the major molecular concepts of the chapter and the other, the impacts of those concepts on society. By putting these summaries side-by-side, the student can clearly see the connections.

A Tour of the Text

GENERAL CHAPTER STRUCTURE

Each chapter opens with a brief paragraph introducing the chapter's main topics and explaining to students why these topics are relevant to their lives. These openers pose questions to help students understand the importance of the topics. For example, the opening paragraph to Chapter 1 states, "As you read these pages, think about the scientific method—its inception just a few hundred years ago has changed human civilization. What are some of those changes? How has the scientific method directly impacted the way you and I live?"

> Each chapter introduces the material with Questions for Thought.

The opening paragraph of each chapter is followed by *Questions for Thought* directly related to chapter content. These questions are answered in the main body of each chapter; presenting them early provides a context for the chapter material.

Most chapters, as appropriate, follow with a description or thought experiment about an everyday experience. The observations of the thought experiment are then explained in molecular terms. For example, a familiar experience may be washing a greasy dish with soapy water. Why does plain water not dissolve the grease? The molecular reason is then given, enhanced by artwork that shows a picture of a soapy dish and a magnification showing what happens with the molecules.

Continuing this theme, the main body of each chapter introduces chemical principles in the context of discovering the molecular causes behind everyday observations. What is it about helium *atoms* that makes it possible to breathe small amounts of helium *gas*—as in a helium balloon—without adverse side effects? What

Preface xxi

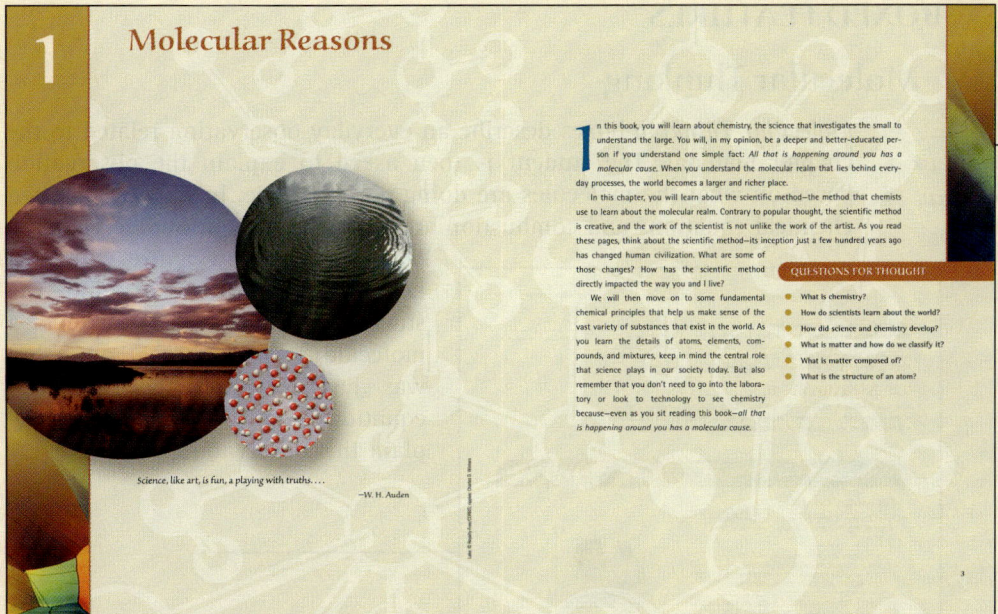

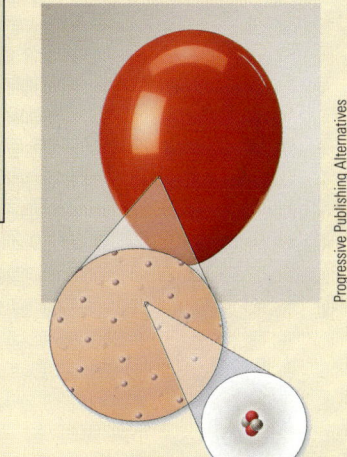

is it about chlorine *atoms* that makes breathing chlorine *gas* dangerous? What happens to water *molecules* when water boils? These questions have molecular answers that teach and illustrate chemical principles. The text develops the chemical principles and concepts involved in a molecular understanding of the macroscopic observations.

Once the student is introduced to basic concepts, consumer applications and environmental problems follow. The text, however, does not separate principles and applications. Early chapters involving basic principles also contain applications, and later chapters with more emphasis on applications build on and expand basic principles.

EXAMPLES AND YOUR TURN EXERCISES

Example problems are included throughout the text, followed by related *Your Turn* exercises for student practice. In designing the text, I made allowances for different instructor preferences on quantitative material. While a course for nonmajors is not usually highly quantitative, some instructors prefer more quantitative material than others. To accommodate individual preferences, many quantitative sections, including some *Examples* and *Your Turn* exercises, can be easily omitted. These are often placed toward the end of chapters for easy omission. Similarly, exercises in the back of each chapter that rely on quantitative material can also be easily omitted. Instructors wishing a more quantitative course should include these sections, while those wanting a more qualitative course can skip them. The answers to the *Your Turn* exercises can be found in Appendix 3.

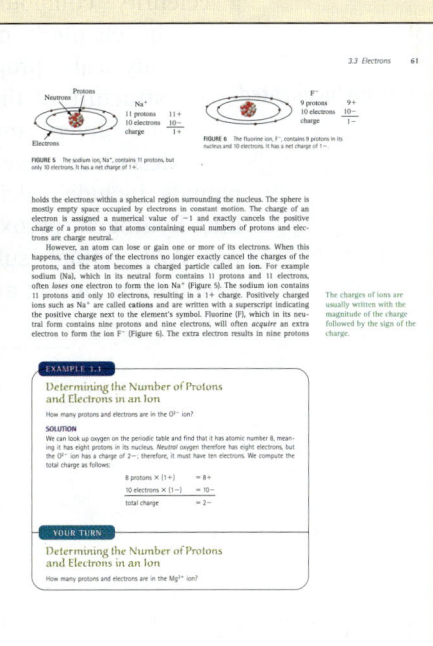

BOXED FEATURES

Molecular Thinking

Boxed features show relevance and ask students to interact with the material.

Molecular Thinking boxes describe an everyday observation related to the chapter material. The student is then asked to explain the observation based on what the molecules are doing. For example, in Chapter 4, when chemical equations and combustion are discussed, the *Molecular Thinking* box describes how a fire will burn hotter in the presence of wind. The student is then asked to give a molecular reason—based on what was just learned about chemical equations and combustion—to explain this observation.

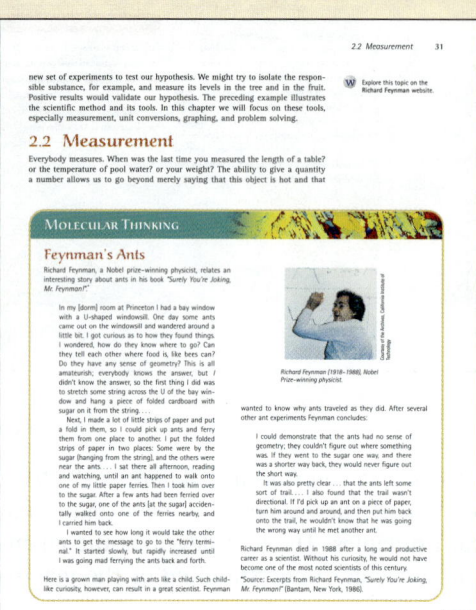

Molecular Focus

Molecular Focus boxes highlight a "celebrity" compound related to the chapter's material. The physical properties and structure of the compound are given and its use(s) described. Featured compounds include calcium carbonate, hydrogen peroxide, ammonia, AZT, retinal, sulfur dioxide, ammonium nitrate, and others.

Celebrity compounds are highlighted.

The Molecular Revolution

Molecular Revolution boxes highlight topics of modern research and recent technology related to the chapter's material. Examples include the measuring of global temperatures, imaging atoms with scanning tunneling microscopy, and the development of fuel cell and hybrid electric vehicles.

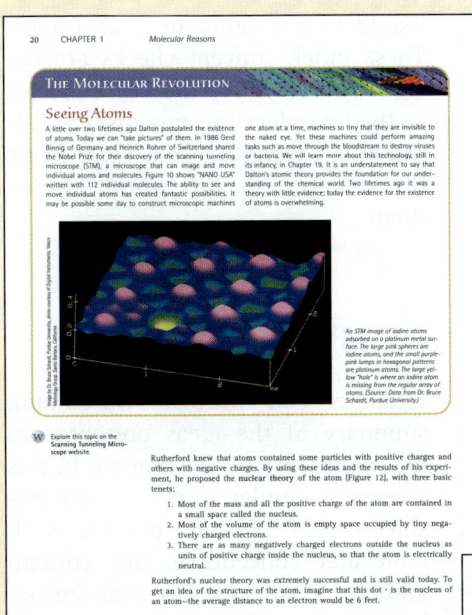

What if . . .

What if . . . boxes discuss topics with societal, political, or ethical implications. At the end of the discussion there are one or more open-ended questions for group discussion. Topics include the Manhattan Project, government subsidies for the development of alternative fuels, stem cell research, and others.

APPLY YOUR KNOWLEDGE   New!

Apply Your Knowledge boxes are a new feature to this edition. The student is asked to use a conceptual idea to answer a practical question. For instance, in Chapter 3, the *Apply Your Knowledge* box presents the situation of a friend who tells you that a tabloid reported the discovery of a new form of carbon that contains eight protons in the nucleus of its atoms and spontaneously turns into diamond. How

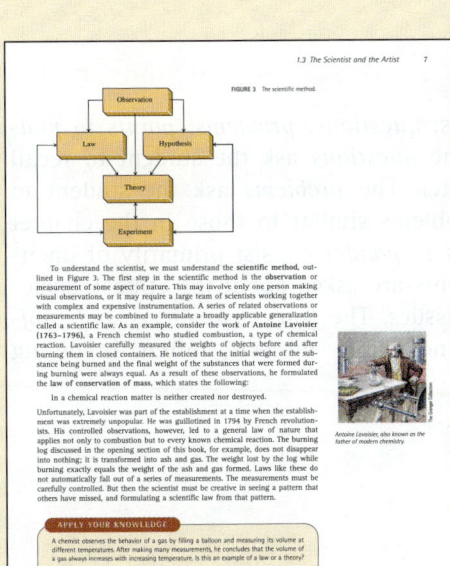

Chapter summaries review main molecular concepts and their societal impacts.

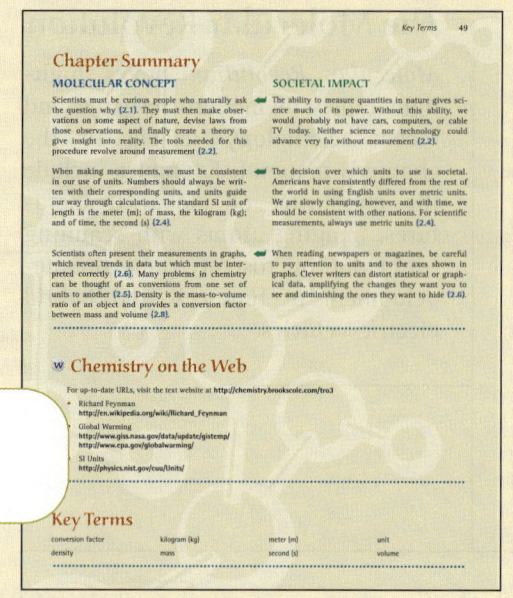

would you respond to your friend? These quick concept checks are designed to reinforce the key concepts in the text, develop students' critical-thinking skills, and help them relate the material to the world around them.

CHAPTER SUMMARIES

Chapters end with a two-column summary of the ideas presented in the main body of the chapter. In this summary, students get a side-by-side review of the chapter with molecular concepts in one column and the coinciding societal impact in the other. The chapter summary allows the student to get an overall picture of the chapter and strengthens the connection between principles and applications.

New! ## CHEMISTRY ON THE WEB

The *Chemistry on the Web* section features a list of URLs for the websites referenced within the chapter. They can easily be assigned for further exploration or research. Weblinks are also provided on the Student Book Companion Website.

KEY TERMS

Each chapter has a set of key terms from within that chapter for review and study. Each of the key terms is defined in the Glossary at the end of the text.

STUDENT EXERCISES

All chapters contain exercises of four types: *questions, problems, points to ponder,* and *feature problems and projects.* The *questions* ask the student to recall many of the key concepts from the chapter. The *problems* ask the student to apply what they have learned to solve problems similar to those in the chapter *Examples* and *Your Turn* boxes. The *points to ponder* consist primarily of open-ended short-essay questions in which students are asked about the ethical, societal, and political implications of scientific issues. The *feature problems and projects* contain problems with graphics and short projects, often involving web-based inquiry.

Accompanying Materials

Online Instructor's Resource Manual
Written by Ann Tro of Westmont College and updated by Richard Jarman of the College of DuPage, this manual contains detailed solutions to all of the end-of-chapter problems in the text.

New! **iLrn Computerized Testbank**
With this easy-to-use software, professors can create, deliver, and customize tests in minutes. It contains questions and problems authored specifically for the text.

Test Bank Online
The Test Bank, revised by Stephen J. Glueckert of the University of Southern Indiana, features more than 700 multiple-choice questions for instructors to use for tests, quizzes, or homework assignments. It can be accessed from the instructors' section of the Book Companion Website.

New! **Microsoft® PowerPoint® Slides**
A presentation tool created by Jeannine Eddleton of Virginia Polytechnic Institute and State University, these slides provide text, art, photos, and tables in an electronic format that is easily exported into other software packages. In addition, you can customize your presentations by importing your own personal lecture slides or notes. The slides can be found on the instructors' section of the Book Companion Website.

New! **JoinIn™ on TurningPoint® CD-ROM**
JoinIn content for Response Systems, authored by Chris Truitt of Texas Tech University, allows you to transform your classroom and assess your students' progress with instant in-class quizzes and polls. Our exclusive agreement to offer TurningPoint software lets you pose book-specific questions and display students' answers seamlessly within the Microsoft PowerPoint slides of your own lecture, in conjunction with the "clicker" hardware of your choice. Enhance how your students interact with you, your lecture, and each other. Please consult your Brooks/Cole representative for further details.

Transparency Acetates
The overhead transparency acetates offer instructors a selection of over 75 full-color images correlated to the textbook content illustrating the most pedagogically important concepts.

New! **Student Book Companion Website**
Organized by chapter, this outstanding site features chapter-by-chapter online quizzes and weblinks from the Chemistry on the Web sections in the textbook.

New! **OWL: Online Web-based Learning System**
Developed over the last several years at the University of Massachusetts, Amherst, and class-tested by thousands of students, OWL is a fully customizable and flexible web-based homework system and assessment tool with course management. With both numerical and chemical parameterization and useful, specific feedback built right in, OWL produces over 100,000 general chemistry questions

correlated to this text. The OWL system also features a database of simulations, guided tutorials, and problems correlated to the textbook content. Instructors are able to customize the OWL program, use the grade book feature, and generate multiple reports. OWL provides an excellent solution for those who wish to place more emphasis on the quantitative aspects of chemistry.

New! **Inquiry-based Laboratories for Liberal Arts Chemistry**
By Vickie Williamson and Larry Peck of Texas A&M University, *Inquiry-based Laboratories for Liberal Arts Chemistry* offers 19 experiments. The focus of the manual is conceptual learning of the chemical phenomena in our everyday lives. It employs the learning cycle approach, which is used as the underlying model for the guided and open inquiry/application laboratories. An online instructor's guide is also available.

Acknowledgments

Although the cover of this book bears only one name, it should rightly bear many. I am grateful to my colleagues at Westmont College, who have given me the space to write this book. I am especially grateful to Shirley Mullen, Allan Nishimura, and David Marten for their support. I am grateful to my editors, Lisa Lockwood, Sarah Lowe, and Lisa Weber, who have been incredibly gracious and helpful to me throughout this revision. I am also grateful to Ronn Jost, from Lachina Publishing Services, who was attentive to every detail and was a wonderful person to work with.

Thanks also to those who supported me personally while writing this book. I am particularly grateful to my wife, Ann, whose love healed a broken man. Thanks to my children, Michael, Ali, Kyle, and Kaden—they are my *raison d' etre*. I come from a large and close extended Cuban family who has stuck by me through all manner of difficult circumstances. I thank my parents, Nivaldo and Sara, and my siblings, Sarita, Mary, and Jorge. Thanks also to Pam—may her spirit rest in peace.

I am greatly indebted to the reviewers of each of the editions of this book who are listed below. They have all left marks on the work you are now holding. Lastly, I thank my students, whose lives energize me and whose eyes continually provide a new way for me to see the world. I am particularly grateful to students Audrey Farkas, Samantha Scheidler, and Daniel Hoss, who helped me in the preparation and proofreading of the manuscript for the third edition.

—Nivaldo J. Tro
Westmont College

THIRD EDITION REVIEWERS

Jeannine Eddleton, Virginia Polytechnic Institute and State University
Stephen J. Glueckert, University of Southern Indiana
Michael Hampton, University of Central Florida
Karen Hanner, Washington State Community College
Eileen Hinks, Virginia Military Institute
Richard H. Jarman, College of DuPage
Gregory A. Oswald, North Dakota State University
Vicki Berger Paulissen, Eastern Michigan University
Albert Plaush, Saginaw Valley State University
Anne Marie Sokol, Buffalo State College
Nhu-Y Stessman, California State University, Stanislaus

SECOND EDITION REVIEWERS

Thomas Goyne, Valparaiso University
Katrina Hartman, Aquinas College

William C. McHarris, Michigan State University
Anne Marie Sokol, Buffalo State College

FIRST EDITION REVIEWERS

Ronald Backus, American River College
Morris Bader, Moravian College
Ronald Baumgarten, University of Illinois at Chicago
Barbara Burke, California State Polytechnic University, Pomona
Marvin Dixon, William Jewell College
Jeff Draves, University of Central Arkansas
Jerry Driscoll, University of Utah
Lawrence Duffy, University of Alaska, Fairbanks
Karen Eichstadt, Ohio University
Seth Elsheimer, University of Central Florida
Gordon Ewing, New Mexico State University
Sharon L. Garlund, Pima Community College
Patrick Garvey, Des Moines Area Community College
James Golen, University of Massachusetts, Dartmouth
Marie Herrmann, University of Cincinnati–Raymond Walters College
Toney Keeney, Southwest Texas Junior College
Keith Kennedy, St. Cloud State University
Leslie N. Kinsland, University of Southwestern Louisiana
David Lippmann, Southwest Texas State University
Kenneth Loach, State University of New York College at Plattsburgh

Lawrence Mack, Bloomsburg University
Joyce Miller, University of Wisconsin, Platteville
Joseph P. Nunes, State University of New York College of Agriculture and Technology at Cobleskill
Gordon Parker, University of Michigan, Dearborn
Alan Pribula, Towson University
Edith Rand, East Carolina University
Martin Salzman, Providence College
Elsa Santos, Colorado State University
George Schenk, Wayne State University
James Schreck, University of Northern Colorado
Kerri Scott, University of Mississippi
Dennis L. Steven, University of Nevada, Las Vegas
Dan M. Sullivan, University of Nebraska at Omaha
Tamar Susskind, Oakland Community College
Joseph Tausta, State University of New York College at Oneonta
Naola VanOrden, Sacramento City College
George Wahl, North Carolina State University
Robert Wallace, Bentley College
Karen Weaver, University of Central Arkansas
Sidney Young, University of Southern Alabama

1 Molecular Reasons

Science, like art, is fun, a playing with truths....

—W. H. Auden

In this book, you will learn about chemistry, the science that investigates the small to understand the large. You will, in my opinion, be a deeper and better-educated person if you understand one simple fact: *All that is happening around you has a molecular cause.* When you understand the molecular realm that lies behind everyday processes, the world becomes a larger and richer place.

In this chapter, you will learn about the scientific method—the method that chemists use to learn about the molecular realm. Contrary to popular thought, the scientific method is creative, and the work of the scientist is not unlike the work of the artist. As you read these pages, think about the modern scientific method—its inception just a few hundred years ago has changed human civilization. What are some of those changes? How has the scientific method directly impacted the way you and I live?

We will then move on to some fundamental chemical principles that help us make sense of the vast variety of substances that exist in the world. As you learn the details of atoms, elements, compounds, and mixtures, keep in mind the central role that science plays in our society today. But also remember that you don't need to go into the laboratory or look to technology to see chemistry because—even as you sit reading this book—*all that is happening around you has a molecular cause.*

QUESTIONS FOR THOUGHT

- What is chemistry?
- How do scientists learn about the world?
- How did science and chemistry develop?
- What is matter and how do we classify it?
- What is matter composed of?
- What is the structure of an atom?

1.1 Firesticks

Flames are fascinating. From the small flicker of a burning candle to the heat and roar of a large campfire, flames captivate us. Children and adults alike will stare at a flame for hours—its beauty and its danger demand our attention. My children have a beloved campfire ritual they call "firesticks." They find dry tree branches, two to three feet long, and ignite the tips in the campfire. They then pull the flaming branches out of the fire and wave them in the air, producing a trail of light and smoke. My reprimands about the danger of this practice work for only several minutes, and then waving wands of fire find their way back into their curious little hands.

As fascinating as flames are, an unseen world—even more fantastic—lies beneath the flame. This unseen world is the world of molecules, the world I hope you see in the pages of this book. We will define molecules more carefully later; for now think of them as tiny particles that make up matter—so tiny that a single flake of ash from a fire contains one million trillion of them. The flame on my children's firesticks and in the campfire is composed of molecules, billions of billions of them rising upward and emitting light (Figure 1).

The molecules in the flame come from an extraordinary transformation—called a **chemical reaction**—in which the molecules within the wood combine with certain molecules in air to form new molecules. The new molecules have excess energy that they shed as heat and light as they escape in the flame. Some of them, hopefully after cooling down, might find their way into your nose, producing the smell of the fire.

Let's suppose for a moment that we could see the molecules within the burning wood—we would witness a frenzy of activity. A bustling city during rush hour would appear calm in comparison. The molecules in the wood, all vibrating and jostling trillions of times every second, would rapidly react with molecules in the air. The reaction of a single molecule with another would occur within a split second, and the new molecules would fly off in a trail of heat and light, only to reveal the next molecule in the wood—ready to react. This process would repeat itself trillions of times every second as the wood burns. Yet on the macroscopic scale—the scale that we see—the process looks calm. The wood disappears slowly, and the flame from a few good logs lasts several hours.

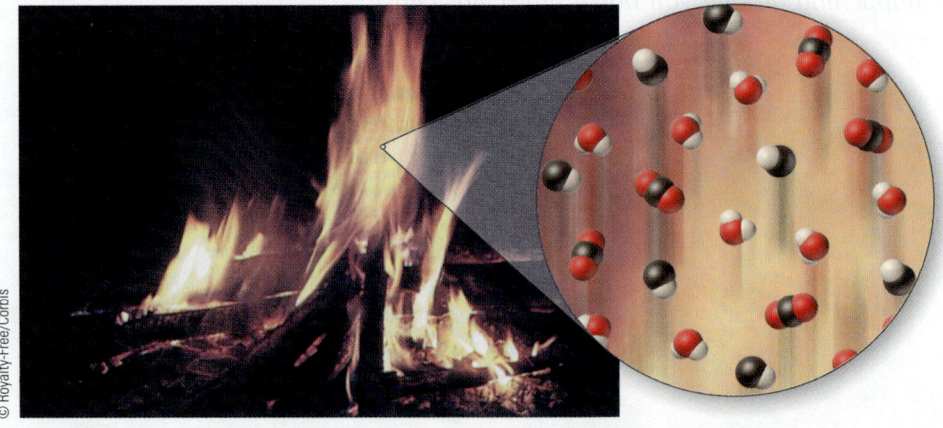

FIGURE 1 The flame you see in a fire is composed of newly created, energetic molecules. They form from the reaction between the molecules within the log and the molecules in the air. They move upward, away from the log, giving off heat and light as they travel.

1.2 Molecular Reasons

All that is happening around us has a molecular cause. When we write, eat, think, move, or breathe, molecules are in action, undergoing changes that make these things happen. The world that we see—that of everyday objects—is determined by the world we cannot see—that of atoms, molecules, and their interactions. **Chemistry** is *the science that investigates the molecular reasons for the processes occurring in our macroscopic world.* Why are leaves green? Why do colored fabrics fade on repeated exposure to sunlight? What happens when water boils? Why does a pencil leave a mark when dragged across a sheet of paper? These basic questions can be answered by considering atoms and molecules and their interactions with each other.

For example, over time we might see a red shirt fade as it is exposed to sunlight. The molecular cause is energy from the sun, which decomposes the molecules that gave the shirt its red color. We may notice that nail polish remover

MOLECULAR THINKING

Why Should Nonscience Majors Study Science?

You may be reading this book only because it is required reading in a required course. You are probably not a science major and might be wondering why you should study science. There are at least three reasons why you should study science, specifically because you are not a science major.

First, modern science influences culture and society in profound ways and raises ethical questions that only society as a whole can answer. For example, in 2001, scientists at a biotechnology company in Massachusetts succeeded for the first time in cloning (making a biological copy of) a human embryo. Their reason for cloning the embryo was *not* human reproduction (they were not trying to make a race of superhumans or clones of themselves) but rather to cure and treat diseases. This kind of cloning, called *therapeutic cloning* (as opposed to reproductive cloning), holds as its goal the creation of specialized cells (called stem cells) to be used, for example, to cure diabetes or to mend damaged spinal cords. The potential benefits of this research are significant, but it also carries some moral risk. Only society as a whole can decide if the benefits are worth the risk. If our society is to make intelligent decisions on issues such as this, we, as citizens of that society, should have a basic understanding of the scientific principles at work.

Second, decisions involving scientific principles are often made by nonscientists. Politicians are generally not trained in science, nor are the people electing the politicians. Yet politicians make decisions concerning science policy, science funding, and environmental regulation. A clever politician could impose unsound scientific policy on an uninformed electorate. For example, Adolf Hitler proposed his own versions of Nazi genetics on the German people. He wrongly proposed that the Aryan race could make itself better by isolating itself from other races. According to Hitler, Aryans should only reproduce with other Aryans to produce superior human beings. However, any person with a general knowledge of genetics would know that Hitler was wrong. Excessive inbreeding actually causes genetic weaknesses in a population. For this reason, purebred dogs have many genetic problems, and societal taboos exist for intrafamily marriages. History demonstrates other examples of this sort of abuse. Agriculture in the former Soviet Union still suffers from years of misdirected policies based on communistic ideas of growing crops, and South America has seen failures in land use policies that were scientifically ill-informed. It behooves any of us interested in making this planet a viable place to live to understand the scientific issues that face our society and to make intelligent decisions that will benefit our children and grandchildren.

Third, science is a fundamental way to understand the world around us and therefore reveals knowledge not attainable by other means. Such knowledge serves to deepen and enrich our lives. For example, an uninformed observer of the night sky may marvel at its beauty but will probably not experience the awe that comes from knowing that even the closest star is trillions of miles away or that stars produce light in a process that could only start at temperatures exceeding millions of degrees. For the uninformed, the world is a two-dimensional, shallow place. For the informed, the world becomes a deeper, richer, and more complex place. In chemistry, we learn about the world that exists behind the world we see, a world present all around us and even inside of us. Through its study we are better able to understand our world and better able to understand ourselves.

FIGURE 2 When sugar dissolves into coffee, the sugar molecules mix with the water molecules.

Chemists investigate the molecular reasons for physical phenomena.

accidentally spilled on our hand makes our skin feel cold as it evaporates. The molecular cause is molecules in our skin colliding with the evaporating molecules in the nail polish remover, losing energy to them, and producing the cold sensation. We may see that a teaspoon of sugar stirred into coffee readily dissolves (Figure 2). The sugar seems to disappear in the coffee. However, when we drink the coffee, we know the sugar is still there because we taste its sweetness. The molecular cause is that a sugar molecule has a strong attraction for water molecules and prefers to leave its neighboring sugar molecules and mingle with the water. We see this as the disappearing of the solid sugar, but it is not disappearing at all, just mixing on the molecular level. Chemists, by using the scientific method, investigate the molecular world; they examine the molecular reasons for our macroscopic observations.

1.3 The Scientist and the Artist

Science and art are often perceived as different disciplines, attracting different types of people. The artist is perceived as a highly creative individual, uninterested in facts or numbers. The scientist, in contrast, is perceived as a stilted, uncreative individual interested only in facts and numbers. Both images are false, however, and the two professions have more in common than is generally imagined.

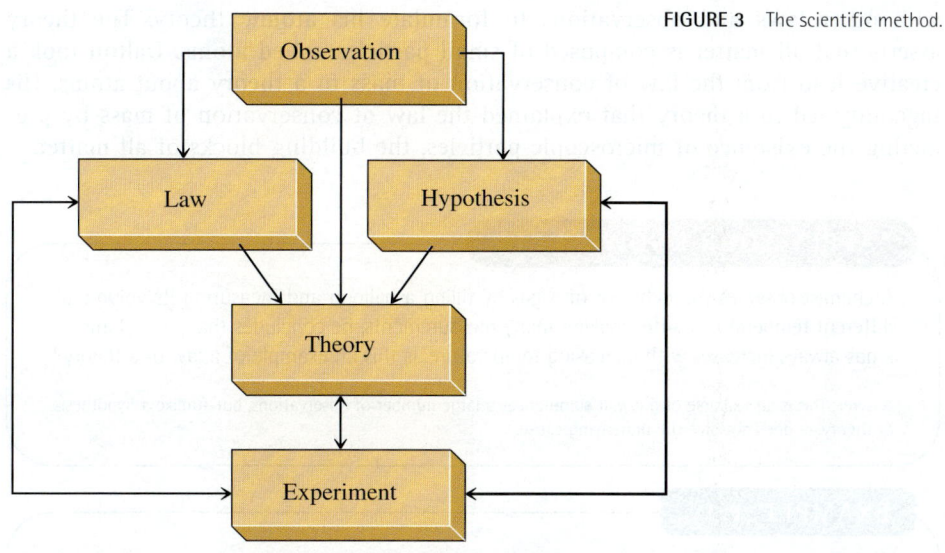

FIGURE 3 The scientific method.

To understand the scientist, we must understand the **scientific method**, outlined in Figure 3. The first step in the scientific method is the **observation** or measurement of some aspect of nature. This may involve only one person making visual observations, or it may require a large team of scientists working together with complex and expensive instrumentation. A series of related observations or measurements may be combined to formulate a broadly applicable generalization called a scientific law. As an example, consider the work of **Antoine Lavoisier (1763–1794)**, a French chemist who studied combustion, a type of chemical reaction. Lavoisier carefully measured the weights of objects before and after burning them in closed containers. He noticed that the initial weight of the substance being burned and the final weight of the substances that were formed during burning were always equal. As a result of these observations, he formulated the **law of conservation of mass**, which states the following:

In a chemical reaction matter is neither created nor destroyed.

Antoine Lavoisier, also known as the father of modern chemistry.

Unfortunately, Lavoisier was part of the establishment at a time when the establishment was extremely unpopular. He was guillotined in 1794 by French revolutionists. His controlled observations, however, led to a general law of nature that applies not only to combustion but to every known chemical reaction. The burning log discussed in the opening section of this book, for example, does not disappear into nothing; it is transformed into ash and gas. The weight lost by the log while burning exactly equals the weight of the ash and gas formed. Laws like these do not automatically fall out of a series of measurements. The measurements must be carefully controlled. But then the scientist must be creative in seeing a pattern that others have missed, and formulating a scientific law from that pattern.

Scientific laws summarize and predict behavior, but they do not explain the underlying cause. A **hypothesis** is an initial attempt to explain the underlying causes of observations and laws. A hypothesis is a tentative model (educated by observation) that is then tested by an **experiment**, a controlled observation specifically designed to test a hypothesis. One or more confirmed hypotheses (possibly with the additional support of observations and laws) may evolve into an overarching model of reality called a **theory**. A good theory often predicts behavior far beyond the observations and laws from which it was formulated. For example, John Dalton, an English chemist, used the law of conservation of mass along

The atomic theory is described in more detail in Section 1.9.

with other laws and observations to formulate his atomic theory. The theory asserts that all matter is composed of small particles called atoms. Dalton took a creative leap from the law of conservation of mass to a theory about atoms. His ingenuity led to a theory that explained the law of conservation of mass by predicting the existence of microscopic particles, the building blocks of all matter.

> **APPLY YOUR KNOWLEDGE**
>
> A chemist observes the behavior of a gas by filling a balloon and measuring its volume at different temperatures. After making many measurements, he concludes that the volume of a gas always increases with increasing temperature. Is this an example of a law or a theory?
>
> Answer: This is an example of a law. It summarizes a large number of observations, but—unlike a hypothesis or theory—it does not give the underlying cause.

EXAMPLE 1.1

The Scientific Method

Suppose you are an astronomer mapping the galaxies in the sky for the very first time. You discover that all galaxies are moving away from Earth at high speeds. As part of your studies, you measure the speed and distance from the Earth of a number of galaxies. Your results are shown here.

Distance from Earth	Speed Relative to Earth
5.0 million light-years	600 miles/second (mi/s)
8.4 million light-years	1000 mi/s
12.3 million light-years	1500 mi/s
20.8 million light-years	2500 mi/s

Formulate a law based on your observations.
Because laws summarize a number of related observations, we can formulate the following law from the tabulated observations:

> The farther away a galaxy is from Earth, the faster its speed.

Devise a hypothesis or theory that might explain the law.
You may devise any number of hypotheses or theories consistent with the preceding law. Your hypotheses must, however, give the underlying reasons behind the law. One possible hypothesis:

> Earth has a slowing effect on all galaxies. Those galaxies close to Earth experience this effect more strongly than those that are farther away and therefore travel more slowly.

Another possible hypothesis:

> Galaxies were formed in an expansion that began sometime in the past and are therefore moving away from each other at speeds that depend on their separation.

What kind of experiments would help validate or disprove these hypotheses?
For the first hypothesis, you might devise experiments that try to measure the nature of the slowing effect that Earth exerts on galaxies. For example, the force responsible for the slowing may also affect the Moon's movement, and its effect might be measured by experiment. For the second hypothesis, experiments that look for other evidence of an expansion would work. For example, you might try to look for remnants of the heat or light given off by the expansion. Experimental confirmation of your hypothesis could result in the evolution of the hypothesis into a theory for how the universe came to exist in its present form.

Finally, like a hypothesis, a theory is subject to experiments. A theory is valid only if all experiments to date are consistent with it. If an experiment is inconsistent with a particular theory, that theory must be revised, and a new set of experiments must be performed to test the revision. A theory is never proved, only validated by experimentation. The constant interplay between theory and experiment gives science its excitement and power.

The process by which a set of observations leads to a model of reality is the scientific method. It is similar, in some ways, to the process by which a series of observations of the world leads to a magnificent painting. Like the artist, the scientist must be creative. Like the artist, the scientist must see order where others have seen only chaos. Like the artist, the scientist must create a finished work that imitates the world. The difference between the scientist and the artist lies in the stringency of the imitation. The scientist must constantly turn to experiment to determine if his or her ideas about the world are valid.

1.4 The First People to Wonder About Molecular Reasons

The Greek philosophers are the first people on record to have thought deeply about the nature of matter. As early as 600 B.C., these scholars wanted to know the *why* of things. However, they were immersed in the philosophical thought of their day that held that physical reality is an imperfect representation of the ideal. As a result, they did not emphasize experiments on the imperfect physical world as a way to understand it. According to **Plato** (428–348 B.C.), *reason alone* was the superior way to unravel the mysteries of nature. Remarkably, Greek ideas about nature led to some ideas similar to modern ones.

Democritus (460–370 B.C.) theorized that matter was ultimately composed of small, indivisible particles he called *atomos* or atoms, meaning "not to cut." Democritus believed that if you divided matter into smaller and smaller pieces, you would eventually end up with very small particles (atoms) that could not be divided any further. He is quoted as saying, "Nothing exists except atoms and empty space; everything else is opinion." Although Democritus was right by modern standards, most Greek thinkers rejected his atomistic viewpoint.

Thales (624–546 B.C.) reasoned that any substance could be converted into any other substance, so that all substances were in reality one basic material. Thales believed the one basic material to be water. He said, "Water is the principle, or the element of things. All things are water." **Empedocles** (490–430 B.C.), on the other hand, suggested that all matter was composed of four basic materials or elements: air, water, fire, and earth. This idea was accepted by **Aristotle** (384–321 B.C.), who added a fifth element—the heavenly ether—perfect, eternal, and incorruptible. In Aristotle's mind, the five basic elements made all matter, and his idea reigned for 2000 years.

1.5 Immortality and Endless Riches

The predecessor of chemistry, called **alchemy**, flourished in Europe during the Middle Ages. Alchemy was a partly empirical, partly magical, and entirely secretive pursuit with two main goals: the transmutation of ordinary materials into gold, and the discovery of the "elixir of

Alchemists sought to turn ordinary materials into gold and to make "the elixir of life," a substance that would grant immortality.

life," a substance that would grant immortality to any who consumed it. In spite of what might today appear as misdirected goals, alchemists made some progress in our understanding of the chemical world. Through their obsession with turning metals into gold, they learned much about metals. They were able to form alloys—mixtures of metals—with unique properties. They also contributed to the development of laboratory separation and purification techniques that are still used today. In addition, alchemists made advances in the area of pharmacology by isolating natural substances and using them to treat ailments. Because of the mystical nature of alchemy and the preoccupation with secrecy, however, knowledge was not efficiently propagated, and up to the 16th century, progress was slow.

1.6 The Beginning of Understanding

The publication of two books in 1543 marks the beginning of what is now called **the scientific revolution**. The first book was written by **Nicholas Copernicus** (1473–1543), a Polish astronomer who claimed that the Sun was the center of the universe. In contrast, the Greeks had reasoned that Earth was the center of the universe, with all heavenly bodies, including the Sun, revolving around Earth. Although complex orbits were required to explain the movement of the stars and planets, the Earth-centered universe put humans in the logical center of the created order. Copernicus, by using elegant mathematical arguments and a growing body of astronomical data, suggested exactly the opposite—the Sun stood still and Earth revolved around it. The second book, written by **Andreas Vesalius** (1514–1564), a Flemish anatomist, portrayed human anatomy with unprecedented accuracy.

Galileo Galilei expanded on Copernicus's ideas of a Sun-centered rather than an Earth-centered universe.

The uniqueness of these books was their overarching emphasis on observation and experiment as the way to learn about the natural world. The books were revolutionary, and Copernicus and Vesalius laid the foundation for a new way to understand the world. Nonetheless, progress was slow. Copernicus's ideas were not popular among the religious establishment. **Galileo Galilei** (1564–1642), who confirmed and expanded on Copernicus's ideas, was chastised by the Roman Catholic Church for his views. Galileo's Sun-centered universe put man outside of the geometric middle of God's created order and seemed to contradict the teachings of Aristotle and the Church. As a result, the Roman Catholic Inquisition

What if...

Observation and Reason

Throughout this text, I will pose a number of open-ended questions that you can ponder and discuss. Some will have better-defined answers than others, but none will have a single correct answer. The first one follows.

The field of science is relatively young compared to other fields such as philosophy, history, or art. It has, however, progressed quickly. In the four and one-half centuries since the scientific revolution, science and its applications have dramatically changed our lives. In contrast, the tens of centuries before 1543 proceeded with comparatively few scientific advances. A major factor in the scarcity of scientific discoveries before 1543 was the Greek emphasis on reason over observation as the key to knowledge. Although some Greek philosophers, such as Aristotle, spent a great deal of time observing and describing the natural world, they did not emphasize experimentation and the modification of ideas based on the outcomes of experiments. What if the Greeks had placed a greater emphasis on experimentation? What if Democritus had set out to prove his atomistic view of matter by performing experiments? Where do you think science might be today?

forced Galileo to recant his views. Galileo was never tortured, but he was subject to house arrest until he died.

The scientific method progressed nonetheless, and alchemy was transformed into chemistry. Chemists began to perform experiments to answer fundamental questions such as these: What are the basic elements? Which substances are pure and which are not? In 1661, **Robert Boyle** (1627–1691) published *The Skeptical Chymist,* in which he criticized Greek ideas concerning a four-element explanation of matter. He proposed that an element must be tested to determine if it was really simple. If a substance could be broken into simpler substances, it was not an element.

1.7 The Classification of Matter

Matter can be classified by its composition (what it's composed of) or by its phase (solid, liquid, or gas). We examine each of these in turn.

CLASSIFYING MATTER BY ITS COMPOSITION

Boyle's approach led to a scheme, shown in Figure 4, that we use to classify matter today. In this scheme, all matter is first classifiable as either a **pure substance** or a **mixture**.

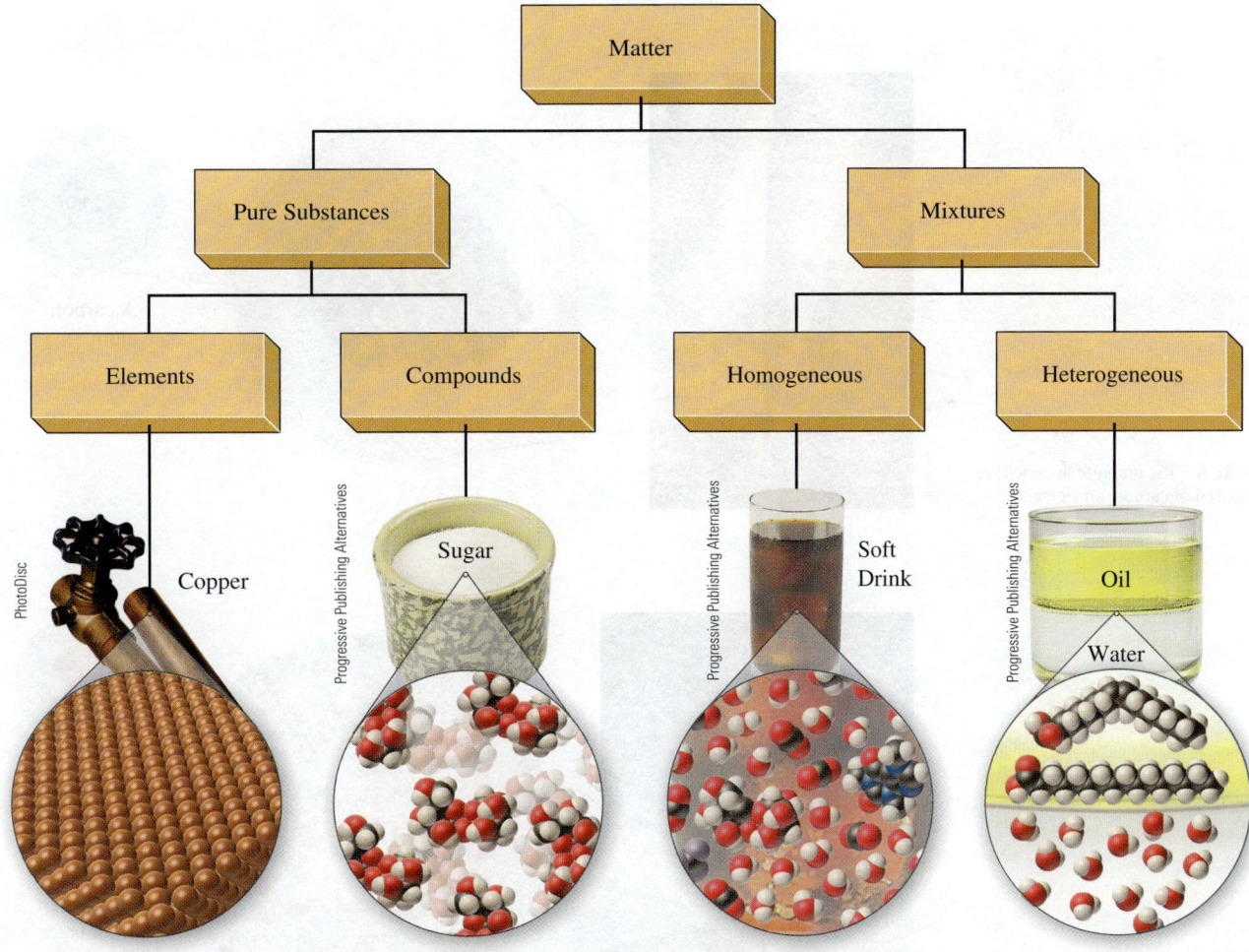

FIGURE 4 Classification scheme for matter. Matter may be either a pure substance or a mixture. If it is a pure substance, it may be either an element or a compound. If it is a mixture, it may be either a homogeneous mixture or a heterogeneous mixture.

PURE SUBSTANCES

W Explore this topic on the **Interactive Periodic Table** website.

A pure substance may be either an element or a compound. An **element** is a substance that cannot be decomposed into simpler substances. The graphite in pencils (Figure 5) is an example of an element—carbon. No amount of chemical transformation can decompose graphite into simpler substances; it is pure carbon. Other examples of elements include oxygen, a component of air; helium, the gas in helium balloons; and copper, used in plumbing and as a coating on pennies. The smallest identifiable unit of an element is an **atom**. There are about 90 different elements in nature and therefore about 90 different kinds of atoms.

A **compound** is a substance composed of two or more elements in fixed, definite proportions. Compounds are more common in nature than elements because most elements tend to combine with other elements to form compounds. Water (Figure 6), table salt, and sugar are examples of compounds; they can all be decomposed into simpler substances. Water and table salt are difficult to decompose, but sugar is easy to decompose. You may have decomposed sugar yourself while cooking. The black substance left on your pan after burning sugar is carbon, one of sugar's constituent elements. The smallest identifiable unit of many compounds is a **molecule**, two or more atoms bonded together.

Ionic compounds, as you will learn in Chapter 4, are not composed of molecules but rather consist of their constituent elements in a repeating three-dimensional array.

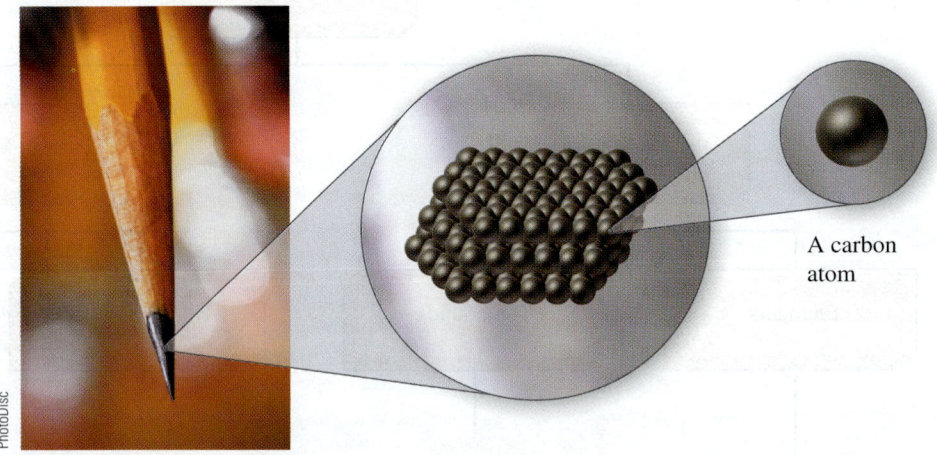

FIGURE 5 The graphite in pencils is composed of carbon, an element.

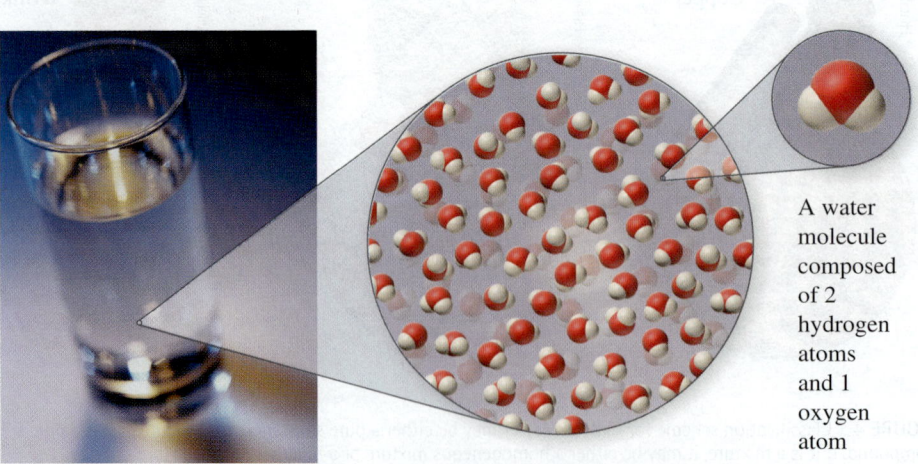

FIGURE 6 Water is a compound whose molecules are composed of 2 hydrogen atoms (white) and 1 oxygen atom (red).

MIXTURES

A mixture is a combination of two or more pure substances in variable proportions. The pure substances may themselves be either elements or compounds. The flame from a burning log is a good example of a mixture. It contains various gases whose proportions vary considerably from one flame to another. A cup of coffee, a can of soda, and ordinary soil are also examples of mixtures. In fact, most of the matter we encounter is in the form of mixtures. The air we breathe is a mixture; seawater is a mixture (Figure 7); food is a mixture; we can even think of ourselves as a very complex mixture.

FIGURE 7 Air is a mixture whose major components are nitrogen (blue) and oxygen (red). Seawater is a mixture whose primary components are water and salt.

EXAMPLE 1.2

Classifying Matter

Determine whether each of the following is an element, a compound, or a mixture. If it is a mixture, classify it as homogeneous or heterogeneous.

   a. copper wire
   b. water
   c. salt water
   d. Italian salad dressing

SOLUTION

Every element is listed in alphabetical order in the table on the back inside cover of this text. Use this table to determine if the substance is an element. If the substance is not listed in the table, but is a pure substance, then it must be a compound. If the substance is not a pure substance, then it is a mixture.

   a. Copper is listed in the table of elements. It is an element.
   b. Water is not listed in the table of elements, but it is a pure substance; therefore, it is a compound.
   c. Salt water is composed of two different substances, salt and water; it is a mixture. Different samples of salt water may have different proportions of salt and water, a property of mixtures. Its composition is uniform throughout; thus, it is a homogeneous mixture.
   d. Italian salad dressing contains a number of substances and is therefore a mixture. It usually separates into at least two distinct regions—each with a different composition—and is therefore a heterogeneous mixture.

> **YOUR TURN**
>
> ## Classifying Matter
>
> Determine whether each of the following is a pure element, a compound, or a mixture. If it is a mixture, classify it as homogeneous or heterogeneous:
>
> a. pure salt
> b. helium gas
> c. chicken noodle soup
> d. coffee

Mixtures may be composed of two or more elements, two or more compounds, or a combination of both. Mixtures can be classified according to how uniformly the substances that compose them mix. A **heterogeneous mixture**, such as oil and water, is separated into two or more regions with different compositions. A **homogeneous mixture**, such as salt water, has the same composition throughout.

CLASSIFYING MATTER BY ITS PHASE

Another way to classify matter is according to its phase. Matter can exist as either a **solid**, a **liquid**, or a **gas** (Figure 8). In solid matter, atoms or molecules are in

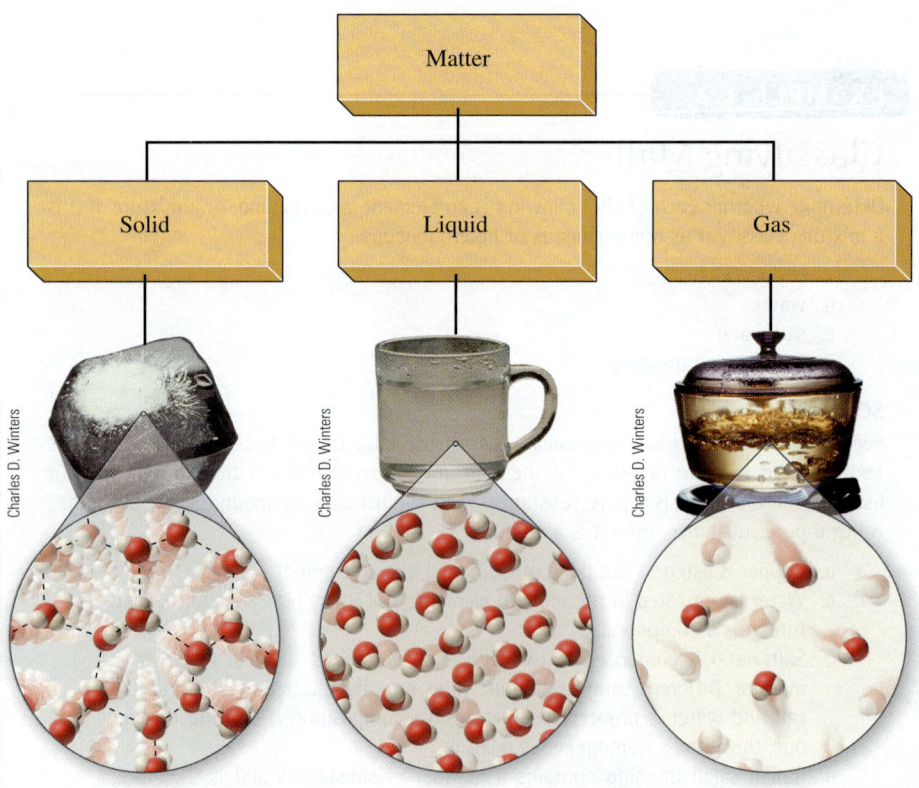

FIGURE 8 The phases of matter. Matter can either be a solid, a liquid, or a gas.

> **EXAMPLE 1.3**
>
> ## The Conservation of Mass
>
> A chemist combines sodium and chlorine, and they react to form sodium chloride. The initial masses of the sodium and chlorine were 11.5 and 17.7 grams (g), respectively. The mass of the sodium chloride was 29.2 g. Show that these results are consistent with the law of conservation of mass.
>
> **SOLUTION**
> The sum of the masses of sodium and chlorine is as follows:
>
> $$\underset{\substack{\text{mass} \\ \text{sodium}}}{11.5 \text{ g}} + \underset{\substack{\text{mass} \\ \text{chlorine}}}{17.7 \text{ g}} = \underset{\substack{\text{total} \\ \text{mass}}}{29.2 \text{ g}}$$
>
> The masses of sodium and chlorine add up to the mass of the sodium chloride; therefore, matter was neither created nor destroyed.
>
> **YOUR TURN**
>
> ## The Conservation of Mass
>
> A match is weighed and then burned. The ashes are found to weigh much less. How can this be consistent with the conservation of mass?

For example, if we decompose an 18.0 g sample of water into its constituent elements, we would obtain 16.0 g of oxygen and 2.0 g of hydrogen, a ratio of oxygen to hydrogen of

$$\frac{\text{oxygen}}{\text{hydrogen}} = \frac{16.0 \text{ g}}{2.0 \text{ g}} = 8.0$$

This ratio would be the same regardless of the size of the water sample or where the water sample was obtained. Similarly, all samples of ammonia contain 14.0 g of nitrogen to every 3.0 g of hydrogen, a ratio of nitrogen to hydrogen of 4.67. The composition of each compound is constant.

> **EXAMPLE 1.4**
>
> ## Constant Composition of Compounds
>
> Two samples of water are obtained from two different sources. When the water is decomposed into its constituent elements, one sample of water produces 24.0 g of oxygen and 3.0 g of hydrogen, while the other sample produces 4.0 g of oxygen and 0.50 g of hydrogen. Show that these results are consistent with the law of constant composition.
> For the first sample:
>
> $$\frac{\text{oxygen}}{\text{hydrogen}} = \frac{24.0 \text{ g}}{3.0 \text{ g}} = 8.0$$
>
> For the second sample:
>
> $$\frac{\text{oxygen}}{\text{hydrogen}} = \frac{4.0 \text{ g}}{0.5 \text{ g}} = 8.0$$

The two samples have the same proportions of their constituent elements and therefore are consistent with the law of constant composition.

YOUR TURN

Constant Composition of Compounds

Two samples of sugar are decomposed into their constituent elements. One sample of sugar produces 18.0 g carbon, 3.0 g hydrogen, and 24.0 g oxygen; the other sample produces 24.0 g carbon, 4.0 g hydrogen, and 32.0 g oxygen. Find the ratio of carbon to hydrogen and the ratio of oxygen to hydrogen for each of the samples and show they are consistent with the law of constant composition.

THE ATOMIC THEORY

In 1808 **John Dalton** (1766–1844), a British scientist, used the laws of Lavoisier and Proust, as well as data from his own experiments, to formulate a fundamental theory of matter. Dalton had an unusual ability to synthesize a theory from a variety of different pieces of information. His **atomic theory** had three parts:

1. All matter is composed of indivisible particles called atoms that cannot be created or destroyed.
2. All atoms of a given element are alike in mass and other properties. These properties are unique characteristics of each element and different from other elements.
3. Atoms of different elements combine to form compounds in simple whole-number ratios. For example, the compound water is formed from 2 hydrogen atoms and 1 oxygen atom. The numbers, 2 and 1, are simple whole numbers.

John Dalton, the British scientist who postulated the atomic theory.

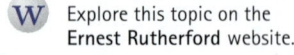

Explore this topic on the **Ernest Rutherford** website.

1.10 The Nuclear Atom

By the late 1800s most scientists were convinced that matter was composed of atoms. However, they did not know what atoms were like. Were they solid spheres like billiard balls, or were they soft and pliable like blueberry muffins? In 1909 the internal structure of the atom was studied by **Ernest Rutherford** (1871–1937). Rutherford examined atomic structure by directing small particles, called alpha particles, at a thin sheet of gold foil (Figure 10). If the atoms in the foil were like soft blueberry muffins, an idea favored by Rutherford, the small particles should shoot right through the gold foil.

Rutherford's experiment produced results that were different from what he expected. The majority of the particles did pass directly through the foil, but some particles experienced a slight deflection as they passed through the foil, and a small fraction bounced right back. These results puzzled Rutherford, who said, "[this is] about as credible as if you had fired a 15 inch shell at a piece of tissue paper and it came back and hit you." What must the structure of the atom be to explain this odd behavior?

Rutherford devised a theory that explained his observed results. He concluded that matter must not be as uniform as it appears; it must contain large regions of empty space accompanied by small regions of very dense material. Rutherford knew that atoms contained some particles with positive charges and others with negative charges. By using these ideas and the results of his experi-

TABLE 1			
The Phases of Matter			
Solid	Incompressible	Fixed volume	Fixed shape
Liquid	Incompressible	Fixed volume	Variable shape
Gas	Compressible	Variable volume	Variable shape

close contact and in fixed locations. Consequently, solid matter is rigid, has a fixed shape, and is incompressible. Good examples of solid matter include ice, copper metal, and diamond. In liquid matter, atoms or molecules are also in close contact, but not in fixed locations—they are free to move around each other. As a result, liquids have a fixed volume and are incompressible, but they don't have a rigid shape. Instead, they flow to assume the shape of their container. Good examples of liquid matter include water, rubbing alcohol, and vegetable oil.

In gaseous matter, atoms or molecules are not in close contact but are separated by large distances. The atoms or molecules are in constant motion and often collide with each and with the walls of their container. Consequently, gaseous matter does not have a fixed shape or a fixed volume but rather assumes the shape and volume of its container. In addition, gaseous matter is compressible. Good examples of gaseous matter include steam, helium, and air. Table 1 summarizes the phases of matter and the properties of each phase.

1.8 The Properties of Matter

Every day, we tell one substance from another based on its properties. For example, we tell the difference between gasoline and water because they smell differently, or between sugar and salt because they taste differently. The characteristics that distinguish a substance and make it unique are its **properties**. In chemistry, we distinguish between **physical properties**, those properties that a substance displays without changing its composition, and **chemical properties**, those properties that a substance displays only when changing its composition. For example, the smell of alcohol is a physical property. When we smell alcohol, it does not change its composition. However, the flammability of alcohol—its tendency to burn—is a chemical property. When alcohol burns, it combines with oxygen in the air to form other substances.

We can also distinguish between two different kinds of changes that occur in matter—physical change and chemical change. When matter undergoes a **physical change**, it changes its appearance but not its composition. For example, when we smell alcohol, some of the alcohol had to vaporize into the air—this is a physical change. When alcohol vaporizes, the alcohol molecules change from the liquid

> **APPLY YOUR KNOWLEDGE**
>
> Water is put on the stove and heated with a natural gas burner. After some time, the water begins to bubble, and steam is given off. Is this a physical or chemical change?
>
> Answer: Physical change. The water changes from liquid to gas, but its composition does not change.

FIGURE 9 The evaporation of alcohol from its container is a physical change. The alcohol does not change its composition upon vaporization.

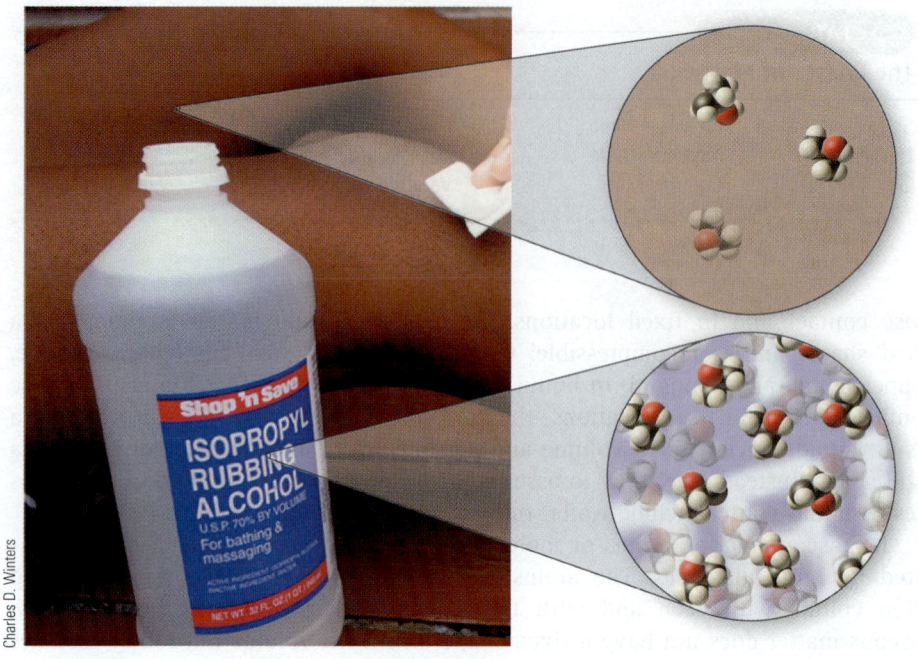

A physical change results in a different form of the same substance; a chemical change results in a completely new substance.

phase to the gas phase, but they remain alcohol (Figure 9). When matter undergoes a **chemical change**, on the other hand, it changes its composition. When alcohol burns, for example, it undergoes a chemical change.

It is not always easy to tell if a change is physical or chemical. In general, changes in the phase of a substance, such as melting or boiling, or changes that are only in appearance, such as cutting or bending, are always physical changes. Chemical changes, on the other hand, often emit or absorb heat or light or result in a color change of the substance.

1.9 The Development of the Atomic Theory

As we have seen, Democritus was the first person to suggest that matter was ultimately composed of atoms. However, the atomic theory was not completely stated and accepted until the early 1800s. The laws of conservation of mass and constant composition both led to the atomic theory.

THE CONSERVATION OF MASS

The mass of something is a measure of the quantity of matter within it. The difference between mass and weight is described in Section 2.3.

In 1789 Antoine Lavoisier published a chemical textbook titled *Elementary Treatise on Chemistry*. Lavoisier is known as the father of modern chemistry because he was among the first to study chemical reactions carefully. As we saw previously, Lavoisier studied combustion, and by burning substances in closed containers, he was able to establish the law of conservation of mass, which states that matter is neither created nor destroyed in a chemical reaction.

A second French chemist, **Joseph Proust** (1754–1826), established the **law of constant composition**, which states the following:

> All samples of a given compound have the same proportions of their constituent elements.

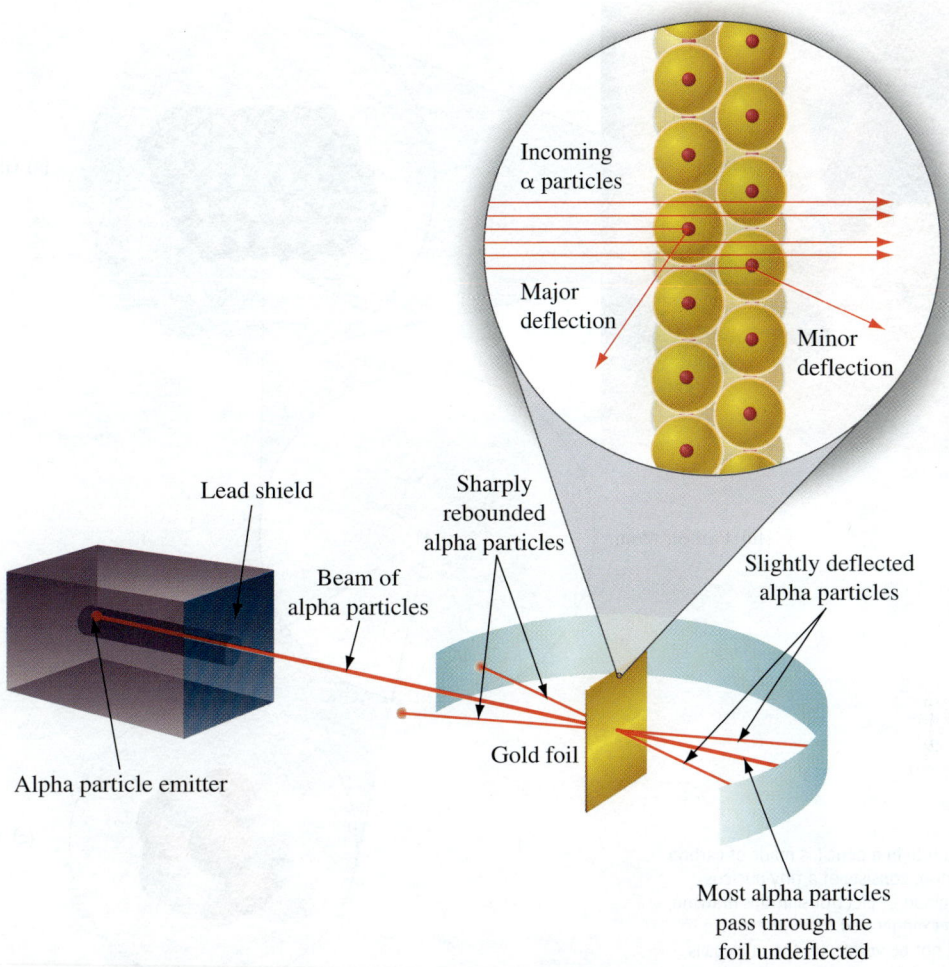

FIGURE 10 Rutherford's gold foil experiment. A beam of alpha particles is directed at a thin sheet of gold foil. Some of the particles pass through the foil undeflected. Others are deflected a small amount. A tiny fraction bounce back in the direction they came from. (Source: Adapted from *General, Organic, and Biological Chemistry*, by David G. Lygre. Copyright © 1995 Brooks/Cole.)

ment, he proposed the **nuclear theory** of the atom (Figure 11), with three basic tenets:

1. Most of the mass and all the positive charge of the atom are contained in a small space called the nucleus.
2. Most of the volume of the atom is empty space occupied by tiny negatively charged electrons.
3. There are as many negatively charged electrons outside the nucleus as units of positive charge inside the nucleus, so that the atom is electrically neutral.

Rutherford's nuclear theory was extremely successful and is still valid today. To get an idea of the structure of the atom, imagine that this dot · is the nucleus of an atom—the average distance to an electron would be 6 feet.

Later work by Rutherford and others demonstrated that the nucleus of the atom consists of two kinds of particles: positively charged particles called **protons** and neutral particles called **neutrons**. The number of protons in the nucleus of a

Protons, neutrons, and electrons are discussed in more detail in Chapter 3.

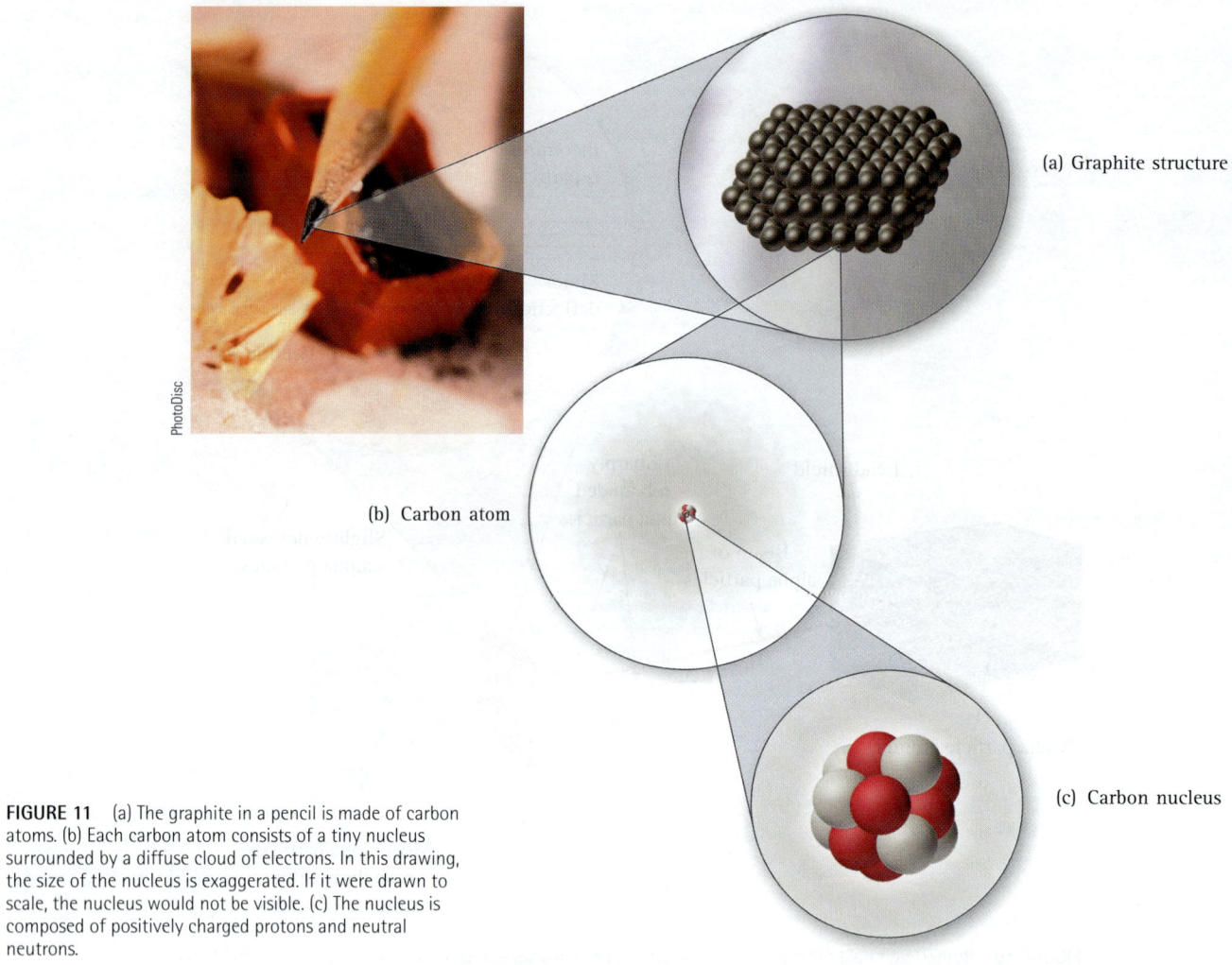

FIGURE 11 (a) The graphite in a pencil is made of carbon atoms. (b) Each carbon atom consists of a tiny nucleus surrounded by a diffuse cloud of electrons. In this drawing, the size of the nucleus is exaggerated. If it were drawn to scale, the nucleus would not be visible. (c) The nucleus is composed of positively charged protons and neutral neutrons.

neutral atom is always equal to the number of electrons outside the nucleus. Although the number of neutrons in the nucleus is often close to the number of protons, no simple rule predicts the relative number of neutrons and protons. The dense nucleus contains over 99% of the mass of the atom. Like water droplets that make up the volume of a cloud, tiny electrons make up most of the volume of the atom. However, the electrons have almost no mass in comparison with the protons and neutrons that compose the nucleus.

Although everyday experience suggests that matter is solid and uniform, experiments since Rutherford's time validate that it is not. If atoms were solid nuclear material and not primarily empty space, a single grain of sand would weigh 10 million pounds. Astronomers believe there are some places in the universe where the structure of the atom has broken down to form "solid" matter. Black holes and neutron stars are two such examples of these very dense forms of matter. The large mass and small size associated with black holes cause intense gravitational fields that attract all things in their general vicinity, including light. As a result, black holes neither reflect nor emit light, making them truly "black."

THE MOLECULAR REVOLUTION

Seeing Atoms

A little over two lifetimes ago the atomic theory was broadly accepted for the first time in history (in part because of John Dalton). Today we can "take pictures" of them. In 1986 Gerd Binnig of Germany and Heinrich Rohrer of Switzerland shared the Nobel Prize for their discovery of the scanning tunneling microscope (STM), a microscope that can image and move individual atoms and molecules. Figure 12 shows "NANO USA" written with 112 individual molecules. The ability to see and move individual atoms has created fantastic possibilities. It may be possible some day to construct microscopic machines one atom at a time, machines so tiny that they are invisible to the naked eye. Yet these machines could perform amazing tasks such as move through the bloodstream to destroy viruses or bacteria. We will learn more about this technology, still in its infancy, in Chapter 19. It is an understatement to say that Dalton's atomic theory provides the foundation for our understanding of the chemical world. Two lifetimes ago it was a theory with little evidence; today the evidence for the existence of atoms is overwhelming.

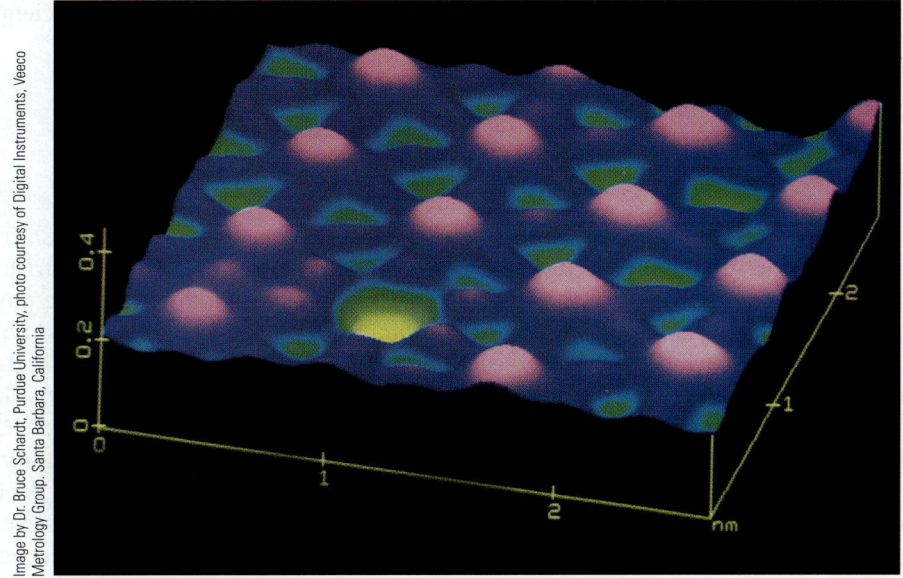

An STM image of iodine atoms adsorbed on a platinum metal surface. The large pink spheres are iodine atoms, and the small purple-pink lumps in hexagonal patterns are platinum atoms. The large yellow "hole" is where an iodine atom is missing from the regular array of atoms. (Source: Data from Dr. Bruce Schardt, Purdue University.)

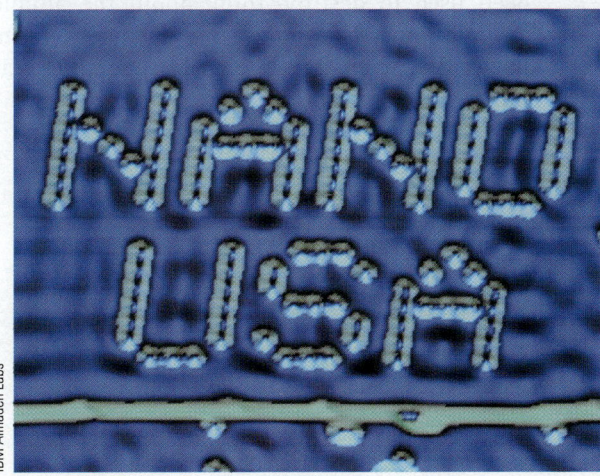

FIGURE 12 "NANO USA" written with 112 individual carbon monoxide molecules and imaged using an STM at IBM Corporation.

 Explore this topic on the **Scanning Tunneling Microscope** website.

Chapter Summary

MOLECULAR CONCEPT

Chemistry is the science that examines the molecular reasons for macroscopic phenomena (1.2). It uses the scientific method, which emphasizes observation and experiment, to explore the link between our everyday world and the world of molecules and atoms (1.3).

As early as 600 B.C., people wondered about the underlying reasons for the world and its behavior. Several Greek philosophers, including Plato, Democritus, Thales, Empedocles, and Aristotle, believed that reason was the primary way to unravel the mysteries of nature. They made some progress in understanding the natural world and introduced fundamental ideas such as atoms and elements (1.4). Alchemy, the predecessor of chemistry, flourished in the Middle Ages and contributed to chemical knowledge, but because of its secretive nature, knowledge was not efficiently propagated, and progress came slowly (1.5).

In the 16th century, scientists began to focus on observation and experiment as the key to understanding the natural world (1.6). Books written by Copernicus and Vesalius exemplify this change of perspective and mark the beginning of the scientific revolution. These books were followed by relatively rapid developments in our understanding of the chemical world. In the 17th century, Boyle initiated a scheme that can be used to classify matter according to its composition (1.7, 1.8). Lavoisier formulated the law of conservation of mass, and Proust, the law of constant composition. In the early 19th century, John Dalton built on these laws to develop the atomic theory (1.9). In the early 20th century, about one hundred years ago, Rutherford examined the internal structure of the atom and found it to be mostly empty space with a very dense nucleus in the center and negatively charged electrons moving around it (1.10). The foundations of modern chemistry were laid.

SOCIETAL IMPACT

Understanding chemistry deepens our understanding of the world and our understanding of ourselves because all matter, even our own brains and bodies, are made of atoms and molecules (1.2).

The Greek philosophers produced ideas about the nature of world that are still valid. However, the Greeks did not emphasize observation and experimentation to the extent that we do today (1.4). Consequently, the development of modern science is a relatively recent phenomena. Our society has profoundly changed after only 450 years of scientific progress.

Major societal changes resulted from this shift in viewpoint (1.7–1.10). Science began to flourish as we learned how to understand and control the physical world. Technology grew as we applied the knowledge acquired through the scientific method to improve human lives. Think about your own life—how would it be different without technology? With these changes, however, came the responsibility to use knowledge and technology wisely—only society as a whole can do that. The power that science bestowed upon us has been used to improve society—think about advances in medicine—but it has also been used to destroy—think about the atomic bomb or pollution. How do we, as a society, harness the power that science has given for good while avoiding the negative effects?

Chemistry on the Web

For up-to-date URLs, visit the text website at **http://chemistry.brookscole.com/tro3**

- Interactive Periodic Table
  http://www.chemistry.org/portal/a/c/s/1/acsdisplay.html?DOC=sitetools%5Cperiodic_table.html#
- Scanning Tunneling Microscope
  http://www.lbl.gov/Science-Articles/Archive/STM-under-pressure.html
  http://www.inano.dk/sw2249.asp
- Ernest Rutherford
  http://www.rutherford.org.nz/

Key Terms

alchemy	Dalton, John	law of conservation of mass	properties
Aristotle	Democritus	law of constant composition	protons
atom	element	liquid	Proust, Joseph
atomic theory	Empedocles	mixture	pure substance
Boyle, Robert	experiment	molecule	Rutherford, Ernest
chemical change	Galilei, Galileo	neutrons	scientific method
chemical properties	gas	nuclear theory	scientific revolution
chemical reaction	heterogeneous mixture	observation	solid
chemistry	homogeneous mixture	physical change	Thales
compound	hypothesis	physical properties	theory
Copernicus, Nicholas	Lavoisier, Antoine	Plato	Vesalius, Andreas

Exercises

Assess your understanding of this chapter's topics with an online chapter quiz at **http://chemistry.brookscole.com/tro3**.

QUESTIONS

1. What is meant by the statement "macroscopic observations have molecular causes"? Give two examples.
2. Why should nonscience majors study science?
3. Explain the following observations on a molecular level:
   a. A brightly colored carpet appears faded in an area near a large window.
   b. A teaspoon of salt dissolves when stirred into water.
4. Define chemistry.
5. Describe the scientific method.
6. What is the difference between a law and a theory?
7. How are science and art similar? How are they different?
8. What is different about the Greek philosophers' approach to scientific knowledge and the approach taken today?

9. Match each of the people in column A with their contribution to scientific knowledge in column B.

A	B
Galileo	conservation of mass
Democritus	all things are water
John Dalton	Inquisition
Andreas Vesalius	the nuclear atom
Empedocles	Sun-centered universe
Joseph Proust	human anatomy
Copernicus	the atomic theory
Ernest Rutherford	constant composition
Thales	*atomos*
Antoine Lavoisier	four basic elements
Robert Boyle	criticized idea of four Greek elements

10. What were the two main pursuits of alchemy? What were the contributions of alchemy to modern chemistry?
11. When did the scientific revolution begin? What events signaled its beginning?
12. What is the difference between an element and a compound? Give two examples of each.
13. What is the difference between a pure substance and a mixture? Give two examples of each.
14. What is the difference between a homogeneous mixture and a heterogeneous mixture? Give one example of each.
15. Explain the differences between a solid, a liquid, and a gas based on the atoms or molecules that compose them.
16. Match each term in the left column with three terms from the right column. You may use terms in the right column more than once.

solid	incompressible
liquid	compressible
gas	fixed volume
	fixed shape
	variable shape
	variable volume

17. What is the atomic theory?
18. Describe Rutherford's gold foil experiment.
19. Describe the structure of the atom as explained by Rutherford. How was this picture consistent with the outcome of his gold foil experiment?
20. What is a black hole?

PROBLEMS

21. Classify each of the following as an observation or a law.
    a. Two grams of hydrogen combine with sixteen grams of oxygen to form eighteen grams of water.
    b. Chlorine and sodium readily combine in a chemical reaction that emits much heat and light.
    c. The properties of elements vary periodically with the mass of their atoms.
22. Classify each of the following as a law or a theory.
    a. All matter is composed of atoms.
    b. When the temperature of a gas is increased, the volume of the gas increases.
    c. Gases are composed of particles in constant motion.
23. To the best of your knowledge, classify each of the following as an element, a compound, or a mixture. If it is a mixture, classify it as homogeneous or heterogeneous.
    a. silver coin
    b. air
    c. coffee
    d. soil
24. To the best of your knowledge, classify each of the following as an element, a compound, or a mixture. If it is a mixture, classify it as homogeneous or heterogeneous.
    a. pure water
    b. copper wire
    c. graphite in a pencil
    d. oil and water
25. Classify each of the following properties as chemical or physical.
    a. the ability of copper metal to bend easily
    b. the tendency of iron to rust
    c. the flammability of paint thinner
    d. the smell of gasoline
26. Classify each of the following properties as chemical or physical.
    a. the tendency of acetone to vaporize easily
    b. the boiling point of water
    c. the tendency of gunpowder to explode upon ignition
    d. the vaporization of dry ice
27. Classify each of the following changes as physical or chemical.
    a. the crushing of salt
    b. the rusting of iron
    c. the burning of natural gas in a stove
    d. the vaporization of gasoline
28. Classify each of the following changes as physical or chemical.
    a. the burning of butane in a lighter
    b. the freezing of water
    c. the bending of a copper rod
    d. the fading of a brightly colored carpet upon excessive exposure to sunlight
29. The burning of gasoline in automobile engines is a chemical reaction. In light of the law of conservation of mass, explain what happens to the gasoline in your car's tank as you drive.
30. A campfire is a chemical reaction involving wood and oxygen from the air. In light of the law of conservation of mass, explain what happens to the mass of the wood during the reaction. How is mass conserved if the wood disappears?

31. Determine if any of the following data sets on chemical reactions are inconsistent with the law of conservation of mass and therefore erroneous.
    a. 6 grams of hydrogen react with 48 grams of oxygen to form 54 grams of water.
    b. 10.0 grams of gasoline react with 4.0 grams of oxygen to form 9.0 grams of carbon dioxide and 5.0 grams of water.
32. Determine if any of the following data sets on chemical reactions are inconsistent with the law of conservation of mass and therefore erroneous.
    a. 8 grams of natural gas react with 38 grams of oxygen gas to form 17 grams of carbon dioxide and 19 grams of water.
    b. 5.7 grams of sodium react with 8.9 grams of chlorine to produce 14.6 grams of sodium chloride.
33. A chemist combines 11 grams of sodium with 14 grams of chlorine. A spectacular reaction occurs and produces sodium chloride. After the reaction, the chemist finds that all the chlorine was used up by the reaction, but 2 grams of sodium remained. How many grams of sodium chloride were formed?
34. A chemist combines 12 grams of hydrogen with 104 grams of oxygen in an explosive reaction that forms water as its sole product. All of the hydrogen reacts but 8 grams of oxygen remain. How many grams of water are formed?
35. Several samples of carbon dioxide are obtained and decomposed into carbon and oxygen. The masses of the carbon and oxygen are then weighed, and the results are tabulated as shown here. One of these results does not follow the law of constant composition and is therefore wrong. Which one?
    a. 12 grams of carbon and 32 grams of oxygen
    b. 4.0 grams of carbon and 16 grams of oxygen
    c. 1.5 grams of carbon and 4.0 grams of oxygen
    d. 22.3 grams of carbon and 59.4 grams of oxygen
36. Several samples of methane gas, the primary component of natural gas, are decomposed into carbon and hydrogen. The masses of the carbon and hydrogen are then weighed, and the results are tabulated as shown here. Which of these does not follow the law of constant composition?
    a. 4.0 grams hydrogen and 12.0 grams carbon
    b. 1.5 grams hydrogen and 4.5 grams carbon
    c. 7.0 grams hydrogen and 17.0 grams carbon
    d. 10 grams hydrogen and 30 grams carbon
37. According to Rutherford's model of the atom, how many electrons would be found in each of the following atoms?
    a. lithium, which has 3 protons in its nucleus
    b. magnesium, which has 12 protons in its nucleus
38. According to Rutherford's model of the atom, how many electrons would be found in each of the following atoms?
    a. chlorine, which has 17 protons in its nucleus
    b. calcium, which has 20 protons in its nucleus

POINTS TO PONDER

39. There are many political issues that involve scientific data and information. Can you think of at least five political issues that involve science?
40. Why was the early Greek idea of only four or five basic elements unsatisfactory?
41. Albert Einstein said, "Imagination is more important than knowledge." What do you think he meant by this?
42. In Section 1.3, I suggested that science and art are equally creative enterprises. Do you agree or disagree? Why?
43. If matter is mostly empty space, what are you touching when your finger touches this book?
44. The nuclei of a limited number of atoms are investigated to determine the number of protons and neutrons in each. The following table summarizes the results:

    | helium | 2 protons, 2 neutrons |
    | carbon | 6 protons, 6 neutrons |
    | nitrogen | 7 protons, 7 neutrons |

    a. Form a scientific law based on these limited measurements.
    b. What other measurements could be carried out to further confirm this law?
    c. Devise a theory that would explain your law.
    d. How would the following results affect your theory and your law?

    | uranium | 92 protons, 143 neutrons |
    | chromium | 24 protons, 28 neutrons |

45. In Section 1.5, I mentioned that the goals of alchemy—turning substances into gold and producing immortality—appear misdirected by today's standards. In what ways have we as a society abandoned these goals today? In what ways have we not?

Feature Problems and Projects

46. Based on the molecular views shown for each of the following substances, classify them as an element, a compound, a homogeneous mixture, or a heterogeneous mixture:

   a.
   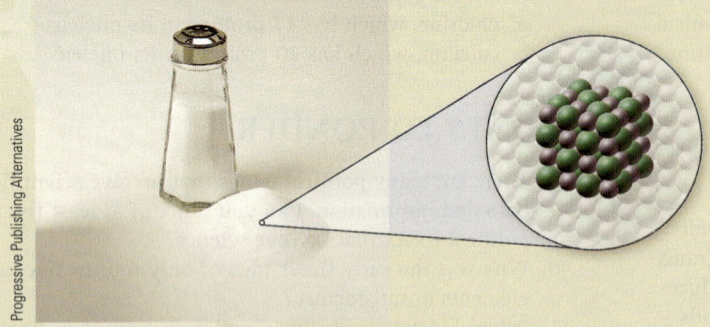

   b.
   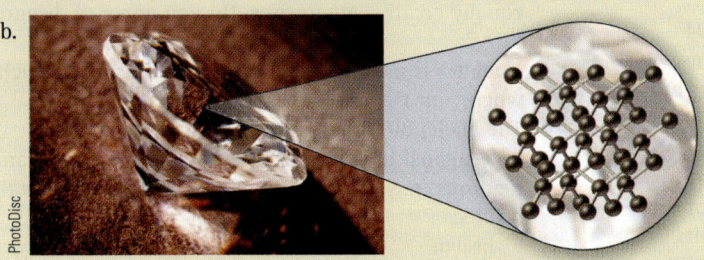

   c.

   d.
   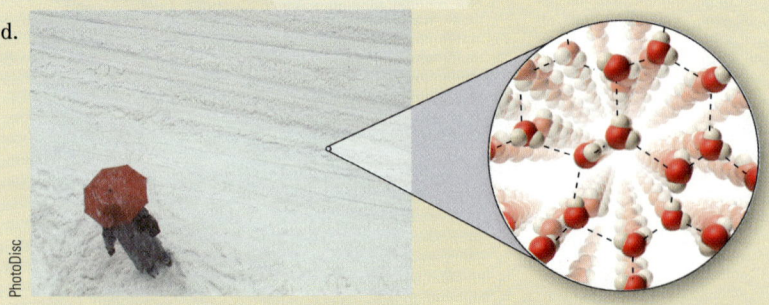

47. Read the section on page 5 entitled "Why Should Nonscience Majors Study Science?" Do you agree or disagree? Your institution probably has a science requirement as part of its general education requirements. Suppose your institution was reevaluating this requirement. Write a brief letter to the chief academic officer of your institution either in favor of the current science requirement or against it. Be persuasive in your reasons.

48. Read The Molecular Revolution box in this chapter on seeing atoms. How is the technology explained in the box a confirmation of the atomic theory? Does the technology "prove" the atomic theory?

 49. Pick an element in the periodic table from the inside front cover of this book. Use a search engine such as Google to find any one of a number internet sites featuring interactive periodic tables (see for example **http://www.chemicalelements.com**). Find out as much as you can about your element and write a paragraph summarizing the information.

2 The Chemist's Toolbox

The language of mathematics reveals itself unreasonably effective in the natural sciences ... a wonderful gift which we neither understand nor deserve.

—Eugene Paul Wigner

1 In this chapter, you will learn how to use some chemists' tools—the hammers, wrenches, and screwdrivers of chemistry. Just as a carpenter learns to use a hammer to drive nails and a screwdriver to build a cabinet, so you must learn to use the tools of measurement and problem solving to understand chemical principles and build chemical knowledge. The ability to be precise and to assign numbers to measurements gives science much of its power. How much you know about a physical or chemical process is often related to how well you can measure some aspect of it.

QUESTIONS FOR THOUGHT

- Why is measurement important?
- How do we write big and small numbers compactly?
- What units should we use in reporting measurements?
- How do we convert between different units?
- How do we read and interpret graphs?
- How do we solve problems in chemistry?
- What is density?

2.1 Curious About Oranges

As children, we probably asked *why* more times than our parents cared to answer. Why is the sky blue? Why do objects fall down and not up? Why are oranges sweet? As we got older, we began to accept things as they are—unfortunately, we often stopped asking such questions. Most good scientists, however, have an insatiable curiosity—they *always* want to know why. The importance of curiosity in science cannot be overstated. If early scientists had not cared about why things were as they were, science would not exist. Therefore, as a student of science, the first tool you must develop is your curiosity. You must relearn what you knew well as a child—how to ask the question *why*.

As an example of scientific thinking and the tools involved in science, consider an orange tree that produces oranges with different levels of sweetness. Suppose we become curious about why some oranges are sweeter than others. To study the problem using the scientific method (Section 1.3), we begin with observation. We must observe or measure the sweetness of oranges as well as other properties that may be related to sweetness. For example, we might begin by measuring the sweetness, size, and color of oranges. How would you measure an orange's size? Level of sweetness? Color? The measurements must be accurate. They must also use a standard *unit of measurement*. We cannot merely classify the oranges as large or small, sweet or sour, orange or green. We must differentiate between small differences in these quantities.

Why are oranges sweet?

After completing our measurements, we might have to perform some calculations, perhaps converting our measurements from one set of units to another. Then, we would tabulate our results in a table or graph, which allows us to see trends and correlations between properties. For example, the graph shown in Figure 1 shows a hypothetical correlation between the level of sweetness and the size of the orange. We might summarize this trend with a general statement such as, "In oranges, sweetness increases with increasing size." If this statement had no exceptions, it would be a scientific law.

The next step in the scientific method is to devise a hypothesis explaining why the bigger oranges are sweeter. For example, we might propose that oranges grow in response to a particular substance secreted by the tree. This substance might determine both sweetness and size. We could then embark on an entire

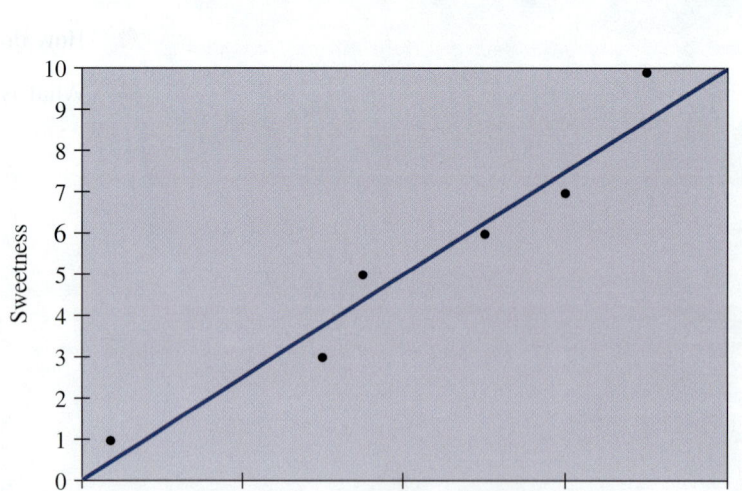

FIGURE 1 In this graph, we have plotted the sweetness of oranges, on a scale from 1 to 10, as a function of their diameters in inches.

new set of experiments to test our hypothesis. We might try to isolate the responsible substance, for example, and measure its levels in the tree and in the fruit. Positive results would validate our hypothesis. The preceding example illustrates the scientific method and its tools. In this chapter we will focus on these tools, especially measurement, unit conversions, graphing, and problem solving.

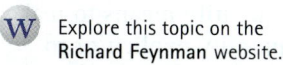
Explore this topic on the **Richard Feynman** website.

2.2 Measurement

Everybody measures. When was the last time you measured the length of a table? or the temperature of pool water? or your weight? The ability to give a quantity a number allows us to go beyond merely saying that this object is hot and that

---

MOLECULAR THINKING

Feynman's Ants

Richard Feynman, a Nobel prize–winning physicist, relates an interesting story about ants in his book *"Surely You're Joking, Mr. Feynman!"*:

> In my [dorm] room at Princeton I had a bay window with a U-shaped windowsill. One day some ants came out on the windowsill and wandered around a little bit. I got curious as to how they found things. I wondered, how do they know where to go? Can they tell each other where food is, like bees can? Do they have any sense of geometry? This is all amateurish; everybody knows the answer, but *I* didn't know the answer, so the first thing I did was to stretch some string across the U of the bay window and hang a piece of folded cardboard with sugar on it from the string. . . .
>
> Next, I made a lot of little strips of paper and put a fold in them, so I could pick up ants and ferry them from one place to another. I put the folded strips of paper in two places: Some were by the sugar (hanging from the string), and the others were near the ants. . . . I sat there all afternoon, reading and watching, until an ant happened to walk onto one of my little paper ferries. Then I took him over to the sugar. After a few ants had been ferried over to the sugar, one of the ants [at the sugar] accidentally walked onto one of the ferries nearby, and I carried him back.
>
> I wanted to see how long it would take the other ants to get the message to go to the "ferry terminal." It started slowly, but rapidly increased until I was going mad ferrying the ants back and forth.

Here is a grown man playing with ants like a child. Such childlike curiosity, however, can result in a great scientist. Feynman

Richard Feynman (1918–1988), Nobel Prize-winning physicist.

wanted to know why ants traveled as they did. After several other ant experiments Feynman concludes:

> I could demonstrate that the ants had no sense of geometry; they couldn't figure out where something was. If they went to the sugar one way, and there was a shorter way back, they would never figure out the short way.
>
> It was also pretty clear . . . that the ants left some sort of trail. . . . I also found that the trail wasn't directional. If I'd pick up an ant on a piece of paper, turn him around and around, and then put him back onto the trail, he wouldn't know that he was going the wrong way until he met another ant.

Richard Feynman died in 1988 after a long and productive career as a scientist. Without his curiosity, he would not have become one of the most noted scientists of this century.

*Source: Excerpts from Richard Feynman, *"Surely You're Joking, Mr. Feynman!"* (Bantam, New York, 1986).

To *quantify* means to make a quantity explicit or to assign a number to it.

one cold or that this one is large and that one small—it allows us to *quantify* the difference. For example, two samples of water may feel equally hot to our hands, but when we measure their temperatures, one sample has a temperature of 78°F and the other 82°F. By assigning a number to the water temperatures, we can differentiate between small differences.

EXPRESSING UNCERTAINTY IN MEASURED QUANTITIES

Because of the limitations of measuring devices, measurements always involve some uncertainty. For example, consider measuring the diameter of a coin with two different rulers (Figure 2). One ruler has marks every one millimeter, and the other ruler, which is much coarser, has marks every one centimeter. Which ruler has more uncertainty associated with it? Obviously, it is the one with the coarser (or less fine) marks.

Scientists report measured quantities in a way that reflects the uncertainty associated with the measuring device. The general rule for reporting measured quantities is as follows:

> Each digit in a reported quantity is taken to be certain, except the last, which is estimated.

For example, the result of measuring the coin with the ruler marked at every one millimeter (Figure 2a) should be reported as 2.33 centimeters. Note that the last digit is estimated. Another person may have reported the measurement as 2.32 or 2.34 centimeters. Each digit is certain except the last. A measurement with the other ruler (Figure 2b) should be reported as 2.3 centimeters. Again, the last digit is estimated. The different number of digits used to report the diameter of the coin reflects the difference in certainty associated with the two rulers.

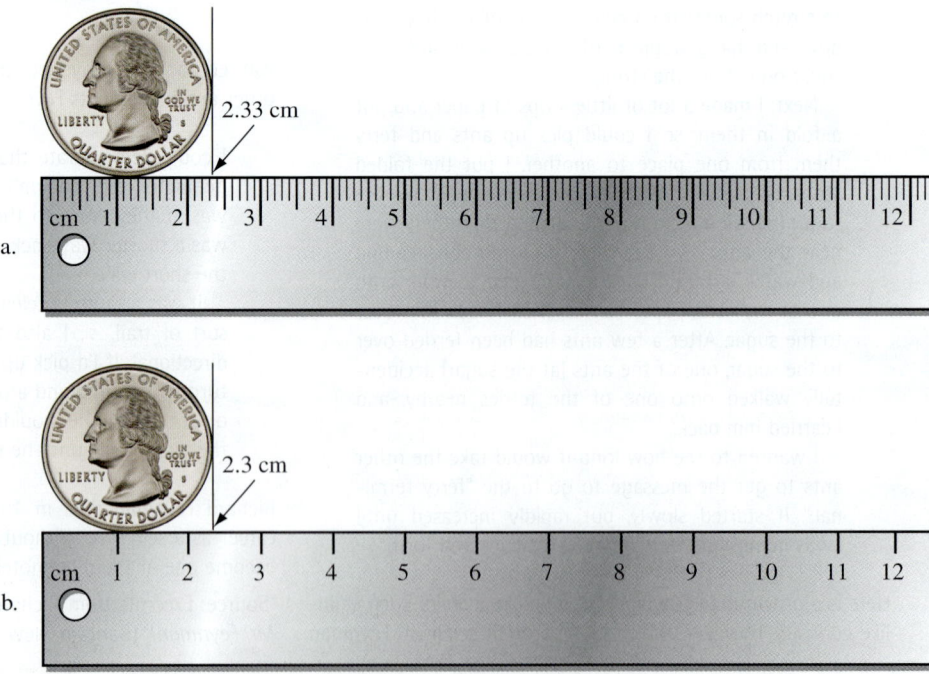

FIGURE 2 The number of digits reported in a measured quantity reflects the uncertainty associated with the measuring device. (Source: PhotoDisc Green/Getty Images)

THE MOLECULAR REVOLUTION

Measuring Average Global Temperatures

The ability to measure quantities carefully and accurately while eliminating systematic error is central to science. Today, careful measurement allows us to track many of the earth's vital signs with unprecedented accuracy. For example, the average temperature of Earth is critical to most forms of agriculture, to determining ocean levels, and to predicting the severity of weather; however, it is difficult to measure. Several years ago, in response to concerns about global warming, two scientists analyzed historical temperature measurements. The task was not simple. The records left behind by previous temperature-measuring stations were often incomplete and inconsistent. For example, a particular temperature-recording station may have changed location, or a town or city may have grown in the vicinity of a measuring station. These changes could lead to an apparent change in temperature that was attributable, not to changes in global climate, but to the changes in location or the surroundings of the measuring station. In addition, some techniques of measuring temperature inherently gave different results than others. Consequently, the researchers had to carefully determine how the measurements were made and discard those measurements where spurious changes occurred.

In a process that took 10 years, these scientists analyzed the temperature records of over 3000 temperature-recording stations throughout the world. Their conclusion was that the average temperature of the planet has increased by 0.6°C in the last century. Note the number of digits in this reported result and remember that the uncertainty is in the last reported digit. The temperature rise could be as much as 0.7°C or as little as 0.5°C, but by reporting their result as 0.6°C, the scientists are saying that the temperature rise is not 1.0°C. Although this temperature increase is small, it is significant. Ocean levels are measurably higher and severe weather—such as tornadoes and hurricanes—seems to be on the rise. In Chapter 9, we examine a theory that could explain the increase in global temperatures.

Global temperatures are critical to agriculture, weather, and ocean levels.

When using measured quantities in calculations, we must be careful to preserve the certainty associated with the measured quantities. For a full treatment of this concept, called *significant figures*, see Appendix 1.

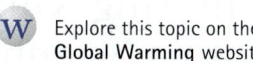 Explore this topic on the **Global Warming** website.

2.3 Scientific Notation

Have you ever had to write a very large or very small number? For example, the radius of the Earth, the distance from its center to its outer edge, is 6,400,000 meters. The radius of a hydrogen atom, on the other hand, is 0.00000000012 meters. These numbers can be written more compactly by using scientific notation:

$$6{,}400{,}000 = 6.4 \times 10^6$$
$$0.00000000012 = 1.2 \times 10^{-10}$$

A number written in scientific notation has two parts: a decimal part, a number between 1 and 10; and an exponential part, 10 raised to an exponent, n.

$$\underbrace{1.2}_{\text{decimal part}} \times \underbrace{10^{-10}}_{\text{exponential part}} \leftarrow \text{exponent } (n)$$

The exponential part means 1 multiplied by 10 n times for *positive n*, and 1 divided by 10 n times for *negative n*.

Adding, subtracting, multiplying, and dividing numbers written in scientific notation are covered in Appendix 4.

Positive n:

$$10^0 = 1$$
$$10^1 = 1 \times 10$$
$$10^2 = 1 \times 10 \times 10 = 100$$
$$10^3 = 1 \times 10 \times 10 \times 10 = 1{,}000$$
$$10^4 = 1 \times 10 \times 10 \times 10 \times 10 = 10{,}000$$

Negative n:

$$10^{-1} = \frac{1}{10} = 0.1$$
$$10^{-2} = \frac{1}{10 \times 10} = 0.01$$
$$10^{-3} = \frac{1}{10 \times 10 \times 10} = 0.001$$
$$10^{-4} = \frac{1}{10 \times 10 \times 10 \times 10} = 0.0001$$

To express a number in scientific notation, follow these steps:

1. Move the decimal point to obtain a number between 1 and 10 (the decimal part).
2. Write the decimal part multiplied by 10 raised to the number of places you moved the decimal.
3. The exponent is *positive* if you moved the decimal point to the left and *negative* if you moved it to the right.

Entering Numbers Written in Scientific Notation into a Scientific Calculator

Most scientific calculators can accommodate numbers in scientific notation. For example, to enter the number 8.9×10^8, use the following keys:

[8] [.] [9] [EE] [8]

On some calculators the "EE" key is labeled "exp" or "EEX". The display will read

8.9   E   8

The number on the left is the decimal part and the number on the right is the exponent. To enter the number 8.9×10^8, use the following keys:

[8] [.] [9] [EE] [+/-] [8]

On some calculators, the "+/-" key may be labeled "CHS" or simply "-". On many calculators "EE" is accessed through the "2nd" function key. Some calculators may differ. See your owner's manual for the instructions on your specific calculator.

A scientific calculator accepts numbers in scientific notation.

For example, suppose we want to convert 37,225 to scientific notation. We move the decimal point to the left four places to obtain a number between 1 and 10. Since we moved the decimal point to the left four places, the exponent is positive 4.

$$37225 = 3.7225 \times 10^4$$

To convert 0.0038 to scientific notation, we move the decimal point to the right three places to obtain a number between 1 and 10. The exponent is then -3.

$$0.0038 = 3.8 \times 10^{-3}$$

EXAMPLE 2.1

Writing Numbers in Scientific Notation

Light travels through space at a speed of 186,000 miles/second. Express this number in scientific notation.

SOLUTION
To obtain a number between 1 and 10, we move the decimal point to the left five decimal places; therefore, the exponent is positive 5.

$$186{,}000 \text{ mi/s} = 1.86 \times 10^5 \text{ mi/s}$$

YOUR TURN

Scientific Notation

Express the number 0.0000023 in scientific notation.

APPLY YOUR KNOWLEDGE

A number written in scientific notation has a negative exponent. In which direction should you move the decimal point to obtain a number a written in decimal notation?

Answer: Move it to the left.

2.4 Units in Measurement

Most measurements require a **unit**, a fixed, agreed-upon quantity by which other quantities are measured. For example, you measure the length of a table in centimeters or inches, and you measure your weight in pounds. Inches, centimeters, and pounds are all examples of units. In *The Hitchhiker's Guide to the Galaxy*, a novel by Douglas Adams, a supercomputer was asked to find the answer to life, the universe, and everything. After many years of computation, the computer replied, "42." Aside from being amused, we are left asking the question, "42 what?" The units were missing from the answer, and therefore much of the meaning was missing as well. When making measurements, we must decide on what *units* to use.

 Explore this topic on the **SI units** website.

TABLE 1
Important SI Standard Units

Quantity	Unit	Symbol
Length	meter	m
Mass	kilogram	kg
Time	second	s
Temperature*	kelvin	K

*Temperature and its units are discussed in Section 9.5.

There are several systems of units in common use. To minimize confusion, scientists around the world agree to use the International System of Units, or SI units, based on the metric system (Table 1).

LENGTH

The SI unit of length is the **meter (m)**. The meter was originally defined as 1/10,000,000 of the distance from the North Pole to the equator along a meridian passing through Paris. In 1983, it was redefined as the distance light travels in a certain period of time, 1/299,792,458 s. When length is expressed in meters, a human has a height of about 2 m, while the size of a dust particle is about 0.0001 m. One meter is slightly longer than a yard (1 meter = 39.4 inches).

PREFIX MULTIPLIERS

The International System of Units has also developed prefix multipliers that are used with the basic SI units. For example, the unit kilometer (km) has the prefix "kilo" meaning 1000 or 10^3.

$$1 \text{ km} = 1000 \text{ m} = 10^3 \text{ m}$$

Similarly, the unit centimeter (cm) has the prefix "centi" meaning 0.01 or 10^{-2}.

$$1 \text{ cm} = 0.01 \text{ m} = 10^{-2} \text{ m}$$

Table 2 summarizes the most commonly used prefix multipliers along with the symbols used to denote them.

TABLE 2
SI Prefix Multipliers

Prefix	Symbol	Multiplier	
giga	G	1,000,000,000	(10^9)
mega	M	1,000,000	(10^6)
kilo	k	1,000	(10^3)
deci	d	0.1	(10^{-1})
centi	c	0.01	(10^{-2})
milli	m	0.001	(10^{-3})
micro	µ	0.000001	(10^{-6})
nano	n	0.000000001	(10^{-9})

MASS

The **mass** of an object is a measure of the quantity of matter within it. The standard SI unit of mass, the **kilogram (kg)**, is the mass of a particular block of platinum and iridium kept at the International Bureau of Weights and Measures, at Sevres, France (Figure 3). Mass is a subtly different quantity than weight. Weight is a measure of the gravitational pull on an object. Our weight on other planets would be different as a result of different gravitational forces. For example, on large planets like Jupiter or Saturn, we would weigh more than on Earth, because the pull of gravity on these planets is greater. However, our mass, the quantity of matter in our bodies, would remain the same. If we express mass in kilograms, an average female adult has a mass of approximately 60 kg, this book has a mass of about 1 kg, and a mosquito has a mass of 0.00001 kg. One kilogram equals 2.20 pounds (lb). A second common unit of mass is the gram (g) defined as 1/1000 of a kilogram. A dime has a mass of about 2 g, and a nickel, about 5 g.

TIME

The standard SI unit of time is the **second (s)**. The second was originally defined as 1/60 of a minute (min), which in turn was defined as 1/60 of an hour (h), which was defined as 1/24 of a day. However, the length of a day varies slightly because the speed of the Earth's rotation is not perfectly constant. As a result, a new definition was required. Today, a second is defined by an atomic standard using a cesium clock (Figure 4).

VOLUME

The **volume** of an object is a measure of the amount of space it occupies. Volume is a *derived unit*—it is composed of other units multiplied together. The units of

FIGURE 3 A duplicate of the international standard kilogram, called kilogram 20, is kept at the National Institute of Standards and Technology near Washington, DC.

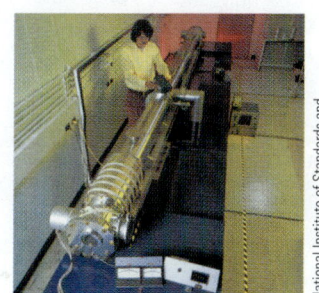

FIGURE 4 The standard of time, the second, is defined using a cesium clock.

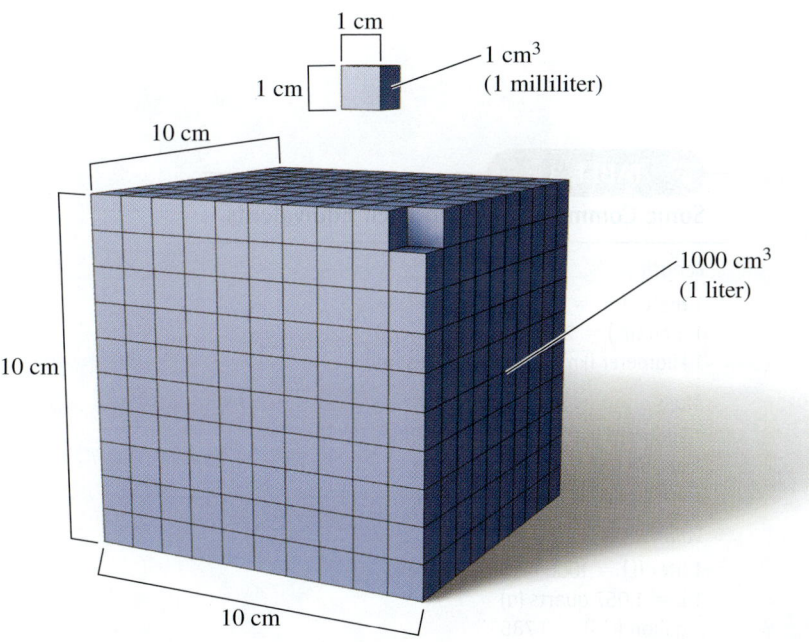

The units of volume are the units of length cubed. This cube measures 10 cm by 10 cm by 10 cm. Its volume is 1000 cm³, which is equivalent to 1 L.

A can of soda holds about 350 mL of liquid.

volume are the units of length raised to the third power. We express volume in units such as cubic meters (m^3) or cubic centimeters (cm^3). A typical bedroom has a volume of 72 m^3, and a can of soda has a volume of 350 cm^3. In chemistry, we also use the units of liter (L) and milliliter (mL). A gallon of milk contains 3.78 L, and a can of soda contains 350 mL. The cubic centimeter and milliliter are equal (1 cm^3 = 1 mL).

2.5 Converting Between Units

When using measured quantities, we often must convert a quantity expressed in one unit to another unit. You probably make simple conversions between commonly used units every day. For example, you might have converted 2 feet to 24 inches, or 1.5 hours to 90 minutes. To perform these calculations, you need to know the numerical relationships between units. For example, to convert feet to inches, you must know that 12 inches = 1 foot. To convert hours to minutes, you must know that 60 minutes = 1 hour. Table 3 shows some common units and their equivalents in other units.

When converting a quantity from one unit to another, the units themselves help to determine the correctness of the calculation. Units are multiplied, divided, and canceled like any other algebraic quantity. Consider the following conversion from yards (yd) to inches (in.):

$$4.00 \text{ yd} \times \frac{36 \text{ in.}}{1 \text{ yd}} = 144 \text{ in.}$$

conversion factor

The quantity $\frac{36 \text{ in.}}{1 \text{ yd}}$ is a **conversion factor**, a fraction with the units we are converting *to* on the top and the units we are converting *from* on the bottom. Notice that the unit we are converting from, *yd*, cancels and we are left with the unit we are converting to, *in*.

TABLE 3

Some Common Units and Their Equivalents

Length

1 meter (m) = 39.37 inches (in.)
1 inch (in.) = 2.54 centimeters (cm)
1 kilometer (km) = 0.6214 miles (mi)

Mass

1 kilogram (kg) = 2.20 pounds (lb)
1 pound (lb) = 453.59 grams (g)
1 ounce (oz) = 28.35 grams (g)

Volume

1 liter (L) = 1000 mL = 1000 cm^3
1 L = 1.057 quarts (q)
1 gallon (gal) = 3.785 L

> **EXAMPLE 2.2**
>
> ## Converting Between Units
>
> Convert 22 cm to inches.
>
> **SOLUTION**
> From Table 2, we find that 1 in. = 2.54 cm. We then start with the given quantity (22 cm) and multiply by the conversion factor (written so that centimeters, the unit we are converting *from*, is on the bottom and inches, the unit we are converting *to*, is on the top) to arrive at inches:
>
> $$22 \text{ cm} \times \frac{1 \text{ in.}}{2.54 \text{ cm}} = 8.7 \text{ in.}$$
>
> **YOUR TURN**
>
> ## Converting Between Units
>
> Convert 34.0 cm to inches.

Conversion factors can be constructed from any two quantities known to be equal. In our example, 36 in. = 1 yd, so we constructed the conversion factor by dividing both sides of this equivalence by 1 yd:

$$\frac{36 \text{ in.}}{1 \text{ yd}} = \frac{1 \text{ yd}}{1 \text{ yd}}$$

$$\frac{36 \text{ in.}}{1 \text{ yd}} = 1$$

The quantity is $\frac{36 \text{ in.}}{1 \text{ yd}}$ equal to 1 and will convert a quantity from yards to inches.

Suppose we want to convert the other way, from inches to yards. If we use the same conversion factor, our units do not cancel correctly:

$$144.0 \text{ in.} \times \frac{36 \text{ in.}}{1 \text{ yd}} = \frac{5184 \text{ in.}^2}{\text{yd}}$$

The units in the answer are nonsense. We should also realize that the value of the answer is wrong. We know that 144.0 in. cannot be equivalent to 5184 in.²/yd. To perform this conversion correctly, we must invert the conversion factor:

$$144.0 \text{ in.} \times \frac{1 \text{ yd}}{36 \text{ in.}} = 4.000 \text{ yd}$$

Conversion factors can be inverted because they are always equal to 1 and 1/1 is equal to 1.

For conversions between units, we are usually given a quantity in some unit and asked to find the quantity in a different unit. These calculations often take the following form:

==(quantity given)== × ==(conversion factor(s))== = ==(quantity sought)==

Always start by writing the quantity that is given and its unit. Then multiply by one or more conversion factors, canceling units, to arrive at the quantity that is sought in the desired units.

> **EXAMPLE 2.3**
>
> ### Converting Between Units When Multiple Steps Are Required
>
> A German recipe for making pastry dough calls for 1.2 L of milk. Your measuring cup only measures in cups. How many cups of milk should you use? 4 cups (cp) = 1 quart (q). (Note: Conversion factors like this, where 4 cp is defined as 1 q, are exact. Therefore, they do not limit the number of significant figures in the answer.)
>
> **SOLUTION**
>
> For this calculation, you must make two conversions, one from liters to quarts and a second from quarts to cups:
>
> $$1.2 \, \cancel{L} \times \frac{1.057 \, \cancel{q}}{1 \, \cancel{L}} \times \frac{4 \, cp}{1 \, \cancel{q}} = 5.1 \, cp$$
>
> **YOUR TURN**
>
> ### Converting Between Units When Multiple Steps Are Required
>
> Suppose the recipe in the preceding example called for 1.5 L of milk, and your measuring cup only measured in ounces. How many ounces of milk should you use? 1 cup (cp) = 8 ounces (oz).

2.6 Reading Graphs

To see trends in numerical data, scientists often display that data in graphs. For example, Figure 1 in this chapter is a graph showing the sweetness of oranges as a function of their diameter. We can clearly see from the graph that orange sweetness increases with increasing orange diameter. Graphs are often used in newspapers and magazines to display scientific, economic, and other types of numerical data. Understanding graphs is an important tool, not just for the science student but for any educated reader.

A set of data points can be graphed to emphasize or draw attention to particular changes. For example, suppose I want to convince my son that his allowance increases over the last ten years have been generous. He started out with a $9.00 allowance in 1995 and has received a $0.25 increase every year. I prepared two graphs of his weekly allowance as a function of time, shown in Figure 5. Which graph should I use to convince my son of my generosity? Both graphs show the exact same data, yet the increase appears larger in the second graph. Why? The y-axis in the first graph begins at $0.00, while the y-axis in the second graph begins at $8.00. Consequently the change—from $9.00 in 1995 to $11.50 in 2005—appears larger in the second graph than in the first, and I appear more generous.

In this case, I wanted to exaggerate my allowance increases in a slightly misleading way. However, data are often graphed in this way, not to be misleading, but to better show small changes in a quantity. The range of the axes in any graph can be chosen to accentuate changes, and the first thing to do when reading a graph is to look at the axes and examine the range of each. Do the axes begin at zero? If not, what is the effect on how the data appear?

FIGURE 5 Graphs of weekly allowance increases. The year is graphed on the x-axis and the amount of allowance is on the y-axis.

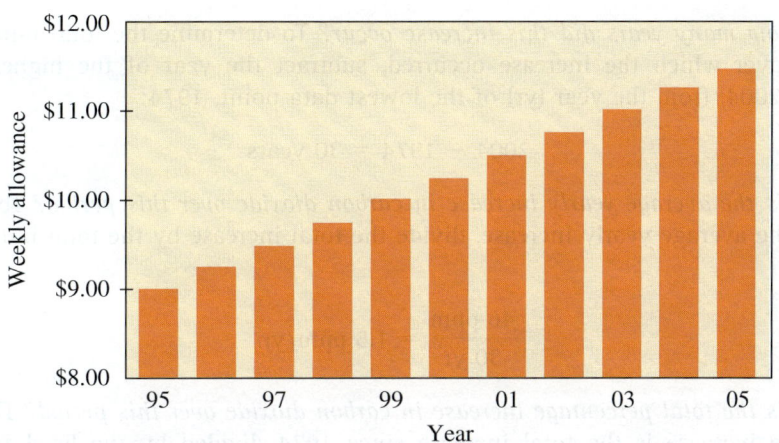

EXTRACTING INFORMATION FROM GRAPHICAL DATA

We can extract important information from a graph. For example, consider the graph in Figure 6 showing atmospheric carbon dioxide levels from 1976 to 2004. Increases in atmospheric carbon dioxide are thought to be at least partially

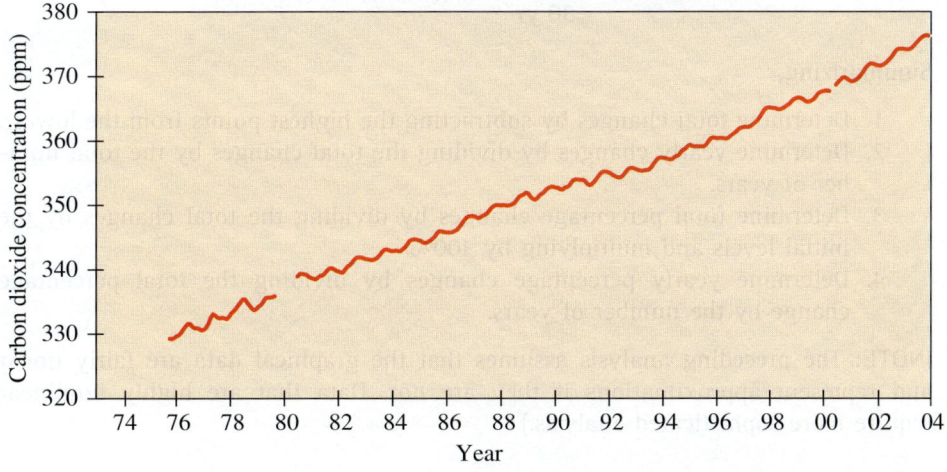

FIGURE 6 Atmospheric carbon dioxide levels as measured at the South Pole, Antarctica. (Source: National Oceanic and Atmospheric Administration, NOAA.)

Global warming and its connection to carbon dioxide levels are discussed in detail in Chapter 9.

ppm is an abbreviation for parts per million, a unit of concentration discussed in detail in Section 12.6.

responsible for increases in global temperatures (global warming), so a number of stations throughout the world monitor the levels of this gas in the atmosphere. The graph shows the measurements from one of these stations. First, look at the axes of the graph. The x-axis is the year, easy enough to understand. The y-axis is concentration or amount of carbon dioxide in the atmosphere. The scale of the y-axis begins at 320 parts per million (ppm) and ranges to 380 ppm. If this scale began at zero, the changes in carbon dioxide concentration would appear smaller. In order to see the increase better, the y-axis begins at 320 ppm, a level near the first data point.

We can extract several pieces of important information from this graph. One is the *total increase in carbon dioxide over the time period*. To determine the total increase, subtract the highest carbon dioxide level—375 ppm in 2004—from the lowest level—329 ppm in 1974:

$$375 - 329 = 46 \text{ ppm}$$

Over how many years did this increase occur? To determine the total number of years over which the increase occurred, subtract the year of the highest data point, 2004, from the year (yr) of the lowest data point, 1974:

$$2004 - 1974 = 30 \text{ years}$$

What is the average yearly increase in carbon dioxide over this period? To determine the average yearly increase, divide the total increase by the total number of years:

$$\frac{46 \text{ ppm}}{30 \text{ yr}} = 1.5 \text{ ppm/yr}$$

What is the total percentage increase in carbon dioxide over this period? The percentage increase is the total increase since 1974 divided by the level in 1974 times 100%:

$$\frac{46 \text{ ppm}}{329 \text{ ppm}} \times 100\% = 14\%$$

What is the average yearly percentage increase in carbon dioxide over this period? The average yearly increase is simply the total percentage increase divided by the total number of years.

$$\frac{14\%}{30 \text{ yr}} = 0.46 \text{ \%/yr}$$

Summarizing,

1. Determine total changes by subtracting the highest points from the lowest.
2. Determine yearly changes by dividing the total changes by the total number of years.
3. Determine total percentage changes by dividing the total changes by the initial levels and multiplying by 100%.
4. Determine yearly percentage changes by dividing the total percentage change by the number of years.

(NOTE: The preceding analysis assumes that the graphical data are fairly linear and represent approximations if they are not. Data that are highly nonlinear require more sophisticated analysis.)

EXAMPLE 2.4

Extracting Information from Graphical Data

The following graph shows the concentration of an atmospheric pollutant, sulfur dioxide, for the period 1988–2000. Sulfur dioxide, a product of fossil fuel combustion, is a main precursor to acid rain. In recent years, because of mandatory reductions legislated by the Clean Air Act, atmospheric sulfur dioxide levels have decreased in the atmosphere as shown by this graph.

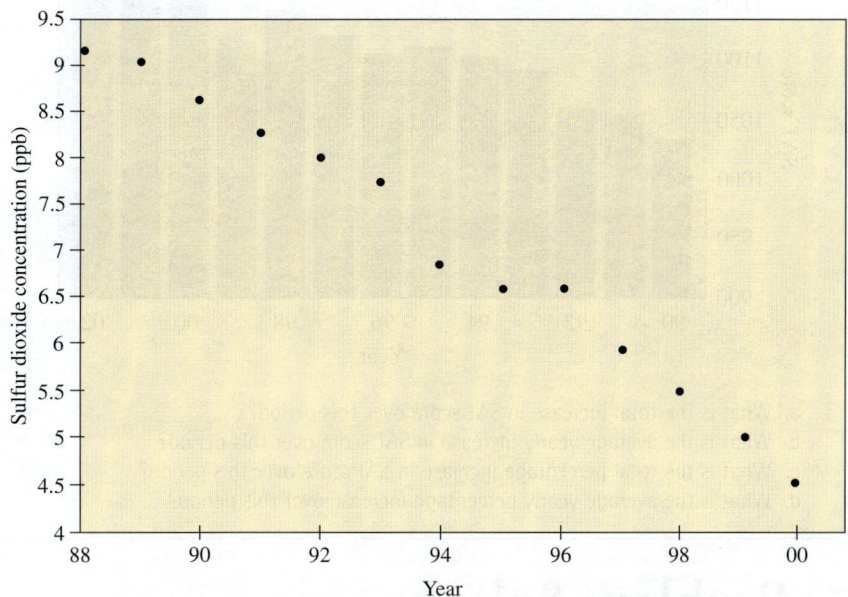

Sulfur dioxide concentration (ppb) (Source: U.S. EPA: 2000 National Air Status and Quality Trends Report.)

a. What is the total decrease in sulfur dioxide concentration over this period?
b. What is the average yearly decrease in sulfur dioxide over this period?
c. What is the total percentage decrease in sulfur dioxide over this period?
d. What is the average yearly percentage decrease over this period?

SOLUTION

a. The total decrease is the difference between the highest sulfur dioxide level on the graph, 9.3 parts per billion (ppb), and the lowest, 4.7 ppb.

$$9.2 \text{ ppb} - 4.6 \text{ ppb} = 4.6 \text{ ppb}$$

b. The average yearly decrease is the total decrease, 4.6 ppb, divided by the total number of years, 12.

$$\frac{4.6 \text{ ppb}}{12 \text{ yr}} = 0.38 \text{ ppb/yr}$$

c. The total percentage decrease is the total decrease, 4.6 ppb, divided by the initial level, 9.3 ppb, and multiplied by 100%.

$$\frac{4.6 \text{ ppb}}{9.3 \text{ ppb}} \times 100\% = 49\%$$

d. The yearly percentage decrease is the total percentage decrease, 49%, divided by the total number of years, 12.

$$\frac{49\%}{12 \text{ yr}} = 4.1\%/\text{yr}$$

> **YOUR TURN**
>
> ### Extracting Information from Graphical Data
>
> The following graph shows the average SAT score for incoming students at Upandcoming University.
>
>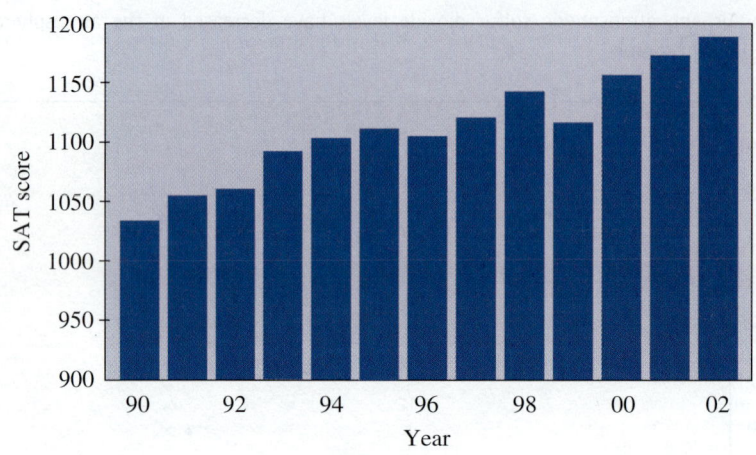
>
> a. What is the total increase in SAT score over this period?
> b. What is the average yearly increase in SAT score over this period?
> c. What is the total percentage increase in SAT score over this period?
> d. What is the average yearly percentage increase over this period?

2.7 Problem Solving

Solving problems requires practice. You can solve many of the problems in this book by paying close attention to units and their conversions (as outlined in Section 2.6). However, sometimes getting started is difficult. How do you interpret a written problem and get started solving it? Many (but not all) problems can be solved with the following strategy:

1. Write all given quantities including their units under the heading **given**.
2. Write the quantity that is sought including its units under the heading **find**.
3. Write all known conversion factors.
4. Start with what is given and multiply by the appropriate conversion factors, canceling units, to obtain the desired units in the result.
5. Round off the answer so that the number of digits in the answer is approximately the same as the number of digits in the given quantities. (For a more accurate treatment of significant figures, see Appendix 1.)

Many problems besides strict unit conversions can be thought of as conversion problems and can therefore be solved with the preceding procedure.

> **EXAMPLE 2.5**
>
> ### Solving Word Problems
>
> A 5.0-km run is being planned through a city. If 8.0 city blocks make up 1.0 mi, how many blocks need to be set apart for the run?

SOLUTION

Begin by setting up the problem as outlined in steps 1 through 3 above.

Given
5.0 km

Find
blocks

Conversion Factors
8.0 blocks = 1.0 mi

Then, following step 4, begin with the given quantity (5.0 km) and multiply by the appropriate conversion factors, canceling units, to obtain the desired units:

$$5.0 \text{ km} \times \frac{0.62 \text{ mi}}{1.0 \text{ km}} \times \frac{8.0 \text{ blocks}}{1.0 \text{ mi}} = 25 \text{ blocks}$$

Your calculator will read 24.8 as the answer; however, round to 25, two digits, because the given quantity, 5.0 km, is expressed to two digits.

YOUR TURN

Solving Word Problems

A swimmer wants to swim 1.50 mi each day in a pool that is 60.0 feet (ft) long. How many pool lengths must she swim? (1.00 miles = 5280 feet)

EXAMPLE 2.6

Solving Word Problems Involving Units Raised to a Power

How many cubic feet are there in a room measuring 80.0 m³?

Given
80.0 m³

Find
Volume in cubic feet (ft³)

Most tables of conversion factors do not include a conversion between m³ and ft³, but they do have a conversion between m and ft (3.28 ft = 1 m). To obtain the correct conversion factor, cube both sides of this equivalence:

$$(3.28 \text{ ft})^3 = (1 \text{ m})^3$$

$$35.3 \text{ ft}^3 = 1 \text{ m}^3$$

Now you can solve the problem:

$$80.0 \text{ m}^3 \times \frac{35.3 \text{ ft}^3}{1 \text{ m}^3} = 2.82 \times 10^3 \text{ ft}^3$$

Your calculator will read 2824 ft; express your answer to three digits because the original quantity was expressed to three digits.

> **YOUR TURN**
>
> ### Solving Word Problems Involving Units Raised to a Power
>
> How many cubic inches are there in 1.00 yd³?

2.8 Density: A Measure of Compactness

An old riddle asks, "Which weighs more, a ton of bricks or a ton of feathers?" The correct answer is neither–they both weigh the same, 1 ton. If you answered bricks, you confused weight with density. The **density** (*d*) of a substance is a measure of how much of its mass is in a given amount of space and is defined as the ratio of its mass (*m*) to its volume (*V*).

$$\text{density } (d) = \frac{\text{mass } (m)}{\text{volume } (V)}$$

Remember that cm^3 and mL are equivalent units.

The units of density are those of mass divided by those of volume and are most commonly expressed in grams per cubic centimeters (g/cm^3) or grams per milliliter (g/mL). The densities of some common substances are shown in Table 4. If a

> **EXAMPLE 2.7**
>
> ### Calculating Density
>
> You find a gold-colored stone in the bottom of a riverbed. To determine if the stone is real gold, you measure its volume (by displacement of water) to be 3.8 cm^3 and its mass to be 29.5 grams. Is the stone real gold?
>
> **SOLUTION**
>
> **Given**
> $V = 3.8\ cm^3$, $m = 29.5$ g
>
> **Find**
> density (to determine if the stone is real gold)
> To find the density, divide the stone's mass by its volume.
>
> $$d = \frac{m}{V} = \frac{29.5\ g}{3.8\ cm^3} = 7.7\ \frac{g}{cm^3}$$
>
> Since the density of gold is 19.3 g/cm^3, the stone is not gold.
>
> **YOUR TURN**
>
> ### Calculating Density
>
> A sample of an unknown liquid has a mass of 5.8 grams and a volume of 6.9 mL. Calculate its density in g/mL.

TABLE 4
Density of Some Common Substances

Substance	Density (g/cm³)
Water	1.0
Ice	0.92
Lead	11.4
Gold	19.3
Silver	10.5
Copper	8.96
Aluminum	2.7
Charcoal, oak	0.57
Glass	2.6
Wood, oak	0.60–0.90
Wood, white pine	0.35–0.50

substance is less dense than water, it will float. If it is denser than water, it will sink. Which of the substances listed in Table 4 will float on water?

DENSITY AS A CONVERSION FACTOR

You can use density as a conversion factor between volume and mass. For example, given that water has a density of 1.00 g/cm³, what is the mass of 350 cm³ of water?

$$350 \text{ cm}^3 \times \frac{1.00 \text{ g}}{1 \text{ cm}^3} = 350 \text{ g}$$

Notice that density converts cm³, a unit of volume, into grams, a unit of mass. Density can also be inverted and used to convert mass to volume.

Pumice is a type of volcanic rock that is less dense than water and therefore floats.

EXAMPLE 2.8

Using Density as a Conversion Factor

A sample of alcohol has a mass of 50.0 kg and its density is 0.789 g/cm³. What is its volume in liters?

Given
50.0 kg

Find
volume in L

Conversion Factors
$d = 0.789$ g/cm³
1000 g = 1 kg
1 mL = 1 cm³
1000 mL = 1 L

In this case, we must convert the mass to grams before we use the density to convert to cm³. Then we convert to mL and finally to L.

$$50.0 \text{ kg} \times \frac{1000 \text{ g}}{1 \text{ kg}} \times \frac{1 \text{ cm}^3}{0.789 \text{ g}} \times \frac{1 \text{ mL}}{1 \text{ cm}^3} \times \frac{1 \text{ L}}{1000 \text{ mL}} = 63.3 \text{ L}$$

YOUR TURN

Density

A sample of iron pellets weighs 23.0 g and has a density of 7.86 g/cm³. How much volume do the pellets occupy in cubic millimeters?

APPLY YOUR KNOWLEDGE

Without doing any calculations, determine if substance A with a density of 1.0 g/cm³ is more or less dense than substance B with a density of 1.0 kg/m³.

Answer: Substance A is denser than substance B. The numerator of the density of substance B is 1000 times greater than the numerator of the density of substance A. However, the denominator is a million times greater (1 m³ = 10⁶ cm³). Therefore, substance B is less dense.

Chapter Summary

MOLECULAR CONCEPT

Scientists must be curious people who naturally ask the question why (2.1). They must then make observations on some aspect of nature, devise laws from those observations, and finally create a theory to give insight into reality. The tools needed for this procedure revolve around measurement (2.2).

When making measurements, we must be consistent in our use of units. Numbers should always be written with their corresponding units, and units guide our way through calculations. The standard SI unit of length is the meter (m); of mass, the kilogram (kg); and of time, the second (s) (2.4).

Scientists often present their measurements in graphs, which reveal trends in data but which must be interpreted correctly (2.6). Many problems in chemistry can be thought of as conversions from one set of units to another (2.5). Density is the mass-to-volume ratio of an object and provides a conversion factor between mass and volume (2.8).

SOCIETAL IMPACT

The ability to measure quantities in nature gives science much of its power. Without this ability, we would probably not have cars, computers, or cable TV today. Neither science nor technology could advance very far without measurement (2.2).

The decision over which units to use is societal. Americans have consistently differed from the rest of the world in using English units over metric units. We are slowly changing, however, and with time, we should be consistent with other nations. For scientific measurements, always use metric units (2.4).

When reading newspapers or magazines, be careful to pay attention to units and to the axes shown in graphs. Clever writers can distort statistical or graphical data, amplifying the changes they want you to see and diminishing the ones they want to hide (2.6).

Chemistry on the Web

For up-to-date URLs, visit the text website at http://chemistry.brookscole.com/tro3

- Richard Feynman
  http://en.wikipedia.org/wiki/Richard_Feynman
- Global Warming
  http://www.giss.nasa.gov/data/update/gistemp/
  http://www.epa.gov/globalwarming/
- SI Units
  http://physics.nist.gov/cuu/Units/

Key Terms

| conversion factor | kilogram (kg) | meter (m) | unit |
| density | mass | second (s) | volume |

Exercises

 Assess your understanding of this chapter's topics with an online chapter quiz at **http://chemistry.brookscole.com/tro3**.

QUESTIONS

1. Why is curiosity an important part of the scientific enterprise?
2. Explain the importance of measurement in science.
3. How is uncertainty reflected in the way that measured quantities are written and reported?
4. What is the difference between reporting the quantity nine inches as 9 inches and 9.00 inches?
5. What is a unit? Why are units important when expressing a measured quantity?
6. List three units for length and give one example for each (i.e., list an object and its length expressed in that unit).
7. List three units for mass (or weight) and give one example for each (i.e., list an object and its mass, or weight, expressed in that unit).
8. List three units for time and give one example for each.
9. List three units for volume and give one example for each.
10. What is a conversion factor? Give two examples of conversion factors and explain what units they convert from and what units they convert to.
11. Why are graphs useful in analyzing and evaluating numerical data? Why is it important to notice the range of the y-axis on a graph?
12. Identify the decimal part, the exponential part, and the exponent of the number 9.66×10^{-5}.
13. What is density? Give two examples of possible units for density.
14. Since oil floats on water, what can you say about the relative densities of oil and water?

PROBLEMS

15. Express each of the following in scientific notation:
    a. 0.0045 g
    b. 23,000 L
    c. 299,790,000 m/s (speed of light)
    d. 0.00000035 m
16. Express each of the following numbers in scientific notation:
    a. 32,667,000 (population of California)
    b. 5,926,467,000 (world population)
    c. 0.00000000007461 m (length of a hydrogen-hydrogen chemical bond)
    d. 0.000015 m (diameter of a human hair)
17. Express each of the following in decimal notation:
    a. 6.4×10^6 m (radius of the Earth)
    b. 7.9×10^{-11} m (radius of a hydrogen atom)
    c. 2.7×10^8 (U.S. population)
    d. 25×10^{-2}
18. Express each of the following in decimal notation:
    a. 6.022×10^{23} (number of carbon atoms in 12.01 grams of carbon)
    b. 2.99×10^8 m/s (speed of light)
    c. 450×10^{-9} m (wavelength of blue light)
    d. 16×10^9 years (approximate age of the universe)
19. The circumference of the Earth at the equator is 40,075 km. Convert this distance into each of the following units:
    a. meters
    b. miles
    c. feet
20. The distance from New York to Los Angeles is 2777 miles. Convert this distance into each of the following units:
    a. kilometers
    b. meters
    c. feet
21. A can of soda contains 12 fluid ounces. What is this volume in mL? (128 fluid ounces = 3.785 L)
22. A laboratory beaker can hold 150 mL. How many fluid ounces can it hold? (128 fluid ounces = 3.785 L)
23. Perform each of the following conversions within the metric system:
    a. 6254 mm to m
    b. 3.28 kg to g
    c. 2566 mg to kg
    d. 0.0256 L to mL
24. Perform each of the following conversions within the metric system:
    a. 4.15 cm to m
    b. 1458.4 g to kg
    c. 18.9 cm to mm
    d. 318 mL to L
25. Perform the following conversions between the English and metric systems:
    a. 25.2 in. to cm
    b. 106 ft to m
    c. 5077 yd to km
    d. 3.1 in to mm
26. Perform the following conversions between the metric and English systems:
    a. 49 cm to in.
    b. 17.2 m to ft
    c. 23.1 km to mi
    d. 1815 m to ft

27. A pond has a surface area of 1244 m². Convert this quantity into each of the following units:
    a. ft²
    b. km²
    c. dm²
28. An orange has a volume of 63 cm³. Convert this quantity into each of the following units:
    a. mm³
    b. in.³
    c. dm³
29. Determine the number of
    a. square meters (m²) in 1 square kilometer (km²)
    b. cubic centimeters (cm³) in 1 cubic foot (ft³)
    c. square feet (ft²) in 1 square yard (yd²)
30. Determine the number of
    a. square centimeters (cm²) in 1 square meter (m²)
    b. cubic inches (in.³) in 1 cubic yard (yd³)
    c. square centimeters (cm²) in 1 square foot (ft²)
31. A sports utility vehicle gets 12 miles per gallon of gas. Calculate this quantity in miles per liter and kilometers per liter.
32. A hybrid electric vehicle (HEV) is a car with both a gasoline powered engine and an electric motor. Some HEVs achieve mileage ratings as high as 75 miles per gallon of gas. Calculate this quantity in miles per liter and kilometers per liter.
33. The following graph shows the concentration of an atmospheric pollutant, carbon monoxide, for the period 1990–2000. As shown by this graph, carbon monoxide levels have decreased in the atmosphere because of mandatory reductions legislated by the Clean Air Act.

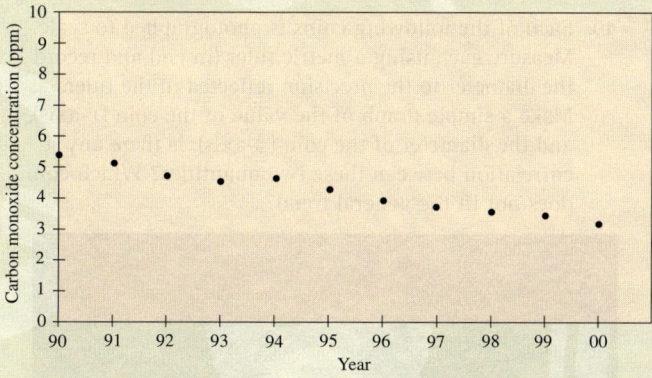

(Source: U.S. EPA: 2000 National Air Status and Quality Trends Report.)

   a. What is the total decrease in carbon monoxide over this period?
   b. What is the average yearly decrease in carbon monoxide over this period?
   c. What is the total percentage decrease in carbon monoxide over this period?
   d. What is the average yearly percentage decrease in carbon monoxide over this period?
34. The following graph shows the historical concentration of atmospheric carbon dioxide in the atmosphere. Carbon dioxide concentrations began increasing in the 1800s as humans started to rely on combustion for energy (carbon dioxide is a product of combustion). The increase in atmospheric carbon dioxide is linked to the increases in global temperatures known as global warming.

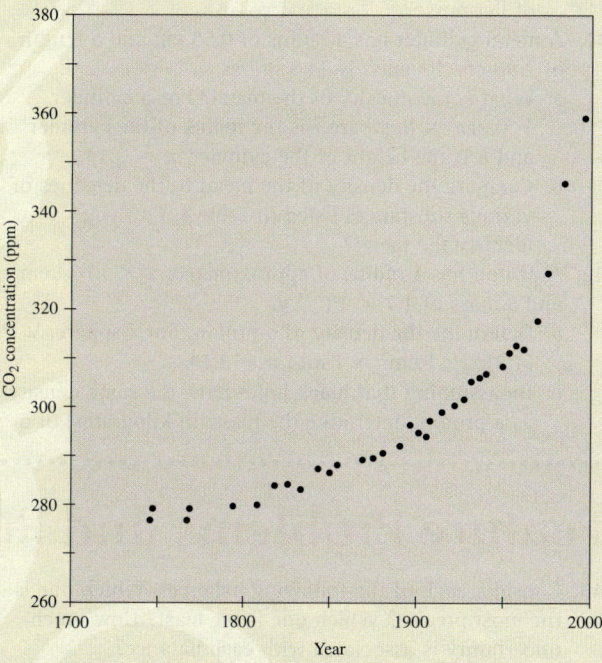

Atmospheric carbon dioxide derived from Siple Station ice core measurements and from levels measured at the South Pole, Antarctica (Source: ORNL/CDIAC65. See credits section at the back of the book.)

   a. What is the total increase in carbon dioxide between 1800 and 2000?
   b. What is the average yearly increase in carbon dioxide over this period?
   c. What is the total percentage increase in carbon dioxide over this period?
   d. What is the average yearly percentage increase in carbon dioxide over this period?
35. A 28.4 cm³ sample of titanium has a mass of 127.8 g. Calculate the density of titanium in g/cm³.
36. A 1.5 cm³ sample of silicon has a mass of 3.5 g. Calculate the density of silicon in g/cm³.

37. A 5.00 L sample of pure glycerol has a mass of 6.30×10^3 g. What is the density of glycerol in g/cm^3?
38. A 3.80 mL sample of mercury has a mass of 51.4 g. What is the density of mercury in g/cm^3?
39. Ethylene glycol (antifreeze) has a density of 1.11 g/cm^3.
    a. What is the mass in g of 417 mL of this liquid?
    b. What is the volume in L of 4.1 kg of this liquid?
40. A thief plans to steal a bar of gold from a woman's purse and replace it with a bag of sand. Assume that the volumes of the gold and the sand are 350 mL each and that the density of sand is 3.00 g/cm^3.
    a. Calculate the mass of each object.
    b. Would the woman notice a difference in the weight of her purse?
41. A metal cylinder has a radius of 0.55 cm and a length of 2.85 cm. Its mass is 24.3 grams.
    a. What is the density of the metal? For a cylinder, $V = \pi r^2 \times h$, where r is the radius of the cylinder and h is the height of the cylinder. $\pi = 3.14$.
    b. Compare the density of the metal to the densities of various substances listed in Table 4. Can you identify the metal?
42. A proton has a radius of approximately 1×10^{-13} cm and a mass of 1.7×10^{-24} g.
    a. Determine the density of a proton. For a sphere, $V = (4/3)\pi r^3$; $1 cm^3 = 1$ mL; $\pi = 3.14$.
    b. By assuming that black holes have the same density as a proton, determine the mass (in kilograms) of a black hole the size of a sand particle. A typical sand particle has a radius of 1×10^{-4} m.

POINTS TO PONDER

43. What did Einstein mean when he said, "The most incomprehensible thing about the world is that it is comprehensible."
44. Which do you think is more valuable in the scientific enterprise, creativity or mathematical ability? Why?
45. Think of something from your everyday life about which you are curious. Design an experiment or set of measurements that might help satisfy your curiosity. What are the possible results? What would they show?
46. Luis Alvarez said, "There is no democracy in [science]. We can't say that some second-guy has as much right to an opinion as Fermi." Fermi was a world-famous scientist who received the 1938 Nobel Prize in physics. What do you think Alvarez meant by this statement? Do you agree?
47. As we saw in this chapter (Section 2.6), numerical data can be presented in ways that call attention to certain changes that the presenter might want emphasized. Can you think of any instances where a politician or other public figure presented data in a way that favored his or her particular viewpoint?

Feature Problems and Projects

48. Consider each of the following balances. Which one is the most precise? Which one is the least? How much uncertainty is associated with each balance?
    a.
    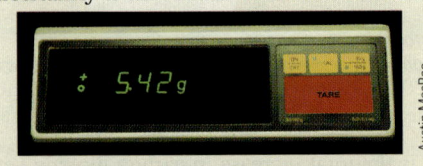
    b.

    c.

49. Each of the following coins is photographed to scale. Measure each using a metric ruler (in cm) and record the diameter to the precision reflected in the ruler. Make a simple graph of the value of the coin (x-axis) and the diameter of the coin (y-axis). Is there any correlation between these two quantities? Which coin does not fit the general trend?

50. Obtain an outdoor thermometer and record the temperature every day for one month. Be sure to record the temperature at the same time every day. Make a graph of your temperature beginning with day 1 and ending at day 30. Are there any trends in the temperature? If there is a trend, calculate the average daily change.

51. Visit the websites listed under global warming in the Chemistry on the Web section at the end of this chapter. Write a short essay explaining the change in global temperatures and the risks associated with that change. Pay special attention to any numerical values given on the websites. How many digits are reported in these values? What does this say about the certainty in the measurements?

3 Atoms and Elements

Nothing exists except atoms and empty space; everything else is opinion.

—Democritus

In this chapter, we will see how everything—you, me, the air we breathe, and the chair you sit on—is ultimately composed of atoms. One substance is different from another because the atoms that compose each substance are different. How are the atoms different? Some substances share similar properties. For example, helium, neon, and argon are all inert (nonreactive) gases. Are their atoms similar? If so, how?

Keep in mind the scientific method and especially the nature of scientific theories as you learn about atoms. You will learn two theories in this chapter—the Bohr theory and the quantum mechanical theory—that model atoms. These models of reality help us to understand the differences among the atoms of various elements, and the properties of the elements themselves. The connection between the microscopic atom and the macroscopic element is the key to understanding the chemical world. Once we understand—based on their atoms—why elements differ from one another, we can begin to understand our world and even ourselves on a different level. For example, we can begin to understand why some atoms are dangerous to the environment or to human life and why others are not.

QUESTIONS FOR THOUGHT

- What composes all matter?
- What makes one element different from another? How do the atoms of different elements differ from one another?
- What are atoms composed of?
- How do we specify a given atom?
- Do similarities between atoms make the elements they compose similar? What are those similarities?
- How do we create a model for the atom that explains similarities and differences among elements? How do we use that model?
- How do we know numbers of atoms in an object? For example, can we calculate the number of atoms in a penny?

3.1 A Walk on the Beach

As we will see in the next chapter, most atoms exist, not as free particles, but as groups of atoms bound together to form molecules.

The exact number of naturally occurring elements is controversial because some elements that were first discovered when they were synthesized were later found in minute amounts in nature. The periodic table shows more than 91 elements. The additional elements do not occur naturally (at least not in any substantial quantities), but have been synthesized by humans.

A walk along the beach on a breezy day provides us with ample opportunity to begin thinking about atoms (Figure 1). As we walk, we feel the wind on our skin and the sand under our feet. We hear the waves crashing, and we smell the salt air. What is the ultimate cause of these sensations? The answer is simple—atoms. When we feel the breeze on our face, we are feeling atoms. When we hear the crash of the waves, we are hearing atoms. When we pick up a handful of sand, we are picking up atoms; and when we smell the air, we are smelling atoms. We eat atoms, we breathe atoms, and we excrete atoms. Atoms are the building blocks of the physical world; they are the tinker toys of nature. They are all around us, and they compose all matter, including our own bodies.

Atoms are unfathomably small. A single sand grain, barely visible to our eye, contains more atoms than we could ever count or imagine. In fact, the number of atoms in a sand grain far exceeds the number of sand grains on the largest of beaches.

If we are to understand the connection between the microscopic world and the macroscopic world, we must begin by understanding the atom. As we learned in Chapter 1, an **atom** is the smallest identifiable unit of an element. There are 91 different elements in nature and consequently 91 different kinds of atoms. As far as we know, these 91 elements are the same ones that make up the entire universe. Figure 2 shows the relative abundances of the most common elements on Earth and in the human body.

Some of the elements in nature, such as helium and neon, are inert; they do not tend to combine with other elements to form compounds. What do their sim-

FIGURE 1 Everything around us, including our own bodies, is ultimately composed of atoms. Most of the atoms are bound together to form molecules, but before we try to understand molecules, we will examine the atoms that compose them.

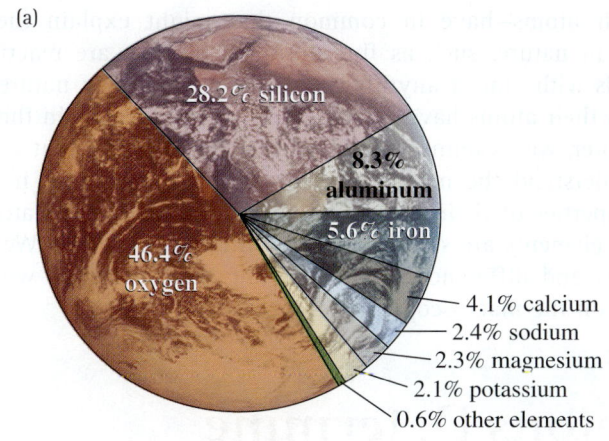

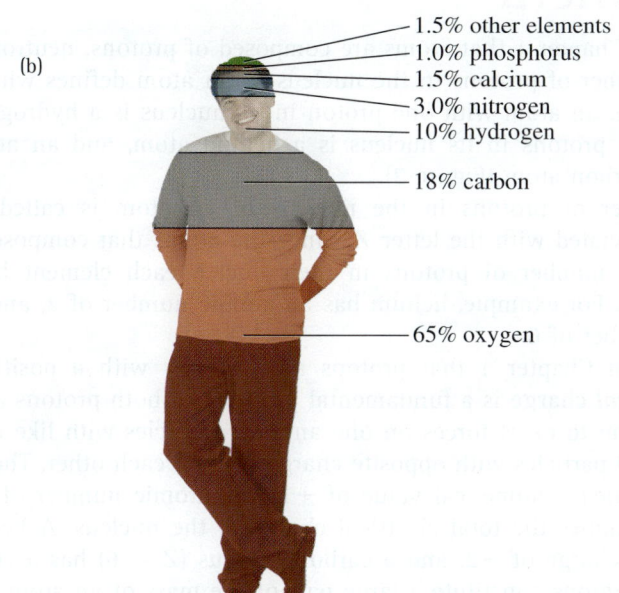

FIGURE 2 The elemental composition by mass of (a) the Earth's crust and (b) the human body.

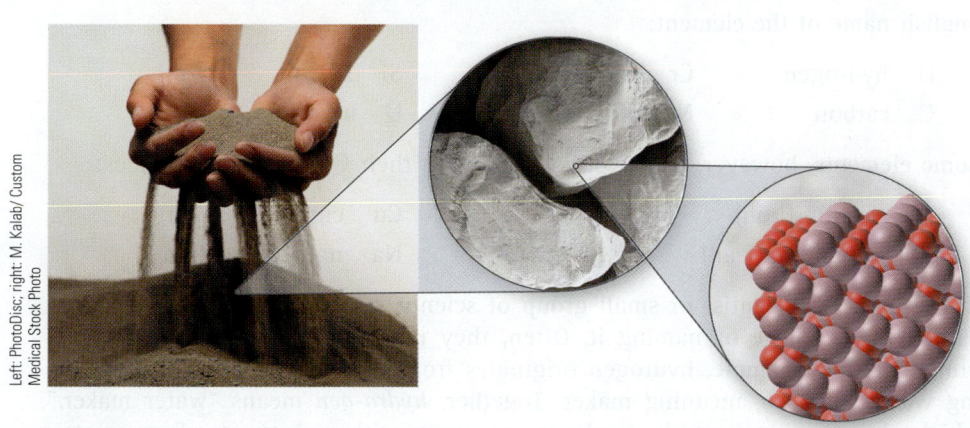

A single grain of sand contains approximately 1×10^{20} atoms.

plest units—their atoms—have in common that might explain their inertness? Other elements in nature, such as fluorine and chlorine, are reactive; they will form compounds with almost anything and are not found in nature as pure elements. What do their atoms have in common that might explain their reactivity?

In this chapter, we examine the elements and the atoms that compose them. We want to understand the macroscopic properties of elements in terms of the microscopic properties of their atoms. We will focus on how the atoms that constitute different elements are similar and how they are different. We will see how these similarities and differences on the atomic scale correlate with similarities and differences on the macroscopic scale.

3.2 Protons Determine the Element

We learned in Chapter 1 that atoms are composed of protons, neutrons, and electrons. The number of protons in the nucleus of an atom defines what element it is. For example, an atom with one proton in its nucleus is a hydrogen atom. An atom with two protons in its nucleus is a helium atom, and an atom with six protons is a carbon atom (Figure 3).

The number of protons in the nucleus of an atom is called the **atomic number**, abbreviated with the letter Z. Since the atoms that compose an element have a unique number of protons in their nuclei, each element has a *unique* atomic number. For example, helium has an atomic number of 2, and carbon has an atomic number of 6.

Recall from Chapter 1 that protons are particles with a positive electrical charge. *Electrical* **charge** is a fundamental property of both protons and electrons that causes them to exert forces on one another. Particles with like charges repel each other, and particles with opposite charges attract each other. The charge of a proton is assigned a numerical value of $+1$. The atomic number of an atom, Z, therefore determines the total electrical charge of the nucleus. A helium nucleus ($Z = 2$) has a charge of $+2$, and a carbon nucleus ($Z = 6$) has a charge of $+6$, for example. Protons constitute a large part of the mass of an atom. The mass of a proton is 1.0 atomic mass units (amu). An **amu** is a unit of mass equivalent to 1/12 the mass of a carbon-12 nucleus and is equivalent to 1.67×10^{-24} g.

We denote elements by their **chemical symbol**, a one- or two-letter abbreviation for the element. For example, we denote hydrogen by the chemical symbol H; helium by He, and carbon by C. Most chemical symbols are based on the English name of the element:

| H | hydrogen | Cr | chromium | Si | silicon |
| C | carbon | N | nitrogen | U | uranium |

Some elements, however, have symbols based on their Greek or Latin names:

| Fe | ferrum (iron) | Cu | cuprum (copper) |
| Pb | plumbum (lead) | Na | natrium (sodium) |

In the past, the scientist or small group of scientists who discovered an element earned the privilege of naming it. Often, they named the element based on its properties. For example, hydrogen originates from the Greek roots *hydro*, meaning water, and *gen*, meaning maker. Together, *hydro-gen* means "water maker," which describes hydrogen's tendency to react with oxygen to form water.

Z = atomic number, the number of protons in the nucleus of an atom. Changing the number of protons changes the element.

Curium was named after Marie Curie, a co-discoverer of radioactivity.

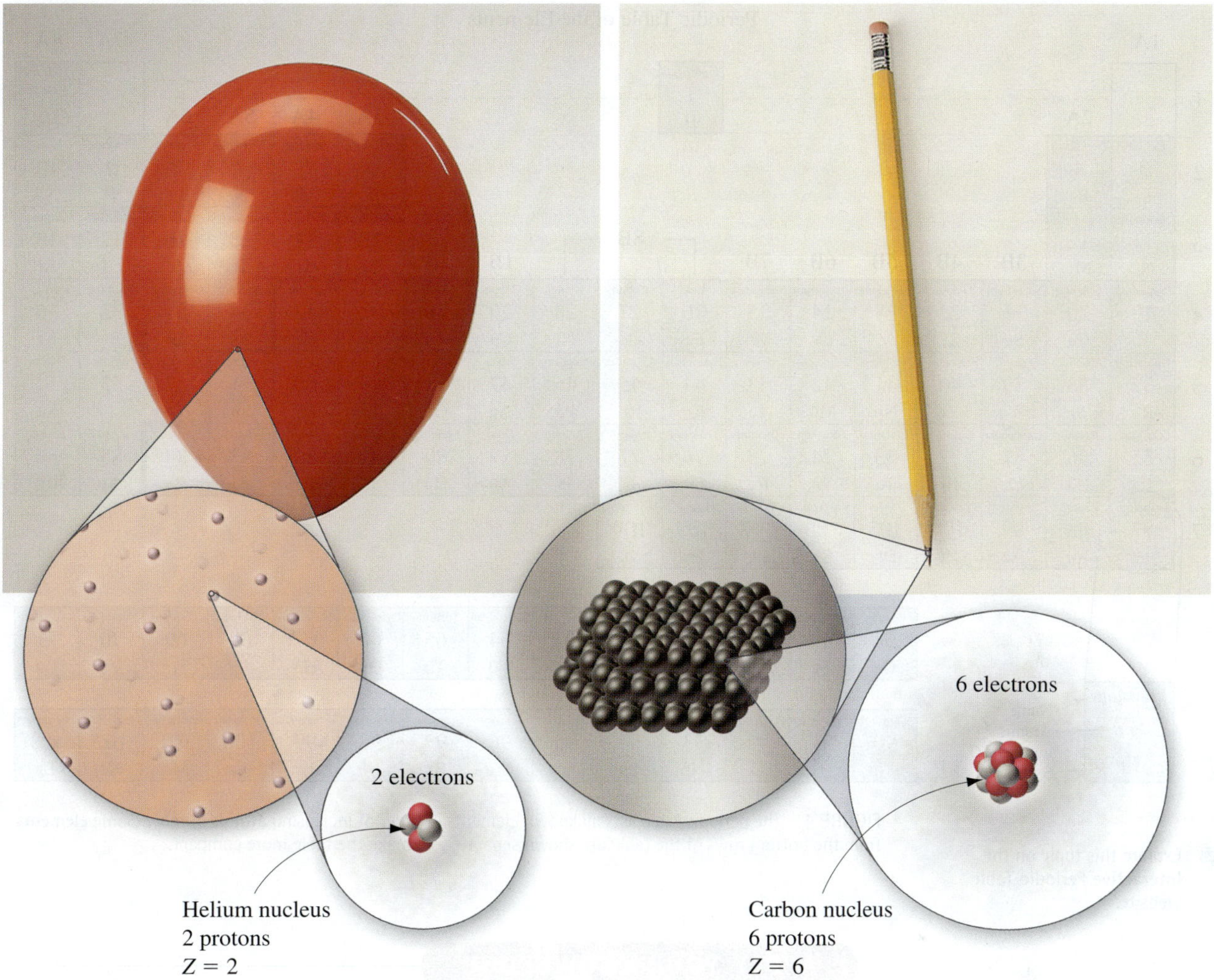

Progressive Publishing Alternatives

FIGURE 3 Helium, which is lighter than air, is often used to cause balloons to float. A helium atom contains two protons in its nucleus ($Z = 2$). Graphite, a form of carbon, is the "lead" in pencils. A carbon atom contains six protons in its nucleus ($Z = 6$).

Chlorine comes from the root *khloros,* meaning greenish yellow, which describes chlorine's color. Other elements were named after their place of discovery:

Eu	europium (Europe)	Am	americium (America)
Ga	gallium (Gallia, France)	Ge	germanium (Germany)
Bk	berkelium (Berkeley, CA)	Cf	californium (California)

Some elements were named after famous scientists:

Cm curium (Marie Curie)
Es einsteinium (Albert Einstein)
No nobelium (Alfred Nobel)

W Explore this topic on the **Marie Curie—Biography** website.

Today, all naturally occurring elements have probably been discovered. In addition, new elements—that don't exist in nature—have been created by accelerating protons or neutrons into the nuclei of naturally occurring elements. The protons and neutrons are incorporated into the nucleus, and elements of higher atomic number form. All naturally occurring elements, as well as those made by

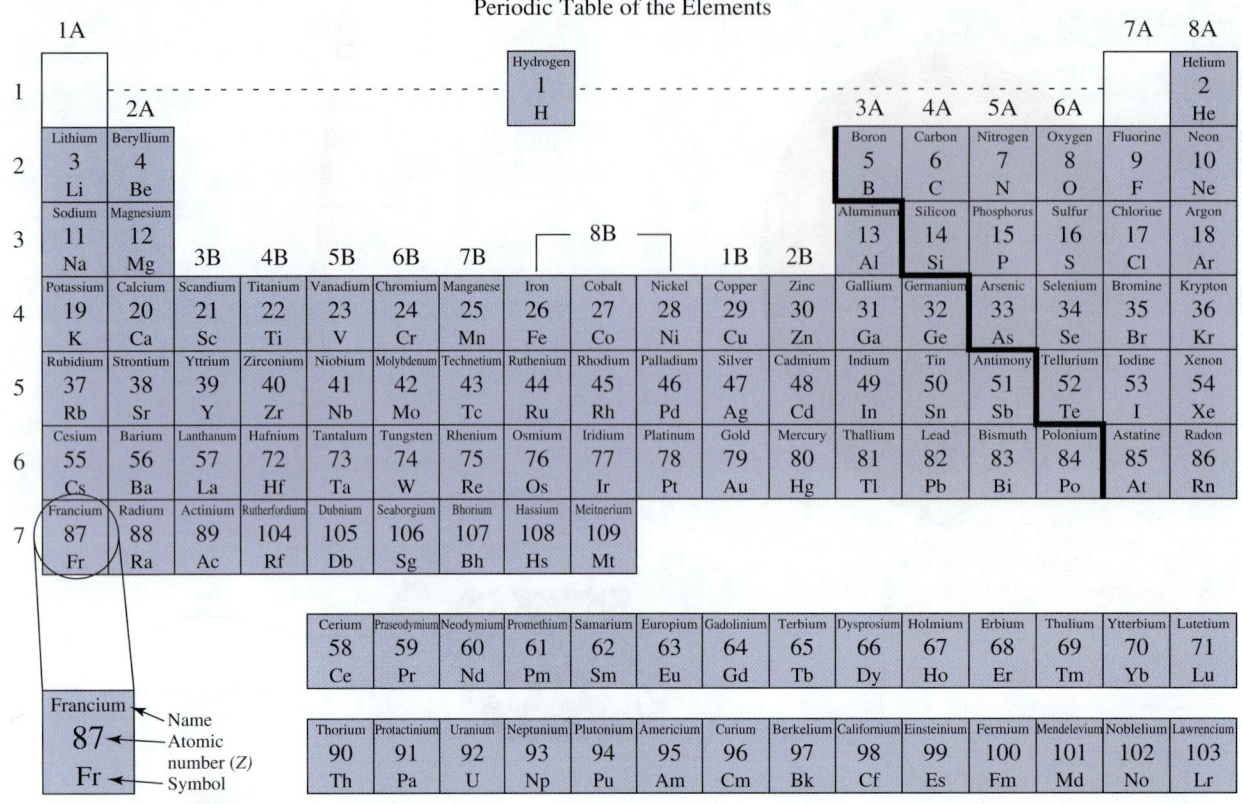

FIGURE 4 The periodic table lists all known elements in order of increasing atomic number. Some elements from the bottom rows of the table are shown separately to make the table more compact.

W Explore this topic on the **Interactive Periodic Table** website.

> **APPLY YOUR KNOWLEDGE**
>
> Your friend tells you about an article that he read in a tabloid that reported the discovery of a new form of carbon containing eight protons in the nucleus of its atoms. According to the article, this form of carbon spontaneously turns into diamond. How would you respond to your friend?
>
> Answer: You should tell your friend that the "form of carbon containing eight protons" was discovered long ago, and it is not carbon at all. We call it oxygen and it does not form diamonds.

humans, are listed in the **periodic table** shown in Figure 4 in order of increasing atomic number. There is also an alphabetical listing of the elements on the inside back cover of this book.

3.3 Electrons

A neutral atom has as many electrons outside of its nucleus as protons within its nucleus. Therefore, a hydrogen atom has one electron, a helium atom has two electrons, and a carbon atom has six electrons. Electrons have a very small mass compared to protons (0.00054 amu) and are negatively charged. Electrons therefore experience a strong attraction to the positively charged nucleus. This attraction

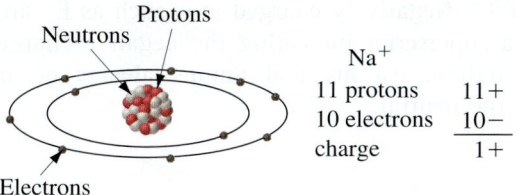

FIGURE 5 The sodium ion, Na⁺, contains 11 protons, but only 10 electrons. It has a net charge of 1+.

FIGURE 6 The fluorine ion, F⁻, contains 9 protons in its nucleus and 10 electrons. It has a net charge of 1−.

holds the electrons within a spherical region surrounding the nucleus. The sphere is mostly empty space occupied by electrons in constant motion. The charge of an electron is assigned a numerical value of −1 and exactly cancels the positive charge of a proton so that atoms containing equal numbers of protons and electrons are charge neutral.

However, an atom can lose or gain one or more of its electrons. When this happens, the charges of the electrons no longer exactly cancel the charges of the protons, and the atom becomes a charged particle called an **ion**. For example sodium (Na), which in its neutral form contains 11 protons and 11 electrons, often *loses* one electron to form the ion Na⁺ (Figure 5). The sodium ion contains 11 protons and only 10 electrons, resulting in a 1+ charge. Positively charged ions such as Na⁺ are called **cations** and are written with a superscript indicating the positive charge next to the element's symbol. Fluorine (F), which in its neutral form contains nine protons and nine electrons, will often *acquire* an extra electron to form the ion F⁻ (Figure 6). The extra electron results in nine protons

The charges of ions are usually written with the magnitude of the charge followed by the sign of the charge.

EXAMPLE 3.1

Determining the Number of Protons and Electrons in an Ion

How many protons and electrons are in the O^{2-} ion?

SOLUTION

We can look up oxygen on the periodic table and find that it has atomic number 8, meaning it has eight protons in its nucleus. *Neutral* oxygen therefore has eight electrons, but the O^{2-} ion has a charge of 2−; therefore, it must have ten electrons. We compute the total charge as follows:

8 protons × (1+)	= 8+
10 electrons × (1−)	= 10−
total charge	= 2−

YOUR TURN

Determining the Number of Protons and Electrons in an Ion

How many protons and electrons are in the Mg^{2+} ion?

and ten electrons for a net charge of 1⁻. Negatively charged ions such as F⁻ are called **anions** and are written with a superscript indicating the negative charge next to the element's symbol. In nature, cations and anions always occur together, so that, again, matter is charge neutral.

3.4 Neutrons

Atoms also contain neutral particles called **neutrons** within their nuclei. Neutrons have almost the same mass as protons but carry no electrical charge. Unlike protons, whose number is constant for a given element, the number of neutrons in the atoms of an element can vary. For example, carbon atoms have 6 protons in their nuclei, but may have 6, 7, or 8 neutrons (Figure 7). Carbon atoms exist in all three forms. Atoms with the same atomic number but different numbers of neutrons are called **isotopes**. Some elements, such as fluorine (F) and sodium (Na), have only one naturally occurring isotope—every fluorine atom in nature contains 10 neutrons and 9 protons in its nucleus and every sodium atom contains 12 neutrons and 11 protons. Other elements, such as carbon and chlorine, have two or more naturally occurring isotopes. Naturally occurring chlorine (Figure 8), for example, is composed of atoms with 18 neutrons in their nuclei (75.77% of all chlorine atoms) and 20 neutrons in their nuclei (24.23% of all chlorine atoms).

Scientists can synthesize some isotopes that are not normally found in nature. For example, almost all naturally occurring hydrogen atoms contain either no neutrons (99.985%) or one neutron (0.015%)—the isotope with one neutron is called deuterium. However, scientists have also made tritium, hydrogen atoms with two neutrons. Tritium is still hydrogen—it contains one proton in its nucleus—but its mass is higher because it has extra neutrons.

The neutrons and protons account for almost the entire mass of an atom. Consequently, the sum of the number of neutrons and protons in the nucleus of a given atom is called its **mass number** and has the symbol A. Naturally occurring sodium atoms, for example, with 11 protons and 12 neutrons in their nuclei have a mass number of 23. The two different isotopes of chlorine have two different mass

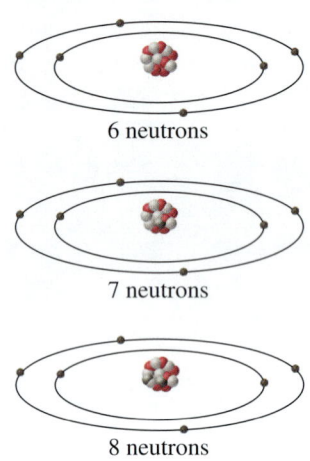

FIGURE 7 While all carbon atoms have 6 protons in their nuclei, the number of neutrons can be 6, 7, or 8. All three types of carbon atoms exist.

> ### EXAMPLE 3.2
> #### Atomic Numbers and Mass Numbers
> What is the atomic number (Z) and mass number (A) of the neon isotope that contains 12 neutrons in its nucleus?
>
> **SOLUTION**
> From the periodic table, we find that neon has atomic number 10, meaning that it contains 10 protons in its nucleus. Its mass number is the sum of the number of protons (10) and the number of neutrons (12) for a total of 22. Therefore,
>
> $$Z = 10, A = 22$$
>
> ##### YOUR TURN
> #### Atomic Numbers and Mass Numbers
> What is the atomic number (Z) and mass number (A) of the silicon isotope that contains 14 neutrons in its nucleus?

numbers: The chlorine isotope with 17 protons and 18 neutrons has a mass number of 35; the isotope with 17 protons and 20 neutrons has a mass number of 37.

3.5 Specifying an Atom

Table 1 summarizes the properties of the three subatomic particles that compose the atom. We can specify any atom, isotope, or ion by specifying the number of each of these particles within it. This is usually done by listing its atomic number (Z), its mass number (A), and its **charge** (C). We combine these with the atomic symbol of the element (X) in the following manner:

$$^A_Z X^C$$

Because every element has a unique atomic number, the chemical symbol, X, and the atomic number, Z, are redundant. For example, if the element is carbon, then the symbol is C and the atomic number is 6. If the atomic number were 7, the symbol would have to be that of nitrogen, N.

A shorter notation, often used when referring to neutral atoms, has the chemical symbol or name followed by a dash and its mass number:

$$X - A$$

Here, X is the symbol or name of the element and A is the mass number. For example, U-235 refers to the uranium isotope with mass number 235, and C-12 to the carbon isotope with mass number 12.

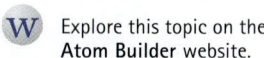
Explore this topic on the **Atom Builder** website.

EXAMPLE 3.3

Determining Protons, Neutrons, and Electrons

How many protons, neutrons, and electrons are in the following ion:

$$^{50}_{22}\text{Ti}^{2+}$$

SOLUTION

protons = Z = 22

neutrons = $A - Z$ = 50 − 22 = 28

electrons = $Z - C$ = 22 − 2 = 20

YOUR TURN

Determining Protons, Neutrons, and Electrons

How many protons, neutrons, and electrons are in the following ion:

$$^{35}_{17}\text{Cl}^-$$

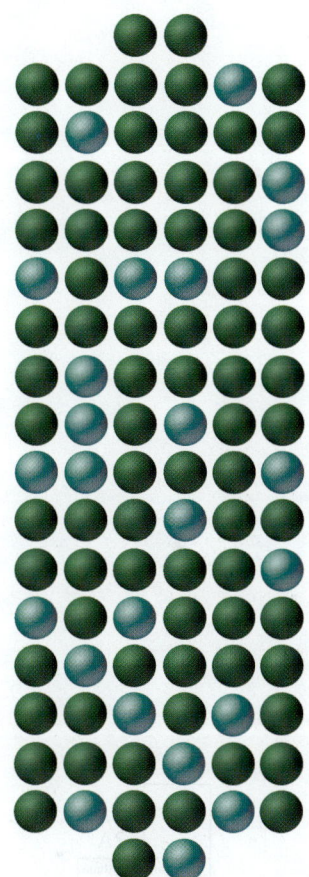

● Cl atoms with 18 neutrons

● Cl atoms with 20 neutrons

FIGURE 8 Out of 100 naturally occurring chlorine atoms, approximately 76 of them have 18 neutrons and 24 of them have 20 neutrons. They all have 17 protons.

TABLE 1

Subatomic Particles

	Mass (g)	Mass (amu)	Charge
Proton	1.6726×10^{-24}	1.0073	1+
Neutron	1.6749×10^{-24}	1.0087	0
Electron	0.000911×10^{-24}	0.000549	1−

> **APPLY YOUR KNOWLEDGE**
>
> What is the difference between an isotope and an ion?
>
> a) An isotope is defined by the relative number of protons and electrons, while an ion is defined by the number of protons and neutrons.
>
> b) An ion is defined by the relative number of protons and electrons, while an isotope is defined by the number of protons and neutrons.
>
> c) Two different ions must always correspond to two different elements, but two different isotopes could correspond to the same element.
>
> Answer: b. An ion is a charged particle whose charge depends on the relative number of protons and electrons. An isotope is an atom with a specified number of neutrons.

3.6 Atomic Mass

A characteristic of an element is the mass of its atoms. Hydrogen, containing only 1 proton in its nucleus, is the lightest element, while uranium, containing 92 protons and over 140 neutrons, is among the heaviest. The difficulty in assigning a mass to a particular element is that each element may exist as a mixture of two or more isotopes with different masses. Consequently, we assign an average mass to each element, called **atomic mass**. Atomic masses are listed in the periodic table and represent a weighted average of the masses of each naturally occurring isotope for that element (Figure 9).

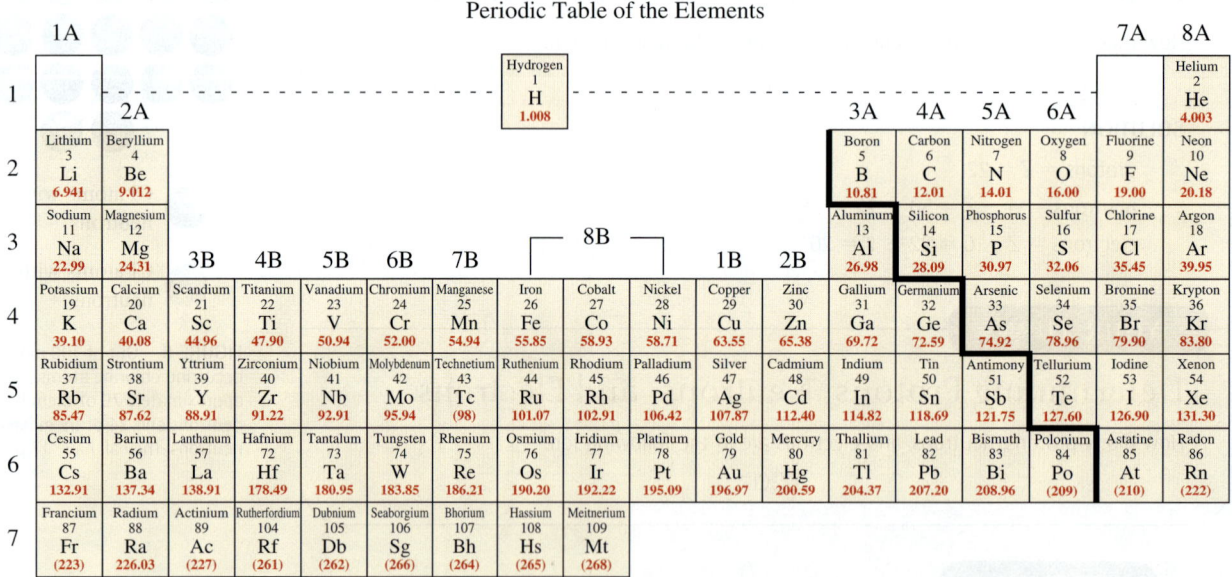

FIGURE 9 The periodic table. The number listed below each element is its atomic mass.

WHAT IF...

Complexity Out of Simplicity

Like a composer who builds an entire symphony on one basic theme by presenting it over and over in slightly different ways, so nature builds tremendous complexity from relatively simple particles by combining them over and over again in slightly different ways. Beethoven's famous "Fifth Symphony" is built on the simplest of themes: three short notes followed by a long one—da da da daaaaaa. Beethoven's brilliance is in how he plays on this theme, repeating it, sometimes slowly, sometimes quickly, sometimes with other notes thrown in. In the end, his simple theme is woven into a masterpiece of beauty and complexity. So it is with nature.

In nature, the simplest theme is the fundamental particle that physicists tell us composes all ordinary matter—the quark. Quarks have a number of odd properties that need not concern us here. What is important about them is that they compose the protons and neutrons that in turn compose atoms. Quarks come in six types, but ordinary matter is composed of only two, called *up* and *down*. An *up* quark has a $2/3+$ charge, and a *down* quark has a $1/3-$ charge.

By combining these quarks in slightly different ways, we get protons and neutrons, the main building blocks of atoms. A proton is composed of three quarks: two up and one down. The charges of these quarks sum to give the net charge of the proton (1+). A neutron is also composed of three quarks: one up and two down, which results in a particle with no net charge.

Protons and neutrons combine with electrons—which, as far as we know, are fundamental particles that are not composed of any others—to form 91 different naturally occurring elements, each with its own properties and its own chemistry, each different from the other. For example, combine one proton and one electron and you get hydrogen—the explosive, buoyant gas that filled the *Hindenberg* airship and resulted in its unfortunate demise in 1937. Combine two protons, two neutrons, and two electrons and you get helium—the inert gas that fills balloons and makes your voice sound funny when inhaled. Combine six protons, six neutrons, and six electrons and you get carbon—the solid substance that composes both graphite (pencil lead) and diamond, depending on how the carbon atoms join together. Diversity out of simplicity—this is nature's way.

This pattern continues beyond elements and into compounds. Combine a handful of elements—carbon, hydrogen, nitrogen, and oxygen—with each other and you get millions of different organic compounds (the compounds that compose living organisms), each one with its own properties, each one with its own chemistry, and each one unique. From a few basic particles, we get millions of compounds—so much diversity from so simple a beginning. And yet, it is exactly this kind of diversity that makes life possible. Life could not exist if elements did not combine to form compounds; it takes molecules in all their different sizes, shapes, and properties to create the complexity necessary for life. Even within living things, the variation of a simple theme to create complexity continues. We will examine this variation more fully in Chapter 16 when we consider the chemistry of life—biochemistry.

QUESTIONS: What if quarks didn't combine to form protons and neutrons? What if protons and neutrons didn't combine to form atoms? What if atoms didn't combine to form molecules? Can you think of anything else where complexity emerges from simplicity?

CALCULATING ATOMIC MASS

The atomic mass of any element is calculated according to the following equation:

$$\text{atomic mass} = (\text{fraction isotope 1}) \times (\text{mass isotope 1}) + (\text{fraction isotope 2}) \times (\text{mass isotope 2}) + \cdots$$

For example, we saw that naturally occurring chlorine has two isotopes: 75.77% of chlorine atoms are chlorine-35 (mass 34.97 amu) and 24.23% are chlorine-37 (mass 36.97 amu). The atomic mass is calculated by summing the atomic masses of each isotope multiplied by its fractional abundance:

Cl atomic mass = 0.7577 (34.97 amu) + 0.2423 (36.97 amu) = 35.45 amu

Notice that the percent abundances must be converted to fractional abundances by dividing them by 100. The atomic mass of chlorine is closer to 35 than

> **EXAMPLE 3.4**
>
> ### Calculating Atomic Mass
>
> Carbon has two naturally occurring isotopes with masses 12.00 and 13.00 amu and abundances 98.90 and 1.10%, respectively. Calculate the atomic mass of carbon.
>
> **SOLUTION**
>
> $$\text{C atomic mass} = 0.9890\,(12.00\text{ amu}) + 0.0110\,(13.00\text{ amu}) = 12.01\text{ amu}$$
>
> **YOUR TURN**
>
> ### Calculating Atomic Mass
>
> Magnesium has three naturally occurring isotopes with masses 23.99, 24.99, and 25.98 amu and natural abundances 78.99, 10.00, and 11.01%. Calculate the atomic mass of magnesium.

37 because naturally occurring chlorine contains more chlorine-35 atoms than chlorine-37 atoms.

3.7 Periodic Law

In the 1860s, a popular Russian professor named **Dmitri Mendeleev** (1834–1907) at the Technological Institute of St. Petersburg wrote a chemistry textbook. Drawing on the growing knowledge of the properties of elements, he noticed that some elements had similar properties and he grouped these together. For example, helium, neon, and argon are all chemically inert gases and could be put into one group. Sodium and potassium are reactive metals and could be put into another group.

Mendeleev found that if he listed the elements in order of increasing atomic mass, these similar properties would recur in a periodic fashion. Mendeleev summarized these observations in the periodic law, which states the following:

> When the elements are arranged in order of increasing atomic mass, certain sets of properties recur periodically.

Russian Professor Dmitri Mendeleev, who arranged the first periodic table.

Mendeleev then organized all the known elements in a table (a precursor to today's periodic table) so that atomic mass increased from left to right and elements with similar properties aligned in the same vertical columns. For this to work, Mendeleev had to leave some gaps in his table. He predicted that elements would be discovered to fill those gaps. He also proposed that some measured atomic masses were wrong. In both cases, Mendeleev was correct. Within 20 years of Mendeleev's proposal, three gaps were filled with the discoveries of gallium (Ga), scandium (Sc), and germanium (Ge). Today's periodic table (Figure 9) shows all the known elements and is foundational to modern chemistry.

Mendeleev did not know *why* the periodic law existed. His law, like all scientific laws, summarized a large number of observations but did not give the underlying reasons for the observed behavior. The next step, following the scientific method, was to devise a theory that explained the law and provided a model for atoms. In Mendeleev's time, there was no theory to explain the periodic law.

Before looking more closely at the periodic table, we briefly explore two theories that explain why the properties of elements recur in a periodic fashion. The first theory is called the Bohr model for the atom, after **Niels Bohr** (1885-1962), and it links the macroscopic observation—that certain elements have similar properties that recur—to the microscopic reason—that the atoms comprising the elements have similarities that recur.

3.8 A Theory That Explains the Periodic Law: The Bohr Model

As we learned in Chapter 1, science gives insight into nature through models or theories that explain laws and predict behavior. Here we look at the **Bohr model**, a theory for the way electrons behave in atoms that explains the periodic law and links the macroscopic properties of elements to the microscopic properties of their atoms. We must keep in mind that this is a *model* for the atom and not the real thing. In fact, a more successful and complete model for the atom, called the *quantum mechanical model,* is significantly different, as we will see in the next section. Nonetheless, the simple model developed here helps us to understand the periodic law and allows us to predict a great deal of chemical behavior.

Niels Bohr (1885-1962)

Although the number of protons in the nucleus of an atom determines the element, the number of electrons outside the nucleus defines its chemical behavior. Niels Bohr focused on those electrons and modeled them as being in orbits around the nucleus, much like planets are in orbits around the Sun (Figure 10). However, in contrast to planets, that could theoretically orbit at any distance whatsoever, Bohr proposed that electrons could only orbit at specific, fixed distances from the nucleus. Those fixed distances then corresponded to specific, fixed energies for the electron. Electrons closer to the nucleus have lower energy than those that are further away.

Bohr specified each orbit with an integer n, called the orbit's **quantum number**. The higher the quantum number is, the greater is the distance between the electron and the nucleus and the higher is the electron's energy. Bohr also stipulated that the orbits could only hold a maximum number of electrons determined by the value of n as follows:

Bohr orbit with $n = 1$ holds a maximum of two electrons
Bohr orbit with $n = 2$ holds a maximum of eight electrons
Bohr orbit with $n = 3$ holds a maximum of eight electrons*

In an atom, an electron occupies the lowest energy orbit that is available. The one electron in the hydrogen atom seeks the lowest energy orbit and therefore

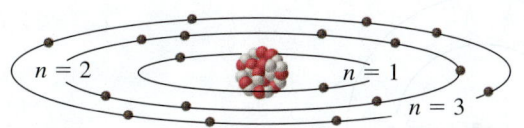

FIGURE 10 In the Bohr model, electrons occupy orbits that are at fixed energies and fixed radii.

---

* This is an oversimplification, but it will suit our purposes. In the quantum mechanical model, Bohr orbits are replaced with quantum mechanical orbitals and quantum shells. The quantum shell with $n = 3$ can actually hold 18 electrons; however, the quantum shell with $n = 4$ begins to fill when there are only 8 electrons in the quantum shell with $n = 3$.

occupies the Bohr orbit with $n = 1$. Because the orbit can hold a maximum of two electrons, it is half full.

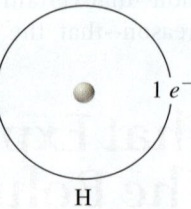
H

Helium's two electrons just fill the $n = 1$ orbit.

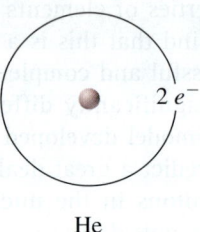
He

The next element, Li, has three electrons; two of those electrons are in the $n = 1$ orbit, but the third has to go into the $n = 2$ orbit.

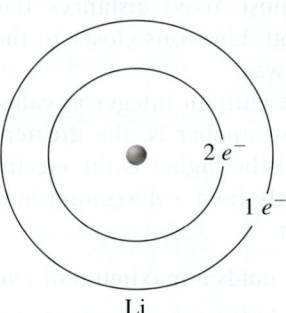
Li

Skipping to fluorine, which has nine electrons, two of those electrons are in the $n = 1$ orbit and the remaining seven are in the $n = 2$ orbit. The outer orbit is almost full.

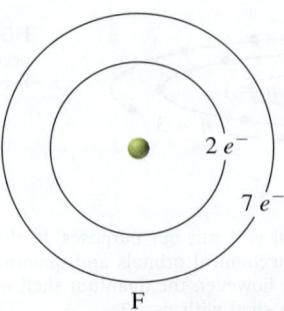
F

The orbit fills in neon, which has ten electrons, two in the $n = 1$ orbit and eight in the $n = 2$ orbit.

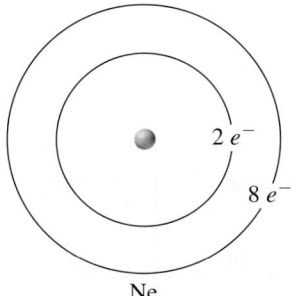

Ne

The next element, sodium, has eleven electrons. Of those, two electrons are in the $n = 1$ orbit, eight are in the $n = 2$ orbit, and one is in the $n = 3$ orbit.

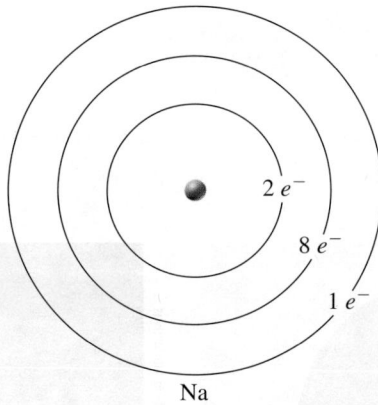

Na

By continuing this process, and remembering that electrons fill the lowest energy orbits available, we can draw similar diagrams for the first 20 elements in the periodic table (Figure 11). These diagrams are called **Bohr diagrams** or **electron configurations**. They show the arrangement of an atom's electrons around the nucleus for each element. Remember that we are examining the Bohr model in order to explain why certain elements have similar properties and why properties recur in a periodic fashion. Helium, neon, and argon are three elements with similar properties—they are chemically inert gases. Do you notice anything similar about their electron configurations in Figure 11? Lithium, sodium, and potassium also have similar properties—they are chemically reactive metals. Do you notice any similarities in their electron configurations?

The connection between the microscopic electron configuration and the macroscopic property of the element lies in the number of electrons in the element's outer Bohr orbit:

- Atoms with a full outer orbit are extremely stable; they are chemically inert and do not react readily with other elements.
- Atoms having outer orbits that are not full are unstable and will undergo chemical reactions with other elements to gain full outer orbits.

70 CHAPTER 3   Atoms and Elements

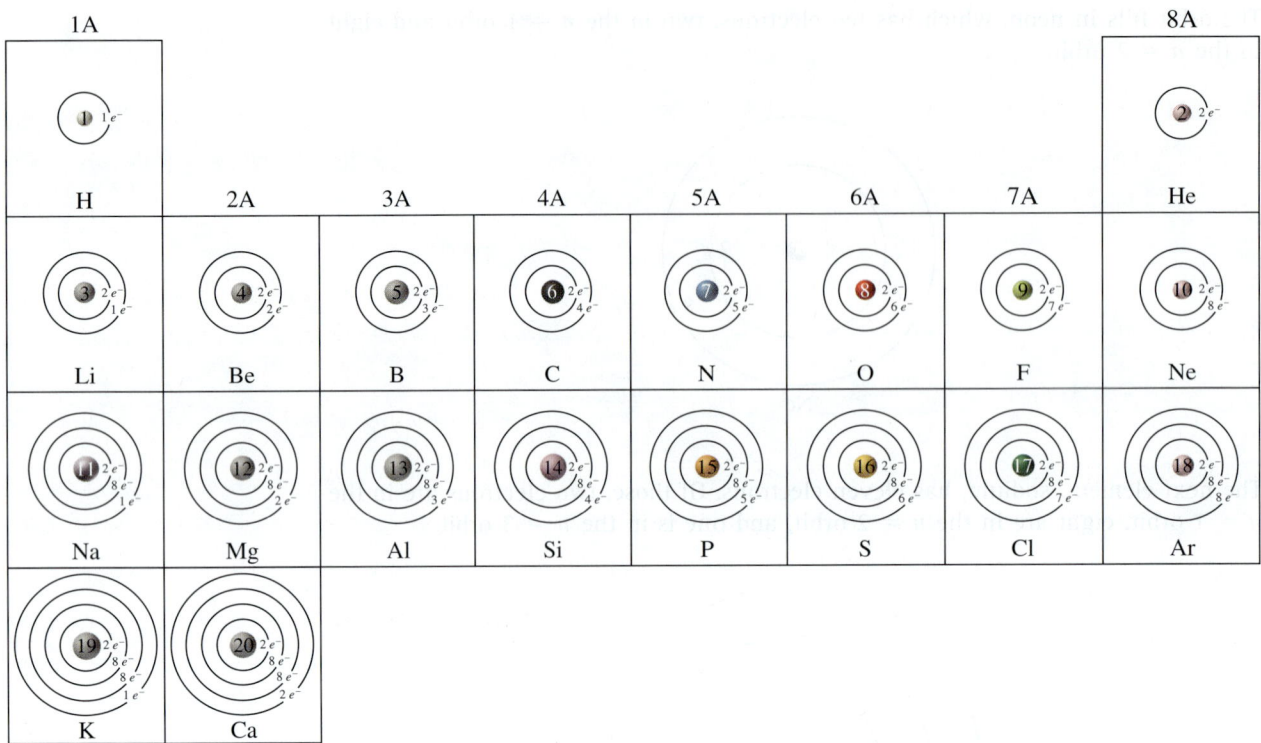

FIGURE 11 Electron configurations for the first 20 elements.

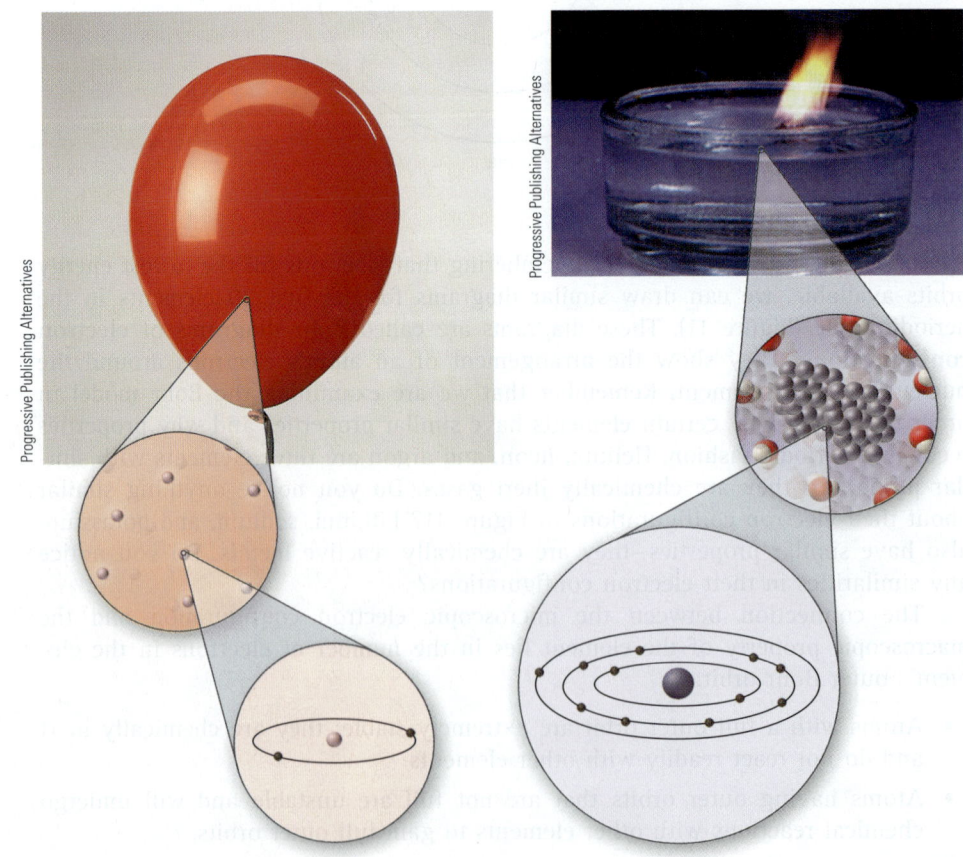

(a) Helium balloon. Helium is an inert element that naturally occurs as a gas. Helium does not react with anything. The atoms of inert elements have full outer electron orbits. (b) Sodium reacting with water: Sodium is a very reactive element, and when it is mixed with water, a reaction takes place. The atoms of reactive elements do not have full outer electron orbits.

> **APPLY YOUR KNOWLEDGE**
>
> Based on their electron configurations, which of these elements—argon, sulfur, or magnesium—do you predict to be most chemically inert (nonreactive)?
>
> Answer: Because argon has a full outer orbit and the others do not, it is most stable.

The kinds of reactions that elements undergo and the kinds of compounds they form depend on the number of electrons in their outer orbit. Therefore, two or more elements whose atoms have the same number of electrons in their outer orbits will undergo similar chemical reactions and form similar compounds. Because the electrons in the outer orbit of an element's atoms are so critical in determining the element's properties, they are singled out and called **valence electrons**. Much of the chemistry you encounter in this book is a reflection of the variation in an element's valence electrons. Neon and argon both have eight valence electrons, while lithium, sodium, and potassium have one valence electron. The reason certain groups of elements have similar properties is that they have a similar number of valence electrons.

3.9 The Quantum Mechanical Model for the Atom

Bohr modeled electrons as negatively charged, orbiting particles restricted to certain distances from the nucleus. However, in the early 1900s, a new model for the atom, called the **quantum mechanical model**, was developed. Although the Bohr model is useful for explaining much of the chemical behavior we encounter in this book, it leaves us with an outdated picture for how electrons exist in atoms. A brief introduction to the quantum mechanical model can give us a more modern picture of the atom.

In the quantum mechanical model, orbits are replaced with **orbitals**. Quantum mechanical orbitals resulted from discoveries showing that the electron—which was thought of as a particle—also displayed properties that we normally associate with waves. The wave nature of the electron meant that its motions around the nucleus were more complex than simple circular orbits. Consequently, a quantum mechanical orbital does not specify an electron's exact path, but shows the *probability* of finding the electron in a certain region. For example, consider the 1s quantum mechanical orbital shown in Figure 12. This orbital is the quantum mechanical equivalent of the $n = 1$ orbit in the Bohr model. We can understand this image of the 1s orbital with a simple analogy. Imagine a moth flying around a bright light bulb. Imagine further a camera that takes a picture of the moth every minute for 2 or 3 hours. Then imagine combining all those pictures into one image. That image would be a probability map—it would show where the moth spent most of its time. The 1s orbital is similar—it shows where an electron in the 1s orbital spends most of its time. The darker regions show higher probability of finding the electron. You can see from Figure 12 that the electron spends more time near the nucleus than far away from it.

Another way to depict the 1s orbital is with a sphere that encloses the 90% probability boundary—the boundary within which the electron spends 90% of its

1s orbital

FIGURE 12 The quantum mechanical 1s orbital.

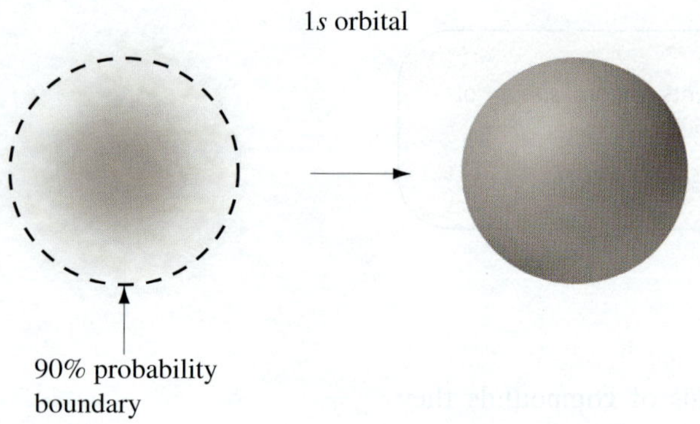

FIGURE 13 The 1s orbital depicted by showing its 90% probability boundary. (Source: Progressive Publishing Alternatives)

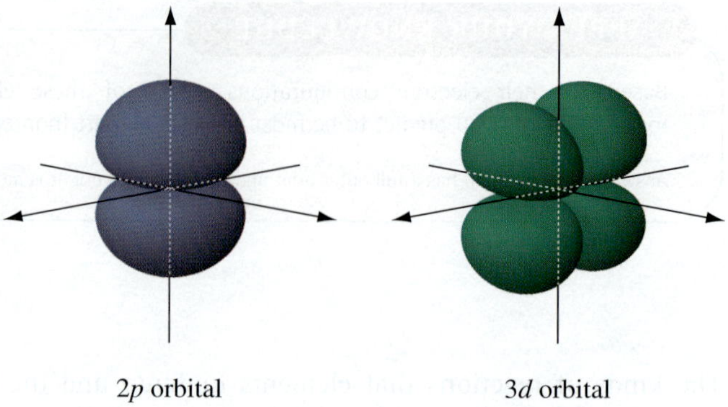

FIGURE 14 The 2p and 3d quantum mechanical orbitals.

time (Figure 13). Other quantum mechanical orbitals, depicted by showing their 90% probability boundaries, are shown in Figure 14. Quantum mechanical orbitals are grouped into different shells. These shells fill in a manner similar to the filling of Bohr orbits, but the exact way they fill is beyond the scope of this book.

You can see that the quantum mechanical model does not exactly locate the electron's path, but rather, predicts where it spends most of its time. Erwin Schrödinger (1887–1961) shook the scientific establishment when he proposed this model in 1926. The most shocking difference was the statistical nature of the quantum mechanical model. According to quantum mechanics, the paths of electrons are *not* like the paths of baseballs flying through the air or of planets orbiting the Sun, both of which are predictable. For example, we can predict where

WHAT IF. . .

Philosophy, Determinism, and Quantum Mechanics

We often think of science in terms of the technology it produces—because of science we have computers, medicines, and CD players, for example. However, science also contributes to basic human knowledge and makes discoveries that affect other academic disciplines. The discovery of quantum mechanics in the 20th century, for example, had a profound effect on our fundamental understanding of reality and on the field of philosophy. At stake was a philosophical question that has been debated for centuries: Is the future predetermined?

The idea that the future *is* predetermined is called *determinism*. In this view, future events are caused by present events that are in turn caused by past events, so that all of history is simply one long chain of causation, each event being caused by the one before it. Before the discovery of quantum mechanics, the case for determinism seemed strong. Newton's laws of motion described the future path of any particle based on its current position (where it was) and its velocity (how fast and what direction it was going). We all have a sense of Newton's laws because we have seen objects such as baseballs or billiard balls behave according to them. For example, an outfielder can predict where a baseball will land by observing its current position and velocity. The outfielder predicts the *future* path of the baseball based on its *current* path—this is determinism.

The discovery of quantum mechanics challenged the idea that our universe behaves deterministically. Electrons, and all other small particles such as protons and neutrons, do not appear to behave deterministically. An outfielder chasing an electron could not predict where it would land. The subatomic world is indeterminate—the present does not determine the future. This was a new idea. Erwin Schrödinger himself once said of quantum mechanics, "I don't like it, and I am sorry I ever had anything to do with it," and Neils Bohr said, "anyone who is not shocked by quantum mechanics has not understood it." To some, an indeterminate universe was threatening. To others, the idea that the future was not predetermined—at least for subatomic particles—came as a pleasant surprise. In philosophy, the debate continues. However, the indeterminate nature of the subatomic world dealt a severe blow to the idea that every event in the universe was determined by the event before it.

the Earth will be in its orbit around the Sun in 2 years, 20 years, or even 200 years. This is not so for an electron. We cannot predict exactly where an electron will be at any given time—we can only predict the probability of finding it in a certain region of space.

So, which model is correct? Is it the Bohr model or the quantum mechanical model? Remember that, in science, we build models (or theories) and then perform experiments in an attempt to validate them. The Bohr model has been shown to be invalid by experiments. The quantum mechanical model is consistent with all experiments to date. Of course, this doesn't make the quantum mechanical model theory "true." Scientific theories are never proven true, only valid. This also does not mean that the Bohr model is not useful. In fact, the Bohr model is sufficient to predict much of the chemical behavior we encounter in this book. However, the quantum mechanical model gives us a better picture of atoms.

W Explore this topic on the **Quantum Philosophy** website.

3.10 Families of Elements

Elements such as He, Ne, and Ar that have similar outer electron configurations (in this case, full outer orbits), have similar properties, and form a **family** or **group of elements**. These groups fall in vertical columns on the periodic table. Each column in the periodic table is assigned a group number, which is shown directly above the column (Figure 15). Some groups are also given a name.

FIGURE 15 The periodic table. Vertical columns are called groups or families of elements. The elements within a group have similar properties.

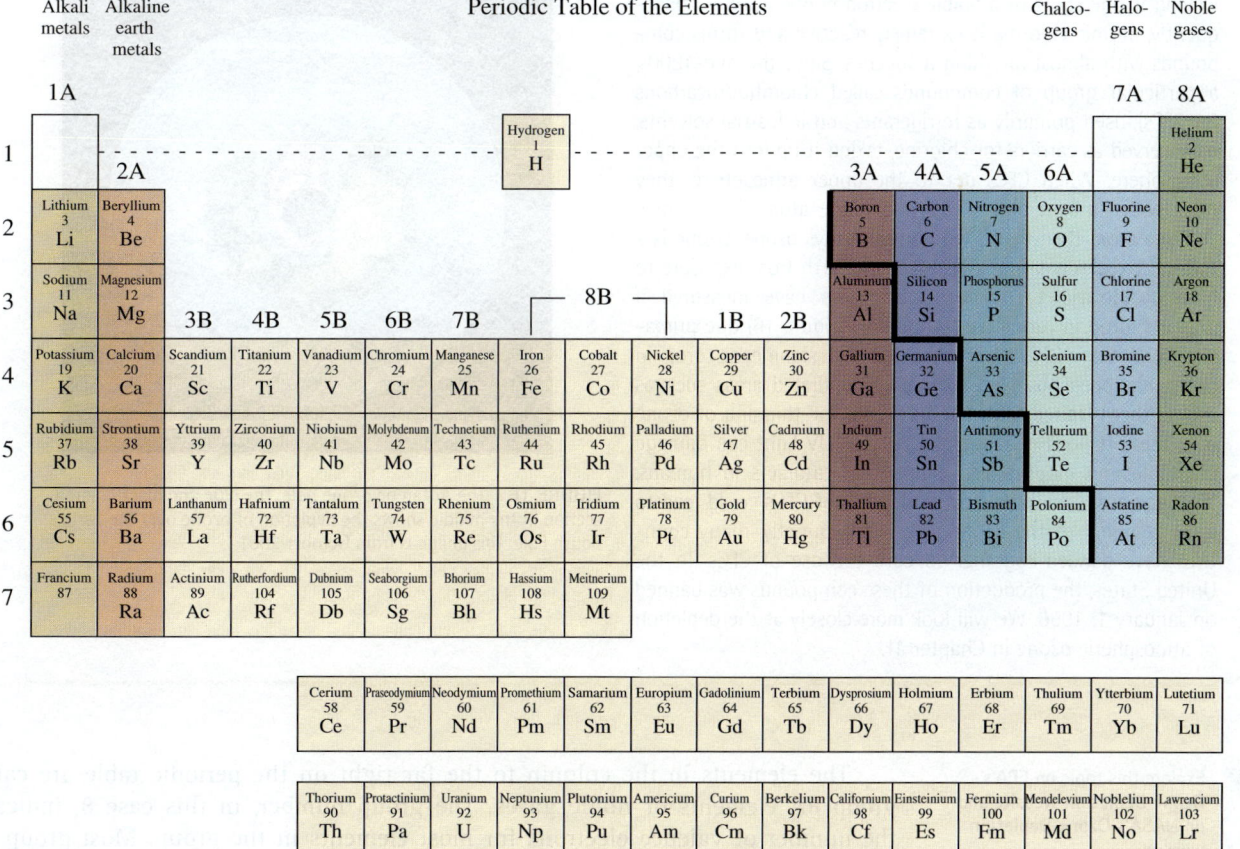

MOLECULAR THINKING

Is Breathing Helium Dangerous?

Helium balloons float because the helium gas within the balloon is less dense than the surrounding air. A balloon rising in air is like a submerged cork rising toward the surface of water: The cork is less dense than the water and thus it floats. Helium balloons are relatively safe because the helium gas within them is inert; it does not react with anything. Highly reactive substances are usually more dangerous to humans because they react with biological molecules that compose the body and can change or damage them. The helium within a balloon is so inert that it can even be inhaled *in small amounts* without any long-term physiological effects. The helium only causes a high, funny-sounding voice. Inhaling moderate to large amounts of helium, on the other hand, will cause suffocation because your body will not get enough oxygen.

QUESTION: What is the *molecular* (or in this case *atomic*) *reason* for the inertness of helium; that is, what is it about helium *atoms* that makes helium gas inert?

THE MOLECULAR REVOLUTION

The Reactivity of Chlorine and the Depletion of the Ozone Layer

As we saw in Section 3.9, chlorine has seven valence electrons, leaving it one short of a stable electron configuration. Consequently, atomic chlorine is extremely reactive and forms compounds with almost anything it touches. Since the mid-1900s, a particular group of compounds called chlorofluorocarbons (or CFCs), used primarily as refrigerants and industrial solvents, have served as carriers for chlorine, taking it up into the upper atmosphere. When CFCs get to the upper atmosphere, they react with sunlight and release a chlorine atom. The reactive chlorine atom then reacts with and destroys ozone. Ozone is a form of oxygen gas that shields life on Earth from exposure to harmful ultraviolet (UV) light. Scientists have measured a dramatic drop in ozone over Antarctica (Figure 16) due primarily to Cl from CFCs. A smaller, but still significant, drop in ozone has been observed over more-populated areas such as the northern United States and Canada. The thinning of ozone over these regions is dangerous because UV light can damage plant life and induces skin cancer and cataracts in humans. Most scientists think that continued use of CFCs could lead to more thinning of the ozone layer. Consequently, many countries have banded together to curb the use of CFCs. In the United States, the production of these compounds was banned on January 1, 1996. We will look more closely at the depletion of atmospheric ozone in Chapter 11.

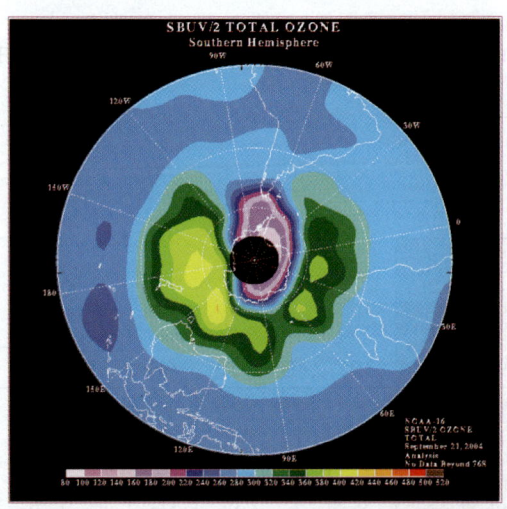

FIGURE 16 The Antarctic ozone hole. The blue and black colored section in the middle shows the depletion of ozone over the Earth's South Pole. This image is from October 2001.

 Explore this topic on **EPA's Ozone Depletion** website and on **NASA's Ozone Depletion** website.

The elements in the column to the far right on the periodic table are called group 8A elements or **noble gases**. The group number, in this case 8, indicates the number of valence electrons for most elements in the group. Most group 8A elements have eight valence electrons, except helium, which has two. All group

8A elements have full outer orbits and are chemically inert. They have stable electron configurations and therefore do not readily react with other elements to form compounds.

The elements in the column to the far left on the periodic table are called group 1A elements or **alkali metals**. They all have one valence electron and are very reactive. In their reactions, they lose that valence electron to acquire an electron configuration like that of a noble gas. The elements in the second column of the periodic table are called group 2A elements or **alkaline earth metals**. They have two valence electrons and lose them in chemical reactions to acquire a noble gas electron configuration.

Elements in the third column, or group 3A elements, have three valence electrons. In certain instances, some group 3A elements will lose three electrons, but they will also share electrons with another element to attain stability (more on this in Chapter 4). Similarly, group 4A and 5A elements have four and five valence electrons, respectively, and usually share these with another element to acquire stable electron configurations.

Group 6A elements are called the **chalcogens** or the **oxygen family** and are two electrons short of having full outer orbits. They will often gain two electrons when they react to attain a stable noble gas electron configuration. Group 7A elements are called the **halogens** and are only one electron short of acquiring a stable configuration. They undergo vigorous chemical reaction to attain an additional electron.

The elements in the periodic table can be classified even more broadly as metals, nonmetals, and metalloids, as shown in Figure 17. **Metals** tend toward the

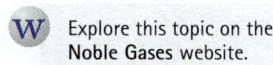

Explore this topic on the **Noble Gases** website.

Sodium is an alkali metal. It is so reactive that it must be protected from air and moisture by storage in a nonaqueous liquid.

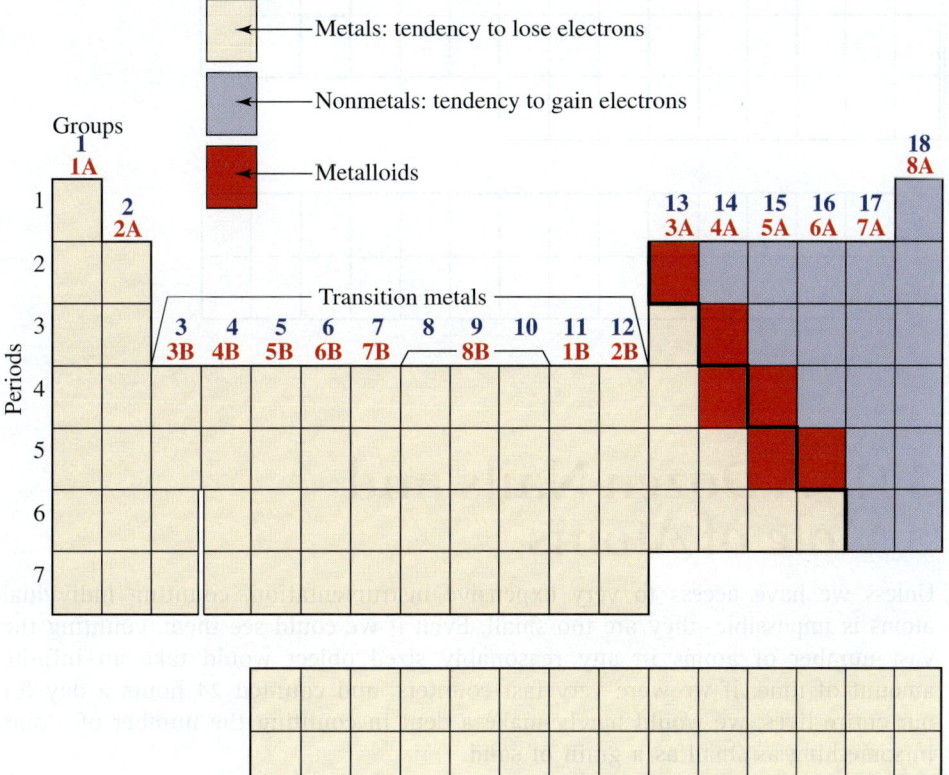

FIGURE 17 The elements can be broadly divided into metals, which primarily lose electrons in chemical reactions; nonmetals, which primarily gain electrons in chemical reactions; and metalloids, which can do either.

Chlorine is a halogen.

left side of the periodic table and tend to lose electrons in their chemical reactions, whereas **nonmetals** tend toward the right side of the periodic table and tend to gain electrons in their chemical reactions. The elements that fall in the middle-right of the periodic table are called **metalloids** and show mixed tendencies. Groups 3B through 2B in the center of the periodic table are also metals and are called the **transition metals**. Transition metals lose electrons in their chemical reactions but do not necessarily acquire noble gas configurations.

MOLECULAR ELEMENTS

Although the simplest identifiable units of an element are its corresponding atoms, a number of elements exist in nature as diatomic molecules—two atoms bonded together—as shown in Figure 18. These elements are hydrogen, nitrogen, oxygen, fluorine, chlorine, bromine, and iodine. In Chapter 5, we will learn a theory that explains why these elements form diatomic molecules.

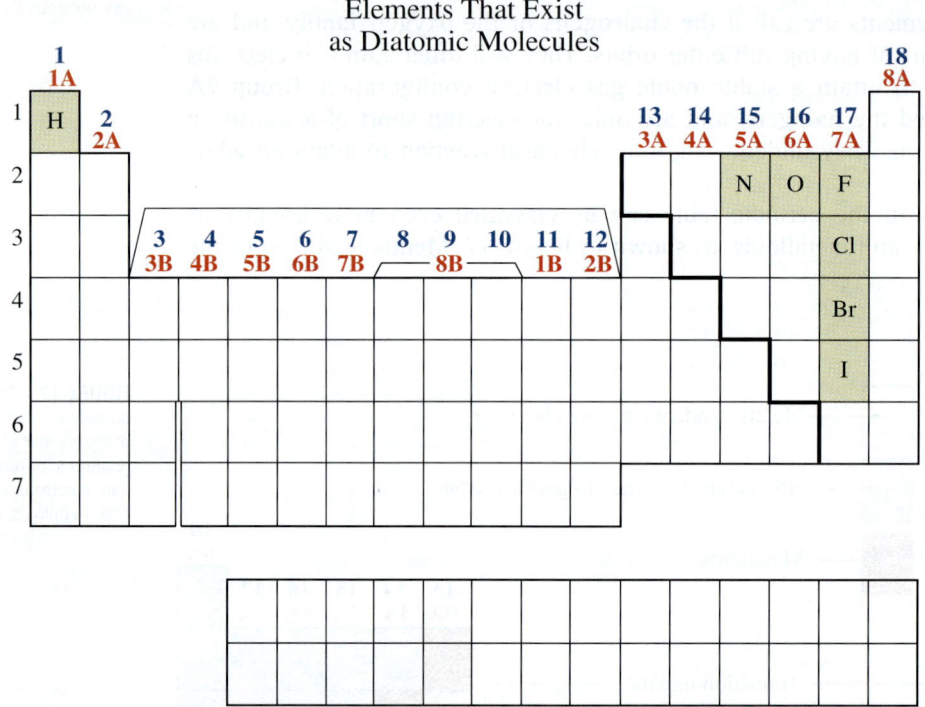

FIGURE 18 The elements whose positions in the periodic table are shaded exist as diatomic elements.

3.11 A Dozen Nails and a Mole of Atoms

Unless we have access to very expensive instrumentation, counting individual atoms is impossible—they are too small. Even if we could see them, counting the vast number of atoms in any reasonably sized object would take an infinite amount of time. If we were very fast counters, and counted 24 hours a day for our entire lives, we would barely make a dent in counting the number of atoms in something as small as a grain of sand.

3.11 A Dozen Nails and a Mole of Atoms

If we want to know the number of atoms in anything of substantial size, we need another method besides direct counting. One such method involves the *mole concept*, which relates the mass of a sample of an element to the number of atoms within it, so that we can determine the number of atoms by weighing.

To understand the mole concept, consider this analogy. Imagine buying nails by the pound at the local hardware store. How do we determine the number of nails in a certain weight of nails? We need a relationship between the weight and the number of nails. For example, suppose we bought 1.50 lb of medium-sized nails. Suppose further that the nails weighed 0.100 pounds per dozen (0.100 lb/doz). To determine the number of nails in our sample, we first convert from the weight to the number of dozens and then from dozens to the number of nails. The two conversion factors are 0.100 lb = 1 doz and 1 doz = 12 nails and the calculation proceeds as follows:

$$\text{weight} \rightarrow \text{dozen} \rightarrow \text{nails}$$

$$1.5 \text{ lb} \times \frac{1 \text{ doz}}{0.10 \text{ lb}} \times \frac{12 \text{ nails}}{1 \text{ doz}} = 180 \text{ nails}$$

With these conversion factors, we can easily count nails by weighing them. What if we had larger nails? smaller ones? The first conversion factor would change because there would be a different weight per dozen nails, but the second conversion factor would remain the same. A dozen corresponds to 12 nails, regardless of their size.

Chemists use a number that is similar in concept to the dozen. However, 12 is too small a number when referring to atoms. We need a much larger number because atoms are so small. Consequently, the chemist's "dozen" is called the **mole (mol)** and corresponds to 6.022×10^{23}. This number is called Avogadro's number, named after **Amadeo Avogadro (1776–1856)**, and is a convenient number to use when speaking of atoms. A mole of atoms contains 6.022×10^{23} atoms and makes up objects of reasonable sizes. For example, 22 pure copper pennies (pre-1983) contain approximately one mole of copper atoms, and a few large helium balloons contain approximately one mole of helium atoms. There is nothing mysterious about a mole; it is just a certain number of objects, 6.022×10^{23}, just like a dozen is a certain number of objects, 12.

For atoms, the quantity that is analogous to the other conversion factor in our nail calculation—the weight of one dozen nails—is the mass of one mole of

Amadeo Avogadro (1776–1856).

(a)

(b)

(a) 22 pennies contain about one mole of copper atoms. (Note: Pennies were made from pure copper until 1983.) (b) Six helium balloons contain about 1 mol of helium atoms.

78   CHAPTER 3   Atoms and Elements

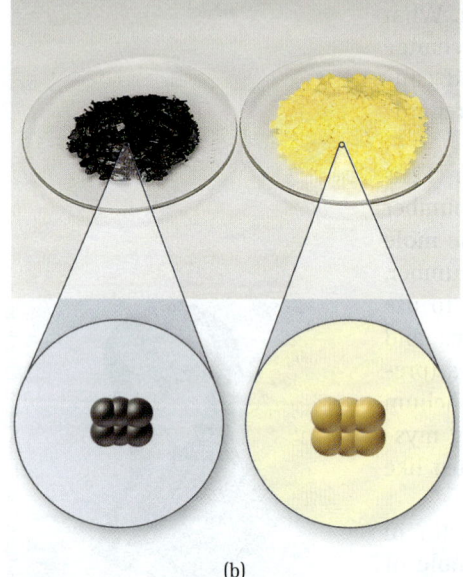

(a)

(b)

FIGURE 19 Just as 1 doz large nails constitutes more matter than 1 doz small nails (a), 1 mol sulfur atoms constitutes more matter than 1 mol carbon atoms (b). Sulfur atoms are bigger and heavier than carbon atoms.

atoms. This quantity is called **molar mass**. Avogadro's number is chosen so that the numerical value of the atomic mass of an element in amu is equal to the molar mass of that element in grams per mole. In other words, the atomic mass of any element gives the mass of one mole of its atoms.

$$\text{atomic mass (in amu)} = \text{molar mass}\left(\text{in } \frac{\text{grams}}{\text{mole}}\right)$$

In analogy to the nail example, where the weight of nails per dozen depends on the nail's size, so the molar mass depends on the atom's size. The bigger the atom, the more mass per mole (Figure 19). For example:

Molar masses from periodic table →

1.008 g hydrogen = 1 mole hydrogen = 6.022×10^{23} H atoms
12.011 g carbon = 1 mole carbon = 6.022×10^{23} C atoms
32.06 g sulfur = 1 mole sulfur = 6.022×10^{23} S atoms

So from the periodic table, we can find the mass of one mole of atoms for any element. Now, how do we use that quantity, in combination with the number of atoms in a mole, to determine the number of atoms in sample of a given mass? Suppose we want to know the number of aluminum atoms in 12.5 grams of aluminum. We first convert from the mass to the number of moles, and then from the number of moles to the number of atoms:

$$\text{mass} \rightarrow \text{moles} \rightarrow \text{atoms}$$

$$12.5 \text{ g} \times \frac{1 \text{ mole}}{26.98 \text{ g}} \times \frac{6.022 \times 10^{23} \text{ atoms}}{\text{mole}} = 2.79 \times 10^{23} \text{ atoms}$$

For more practice with the mole concept, consider the following examples.

EXAMPLE 3.5

The Mole Concept

The helium gas within a medium-sized helium balloon weighs 0.55 g. How many moles of helium are in the balloon?

SOLUTION

Following our problem-solving procedure from Chapter 2, we first write down the information the problem gives and what it asks us to find.

Given
0.55 g of He

Find

moles of helium

We need a conversion factor between grams of helium and moles, which is helium's molar mass, 4.00 g/mol. We then start with the mass and convert to the number of moles. (The molar mass of helium and other elements can be obtained from the periodic table on the inside front cover of this book.)

$$0.55 \text{ g} \times \frac{1 \text{ mol}}{4.00 \text{ g}} = 0.14 \text{ mol}$$

YOUR TURN

The Mole Concept

A diamond, which is pure carbon, contains 0.020 mol of carbon. What is the mass, in grams, of the diamond?

EXAMPLE 3.6

The Mole Concept II

Five pennies are found to weigh 15.3 g. If the pennies are pure copper, how many copper atoms are in the five pennies?

SOLUTION

Given

15.3 g

Find

number of copper atoms

We need two conversion factors. From the periodic table, we find the molar mass of copper to be 63.55 g/mol. We also need Avogadro's number, 6.022×10^{23} atoms = 1 mol. Starting with the mass, we first convert to moles and then to number of atoms:

$$15.3 \text{ g} \times \frac{1 \text{ mole}}{63.55 \text{ g}} \times \frac{6.022 \times 10^{23} \text{ atoms}}{1 \text{ mole}} = 1.45 \times 10^{23} \text{ atoms}$$

YOUR TURN

The Mole Concept II

Calculate the number of atoms in a pure gold ring weighing 17 g.

Chapter Summary

MOLECULAR CONCEPT

We have seen that all things, including ourselves, are ultimately composed of atoms and that the macroscopic properties of substances ultimately depend on the microscopic properties of the atoms that compose them **(3.1)**. We completely specify an atom by indicating each of the following **(3.2–3.5)**:

- its atomic number (Z), which is the number of protons in its nucleus
- its mass number (A), which is the sum of the number of protons and neutrons in its nucleus.
- its charge (C), which depends on the relative number of protons and electrons.

The mass number and charge can vary for a given element, but the atomic number defines the element and is therefore always the same for a given element. Atoms that have the same atomic number but different mass numbers are called isotopes, and atoms that have lost or gained electrons to acquire a charge are called ions. A positive ion is called a cation, and a negative one is called an anion.

A characteristic of an element is its atomic mass, a weighted average of the masses of the isotopes that naturally compose that element **(3.6)**. The atomic mass is numerically equivalent to molar mass, the mass of one mole of that element in grams. The molar mass provides a conversion factor between grams and moles.

In the Bohr model for the atom, electrons orbit the nucleus much like planets orbit the Sun **(3.8)**. The electrons in the outermost Bohr orbit are called the valence electrons and are key in determining an element's properties. Elements with full outer orbits are chemically stable, whereas those with partially filled outer orbits are not. Elements with the same number of valence electrons form families or groups and occur in vertical columns in the periodic table **(3.7)**. Elements with small numbers of valence electrons tend to lose them in chemical reactions and are called metals. Elements with large numbers of valence electrons tend to gain additional electrons in their reactions and are called nonmetals **(3.10)**.

In the quantum mechanical model for the atom, electrons exist in orbitals **(3.9)**. Orbitals show the relative probabilities of finding the electron in the space around the nucleus.

SOCIETAL IMPACT

Because all matter is made of atoms, we can better understand matter if we understand atoms. The processes that occur around us at any time are caused by changes in the atoms that compose matter **(3.1)**. Except in special cases—specifically, nuclear reactions—elements don't change. A carbon atom remains a carbon atom for as long a time as we can imagine. Pollution, then, is simply misplaced atoms—atoms that, because of human activity, have found their way into places that they do not belong. However, because atoms don't change, pollution is not an easy problem to solve. The atoms that cause pollution must somehow be brought back to their original place, or at least to a place where they won't do any harm.

Molar masses help us to calculate the number of atoms in a given object simply by weighing it **(3.11)**.

The microscopic models developed in this chapter will be directly applicable in explaining why elements form the compounds that they do **(3.8, 3.9)**. Reactive atoms, such as chlorine, are reactive because they have seven valence electrons when eight are required for stability **(3.7)**. Consequently, chlorine reacts with other elements in an attempt to gain an additional electron. Reactive atoms like chlorine can become environmental problems, especially when human activity moves them to places where they are usually not found. For example, chlorine atoms are transported into the upper atmosphere by synthetic compounds called chlorofluorocarbons. Once there, they react with the ozone layer and destroy it **(3.10)**.

Chemistry on the Web

For up-to-date URLs, visit the text website at **http://chemistry.brookscole.com/tro3**

- Marie Curie - Biography
  http://nobelprize.org/physics/laureates/1903/marie-curie-bio.html
- Interactive Periodic Table
  http://www.chemicalelements.com/
- Atom Builder
  http://www.pbs.org/wgbh/aso/tryit/ (select atom builder)
- Quantum Philosophy
  http://www.fortunecity.com/emachines/e11/86/qphil.html
- The Noble Gases
  http://pubs.acs.org/cen/80th/noblegases.html
- EPA's Ozone Depletion Website
  http://www.epa.gov/docs/ozone/
- NASA's Ozone Depletion Website
  http://www.nas.nasa.gov/About/Education/Ozone/

Key Terms

alkali metal	Bohr model	ion	noble gas
alkaline earth metal	Bohr, Niels	isotope	nonmetal
amu	cation	mass number (*A*)	orbitals
anions	chalcogens	Mendeleev, Dmitri	oxygen family
atom	charge	metal	periodic table
atomic mass	chemical symbol	metalloid	quantum mechanical model
atomic number (*Z*)	electron configuration	molar mass	quantum number (*n*)
Avogadro, Amadeo	family or group of elements	mole	transition metal
Bohr diagram	halogen	neutrons	valence electron

Exercises

Assess your understanding of this chapter's topics with an online chapter quiz at **http://chemistry.brookscole.com/tro3**

QUESTIONS

1. Why is it important to understand atoms?
2. Describe an atom.
3. What defines an element? How many naturally occurring elements exist?
4. List three different ways that chemical elements were named.
5. Use the periodic table to write the name and the atomic number of the element that corresponds to each of the following symbols:

H	He	Li	Be	B	C	N	O	F	Ne
Na	Mg	Al	Si	P	S	Cl	Ar	Fe	Cu	Br
Kr	Ag	I	Xe	W	Au	Hg	Pb	Rn	U

6. Define each of the following terms:
   a. atomic number
   b. mass number
   c. isotope
   d. ion
   e. atomic mass
   f. molar mass
   g. Avogadro's number
   h. chemical symbol

7. Write the mass and charge of the proton, neutron, and electron.
8. What is the periodic law?
9. What was Mendeleev's largest contribution to the development of modern chemistry?
10. Explain the Bohr model for the atom. How does the model explain the periodic law?
11. Explain the quantum mechanical model for the atom. How is it different from the Bohr model?
12. Give two examples of each:
    a. alkali metal
    b. alkaline earth metal
    c. halogen
    d. noble gas
    e. metal
    f. nonmetal
    g. transition metal
    h. metalloid
13. Which elements exist as diatomic molecules?
14. Explain the difference and similarity between atomic mass and molar mass.

PROBLEMS

15. Determine the number of protons and electrons in each of the following ions:
    a. K^+
    b. S^{2-}
    c. Ca^{2+}
    d. Br^-
    e. Al^{3+}
16. Determine the number of protons and electrons in each of the following ions:
    a. Na^+
    b. O^{2-}
    c. Cr^{2+}
    d. I^-
    e. Fe^{3+}
17. Give the atomic number (Z) and the mass number (A) for each of the following:
    a. a carbon atom with 8 neutrons
    b. an aluminum atom with 14 neutrons
    c. an argon atom with 20 neutrons
    d. a copper atom with 36 neutrons
18. Give the atomic number (Z) and mass number (A) for each of the following:
    a. an oxygen atom with 8 neutrons
    b. a chlorine atom with 18 neutrons
    c. a sodium atom with 12 neutrons
    d. a uranium atom with 143 neutrons
19. The following isotopes have applications in medicine. Write their symbols in the form $^A_Z X$.
    a. cobalt-60
    b. phosphorus-32
    c. iodine-131
    d. sulfur-35
20. The following isotopes are important in nuclear power. Write their symbols in the form $^A_Z X$.
    a. U-235
    b. U-238
    c. Pu-239
    d. Xe-144
21. Determine the number of protons, neutrons, and electrons in each of the following:
    a. ^{238}U (used in nuclear reactors)
    b. ^{14}C (used in carbon dating of fossils)
    c. $^{23}Na^+$
    d. $^{81}Br^-$
    e. $^{16}O^{2-}$
22. Determine the number of protons, neutrons, and electrons in each of the following:
    a. ^{239}Pu
    b. $^{52}Cr^{3+}$
    c. $^{16}O^{2-}$
    d. $^{40}Ca^{2+}$
23. Give electron configurations according to the Bohr model for each of the following elements. Try to not use Figure 11, but instead determine the configuration based on your knowledge of the number of electrons in each atom and the maximum number of electrons in each Bohr orbit. Indicate which of these elements you expect to be the most reactive and the least reactive.
    a. B
    b. Si
    c. Ca
    d. F
    e. Ar
24. Give electron configurations according to the Bohr model for each of the following elements. Indicate which of these elements you expect to be the most reactive and the least reactive.
    a. He
    b. Al
    c. Be
    d. Ne
    e. O

25. How many valence electrons are in each of the elements in problem 23?
26. How many valence electrons are in each element of problem 24?
27. Draw electron configurations for each of the following elements according to the Bohr model. Indicate which electrons are valence electrons.
    a. Li
    b. C
    c. F
    d. P
28. Draw electron configurations for each of the following elements according to the Bohr model. Indicate which electrons are valence electrons.
    a. Mg
    b. S
    c. He
    d. K
29. Which two of the following elements would you expect to be most similar to each other? Why?
    Mg, N, F, S, Ne, Ca
30. Group the following elements into three similar groups of two each:
    Na, O, Ne, Li, Ar, S
31. We have seen that the reactivity of an element is determined by its electron configuration. What is the electron configuration of the ion Cl^-? (Hint: You must add one additional electron beyond the number of electrons that chlorine would normally have.) How does its reactivity compare with neutral Cl? How would the reactivities of Na and Na^+ compare?
32. What is the electron configuration of Mg^{2+}? How does its reactivity compare with neutral Mg? How do the reactivities of F^- and F compare?
33. Classify each of the following elements as a metal, a nonmetal, or a metalloid:
    a. Cr
    b. N
    c. Ca
    d. Ge
    e. Si
34. Classify each of the following as a metal, a nonmetal, or a metalloid:
    a. Mn
    b. I
    c. Te
    d. Sb
    e. O
35. Calculate the atomic weight of neon (Ne), composed of three naturally occurring isotopes with the following natural abundances and masses:
    a. 90.51% Ne-20 (mass = 19.992 amu)
    b. 0.27% Ne-21 (mass = 20.993 amu)
    c. 9.22% Ne-22 (mass = 21.991 amu)
36. An element has two naturally occurring isotopes. Isotope 1 has a mass of 106.905 amu and a relative abundance of 51.8%, and isotope 2 has a mass of 108.904 amu and a relative abundance of 48.2%. Find the atomic weight of this element and, by comparison to the periodic table, identify it.
37. A fictitious element has two naturally occurring isotopes and has an atomic weight of 29.5 amu.
    a. If the natural abundance of isotope 1 is 33.7%, what is the natual abundance of isotope 2?
    b. If the mass of isotope 2 is 30.0 amu, what is the mass of isotope 1?
38. Copper has two naturally occurring isotopes. Cu-63 has a mass of 62.939 amu and a relative abundance of 69.17%. Use the atomic weight of copper to determine the mass of the second copper isotope.
39. How many moles of copper are there in 138 g of copper?
40. How many moles of Al are there in an aircraft part weighing 2.6 kg?
41. How many moles are there in each of the following?
    a. 4.8 kg of Au
    b. 56.8 g of Na
    c. 7.9×10^{24} C atoms
    d. 4.5×10^2 He atoms
42. What is the mass of each of the following?
    a. 7.30 mol of P
    b. 14.8 mol of C
    c. 5.00×10^{23} Cr atoms
    d. 2.80×10^{24} I atoms

43. How many Ag atoms are present in a piece of pure silver jewelry weighing 38.7 g?
44. How many platinum atoms are in a pure platinum ring weighing 4.8 g?
45. A pure gold necklace has a volume of 1.8 cm³. How many gold atoms are in the necklace? The density of gold is 19.3 g/cm³.
46. A titanium bicycle component has a volume of 32 cm³. How many titanium atoms are in the bicycle component? Titanium density = 4.50 g/cm³.
47. An iron sphere has a radius of 3.4 cm. How many iron atoms are in the sphere? Iron has a density of 7.86 g/cm³ and a sphere has a volume of $V = \left(\frac{4}{3}\right) \times \pi r^3$.
48. Calculate the number of atoms in the universe. The following steps will guide you through this calculation:
    a. Planets constitute less than 1% of the total mass of the universe and can therefore be neglected. Stars make up most of the visible mass of the universe, so we need to determine how many atoms are in a star. Stars are primarily composed of hydrogen atoms and our Sun is an average-sized star. Calculate the number of hydrogen atoms in our Sun given that the radius of the Sun is 7×10^8 m and its density is 1.4 g/cm³. The volume of a sphere is given by
    $$V = \left(\frac{4}{3}\right) \times \pi r^3$$
    (Hint: Use the volume and the density to get the mass of the Sun.)
    b. The average galaxy (like our own Milky Way galaxy) contains 1×10^{11} stars, and the universe contains 1×10^9 galaxies. Calculate the number of atoms in an average galaxy and finally the number of atoms in the entire universe.
    c. You can hold 1×10^{23} atoms in your hand (five copper pennies constitute 1.4×10^{23} copper atoms.) How does this number compare with the number of atoms in the universe?

POINTS TO PONDER

49. The introduction to this chapter states that everything is made of atoms, including ourselves. Does that affect the way you view human life? Do the atoms in our bodies follow the same physical laws as atoms in soil or rocks or water? If so, does this make human life any less unique?
50. Suppose the absolute value of the charge of the electron was slightly greater than the charge of a proton. How would matter behave? What does this tell you about how small changes in the atomic realm affect the macroscopic realm?
51. When we refer to doughnuts or cookies, we often refer to 1 doz of them, which corresponds to 12. Why is the dozen an inconvenient number when referring to atoms? Why is Avogadro's number, 6.022×10^{23}, more convenient?
52. Draw 10^1 squares on a piece of paper and time yourself as you do it. How long would it take you to draw 10^2 squares? 10^4 squares? Estimate how long it would take you to draw 10^{15} squares (60 seconds = 1 minute, 5.3×10^5 minutes = 1 year). Comment on expressing numbers in an exponential fashion.
53. Why does Avogadro's number have such an odd value? Why not pick a nice round number like 1.0×10^{24}?
54. Can you think of anything that is not composed of atoms?
55. Explain the connection between the properties of an element and the atoms that compose it.
56. Read the interest box in this chapter entitled, *What If . . . Philosophy, Determinism, and Quantum Mechanics*. Explain the concept of determinism. Do you think the universe is deterministic? Can you think of other arguments, besides quantum mechanics, against determinism?

Feature Problems and Projects

57. Here are three fictitious elements and a molecular view of the atoms that compose them. The molar mass of the middle element, (b), is 25 grams per dozen (g/doz). (The atoms of these fictitious elements are much larger than ordinary atoms.) Based on the size of the atoms, do you expect the atomic masses of elements (a) and (c) to be greater than or less than (b)? How many atoms are present in 175 g of element (b)?

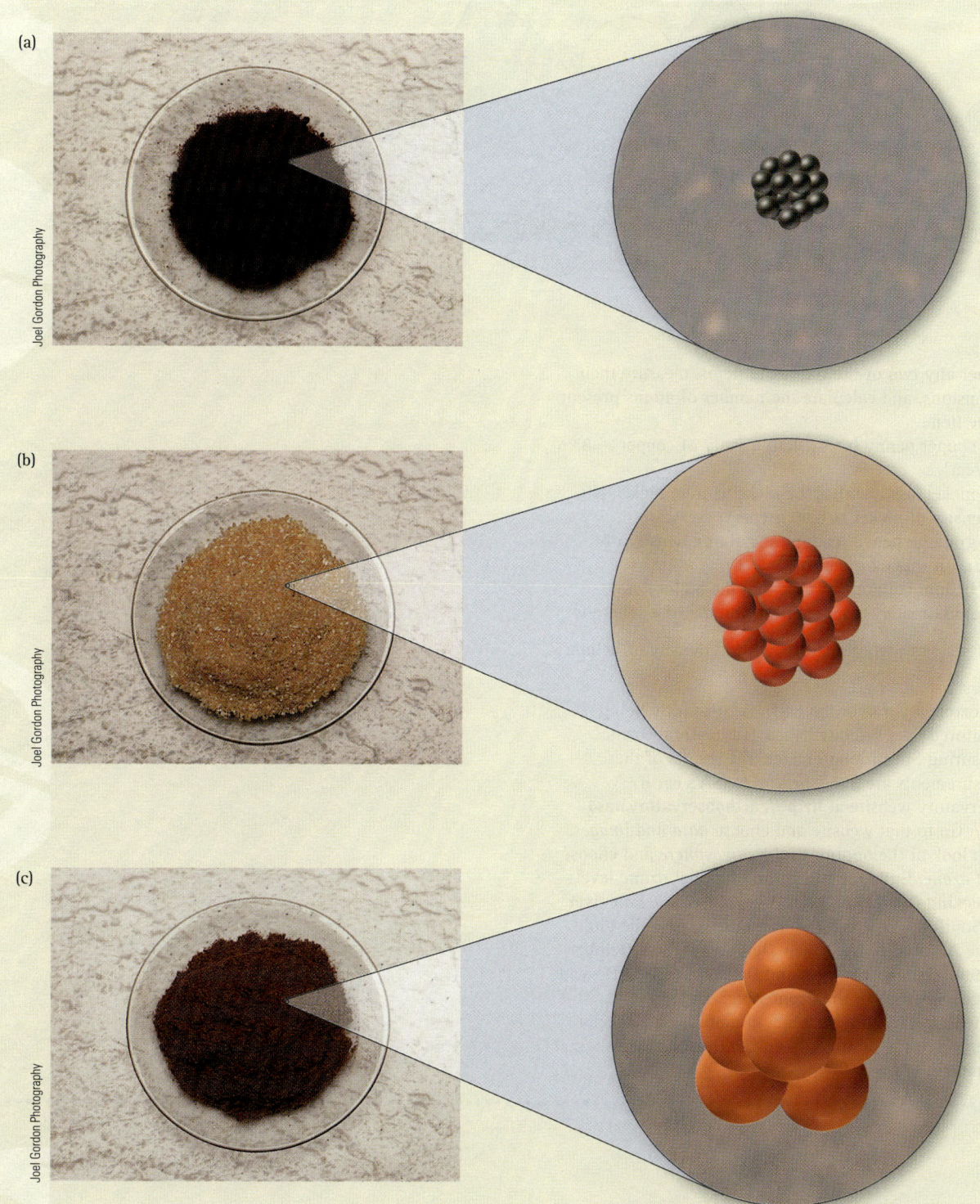

58. Read the box entitled, *What If . . . Complexity Out of Simplicity*. Explain the comment in the following cartoon.

59. Gather any two of the following items, measure their dimensions, and calculate the number of atoms present in the item.
    a. a copper penny (pre-1983), density of copper = 8.96 g/cm³
    b. a nickel (assume that the nickel is pure nickel, Ni), density of nickel = 8.90 g/cm³
    c. a graphite pencil lead (pure carbon), density of carbon = 2.62 g/cm³
    d. a helium balloon (assume that the balloon is approximately spherical; the volume of a sphere is $V = \frac{4}{3}\pi r^3$, where r is the radius), density of helium gas = 0.0899 g/L

60. The ozone layer over the Earth is constantly being monitored by a satellite called TOMS, Total Ozone Measuring Spectrophotometer. The results of these measurements are published on NASA's earth observatory website at **http://earthobservatory.nasa.gov/**. Go to that website and choose *data and images*. Then look at the options under *atmosphere* and choose *total ozone*. Build an animation of global ozone levels by picking October of 1980. View the ozone hole over the South Pole by using your mouse to turn the globe to see Antarctica. Next view October of 1985, October of 1990, October of 1996, and October of 2001. Do you notice any trend in the ozone hole? Staying in a particular year, view subsequent months. Pay special attention to the South Pole during October. What happens?

4 Molecules, Compounds, and Chemical Reactions

Knowledge is an attitude, a passion, actually an illicit attitude. For the compulsion to know is just like dipsomania, erotomania, and homicidal mania, in producing a character that is out of balance. It is not true that the scientist goes after truth. It goes after him.

—Robert Musil, 1880–1942

All matter is ultimately composed of atoms, but those atoms are usually bonded together to form molecules. Molecules are at the core of chemical processes; they compose most substances, and they form and transform in chemical reactions. These topics are the focus of this chapter. What are molecules? How do we represent them on paper, and how do we name them? How do we represent the reactions they undergo?

Molecules and their reactions are central to our daily existence—in our bodies when we breathe, on our stoves when we cook, and in our cars when we drive. The changes we see in the substances around us always involve some sort of change at the molecular level. As you learn about molecules, remember that our understanding of the molecule has transformed society. We can make molecules to serve specific purposes that we design. For example, we can make molecules that form durable containers (plastics), or we can make molecules that kill unwanted organisms within the human body (antibiotics), or we can make molecules that prevent pregnancy (the birth-control pill). The power to understand and manipulate molecules has changed the way we live and will continue to affect our lives, even in ways we may not yet imagine, for the foreseeable future.

QUESTIONS FOR THOUGHT

- Are most substances composed of isolated atoms? How do atoms group together in compounds?

- How do the properties of substances depend on the molecules that compose them? How sensitive are those properties to changes within these molecules?

- How do we represent compounds? How do we name them?

- How do compounds that form from a metal and a nonmetal differ from compounds that form from two or more nonmetals?

- What is the characteristic mass of a molecule?

- How much of a particular element is present in a particular compound?

- How do molecules change in chemical reactions? How do we represent those changes?

4.1 Molecules Cause the Behavior of Matter

If someone were to ask me what the most important piece of information about the physical world is, my answer would be this: Molecules cause the behavior of matter. In this simple, six-word sentence lays the key that explains the physical world and our experience of it. If the rest of scientific information were to vanish, this one idea would be the most helpful in recreating what we currently know. Matter is the way it is because molecules are the way they are. Water does what water molecules do, air does what air molecules do, and humans do what the molecules that compose us do. It is that simple, and it is always true. In all of science history, an exception to this rule has not yet been found.

As you sit wherever you are sitting staring at this page of light and dark spots, molecules are interacting in a complex dance. Molecules within ink interact with molecules in paper to create the light and dark spots on the page. Those molecules interact with light, causing the image of light and dark spots to be transmitted to your eyes. The light then strikes certain molecules within your eyes that change shape and cause a signal to be sent to your brain. Molecules within your brain then interpret the code to produce the ideas, concepts, and images that constitute reading. Although we don't know enough to explain this complex process in detail, we can map portions of it. For simpler processes, we can map the entire thing. In this chapter, we begin our journey of understanding molecules and how they form the macroscopic substances we experience every day.

4.2 Chemical Compounds and Chemical Formulas

Recall from Section 1.7 that a compound is a substance composed of two or more elements in fixed, definite proportions.

In Chapter 3, we learned that all matter is composed of atoms. In this chapter, we learn that those atoms are usually bound together to form compounds. A survey of some common substances reveals that most of them are compounds or mixtures of compounds. For example, one of the most common substances we encounter is water with molecules composed of two hydrogen atoms bonded to an oxygen atom. Most samples of water are not pure water, but a mixture of

Water is a compound whose molecules are composed of two hydrogen atoms for every oxygen atom.

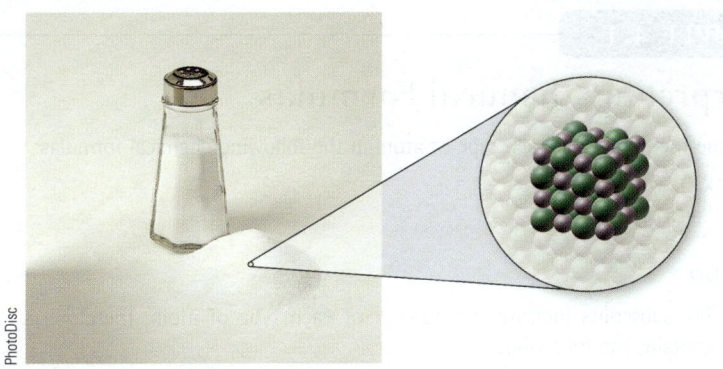

Salt is a compound composed of sodium and chlorine ions in a 1:1 ratio. Unlike water, which is composed of discrete molecules, sodium chloride exists in a three-dimensional crystalline array in which sodium and chloride ions alternate.

water with other compounds. Seawater contains, among other substances, table salt (sodium chloride). Table salt is a compound composed of sodium and chlorine atoms bonded together in a one-to-one (1:1) ratio. Some other common compounds include sugar, natural gas, and ammonia.

CHEMICAL FORMULAS

We represent a compound with a **chemical formula**, which, at minimum, indicates the elements present in the compound and the relative number of atoms of each. For example, NaCl is the chemical formula for table salt. It shows that table salt is composed of sodium and chlorine atoms in a 1:1 ratio. H_2O is the chemical formula for water, showing that water is composed of hydrogen and oxygen in a 2:1 ratio. Compounds such as H_2O should never be confused with mixtures of hydrogen and oxygen. Hydrogen and oxygen are both gases and a mixture of the two is also a gas. In contrast, the compound H_2O is a liquid with vastly different properties with which we are all familiar. Because the subscripts in a chemical formula indicate the relative number of each atom in the compound, they do not change for a given compound. Water is always H_2O and never anything else. If we change the subscripts in a formula, we change the compound itself. For example, adding a 2 on the end of H_2O results in H_2O_2, which is hydrogen peroxide, a very different substance. The chemical formulas of some other common compounds are CO_2 for carbon dioxide, CO for carbon monoxide, and $C_{12}H_{22}O_{11}$ for table sugar (Figure 1).

We discuss chemical bonds in more detail in the following sections and in Chapter 5. For now, simply think of bonds as links that hold atoms together.

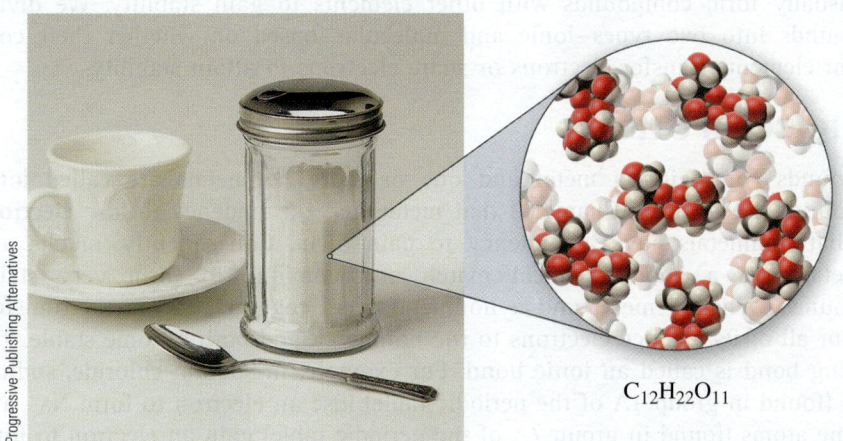

$C_{12}H_{22}O_{11}$

FIGURE 1 The chemical formula for table sugar tells us the number of atoms in each molecule in the compound. What elements, and how many atoms of each, is a sugar molecule composed of?

EXAMPLE 4.1

Interpreting Chemical Formulas

Determine the number of each type of atom in the following chemical formulas:

a. K_2O
b. $Ca(NO_3)_2$

SOLUTION

a. The subscripts indicate the number of each type of atom. Therefore, K_2O contains the following:

    2 K
    1 O

b. When a group of atoms is placed in parentheses with a subscript outside of the parentheses, that subscript applies to the entire group of atoms. Therefore, $Ca(NO_3)_2$ contains the following:

    1 Ca
    2 N
    6 O

YOUR TURN

Interpreting Chemical Formulas

Determine the number of each type of atom in the following chemical formulas:

a. CCl_4
b. $Al_2(SO_4)_3$

4.3 Ionic and Molecular Compounds

Elements combine to form compounds because, as we learned from the Bohr model, only a few elements—namely the noble gases—exist as isolated atoms with stable electron configurations. Elements with unstable electron configurations will usually form compounds with other elements to gain stability. We divide compounds into two types—ionic and molecular—based on whether their constituent elements transfer electrons or share electrons to attain stability.

IONIC COMPOUNDS

Compounds containing a metal and one or more nonmetals are called **ionic compounds**. Recall from Chapter 3 that metals have a tendency to lose electrons and that nonmetals have a tendency to gain them. Consequently, metals and nonmetals are a good chemical match and combine to form very stable compounds. When a metal and a nonmetal bond together, the metal transfers some or all of its valence electrons to the nonmetal, and both become stable. The resulting bond is called an **ionic bond**. For example, in sodium chloride, sodium atoms (found in group 1A of the periodic table) lose an electron to form Na^+ and chlorine atoms (found in group 7A of the periodic table) gain an electron to form Cl^-. The compound, NaCl, is then held together by the attraction between

FIGURE 2 The Na⁺ and Cl⁻ ions in sodium chloride alternate in a three-dimensional cubic pattern. When table salt is viewed with a microscope we can see the cubic-shaped crystals, which result from the cubic arrangement of the sodium and chloride ions.

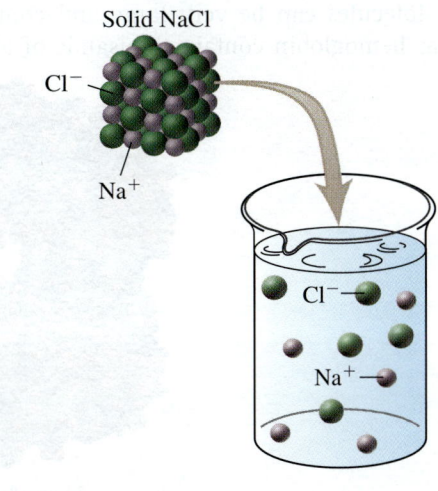

FIGURE 3 When sodium chloride dissolves in water, it dissociates into its component ions.

positively charged sodium ions and negatively charged chlorine ions. Both Na⁺ and Cl⁻ have full outer Bohr orbits, and the compound is stable. In NaCl crystals, the sodium ions and chlorine ions alternate in three dimensions to form a crystalline lattice, as shown in Figure 2. The crystal is held together by ionic bonds between positively charged sodium ions and negatively charged chlorine ions. The well-ordered cubic pattern of atoms in the lattice results in well-ordered cubic-shaped salt crystals.

When ionic compounds dissolve in water, they dissociate to form ions (Figure 3). For example, salt water does not contain any NaCl units, but it contains Na⁺ ions and Cl⁻ ions instead. Solutions with dissolved ions are called **electrolyte solutions** and are good conductors of electricity due to the mobility of the charged ions. (Electricity is the flow of charged particles.)

MOLECULAR COMPOUNDS

Compounds containing only nonmetals are called **molecular compounds**. In a molecular compound, the atoms share their electrons—although not always equally, as we will see in Chapter 5—to gain stability. The resulting bond is called a **covalent bond**. Unlike ionic compounds, which are composed of repeating positive and negative ions in large arrays, molecular compounds are composed of clusters of two or more atoms bonded together to form **molecules**. Molecules are represented by a more specific kind of chemical formula called a **molecular formula**. A molecular formula is a chemical formula that specifies the actual number of each kind of atom in the molecule, not just the simplest ratio. For example, benzene, a minority component of gasoline and a carcinogen, has the molecular formula C_6H_6. Even though the simplest ratio of carbon to hydrogen in benzene is 1:1, its molecular formula is C_6H_6, not CH, as benzene *molecules* contain 6 carbon atoms and 6 hydrogen atoms, all bonded together.

benzene molecule (C_6H_6)

Molecules can be very large and complex. For example, protein molecules such as hemoglobin contain thousands of atoms.

hemoglobin molecule

The shapes of molecules are covered in detail in Section 5.6. The properties of water are covered in detail in Chapter 12.

Molecules govern macroscopic behavior. The bulk properties of molecular compounds depend on the molecules that compose them; a slight change in the molecule has profound consequences. For example, changing the shape of the water molecule from bent to straight would cause water, which normally boils at 100°C (212°F), to boil at temperatures well below 0°C (32°F).

water molecule (H_2O)

The result of this change would be devastating. The Earth's oceans would boil away, as would much of our body because we are made up of large amounts of water.

Consider the hemoglobin molecule shown previously. The 10,000 atoms that compose it, and the way they are joined together, govern its function: to transport oxygen in our blood. Sickle cell anemia, a serious disease, is the result of just a few misplaced atoms in the hemoglobin molecule. For this reason, we emphasize *molecular* reasons for macroscopic behavior. The molecule—its atoms, its shape, its structure, and its bonds—is responsible for most of what we observe and experience. A small change in any of these, substituting an atom here or changing a shape there, dramatically changes the properties of the compound that the molecule composes. Take a moment to examine the forms of the molecules shown below, remembering that every atom and every shape influence the properties we usually associate with these compounds.

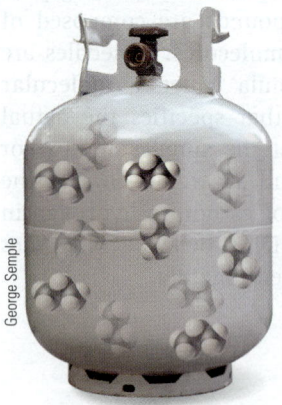

Propane (C_3H_8), used as a fuel

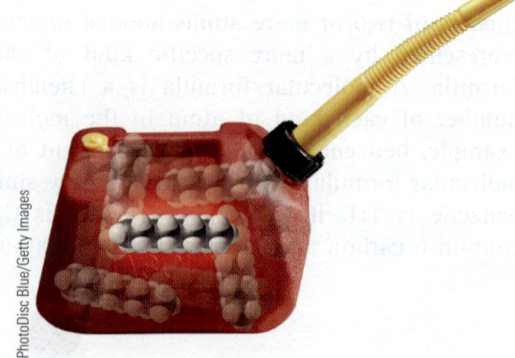

Octane (C_8H_{18}), a major component of gasoline

4.3 Ionic and Molecular Compounds

Cholesterol ($C_{27}H_{46}O$), a partial cause of arteriosclerosis, or clogging of the arteries

Ammonia (NH_3), a pungent gas used as a household cleaner

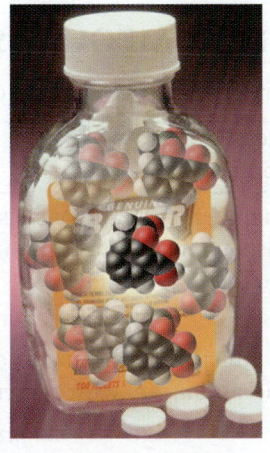

Acetylsalicylic acid ($C_9H_8O_4$), or aspirin

Sucrose ($C_{12}H_{22}O_{11}$), or sugar

WHAT IF...

Problem Molecules

On one of my daily runs, I found myself behind two large, old trucks going so slowly that I couldn't help but keep up with them. The combination of their being old and being trucks made them particularly good polluters. I coughed and felt dizzy as they spewed black smoke into the air that I was trying to breathe. Some of the molecules in that smoke were problem molecules—molecules that, because of their particular combination of atoms and because of their particular shape, cause problems in the environment, and, in this case, in my respiratory system. One of the problem molecules in truck and automobile exhaust is carbon monoxide (CO). Carbon monoxide is deceptively similar, both in size and in shape, to oxygen (O_2).

Consequently, carbon monoxide fits snugly into the spot within hemoglobin—the oxygen carrier in blood—that is normally reserved for oxygen, and this hinders the body's ability to get oxygen to the muscles and to the brain where it is most needed, hence my dizziness. Problem molecules create problems because their particular characteristics produce an undesirable effect either on humans or on our environment. Some of these problem molecules are naturally present in the world—the poison in poisonous mushrooms, for example—and others are put into the environment as byproducts of societal processes such as the combustion of fuel for transportation.

QUESTIONS: How do problem molecules affect your daily life? Can you think of any specific examples? Why is it important to study and understand the chemistry of problem molecules? Is this important for the average citizen or just for the scientist? Why or why not?

O_2 CO

4.4 Naming Compounds

The vast numbers of compounds that exist require us to devise a systematic way to name them. Many compounds, however, also have common names that can only be learned through familiarity or memorization. For example, H_2O has the common name *water* and the systematic name *dihydrogen monoxide*. NH_3 has the common name *ammonia* and the systematic name *nitrogen trihydride*. A common name is a sort of nickname for a compound, used by those who are familiar with it. Some compounds, such as carbon dioxide for example, are known only by their systematic names. Others, such as water, are known only by their common names. The systematic name of a compound can be assigned based on its chemical formula. In this section, we learn how to assign systematic names to simple ionic and molecular compounds.

NAMING IONIC COMPOUNDS

The names for binary (two-element) ionic compounds have the following form:

| name of cation (metal) | base name of anion (nonmetal) + *ide* |

For example, the name for NaCl consists of the name of the cation, *sodium*, followed by the base name of the anion, *chlor*, with the ending *ide*. The full name is as follows:

NaCl   *sodium chloride*

The name for MgO consists of the name of the cation, *magnesium*, followed by the base name of the anion, *ox*, with the ending *ide*. The full name is as follows:

MgO   *magnesium oxide*

The name for Li_2S consists of the name of the cation, *lithium*, followed by the base name of the anion, *sulf*, with the ending *ide*. The full name is as follows:

Li_2S *lithium sulfide*

Note that the names of ionic compounds DO NOT contain prefixes such as *di* or *tri* to indicate the number of each type of atom. However, ionic compounds containing a *transition* metal (see the periodic table on the inside front cover of this book) usually include a roman numeral indicating the charge of the metal. For example, $FeCl_3$ is named as follows:

$FeCl_3$ *iron(III) chloride*

The base names for various nonmetals and their most common charges in ionic compounds are shown in Table 1.

> In a more comprehensive treatment of nomenclature, compounds containing ions that exhibit more than one charge have the ionic charge indicated by the Roman numeral. For example, Sn and Pb are not transition metals, but their names require a Roman numeral because their charge can vary from one compound to another.

APPLY YOUR KNOWLEDGE

Locate each of the ions in Table 1 on the periodic table (on the inside cover of this book). Why do F, Cl, Br, and I form ions with 1− charges? Why do oxygen and sulfur form ions with a 2− charge? Why does N form ions with a 3− charge?

Answer: Recall from Chapter 3 that F, Cl, Br, and I are halogens, and each has seven valence electrons. To attain a full outer orbit, each atom must gain one electron; therefore, these atoms tend to form ions with 1− charges. Oxygen and sulfur have six valence electrons and must gain two electrons to attain a full outer orbit; therefore, they tend to form ions with 2− charges. Nitrogen has five valence electrons and must gain three electrons to attain a full outer orbit; therefore, nitrogen tends to form an ion with a 3− charge.

TABLE 1
Some Common Anions

Nonmetal	Symbol for Ion	Base Name	Anion Name
Fluorine	F^-	Fluor	Fluoride
Chlorine	Cl^-	Chlor	Chloride
Bromine	Br^-	Brom	Bromide
Iodine	I^-	Iod	Iodide
Oxygen	O^{2-}	Ox	Oxide
Sulfur	S^{2-}	Sulf	Sulfide
Nitrogen	N^{3-}	Nitr	Nitride

Many ionic compounds contain anions with more than one atom. These ions are called **polyatomic ions** and are tabulated in Table 2. In naming compounds that contain these polyatomic ions, simply use the name of the polyatomic ion as the name of the anion. For example, KNO_3 is named according to its cation, *potassium,* and its polyatomic anion, *nitrate*. The full name is as follows:

KNO_3 *potassium nitrate*

TABLE 2
Some Common Polyatomic Ions

Name	Formula
Carbonate	CO_3^{2-}
Bicarbonate	HCO_3^-
Hydroxide	OH^-
Nitrate	NO_3^-
Phosphate	PO_4^{3-}
Sulfate	SO_4^{2-}

MOLECULAR FOCUS

Calcium Carbonate

Within most chapters of this text, we will highlight a "celebrity" compound in a *Molecular Focus* box. You have probably encountered these compounds in some way or another. We begin with calcium carbonate, an ionic compound that is abundant in nature.

Formula: $CaCO_3$
Molar Mass: 100.09 g/mol
Melting point: 1339°C (calcite form)

Calcium carbonate is an example of an ionic compound containing a polyatomic ion (CO_3^{2-}). Calcium carbonate is common in nature, occurring in eggshells, seashells, limestone, and marine sediments. It occurs most dramatically in stalactites and stalagmites in limestone caves. These formations develop over time because rainwater, containing atmospheric CO_2 that makes it acidic (more on this in Chapter 13), dissolves calcium carbonate from soils and rocks. As the calcium carbonate-saturated water seeps into the ground, some of the CO_2 escapes, lowering the acidity of the rainwater and causing the calcium carbonate to deposit as a solid. When this occurs in an underground cave, the dripping water forms structures called stalactites, which hang down from the ceiling of a cave, and stalagmites, which protrude up from the floor of a cave.

Calcium carbonate is used in many consumer products because of its low toxicity, structural stability, and tendency to

The stalactites and stalagmites of limestone caves are composed of calcium carbonate.

neutralize acids. It is the main ingredient in a number of building materials, including cement and marble. It also is the main component of popular over-the-counter antacids such as Tums and is commonly used to remove excess acidity from wines.

> **EXAMPLE 4.2**
>
> ### Naming Ionic Compounds
>
> Give the name for the compound MgF_2.
>
> **SOLUTION**
> The cation is *magnesium*. The anion is fluorine, which becomes *fluoride*. The correct name is *magnesium fluoride*.
>
> **YOUR TURN**
>
> ### Naming Ionic Compounds
>
> Give the name for the compound KBr.

> **EXAMPLE 4.3**
>
> ### Naming Ionic Compounds That Contain a Polyatomic Ion
>
> Give the name for the compound NaOH.
>
> **SOLUTION**
> The cation is *sodium*. The anion is a polyatomic ion whose name we can find in Table 2. The name for OH is *hydroxide*. The correct name for the compound is *sodium hydroxide*.
>
> **YOUR TURN**
>
> ### Naming Ionic Compounds That Contain a Polyatomic Ion
>
> Give the name for the compound $CaCO_3$.

NAMING MOLECULAR COMPOUNDS

The names for binary (two-element) molecular compounds are similar to those of ionic compounds. Even though both elements are nonmetals in a molecular compound, the more metallic element (that closest to the left bottom side of the periodic table) is written first and the less metallic element is written second. In addition, prefixes are used to show how many atoms of each element are present in each molecule. The overall form is as follows:

| (prefix) name of more metallic element | (prefix) name of less metallic element + *ide* |

The prefixes given to each element indicate the number of atoms of that element that are present. If the first element consists of only one atom, the prefix *mono* is omitted. For example, the compound CO_2 is named according to the first element, carbon, with no prefix because mono is omitted for the first element; and the second element, oxygen, which becomes *oxide* with the prefix *di* to indicate two oxygen atoms. The full name is as follows:

CO_2 carbon *dioxide*

The compound N_2O, also called laughing gas, is named according to the first element, nitrogen, with the prefix *di* to indicate that there are two nitrogen atoms; and the second element, *oxide*, with the prefix *mono* to indicate one oxygen atom. Because *mono* ends with a vowel and *oxide* begins with one, the *o* is dropped on *mono* and the two are combined as *monoxide*. The entire name for N_2O is as follows:

N_2O *dinitrogen monoxide*

Prefixes

mono = 1   *di* = 2
tri = 3   *tetra* = 4
penta = 5   *hexa* = 6

EXAMPLE 4.4

Naming Molecular Compounds

Name the compounds CCl_4, BCl_3, and SF_6.

SOLUTION

CCl_4	carbon tetrachloride
BCl_3	boron trichloride
SF_6	sulfur hexafluoride

YOUR TURN

Naming Molecular Compounds

Name the compound N_2O_4.

APPLY YOUR KNOWLEDGE

What is wrong with the following compound name: *monoselenium dioxide*?

Answer: *Mono* is normally dropped from the first element. The correct name is *selenium dioxide* (SeO_2).

4.5 Formula Mass and Molar Mass of Compounds

We have learned how compounds are represented with chemical formulas and how they are named. We now turn to their characteristic mass. Just as an element has a characteristic mass called its *atomic mass*, so a compound has a characteristic mass called its **formula mass**. The formula mass of a compound is computed by summing the atomic masses of all the atoms in its formula. For example, the formula mass of water, H_2O, is computed as follows:

$$H_2O \text{ formula mass} = 2(1.01 \text{ amu}) + 16.00 \text{ amu}$$
$$= 18.02 \text{ amu}$$

where 2(1.01 amu) comes from the atomic mass of hydrogen and 16.00 amu is the atomic mass of oxygen.

EXAMPLE 4.5

Calculating Formula Mass

Calculate the formula mass of carbon tetrachloride, CCl_4.

SOLUTION
To find the formula mass, we sum the atomic masses of each atom in the molecule. Because the molecule contains one C atom and four Cl atoms, its formula mass is as follows:

$$CCl_4 \text{ formula mass} = 12.01 \text{ amu} + 4(35.45 \text{ amu})$$
$$= 153.81 \text{ amu}$$

where 12.01 amu is the atomic mass of carbon and 35.45 amu is the atomic mass of chlorine.

YOUR TURN

Calculating Formula Mass

Calculate the formula mass of dinitrogen monoxide (N_2O), an anaesthetic gas also called laughing gas.

MOLAR MASS

Recall from Chapter 3 that the atomic mass of an element in amu is numerically equivalent to its molar mass in g/mol. The same relationship is true for compounds—the formula mass of a compound in amu is numerically equivalent to its **molar mass** in g/mol.

$$\text{formula mass (amu)} \leftrightarrow \text{molar mass (g/mol)}$$

For example, H_2O has a formula mass of 18.02 amu; therefore, H_2O has a molar mass of 18.02 g/mol—one mole of water molecules has a mass of 18.02 grams. Just as the molar mass of an element is a conversion factor between grams of the element and moles of the element, so the molar mass of a compound is a conversion factor between grams of the compound and moles of the molecule.

EXAMPLE 4.6

Using the Molar Mass to Find the Number of Molecules in a Sample of a Compound

Calculate the number of water molecules in a raindrop with a mass of 0.100 g.

SOLUTION

Begin by writing down the quantities you are given and the quantity you are asked to find.

Given

0.100 g H$_2$O

Find

number of water molecules

Use the molar mass of water (calculated previously) as a conversion factor between grams of H$_2$O and moles of H$_2$O. Then use Avogadro's number to find the number of water molecules.

$$0.100 \cancel{g} \times \frac{1 \cancel{\text{mole}}}{18.01 \cancel{g}} \times \frac{6.022 \times 10^{23} \text{ molecules}}{\cancel{\text{mole}}} = 3.34 \times 10^{21} \text{ molecules}$$

YOUR TURN

Using the Molar Mass to Find the Number of Molecules in a Sample of a Compound

Calculate the number of carbon tetrachloride (CCl$_4$) molecules in 3.82 g of carbon tetrachloride.

4.6 Composition of Compounds: Chemical Formulas as Conversion Factors

We often want to know how much of a particular element is present in a particular compound. For example, an iron manufacturer may want to know how much iron (Fe) is present in a ton of iron oxide (Fe$_2$O$_3$), or an estimate of the threat of ozone depletion may require knowing how much chlorine (Cl) is in a ton of a particular chlorofluorocarbon such as Freon-12 (CF$_2$Cl$_2$). The information necessary for these types of calculations are inherent in chemical formulas.

We can understand the concept behind these calculations with a simple analogy. Asking how much iron is in a ton of iron oxide is much like asking how many tires are in 121 cars. We need a conversion factor between tires and cars. For cars, the conversion factor comes from our knowledge about cars; we know that each car has four tires (Figure 4).

We can write:

$$4 \text{ tires} \equiv 1 \text{ car}$$

The ≡ sign means "equivalent to." Although four tires do not equal one car—a car obviously has many other components—four tires are equivalent to one car, meaning that each car must have four tires to be a complete car. With this equivalence, we can build a conversion factor to find the number of tires in 121 cars:

$$121 \cancel{\text{cars}} \times \frac{4 \text{ tires}}{1 \cancel{\text{car}}} = 484 \text{ tires}$$

Chlorine within chlorofluorocarbons depletes atmospheric ozone, a shield against harmful ultraviolet light. This topic is covered in detail in Chapter 11.

4 tires ≡ 1 car

2 H atoms ≡ 1 H$_2$O molecule

FIGURE 4 The conversion factors inherent in a chemical formula are much like the conversion factors inherent in the "formula" that specifies a car.

Similarly, a chemical formula gives us equivalences between the elements in a particular compound and the compound itself. For example, the formula for water (H_2O) means there are 2 H atoms per O atom, or 2 H atoms per H_2O molecule. We write this as:

$$2 \text{ H atoms} \equiv 1 \text{ O atom} \equiv 1 \text{ } H_2O \text{ molecule}$$

or

$$2 \text{ mol H atoms} \equiv 1 \text{ mol O atoms} \equiv 1 \text{ mol } H_2O \text{ molecules}$$

These equivalences allow us to determine the amounts of the constituent elements present in a given amount of a compound. For example, suppose we want to know the number of moles of H in 12 mol of H_2O. We begin the calculation with the 12 moles of H_2O and use the conversion factor obtained from the chemical formula to calculate the moles of H.

$$12 \text{ mol } H_2O \times \underset{\text{conversion factor from chemical formula}}{\frac{2 \text{ mol H}}{1 \text{ mol } H_2O}} = 24 \text{ mol H}$$

Therefore, 12 moles of H_2O molecules contain 24 moles of H atoms. If the amount of H_2O were given in grams rather than moles, we would simply convert from grams to moles first, and then proceed as illustrated earlier.

EXAMPLE 4.7

Chemical Formulas as Conversion Factors (Moles to Moles)

Determine the number of moles of chlorine atoms in 4.38 moles of Freon-12, a chlorofluorocarbon that has the chemical formula CF_2Cl_2.

SOLUTION

Given
4.38 mol CF_2Cl_2

Find
moles of Cl

The equivalence between moles of Freon-12 and moles of Cl is obtained from the chemical formula.

$$2 \text{ mol Cl} \equiv 1 \text{ mol } CF_2Cl_2$$

Begin with moles of CF_2Cl_2 and multiply by the conversion factor to obtain moles of Cl:

$$4.38 \text{ mol } CF_2Cl_2 \times \frac{2 \text{ mol Cl}}{1 \text{ mol } CF_2Cl_2} = 8.76 \text{ mol Cl}$$

YOUR TURN

Chemical Formulas as Conversion Factors (Moles to Moles)

How many moles of Cl are in 2.43 moles of CCl_4?

> **EXAMPLE 4.8**
>
> ### Chemical Formulas as Conversion Factors (Mass to Mass)
>
> How many grams of oxygen are contained in 13.5 grams of glucose ($C_6H_{12}O_6$)?
>
> **SOLUTION**
>
> **Given**
> 13.5 g $C_6H_{12}O_6$
>
> **Find**
> grams of O
>
> The equivalence between moles of oxygen and moles of glucose is obtained from the chemical formula:
>
> $$6 \text{ mol O} \equiv 1 \text{ mol } C_6H_{12}O_6$$
>
> However, we are given grams of glucose, not moles. We need the following molar masses to convert between grams and moles.
>
> $$\text{molar mass of O} = 16.00 \text{ g/mol}$$
>
> $$\text{molar mass of } C_6H_{12}O_6 = 6\,(12.01) + 12\,(1.01) + 6\,(16.00)$$
> $$= 180.18 \text{ g/mol}$$
>
> We must first convert grams of glucose to moles of glucose using the molar mass of glucose. Then we can convert to moles of oxygen and finally to grams of oxygen using the molar mass of oxygen. The entire calculation has the following form:
>
> $$\text{g } C_6H_{12}O_6 \rightarrow \text{mol } C_6H_{12}O_6 \rightarrow \text{mol O} \rightarrow \text{g O}$$
>
> use glucose molar mass as conversion factor | use conversion factor from chemical formula | use oxygen molar mass as conversion factor
>
> Beginning with grams of glucose, the calculation proceeds as follows:
>
> $$13.5 \text{ g } C_6H_{12}O_6 \times \frac{\text{mol } C_6H_{12}O_6}{180.18 \text{ g } C_6H_{12}O_6} \times \frac{6 \text{ mol O}}{\text{mol } C_6H_{12}O_6} \times \frac{16.00 \text{ g O}}{\text{mol O}} = 7.19 \text{ g O}$$
>
> **YOUR TURN**
>
> ### Chemical Formulas as Conversion Factors (Mass to Mass)
>
> How many grams of O are in 2.5 grams of CO_2?

4.7 Forming and Transforming Compounds: Chemical Reactions

Compounds are made in chemical reactions and can be transformed into other compounds by chemical reactions. For example, methane (CH_4) is a colorless and odorless gas. It is the major component of natural gas and burns in air, reacting with oxygen and producing heat. We are all familiar with the flame from a natural gas burner, but what happens at the molecular level? Methane molecules stream

FIGURE 5 A chemical reaction occurs in a natural gas flame.

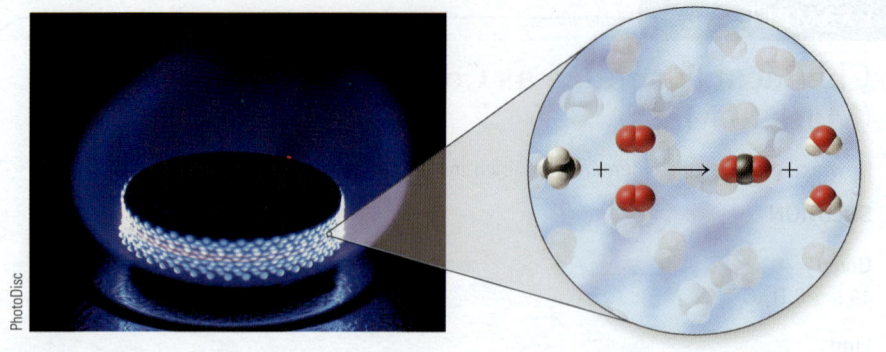

from a gas line and mix with oxygen (O_2) molecules present in air (Figure 5). The reaction begins with collisions between methane and oxygen. If the temperature is high enough, methane and oxygen molecules react to produce different molecules, carbon dioxide (CO_2) and water (H_2O). In a typical stove-top flame, trillions upon trillions of molecules undergo this reaction every second to produce the heat and light of the flame. The air above the flame contains the reaction products, carbon dioxide (CO_2) and water (H_2O), in the form of water vapor.

THE CHEMICAL EQUATION

Chemical reactions are represented by **chemical equations**. The reaction between methane and oxygen, for example, is written as follows:

$$\underbrace{CH_4 + O_2}_{\text{reactants}} \rightarrow \underbrace{CO_2 + H_2O}_{\text{products}} \quad \text{(unbalanced)}$$

The starting substances on the left side of the equation are called the **reactants**, and the new substances on the right side are called the **products**. Because chemical equations represent real chemical reactions, the number of atoms of each element on both sides of the equation must be equal—the equation must be *balanced*. New atoms do not form during a reaction, nor do atoms vanish; matter is conserved. The preceding equation is not balanced, as there are two oxygen atoms on the left side of the equation and three on the right. Also, there are four hydrogen atoms on the left and only two on the right.

$$CH_4 + O_2 \rightarrow CO_2 + H_2O$$

Reactants	Products
1 C atom	1 C atom
4 H atoms	2 H atoms
2 O atoms	3 O atoms

To balance the equation, we add coefficients to the reactants and products to make the number of atoms of each type of element on both sides of the equation equal. This changes the *number of* molecules involved in the reaction, but it does not change the *kind of* molecules. In this case, we put the coefficient 2 before O_2 in the reactants side of the equation, and the coefficient 2 before H_2O in the products side. Now the reaction is balanced.

$$CH_4 + 2\ O_2 \rightarrow CO_2 + 2\ H_2O$$

Reactants	Products
1 C atom	1 C atom
4 H atoms	4 H atoms (2 × H_2O)
4 O atoms (2 × O_2)	4 O atoms (1 × CO_2 + 2 × H_2O)

The total number of atoms of a given element in a compound is obtained by multiplying the subscript for the element by the coefficient for the compound. As you can see, the equation is now balanced. In general, chemical equations are balanced by following these guidelines:

1. If an element occurs in only one compound on both sides of the equation, balance that element first.
2. If an element occurs as a free element on either side of the chemical equation, balance that element last.
3. Change only the coefficients to balance a chemical equation, never the subscripts. Changing the subscripts changes the compound.
4. Clear coefficient fractions by multiplying the entire equation by the appropriate factor.

EXAMPLE 4.9

Balancing Chemical Equations

Balance the following chemical equation:

$$P_2H_4 \rightarrow PH_3 + P_4$$

SOLUTION

Because H occurs in only one compound on each side of the equation, we balance it first:

$$3\ P_2H_4 \rightarrow 4\ PH_3 + P_4$$

Next we balance P:

$$3\ P_2H_4 \rightarrow 4\ PH_3 + \frac{1}{2} P_4$$

Although the equation is technically balanced as written, we usually clear coefficient fractions such as 1/2 by multiplying both sides of the equation by 2:

$$\left[3\ P_2H_4 \rightarrow 4\ PH_3 + \frac{1}{2} P_4 \right] \times 2$$

$$6\ P_2H_4 \rightarrow 8\ PH_3 + P_4$$

We can always check if an equation is balanced by counting the number of each type of atom on each side:

Reactants	Products
12 P atoms (6 × P_2H_4)	12 P atoms (8 × PH_3 + 1 × P_4)
24 H atoms (6 × P_2H_4)	24 H atoms (8 × PH_3)

The equation is balanced.

> **YOUR TURN**
>
> ### Balancing Chemical Equations
>
> Balance the following chemical equation:
>
> $$HCl + O_2 \rightarrow H_2O + Cl_2$$

4.8 Reaction Stoichiometry: Chemical Equations as Conversion Factors

The coefficients in a chemical equation can be used as conversion factors in calculations much as the subscripts in a chemical formula were used previously. These calculations are important because they allow us to predict how much of a particular reactant might be needed in a particular reaction, or how much of a particular product will be formed. For example, one of the gases that contribute to global warming is carbon dioxide (CO_2). Carbon dioxide is a product of the combustion of fossil fuels such as methane, the primary component of natural gas. From the previous section, the chemical equation for the combustion of methane is as follows:

$$CH_4 + 2\ O_2 \rightarrow CO_2 + 2\ H_2O$$

We may want to calculate how much CO_2 is emitted into the atmosphere as a result of methane combustion. The total amount of methane used is known because gas companies keep accurate records. What we need is a conversion factor between the amount of methane used and the amount of carbon dioxide produced. The chemical equation gives us this conversion factor.

We can understand the concept behind these calculations with a simple analogy. Asking how much carbon dioxide is produced from the combustion of a certain amount of methane is much like asking how many pizzas can be made from a certain amount of cheese.

MAKING PIZZA AND MAKING MOLECULES

My recipe for pizza is simple:

$$1\ \text{pizza crust} + 8\ \text{oz cheese} + 4\ \text{oz tomato} \rightarrow 1\ \text{pizza}$$

Suppose that I have 32 oz of cheese on hand. Assuming I have enough of everything else, how many pizzas can I make from this cheese? From my recipe, I have the following equivalence:

$$8\ \text{oz cheese} \equiv 1\ \text{pizza}$$

The recipe for making pizza gives us numerical relationships among the different ingredients. Similarly, a chemical equation gives us the numerical relationships among the reactants and products in a chemical reaction.

The Molecular Revolution

Engineering Animals to Do Chemistry

In the simplest case, a chemist synthesizes a compound in the laboratory via a chemical reaction whose product is the desired compound. As our knowledge of chemistry has grown, however, so has the complexity of the molecules we wish to synthesize. Modern chemical synthesis often requires many reaction steps as well as numerous identification, separation, and purification steps along the way.

Some compounds, especially biological ones such as proteins, are so complex that they cannot be synthesized in the laboratory. However, these compounds are often needed to treat disease. Until recently, human proteins were available only from human sources, making them scarce and expensive. Advances in genetics have allowed scientists to transplant the human "blueprint" for making certain proteins into animals (more on this in Chapter 16). The animals then become factories for the desired protein. For example, scientists have altered the genes of a pig named Genie to produce human protein C, a blood protein central to blood clotting and absent in hemophiliacs. Now Genie's milk is rich in this critically needed human protein. Although still in experimental stages, such novel ways of synthesizing medical compounds could be extremely beneficial. For example, hemophiliacs often need blood proteins such as human protein C to cause their blood to clot. Without them, they can bleed to death. Currently, treatments are limited to patients who are actually bleeding, and each treatment costs tens of thousands of dollars.

Geneticists have transferred the codes for making human proteins into animals, such as this pig. The pig then becomes a factory for the protein.

Hemophiliacs could benefit from regular infusions of the necessary blood proteins, but the cost—over $100,000 per year—and the scarcity of the proteins make this almost impossible. The wide-scale use of genetically altered animals to synthesize medical compounds could produce ample amounts of the necessary proteins and bring the cost down dramatically.

As before, the ≡ sign means "equivalent to." Although 8 oz of cheese does not equal 1 pizza—good pizza requires much more than just cheese—8 oz of cheese is equivalent to one pizza, meaning that each pizza must have 8 oz of cheese to be complete. With this equivalence, we can construct the correct conversion factor to find the number of pizzas that are possible with 8 oz of cheese:

$$32 \text{ oz cheese} \times \frac{1 \text{ pizza}}{8 \text{ oz cheese}} = 4 \text{ pizzas}$$

↑
conversion factor from equivalences in recipe

Similarly, in a chemical reaction, we have a "recipe" for how molecules combine to form other molecules. For example, water is formed by the reaction of hydrogen and oxygen:

$$2\,H_2 + O_2 \rightarrow 2\,H_2O$$

The balanced reaction tells us that two H_2 molecules react with one O_2 molecule to form two H_2O molecules, or that two moles of H_2 react with one mole of O_2 to form two moles of H_2O. Because it is most useful to work with moles, we write our conversion factors as

$$2 \text{ mol } H_2 \equiv 1 \text{ mol } O_2 \equiv 2 \text{ mol } H_2O$$

If we start with 2 moles of H_2, and we have as much oxygen as we need, how much water can we make? We use the equivalence that we have already established to build a conversion factor that converts from moles of H_2 to moles of H_2O.

$$2 \text{ mol } H_2 \times \frac{2 \text{ mol } H_2O}{2 \text{ mol } H_2} = 2 \text{ mol } H_2O$$

conversion factor from equivalences in balanced chemical equation

We therefore have enough H_2 to produce 2 moles of H_2O. If the initial amounts of reactants are given in grams rather than moles, we simply convert from grams to moles first, and then proceed as illustrated in Example 4.11.

> **APPLY YOUR KNOWLEDGE**
>
> The coefficients in the chemical reaction $2A + 3B \rightarrow$ products gives us a relationship between which of the following:
>
> a. grams of A and grams of B
> b. density of A and density of B
> c. moles of A and moles of B
>
> Answer: c. The coefficients in a chemical reaction always give relationships between moles, never mass.

EXAMPLE 4.10

Using Chemical Equation Coefficients as Conversion Factors (Moles to Moles)

Aluminum oxide, Al_2O_3, can be made by the following reaction:

$$4 \text{ Al} + 3 \text{ O}_2 \rightarrow 2 \text{ Al}_2O_3$$

How many moles of aluminum oxide can be produced from 2.2 moles of Al? Assume that there is more than enough O_2 present.

SOLUTION
Begin by setting up the problem in the standard way.

Given
2.2 moles of Al

Find
moles of Al_2O_3
Equivalence (from balanced chemical equation):

$$4 \text{ mol Al} \equiv 2 \text{ mol Al}_2O_3$$

We use this equivalence to find how many moles of Al_2O_3 can be produced from the initial amount of Al:

$$2.2 \text{ mol Al} \times \frac{2 \text{ mol Al}_2O_3}{4 \text{ mol Al}} = 1.1 \text{ mol Al}_2O_3$$

There is enough Al to produce 1.1 mol Al_2O_3.

> **YOUR TURN**
>
> ### Using Chemical Equation Coefficients as Conversion Factors (Moles to Moles)
>
> An industrially important reaction is the production of ammonia, NH_3, from N_2 and H_2:
>
> $$N_2 + 3\,H_2 \rightarrow 2\,NH_3$$
>
> How many moles of NH_3 can be produced from 19.3 mol of H_2? Assume that there is more than enough N_2 present.

EXAMPLE 4.11

Using Chemical Equation Coefficients as Conversion Factors (Mass to Mass)

Suppose the city in which you live burns 2.1×10^8 grams of methane per day (approximate usage for a city with 50,000 inhabitants). How many grams of CO_2 are produced? The balanced reaction for the combustion of methane is as follows:

$$CH_4 + 2\,O_2 \rightarrow CO_2 + 2\,H_2O$$

SOLUTION

Given
2.1×10^8 g CH_4

Find
grams of CO_2

The necessary equivalence comes from the chemical equation:

$$1 \text{ mol } CH_4 \equiv 1 \text{ mol } CO_2$$

However, we are given grams of CH_4, not moles. We need the following molar masses to convert between grams and moles.

$$\text{molar mass of } CH_4 = 12.01 + 4(1.01) = 16.05 \text{ g/mol}$$

$$\text{molar mass of } CO_2 = 12.01 + 2(16.00) = 44.01 \text{ g/mol}$$

We must first convert grams of CH_4 to moles of CH_4 using the molar mass of CH_4. Then we can convert to moles of CO_2 and finally to grams of CO_2 using the molar mass of CO_2. The entire calculation has the following form:

$$g\ CH_4 \rightarrow mol\ CH_4 \rightarrow mol\ CO_2 \rightarrow g\ CO_2$$

- use CH_4 molar mass as conversion factor
- use conversion factor from chemical equation
- use CO_2 molar mass as conversion factor

Beginning with grams of CH_4, the calculation proceeds as follows:

$$2.1 \times 10^8 \text{ g } CH_4 \times \frac{\text{mol } CH_4}{16.05 \text{ g } CH_4} \times \frac{1 \text{ mol } CO_2}{1 \text{ mol } CH_4} \times \frac{44.01 \text{ g } CO_2}{\text{mol } CO_2} = 5.8 \times 10^8 \text{ g } CO_2$$

Consequently, 5.8×10^8 g of CO_2 are produced.

> **YOUR TURN**
>
> ### Using Chemical Equation Coefficients as Conversion Factors (Mass to Mass)
>
> One of the reactions occurring in automobile engines is the combustion of octane (C_8H_{18}):
>
> $$2\ C_8H_{18} + 25\ O_2 \rightarrow 16\ CO_2 + 18\ H_2O$$
>
> Assume that gasoline is pure octane and that you burn approximately 6.44×10^4 grams of octane per week (about 20 gal). How many grams of CO_2 are produced?

> **APPLY YOUR KNOWLEDGE**
>
> Consider the following reaction: $2A + 3B \rightarrow 2C$
> If you have 2 moles of A and 6 moles of B, what is the maximum number of moles of C that can be made by the reaction?
>
> Answer: 2 moles of C. Even though you have enough of B to make 4 moles of C, you only have enough of A to make 2 moles of C. The moles of A limit the amount of product that you can make.

MOLECULAR THINKING

Campfires

A campfire is a good example of a chemical reaction. As we saw in Chapter 1, a campfire consists of molecules from wood combining with oxygen from air to form carbon dioxide, water, and heat. Have you ever noticed that it is easier to build a good fire if there is a breeze? It takes some extra effort to get the fire going in the breeze, but once it ignites, the breeze causes the fire to burn more intensely than if the air were still. Why?

Answer: The two reactants in the campfire are the wood and oxygen from air. In still air, the oxygen around the wood is used up as the wood burns. In a breeze, the fire is constantly fed more oxygen by the moving air.

Why do fires burn more intensely in windy conditions?

Chapter Summary

MOLECULAR CONCEPT

In this chapter, we saw that most of the substances in nature are compounds, combinations of elements in fixed ratios (4.1). Compounds are represented by their chemical formulas, which identify the type and relative amounts of each element present (4.2). For molecular compounds, a molecular formula indicates the number and type of atoms in each molecule.

Chemical compounds are divided into two types, ionic and molecular, each with its own naming system (4.3, 4.4). An ionic compound is a metal bonded to a nonmetal via an ionic bond. In an ionic bond, an electron is transferred from the metal to the nonmetal, making the metal a cation (positively charged) and the nonmetal an anion (negatively charged). In its solid form, an ionic compound consists of a three-dimensional lattice of alternating positive and negative ions. A molecular compound is a nonmetal bonded to a nonmetal via a covalent bond. In a covalent bond, electrons are shared between the two atoms. Molecular compounds contain identifiable clusters of atoms called molecules.

The sum of the atomic masses of all the atoms in a chemical formula is called the formula mass (4.5). It is a conversion factor between mass of the compound and moles of its molecules. The numerical relationships inherent in chemical formulas can help us determine the amount of a given element within a given compound (4.6).

Chemical reactions, in which compounds are formed or transformed, are represented by chemical equations (4.7). The substances on the left side of a chemical equation are called the reactants, and the substances on the right side are called the products. The number of atoms of each type on each side of the chemical equation must be equal in order for the equation to be balanced. The coefficients in the chemical equation help us determine numerical relationships between the amounts of reactants and products (4.8).

SOCIETAL IMPACT

Because most common substances are compounds, we must understand compounds to understand what is happening in the world around us (4.1). Compounds, either natural or man-made, are important in everything from the materials we use every day—plastics, detergents, and antiperspirants—to the environmental problems that we face as a society—ozone depletion, air pollution, and global warming.

Common ionic compounds include table salt, calcium carbonate (the active ingredient in some antacids), and sodium bicarbonate (also called baking powder). Common molecular compounds include water, sugar, and carbon dioxide. The properties of molecules determine the properties of the molecular compound they compose (4.3).

We often want to know how much of an element is present in a given compound (4.5). For example, those watching their sodium intake might want to know how much sodium is in a given amount of sodium chloride (table salt).

Chemical reactions keep both our society and our own bodies going (4.7). Ninety percent of our society's energy is derived from chemical reactions, primarily the combustion of fossil fuels. Our own bodies derive energy from the foods we eat by orchestrating a slow combustion of the molecules contained in food. The products of useful chemical reactions can sometimes present environmental problems. For example, carbon dioxide, one of the products of fossil-fuel combustion, may be causing the planet to warm through a process called the greenhouse effect.

Chemistry on the Web

For up-to-date URLs, visit the text website at http://chemistry.brookscole.com/tro3

- Compound Data Base
  http://chemfinder.cambridgesoft.com/
- Carbon Monoxide Poisoning
  http://www.cpsc.gov/cpscpub/pubs/466.html
- Everyday Chemical Reactions
  http://www.museum.vic.gov.au/scidiscovery/reactions/index.asp
- Greenhouse Effect
  http://www.enchantedlearning.com/subjects/astronomy/planets/earth/Greenhouse.shtml

Key Terms

chemical equation
chemical formula
covalent bond
electrolyte solution
formula mass
ionic bond
ionic compound
molar mass
molecular compound
molecular formula
molecule
polyatomic ion
product
reactant

Exercises

 Assess your understanding of this chapter's topics with an online chapter quiz at http://chemistry.brookscole.com/tro3

QUESTIONS

1. In Chapter 3, we learned that all matter is composed of atoms. In this chapter, we learned that most common substances are either compounds or mixtures of compounds. How can these both be true? Explain.
2. Name some common everyday compounds.
3. What is a chemical formula?
4. What kinds of elements form ionic compounds? What kinds of elements form molecular compounds?
5. Explain the difference between an ionic bond and a covalent bond.
6. What is a molecule? Give an example.
7. What determines the properties of molecular compounds?
8. What is the difference between a common name for a compound and a systematic name?
9. Give the general form for naming ionic compounds.
10. What is a polyatomic ion? Give some examples.
11. Give the general form for naming molecular compounds.
12. What is formula mass? What is the relationship between formula mass and molar mass?
13. Explain the numerical relationships inherent in a chemical formula.
14. What is a chemical equation? What are reactants? What are products?
15. Why must chemical equations be balanced?
16. Explain the numerical relationships inherent in a chemical equation.

PROBLEMS

17. Determine the number of each type of atom in each of the following chemical formulas:
    a. CaF_2
    b. CH_2Cl_2
    c. $MgSO_4$
    d. $Sr(NO_3)_2$
18. Determine the number of each type of atom in each of the following chemical formulas:
    a. Na_2O
    b. CF_2Cl_2
    c. K_2CO_3
    d. $Mg(HCO_3)_2$
19. Classify each of the following compounds as ionic or molecular:
    a. KCl
    b. CO_2
    c. N_2O
    d. $NaNO_3$

20. Classify each of the following compounds as ionic or molecular:
    a. CO
    b. $ZnBr_2$
    c. CH_4
    d. NaF
21. Name each of the following compounds:
    a. NaF
    b. $MgCl_2$
    c. Li_2O
    d. Al_2O_3
    e. $CaCO_3$
22. Name each of the following compounds:
    a. SrO
    b. $BeCl_2$
    c. $MgSO_4$
    d. $CaCl_2$
    e. NaOH
23. Name each of the following compounds:
    a. BCl_3
    b. CO_2
    c. N_2O
24. Name each of the following compounds:
    a. SF_6
    b. N_2O_4
    c. SO_2
25. Give chemical formulas for each of the following:
    a. nitrogen dioxide
    b. calcium chloride
    c. carbon monoxide
    d. calcium sulfate
    e. sodium bicarbonate
26. Give chemical formulas for each of the following:
    a. sulfur trioxide
    b. phosphorus pentachloride
    c. carbon disulfide
    d. silicon tetrachloride
    e. potassium hydroxide
27. Calculate the formula mass for each of the following compounds:
    a. CO
    b. CO_2
    c. C_6H_{14}
    d. HCl
28. Calculate the formula mass for each of the following compounds:
    a. NaCl (table salt)
    b. CH_4 (natural gas)
    c. NH_4Cl
    d. $C_{12}H_{22}O_{11}$ (sugar)
29. The formula mass of an unknown compound containing only C and H is 44 amu. What is its molecular formula?
30. The formula mass of an unknown compound containing only C and H is 70.1 amu. What is its molecular formula?
31. Calculate the number of water molecules (H_2O) in 10.0 g of water.
32. Acetone (C_3H_6O) is used as nail polish remover. If the amount of acetone used to remove the polish from one fingernail is 0.5 g, how many molecules were used?
33. Determine the number of sugar molecules in 5.0 g of sugar ($C_{12}H_{22}O_{11}$).
34. One drop of water from a medicine dropper has a volume of approximately 0.050 mL (1 mL = 1 cm^3).
    a. Determine the number of water molecules in a drop of water. (The density of water is 1.0 g/cm^3).
    b. How many H atoms would be present in these molecules?
35. How many chlorine atoms are in each of the following?
    a. 124 CCl_4 molecules
    b. 38 HCl molecules
    c. 89 CF_2Cl_2 molecules
    d. 1368 $CHCl_3$ molecules
36. How many hydrogen atoms are in each of the following?
    a. 5 H_2O molecules
    b. 58 CH_4 molecules
    c. 1.4×10^{22} $C_{12}H_{22}O_{11}$ molecules
    d. 14 dozen NH_3 molecules
37. Find the number of moles of oxygen in each of the following:
    a. 0.20 mol H_2O
    b. 12 mol CO
    c. 15 mol CO_2
    d. 25 g NO_2
38. Find the number of moles of nitrogen in each of the following:
    a. 0.58 mol N_2
    b. 1.7 mol NO_2
    c. 0.12 mol N_2O
    d. 1.3 mol N_2H_4
39. Determine the mass of sodium (in grams) contained in 5.8 grams of table salt (NaCl).
40. Determine the mass of Cl (in grams) contained within 155 grams of CF_2Cl_2.
41. Determine the mass of iron (in kilograms) contained in 15.8 kilograms of Fe_2O_3.
42. Determine the mass of carbon (in kilograms) contained in 178 kilograms of CO_2.
43. Balance each of the following chemical equations:
    a. $HCl + O_2 \rightarrow H_2O + Cl_2$
    b. $NO_2 + H_2O \rightarrow HNO_3 + NO$
    c. $CH_4 + O_2 \rightarrow CO_2 + H_2O$
44. Balance each of the following chemical equations:
    a. $PbO + NH_3 \rightarrow Pb + N_2 + H_2O$
    b. $Mg_3N_2 + H_2O \rightarrow Mg(OH)_2 + NH_3$
    c. $C_3H_8 + O_2 \rightarrow CO_2 + H_2O$
45. Balance each of the following chemical equations:
    a. $Al + O_2 \rightarrow Al_2O_3$
    b. $NO + O_2 \rightarrow NO_2$
    c. $H_2 + Fe_2O_3 \rightarrow Fe + H_2O$
46. Balance each of the following chemical equations:
    a. $C_2H_6 + O_2 \rightarrow CO_2 + H_2O$
    b. $H_2S + SO_2 \rightarrow S + H_2O$
    c. $CH_3OH + O_2 \rightarrow CO_2 + H_2O$

47. Water can be synthesized according to the following unbalanced chemical equation:
$$H_2 + O_2 \rightarrow H_2O$$
   a. Balance the reaction.
   b. How many moles of H_2O can be produced if you have 2.72 mol of O_2 and an unlimited amount of H_2?
   c. How many moles of H_2 would be necessary to produce 10.0 g of water?

48. Billions of pounds of urea, $CO(NH_2)_2$, are produced annually for use as a fertilizer. The principal reaction employed is
$$2NH_3 + CO_2 \rightarrow CO(NH_2)_2 + H_2O$$
By assuming unlimited amounts of CO_2, how many moles of urea can be produced from each of the following amounts of NH_3?
   a. 2 mol NH_3
   b. 0.45 mol NH_3
   c. 10 g NH_3
   d. 2.0 kg NH_3

49. Natural gas is burned in homes for heat according to the following reaction:
$$CH_4 + 2O_2 \rightarrow 2H_2O + CO_2$$
Determine how many grams of CO_2 would be formed for each of the following amounts of methane (assume unlimited amounts of oxygen):
   a. 2.3 mol CH_4
   b. 0.52 mol CH_4
   c. 11 g CH_4
   d. 1.3 kg CH_4

50. A component of acid rain, sulfuric acid, forms by the combination of water with sulfur oxide pollutants that are released into the atmosphere as a byproduct of fossil fuel combustion:
$$SO_3 + H_2O \rightarrow H_2SO_4$$
If you have an unlimited amount of H_2O, determine the number of kilograms of H_2SO_4 that can be formed from 1.0×10^3 kg of SO_3.

POINTS TO PONDER

51. Sugar is a compound that contains C, H, and O. If sugar is heated in a flame, it will smoke and turn black. What has happened at the molecular level? What is the black substance?

52. The Molecular Revolution box in this chapter describes a pig whose genes have been modified to contain human genetic material. Some have criticized the genetic engineering of animals, accusing geneticists of trying to play God. What do you think? Should we tinker with an animal's genetic makeup if the potential outcome is the saving of human lives and a reduction of human suffering? Why or why not?

53. Albert Einstein is quoted as saying, "Not only to know how nature is and how her transactions are carried through, but also to reach as far as possible the utopian and seemingly arrogant aim of knowing why nature is thus and not otherwise." What do you think Einstein meant by this?

54. Let us suppose for a moment that humans had not learned anything about how to carry out or control any chemical reactions. How would your life be different? Start by thinking of waking in the morning. Go through your morning routine and identify the chemical reactions that you carry out directly or products that you use that were made from chemical reactions. How would your morning routine change if humans lacked chemical knowledge?

55. Read the What If . . . Problem Molecules box in this chapter. One such "problem molecule" when it occurs at ground level is ozone, three oxygen atoms bonded together. Ground-level ozone is formed from the action of sunlight on motor vehicle exhaust, and it irritates the eyes and lungs. Ozone occurring in the upper atmosphere (stratospheric ozone), however, is a natural and essential part of the Earth's ecosystem. Stratospheric ozone protects life on Earth from harmful ultraviolet sunlight. How can the same molecule be a problem at ground level but beneficial in the upper atmosphere? Is there anything inherent about the ozone molecule itself that makes it a problem or a benefit?

Feature Problems and Projects

56. For each of the following space-filling molecular models, write a chemical formula. Carbon atoms are black, hydrogen atoms are white, oxygen atoms are red, and nitrogen atoms are blue.

    a.

    b.

    c.

    d.

57. For each of the following chemical reactions, draw in the missing molecules necessary to balance their equations. Carbon atoms are black, hydrogen atoms are white, oxygen atoms are red, and nitrogen atoms are blue.

    a.

    b.

    c.

    d.

58. Read the What If . . . Problem Molecules box in this chapter. The Environmental Protection Agency (EPA) monitors problem molecules and publishes the results on their website **http://www.epa.gov/air/data/index.html**. Go to that website and view the *reports and maps* section. Use the menus to request a map of the U.S. nonattainment areas for carbon monoxide (CO) and print the map. This map shows regions where carbon monoxide air pollution levels persistently exceed national air quality standards. Do you live in a region where the levels are exceeded? Try this for other pollutants such as ozone (O_3).

59. Read The Molecular Revolution box in this chapter. Write a one-page response to the following statement: Altering the genetic make-up of any organism is like playing God and should not be done under any circumstances. Do you agree or disagree with the statement? Why?

5 Chemical Bonding

Man masters nature not by force but by understanding. That is why science has succeeded where magic failed: because it has looked for no spell to cast.

—Jacob Bronowski

In this chapter, we explore chemical bonding. Chemical bonding is the reason that there are millions of different substances in the universe instead of just the 91 naturally occurring elements. As you read, think about why certain elements combine with certain other elements. For example, why does hydrogen readily combine with oxygen to form water? Why are there two hydrogen atoms to every one oxygen atom in a water molecule? We will examine a model for chemical bonding, called Lewis theory, that explains why atoms combine as they do. We will also discuss another model, called VSEPR theory, that predicts the geometry of molecules.

These relatively simple models—just lines and dots on a piece of paper—predict important properties of the substances around you every day. Our understanding of chemical bonding has allowed us to put together molecules that never existed before. Many of these molecules have impacted society and changed the way we live. For example, nylon, plastic, latex, and AIDS drugs were all synthesized by chemists who understood chemical bonding and knew how to put molecules together to achieve specific purposes. How would your life be different without them?

QUESTIONS FOR THOUGHT

- Are the properties of compounds similar to or different from the properties of the elements that compose them?
- Why do elements combine to form compounds?
- How can we predict the observed ratios of elements in compounds?
- Why is the shape of a molecule important?
- How can we predict the shapes of molecules?
- How can we use the shape of a molecule to predict the properties of the compound that the molecule composes?

Explore this topic on the **Salt** website.

Recall from Chapter 3 that chlorine is a halogen (group 7A) and therefore has seven electrons in its outer Bohr orbit. Sodium is an alkali metal (group 1A) and therefore has one electron in its outer Bohr orbit. Recall also that a stable configuration consists of having eight electrons in the outer Bohr orbit.

5.1 From Poison to Seasoning

Chlorine, a pale yellow gas, reacts with almost everything. If inhaled, chlorine is toxic and can cause death. It was even used as a chemical weapon during World War I. Sodium, a shiny metal, is equally reactive and toxic. A small piece of sodium explodes violently when dropped in water. However, when chlorine and sodium combine in a chemical reaction, they produce the relatively harmless seasoning we call table salt. How can two poisons combine to form a flavor enhancer that tastes great on steak? The answer is in chemical bonding.

Chlorine is reactive because chlorine atoms have seven valence electrons; they are one electron short of a stable configuration. Sodium is reactive because sodium atoms have one valence electron; they are one electron beyond a stable configuration.

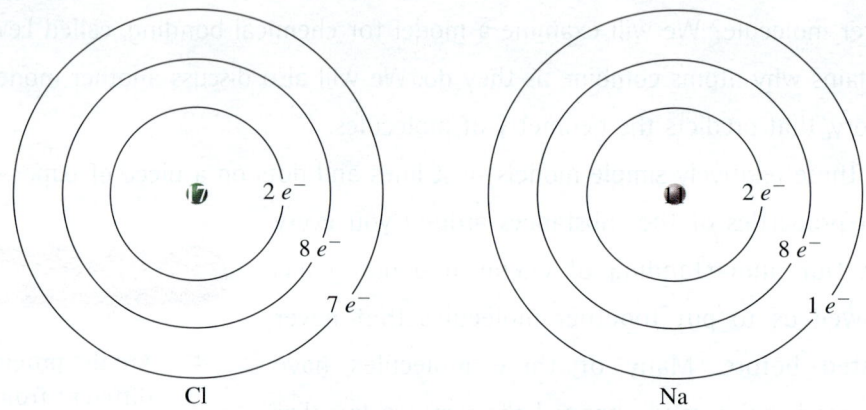

Both elements are toxic because they can alter a biological molecule by forcefully exchanging electrons with it. Sodium transfers electrons *to* the molecule, and chlorine transfers electrons *away* from the molecule.

When sodium and chlorine combine, they undergo a chemical reaction in which sodium atoms transfer electrons to chlorine atoms in a frenzy of activity that would be a marvel to witness (Figure 1). Both atoms are stabilized by the

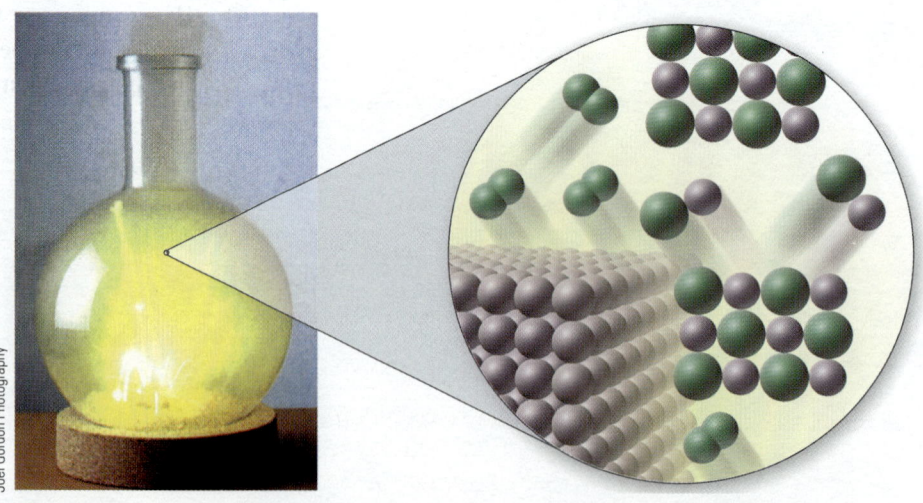

FIGURE 1 Sodium and chlorine react violently.

exchange, and a chemical bond forms. The resulting substance, sodium chloride, is a stable, nontoxic compound because both the sodium cation (Na^+) and chlorine anion (Cl^-) now have stable electron configurations; they both have eight electrons in their outer orbits.

In this chapter, we examine chemical bonding. We develop a simple theory, called **Lewis theory**, that gives insight into chemical bonding and explains why particular elements combine in particular proportions. For example, Lewis theory predicts that hydrogen and oxygen should form a compound in which two hydrogen atoms combine with one oxygen atom. In nature, we indeed find such a compound—water. The theoretical prediction of Lewis theory has physical meaning in the vast oceans, lakes, and ice caps of our planet.

We will also begin to correlate the macroscopic properties of molecular compounds with the microscopic properties of their smallest identifiable units, molecules. To this end, we study another model—called **valence shell electron pair repulsion (VSEPR)** theory—that predicts the shapes of molecules. For example, VSEPR theory predicts that the two hydrogen atoms and one oxygen atom in the water molecule should have a shape resembling a boomerang. When we examine water in nature, we indeed find that water molecules are shaped like boomerangs.

H_2O

Again, our theoretical predictions have real physical meaning. The shape of water molecules determines the properties of the water that we drink, wash with, and bathe in. Water would be a very different substance if its molecules had a different shape.

5.2 Chemical Bonding and Professor G. N. Lewis

Imagine that one of your classmates takes copious notes each time your weekly chemistry test scores are announced—you may think it a little strange. She then asks everyone how much time they spend studying and how they study. Your classmate is making observations on the behavior of the class. She soon announces that she has a created a model that can predict test scores, and nine times out of ten, her predictions are correct. Her theoretical model—that she created in her mind based on her observations—predicts something about the real world. She is delighted.

In a similar way, in the early 1900s, an American chemist named **G. N. Lewis (1875–1946)** at the University of California, Berkeley, made observations and constructed a theoretical model. However, his model predicted something more important than exam scores—it predicted the molecules that would form from certain elements, and, nine times out of ten, he was correct.

Lewis's model focused on the following fundamental ideas:

1. The valence electrons are most important in chemical bonding.
2. Valence electrons are transferred from one atom to another to form ionic bonds, or shared between atoms to form covalent bonds.

Recall from Section 1.3 that the scientific method begins with observations and then moves on to laws, hypotheses, and theories.

 Explore this topic on the G. N. Lewis website.

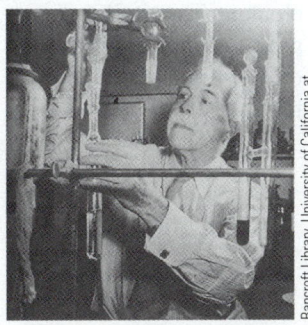

Professor G. N. Lewis taught at the University of California at Berkeley. A chemistry building at the university is named in his honor.

3. When valence electrons are transferred or shared, the atoms in the compound form full outer Bohr orbits and therefore gain stable electron configurations. Because a stable electron configuration usually involves eight electrons, this is known as the **octet rule**. (Significant exceptions include helium and hydrogen, for which a stable configuration is only two electrons, called a duet.)

In Lewis theory, we use dots to represent valence electrons. The **Lewis structure** of an element is its chemical symbol surrounded by a number of dots equal to the number of valence electrons for that element. Each dot represents one valence electron. For example, the electron configuration for silicon according to the Bohr model, and its Lewis structure, are shown below:

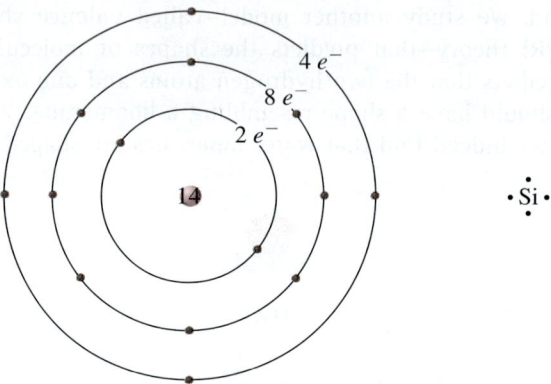

In Lewis theory, we ignore the inner or *core* electrons and represent the four valence electrons with four dots surrounding the chemical symbol for silicon. Because carbon also has four valence electrons, its Lewis structure also has four dots:

The Lewis structures for several other elements are shown below.

As we learned in Section 3.10, the number of valence electrons (and therefore the number of dots) for an element is given by the group number of that element in the periodic table.

The dots, which represent the valence electrons, are placed around the element's symbol with a maximum of two dots per side. Neither the exact location of the dots, nor the order in which they are drawn, is critical. Stable configurations are easily spotted because their Lewis structures contain eight dots (an octet), or two dots (a duet) in the case of helium. Elements without an octet tend to react with other elements to form an octet, except for hydrogen, which reacts to form a duet. According to Lewis theory, chemical bonding brings elements together in the correct ratios so that, either by transferring or by sharing electrons, all the atoms involved form an octet. This is a simple, powerful theory of chemical bonding. When we look at compounds in nature, such as sodium chloride or water, we find that Lewis theory is correct in many of its predictions.

MOLECULAR THINKING

Fluoride

The fluoride ion, F⁻, helps protect teeth against decay. Studies have shown that when sodium fluoride—an ionic compound containing the fluoride ion—is added to drinking water, tooth decay decreases by 65%. Elemental fluorine, on the other hand, is the most reactive element in the universe, reacting with virtually anything it contacts, making it extremely toxic.

QUESTION: What is the molecular reason for the reactivity of elemental fluorine and the stability of sodium fluoride?

Adding fluoride to drinking water reduces tooth decay by 65%.

George Semple

APPLY YOUR KNOWLEDGE

Based on the Lewis structures for hydrogen and helium, explain why buoyant balloons are filled with helium instead of hydrogen gas, even though hydrogen gas is cheaper and more buoyant.

$$H\cdot \qquad He:$$

Answer: Helium has a duet and is therefore chemically stable. It will not react with other elements and, therefore, should the balloon pop, would not represent a danger to humans. Hydrogen, on the other hand, is one electron short of a duet. Elements that are very close to having an octet, or a duet in the case of hydrogen, are among the most reactive. Hydrogen atoms are so reactive that they react with each other to form hydrogen molecules, but hydrogen molecules themselves are also reactive and combine explosively with oxygen to form water.

5.3 Ionic Lewis Structures

In Lewis theory, we combine the Lewis structures of elements to form Lewis structures for compounds. We represent ionic bonding—the bonding between a metal and a nonmetal through the transfer of electrons—by moving dots from the Lewis structure of the metal to the Lewis structure of the nonmetal. The metal becomes a cation and the nonmetal becomes an anion (see Section 3.3). For example, sodium and chlorine have the following Lewis structures:

$$Na\cdot \qquad \cdot \ddot{\underset{..}{Cl}}:$$

In forming the Lewis structure for NaCl, we move sodium's valence electron to chlorine:

$$Na\cdot \quad \cdot \ddot{\underset{..}{Cl}}: \longrightarrow Na^+[:\ddot{\underset{..}{Cl}}:]^-$$

Because sodium loses its valence electron, it has no dots. This corresponds to a stable configuration because the outermost *occupied* Bohr orbit (not shown in the

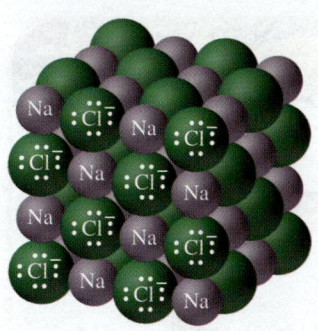

FIGURE 2 Sodium chloride lattice.

Lewis diagram) contains an octet. Look back at the sodium atom in Section 5.1 to confirm this. Chlorine has an octet as indicated by the eight electron dots. Because the metal and nonmetal each acquires a charge, we indicate the magnitude of the charge on the upper right corner of the symbol. We also enclose the anion in brackets to show that the charge belongs to the entire chlorine assembly, including the electron "dots." Recall from Section 4.3 that in sodium chloride crystals, the sodium cations and chloride anions alternate in a three-dimensional crystalline lattice held together by the attraction between cations and anions (Figure 2). We can now see how each of these ions has a stable electron configuration.

Consider magnesium fluoride, MgF_2. Magnesium and fluorine have the following Lewis structures:

$$Mg: \quad \cdot \ddot{F}:$$

For magnesium to obtain an octet, it must lose both of its valence electrons. However, fluorine needs only one electron to fill its octet; therefore, two fluorine atoms are needed for every magnesium atom:

$$[:\ddot{F}:]^- \quad Mg^{2+} \quad [:\ddot{F}:]^-$$

Magnesium has lost all its valence electrons and appears without any dots, while the fluorine atoms both acquire octets. When we look at magnesium fluoride in nature, we indeed find that it is composed of two fluoride ions to every magnesium ion, just as Lewis theory predicts.

EXAMPLE 5.1

Drawing Lewis Structures for Ionic Compounds

Draw a Lewis structure for MgO.

SOLUTION

The Lewis structures for Mg and O are as follows:

$$Mg: \quad \cdot \ddot{O}:$$

Mg must lose two electrons to form an octet, and O must gain two:

$$Mg^{2+} \quad [:\ddot{O}:]^{2-}$$

YOUR TURN

Drawing Lewis Structures for Ionic Compounds

Draw a Lewis structure for $BeCl_2$.

EXAMPLE 5.2

Using Lewis Structures to Determine the Correct Chemical Formula for Ionic Compounds

Use Lewis structures to determine the correct chemical formula for the compound formed between calcium and fluorine.

SOLUTION

The Lewis structures for Ca and F are as follows:

$$\text{Ca:} \quad \cdot \ddot{\text{F}}:$$

Calcium must lose two electrons, while fluorine must gain one electron to acquire an octet. Therefore, two fluorine atoms are required for every calcium atom. The correct Lewis structure is as follows:

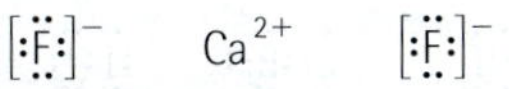

$$[:\ddot{\text{F}}:]^- \quad \text{Ca}^{2+} \quad [:\ddot{\text{F}}:]^-$$

The correct chemical formula is, therefore, CaF_2.

> **YOUR TURN**
>
> ### Using Lewis Structures to Determine the Correct Chemical Formula for Ionic Compounds
>
> Use Lewis structures to determine the correct chemical formula for the compound formed between Li and O.

5.4 Covalent Lewis Structures

In covalent bonds, atoms share their electrons. We represent covalent bonding in Lewis theory by letting atoms share their dots such that some dots count for the octet of more than one atom. For example, we learned in Chapter 3 that elemental chlorine exists as the diatomic (two atom) molecule, Cl_2. Lewis theory explains why. Consider the Lewis structure of chlorine:

Two chlorine atoms can complete their octets by joining to form Cl_2.

$$:\ddot{\text{Cl}}:\ddot{\text{Cl}}:$$

The two electrons in the middle are shared by both atoms and count toward the octet of each so that both chlorine atoms now have octets. Consequently, the chlorine molecule is more stable than the two isolated chlorine atoms, and elemental chlorine exists as Cl_2.

Octet   Octet

We differentiate between two types of electron pairs in Lewis structures. The electrons *between two atoms* are called **bonding pair electrons**, while those *on a single atom* are called **lone pair electrons**.

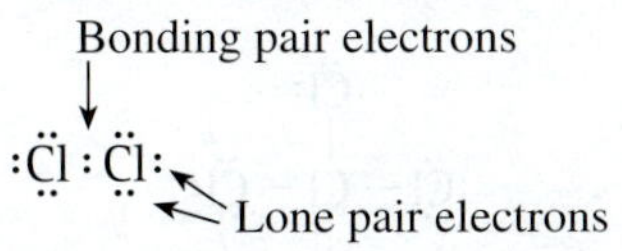

Bonding pair electrons
↓
$:\ddot{\text{Cl}}:\ddot{\text{Cl}}:$ ← Lone pair electrons

Only bonding pair electrons count toward the octet of both atoms. Lone pair electrons only count toward the octet of the atom they are on.

> **APPLY YOUR KNOWLEDGE**
>
> In Chapter 3, we also learned that fluorine, bromine, and iodine each exist as diatomic molecules in nature. Why?
>
> **Answer:** Like chlorine, these elements are all in group 7A (the halogens) on the periodic table and therefore have seven valence electrons. Consequently, they form diatomic molecules to attain octets as shown in these Lewis structures:
>
> :F̈:F̈:   :B̈r:B̈r: :Ï:Ï:

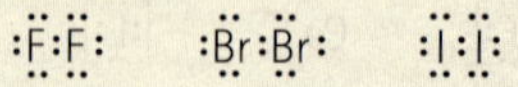

Next consider water. The Lewis structures of hydrogen and oxygen are as follows:

H·   ·Ö:

When these two elements combine to form water, H_2O, the two hydrogen atoms each share their one electron with oxygen. The hydrogen atoms then each have a duet, oxygen has an octet, and a stable water molecule results:

H:Ö:H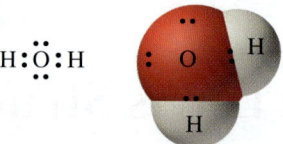

The Lewis structure shows that the ratio of H to O must be 2:1 to form a stable compound. The prediction of Lewis theory is correct. When we examine water in nature, we find that it is composed of two hydrogen atoms to every oxygen atom.

Consider the compound carbon tetrachloride, CCl_4. The Lewis structures for C and Cl are

·Ċ·   ·C̈l:

Carbon needs four additional electrons to complete its octet, while chlorine needs one. The correct Lewis structure is:

:C̈l:
:C̈l:C:C̈l:
:C̈l:

Each atom in this structure has an octet. The carbon atom is surrounded by four bonding electron pairs (an octet), and each chlorine atom has three lone pairs and one bonding pair (an octet). We can simplify our notation by representing bonding pairs of electrons with dashes reinforcing the idea that in Lewis theory a bonding electron pair is a chemical bond:

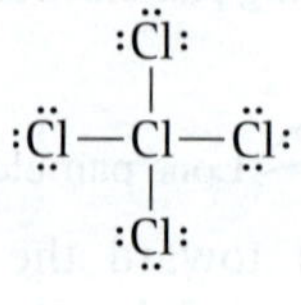

:C̈l:
|
:C̈l—Cl—C̈l:
|
:C̈l:

MULTIPLE BONDS

Atoms often share more than one electron pair to form complete octets. For example, consider elemental oxygen, which exists as the diatomic molecule O_2. The Lewis structure for each isolated oxygen atom has six electrons (because oxygen is in column 6A of the periodic table):

$$\cdot \ddot{\text{O}}: \quad \cdot \ddot{\text{O}}:$$

When we bring the two oxygen atoms together to form O_2, we might initially try to write the following Lewis structure:

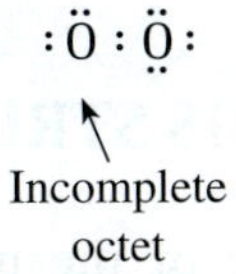

$$:\ddot{\text{O}}:\ddot{\text{O}}:$$

↑
Incomplete octet

However, one oxygen atom in this Lewis structure has an incomplete octet—it is two electrons short of an octet. We can complete its octet by converting one of the lone pairs into a second bonding pair:

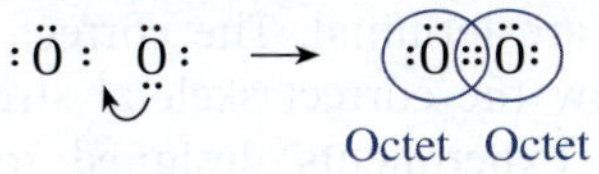

$$:\ddot{\text{O}}:\ddot{\text{O}}: \quad \rightarrow \quad (:\ddot{\text{O}}::\ddot{\text{O}}:)$$

Octet   Octet

This Lewis structure has two bonding pairs, called a **double bond**. The four electrons that constitute the double bond count toward the octet of both oxygen atoms, so both atoms have an octet.

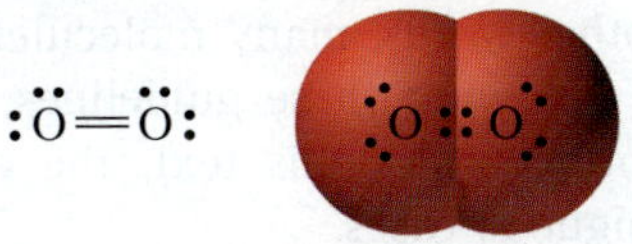

$$:\ddot{\text{O}}=\ddot{\text{O}}:$$

Double bonds are shorter (the bonding atoms are closer together) and stronger than single bonds because there are twice as many electrons in the bond.

Atoms can also share three electron pairs to form complete octets. For example, elemental nitrogen, which exists as the diatomic molecule N_2, shares three electron pairs in its Lewis structure. The Lewis structures for the isolated nitrogen atoms are as follows:

$$\cdot \ddot{\text{N}}: \quad \cdot \ddot{\text{N}}:$$

Again, when we try to write a Lewis structure for N_2 with only one bonding pair, we do not have enough electrons to give each atom an octet:

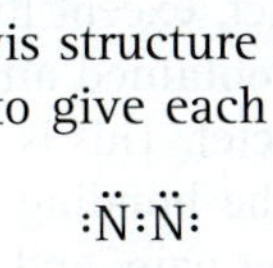

$$:\ddot{\text{N}}:\ddot{\text{N}}:$$

↑↑
Incomplete octets

We can complete the octets on both atoms by moving two lone pairs into the bonding region, converting them to bonding pairs:

$$:\ddot{\text{N}}:\ddot{\text{N}}: \quad \rightarrow \quad (:\text{N}:::\text{N}:)$$

Octet   Octet

This Lewis structure has three bonding pairs, called a **triple bond**. The six electrons that constitute the triple bond count toward the octet of both nitrogen atoms, so both atoms have an octet.

:N≡N:

Triple bonds are shorter and stronger than either single or double bonds. The N_2 molecule is a very stable molecule—it is difficult to break N_2 into its constituent atoms.

STEPS FOR WRITING LEWIS STRUCTURES

1. **Write the skeletal structure of the molecule.** To write a correct Lewis structure for a molecule, the atoms must be in the right positions relative to one another—they must have the correct *skeletal* structure. For example, if we tried to write the skeletal structure of water as H H O, we could not construct a good Lewis structure. In nature, oxygen is the central atom and the hydrogen atoms are terminal. The correct skeletal structure is H O H. We can't always know the correct skeletal structures for molecules unless we perform specific experiments designed to determine their structure. However, we can follow two simple guidelines to help us write reasonable skeletal structures. First, because hydrogen has only one electron and only needs a duet, *it will never be a central atom.* Second, when a molecule contains several atoms of the same type, those atoms will often be in terminal positions. In other words, many molecules tend to be symmetrical, if possible. Even after following these guidelines, however, the best skeletal structure may not be clear. In this text, the correct skeletal structure is specified for all ambiguous cases.
2. **Determine the total number of electrons for the molecule.** Do this by adding together the valence electrons—obtained from the group number in the periodic table—for each atom in the molecule. The final Lewis structure must contain *exactly this total number of electrons.*
3. **Place the electrons as dots to give octets to as many atoms as possible.** Begin by putting a single bond—two dots or a single line—between all bonded atoms. Distribute the remaining electrons beginning with the terminal atoms and ending with the central atom. By using only single bonds, try to give each atom an octet, except hydrogen, which gets a duet.
4. **If the central atom has not obtained an octet, form double or triple bonds as necessary to give it an octet.** This is done by moving lone pair electrons from the outer atoms into the bonding region with the central atom, thus converting them into bonding pairs and forming multiple bonds.

EXAMPLE 5.3

Drawing Lewis Structures for Covalent Compounds

Draw a Lewis structure for NH_3.

SOLUTION

1. **Write the skeletal structure of the molecule.** Nitrogen is the central atom, and the hydrogen atoms are arranged around it. Since hydrogen atoms must be terminal, the correct skeletal structure is as follows:

   H   N   H

   H

2. **Determine the total number of electrons for the molecule.** The total number of valence electrons is computed by adding together the valence electrons of each of the atoms in the molecule.

   # valence electrons for H   # valence electrons for N

   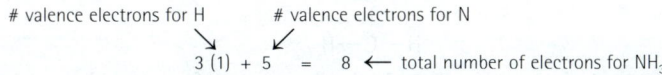

   3 (1) + 5 = 8 ← total number of electrons for NH₃

3. **Place the electrons as dots to give octets to as many atoms as possible.** Put a single bond between each N and H atom, using six of the eight electrons. Place the last two electrons on nitrogen to complete its octet.

   H—N̈—H
       |
       H

Since all of the atoms in the Lewis structure have an octet (or duet for H), the structure is complete.

YOUR TURN

Drawing Lewis Structures for Covalent Compounds

Draw a Lewis structure for PCl₃.

EXAMPLE 5.4

Drawing Lewis Structures for Covalent Compounds II

Draw a Lewis structure for H₂CO. (Carbon is the central atom.)

SOLUTION

1. **Write the skeletal structure of the molecule.** The correct skeletal structure is as follows:

   O

   H   C   H

2. **Determine the total number of electrons for the molecule.** The total number of valence electrons is computed by adding together the valence electrons of each of the atoms in the molecule.

   # valence   # valence   # valence
   electrons for H electrons for C electrons for O

   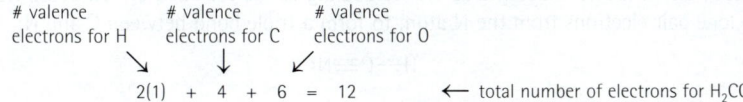

   2(1) + 4 + 6 = 12   ← total number of electrons for H₂CO

3. **Place the electrons as dots to give octets to as many atoms as possible.** Put a single bond between each C and H atom, and between the C and O atom, using six of the twelve electrons:

$$\begin{array}{c} O \\ | \\ H\!\!-\!\!C\!\!-\!\!H \end{array}$$

Arrange the remaining six electrons to give octets (duets to H) to as many atoms as possible beginning with terminal atoms and ending with the central atom:

$$\begin{array}{c} :\!\ddot{O}\!: \\ | \\ H\!\!-\!\!C\!\!-\!\!H \end{array}$$

4. **If the central atom has not obtained an octet, form double or triple bonds as necessary to give it an octet.** We move a lone pair from the O atom to form a double bond between C and O:

$$\begin{array}{c} :\!\ddot{O}\!: \\ \| \\ H\!\!-\!\!C\!\!-\!\!H \end{array}$$

> **YOUR TURN**
>
> ## Drawing Lewis Structures for Covalent Compounds II
>
> Draw a Lewis structure for CO_2.

> **EXAMPLE 5.5**
>
> ## Drawing Lewis Structures for Covalent Compounds III
>
> Draw a Lewis structure for HCN. (Carbon is the central atom.)
>
> **SOLUTION**
> The correct skeletal structure is
>
> $$H \quad C \quad N$$
>
> The total number of valence electrons is
>
> $$1 + 4 + 5 = 10$$
>
> Place a single bond between C and each of the terminal atoms, using four of the ten electrons:
>
> $$H\!\!-\!\!C\!\!-\!\!N$$
>
> Arrange the remaining six electrons to give octets (duets to H) to as many atoms as possible, beginning with terminal atoms and ending with the central atom:
>
> $$H\!\!-\!\!C\!\!-\!\!\ddot{N}\!:$$
>
> Because we ran out of electrons before we could give the central atom an octet, we move two lone pair electrons from the N atom to form a triple bond between C and N:
>
> $$H\!\!-\!\!C\!\!\equiv\!\!N\!:$$

> **YOUR TURN**
>
> ### Drawing Lewis Structures for Covalent Compounds III
>
> Draw a Lewis structure for HC₂H. (The two carbon atoms are in the middle with one hydrogen atom attached to each.)

MOLECULAR FOCUS

Ammonia

Formula: NH_3
Molar Mass: 17.03 g/mol
Melting point: −77.7°C
Boiling point: −33.35°C
Lewis structure:

Three-dimensional structure:

Ammonia is a colorless gas with a pungent odor that smells like urine. Ammonia is usually marketed as ammonia water, a mixture of ammonia and water, which is the form we are most familiar in consumer products such as household cleansers. Its chief use, however, is as a fertilizer. Plants require nitrogen to grow; and even though they are surrounded by nitrogen in the form of atmospheric N_2, the strong triple bond in N_2 prevents plants from using it in this form:

:N≡N:

Many plants obtain nitrogen compounds from bacteria that grow on their roots. These bacteria convert, or "fix," nitrogen from its elemental form into nitrogen compounds that can be used by the plants. However, plants grow better if additional nitrogen compounds are added to the soil, and ammonia is often used for this purpose.

5.5 Chemical Bonding in Ozone

As we learned in Chapter 3, ozone (O_3) is an atmospheric gas that protects life on Earth from excessive exposure to ultraviolet light. Ozone reacts with incoming UV light according to the following reaction:

$$O_3 + UV\ light \rightarrow O_2 + O$$

The ozone molecule absorbs the UV light, causing one of the oxygen-oxygen bonds to break. As a result, the UV light is prevented from reaching the Earth. To understand the uniqueness of ozone, consider its Lewis structure:

:Ö—Ö=Ö:

However, an equally valid Lewis structure has the double bond between the other two oxygen atoms:

:Ö=Ö—Ö:

Initially, it seems that Lewis theory predicts two different types of bonds in the ozone molecule, one double bond and one single bond. However, real ozone molecules have two identical bonds (equal length and strength). Each bond is shorter

 Explore this topic on the **Ozone** website.

than a single bond, but longer than a double bond. In other words, each bond in ozone is like a bond and a half.

In cases such as this, where we can draw two satisfactory Lewis structures for the same molecule, the actual structure is an average between the two:

$$:\!\ddot{\text{O}}\!=\!\ddot{\text{O}}\!=\!\ddot{\text{O}}$$

This averaging of two identical Lewis structures is called **resonance**, and is usually represented by drawing both structures—called **resonance structures**—with a double-headed arrow between them:

$$:\!\ddot{\text{O}}\!-\!\ddot{\text{O}}\!=\!\ddot{\text{O}}\!: \longleftrightarrow :\!\ddot{\text{O}}\!=\!\ddot{\text{O}}\!-\!\ddot{\text{O}}\!:$$

Each bond in ozone is midway between a single and a double bond.

In contrast, oxygen's most common form, O_2, has the following Lewis structure:

$$:\!\ddot{\text{O}}\!=\!\ddot{\text{O}}\!:$$

Notice that O_2 has a double bond, which is stronger than O_3's "bond and a half." This difference in bonding between O_2 and O_3 results in a significant difference in the ability of these molecules to absorb UV light. The UV light does not have enough energy to break the strong bond in O_2, so the light is not absorbed and passes through the O_2 gas in the atmosphere. The bonds in O_3, on the other hand, are not as strong, and the energy of the UV light is sufficient to break one of the two bonds. This match between the strength of O_3 bonds and the energy of UV light accounts for ozone's ability to absorb the most harmful UV light emitted by the Sun and act as a shield to life on the Earth (Figure 3).

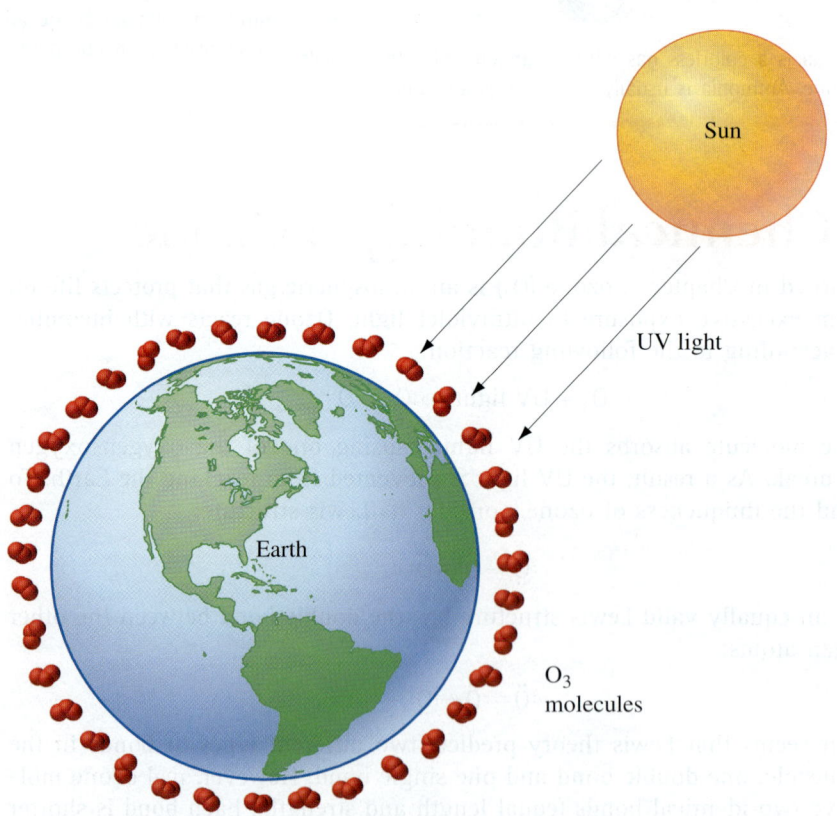

FIGURE 3 Ozone molecules act as a shield against UV light.

5.6 The Shapes of Molecules

In Section 4.3, we learned that the shape of a molecule is an important factor in determining the properties of the substances that it composes. For example, we learned that water would boil away at room temperature if it had a straight shape instead of a bent one. We now develop a simple model called valence shell electron pair repulsion theory that allows us to predict the shapes of molecules from their Lewis structures.

VSEPR theory is based on the idea that the negative charges of bonding electrons and lone pair electrons in a molecule cause them to repel each other, and that these repulsions determine the shape of the molecule. According to VSEPR theory, the geometry of a molecule is determined by maximizing the distance—and therefore minimizing the repulsions—between all electron groups (bonding groups and lone pairs) on the central atom(s) of the molecule. For example, consider methane (CH_4), which has the following Lewis structure:

$$\begin{array}{c} H \\ | \\ H-C-H \\ | \\ H \end{array}$$

Recall from Section 3.2 that particles with like charges repel each other.

On the printed page, we are restricted to two dimensions, and we might think that the electron pairs in each bond can minimize their repulsions with electron pairs in the other bonds by assuming a square-shaped geometry; however, molecules can adopt a three-dimensional shape. The three-dimensional geometry that puts the greatest angle between the four bonding electron pairs on the central carbon atom is *tetrahedral*. A tetrahedron has four identical equilateral triangles as its four sides. The carbon atom is in the center of the tetrahedron and hydrogen atoms are at the four corners.

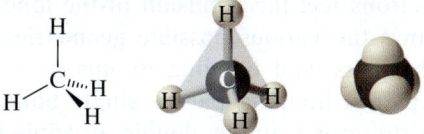

Because Lewis theory requires all atoms to have octets, we might expect all molecules to be tetrahedral. However, because central atoms often have lone pairs as well as bonding pairs, and because double and triple bonds force electrons to group together, a number of different geometries are also possible. For example, the Lewis structure for water has four electron groups on the central atom, two bonding pairs and two lone pairs:

$$H-\ddot{O}-H$$

These four groups of electrons get as far away from each other as possible. The **electron geometry**—the shape assumed by bonding pairs and lone pairs—is, therefore, tetrahedral as it was in CH_4.

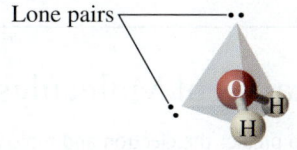

Lone pairs

However, the lone pairs—even though they influence the shape of the molecule—are not really part of its shape. The **molecular geometry**—the shape of the molecule's atoms—is **bent**.

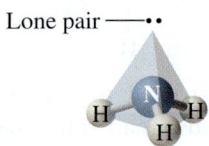

Notice that the bonding electrons feel the repulsion of the lone pairs—this causes the hydrogen atoms to bend toward each other. As we discuss in Section 5.7, this bent geometry is critical to the properties of water.

Consider the Lewis structure of ammonia, NH_3, which has three bonding electron groups and one lone pair:

$$H-\underset{H}{\overset{}{\underset{|}{N}}}-H$$

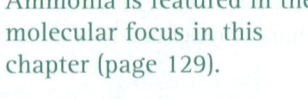

Ammonia is featured in the molecular focus in this chapter (page 129).

Since ammonia has a total of four electron groups around its central atom, the *electron* geometry is again tetrahedral.

However, since one of these electron groups is a lone pair, the resulting *molecular* geometry is **pyramidal**:

Again, the bonding electrons feel the repulsion of the lone pair and bend toward each other. Table 1 shows the various possible geometries based on the number of electron groups, lone pairs, and bonding groups in a molecule. An **electron group** may be a lone pair, a bonding pair, a single bond, a double bond, or a triple bond. A **bonding group** is a single, double, or triple bond.

W Explore this topic on the **VSEPR Shapes** website.

EXAMPLE 5.6

Predicting the Shapes of Molecules

Use Lewis theory and VSEPR to predict the electron and molecular geometry of H_2S.

Solution
The correct Lewis structure is as follows:

$$H-\ddot{\underset{..}{S}}-H$$

The total number of electron pairs is four. Therefore, from Table 1, the *electron* geometry is tetrahedral. The number of bonding pairs is two, so the *molecular* geometry is *bent*.

YOUR TURN

Predicting the Shapes of Molecules

Use Lewis theory and VSEPR to predict the electron and molecular geometry of PCl_3.

5.6 The Shapes of Molecules   133

TABLE 1
VSEPR Geometries

Total Electron Groups	Bonding Groups	Lone Pairs	Electron Geometry	Molecular Geometry	Example	
2	2	0	Linear	Linear	$:\ddot{O}=C=\ddot{O}:$	
3	3	0	Trigonal planar	Trigonal planar	$\begin{array}{c}:\!O\!:\\ \|\|\\ H-C-H\end{array}$	
3	2	1	Trigonal planar	Bent	$:\ddot{O}-\ddot{S}=\ddot{O}:$	
4	4	0	Tetrahedral	Tetrahedral	$\begin{array}{c}H\\ \|\\ H-C-H\\ \|\\ H\end{array}$	
4	4	1	Tetrahedral	Pyramidal	$\begin{array}{c}H-\ddot{N}-H\\ \|\\ H\end{array}$	
4	2	2	Tetrahedral	Bent	$H-\ddot{\underset{..}{O}}-H$	

EXAMPLE 5.7

Predicting the Shapes of Molecules II

Use Lewis theory and VSEPR to predict the molecular geometry of SiH_4.

SOLUTION
The Lewis structure is as follows:

$$\begin{array}{c}H\\ \|\\ H-Si-H\\ \|\\ H\end{array}$$

There are four groups of electrons around the central atom, and they are all bonding pairs; therefore, the correct geometry is tetrahedral.

> **YOUR TURN**
>
> ## Predicting the Shapes of Molecules II
>
> Use Lewis theory and VSEPR to predict the correct molecular geometry of CF_2Cl_2. (Carbon is the central atom in the skeletal structure.)

MULTIPLE BONDS AND MOLECULAR SHAPES

A double or triple bond acts exactly like a bonding electron pair (or a single bond) in determining molecular geometry. For example, CO_2 has the following Lewis structure:

$$:\ddot{O}=C=\ddot{O}:$$

The two pairs of electrons in each bond must stay together and, therefore, act as a group. The two groups minimize their repulsion by getting as far away from each other as possible, resulting in a linear geometry for CO_2. Unlike water or ammonia, there are no lone pairs on the central atom to repel the bonding groups.

> **EXAMPLE 5.8**
>
> ## Predicting the Shapes of Molecules Containing Multiple Bonds
>
> Use Lewis structures and VSEPR to predict the correct molecular geometry for SO_2.
>
> **SOLUTION**
> The Lewis structure is as follows:
>
> $$:\ddot{O}-\ddot{S}=\ddot{O}:$$
>
> There are three groups of electrons around the central atom and one is a lone pair. From Table 1, we see that the correct molecular geometry is bent.
>
> **YOUR TURN**
>
> ## Predicting the Shapes of Molecules Containing Multiple Bonds
>
> Use Lewis structures and VSEPR to predict the correct molecular geometry for H_2CO. (Carbon is the central atom.)

> **EXAMPLE 5.9**
>
> ### Predicting the Shapes of Molecules Containing Multiple Bonds II
>
> Use Lewis structures and VSEPR to predict the correct molecular geometry for C_2H_4. (The two carbon atoms are in the middle with two hydrogen atoms bonded to each carbon.) The Lewis structure is as follows:
>
> $$H-\underset{|}{\overset{H}{C}}=\underset{|}{\overset{H}{C}}-H$$
>
> Because there are two central atoms, we must consider the geometry about each one. Each carbon atom has three groups of electrons and no lone pairs. Therefore, the geometry is trigonal planar about each carbon, resulting in the following molecular geometry:
>
>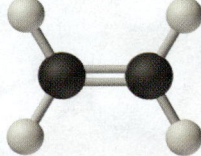
>
> **YOUR TURN**
>
> ### Predicting the Shapes of Molecules Containing Multiple Bonds II
>
> Use Lewis structures and VSEPR theory to predict the molecular geometry of C_2H_2. (The two carbon atoms are in the middle and the hydrogen atoms are on the ends.)

5.7 Water: Polar Bonds and Polar Molecules

Water is among the most important compounds on Earth; without it, life in any form is unimaginable. The unique properties of water—its tendency to be a liquid at room temperature, its ability to dissolve other compounds, its expansion on freezing, and its abundance—are all related to its importance. Like all compounds, the properties of water arise from the molecules that compose it. To see how the properties of water emerge from the properties of water molecules, we must discuss one other important concept—bond polarity.

Consider one of water's two identical covalent bonds. The electrons in this bond are not as evenly shared as they appear in the Lewis structure of water. Within the covalent bond, oxygen attracts the shared electron pair more than hydrogen does. The result is that the electrons are pulled closer to the oxygen atom, causing the electron distribution to be uneven. Covalent bonds with uneven electron distributions are called **polar bonds** because one side of the bond, the side that is electron deficient, develops a slight positive charge (δ^+) while the other side of the bond, the side that is electron rich, develops a slight

The Molecular Revolution

AIDS Drugs

Many patients infected with the human immunodeficiency virus (HIV) that causes acquired immune deficiency syndrome (AIDS) are treated with a class of drugs called *protease inhibitors*. Before these drugs were developed in the early 1990s, HIV-infected individuals would die within a few years of diagnosis. Today, protease inhibitors, when given in combination with other drugs, can reduce HIV to undetectable levels and keep AIDS patients alive.

Protease inhibitors were developed with the help of bonding theories such as those in this chapter. Scientists used these theories to design a molecule that would block the action of HIV-protease, a molecule that HIV requires to reproduce. HIV-protease is an example of a protein, a class of large biological molecules that carry out many important functions in living organisms. The shape of HIV-protease was discovered in 1989. Researchers in drug companies then used the newly discovered shape to design a drug molecule that would fit into the working part of HIV-protease and therefore disable it. Within a couple of years, they had several candidates. Further testing and subsequent human trials resulted in the final development of protease inhibitors, the drugs that treat HIV. Although these drugs haven't cured AIDS, they have turned it into a long-term manageable disease.

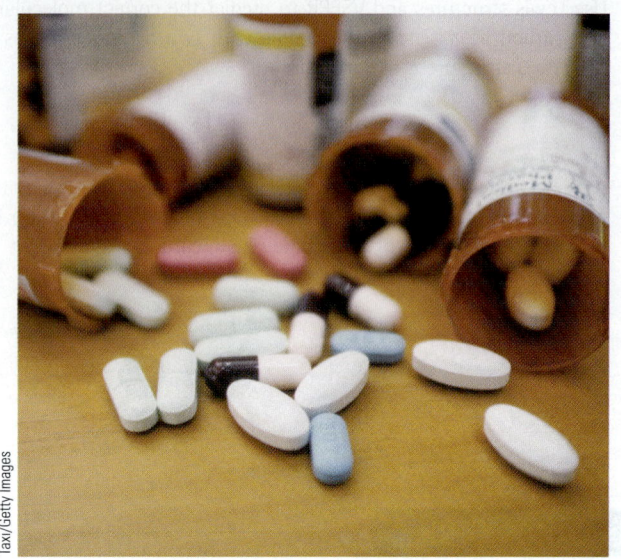

Protease inhibitors are used to treat HIV.

 Explore this topic on the **HIV/AIDS** website.

 Explore this topic on the **Protese Inhibitors** website.

negative charge (δ^-). In other words, the bond develops two poles, one positive and one negative. This is called a **dipole**.

$$\overset{\delta^+}{\text{H}}{-}\overset{\delta^-}{\text{O}}$$

Positive   Negative
pole   pole

Polar bonds are fairly common in molecular compounds and arise whenever elements with different electron-attracting abilities form a bond. The ability of an atom to attract electrons in a covalent bond is called **electronegativity**, and Figure 4 shows how electronegativities vary for different elements across the periodic table. In general, electronegativity increases as you go to the right across a row on the periodic table and decreases as you go down a column. Therefore, the elements at the top right of the periodic table have the highest electronegativities and those at the bottom left have the lowest. Fluorine is the most electronegative element. A covalent bond between two atoms of differing electronegativities will be polar. The greater the difference in electronegativity is, the more polar the bond will be. Identical atoms, or atoms of equal electronegativity, do not form polar bonds—they share their electrons equally.

In analogy to a magnet, a polar bond has two poles, a positive pole and a negative pole. Similarly, an entire molecule may be a **polar molecule** if it has an

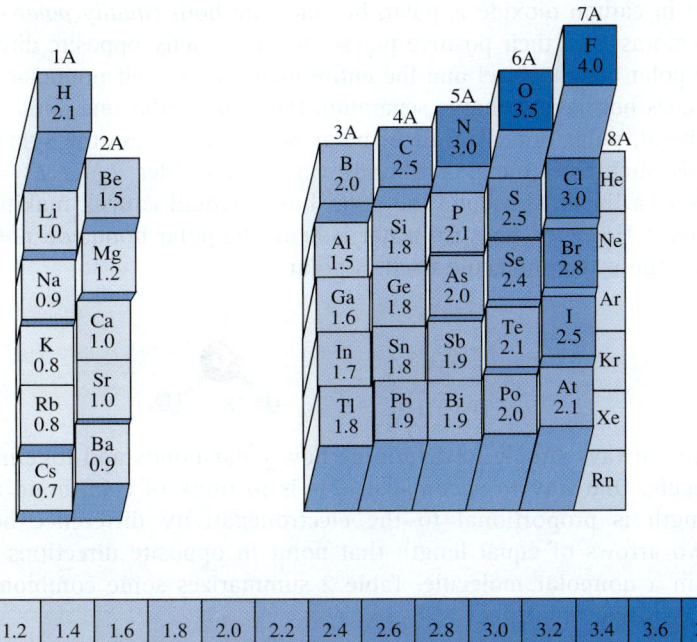

FIGURE 4 Electronegativity of the elements.

uneven electron distribution that results in a negative pole and a positive pole. In a diatomic molecule—one composed of only two atoms—a polar bond always results in a polar molecule. For example, consider carbon monoxide:

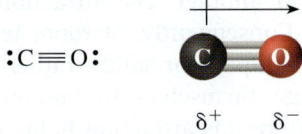

Since carbon and oxygen have different electronegativies, the bond is polar and so is the molecule. One side of the molecule will have a slightly negative charge, and the other side will have a slightly positive charge.

Polar bonds will not always, however, give rise to polar molecules. Consider for example, carbon dioxide:

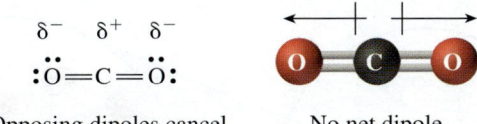

Opposing dipoles cancel   No net dipole

APPLY YOUR KNOWLEDGE

Does a polar bond form between two atoms of the same element? Why or why not?

Answer: No. Two atoms of the same element have identical electronegativities and, therefore, share electrons in the bond equally.

Each bond in carbon dioxide is polar, but they are both *equally polar*—between the same two atoms—and their positive poles point in exactly opposite directions. As a result, the polar bonds cancel and the entire molecule is itself nonpolar. A **nonpolar molecule** does not have a charge separation the way a polar one does.

In general, polar bonds result in polar molecules unless the symmetry of the molecule is such that the polar bonds cancel. Consider water as an example. If water were a linear molecule, the polar bonds would cancel, making the molecule nonpolar. However, because water is bent, the polar bonds do not completely cancel, and the water molecule itself is polar.

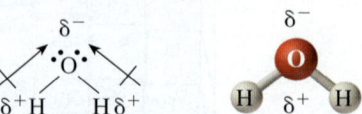

It is not always simple to determine how polar bonds add together or cancel in a molecule. One way to accomplish this is to think of each bond as an arrow whose length is proportional to the electronegativity difference between two atoms. Two arrows of equal length that point in opposite directions will cancel, resulting in a nonpolar molecule. Table 2 summarizes some common geometries and their resulting polarities.

As you already know, negative and positive charges attract one another. In a polar molecule, the positive end of the molecule attracts the negative end of its neighbor.

This attraction is analogous to the attraction between the south pole of one magnet and the north pole of another. The attractions between polar molecules tend to hold them together. Consequently, at room temperature, polar molecules have a greater tendency to be liquids or solids, rather than gases. Figure 5 shows how water molecules arrange themselves to maximize the attraction between their poles. At room temperature this attraction holds water molecules together as a liquid. Without these attractions, water would be a gas at room temperature; therefore, most of the water on Earth would be vapor.

When water freezes to form ice, this structure extends in a three-dimensional lattice (Figure 6) containing water molecules in layers of hexagonal rings. The

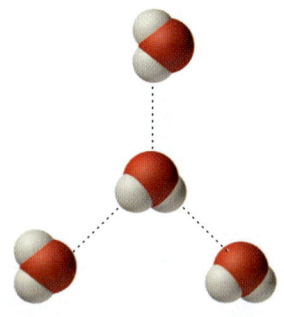

FIGURE 5 Water molecules arrange themselves so that the positive end of one molecule is close to the negative end of another.

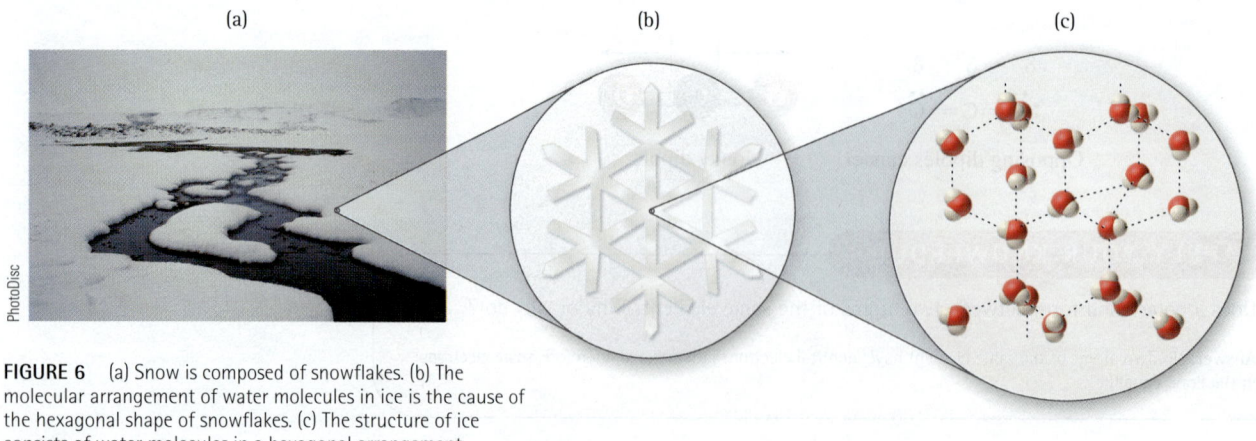

FIGURE 6 (a) Snow is composed of snowflakes. (b) The molecular arrangement of water molecules in ice is the cause of the hexagonal shape of snowflakes. (c) The structure of ice consists of water molecules in a hexagonal arrangement.

TABLE 2
Common Molecular Geometries and Their Resulting Polarities

Nonpolar

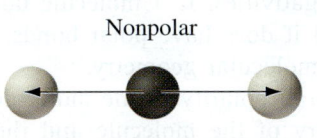

Linear

Two identical polar bonds pointing in opposite directions cancel. The molecule is nonpolar.

Polar

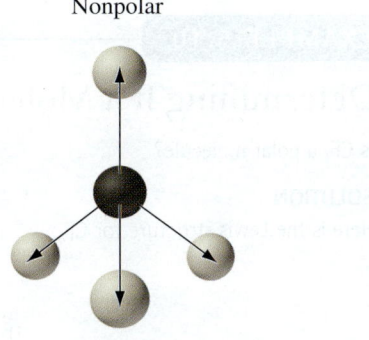

Bent

Two polar bonds at an angle of less than 180° between them will not cancel. The molecule is polar.

Nonpolar

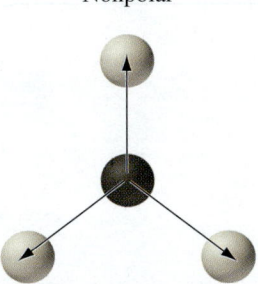

Trigonal planar

Three identical polar bonds at 120° from each other will cancel. The molecule is nonpolar.

Nonpolar

Tetrahedral

Four identical polar bonds in a tetrahedral arrangement will cancel. The molecule is nonpolar.

Polar

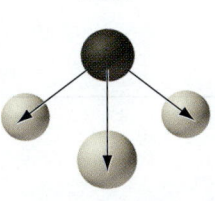

Pyramidal

Three polar bonds in pyramidal arrangement will not cancel. The molecule is polar.

In all cases where the bonds cancel, they are assumed to be identical bonds (i.e., between the same two elements). If the bonds are not identical, they most likely do not cancel, and the molecule will be polar.

•••

hexagonal shape of snowflakes is a direct result of this molecular arrangement. The open spaces in the ice structure result in ice being less dense than liquid water, a unique property of water that makes ice float.

In addition, because polar molecules are attracted to one another, they do not mix well with nonpolar molecules. Oil and water, for example, do not mix because water is polar and oil is nonpolar. We will look at the consequences of polarity more fully in Chapter 12. For now, be able to tell whether simple molecules are polar or not by looking at their Lewis structure and asking two questions:

1. Does the molecule contain polar bonds? A bond is polar if it occurs between two elements of different electronegativities. If a molecule does not have polar bonds, it will not be polar. If it does have polar bonds, it may or may not be polar, depending on the molecular geometry.
2. Do the polar bonds add together to give overall polarity to the molecule? Use VSEPR theory to determine the geometry of the molecule and then determine if the polar bonds cancel. If they do not cancel, the molecule will be polar. Use Table 2 as a guide in determining if a molecule is polar.

EXAMPLE 5.10

Determining If a Molecule Is Polar

Is CF_4 a polar molecule?

SOLUTION

Here is the Lewis structure for CF_4:

$$\ddot{\text{F}}-\underset{\underset{\ddot{\text{F}}}{|}}{\overset{\overset{\ddot{\text{F}}}{|}}{\text{C}}}-\ddot{\text{F}}$$

Because carbon and fluorine have different electronegativities, the bonds are polar. VSEPR theory predicts that CF_4 is tetrahedral. If we think of each bond as an arrow, then the four equal arrows in a tetrahedral arrangement cancel as shown in Table 2. Consequently, the molecule is nonpolar.

YOUR TURN

Determining If a Molecule Is Polar

Is H_2S polar or nonpolar?

APPLY YOUR KNOWLEDGE

We have seen how the shape of water molecules determines the properties of water. If the molecule had a different shape, water would be a different substance. Suppose the shape of water was different in a way that resulted in ice being *more dense* than liquid water? How would a cup of ice water look? Would icebergs exist? What would happen to marine life in winter as ice formed in lakes?

Answer: If ice were denser than liquid water, it would sink. Icebergs would not exist. As water froze in a lake or ocean, it would sink to the bottom, probably killing marine life.

Chapter Summary

MOLECULAR CONCEPT

Most elements do not exist as pure elements in nature. Instead, they bond together to form compounds (5.1). An element's electron configuration determines its chemical reactivity—those elements having full outer Bohr orbits are most stable (5.2). Most elements do not have full outer orbits; therefore, they exchange or share electrons with other elements to obtain a stable configuration. If the electrons are transferred, as between a metal and a nonmetal, the resulting bond is called an ionic bond (5.3). If the electrons are shared, as between two nonmetals, the resulting bond is called a covalent bond (5.4).

We represent chemical bonding with Lewis theory. In this model, valence electrons are represented as dots surrounding the chemical symbol for the element. Compounds are formed by allowing the electrons to be *transferred* between atoms (ionic bonding) or *shared* between atoms (covalent bonding) so that all atoms acquire an octet.

Lewis theory is used in combination with valence shell electron pair repulsion (VSEPR) theory to predict the shapes of molecules (5.6). In this model, electron groups—lone pairs, single bonds, double bonds, or triple bonds—minimize their repulsions by getting as far away from each other as possible. This, in turn, determines the geometry of the molecule.

Covalent bonds are polar if they form between elements of differing electronegativities (5.7). Polar bonds can cancel in a molecule, producing a nonpolar molecule, or they can sum to produce a polar molecule.

SOCIETAL IMPACT

If the 91 naturally occurring elements were chemically stable and did not combine to form compounds, the world would be a vastly different place. There would be only 91 different kinds of substances in the world and life would be impossible (5.1).

Our knowledge of chemical bonding has allowed us to bond atoms together in new ways to form compounds that never existed before. Plastics, Teflon, synthetic fibers, and many pharmaceuticals are among some of the compounds synthesized by chemists that have significantly impacted society (5.3, 5.4).

The shapes of molecules determine many of their properties. Water, for example, would be a vastly different substance if it were shaped differently. Some of our senses depend on molecular shape (5.6). For example, we smell molecules based at least partly on their shape. Receptors within our noses are like molds into which certain molecules can fit. When the mold is occupied by one of these molecules, the brain receives nerve impulses that we interpret as a certain smell.

Chemistry on the Web

For up-to-date URLs, visit the text website at **http://chemistry.brookscole.com/tro3**

- Salt
  http://www.saltinstitute.org/15.html
- G.N. Lewis
  http://www.woodrow.org/teachers/chemistry/institutes/1992/Lewis.html
- Ozone
  http://www.epa.gov/ozone/

- VSEPR Shapes
  http://www.molecules.org/VSEPR_table.html
- HIV/AIDS
  http://aidsinfo.nih.gov/
- Protease Inhibitors
  http://www.thebody.com/treat/protinh.html
  http://www.fda.gov/fdac/features/1999/499_aids.html

Key Terms

bent
bonding group
bonding pair electrons
dipole
double bond
electronegativity
electron geometry
electron group
Lewis, G. N.
Lewis structure
Lewis theory
lone pair electrons
molecular geometry
nonpolar molecule
octet rule
polar bond
polar molecule
pyramidal
resonance
resonance structures
triple bond
valence shell electron pair repulsion (VSEPR)

Exercises

 Assess your understanding of this chapter's topics with an online chapter quiz at **http://chemistry.brookscole.com/tro3**.

QUESTIONS

1. Why is salt, NaCl, relatively harmless even though the elements that compose it, sodium and chlorine, are toxic?
2. If sodium is dropped into water, a loud fizzing noise is heard and sparks fly from the sodium. What do you suppose is occurring?
3. Explain ionic bonding according to Lewis theory.
4. Explain covalent bonding according to Lewis theory.
5. Why is Lewis theory useful? Give some examples.
6. Give the Bohr electron configuration and Lewis electron dot structure for sulfur. Circle the electrons in the Bohr electron configuration that are also in the Lewis structure.
7. Draw electron dot structures for the following elements: Na, Al, P, Cl, and Ar. Which are most chemically reactive? Which are least chemically reactive?
8. Explain VSEPR theory. According to this theory, what determines the shapes of molecules?
9. Why are the shapes of molecules important?
10. Describe a polar covalent bond.
11. In what ways is water unique? What about the water molecule causes the unique properties of water?
12. What is the difference between a polar and a nonpolar bond? What is the difference between a polar and nonpolar molecule?
13. Why do polar molecules have a greater tendency to remain a liquid or a solid at room temperature.
14. Explain, in molecular terms, why oil and water do not mix.

PROBLEMS

15. Draw Lewis structures for each of the following elements. Which elements are chemically stable?
    a. C
    b. Ne
    c. Ca
    d. F

16. Draw Lewis structures for each of the following elements. Which elements are chemically stable?
    a. Br
    b. S
    c. Kr
    d. He
17. Draw a Lewis structure for each of the following ionic compounds:
    a. KI
    b. CaBr$_2$
    c. K$_2$S
    d. MgS
18. Draw a Lewis structure for each of the following ionic compounds:
    a. LiF
    b. Li$_2$O
    c. SrO
    d. SrI$_2$
19. Draw a Lewis structure for each of the following ionic compounds. What chemical formula does Lewis theory predict?
    a. sodium fluoride
    b. calcium chloride
    c. calcium oxide
    d. aluminum chloride
20. Draw a Lewis structure for each of the following ionic compounds. What chemical formula does Lewis theory predict?
    a. sodium oxide
    b. aluminum sulfide
    c. magnesium chloride
    d. beryllium oxide
21. Draw a Lewis structure for each of the following covalent compounds:
    a. I$_2$
    b. NF$_3$
    c. PCl$_3$
    d. SCl$_2$
22. Draw a Lewis structure for each of the following covalent compounds:
    a. OF$_2$
    b. NI$_3$
    c. CS$_2$
    d. Cl$_2$CO (carbon is the central atom)
23. What is wrong with each of the following Lewis structures? Fix the problem and write a correct Lewis structure.
    a. Ca—Ö:
    b. :C̈l=Ö—C̈l:
    c. :F̈—P—F̈:
         |
        :F̈:
    d. :N̈=N̈:
24. What is wrong with each of the following Lewis structures? Fix the problem and write a correct Lewis structure.
    a. :N̈=N=Ö:
    b. :Ö—S̈—Ö:
    c. :B̈r—B̈r:
    d. :Ö=Si—Ö:
25. Use VSEPR to determine the geometry of the molecules in problem 21.
26. Use VSEPR to determine the geometry of the molecules in problem 22.
27. Draw a Lewis structure and use VSEPR to determine the geometry of each of the following molecules. If the molecule has more than one central atom, indicate the geometry about each of these and draw the three-dimensional structure.
    a. ClNO (nitrogen is the central atom)
    b. H$_3$CCH$_3$ (two carbon atoms in the middle, each with three hydrogen atoms attached)
    c. N$_2$F$_2$ (nitrogen atoms in the center and fluorine atoms on the ends)
    d. N$_2$H$_4$ (nitrogen atoms in the center and two hydrogen atoms attached to each nitrogen)
28. Draw a Lewis structure and use VSEPR to determine the geometry of each of the following molecules. If the molecule has more than one central atom, indicate the geometry about each of these and draw the three-dimensional structure.
    a. HCCH (two carbon atoms in the middle, each with one hydrogen atom attached)
    b. CCl$_4$
    c. PH$_3$
    d. HOOH (two oxygen atoms in the middle, each with one hydrogen atom attached)

29. Determine whether each of the following molecules is polar:
    a. HCl
    b. N_2
    c. O_2
    d. CO
30. Determine whether each of the following molecules is polar:
    a. NH_3
    b. CCl_4
    c. SO_2
    d. CH_4
    e. CH_3OH

POINTS TO PONDER

31. Explain why water would be a gas at room temperature if it had a linear rather than a bent geometry.
32. What is the molecular reason behind the hexagonal shape of snowflakes?
33. What would happen to the amount of UV light hitting the Earth if ozone had two double bonds rather than two bonds that are midway between single and double? Explain.
34. One of the observations that led G. N. Lewis to propose his theory was the chemical inertness of the noble gases. Suppose elements were different, such that observations demonstrated the halogens were chemically inert. How would Lewis theory change? Which elements would then be the most reactive?
35. G. N. Lewis developed a model for chemical bonding that you have learned in this chapter. His theory was extremely successful and is used today at all levels of chemistry, from the introductory class to the research laboratory. Why was Lewis's theory so successful?
36. The opening quote of this chapter states that "Man masters nature not by force, but by understanding. That is why science has succeeded where magic failed: because it has looked for no spell to cast." What do you think Bronowski meant by this? How does this apply to chemical bonding?
37. Draw a Lewis structure of the H_2 molecule. If you could somehow see H_2 molecules, would they look like this Lewis structure? In what ways are real H_2 molecules different than their Lewis structure?
38. Since the time of G.N. Lewis, other, more powerful theories have been developed to explain chemical bonding, yet Lewis theory is still incredibly useful. Use what you know about the scientific method and the nature of scientific theories to explain how this can be so.

Feature Problems and Projects

39. The Lewis structures for CH_4, N_2, and CO_2, along with the corresponding space-filling molecular images, are shown here:

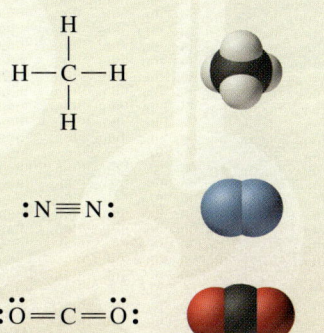

Use Lewis theory and VSEPR to draw similar space-filling molecular images of
   a. CO
   b. NH_3
   c. H_2S
   d. SiH_4

40. CH_3COCH_3 (acetone) is a common laboratory solvent that is often used in nail polish remover. Its Lewis structure and space-filling molecular image are shown here:

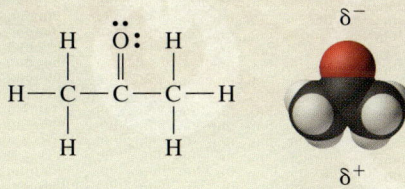

Acetone is a polar molecule; the oxygen end has a slightly negative charge (oxygen is more electronegative), while the carbon and hydrogen end has a slightly positive charge. In liquid acetone, the molecules are attracted to each other via these polar ends—the positive end of one molecule is attracted to the negative end of its neighbor—and therefore align as shown here:

Draw a Lewis structure and space-filling molecular image for CH_2Cl_2 (dichloromethane), another common laboratory solvent. Is the molecule polar? Which end of the molecule has a slightly negative charge? Which end has a slightly positive charge? Draw several space-filling molecular images of CH_2Cl_2 showing how they align in liquid dichloromethane.

41. In this chapter, we learned about chemical bonding, the way atoms join together to form molecules. Human understanding of chemical bonding, and our ability to bond atoms together in controlled ways, has changed our society; it has caused a molecular revolution. How many new molecules can you think of that have affected society in significant ways? Write a short essay describing at least two such impacts on society.

6 Organic Chemistry

Geraniol

2-phenylethanol

Both science and art have to do with ordered complexity.
—Lancelot L. Whyte

In this chapter, we examine a branch of chemistry known as organic chemistry, the study of carbon compounds and their reactions. Carbon is unique because it forms more compounds than any other element and is the backbone of the molecules that compose living organisms. Consequently, many of the molecules in foods, in natural aromas, and in many daily products are organic molecules. Organic molecules also compose most of our society's fuels and are the starting materials from which we make many consumer products such as plastic, Teflon, and nylon. What do organic molecules look like? How can we begin to categorize and understand the many different kinds of organic compounds? As you read this chapter, remember that organic compounds are all around you and even inside you. Most of the molecules in our bodies are organic compounds and they follow chemical laws just like any other substances. This nineteenth-century realization had a profound impact on how humans view themselves. How can it be that our senses, thoughts, and emotions are embodied in organic molecules and their interactions? What does that mean to you?

QUESTIONS FOR THOUGHT

- What are the differences between molecules obtained from living things and those obtained from the earth or mineral sources?

- What are fuels such as natural gas and petroleum composed of?

- Can the same set of atoms be put together in different ways to form different molecules?

- How do we name simple organic molecules?

- What are the molecules used in pesticides?

- What is the "alcohol" in alcoholic beverages? How does it differ from other alcohols such as rubbing alcohol?

- What molecules are responsible for the flavors and aromas in fruits such as raspberries, apples, and pineapple?

- What molecules are responsible for the pleasant aromas in flowers? For the unpleasant odors in rotting fish or decaying flesh?

6.1 Carbon

Look around you. What do you see? Maybe a plastic bottle, a desk, or some plants. Or perhaps your dresser is full of perfume bottles and lotions. There are millions of known compounds in our world and 95% of them have a single element in common—carbon. Carbon forms the backbone of the molecules that compose the plastic bottle, the wood in your desk, the aromas of perfume, and the oils in lotion. Carbon also forms the backbone of the molecules that compose living organisms such as plants, animals, and humans. The number of compounds formed by carbon far outnumbers those formed by all other elements combined. What is so special about carbon?

Carbon is the smallest member and the only nonmetal in group 4A of the periodic table (Figure 1). It has four valence electrons and, in its compounds,

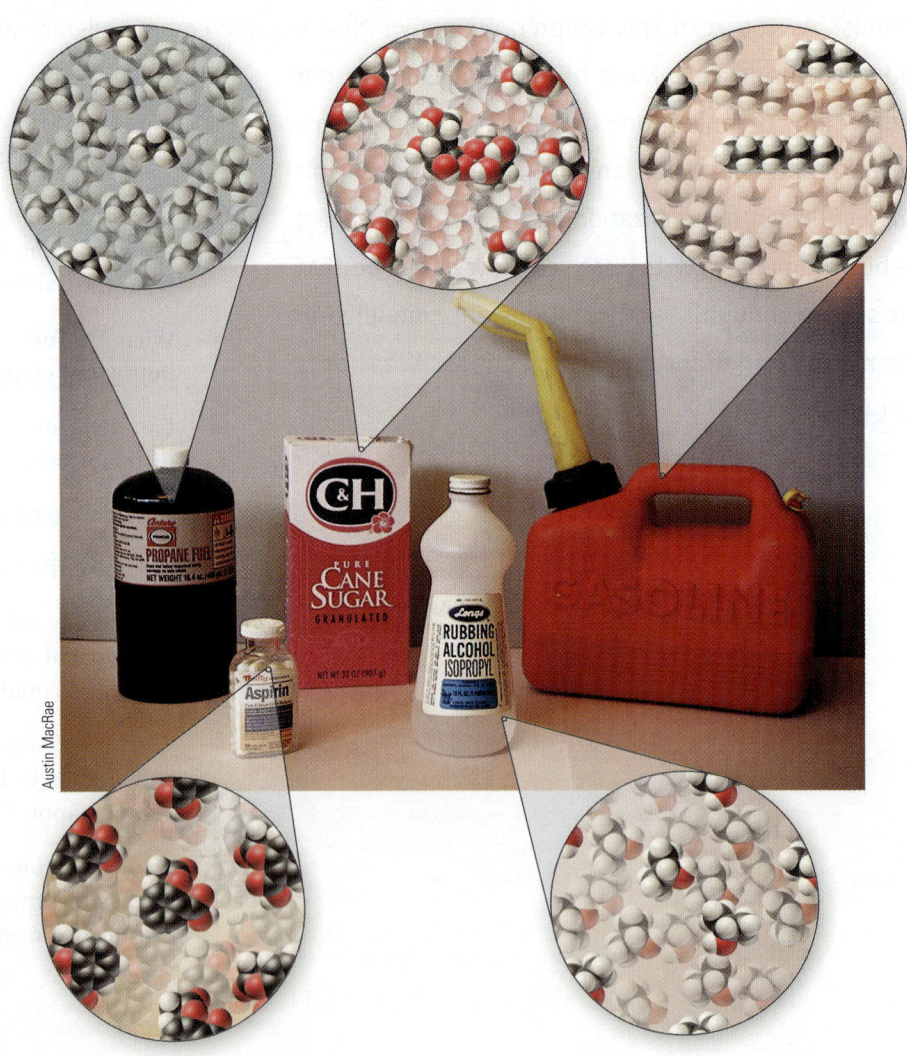

Many common substances are composed of organic compounds.

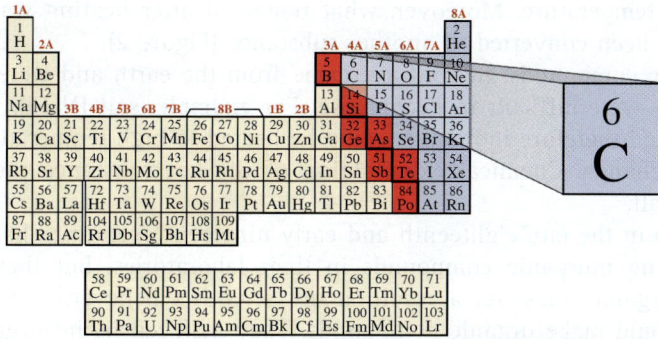

FIGURE 1 Carbon is the smallest member of group 4A in the periodic table.

forms four covalent bonds. Consider the following Lewis structures of carbon-containing compounds:

$$H-\underset{\underset{H}{|}}{\overset{H}{\underset{|}{C}}}-H \qquad H-\underset{\underset{H}{|}}{\overset{\overset{\ddot{O}}{\|}}{C}}-H \qquad H-C\equiv N\colon$$

Notice that in each case, carbon forms four bonds. Notice also that carbon can form single, double, or triple bonds. Carbon can also bond to itself, forming chains, rings, and branched structures.

This great versatility in bonding allows carbon to form a molecular framework with almost endless possibilities. Other elements—especially hydrogen, oxygen, nitrogen, sulfur, and chlorine—can incorporate into this framework to add even more variety.

The study of carbon-containing compounds and their chemistry is generally known as **organic chemistry**. Many organic compounds are found in living—or once living—organisms, but chemists have synthesized many more. Although a complete introductory study of organic chemistry would be the topic of an entire textbook, we can learn the basics of organic chemistry in just a few pages. Although organic molecules are complex, they are also incredibly intricate and even beautiful, and just like all molecules, they determine the properties of the substances they compose, substances that we encounter every day.

6.2 A Vital Force

In the early 1800s, scientists knew that chemical compounds were of two types, organic and inorganic. Organic compounds came from living organisms and were chemically fragile; they were easily decomposed with heat. For example, sugar—derived from plants and therefore organic—would smoke and burn if heated much

FIGURE 2 Burning sugar decomposes it. The black residue left in the pan is carbon.

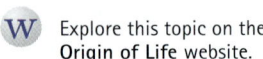
Explore this topic on the **Origin of Life** website.

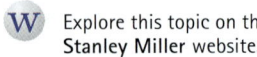
Explore this topic on the **Stanley Miller** website.

above room temperature. Moreover, what remained after heating was no longer sugar; it had been converted to another substance (Figure 2).

Inorganic compounds, in contrast, came from the earth and were chemically durable—they were difficult to decompose. For example, salt (NaCl), mined from the ground and therefore inorganic, could be heated to high temperatures without any visible change. Chemical examination of the hot substance proved it was merely hot salt.

Chemists in the late eighteenth and early nineteenth centuries had succeeded in synthesizing inorganic compounds in their laboratories, but they could not synthesize organic ones. As a result, many scientists believed that only living organisms could make organic compounds. They believed living organisms contained a **vital force** that allowed them to overcome ordinary physical laws and produce organic compounds. The chemist, in his or her laboratory with beakers and test tubes, had to follow the ordinary laws of nature, and therefore would never succeed in making an organic compound. This belief, termed **vitalism**, prevailed through much of the nineteenth century.

In 1828, a German chemist named **Friedrich Wohler (1800–1882)** synthesized an organic compound from an inorganic one in his laboratory. The compound was **urea**, a waste product normally found in the urine of humans and animals (Figure 3). Urea was an organic compound, and according to vitalism, could only be produced within living organisms. The properties of the urea Wohler synthesized in his laboratory were identical in every way to urea obtained from living organisms, but no vital force was necessary to make it. Wohler's synthesis was the beginning of the end for vitalism. The idea that life was somehow beyond physical laws was proved wrong.

THE MOLECULAR REVOLUTION

The Origin of Life

The death of vitalism left every aspect of the natural world open to scientific investigation. Scientists began to understand life itself in terms of natural processes. This stood in stark contrast to previous beliefs in which life had—*at its core*—mystical or supernatural elements exempt from natural law. Eventually, scientists began to examine the *origin* of life—previously the exclusive realm of religion—in terms of natural processes. About 50 years ago, a young graduate student named Stanley Miller at the University of Chicago attempted to re-create primordial Earth in the laboratory. He mixed several different compounds believed to be in the Earth's early atmosphere in a warm water bath and simulated lightning with an electrical spark. After a few days, Miller found that the flask contained not only organic compounds, but organic compounds *central to life*—amino acids. The stunning experiment left many speculating that in time life would be created in the laboratory. It has not turned out that way. We are still trying to solve one of the greatest puzzles of all—how life got its start.

Stanley Miller, now a chemistry professor at the University of California at San Diego, says, "The problem of the origin of life has turned out to be much more difficult than I, and most other people, envisioned." Most origin-of-life researchers have an idea of what a viable theory should look like. Somehow, a group of molecules developed the ability to copy themselves imperfectly; that is, their "offspring" were copies of their predecessors, except for small, inheritable mistakes. Some of these mistakes gave the offspring the ability to replicate better, making them more likely to replicate and therefore pass on the favorable mistake to the next generation. In this way, chemical evolution got its start, producing generations of molecules that slowly got better at replicating themselves. Eventually, living organisms—very good at replicating themselves—evolved.

The main controversy today, however, is just what these first compounds were, how they formed, and how they replicated. The older origin-of-life theories suggest that these compounds were the same ones present in living organisms today, proteins and DNA (more on these in Chapter 16). However, the complexity of proteins and DNA, and the difficulty in getting them to replicate independently, have led some researchers to propose other candidates such as clays, sulfur-based compounds, or pyrite (fool's gold). None of these theories has gained widespread acceptance, and the origin of life continues to be a puzzle with which scientists grapple.

Living organisms must follow the physical laws of the universe. The laws that govern molecules in our brains are no different from the laws that govern molecules in common soil. Since Wohler's time, hundreds of thousands of organic compounds have been synthesized in laboratories all over the world. Even compounds that are central to life, such as proteins and DNA, have been successfully made. The chemistry of living organisms is known as **biochemistry**, covered in Chapter 16. However, organic chemistry and biochemistry are closely related. We need to understand the basics of organic compounds before we can begin to understand biochemistry. In addition, many organic compounds have uses that are not necessarily related to their biological origins. For example, most of our fuels are composed of organic compounds that originated from prehistoric plant and animal life.

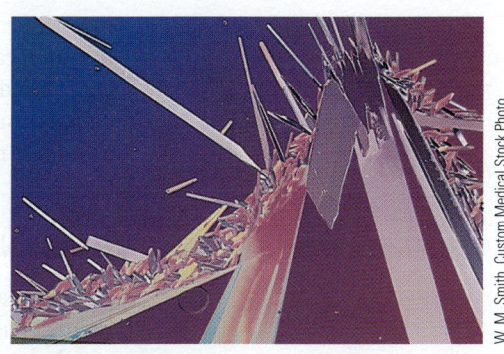

FIGURE 3 Urea crystals seen through polarized light.

6.3 The Simplest Organic Compounds: Hydrocarbons

We can organize our study of organic compounds by grouping them into families and subfamilies with similar characteristics. All organic compounds can be broadly divided into two major types as shown in Table 1. **Hydrocarbons** are compounds containing only carbon and hydrogen. **Functionalized hydrocarbons** are hydrocarbons with additional atoms or groups of atoms—called **functional groups**—incorporated into their structure. We will examine functionalized hydrocarbons later in this chapter. The hydrocarbon family can itself be further divided

TABLE 1

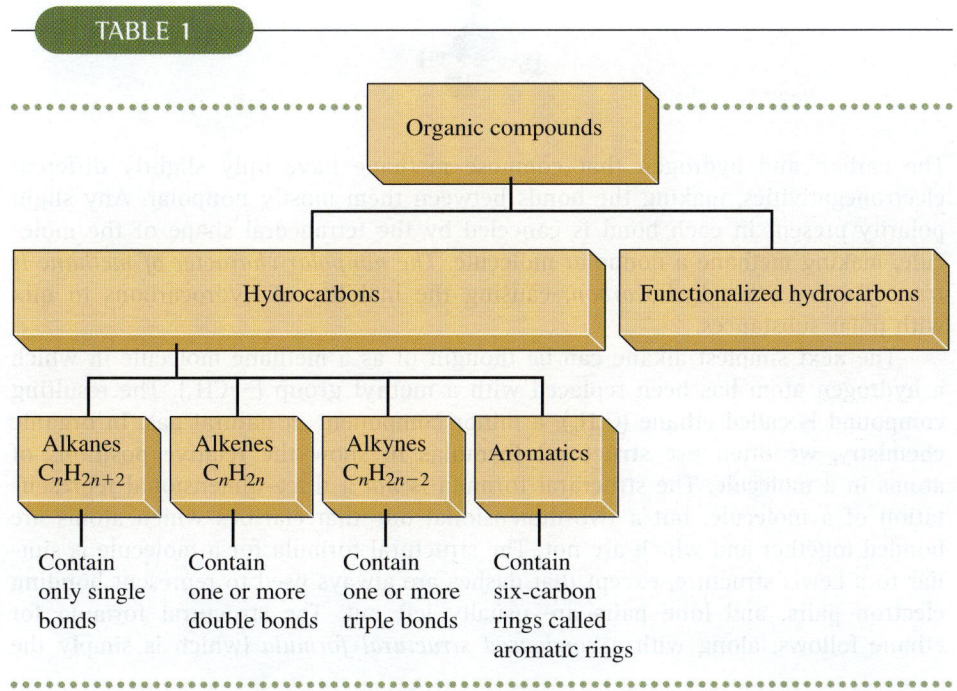

into four subfamilies, three of which depend on whether carbon is bonded to other carbon atoms by single, double, or triple bonds. These three subfamilies are called alkanes, alkenes, and alkynes, respectively. The fourth subfamily, called aromatic hydrocarbons, contains a six-carbon ring structure called an aromatic ring.

ALKANES: GASOLINE AND OTHER FUELS

Alkanes are hydrocarbons in which all carbon atoms are connected by single bonds. They have the general molecular formula C_nH_{2n+2}, where n represents the number of carbon atoms in the molecule. The simplest alkane is methane (CH_4), the major component in natural gas. Since methane is a prototype for all hydrocarbons, many of the properties of hydrocarbons can be understood by understanding methane. Let's use some of what we have learned in Chapter 5 to determine the structure and properties of methane. We begin with the Lewis structure of methane.

$$H-\underset{\underset{H}{|}}{\overset{\overset{H}{|}}{C}}-H$$

Covalent Lewis structures are covered in Section 5.4.

How to determine the shapes of molecules with VSEPR theory is covered in Section 5.6.

Carbon forms four single covalent bonds with four hydrogen atoms. The carbon atom attains an octet, and each hydrogen atom attains a duet—the methane molecule is stable. The molecular geometry of methane can be determined from its Lewis structure using VSEPR theory. Since carbon has four electron groups and no lone pairs, the geometry is tetrahedral.

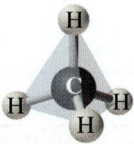

The carbon and hydrogen that compose methane have only slightly different electronegativities, making the bonds between them mostly nonpolar. Any slight polarity present in each bond is canceled by the tetrahedral shape of the molecule, making methane a nonpolar molecule. *The nonpolar character of methane is true of every other hydrocarbon*, causing the inability of hydrocarbons to mix with polar substances.

The next simplest alkane can be thought of as a methane molecule in which a hydrogen atom has been replaced with a **methyl group** (—CH_3). The resulting compound is called ethane (C_2H_6), a minor component in natural gas. In organic chemistry, we often use **structural formulas** to show the relative positions of atoms in a molecule. The structural formula is *not* a three-dimensional representation of a molecule, but a two-dimensional one that clarifies which atoms are bonded together and which are not. The structural formula for a molecule is similar to a Lewis structure, except that dashes are always used to represent bonding electron pairs, and lone pairs are usually left out. The structural formula for ethane follows, along with a *condensed structural formula* (which is simply the structural formula written more compactly) and a three-dimensional representation of the molecule.

Ethane H—C(H)(H)—C(H)(H)—H   CH₃CH₃

Structural formula | Condensed structural formula | Three-dimensional space-filling model

The next simplest alkane can be derived from ethane by inserting a **methylene group** (—CH$_2$—) between the two carbon atoms of ethane. The resulting compound is called propane (C$_3$H$_8$):

Propane H—C(H)(H)—C(H)(H)—C(H)(H)—H   CH₃CH₂CH₃

Propane is the main component in liquid propane (LP) gas, used as fuel in gas barbecues, portable stoves, and camping vehicles (Figure 4). Addition of another methylene group results in butane (C$_4$H$_{10}$), found in butane lighters:

Butane H—C(H)(H)—C(H)(H)—C(H)(H)—C(H)(H)—H   CH₃CH₂CH₂CH₃

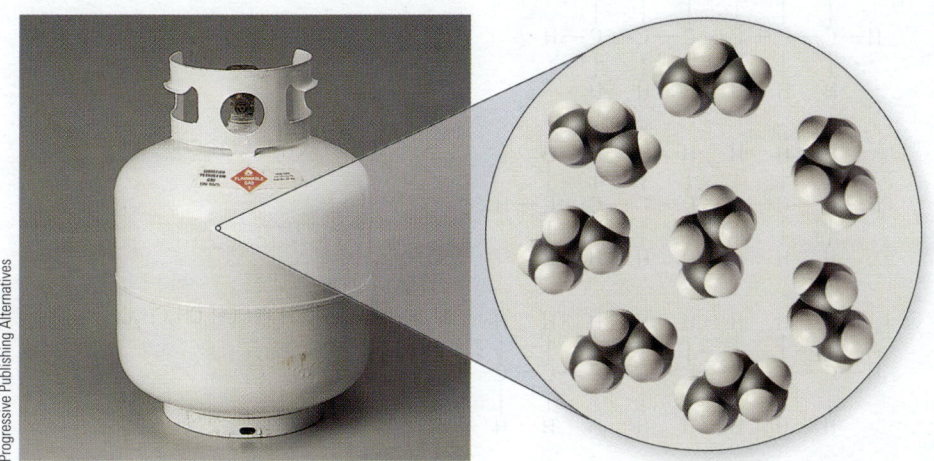

FIGURE 4 Propane is the main component of LP gas.

By successively adding methylene groups, we can form each of the alkanes shown in Table 2.

TABLE 2
Alkanes

n	Name	Molecular Formula	Structural Formula	Condensed Structural Formula
$n = 1$	Methane	CH_4	H–C(H)(H)–H	CH_4
$n = 2$	Ethane	C_2H_6	H–C(H)(H)–C(H)(H)–H	CH_3CH_3
$n = 3$	Propane	C_3H_8	H–C(H)(H)–C(H)(H)–C(H)(H)–H	$CH_3CH_2CH_3$
$n = 4$	Butane	C_4H_{10}	H–C(H)(H)–C(H)(H)–C(H)(H)–C(H)(H)–H	$CH_3CH_2CH_2CH_3$
$n = 5$	Pentane	C_5H_{12}	H–C(H)(H)–C(H)(H)–C(H)(H)–C(H)(H)–C(H)(H)–H	$CH_3CH_2CH_2CH_2CH_3$
$n = 6$	Hexane	C_6H_{14}	H–C(H)(H)–C(H)(H)–C(H)(H)–C(H)(H)–C(H)(H)–C(H)(H)–H	$CH_3CH_2CH_2CH_2CH_2CH_3$
$n = 7$	Heptane	C_7H_{16}	H–C(H)(H)–C(H)(H)–C(H)(H)–C(H)(H)–C(H)(H)–C(H)(H)–C(H)(H)–H	$CH_3CH_2CH_2CH_2CH_2CH_2CH_3$
$n = 8$	Octane	C_8H_{18}	H–C(H)(H)–C(H)(H)–C(H)(H)–C(H)(H)–C(H)(H)–C(H)(H)–C(H)(H)–C(H)(H)–H	$CH_3CH_2CH_2CH_2CH_2CH_2CH_2CH_3$

The alkanes in Table 2 are each represented with a molecular formula, a structural formula, and a condensed structural formula. The condensed structural formula is a way to write the structural formula more compactly. Note that a condensed structural formula, such as $CH_3CH_2CH_3$, *does not* mean C—H—H—H—C—H—H—C—H—H—H but instead represents the following structure:

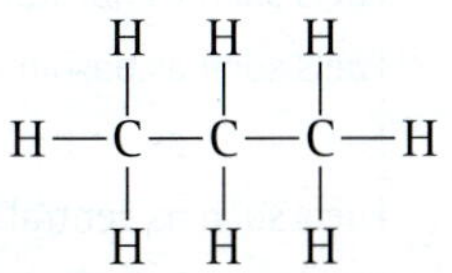

$$\begin{array}{c} H\ \ H\ \ H \\ |\ \ \ |\ \ \ | \\ H-C-C-C-H \\ |\ \ \ |\ \ \ | \\ H\ \ H\ \ H \end{array}$$

Carbon must always form four bonds and hydrogen must always form only one; therefore, hydrogen atoms are always at terminal positions. For longer alkanes, condensed structural formulas may be condensed further by writing $CH_3(CH_2)_nCH_3$ where n represents the number of methylene groups between the two terminal methyl groups. For example, $CH_3(CH_2)_4CH_3$ represents $CH_3CH_2CH_2CH_2CH_2CH_3$.

Straight chain alkanes—those with no branches—are named with a base name, which depends on the number of atoms in the chain as specified in Table 3 and the suffix *-ane*. For example, $CH_3CH_2CH_2CH_2CH_3$ has five carbon atoms and therefore has the base name *pent* followed by the ending *ane*. Its name is pentane.

TABLE 3
Base Names for Hydrocarbons

No. C atoms	Base Name
1	meth
2	eth
3	pro
4	but
5	pent
6	hex
7	hept
8	oct
9	non
10	dec

EXAMPLE 6.1

Drawing Structural and Condensed Structural Formulas

Give the molecular formula and draw the structural formula and condensed structural formula for the alkane with nine carbon atoms.

SOLUTION

The molecule contains 9 carbon atoms and therefore $2(9) + 2 = 20$ hydrogen atoms. Its molecular formula is C_9H_{20}. Its structural formula is as follows:

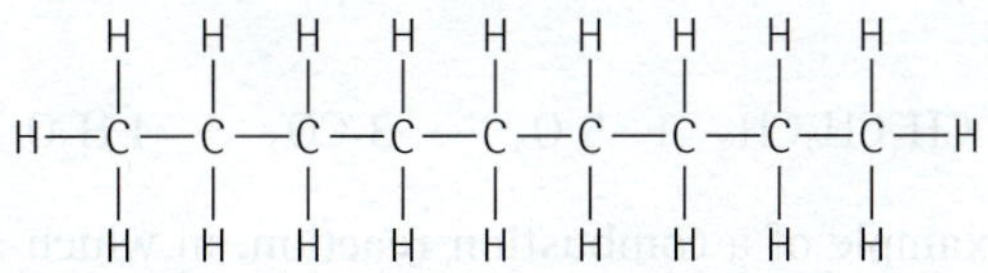

$$\begin{array}{c} H\ \ H\ \ H\ \ H\ \ H\ \ H\ \ H\ \ H\ \ H \\ |\ \ \ |\ \ \ |\ \ \ |\ \ \ |\ \ \ |\ \ \ |\ \ \ |\ \ \ | \\ H-C-C-C-C-C-C-C-C-C-H \\ |\ \ \ |\ \ \ |\ \ \ |\ \ \ |\ \ \ |\ \ \ |\ \ \ |\ \ \ | \\ H\ \ H\ \ H\ \ H\ \ H\ \ H\ \ H\ \ H\ \ H \end{array}$$

Its condensed structural formula is $CH_3CH_2CH_2CH_2CH_2CH_2CH_2CH_2CH_3$ or $CH_3(CH_2)_7CH_3$.

YOUR TURN

Drawing Structural and Condensed Structural Formulas

Give the molecular formula and draw the structural formula and condensed structural formula for the alkane with ten carbon atoms.

TABLE 4
Alkanes and Their Uses

Length of Carbon Chain		Use
$C_1 - C_4$	↓ Increasing boiling point	Fuels such as natural gas, propane, and butane
$C_5 - C_{12}$		Fuels such as gasoline
$C_{12} - C_{18}$		Fuels such as jet fuel
$C_{18} - C_{20}$		Fuels such as central heating fuel
$C_{20} - C_{30}$		Lubricating oils such as engine oil
$C_{30} - C_{40}$		Fuel oils such as ship fuel
$C_{40} - C_{50}$		Waxes and thick oils such as paraffin and petroleum jelly
$> C_{50}$		Tars used in road surfacing

APPLY YOUR KNOWLEDGE

What are the names of the straight chain alkanes with nine and ten carbon atoms?

Answer: nonane and decane

PROPERTIES AND USES OF ALKANES

Table 4 lists the alkanes according to carbon chain length and some of their uses. The boiling point of alkanes increases with increasing carbon chain length. Alkanes with 1 to 4 carbon atoms are gases at room temperature and are used as fuels, as discussed earlier. Alkanes with 5 to 20 carbon atoms are liquids, many of which form the major components of gasoline and other liquid fuels. Octane, with eight carbon atoms, is typical. Alkanes with 20 to 40 carbon atoms are thick liquids that compose lubricants and oils, and alkanes with higher numbers of carbon atoms compose substances such as petroleum jelly and candle wax.

The most important property of the alkanes is their flammability—they burn in oxygen and give off heat—making them good energy sources. For example, propane, the substance that composes LP gas, burns according to the following reaction:

$$CH_3CH_2CH_3 + 5\,O_2 \rightarrow 3\,CO_2 + 4\,H_2O$$

This reaction is an example of a **combustion reaction**, in which a carbon-containing compound reacts with oxygen to form carbon dioxide and water.

Another important property of alkanes is that, like all hydrocarbons, they are nonpolar. This prevents them from mixing with polar substances such as water. For example, if we spill gasoline onto water, it sits on top of the water, and no amount of stirring produces permanent mixing (Figure 5).

ALKENES AND ALKYNES

Alkenes and **alkynes** are hydrocarbons that contain at least one double or triple bond, respectively. The simplest members have the general formulas C_nH_{2n} (alkenes) and C_nH_{2n-2} (alkynes). They have fewer hydrogen atoms per carbon atom

FIGURE 5 Gasoline and water will not mix because alkanes are nonpolar and water is polar.

than alkanes and are therefore referred to as **unsaturated hydrocarbons**. In contrast, alkanes are called **saturated hydrocarbons**, meaning they are *saturated* with H atoms; they have the maximum number of hydrogen atoms per carbon atom. The simplest alkene is ethene, C_2H_4, with the following structure:

Ethene $\quad$ H\C=C/H $\quad$ $CH_2=CH_2$

The geometry of ethene is trigonal planar about each carbon atom; it is a flat, rigid molecule. Each carbon atom forms four bonds, but there is a double bond between the two carbon atoms. An alkene can be derived from an alkane

---

APPLY YOUR KNOWLEDGE

Which of the following structures corresponds to $CH_2=CHCH_3$?

a. C–H–H=C–H–C–H–H–H

b. H–H–C=C–H–C–H–H–H

c.
$$H\!\!-\!\!C=C\!\!-\!\!C\!\!-\!\!H$$
(with appropriate H atoms)

Answer: c. Carbon must always form four bonds, whereas hydrogen forms only one. The double bond between the carbon atoms counts as two bonds for each carbon.

TABLE 5
Alkenes

n	Name	Molecular Formula	Structural Formula	Condensed Structural Formula
$n = 2$	Ethene	C_2H_4	H₂C=CH₂ (structural)	$CH_2{=}CH_2$
$n = 3$	Propene	C_3H_6	(structural)	$CH_2{=}CHCH_3$
$n = 4$	1-Butene	C_4H_8	(structural)	$CH_2{=}CHCH_2CH_3$
$n = 5$	1-Pentene	C_5H_{10}	(structural)	$CH_2{=}CHCH_2CH_2CH_3$

simply by removing two hydrogen atoms on adjacent carbon atoms and inserting a double bond. The four simplest alkenes are shown in Table 5.

The name of an alkene is derived from the corresponding alkane by changing the -*ane* ending to -*ene*. The alkenes have properties similar to alkanes—they are flammable and nonpolar. Their double bond, however, makes them susceptible to the addition of other atoms. A common reaction of alkenes is the addition of substituents across the double bond. For example, alkenes react with hydrogen gas to form alkanes.

$$CH_2{=}CH_2 + H_2 \rightarrow CH_3CH_3$$

The first alkene in Table 5, ethene (C_2H_4), has the common name ethylene and is a natural ripening agent for fruit. Bananas are often picked green and then "gassed," or exposed to ethylene, just before being sold. The banana farmer can pick and ship the bananas while they are green and more durable and still sell the fruit ripe. Most of the other alkenes are seldom encountered directly except as minority components of gasoline.

Alkynes can be derived from alkenes by removing two additional hydrogen atoms from the two carbon atoms on either side of the double bond and placing a triple bond between them. The four simplest alkynes are shown in Table 6. The name of each of these alkynes is derived from the corre-

TABLE 6
Alkynes

n	Name	Molecular Formula	Structural Formula	Condensed Structural Formula
$n = 2$	Ethyne	C_2H_2	H—C≡C—H	CH≡CH
$n = 3$	Propyne	C_3H_4	H—C≡C—C(H)(H)—H	CH≡CCH$_3$
$n = 4$	1-Butyne	C_4H_6	H—C≡C—C(H)(H)—C(H)(H)—H	CH≡CCH$_2$CH$_3$
$n = 5$	1-Pentyne	C_5H_8	H—C≡C—C(H)(H)—C(H)(H)—C(H)(H)—H	CH≡CCH$_2$CH$_2$CH$_3$

sponding alkane by changing the *-ane* ending to *-yne*. The first alkyne in Table 6, ethyne, has the common name acetylene and is used in welding torches. Like alkenes, most of the other alkynes are not commonly encountered directly except as minor components of gasoline.

EXAMPLE 6.2

Drawing Structural Formulas

Draw a structural formula for propyne.

SOLUTION
The *pro-* in propyne indicates three carbon atoms, and the *-yne* indicates a triple bond. We draw the structure by writing three carbon atoms next to each other and placing a triple bond between two of them. We then fill in the hydrogen atoms to give each carbon four bonds:

$$H-C\equiv C-\underset{H}{\overset{H}{C}}-H$$

YOUR TURN

Drawing Structural Formulas

Draw a structural formula for butyne.

6.4 Isomers

In addition to linking in long straight chains, carbon atoms also form branched structures. For example, we saw earlier that butane (C_4H_{10}) has the following structure:

```
              H   H   H   H
              |   |   |   |
n-Butane  H — C — C — C — C — H      CH_3CH_2CH_2CH_3
              |   |   |   |
              H   H   H   H
```

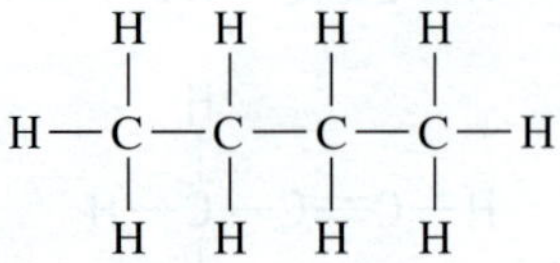

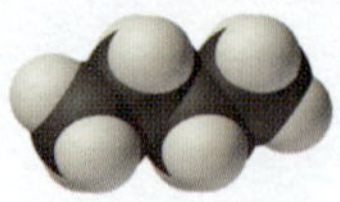

This straight-chain form of butane is called *n*-butane (*n* stands for *normal*). However, another form of butane called isobutane has a different structure:

```
                      H
                      |
                  H — C — H
              H       |       H
              |       |       |            CH_3
Isobutane  H — C —————C —————C — H      CH_3 — CH — CH_3
              |       |       |
              H       H       H
```

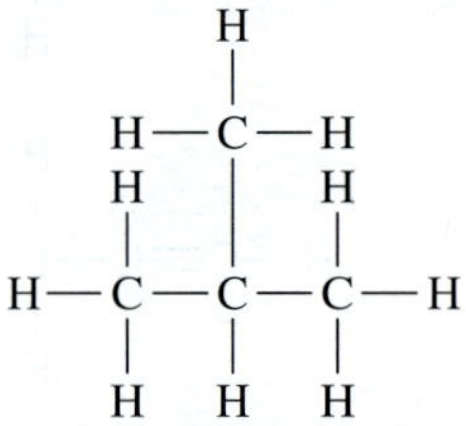

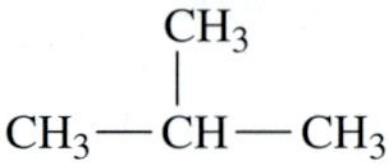

Isobutane has the same molecular formula, C_4H_{10}, as *n*-butane, but a different structure. Molecules having the same molecular formula but different structures are called **isomers**. Isomers have different properties from each other and can be distinguished by inspecting their *structural* formulas. For alkanes, the number of possible isomers increases with the number of carbon atoms in the alkane. Pentane has 3 isomers, hexane has 5, octane has 18, and an alkane with 20 carbon atoms has 366,319.

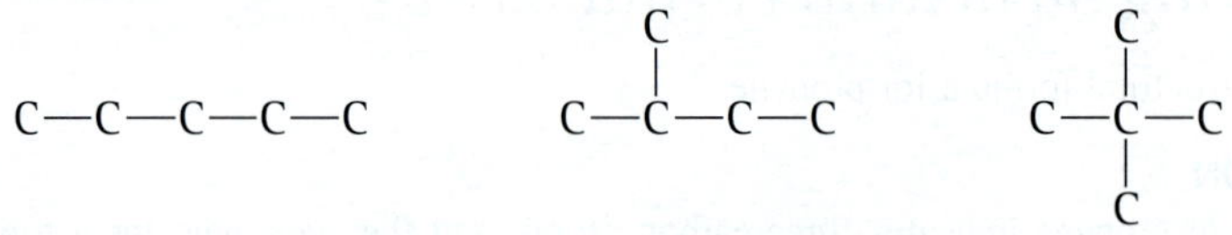

```
                              C                     C
                              |                     |
C — C — C — C — C      C — C — C — C          C — C — C
                                                    |
                                                    C
```

Pentane isomers (H atoms omitted for clarity)

```
                                  C                       C
                                  |                       |
C — C — C — C — C — C      C — C — C — C — C       C — C — C — C
                                                          |
                                                          C

          C                                    C
          |                                    |
    C — C — C — C                        C — C — C — C — C
          |
          C
```

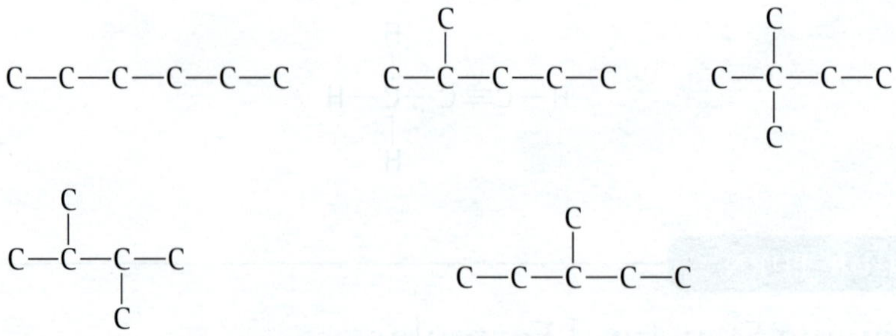

Hexane isomers (H atoms omitted for clarity)

Alkenes and alkynes exhibit another kind of isomerism based on the position of the double or triple bond. For example, butene has two possible structures:

$$\begin{array}{c} H \\ \backslash \\ C=C-C-C-H \\ / | | \\ H H H \end{array} \begin{array}{c} H H H H \\ | | | | \\ H-C-C=C-C-H \\ | | \\ H H \end{array}$$

Since the double bond is in a different position, the molecule is different; however, the two still have the same molecular formula, C_4H_8, and they are isomers.

Isomers can have radically different properties from each other. For example, progesterone, a female reproductive hormone secreted in the latter half of the menstrual cycle and during pregnancy, has the molecular formula $C_{21}H_{30}O_2$

EXAMPLE 6.3

Drawing Structural Formulas for Isomers

Draw structural formulas for two isomers of heptane.

SOLUTION

Heptane isomers must have the same formula but different structures. When drawing structures on paper, it is easiest to draw the carbon chain first and then add hydrogen atoms so that each carbon forms four bonds. We also have to remember that the real structures are three-dimensional and that drawing them on two-dimensional paper can be deceiving. For example, we might try to draw the isomers as follows:

$$CH_3-CH_2-CH_2-CH_2-CH_2-\overset{\overset{\displaystyle CH_3}{|}}{CH_2} \quad \text{and}$$
$$CH_3-CH_2-CH_2-CH_2-CH_2-CH_2-CH_3$$

However, these two structures are identical; the carbon chain has no branches in either structure.

The following two structures are different yet have the same formula; they are isomers.

$$CH_3-\overset{\overset{\displaystyle CH_3}{|}}{CH}-CH_2-CH_2-CH_2-CH_3$$
$$CH_3-CH_2-CH_2-CH_2-CH_2-CH_2-CH_3$$

Notice that one chain has a branch in the carbon backbone, whereas the other does not.

YOUR TURN

Drawing Structural Formulas for Isomers

Draw structural formulas for two isomers of octane.

> **EXAMPLE 6.4**
>
> ### Drawing Structural Formulas for Isomers II
>
> Draw structural formulas for two isomers of pentyne that differ only in the position of the triple bond.
>
> **SOLUTION**
>
> $$CH\equiv C-CH_2-CH_2-CH_3 \quad CH_3-C\equiv C-CH_2-CH_3$$
>
> **YOUR TURN**
>
> ### Drawing Structural Formulas for Isomers II
>
> Draw structural formulas for two isomers of hexene that differ only in the position of the double bond.

FIGURE 6 Progesterone (a) and Tetrahydrocannabinol (THC) (b) are isomers.

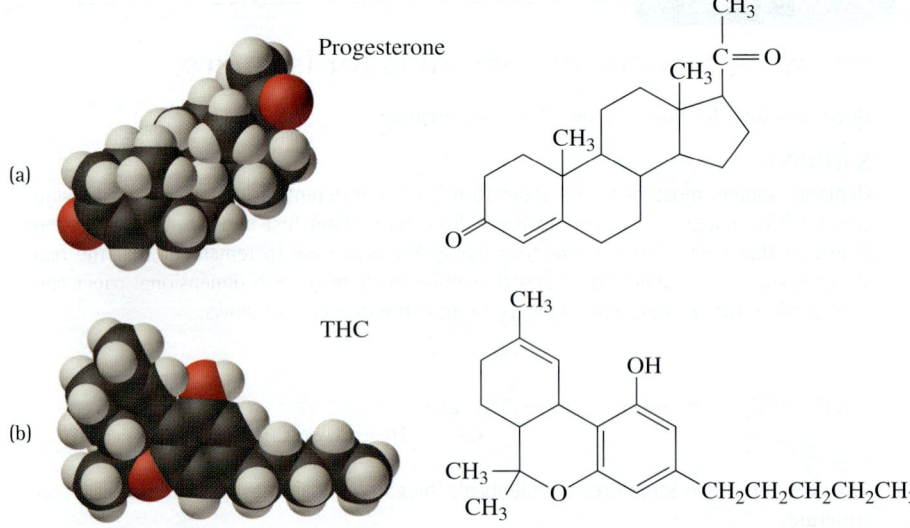

(Figure 6a). Tetrahydrocannabinol (THC), the active constituent of marijuana, also has the molecular formula $C_{21}H_{30}O_2$ (Figure 6b). Progesterone and THC are isomers; they have the same chemical formula but very different structures. Does this mean that they have similar effects on the human body? No! The different structures result in different effects—structure is everything. Nature can take exactly the same atoms in exactly the same numbers and make two different molecules with drastically different properties and functions. One of them, progesterone, makes pregnancy possible; the other, THC, alters the way the brain processes stimuli.

6.5 Naming Hydrocarbons

The large number of organic compounds requires a systematic method of nomenclature that can differentiate between subtle structural features. We will use the nomenclature system recommended by the International Union of Pure and Applied Chemistry (IUPAC), which has become standard.

ALKANES

The following rules allow us to systematically name virtually any alkane. As we go through the rules we will use the following compound as an example.

$$CH_3CH_2CHCH_2CH_2CH_2CH_3$$
$$|$$
$$CH_2$$
$$|$$
$$CH_3$$

1. Select the longest continuous chain, count the number of carbon atoms in that chain, and use Table 1 to determine the base name. The compound in our example has seven carbon atoms in its longest continuous chain and therefore has the base name heptane.

$$CH_3CH_2CHCH_2CH_2CH_2CH_3$$
$$|$$
$$CH_2$$
$$|$$
$$CH_3$$

2. Consider every branch from the longest continuous chain to be a substituent of the base chain. Name each substituent according to the number of carbon atoms in the substituent; however, change the -ane ending to -yl. The compound in our example has one substituent with two carbon atoms. The name of the substituent is ethyl.

$$CH_3CH_2CHCH_2CH_2CH_2CH_3$$
$$|$$
$$CH_2$$
$$|$$
$$CH_3 \quad \text{Ethyl}$$

3. Number the base chain beginning with the end closest to the branching and assign a number to each substituent. The ethyl substituent is in the 3 position.

$$\overset{1}{C}H_3\overset{2}{C}H_2\overset{3}{C}H\overset{4}{C}H_2\overset{5}{C}H_2\overset{6}{C}H_2\overset{7}{C}H_3$$
$$|$$
$$CH_2$$
$$|$$
$$CH_3$$

4. Finally, write the name of the compound in the following order: the number of the substituent followed by a dash, the name of the substituent, and then the base name of the compound. The name of this compound is 3-ethylheptane.

$$CH_3CH_2CHCH_2CH_2CH_2CH_3$$
$$|$$
$$CH_2$$
$$|$$
$$CH_3$$

3-Ethylheptane

Some additional rules may be required in naming more complex compounds.

5. When two or more substituents are present, give each one a number and list them alphabetically. If two or more numberings are possible, choose the set that contains the lowest number. Consider the following compound as an example:

$$\overset{1}{C}H_3\overset{2}{C}H\overset{3}{C}H_2\overset{4}{C}H\overset{5}{C}H_2\overset{6}{C}H_3$$
$$\qquad\;\;|\qquad\;\;|$$
$$\qquad CH_3\quad CH_3$$
$$\qquad\qquad\quad\;\;|$$
$$\qquad\qquad\quad CH_3$$

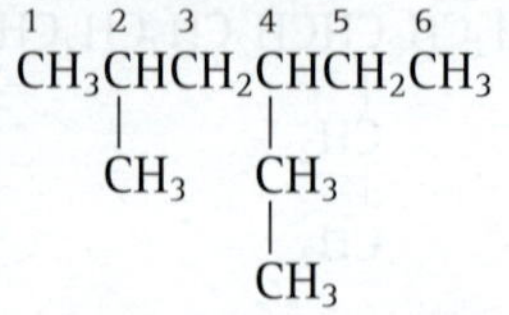

4-Ethyl-2-methylhexane

6. When two or more substituents are on the same carbon, use that number twice. Consider the following compound as an example:

$$\qquad\qquad CH_3$$
$$\qquad\qquad\;|$$
$$\overset{1}{C}H_3\overset{2}{C}H_2\overset{3}{C}\overset{4}{C}H\overset{5}{C}H_2\overset{6}{C}H_3$$
$$\qquad\qquad\;|$$
$$\qquad\qquad CH_2$$
$$\qquad\qquad\;|$$
$$\qquad\qquad CH_3$$

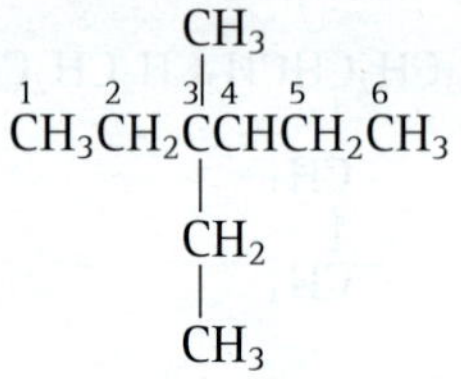

3-Ethyl-3-methylhexane

7. When two or more substituents are identical, so indicate by using the prefixes di-, tri-, and tetra-. Separate the numbers indicating their positions from each other using a comma. Consider the following compound as an example:

$$\qquad\qquad CH_3$$
$$\qquad\qquad\;|$$
$$\overset{1}{C}H_3\overset{2}{C}H\overset{3}{C}H\overset{4}{C}H_3$$
$$\qquad\quad\;|$$
$$\qquad\quad CH_2$$

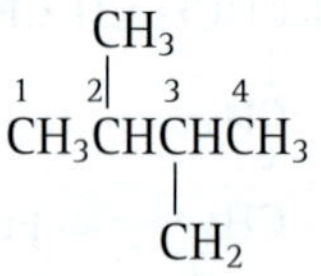

2,3-Dimethylbutane

> ### EXAMPLE 6.5
>
> ## Naming Hydrocarbons
>
> Name the following compounds.
>
> a. 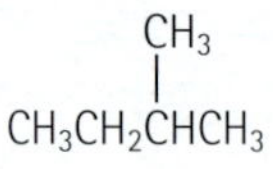
> $$\qquad CH_3$$
> $$\qquad\;|$$
> $$CH_3CH_2CHCH_3$$
>
> b. 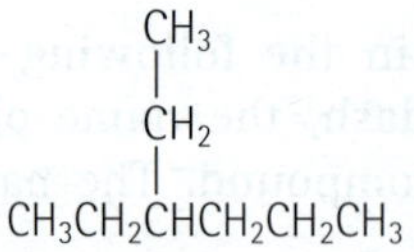
> $$\qquad CH_3$$
> $$\qquad\;|$$
> $$\qquad CH_2$$
> $$\qquad\;|$$
> $$CH_3CH_2CHCH_2CH_2CH_3$$
>
> c. 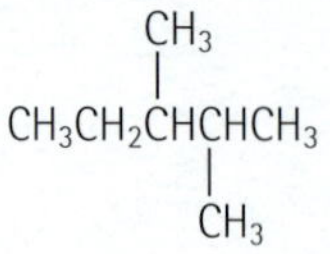
> $$\qquad\quad CH_3$$
> $$\qquad\quad\;|$$
> $$CH_3CH_2CHCHCH_3$$
> $$\qquad\qquad\;|$$
> $$\qquad\qquad CH_3$$
>
> d.
> $$\qquad\qquad CH_3$$
> $$\qquad\qquad\;|$$
> $$\quad CH_3\;\; CH_2$$
> $$\quad\;|\qquad\;|$$
> $$CH_3CHCH_2CHCH_2CH_3$$

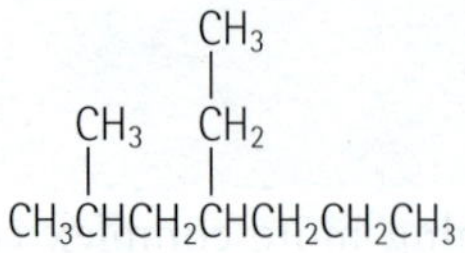

SOLUTION

a. 2-methylbutane

b. 3-ethylhexane

c. 2,3-dimethylpentane

d. 4-ethyl-2-methylheptane

ALKENES AND ALKYNES

Alkenes and alkynes are named in a similar way to alkanes, except for the addition of a number at the beginning to indicate the position of the double or triple bond and a change from *-ane* to *-ene* or *-yne* in the base name. Number the base chain to give the multiple bond the lowest possible number.

EXAMPLE 6.6

Naming Alkenes and Alkynes

Name the following compounds:

a. $CH_2=CHCH_2CH_3$

b. 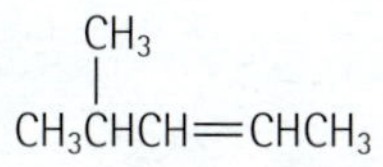
$$\begin{array}{c} CH_3 \\ | \\ CH_3CHCH=CHCH_3 \end{array}$$

c. $CH_3C\equiv CCH_2CH_3$

d. 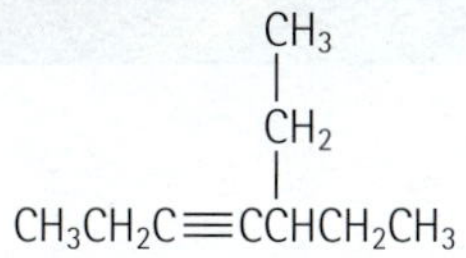
$$\begin{array}{c} CH_3 \\ | \\ CH_2 \\ | \\ CH_3CH_2C\equiv CCHCH_2CH_3 \end{array}$$

SOLUTION

a. 1-butene

b. 4-methyl-2-pentene

c. 2-pentyne

d. 5-ethyl-3-heptyne

6.6 Aromatic Hydrocarbons and Kekule's Dream

In the mid-1800s, organic chemists were just beginning to learn the structures of organic compounds. **Friedrich August Kekule (1829–1896)**, a German chemist, played a central role in elucidating many early structures and is credited with discovering carbon's tendency to form four bonds. One afternoon, Kekule was trying to determine the structure of a particularly stable organic compound called benzene (C_6H_6). Legend has it that while dozing in front of his fireplace, Kekule

 Explore this topic on the Kekule Biography website.

had a dream in which he saw chains of six carbon atoms as snakes. The snakes twisted and writhed until one of them circled around and bit its own tail. Kekule says, "One of the snakes seized its own tail and the form whirled mockingly before my eyes." Stimulated by his dream, Kekule proposed the structure of benzene as a ring of carbon atoms with alternating single and double bonds:

This structure is still accepted today with a minor modification; rather than alternating single and double bonds, each bond is midway between a single and double bond. This change was required because each of the bonds in benzene was found to have the same length, that of a bond intermediate between a single bond and a double bond. As a result, the structure of benzene is often drawn as a hexagon with a circle inside it.

Benzene

 Explore this topic on IBM's Scanning Tunneling Microscope Image Gallery website.

The Molecular Revolution

Determining Organic Chemical Structures

The determination of the structure of organic compounds is an active area of research. Although chemical bonding theories, such as Lewis theory, make good predictions about expected stable structures, the ultimate determination of a structure requires experiment. In the time of Kekule, there were no experimental ways to determine chemical structures with certainty. Even Kekule's proposed ring structure for benzene had little experimental evidence to support it. Today, however, some powerful techniques exist for determining the structure of molecules.

X-ray crystallography determines absolute chemical structures based on the patterns that are formed when X-ray light interacts with a crystal of the compound in question. Nuclear magnetic resonance (NMR) can determine chemical structures based on the interaction of the compound with a magnetic field and radio waves. NMR has become so routine that chemical structures for newly formed compounds can often be determined within several hours of their synthesis. A third technique, called scanning tunneling microscopy, can produce actual images of the molecules themselves. Even though this technique is still in its infancy, it is allowing scientists to get breathtaking visual images of molecules and the atoms that compose them (Figure 7).

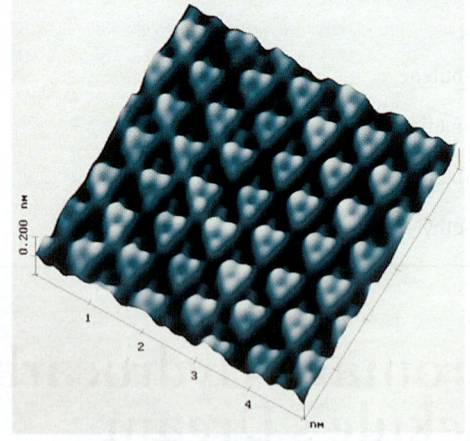

FIGURE 7 STM image of benzene.

Each point of the hexagon represents a carbon atom with a hydrogen atom attached, and the ring represents the bonds that are between double and single.

The ring structure of benzene is particularly stable and is found in many organic molecules. A benzene ring with other substituents attached to it is often called a **phenyl ring**. Benzene rings are also called **aromatic rings** because many of the compounds that contain these rings have notable aromas. Aromatic compounds with two or more aromatic rings fused together are called **polycyclic aromatic hydrocarbons**. A common polycyclic aromatic hydrocarbon is naphthalene ($C_{10}H_8$), used as moth balls.

 Explore this topic on the Polycyclic Aromatic Hydrocarbon Chemicals in Air website.

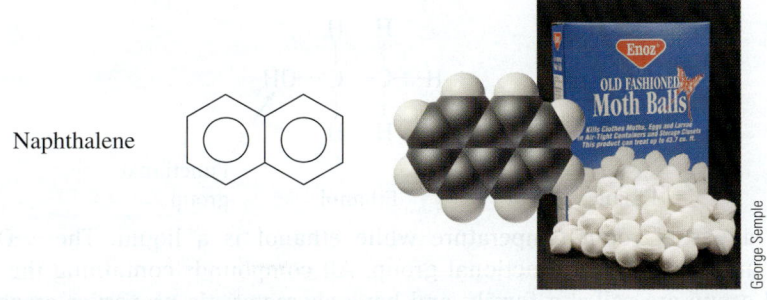

Naphthalene

Many polycyclic aromatic hydrocarbons are formed in the combustion of organic compounds and are dispersed into soils and sediments. Some, like benzene itself and pyrene, are carcinogens that are present in cigarette smoke.

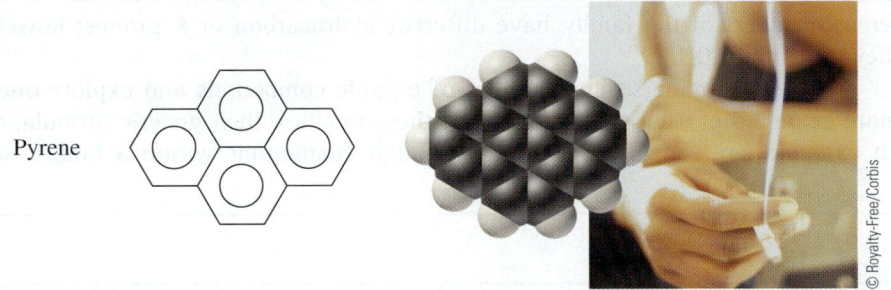

Pyrene

Many compounds containing polycyclic aromatic rings make outstanding dyes. For example, alizarin was used as a red dye for uniforms in Napoleon's army.

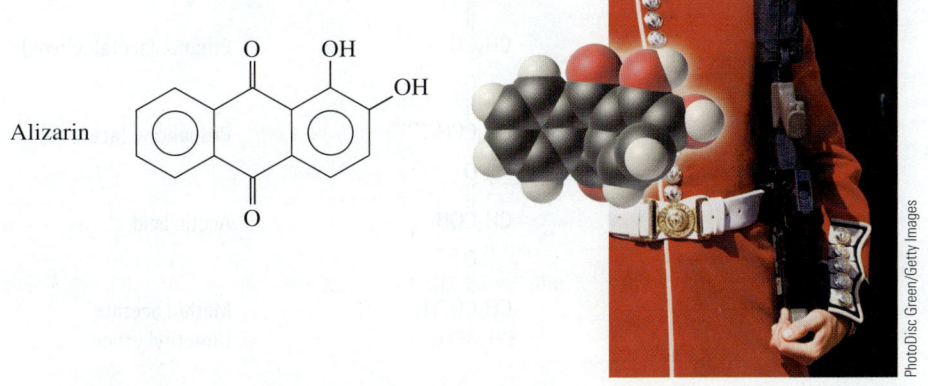

Alizarin

It still exhibits its original brilliance in museum collections, where other dyes have faded.

6.7 Functionalized Hydrocarbons

We have now learned the basic structures of hydrocarbons. These structures form the foundation of the other major grouping of organic compounds, *functionalized hydrocarbons*. Recall from Section 6.3 that functionalized hydrocarbons are hydrocarbons with additional atoms or groups of atoms—called *functional groups*—incorporated into their structure. The insertion of a functional group into a hydrocarbon dramatically alters its properties. For example, consider the difference between ethane, a hydrocarbon, and ethanol, which contains the —OH functional group.

$$\begin{array}{cc} \text{H H} & \text{H H} \\ | \ | & | \ | \\ \text{H—C—C—H} & \text{H—C—C—OH} \\ | \ | & | \ | \\ \text{H H} & \text{H H} \\ \text{Ethane} & \text{Ethanol} \end{array}$$

Functional group

Ethane is a gas at room temperature while ethanol is a liquid. The —OH in ethanol is an example of a functional group. All compounds containing the same functional group are called a **family** and have characteristic properties associated with that functional group. For example, the family of organic compounds containing the —OH functional group are called alcohols, and they tend to be polar because of the presence of the —OH group. Most families can be generically symbolized by R—FG, where R represents the hydrocarbon part of the molecule, and FG a functional group. For example, R—OH symbolizes the alcohols. The different members of the family have different hydrocarbon or R groups; however, they all contain OH.

Next we examine several families of organic compounds and explore one or more examples of each of them. Each of these families, their generic formula, and an example, are shown in Table 7. Although compounds within a family have

TABLE 7
Functionalized Hydrocarbons

Family	General Formula	Example	Name
Chlorinated hydrocarbons	R—Cl	CH_2Cl_2	Dichloromethane
Alcohols	R—OH	CH_3CH_2OH	Ethanol (ethyl alcohol)
Aldehydes	$R-\overset{\overset{\displaystyle O}{\|}}{C}-H$	$CH_3\overset{\overset{\displaystyle O}{\|}}{C}H$	Ethanal (acetaldehyde)
Ketones	$R-\overset{\overset{\displaystyle O}{\|}}{C}-R'$	$CH_3\overset{\overset{\displaystyle O}{\|}}{C}CH_3$	Propanone (acetone)
Carboxylic acids	$R-\overset{\overset{\displaystyle O}{\|}}{C}-OH$	$CH_3\overset{\overset{\displaystyle O}{\|}}{C}OH$	Acetic acid
Esters	$R-\overset{\overset{\displaystyle O}{\|}}{C}O-R'$	$CH_3\overset{\overset{\displaystyle O}{\|}}{C}OCH_3$	Methyl acetate
Ethers	R—O—R'	CH_3OCH_3	Dimethyl ether
Amines	$R-\underset{\underset{\displaystyle}{}}{\overset{\overset{\displaystyle R}{\|}}{N}}-R$	$CH_3CH_2NH_2$	Ethyl amine

many properties in common, they can also be quite different. In addition, many compounds cross family barriers by containing more than one functional group. Nonetheless, the idea of functional groups allows us to organize and classify organic compounds in a systematic way.

6.8 Chlorinated Hydrocarbons: Pesticides and Solvents

Chlorinated hydrocarbons are often found in pesticides, solvents, and refrigerant liquids. They have one or more chlorine atoms substituted for one or more hydrogen atoms on a hydrocarbon and have the general formula R–Cl. The addition of chlorine to a hydrocarbon tends to lower its flammability and chemical reactivity. A simple chlorinated hydrocarbon is dichloromethane (CH_2Cl_2), where two chlorine atoms have taken the place of two hydrogen atoms on methane. Dichloromethane is a liquid that can dissolve many other organic compounds and is often used as a solvent. Dichloromethane was used to extract caffeine, an organic compound, from coffee. A drawback of decaffeinating coffee this way was the residual dichloromethane left in the coffee. Today, most coffees are decaffeinated by other methods.

Chlorinated hydrocarbons are also used as insecticides. Some types of insects have plagued humans because they damage crops and carry serious diseases such as malaria. Shortly before World War II, **Paul Muller (1899–1965)**, a Swiss chemist, demonstrated that a chlorinated hydrocarbon—dichlorodiphenyltrichloroethane (DDT) (Figure 8)—was unusually effective in killing insects, and relatively nontoxic toward humans, a perfect combination. DDT attacked insects, killing mosquitoes, flies, beetles, and almost all other pests. Its stability allowed it to stay on plants and soils for a long time, making it very efficient. Allied troops used DDT during World War II to control lice; farmers used it to increase crop yield; and public health officials used it to kill disease-carrying mosquitoes. DDT is credited with dramatic decreases in the rate of malaria in many countries. For example, in India, malaria cases decreased from 75 million to 5 million per year due to mosquito eradication using DDT. DDT was so successful that Paul Muller was awarded the Nobel Prize in 1948.

The American bald eagle was placed on the endangered species list because of DDT. It has made a dramatic comeback since the ban on DDT.

Although DDT seemed like the dream insecticide, problems began to develop. Some insects became resistant to DDT; the compound was no longer effective in killing them. One of its advantages, its long life in the soil, became a liability. DDT began to accumulate in the soil and in water supplies. Aquatic plants began to have traces of DDT in their cells. The fish that ate the plants began harboring

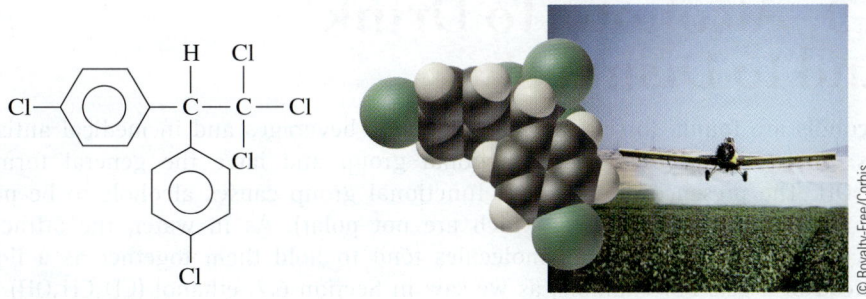

FIGURE 8 Dichlorodiphenyl-trichloroethane (DDT).

FIGURE 9 Freon-114, used in refrigeration and air-conditioning units.

DDT, and so did the birds that ate the fish. Even worse, the levels of DDT became more concentrated as it moved up the food chain—a phenomenon called **bioamplification**. Fish and birds began to die, and our national bird, the American bald eagle, was nearly driven to extinction.

Today, DDT is banned in most developed countries, but other insecticides have replaced it. These replacements are more toxic and less chemically stable than DDT. The high toxicity quickly eliminates pests, and the low stability makes them break down before they contaminate water supplies and accumulate in the food chain.

CHLOROFLUOROCARBONS: OZONE EATERS

A subfamily of chlorinated hydrocarbons contains fluorine as well as chlorine and is called chlorofluorocarbons or CFCs. Two examples of CFCs are dichlorodifluoromethane and trichlorofluoromethane.

Like other chlorinated hydrocarbons, CFCs are chemically stable. This and other properties make them ideal for industrial aerosols, solvents, and air-conditioning and refrigeration coolants such as Freon-114 (Figure 9). However, CFCs pose a danger to our protective ozone layer (see Section 5.5) in the upper atmosphere because, like DDT, CFCs do not decompose in the atmosphere. They eventually drift to the upper atmosphere where high-energy sunlight breaks a chemical bond and releases a reactive chlorine atom. The chlorine atom then reacts with and destroys ozone. This danger to the ozone layer has caused the governments of many developed nations to ban the use of CFCs since January 1, 1996. We will look at ozone depletion in more detail in Chapter 11.

Explore this topic on the **CFCs and the Ozone Hole** website.

6.9 Alcohols: To Drink and to Disinfect

Alcohols are found, for example, in alcoholic beverages and in medical antiseptics. They contain the —OH functional group and have the general formula R—OH. The presence of the —OH functional group causes alcohols to be polar (in contrast to hydrocarbons, which are not polar). As in water, the attractive forces between polar alcohol molecules tend to hold them together as a liquid instead of a gas. For example, as we saw in Section 6.7, ethanol (CH_3CH_2OH) is a liquid at room temperature, whereas ethane (CH_3CH_3) is a gas. We name alcohols according to the name of the hydrocarbon chain modified to end in *-ol*.

ETHANOL

Ethanol is the alcohol present in alcoholic beverages and is also used as an additive to gasoline.

Ethanol

$$\text{H}-\underset{\underset{\text{H}}{|}}{\overset{\overset{\text{H}}{|}}{\text{C}}}-\underset{\underset{\text{H}}{|}}{\overset{\overset{\text{H}}{|}}{\text{C}}}-\text{OH}$$

It is made by the fermentation of sugars normally present in fruits or grains. In this process, living yeasts convert sugars such as glucose ($C_6H_{12}O_6$) and fructose ($C_6H_{12}O_6$)—they are isomers—to ethanol and carbon dioxide.

$$C_6H_{12}O_6 \rightarrow 2\ CH_3CH_2OH + 2\ CO_2$$

Ethanol mixes with water in all proportions because of its polar group. All alcoholic beverages contain primarily ethanol and water, with a few other minor components that give flavor and color. The **proof** of a liquor is twice its alcohol percentage by volume. A 200-proof liquor is pure ethanol; 90-proof whiskey contains 45% ethanol. The alcohol in beer and wine is usually reported as a direct percentage rather than as a proof. Beers contain between 3 and 6% ethanol and wines contain about 12%.

Ethanol is a central nervous system depressant—it slows the rate at which nerve impulses travel. One or two alcoholic drinks lead to mild sedation. Four or more drinks can cause a loss of coordination and even unconsciousness. Overly excessive drinking—20 shots of a 90-proof spirit in a short time—is lethal.

Chronic alcohol abuse is a common problem. There are over 9 million alcoholics in the United States. Alcoholics get over 50% of their daily caloric intake from ethanol; they drink more than 1 L of wine or eight shots of 90-proof vodka per day. Alcoholics are two to three times more likely to develop heart disease, cirrhosis of the liver, and respiratory diseases. On the average, alcoholics die 8 to 12 years earlier than nonalcoholics.

OTHER ALCOHOLS

The addition of a methyl group to ethanol produces isopropyl alcohol, commonly known as rubbing alcohol.

Isopropyl alcohol

$$\text{H}-\underset{\underset{\text{H}}{|}}{\overset{\overset{\overset{\text{H}}{|}}{\overset{\text{H}-\text{C}-\text{H}}{|}}}{\text{C}}}-\text{OH}$$

$$\underset{|}{\overset{CH_3}{CH_3CHOH}}$$

What if...

Alcohol and Society

Our society tends to have conflicting feelings about drinking ethanol. Our government prohibited it with a constitutional amendment in 1920 and then legalized it 14 years later with another constitutional amendment. Some counties in Midwestern states still prohibit the sale of alcohol. Some religions prohibit alcohol consumption, while others allow it. Although excessive intake of alcohol is certainly detrimental, some recent studies have shown that moderate amounts, one to two drinks daily, may lower the risk of heart disease.

QUESTION: What if alcohol had remained prohibited in the United States? Do you think we would have more or fewer alcoholics? Should government take a more or less active role in regulating alcohol consumption?

Isopropyl alcohol is useful for disinfecting cuts and scratches but is more toxic than ethanol—four ounces of isopropyl alcohol can be fatal. Simply substituting a methyl group for a hydrogen on ethanol changes its properties. Such is the molecular world; small changes on the molecular scale result in significant changes in the properties of a compound.

Another common alcohol, often used as a solvent, is methanol.

$$\text{Methanol} \quad H-\underset{H}{\overset{H}{\underset{|}{\overset{|}{C}}}}-OH \quad CH_3OH$$

It, too, is toxic if ingested. Methanol's toxicity results from the breakdown of methanol to formic acid in the liver. Formic acid gets into the bloodstream and causes acidosis (acidification of the blood), which can be lethal. One treatment for methanol poisoning is the consumption of ethanol. Emergency room doctors will often administer ethanol intravenously to patients who have accidentally drunk methanol. The liver preferentially breaks down the ethanol, leaving the methanol to escape in the urine. The patient usually wakes with a terrible hangover, but considering the consequences of methanol poisoning—blindness or death—it is worth the headache.

6.10 Aldehydes and Ketones: Smoke and Raspberries

Aldehydes and **ketones** are commonly found in many pleasant flavors and aromas. Aldehydes have the general formula RCHO where the oxygen is joined to the carbon via a double bond as shown here.

$$\text{Aldehydes} \quad \overset{O}{\underset{}{\overset{\parallel}{RCH}}}$$

The C=O group is called a **carbonyl group**. Notice that the condensed structural formula RCHO *does not* mean R—C—H—O, but instead the structure just shown. Again, carbon must form four bonds, but hydrogen must form only one. Oxygen normally forms two bonds in organic compounds. Aldehydes are named according to the number of carbon atoms and given the ending *-al*. Their common names often end with *-aldehyde*.

The simplest aldehyde has the common name formaldehyde (H_2CO) and formal name methanal.

Formaldehyde

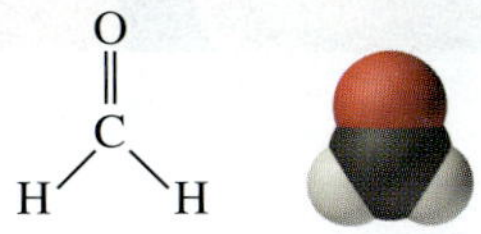

It is a liquid with an acrid smell. Its toxicity to bacteria makes it useful as a preservative of biological specimens. Formaldehyde is also present in wood smoke and aids in the preservation of smoked meats. Formaldehyde's presence in smoke is partially responsible for the burning and tearing of eyes that occur around a campfire.

Acrolein is another commonly encountered aldehyde.

Acrolein

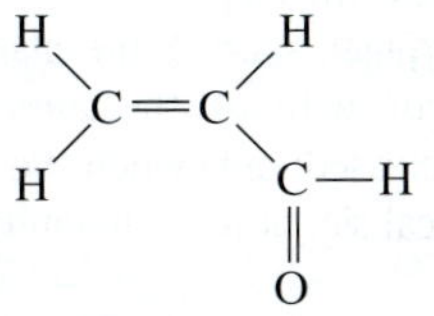

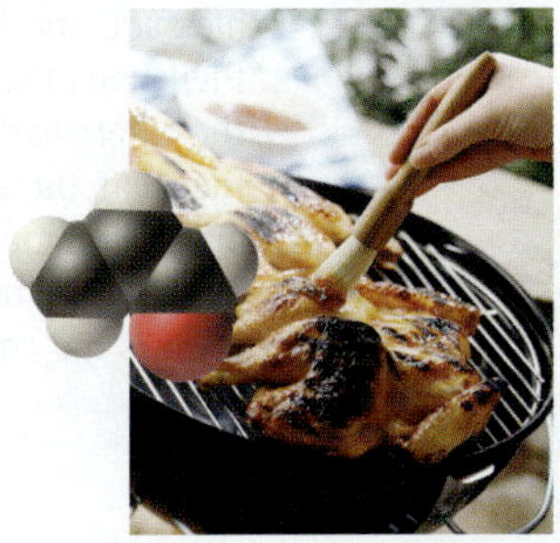

© Royalty-Free/Corbis

It is formed when molecules in some foods decompose during heating. Its acrid smell is enjoyed when barbecuing meats, and its flavor is found in caramelized sugar. Other aldehydes of interest include cinnamaldehyde, which gives cinnamon its distinctive smell and flavor, and benzaldehyde, also called oil of almond.

Cinnamaldehyde

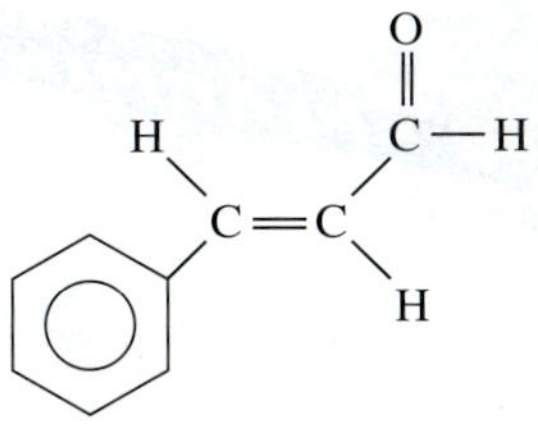

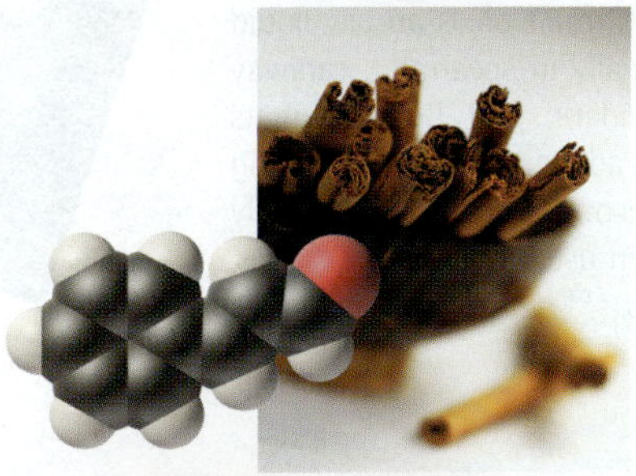

© Royalty-Free/Corbis

Benzaldehyde

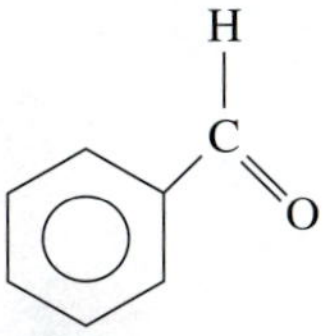

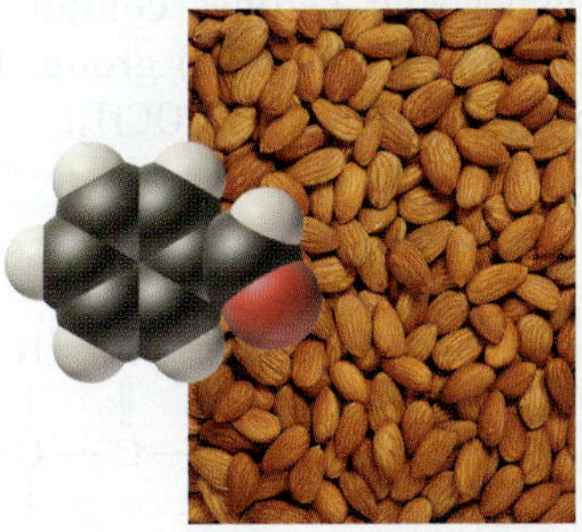

© Royalty-Free/Corbis

Ketones are similar to aldehydes, but instead of one R group, they have two. Ketones have the general formula RCOR', where R' simply means that the second R group could be different than the first.

Ketones

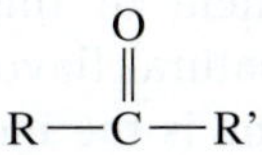

Molecular Focus

Carvone

Formula: $C_{10}H_{14}O$
Molar Mass: 150.2 g/mol
Boiling point: 230°C
Lewis structure:

Three-dimensional structure:

Carvone is a ketone, but it also belongs to a class of naturally occurring organic compounds called terpenes, which usually contain carbon atoms in multiples of five. Terpenes are thought to act as chemical messengers in the communications between plants and insects. For example, certain plants actively produce and release terpenes into the air when they are attacked by caterpillars. Parasitic wasps recognize the chemical message and hone in on its source, the ailing plant. The wasp then finds the caterpillar, paralyzes it, and injects its eggs into the caterpillar's body. The eggs develop into wasp larvae, which then eat the caterpillar, sparing the plant. Surprisingly, if the plant is simply cut without the presence of a caterpillar, terpenes are not released; only when the caterpillar attacks the plant is the chemical signal for help emitted.

Carvone, the main component of spearmint oil, is an oily liquid at room temperature. It occurs naturally in spearmint, caraway seed, dill seed, gingergrass, and mandarin peel. Because of its pleasant aroma and flavor, carvone is added to liqueurs, chewing gum, soaps, and perfumes. It is one of a number of naturally occurring essential oils that have been used as spices, perfumes, and medicines for thousands of years.

Both aldehydes and ketones contain the carbonyl group, but ketones have the additional hydrocarbon (or R) group. The names of ketones end in *-one*. The simplest ketone is acetone (CH_3COCH_3).

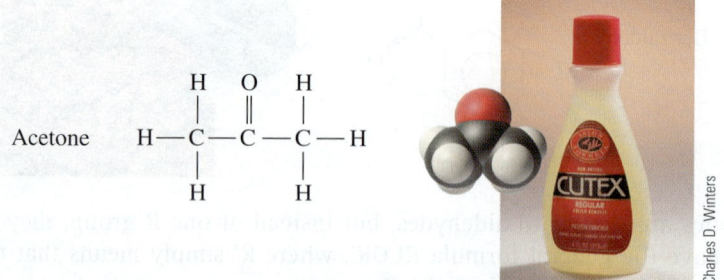

Acetone is the major component of fingernail polish remover. Like aldehydes, ketones are found in several natural flavors and aromas. 2-Heptanone adds to the pleasant smell of cloves. Ionone is the molecule you smell when picking and eating raspberries.

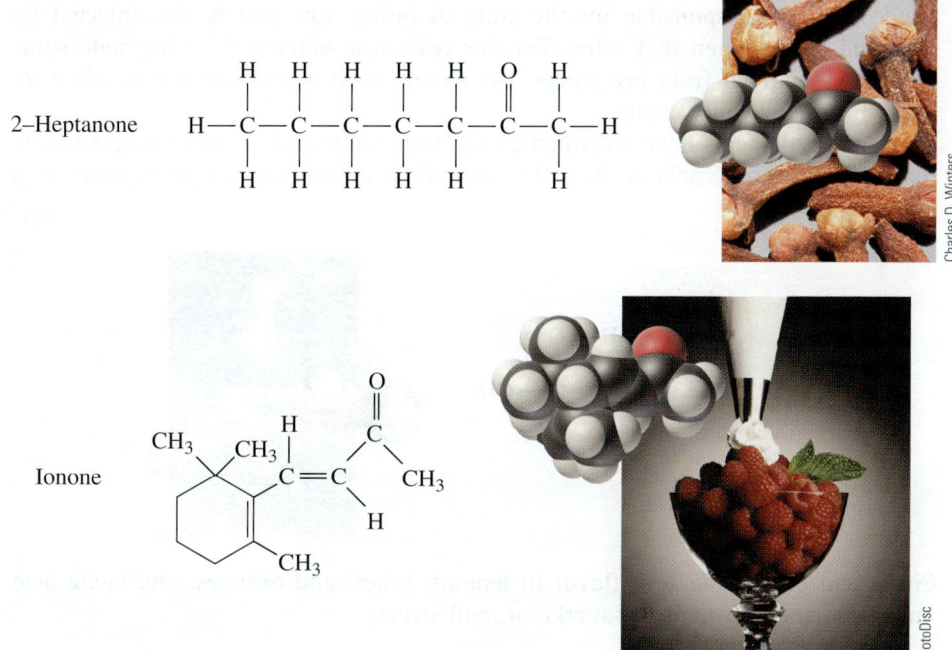

2–Heptanone

Ionone

Butanedione, a diketone, has a cheesy smell and is found in butter and body odor.

Butanedione

Skin-borne bacteria feast on fresh sweat and emit butanedione as a waste product. Deodorants slow this process by killing the bacteria and adding more pleasant-smelling compounds to the mixture.

6.11 Carboxylic Acids: Vinegar and Bee Stings

Carboxylic acids are commonly found in citrus fruits, vinegar, and other sour foods. They contain –COOH and have the general formula RCOOH.

Carboxylic acids

$$\underset{\text{RCOH}}{\overset{\overset{\displaystyle O}{\|}}{}}$$

The simplest carboxylic acid is formic acid (HCOOH).

Formic acid

Formic acid gets its name from the Latin *formica*, meaning "ant." Formic acid is one component responsible for the sting of biting ants and is also injected by bees and wasps when they sting. For this reason, a water and baking soda solution provides relief from bee stings. The baking soda neutralizes the formic acid, removing the chemical sting.

Carboxylic acids have a distinctly sour taste we associate with vinegar or citrus products. Acetic acid is the active ingredient in vinegar and gives sourdough bread its bite.

Acetic acid

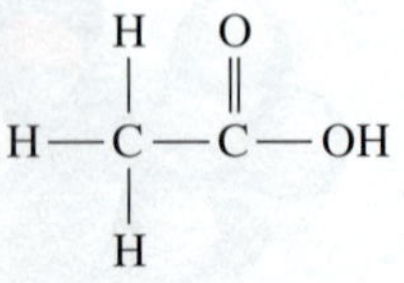

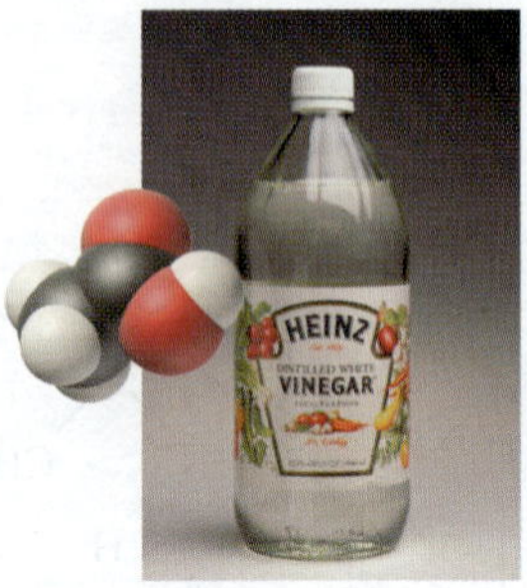

Charles D. Winters

Citric acid causes the sour flavor in lemons, limes, and oranges, and lactic acid imparts tartness to pickles, sauerkraut, and sweat.

Citric acid

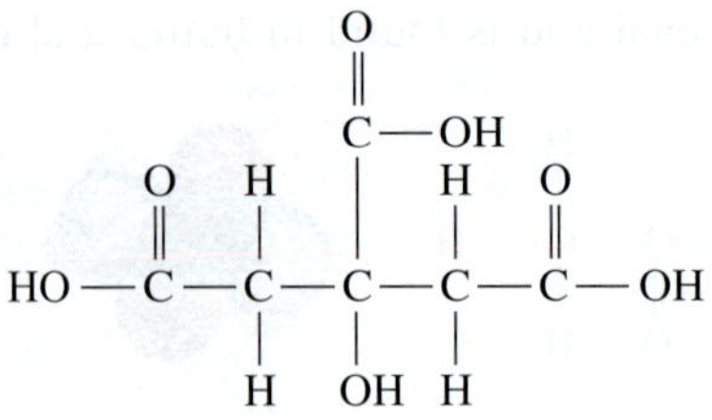

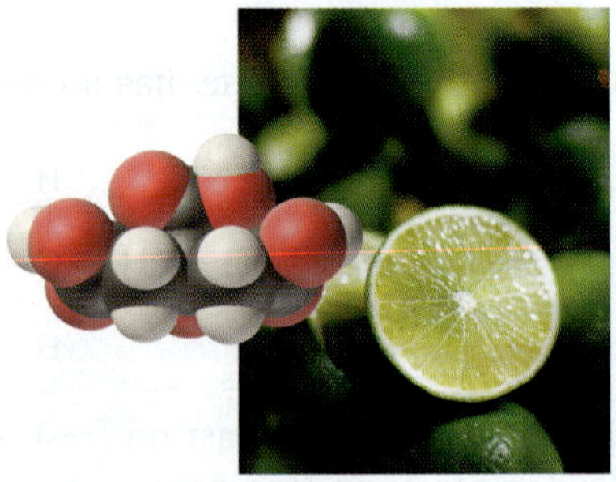

© Royalty-Free/Corbis

Lactic acid

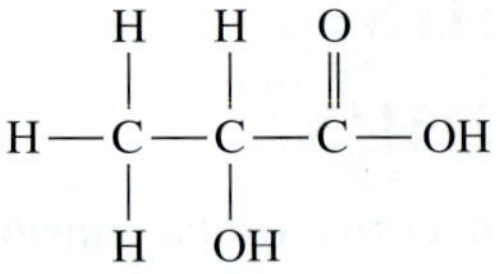

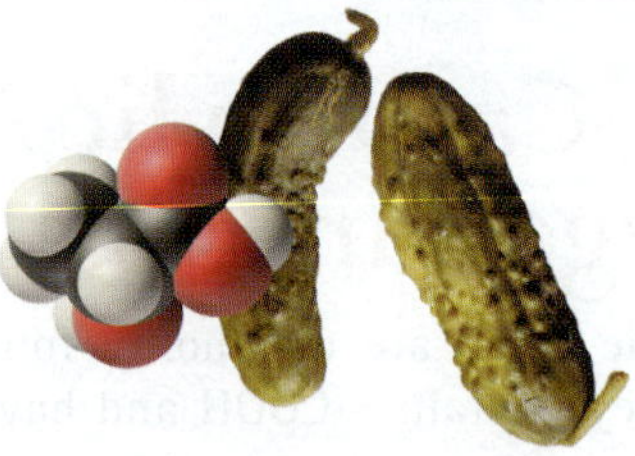

PhotoDisc

6.12 Esters and Ethers: Fruits and Anesthesia

Esters have pleasant aromas and are responsible for many natural flavors and smells. They contain the functional group —COO— and have the general formula R—COO—R'.

Esters $\mathrm{R{-}\overset{\displaystyle O}{\overset{\|}{C}}{-}O{-}R'}$

Esters are named according to the two hydrocarbon groups and end with –*ate*. A common ester is ethyl butyrate ($CH_3CH_2CH_2COOCH_2CH_3$).

Ethyl butyrate

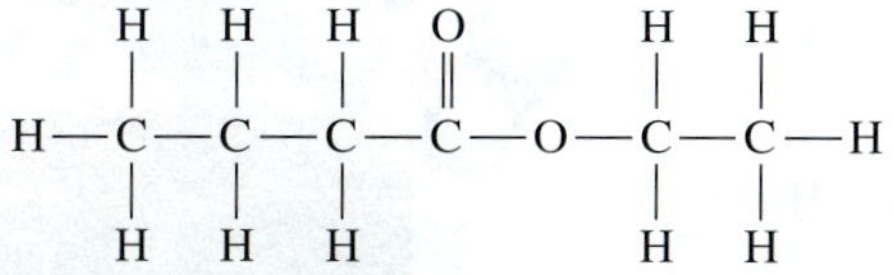

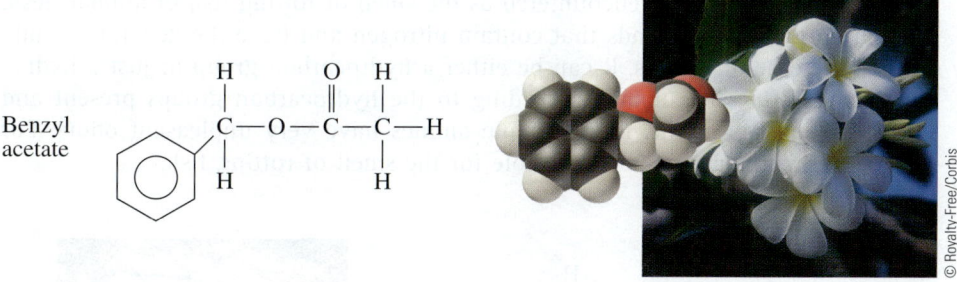

Ethyl butyrate is responsible for the sweet aroma of pineapples. Other common esters include methyl butyrate, found in apples, and ethyl formate, an artificial rum flavor. Several perfumes use esters. For example, oil of jasmine contains benzyl acetate, a molecule found in the jasmine flower.

Benzyl acetate

It is easier and cheaper to synthesize benzyl acetate rather than extract it from jasmine. Most perfumes use synthetic benzyl acetate, but molecularly there is no difference from the "natural" product.

Polyester, a once fashionable cloth, is an example of a **polymer,** a molecule with many similar units bonded together in a long chain. The similar units in polyester are ester groups.

Ethers contain the functional group —O— and have the general formula R-O-R'. Ethers are named according to the two hydrocarbon groups and given the ending *ether*. The simplest ether is dimethyl ether (CH_3—O—CH_3), two methyl groups attached to an oxygen atom.

Dimethyl ether

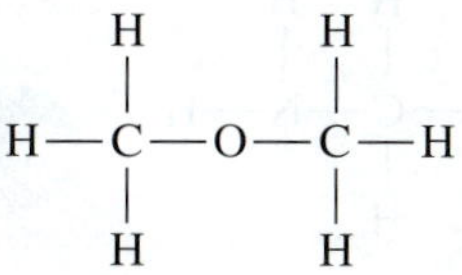

We will learn more about polymers in Chapter 15.

Bugs Bunny cartoons occasionally show the scurrilous rabbit breaking open a container labeled *ether*. As the gas emanates from the jar, Bugs begins moving and talking in slow motion. Poor Elmer Fudd is slowed down by the ether as well and can never catch Bugs. The substance in the jar was diethyl ether, which was used as an anesthetic for some time.

Diethyl ether structure:

```
     H H   H H
     | |   | |
H—C—C—O—C—C—H
     | |   | |
     H H   H H
```

However, this use was terminated because of side effects such as severe nausea.

6.13 Amines: The Smell of Rotten Fish

Amines are most notably encountered as the smell of rotting fish or animal flesh. They are organic compounds that contain nitrogen and have the general formula NR_3. In the case of amines, R can be either a hydrocarbon group or just a hydrogen atom. Amines are named according to the hydrocarbon groups present and given the ending *-amine*. Many simple amines have very unpleasant odors. For example, trimethylamine is responsible for the smell of rotting fish.

Trimethylamine:

```
          H
          |
       H—C—H
       H  |  H
       |  |  |
   H—C—N—C—H
       |     |
       H     H
```

Putrescine is one of the components of decaying animal flesh.

Several illicit drugs such as amphetamine and methamphetamine are also amines.

Amphetamine

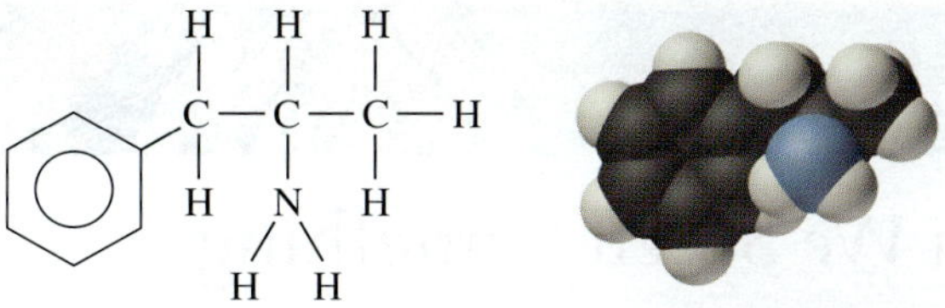

Methamphetamine

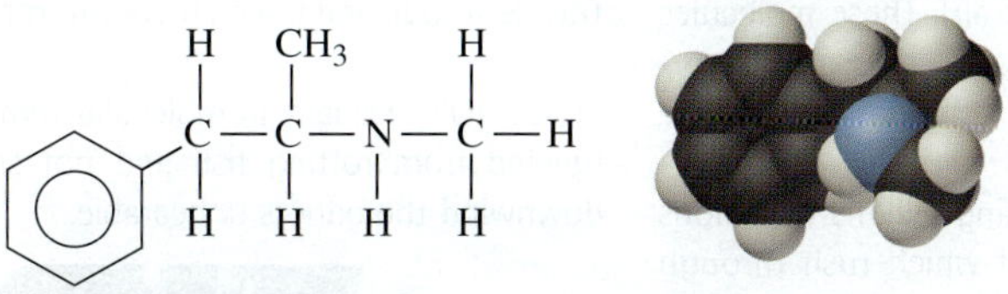

As we will see in Chapter 16, amines play an important role in protein chemistry.

> **APPLY YOUR KNOWLEDGE**
>
> To what family does the molecule CH$_3$COOCH$_3$ belong?
>
> Answer: Since this molecule has the general structure, R–COO–R, it belongs to the ester family.

6.14 A Look at a Label

Although we have invested only a small amount of time in our study of organic chemistry, we can now identify several important kinds of organic compounds. For example, the shaving cream Edge Gel lists as its contents deionized water, palmitic acid, triethanolamine, pentane, fatty acid esters, sorbitol, and isobutane.

We can identify most of these:

deionized water—water treated to remove ions

palmitic acid—a carboxylic acid

triethanolamine—an amine, probably containing three ethanol groups

pentane—a saturated hydrocarbon with five carbon atoms

fatty acid esters—an assortment of esters

sorbitol—an alcohol

isobutane—a saturated hydrocarbon with four carbon atoms

MOLECULAR THINKING

What Happens When We Smell Something

Air contains primarily two kinds of molecules, oxygen (about 20% of air) and nitrogen (about 80% of air). These molecules move at high speeds and bounce off each other and everything else. The collective effect of these collisions is what we call pressure.

We are constantly inhaling and exhaling billions of billions of nitrogen and oxygen molecules, all of which rush through our nose and into our lungs, and most of which rush back out again when we exhale.

If we walk into a blooming rose garden, however, we immediately notice something different when we inhale—a pleasant smell. What causes it? The molecules in the rose garden are not much different than those in ordinary air—20% oxygen and 80% nitrogen. However, there is a small difference—about 1 molecule in every 100 million is geraniol or 2-phenylethanol, the molecules responsible for the smell of roses.

When we inhale these molecules, even in concentrations as small as 1 in 100 million, receptors in our noses grab them. Olfactory receptors are extremely sensitive to molecular shapes and can pick out the 1 geraniol molecule out of the 100 million nitrogen and oxygen molecules (Figure 10). When the geraniol interacts with the receptor in our nose, a nerve signal travels to our brain, which we interpret as *the smell of roses*.

QUESTION: Explain, in molecular terms, why you can stand 2 ft upwind from rotting fish and not smell a thing, while 20 ft downwind the odor is unbearable.

FIGURE 10 Geraniol and 2-phenylethanol are the main components of rose scent. The flowers emit these molecules into the air, which is inhaled through the nose.

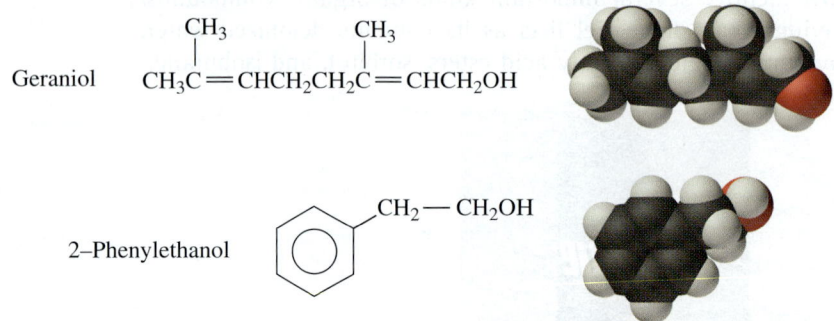

Chapter Summary

MOLECULAR CONCEPT

Many of the substances we encounter, ranging from gasoline to perfumes to our own bodies, are composed of organic compounds. Organic compounds all contain carbon, a unique element that forms four bonds and can bond to itself to form branched and ring structures (6.1). Organic compounds are fragile in comparison to inorganic compounds, and scientists once believed that a vital force was necessary to produce them. This idea was eradicated by Wohler's synthesis of urea, and today, organic compounds are routinely synthesized (6.2).

The simplest organic compounds, containing only carbon and hydrogen, are called hydrocarbons and are used extensively as energy sources. Hydrocarbons can be divided into three subfamilies: alkanes, alkenes, and alkynes. Alkanes are saturated with hydrogen and contain no double bonds, while alkenes and alkynes are unsaturated and contain double or triple bonds, respectively. Alkanes and many other organic compounds form isomers, molecules with the same molecular formula but different structures. Hydrocarbons can be systematically named according to guidelines established by the International Union of Pure and Applied Chemistry (IUPAC) (6.3).

Other families of organic compounds are classified according to their functional groups. These functional groups have common characteristics giving the members of a family common properties (6.4–6.13).

SOCIETAL IMPACT

The synthesis of the first organic compound by Wohler in 1828, and the synthesis of many organic compounds since then, including compounds central to life, changed the idea that living things were somehow outside of physical law (6.2). Since that time, scientists have continued to describe living things in terms of the molecules that compose them with tremendous success. That success, however, has at times left humans wondering how they are unique.

Hydrocarbons are the primary molecules that our society uses for fuel. Natural gas, gasoline, and oil are all mixtures of hydrocarbons with progressively longer carbon chains in going from gas to oil. All of these fuels, called fossil fuels, are of limited supply and will eventually need to be replaced by other energy sources. Their use also causes environmental problems such as air pollution, acid rain, and global warming (6.3).

We can recognize many organic functional groups in compounds we use every day. Pesticides and Freons often contain chlorinated hydrocarbons. Many fruits get their aromas and flavors from aldehydes, ketones, and esters. The sour taste in foods is because of carboxylic acids, and the many rotting smells are due to amines (6.4–6.13).

Chemistry on the Web

For up-to-date URLs, visit the text website at http://chemistry.brookscole.com/tro3

- Origin of Life
  http://www.geocities.com/CapeCanaveral/Lab/2948/orgel.html
- Interview with Stanley Miller
  http://www.accessexcellence.org/WN/NM/miller.html
- Kekule Biography
  http://www.woodrow.org/teachers/chemistry/institutes/1992/Kekule.html
- Polycyclic Aromatic Hydrocarbon Chemicals in Air
  http://www.epa.gov/wtc/pah/

- IBM's Scanning Tunneling Microscope Image Gallery
  http://www.almaden.ibm.com/vis/stm/gallery.html
- CFCs and the Ozone Hole
  http://www.atmosphere.mpg.de/enid/1z2.html

Key Terms

alcohol	carbonyl group	isomers	polymer
aldehyde	carboxylic acid	Kekule, Friedrich August	proof
alkane	combustion reaction	ketone	saturated hydrocarbon
alkene	ester	methylene group	structural formula
alkyne	ether	methyl group	unsaturated hydrocarbon
amine	family (of organic compounds)	Muller, Paul	urea
aromatic ring		organic chemistry	vital force
bioamplification	functional group	phenyl ring	vitalism
biochemistry	functionalized hydrocarbon	polycyclic aromatic hydrocarbon	Wohler, Friedrich
	hydrocarbon		

Exercises

Assess your understanding of this chapter's topics with an online chapter quiz at http://chemistry.brookscole.com/tro3.

QUESTIONS

1. Define organic chemistry.
2. Why is carbon unique?
3. What is the geometry about a carbon atom that is bonded to four other atoms through single bonds?
4. Classify each of the following as organic or inorganic (Hint: Consider its source.):
   a. table salt
   b. sugar
   c. copper
   d. diamond
   e. gold
   f. vegetable oil
5. What is vitalism? Why did vitalism become a popular belief?
6. How did the belief in vitalism end?
7. What are hydrocarbons? functionalized hydrocarbons?
8. What are the general formulas for alkanes, alkenes, and alkynes?
9. Draw the structural formula and condensed structural formula for pentane, and identify the methyl groups and the methylene groups.
10. List four common fuels used by our society, and draw structural formulas for the corresponding hydrocarbons that compose them.
11. How does the boiling point of alkanes vary with carbon chain length?
12. Why are alkenes and alkynes called *unsaturated* hydrocarbons? Why are alkanes called *saturated* hydrocarbons?
13. What are the main properties of alkanes?
14. If water is combined with motor oil, the two substances will not mix. Why?
15. Write an equation for the combustion of butane.
16. Draw structural formulas for propene and propyne.
17. What is ethylene (ethene) used for? What is acetylene (ethyne) used for?
18. What are isomers? Give an example.
19. Do isomers have identical properties?
20. Give the general formula, including the functional group, and draw condensed structural formulas for two examples of each type of the following organic compounds:
    a. alkane
    b. alkene
    c. alkyne
    d. aromatic
    e. chlorinated hydrocarbon
    f. alcohol

21. Give the general formula, including the functional group, and draw condensed structural formulas for two examples of each type of the following organic compounds:
    a. aldehyde
    b. ketone
    c. carboxylic acid
    d. ester
    e. ether
    f. amine
22. What family of organic compound would most likely be used as a
    a. dye
    b. fuel
    c. perfume
    d. refrigerant
    e. insecticide
    f. illicit drug
23. What was DDT used for? What were the problems associated with its use?
24. What are chlorofluorcarbons? What are their uses? Why do they pose a threat to the environment?
25. What are the effects of ethanol on the body?
26. What is used as a treatment for the ingestion of methanol?
27. What is formaldehyde used for?
28. Where is acrolein found?
29. Where are benzaldehyde and cinnamaldehyde found?
30. What is a common use for acetone?
31. Where might you find 2-heptanone, ionone, and butanedione?
32. In what kinds of foods might you find carboxylic acids? What kind of flavor is associated with carboxylic acids?
33. Where might you find ethyl butyrate, methyl butyrate, ethyl formate, and benzyl acetate?
34. What kind of odor is associated with amines?

PROBLEMS

35. Draw structural formulas for three isomers of pentane.
36. Draw structural formulas for any three isomers of octane.
37. Draw the structural formula for *n*-hexene. How many isomers are possible by moving only the position of the double bond?
38. Draw the structural formula for *n*-heptyne. How many isomers are possible by moving only the position of the triple bond?
39. Name each of the following alkanes according to their IUPAC names:

    a. $CH_3CH_2\underset{\underset{CH_3}{|}}{C}HCH_3$

    b. $CH_3\underset{\underset{CH_2}{|}}{C}HCH_2\underset{\underset{CH_3}{|}}{C}HCH_2CH_3$

    c. $CH_3CH_2\underset{\underset{CH_3}{|}}{C}HCH_2\underset{\underset{CH_3}{|}}{C}HCH_3$

    d. $CH_3\underset{\underset{CH_3}{|}}{\overset{\overset{CH_3}{|}}{C}}CH_2CH_2CH_3$

40. Name each of the following alkanes according to their IUPAC names:

    a. $CH_3CH_2\underset{\underset{CH_3}{|}}{\overset{\overset{CH_2CH_3}{|}}{C}H}CH_2CH_2CH_3$

    b. $CH_3\underset{\underset{CH_3}{|}}{C}H\underset{\underset{CH_2CH_3}{|}}{C}HCH_2CH_2CH_3$

    c. $CH_3CH_2\underset{\underset{CH_3}{|}}{C}HCH_2CH_3$

    d. $CH_3\underset{\underset{CH_3}{|}}{C}H\underset{\underset{CH_3}{|}}{C}HCH_3$

41. Name each of the following alkenes:
    a. $CH_3CH=CHCH_3$
    b. $CH_3CH=CH\underset{\underset{CH_3}{|}}{C}HCH_3$
    c. $CH_3\underset{\underset{CH_3}{|}}{C}HCH=CHCH_2CH_3$

42. Name each of the following alkenes:
    a. $CH_3CH=CHCH_2CH_2CH_3$
    b. $CH_3CH=CH_2$
    c. $CH_2=CH\underset{\underset{CH_3}{\overset{\overset{CH_2CH_3}{|}}{}}}{}CH_2CH_3$
       $CH_2=CHCH_2CH_3$ with CH_2CH_3 substituent

43. Name each of the following alkynes:
    a. $CH\equiv CCH_3$
    b. $CH_3CH_2C\equiv CCH_2CH_3$
    c. $CH_3C\equiv C\underset{\underset{CH_2CH_3}{|}}{C}HCH_2CH_3$

44. Name each of the following alkynes:
    a. $CH_3C{\equiv}CCH_2CH_3$
    b. $CH{\equiv}CH$
45. Draw the condensed structural formula for each of the following alkanes:
    a. 2-methylpentane
    b. 3-methylhexane
    c. 2,3-dimethylbutane
    d. 3-ethyl-2-methylhexane
46. Draw the condensed structural formula for each of the following alkanes:
    a. 4-ethyloctane
    b. 3-methylpentane
    c. 4-ethyl-2,3-dimethylheptane
    d. 2,5-dimethylheptane
47. Draw the condensed structural formula for each of the following alkenes and alkynes:
    a. 2-butene
    b. 3-hexyne
    c. 3-methyl-1-pentene
    d. 4-methyl-2-hexyne
48. Draw the condensed structural formula for each of the following alkenes and alkynes:
    a. 2-butyne
    b. 1-heptene
    c. 3-ethyl-2-hexene
    d. 3-methyl-1-pentyne
49. Identify each of the following compounds according to their functional group (e.g., amine, ester, etc.):
    a. $CH_3{-}O{-}CH_2CH_3$
    b. CH_2ClCH_2Cl
    c. CH_3NHCH_3
    d. $CH_3CH_2CH_2\overset{\displaystyle O}{\overset{\displaystyle \|}{C}}H$
50. Identify each of the following compounds according to their functional group (e.g., amine, ester, etc.):
    a. $CH_3CH_2\overset{\displaystyle O}{\overset{\displaystyle \|}{C}}CH_3$
    b. $CH_3\overset{\displaystyle CH_3}{\overset{\displaystyle |}{C}}HOH$
    c. $CH_3CH_2CH_2\overset{\displaystyle O}{\overset{\displaystyle \|}{C}}OCH_3$
    d. $CH_3CH{=}CH_2$
51. Identify each of the following compounds according to their functional group:
    a. $CH_3\overset{\displaystyle O}{\overset{\displaystyle \|}{C}}OH$
    b. (benzene ring)–CH_2CH_3
    c. CH_3CH_2OH
    d. $CH_3\overset{\displaystyle CH_3}{\overset{\displaystyle |}{N}}CH_3$
52. Identify each of the following compounds according to their functional group:
    a. CH_3CH_2Cl
    b. $CH_3CH_2\overset{\displaystyle O}{\overset{\displaystyle \|}{C}}{-}O{-}CH_2CH_3$
    c. $CH_3CH_2CH_2{-}O{-}CH_2CH_3$
    d. $CH_3CH_2\overset{\displaystyle O}{\overset{\displaystyle \|}{C}}{-}OH$
53. Propane, $CH_3CH_2CH_3$, is a gas at room temperature, while propanol, $CH_3CH_2CH_2OH$, is a liquid. Why the difference?
54. Ethane, CH_3CH_3, is a gas at room temperature, while ethanol, CH_3CH_2OH, is a liquid. Why the difference?

POINTS TO PONDER

55. What was the impact of vitalism's downfall on mankind's view of itself?
56. Why do you think our society has mixed feelings about ethanol consumption? The legal drinking age in the United States is 21 years old. Should it be changed? Why or why not?
57. Devise a molecular theory to explain why many esters have pleasant odors, while many amines have foul smells. (Hint: Think about the interactions between these molecules and those in your nose.)
58. Wohler's synthesis of an organic compound demonstrated that physical laws were followed by molecules in living organisms including humans. The belief that life is only molecules and nothing else is termed reductionism—all can be reduced to atoms and molecules. Do you agree or disagree with reductionism? Why?
59. Our understanding of organic molecules and especially biological molecules has given humans the power to manipulate the molecules of life with unprecedented control. Name some ways in which that control has benefited society. Are there any risks associated with such control?
60. Any one molecule can be represented many ways. For example, pentane can be represented as:

    C_5H_{12}

    $CH_3CH_2CH_2CH_2CH_3$

    $$\begin{array}{c}\text{H H H H H}\\ | \; | \; | \; | \; |\\ \text{H}{-}\text{C}{-}\text{C}{-}\text{C}{-}\text{C}{-}\text{C}{-}\text{H}\\ | \; | \; | \; | \; |\\ \text{H H H H H}\end{array}$$

    Why do chemists insist on so many different ways to represent the same molecule? What are the advantages and disadvantages of each?
61. Explain why the formula $CH_3CH_2CH_3$ cannot mean:
    C—H—H—H—C—H—H—C—H—H—H

Feature Problems and Projects

62. Look at each of the following space-filling molecular images of various compounds and decide which, if any, are soluble in water. Remember, carbon is black, hydrogen is gray, oxygen is red, and chlorine is green.

63. Look at the following photo of the label for Dial Instant Hand Sanitizer and identify as many of the contents as possible. Draw structures for those compounds whose structure you know from this chapter.

- From Dial, the makers of America's #1 anti-bacterial soap.
Dermatologist Tested. Hypoallergenic.
Directions: Put a dime-sized amount in your hands. Rub hands briskly until dry.
Children under 6 years of age should be supervised by an adult when using this product.
Active Ingredient: Ethyl Alcohol, 62%
Other Ingredients: Water, Carbomer, Diisopropylamine, Polysorbate 80 (and) Cetyl Acetate (and) Acetylated Lanolin Alcohol, Propylene Glycol (and) Diazolidinyl Urea (and) Methylparaben (and) Propylparaben, Fragrance.
Warning: FLAMMABLE UNTIL DRY. Do not use near fire or flame. FOR EXTERNAL USE ONLY. If swallowed, seek medical advice immediately. Avoid contact with broken skin, face and eyes. In case of eye contact, flush thoroughly with water then seek medical advice. Discontinue use if

64. Obtain one of the following household items, copy its contents onto paper, and identify as many as possible:
    a. hair spray
    b. aerosol deodorant
    c. mouthwash
    d. shaving cream

65. Read The Molecular Revolution box in this chapter entitled "The Origin of Life." Some have argued that the scientific explanation for the origin of life is incorrect and that science's inability to formulate a viable theory for just how life began is evidence that it happened miraculously. What do you think? What happens to this perspective when a viable theory is formulated?

7 Light and Color

I seem to have been only like a boy playing on the seashore, and diverting myself in now and then finding a smoother pebble or a prettier shell than ordinary, whilst the great ocean of truth lay all undiscovered before me.

—Isaac Newton

In this chapter, we examine light by which we all see the world. What is light? How does it propagate? How does it interact with matter to produce color? What we normally call light is actually just a small portion of the electromagnetic spectrum. The entire spectrum contains many different kinds of *light* that we can't see. The radio waves that carry music from transmitters to your home radio, the microwaves that heat food in a microwave oven, and the X-rays used in medicine are all forms of *light*. What are the differences between these forms of light and the light that we see with our eyes?

We will also see how light interacts with matter, or, rather, how it interacts with the atoms and molecules within matter. It is this interaction that produces color.

Keep in mind as you read this chapter that light is constantly streaming into your own eyes all day long. The light causes a chemical reaction in molecules within your retina that sends nerve impulses to your brain that you interpret as sight. In fact, this very reaction allows you to see the words and images on this page.

Keep in mind also that the interaction of light with matter, also known as spectroscopy, is one of the most important tools available to the scientist. By using this tool, scientists can determine the composition of distant stars or the structure of complex molecules. A specific type of spectroscopy—called magnetic resonance imaging, or MRI—has become the premier imaging tool in medicine, allowing doctors to examine soft body tissues with unprecedented accuracy and detail.

QUESTIONS FOR THOUGHT

- What is light?
- What causes color?
- What is infrared light? UV light? Why is UV light dangerous?
- What are X-rays, gamma rays, microwaves, and radio waves?
- How does light interact with molecules and atoms?
- How can light be used to identify molecules and atoms?
- What is a laser and how does it work?

The orange color of leaves in the fall is caused by the interaction of light with β-carotene: molecules (shown here) in the leaves.

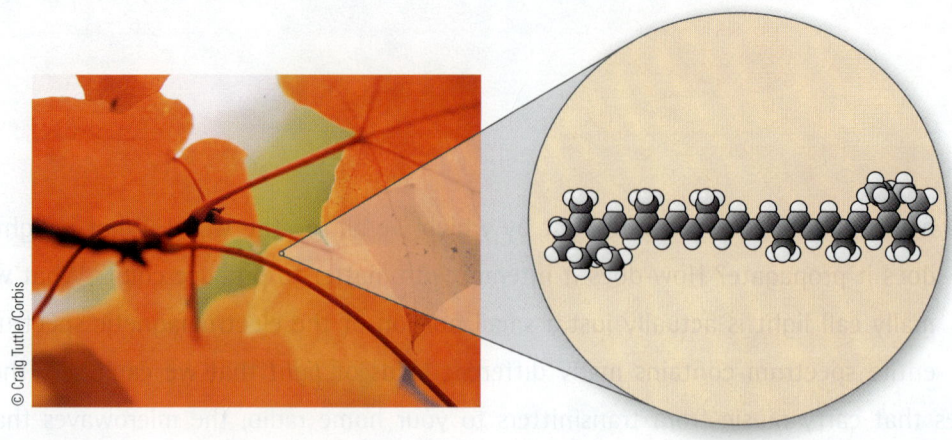

7.1 A New England Fall

The New England fall has captured the imagination of artists, poets, and writers. Trees change from a uniform green into a tapestry of color. What causes the color? Why are leaves green in the spring and summer and red, orange, and brown in the autumn? The answer lies in the molecules within the leaves and their interaction with light.

Sunlight, although appearing white, contains a spectrum of color. We see the colors contained in ordinary sunlight—red, orange, yellow, green, blue, indigo, and violet—in a rainbow. Under the right conditions, airborne water separates white light into its constituent colors, allowing us to see them. Airborne water does not produce the color; it was there in the white light, just mixed. The colors within white light were discovered by Isaac Newton, who used a glass prism to split light into its component colors and a second prism to recombine them into white light again (Figure 1).

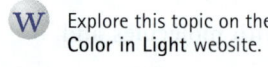
Explore this topic on the **Color in Light** website.

The color of any object depends on how its molecules interact with white light (Figure 2). If the molecules do not absorb any light, the reflected light is white, and the object appears white. If the molecules absorb all light, there is no reflected light, and the object appears black. If the molecules in an object selectively absorb some of light's colors and reflect others, the object appears colored.

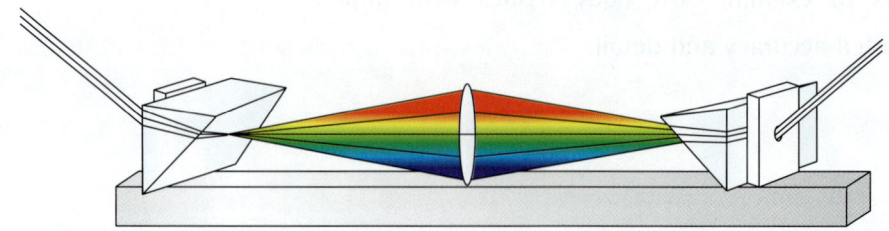

FIGURE 1 Isaac Newton used two prisms and a lens to split light into its constituent colors and then recombine them into white light.

(a) (b) (c)

FIGURE 2 (a) A substance that appears white reflects all colors of light. (b) A substance that appears black absorbs all colors of light. (c) A substance that appears colored absorbs some colors and reflects others.

The molecules present within the colored object determine which colors are reflected and which are absorbed.

The green-blue-yellow light reflected from a green leaf makes the leaf appear a dark shade of green (Figure 3). The molecules responsible for this green color are **chlorophylls** (Figure 4). Chlorophylls absorb light of all colors except green, blue, and some yellow. In the fall, chemical reactions within leaves destroy the chlorophyll molecules and a different class of molecules, called **carotenes** (Figure 5), dominates the leaf's color. Carotenes, also responsible for the color of carrots, absorb all colors except red and orange. As a result, the light that reflects from a leaf rich in carotenes is red-orange, and the leaf appears red-orange. Slight variations in the molecular structure of carotenes are responsible for the range of red colors we see in fall leaves.

FIGURE 3 A green leaf absorbs red, orange, and violet light. Green, yellow, and blue light are reflected, resulting in the dark shade of green that we see.

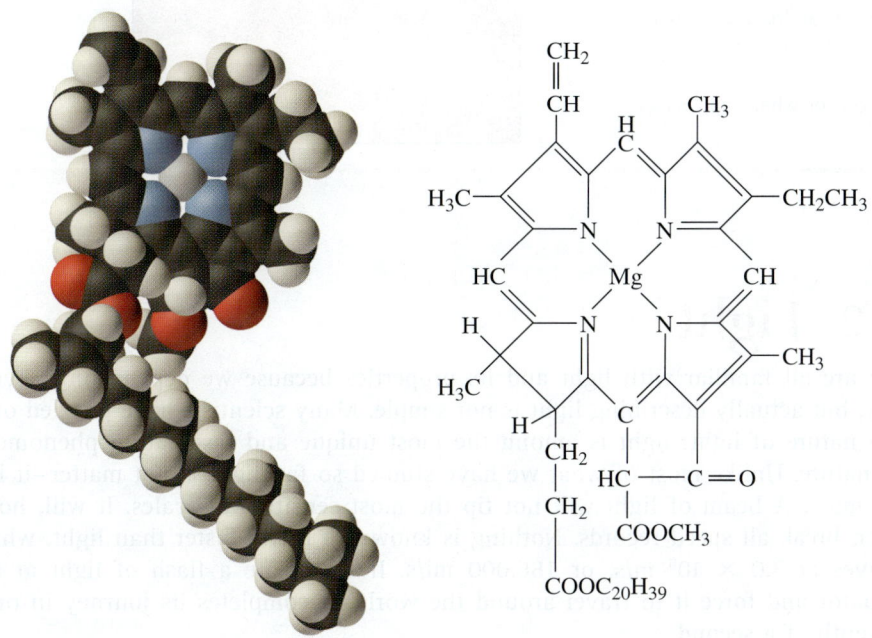

FIGURE 4 A chlorophyll molecule. Chlorophyll is responsible for the green color of leaves.

FIGURE 5 β-Carotene is the molecule responsible for the color of fall leaves.

MOLECULAR THINKING

Changing Colors

Several toys and even some children's clothing change color with temperature. For example, a particular raincoat is purple at room temperature but changes to pink when cooled. When it first gets wet, the drops of cool water produce a splatter of pink color on the otherwise purple coat. Kids love the patterns that form as a result of the interaction between the cold water and the warmth of their body. Other toys change color at the point where they are touched because heat from a hand or finger changes the temperature of the surface of the toy.

Question: Give a molecular explanation for what happens as the shirt changes color.

7.2 Light

We are all familiar with light and its properties because we depend on it every day, but actually describing light is not simple. Many scientists have puzzled over the nature of light; light is among the most unique and fascinating phenomena in nature. Unlike most of what we have studied so far, light is not matter—it has no mass. A beam of light will not tip the most sensitive of scales. It will, however, break all speed records. Nothing is known to travel faster than light, which moves at 3.0×10^8 m/s, or 186,000 mi/s. If we create a flash of light at the equator and force it to travel around the world, it completes its journey in one-seventh of a second.

We experience the high speed of light at a fireworks display or a baseball stadium. We see the action, the explosion of the fireworks, or the swing of the

W Explore this topic on the **Wave-Particle Duality** website.

bat; then a few moments later we hear the sound. The light and sound are formed simultaneously, but the light reaches us much faster than the sound.

One of the complexities of light is related to its dual nature; it has both wave and particle properties. We can think of particles of light, called **photons**, as tiny packets of energy traveling at the speed of light. During the day, billions of photons enter our eyes every second, producing the images we see. However, in the dark, only 10–20 photons reach our eyes each second. Experiments have shown that human eyes are sensitive to as few as 5–10 photons, making them one of the most versatile photon detectors known.

The wave nature of light is embodied in magnetic and electric fields. Because most of us have played with magnets, we know what a **magnetic field** is: It is the area around a magnet where forces are experienced. Hold a magnet between your fingers and bring a paper clip close to it. The clip will experience an attraction to the magnet as it enters the magnet's magnetic field. Similarly, an **electric field** is the area around a charged particle where forces are experienced. Although electric fields may not be as intuitive as magnetic fields, most of us have experienced them as "static electricity." Nonmetallic objects such as combs, brushes, sweaters, or even our own bodies can acquire charge, especially during dry weather. The result of that charge is an electric field that causes clothes to cling and hair to stand on end.

Iron filings around a magnet reveal the pattern of its magnetic field.

The wave nature of light, shown in Figure 6, is an embodied oscillating wave of electric and magnetic fields. The distance between wave crests is called the **wavelength** (represented by the Greek letter λ) and determines the color of the light. Visible light has very short wavelengths. For example, green light has a wavelength of approximately 540 nanometers (nm; 1 nm = 10^{-9} m), and blue light has a wavelength of approximately 450 nm (Figure 7).

The wavelength of light also determines how much energy one of its photons carries. Just as ocean waves carry more energy if their crests are closer together—think about being hit by surf on a beach—light carries more energy if its wavelength is shorter. The photons of blue light are more energetic than those of green light. The relationship between the wavelength and energy of light is an *inverse* relationship, meaning that as the wavelength increases, the energy decreases. We can

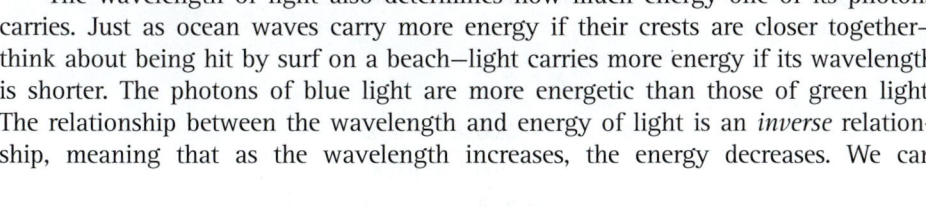

FIGURE 6 A representation of the wave nature of light. The distance between adjacent crests is the wavelength (λ) of the light.

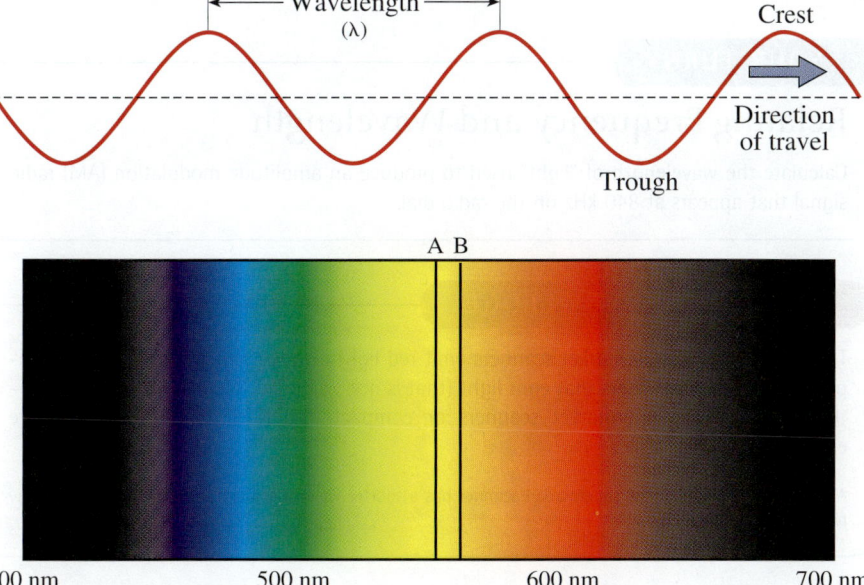

FIGURE 7 Every single point along the spectrum corresponds to a different wavelength of light. Even points A and B, which may appear the same color, correspond to different wavelengths.

express the inverse relationship between the energy of a photon and the wavelength of light as follows:

$$E_{photon} \propto \frac{1}{\lambda}$$

Where E_{photon} is the energy carried by the photon, λ is the wavelength of the light, and $\propto$ means "proportional to."

Another quantity related to wavelength is **frequency** (represented by the Greek letter ν). The frequency of light is the number of cycles or crests that pass through a stationary point in 1 s. The unit of frequency is 1/s and is also called the hertz (Hz). Like energy and wavelength, frequency and wavelength have an inverse relationship: The shorter the wavelength, the higher the frequency. The two are related by the following equation:

$$\nu = c/\lambda$$

where ν is the frequency in units of 1/s, c is the speed of light (3.0×10^8 m/s), and λ is the wavelength, usually in meters.

EXAMPLE 7.1

Relating Frequency and Wavelength

Radio waves are a form of "light" with long wavelengths. Calculate the wavelength used to produce a frequency modulation (FM) radio signal that appears at 100.7 megahertz (MHz; mega = 10^6) on the radio dial.

SOLUTION

We solve the preceding equation for wavelength (λ) and use the given frequency to find the wavelength:

$$\lambda = \frac{c}{\nu}$$

$$= \frac{3.00 \times 10^8 \text{ m/s}}{100.7 \times 10^6 \text{/s}}$$

$$= 2.98 \text{ m}$$

YOUR TURN

Relating Frequency and Wavelength

Calculate the wavelength of "light" used to produce an amplitude modulation (AM) radio signal that appears at 840 kHz on the radio dial.

APPLY YOUR KNOWLEDGE

The lasers used in supermarket scanners emit red light at a wavelength of 633 nm. Compact disc players use lasers that emit light (that is not visible) at 840 nm. Which photons—those emitted by supermarket scanners or compact disc (CD) players—contain more energy per photon?

Answer: The red light of the supermarket scanner has a shorter wavelength and therefore a higher energy per photon.

7.3 The Electromagnetic Spectrum

The word *light* is often used to mean **visible light**—that is, light visible to the human eye. However, this is too narrow a definition. There is no fundamental difference between visible light and light of other wavelengths, except that the human eye can see one but not the other. A general term for all forms of light is **electromagnetic radiation.** Electromagnetic radiation (Figure 8) ranges in wavelength from 10^{-15} m (gamma rays) to 10^4 m (radio waves). Only a small fraction of all electromagnetic radiation, called the *visible region,* can be seen by human eyes. The visible region is bracketed by wavelengths of 400 nm (violet) to 780 nm (red).

Electromagnetic radiation of slightly shorter wavelengths than violet light is called **ultraviolet (UV) light.** UV light is invisible to our eyes. Because of its shorter wavelength, its photons carry more energy than visible light. Consequently, UV light has enough energy in its photons to break chemical bonds and damage biological molecules. The Sun produces substantial amounts of UV light; fortunately, most of it does not reach the Earth's surface due to the absorption of UV light by oxygen and ozone in our atmosphere (see Section 5.5). The UV light that does get through produces sunburns, and excessive exposure may lead to skin cancer and eye cataracts. We protect ourselves against the harmful effects of UV light by wearing UV-rated sunglasses and using sunscreen. Sunscreen contains compounds such as *p*-aminobenzoic acid (PABA) that efficiently absorb UV light, acting like an ultrathin ozone layer right at the surface of our skin.

At shorter wavelengths than UV light are **X-rays,** discovered in 1895 by the German physicist **Wilhelm Roentgen** (1845–1923). Roentgen found that this

 Explore this topic on The **Electromagnetic Spectrum** website.

 Explore this topic on the **Ultraviolet Light** website.

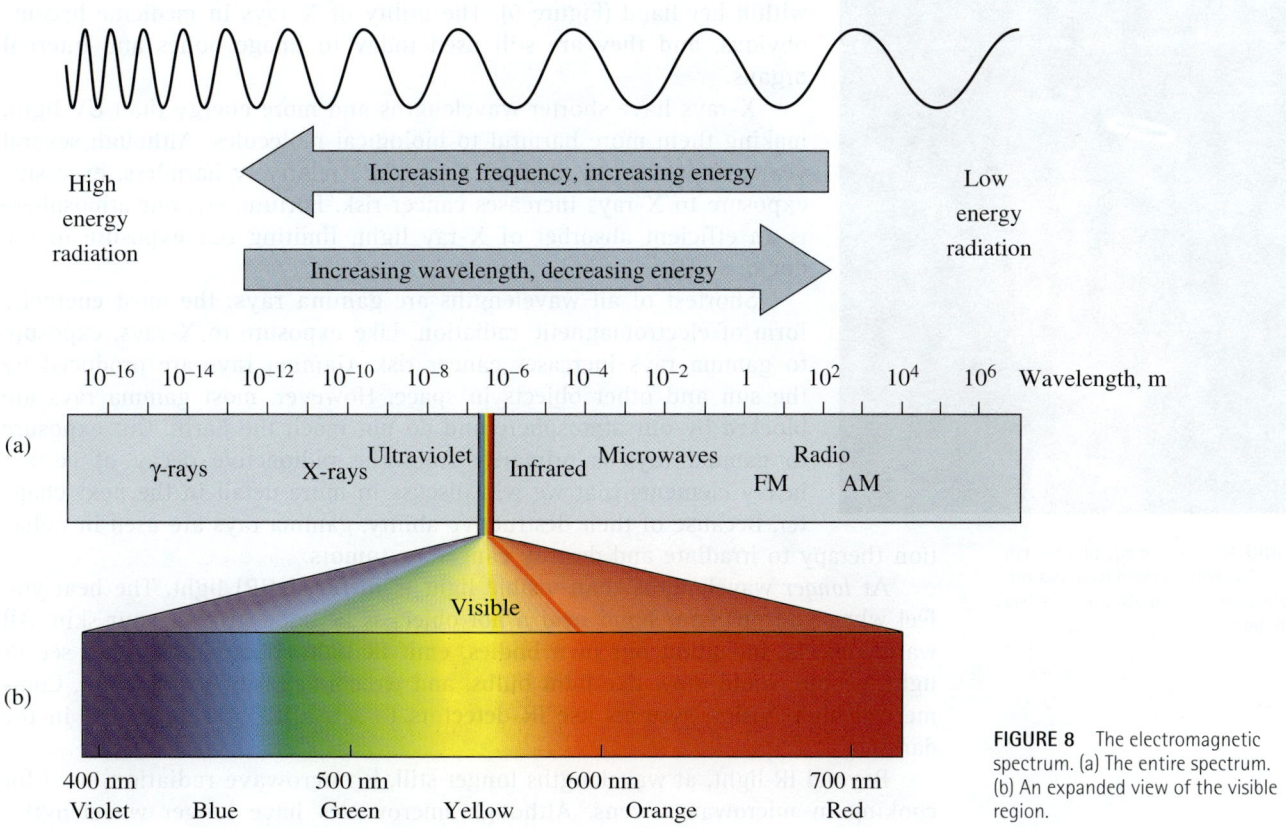

FIGURE 8 The electromagnetic spectrum. (a) The entire spectrum. (b) An expanded view of the visible region.

Sunscreen contains molecules such as p-aminobenzoic acid (PABA) that absorb UV light, preventing UV light from reaching our skin where it can produce sunburn.

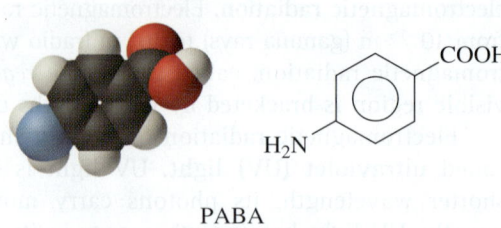

PABA

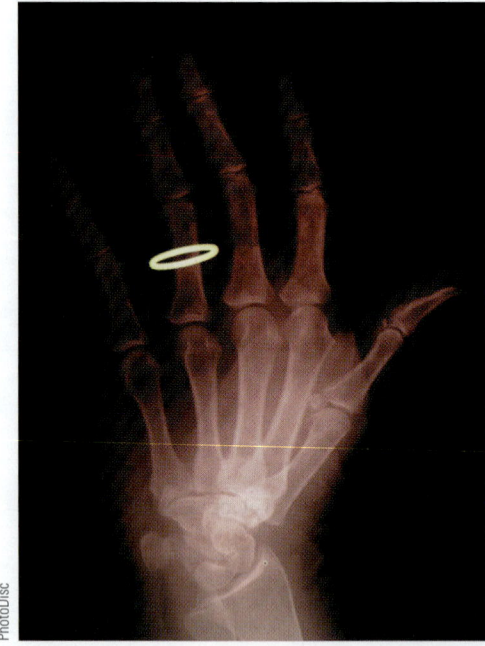

FIGURE 9 X-ray image of a human hand. The light-colored band is a ring. Metals are efficient absorbers of X-ray radiation.

unique form of light penetrated substances that normally blocked light. He used X-rays to photograph coins inside a box and metal objects on the other side of a solid wooden door. Even more remarkably, Roentgen used X-rays to obtain an image of his wife's bones within her hand (Figure 9). The utility of X-rays in medicine became obvious, and they are still used today to image bones and internal organs.

X-rays have shorter wavelengths and more energy than UV light, making them more harmful to biological molecules. Although several yearly exposures to medical X-rays are relatively harmless, excessive exposure to X-rays increases cancer risk. Fortunately, our atmosphere is an efficient absorber of X-ray light, limiting our exposure to the doctor's office.

Shortest of all wavelengths are **gamma rays,** the most energetic form of electromagnetic radiation. Like exposure to X-rays, exposure to gamma rays increases cancer risk. Gamma rays are produced by the sun and other objects in space. However, most gamma rays are blocked by our atmosphere and do not reach the Earth. Our exposure to gamma rays is primarily from the radioactive decay of certain heavy elements that we will discuss in more detail in the next chapter. Because of their destructive ability, gamma rays are used in radiation therapy to irradiate and destroy cancerous tumors.

At *longer* wavelengths than visible light is **infrared (IR)** light. The heat you feel when you put your hand near a hot object is IR light striking your skin. All warm objects, including our own bodies, emit IR light. If our eyes could see IR light, people would glow like light bulbs, and we could easily see at night. Commercial night vision systems use IR detectors to sense IR light and "see" in the dark.

Beyond IR light, at wavelengths longer still, is **microwave radiation,** used for cooking in microwave ovens. Although microwaves have longer wavelengths, and therefore lower energy than visible or IR, microwave energy is efficiently absorbed by water molecules. As a result, microwave radiation efficiently heats substances that contain water while not affecting substances that do not. For example, microwave ovens heat food, high in water, but do not heat plates, where water is absent.

At the longest of all wavelengths are **radio waves** with wavelengths as long as football fields. Radio waves were discovered in 1888 by the German physicist **Heinrich Hertz** (1857–1894). His experiments were continued throughout Europe and North America, and scientists soon discovered that radio waves were useful to transmit communications signals. Today, radio waves transmit the signals responsible for AM and FM radio, cellular telephones, television, satellite links, and other forms of communication.

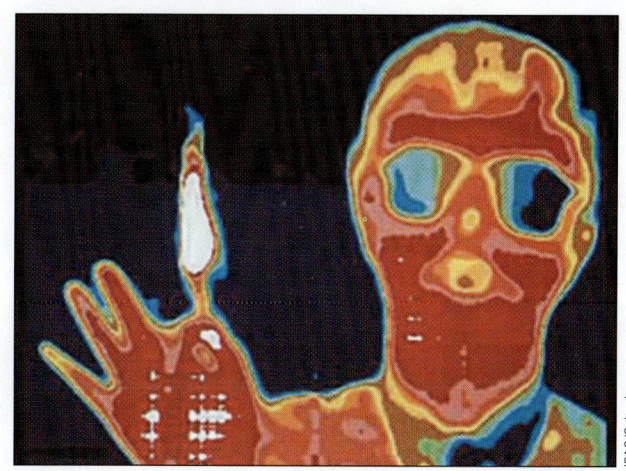

Infrared image of a person lighting a match in the dark. The infrared light emitted by all warm objects can be "seen" with an infrared camera.

What if...

X-Rays—Dangerous or Helpful?

Physicians routinely use X-rays to obtain images of bones, teeth, and internal organs. Yet studies show that exposure to X-rays increases cancer risk. Why do doctors continue to use X-rays? Is not the basic tenet of good medicine to "first do no harm"?

To answer questions like these, we must consider the benefits of X-rays in diagnosis versus their risk. We take risks every day if we perceive the benefit to be worth the risk. For example, every time we drive a car, we are taking a substantial risk—much more substantial, by the way, than the risk associated with a medical X-ray. We take the risk of driving a car because we believe the benefit—getting to the store to buy food, for example—to be worth the risk. A life free of risk is probably not possible. The appropriate question to ask then is, "Is the benefit worth the risk?"

In the case of X-rays, the answer is yes. Only in very large doses are X-rays known to cause cancer. In the much smaller doses acquired during a medical X-ray, the increase in cancer risk is negligible. The benefit, however, is large. The ability for a doctor to see the fractures in a broken bone allows her to diagnose and treat the injury effectively. The risk of *not having* the X-ray—perhaps a permanent disability due to misdiagnosis—is probably much worse.

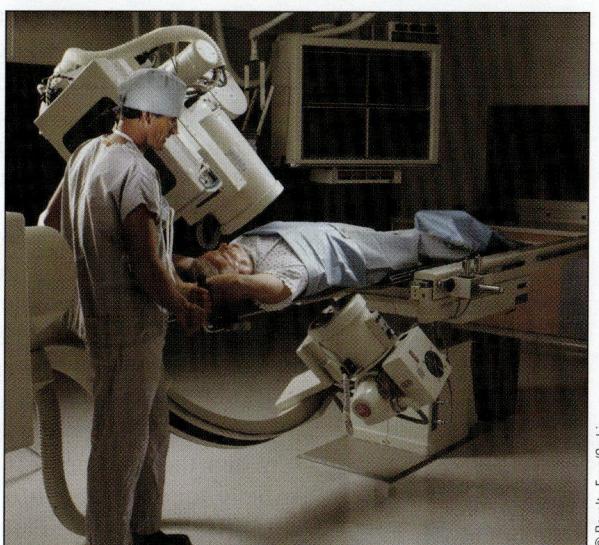

The benefit of getting an X-ray (diagnosing an injury or disease) is worth the risk (a negligible rise in cancer risk).

Question: What if a friend calls you for advice concerning her decision to not visit her doctor after a recent fall? She has a sharp pain whenever she moves her left arm but has grown suspicious of the medical industry and fears that the doctor will take an X-ray. She subscribes to a natural-healing magazine that talks at length about how X-rays increase cancer risk. What advice would you give your friend?

FIGURE 10 If the energy of a photon of light, as determined by its wavelength, exactly matches the energy difference between two electron orbits in an atom or molecule, the light can cause a transition from the lower-energy orbit to the higher-energy orbit. Light of the color corresponding to the energy match will be absorbed.

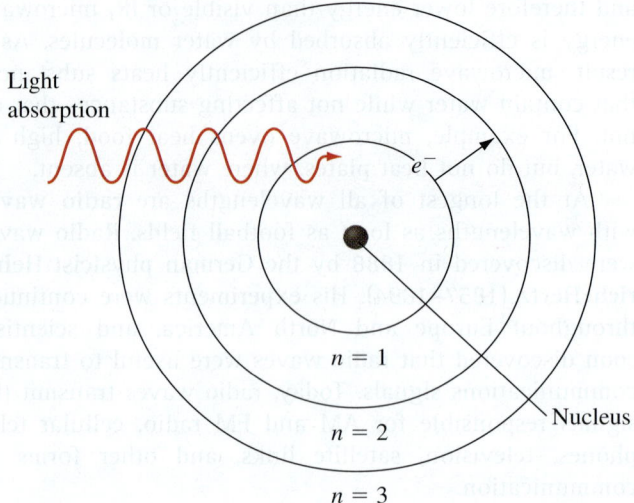

The orbits within atoms were first introduced in Section 3.8.

7.4 Excited Electrons

In Section 7.1 we learned that an object's color depends on the color of the light absorbed by its molecules or atoms. What happens within a molecule or atom when it absorbs light? Because light is a form of energy, a molecule or atom gains energy upon the absorption of light. This energy is captured by electrons that are *excited* from lower-energy orbits to higher-energy ones (Figure 10). Just as energy is required to pick an object off the floor and place it on a table, so energy is required to move an electron from a lower-energy orbit to a higher-energy one. The required energy can be provided by light if the energy of the photon just matches the energy required to move the electron from one orbit to the next. In other words, the energy of the photon (which depends on its wavelength) must be exactly equal to the energy *difference* between the two orbits.

The electron configuration of a molecule or atom with electrons in particular orbits is called an **energy state**. The total energy, or energy state, depends on which orbits are occupied. If all electrons are in the lowest-energy orbits possible, the atom or molecule is in the ground energy state. Light causes an **electronic transition** from one energy state of the molecule or atom to another, higher-energy state. It does this by promoting an electron to a higher-energy orbit. When the electron is in the higher-energy orbit, the atom or molecule is said to be in an **excited state**.

If the energy required to make the transition from one energy state to another corresponds to red light, for example, then red light is absorbed by the molecule, and other colors are not. Some molecules absorb many colors, whereas others absorb only one, and still others absorb none; it depends on the electrons and their orbits.

After an electron in a molecule is excited to a higher-energy orbit with light, the molecule is left in an unstable excited state. The excess energy of this excited state can be dissipated in several ways (Figure 11). If the energy of the absorbed photon were high enough, usually in the UV region or higher, bonds within the

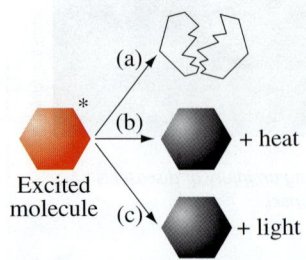

FIGURE 11 An excited molecule can undergo (a) photodecomposition, (b) emission of heat, and (c) emission of light.

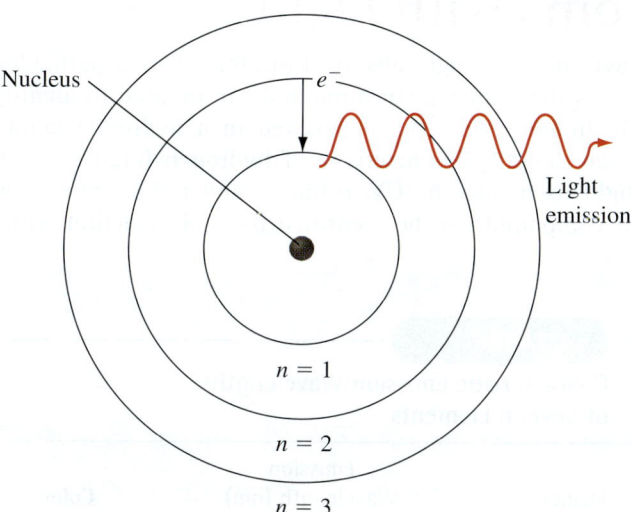

FIGURE 12 As an electron relaxes from an excited state, it can emit light.

molecule can break, and the molecule may fall apart. We call this process **photodecomposition**, and it is partly responsible for the dangers of UV light, X-rays, and gamma rays. These short wavelengths carry enough energy to break chemical bonds in molecules. The fading of fabrics on repeated exposure to sunlight is due to photodecomposition. The molecules within the fabric that gave it the characteristic color are destroyed by the UV component of the sun's rays. As sunlight destroys the color-giving molecules within the fabric, the color of the fabric fades.

Another process that can occur when an atom or molecule is in an excited state is called an **electronic relaxation** (Figure 12). In relaxation, the electron goes back to its original orbit, producing heat or light in the process. We experience the production of heat due to relaxation when we wear a dark-colored shirt on a sunny day. After the molecules within the shirt absorb the sun's light, they relax back to their lower-energy state, emitting heat. The shirt warms and we feel hot. We see the emission of light by the relaxation of excited electrons in **phosphorescence** as occurs in glow-in-the-dark toys. When exposed to light, molecules within the toys absorb light, moving electrons to higher-energy orbits. As the electrons relax back to lower-energy orbits, they emit light, causing the familiar green glow. Another example of relaxation through light emission is the glow of a white T-shirt under black light. Black lights emit UV light, invisible to our eyes but exciting electrons in the white T-shirt. As the electrons relax, they emit visible light in a process called **fluorescence**. The difference between fluorescence and phosphorescence is the time required for electrons to relax—fluorescence is relatively fast, and phosphorescence is slow. Fluorescing objects stop emitting light the instant you remove them from the UV source, whereas phosphorescing objects continue to glow for an extended time in the dark.

In summary, excited electrons can result in photodecomposition—in which molecules break down to other substances—or they can simply undergo relaxation. Relaxation can occur either through the emission of heat or through the emission of light. The emission of light can occur either through phosphorescence, a long-lived process, or through fluorescence, a short-lived process.

Objects will glow under black light because the invisible UV light emitted by these lamps will excite electrons in objects around them to higher-energy orbits. The electrons then relax to lower-energy orbits, emitting light in the process.

7.5 Identifying Molecules and Atoms with Light

The specific wavelengths of light absorbed or emitted by a particular molecule or atom are unique to that molecule or atom and can be used to identify it (Table 1). For example, hydrogen atoms can be excited in a hydrogen lamp. The emitted light contains wavelengths characteristic of hydrogen (Figure 13). By extending the range of light used into the UV, infrared, and radio wave regions, virtually any element or compound can be identified by its interaction with light. Scien-

TABLE 1
Characteristic Emission Wavelengths of Several Elements

Element	Emission Wavelength (nm)	Color
H	656	Red
	486	Green
	434	Blue
	410	Violet
He	706	Red
	587	Yellow
	502	Green
	447	Blue
Li	671	Red
Na	589	Yellow
Hg	579	Yellow
	546	Green
	436	Blue
	405	Violet

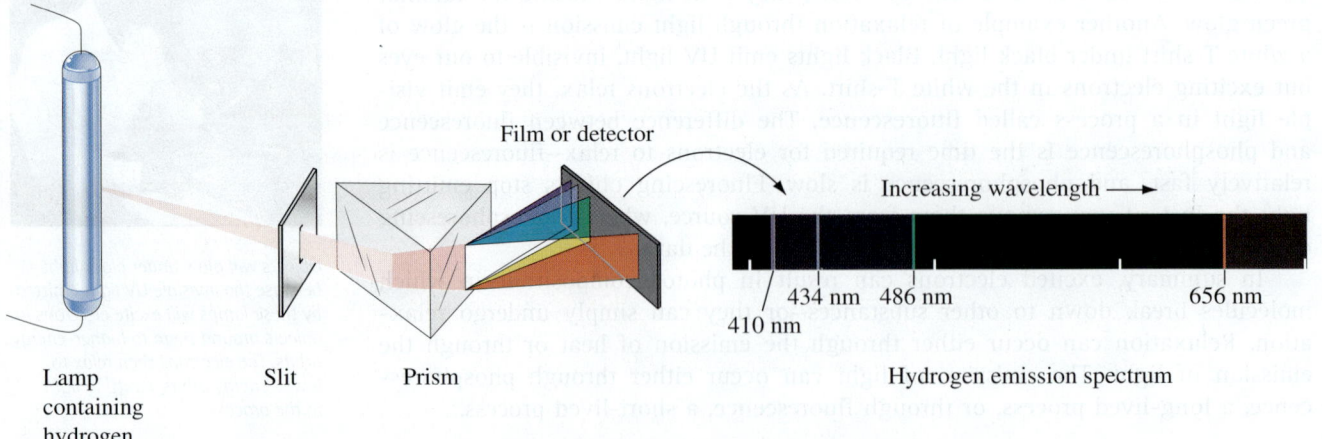

FIGURE 13 Excited hydrogen atoms emit light of specific wavelengths. If you pass the light through a prism, the individual wavelengths can be spatially separated from one another and detected on a photographic film or digital light detector.

tists use the interaction of light with matter, called **spectroscopy,** to identify unknown substances. Astronomers, for example, identify the elements present in stars by examining the light emitted by the star. Chemists can determine the composition of a particular substance by studying the wavelengths of light absorbed by the substance. Atmospheric scientists study slight variations in the intensities of the colors present in sunlight to identify the molecules present in our atmosphere. Spectroscopy is one of the most versatile tools a scientist has at his or her disposal for identifying and quantifying matter.

7.6 Magnetic Resonance Imaging: Spectroscopy of the Human Body

Medical use of X-rays to see into the human body without cutting through skin and tissue revolutionized medicine early in the 20th century. Today, doctors have a more powerful technology to look into the human body. This technology is based on the spectroscopy of hydrogen atoms in a magnetic field and is called **magnetic resonance imaging (MRI).** Because hydrogen atoms are abundant in the human body, both in water (75% of body mass) and in organic molecules, this form of spectroscopy can be used to examine biological tissues. Unlike the spectroscopy discussed previously that involves electrons, MRI involves the *nuclei* of hydrogen atoms and was originally termed **nuclear magnetic resonance (NMR).** Although this name is still used today by many scientists, the medical applications of NMR have been termed MRI to avoid public confusion over the safety of a technique containing the word *nuclear.*

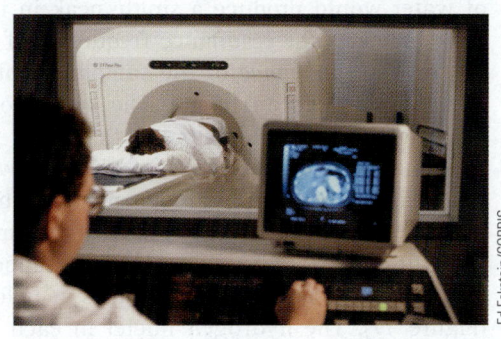

W Explore this topic on the MRI website.

Magnetic resonance imaging allows doctors to see internal organs with exceptional clarity.

Felix Bloch of Stanford University and Edward Purcell of Harvard developed NMR spectroscopy in the 1940s. The pair of scientists shared the Nobel Prize in 1952 for their work. The concepts in NMR can be explained by imagining hydrogen nuclei as tiny magnets (Figure 14). Much like a compass needle aligns in the Earth's magnetic field, these tiny magnets align when placed in an external mag-

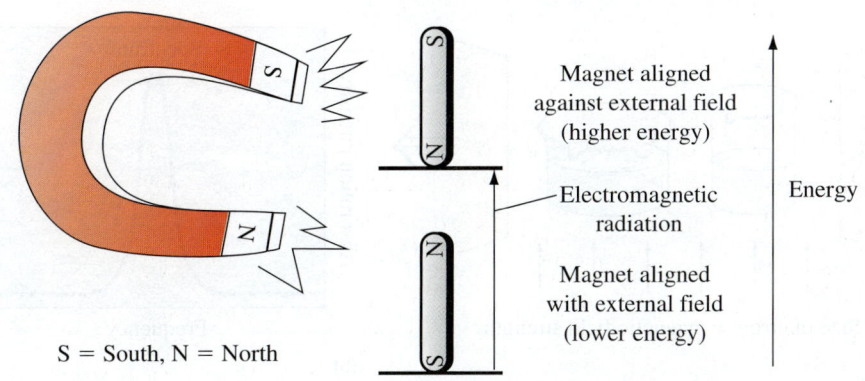

FIGURE 14 A magnet aligned with an external field is in a lower-energy state than one aligned against the external field. Light of the correct wavelength can cause the transition from one state to the other.

netic field, usually created by a large electromagnet. Just as you can change the preferred orientation of a magnet in a magnetic field by pushing it, you can change the orientation of the hydrogen nuclei in the external field by *"pushing"* on them; the pushing in NMR is done by electromagnetic radiation of the right frequency. We say that the electromagnetic radiation causes a *transition* (analogous to transitions involving electrons) from one orientation to another. The energy from the electromagnetic radiation changes the orientation of the tiny magnet in the external magnetic field.

The exact electromagnetic frequency that causes a transition is called the **resonance frequency**. For NMR, the resonance frequency occurs in the radio wave region of the electromagnetic spectrum. The *exact* frequency of radio waves required depends primarily on the strength of the external magnetic field. The stronger the external magnetic field, the more energy required to flip the magnet and the higher the frequency (higher frequency means more energy) of radio waves required to make the transition.

In conventional NMR spectroscopy, a sample is held in a uniform magnetic field while the radio wave frequency is varied. When the frequency becomes resonant, it causes a transition in the nucleus and changes in the magnetism of the sample are observed. A graph that shows the intensity of these changes as a function of the radio wave frequency is called an **absorption spectrum**. A sample of water would produce a single peak in its absorption spectrum (Figure 15); the peak indicates the resonance frequency of hydrogen nuclei in water molecules.

In MRI, an *image* of the sample is obtained. The spectrum in Figure 15 does not contain such an image. The key to obtaining an *image* of a sample lies in creating an external magnetic field that varies in space. For example, suppose two small cylinders of water of different widths are placed in a uniform magnetic field (Figure 16). The hydrogen nuclei in both samples experience the same magnetic field and consequently have the same resonance frequency; the absorption spectrum shows just one peak. Now suppose the two water cylinders are placed in a varying magnetic field—that is, a field whose intensity changes across space (Figure 17). The hydrogen nuclei in each of the cylinders experience a different magnetic field and therefore have a different resonance frequency; the absorption spectrum displays two peaks. Even within one of the cylinders, the nuclei on one side of the cylinder experience a different magnetic field than those on the other side of the cylinder and consequently have a slightly different resonance frequency. The shape of each peak in the absorption spectrum then reflects the shape of the container holding the water sample, and an image is obtained.

By skewing the intensity of the magnetic field in space, the image of any object containing hydrogen atoms is encoded in the frequency of the resonant

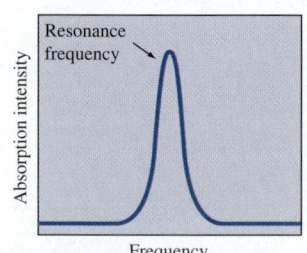

FIGURE 15 A simplified NMR spectrum of water. The peak would shift to a different frequency in a magnetic field of a different strength.

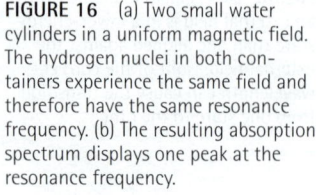

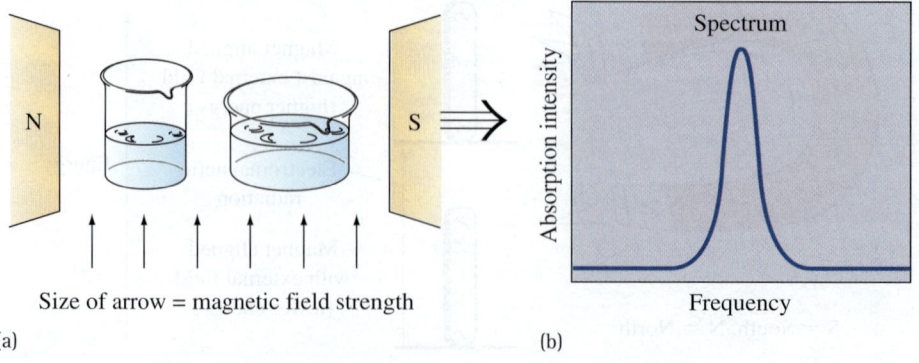

FIGURE 16 (a) Two small water cylinders in a uniform magnetic field. The hydrogen nuclei in both containers experience the same field and therefore have the same resonance frequency. (b) The resulting absorption spectrum displays one peak at the resonance frequency.

7.6 Magnetic Resonance Imaging: Spectroscopy of the Human Body

> ### What if...
>
> ### The Cost of Technology
>
> The cost of medical care has soared in recent years. Part of that increase is because of the high cost of new technologies such as magnetic resonance imaging. These new technologies have, in many cases, proven to be extremely useful in diagnosing and treating diseases. However, they have also brought a dilemma to our society: How much will we pay for health care? Is there a limit? What if new technologies promised therapy for certain diseases but came at tremendous cost? You may be tempted to say that money is no object when dealing with health care; however, in a world with limited resources, money counts. As an extreme example, consider what would happen if all our resources were spent on health care—there would be no resources left for food, clothing, or housing, all basic necessities.
>
> **Question:** On a personal level, how much are you willing to pay for health care—10, 20, or 30% of your income? Why so much? Why not more? We as a society must grapple with this issue as the health care industry continues to change.

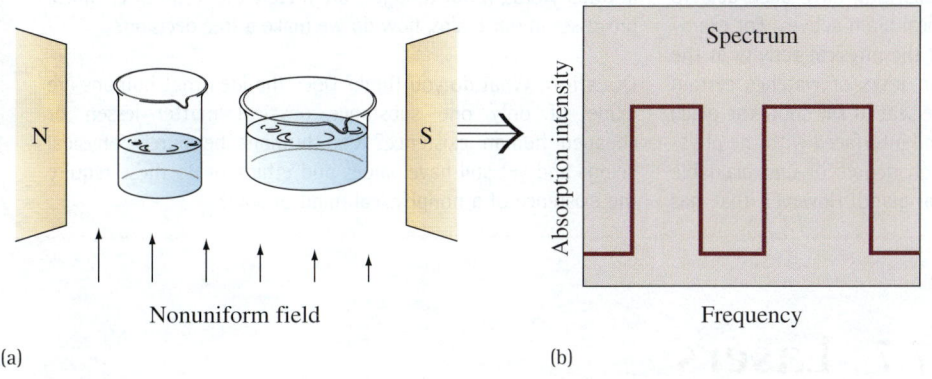

FIGURE 17 (a) Two small water cylinders in a varying magnetic field. The hydrogen nuclei in each container experience a different field and therefore have different resonance frequencies. (b) The resulting absorption spectrum displays two peaks, which match the shape of the cylinders. The shape of the containers is encoded in the frequency of absorption.

radio waves. When the object is a human brain, or knee, or spine, the result is a remarkably clear medical image (Figure 18).

In addition, MRI can measure the time required for nuclei to return to their original orientation after being pushed by electromagnetic radiation. Imagine pushing the needle on a compass and timing how long it takes to realign. The time required for realignment—or more correctly, relaxation—is called the **relaxation time** and is very sensitive to the local environment of the hydrogen atoms. Because different biological tissues contain nuclei in different environments, MRI can distinguish between these different types of tissue.

The ability of MRI to obtain clear, well-resolved images of the internal structures of human patients, in combination with its ability to distinguish between different kinds of tissue—all without the biological risk associated with X-rays—has made it the premiere tool for medical imaging today. It is particularly useful in detecting diseased tissue, cancerous tumors, spinal cord lesions, and other abnormalities of the central nervous system. MRI has also been used to map brain function. Particularly fascinating is the correlation between specific types of brain tasks, such as memorizing or mathematical reasoning, and the area of the brain that is active during the task. Because brain activity produces an increase in oxygen around the active area, which in turn produces a different environment for hydrogen atoms, MRI is used to detect what part of the brain is working when a person is involved in a particular task. Such information has led to remarkable advances in our understanding of brain function (see the What If . . . The Mind–Body Problem box).

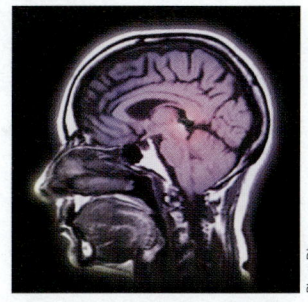

FIGURE 18 MRI images of a human head.

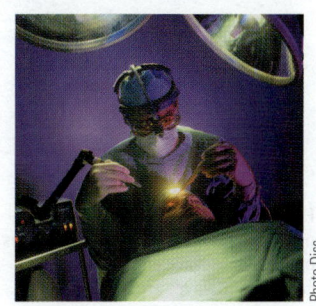

Applications of laser technology include, for example, laser surgery.

What if...

The Mind–Body Problem

In the 17th century, the French philosopher René Descartes uttered his famous words, "I think; therefore, I am." Descartes was a *dualist*, a person who believes that humans are composed of two different substances—a physical body and a nonphysical mind (which can think). Dualism originated with Greek thinkers, especially Plato. Descartes, however, realized that dualism had a fundamental problem: How does a nonphysical mind interact with a physical body? Descartes suggested that the interaction occurred at the pineal gland within the brain. The pineal gland, in Descartes's view, was the seat of the nonphysical soul.

Advances in the study of the brain (known as neuroscience), through techniques such as MRI, have been able to correlate mental activity with physical brain activity. For example, MRI has been used to monitor the physical activity in the brain as a person performs certain tasks or watches certain images. If the pineal gland were the seat of the soul—the place where the nonphysical thinking mind interfaced with the physical body—MRI should detect a high degree of unexplainable activity originating from the pineal gland. However, that has not happened. Instead, researchers find that certain kinds of thoughts correlate with certain areas of the physical brain. In other words, our thoughts seem to originate out of physical matter, not out of a nonphysical mind (or soul).

Because of this type of evidence, most neuroscientists have abandoned mind-body dualism and are working at ways to explain human consciousness through purely physical means. However, many problems remain for such a description ultimately to succeed. For example, explaining consciousness and free will—which are relatively easy to explain under a dualist picture—become difficult to explain under a purely physical picture. In other words, if our thoughts are merely the result of chemical processes in our brains, how do we make a free decision?

Question: What do you think? Does the idea that humans are made of only one substance—physical matter—lessen or cheapen human existence? Can humans be purely physical beings and yet still have values and ethics, or do these require the existence of a nonphysical mind or soul?

7.7 Lasers

Another technology related to light is the **laser**. Discovered in 1960, lasers have evolved from scientific novelties to practical tools with many applications. Lasers are used for visual effects in rock concerts, as precision drills in manufacturing, as aiming devices in laser-sighted weapons, and as invisible scalpels in surgery. They are also integral parts of numerous devices, including supermarket scanners, compact disc players, laser printers, and surveying equipment. The use of lasers in consumer products is well established, and new uses will continue to develop. Science fiction has long played on the idea of laser weapons, including handheld versions that zap an enemy into oblivion or the much larger systems such as those that produce "photon torpedoes" on the Starship *Enterprise*.

The word *laser* is an acronym for light amplification by stimulated emission of radiation. The differences between laser light and ordinary light are twofold (Figure 19). In contrast to white light, which contains many wavelengths, laser light contains only one. For example, the red helium–neon lasers, often seen in laser demonstrations, emit red light of 632.7 nm. Not only is this light red (and not combinations with other colors), but it is a *single shade* of red—the shade that

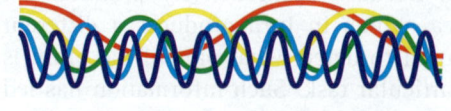

White light   Laser light

FIGURE 19 Laser light differs from ordinary white light in that it contains only a single wavelength and all its component waves are in phase.

is characteristic of 632.7 nm. The second difference between laser light and ordinary light is the alignment of the electromagnetic waves. In contrast to ordinary light, whose electromagnetic waves are randomly oriented, laser light contains waves whose troughs and crests are aligned, or *in phase*. The result is a very pure, intense form of light that does not spread much as it travels through space.

LASER CAVITIES

In a laser, molecules or atoms within the **lasing medium** are excited with light or electrical energy. Electrons jump to higher energy states; when these electrons relax back to their lower-energy states, they emit light. This part of the process is like the fluorescence we discussed earlier (see Section 7.4). In a laser, however, the lasing medium is placed inside a **laser cavity** consisting of two mirrors, one of which is only partially reflecting (Figure 20). The first photon emitted in the direction of one of the mirrors bounces back and passes through the lasing medium again, stimulating the emission of other photons, each one at exactly the same wavelength and with exactly the same wave alignment (or phase) as the original one. These photons will then travel down the cavity, bounce off the other mirror, and pass through the lasing medium again, causing the emission of even more photons. The result is a chain reaction that produces large numbers of photons circulating within the laser cavity. A small fraction of these leaks out of the cavity through the partially reflecting mirror, producing the laser beam.

The primary difference among lasers is the lasing medium. Lasers can be broadly divided into four types: solid-state lasers, gas lasers, dye lasers, and semiconductor lasers.

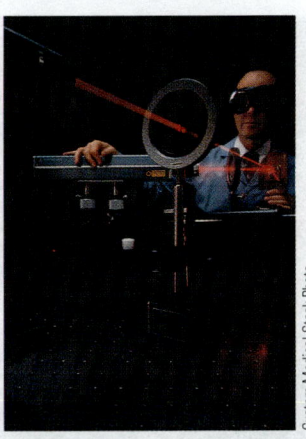

Helium–neon lasers, such as the one in this picture, are often found in undergraduate laboratories and laser displays.

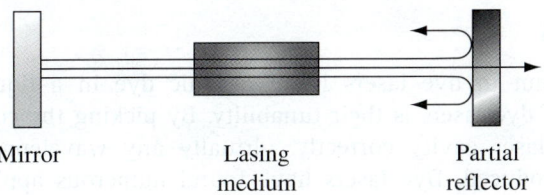

Mirror   Lasing medium   Partial reflector

FIGURE 20 A laser cavity consists of a fluorescent material sandwiched between two mirrors, one of which is only partially reflecting.

SOLID-STATE LASERS

The lasing medium in a solid-state laser is usually a metal ion distributed in a solid crystal. The first working laser was a solid-state ruby laser that used chromium ions dispersed in a sapphire crystal. Ruby lasers produce red light at 694 nm. A more common solid-state laser today is the neodymium-yttrium-aluminum-garnet (Nd:YAG) laser, which employs neodymium ions dispersed in a YAG crystal. Nd:YAG lasers produce infrared light at 1064 nm with moderate to high power (several watts). They can be made to produce a continuous beam of light or can be pulsed to produce short bursts of light. Nd:YAG lasers are used in manufacturing, medicine, and basic scientific research.

GAS LASERS

The lasing medium in gas lasers is a gas or a mixture of gases contained in a tube. The most common gas laser is the helium–neon laser named after the gases present in its cavity. Helium–neon lasers emit a pencil-thin beam of red light at 632.7 nm with relatively low intensity (milliwatts). They are relatively inexpensive and

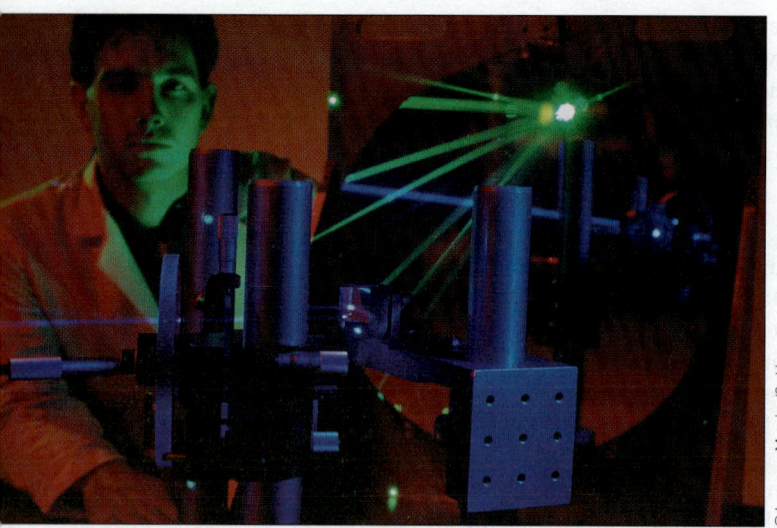

Argon-ion lasers are often used to produce striking visual effects in concerts.

are used in surveying, as a cutting guide in manufacturing, as an alignment tool for optics, and as a light source for holography. Helium–neon lasers are also found in many undergraduate laboratories and laser displays.

Another common gas laser is the argon ion laser, which uses gaseous argon ions as the lasing medium. Argon ion lasers usually produce moderate to high powers (several watts) of green light at 514 nm but can produce light at several other wavelengths in the visible region as well. Argon ion lasers are used in rock concerts to produce spectacular effects. Because laser beams are invisible when propagated through clean air, most visual effects are obtained by filling the air in a concert hall with smoke or water vapor. When the intense, green, argon ion laser beam is propagated through the haze, it appears as a brilliant, pencil-thin beam. The operator often splits a single beam into several and reflects them off moving mirrors to produce complex patterns. The operator can also rapidly scan the beam back and forth to produce what appears to be a shimmering sheet of green light.

A third common type of gas laser is the carbon dioxide (CO_2) laser, which produces infrared light at 10.6 micrometers (μm). The carbon dioxide laser is the most efficient and powerful of all gas lasers producing high powers—up to 1000 watts (W). Such a beam can easily cut through steel and is used in welding, drilling, and cutting.

DYE LASERS

The lasing medium in dye lasers is an organic dye in a liquid solution. The unique feature of dye lasers is their tunability. By picking the correct dye and by configuring the laser cavity correctly, virtually any wavelength in the visible region can be produced. Dye lasers have found numerous applications in basic scientific research and in medicine. An unusual dye laser was made some years ago by A. L. Schawlow of Stanford University. Schawlow built and consumed the first edible laser—an organic dye in Knox gelatin. The lasing medium was said to have been prepared "in accordance with the directions on the package."

SEMICONDUCTOR LASERS (DIODE LASERS)

The very smallest, and in many cases the most inexpensive, of all lasers are semiconductor or diode lasers. The lasing medium consists of two semiconducting materials sandwiched together. Although they produce only modest power, these lasers can be made cheaply and have found utility in supermarket scanners, laser pointers, CD players, and other electronic equipment.

7.8 Lasers in Medicine

Lasers have proved to be useful tools in medicine. Either by efficient heating of tissue with infrared and visible lasers or by the direct scission of chemical bonds within tissue with UV beams, lasers have found a home in the surgeon's arsenal. A laser beam has several advantages over a scalpel: It can make precise cuts through skin and tissue with minimum damage to surrounding areas; it can be

THE MOLECULAR REVOLUTION

Watching Molecules Dance

At 6:00 A.M. on Tuesday, October 12, 1999, Professor Ahmed Zewail of the California Institute of Technology was awakened by a phone call from the Nobel Prize Committee in Sweden informing him that he was the recipient of the 1999 Nobel Prize in chemistry. Professor Zewail's reaction? "I kissed my wife and kids," he says, "I just love science."

Professor Zewail received the Nobel Prize for pioneering a new field of science that he calls "femtochemistry." All of life and all of chemistry depend on the breaking and forming of chemical bonds. Professor Zewail devised a way of watching the breaking and forming of chemical bonds as they occur. The challenge was that these processes occur very quickly, on a time scale of one millionth of one billionth of a second. For example, the chemical change that occurs in the human eye when a photon strikes the retina—a twist about a carbon–carbon double bond—occurs in about 200 femtoseconds (femto = 10^{-15}). Before Professor Zewail's work, no one imagined that you could ever see that twist in real time—now we can. In effect, Professor Zewail has developed a movie camera with a very fast shutter speed—so fast that it can watch molecular movements as they happen. Professor Zewail watches molecules dance, and dance they do. The breaking of the Na–I bond in sodium iodide, for example, takes about ten in-and-out motions before the bond actually breaks.

Professor Zewail does his work with the use of ultrashort pulsed lasers. He creates flashes of light so short that they can resolve atomic motions. His flashing lasers are like strobe lights that catch the atomic dancers in action, a flash of the strobe catching one part of the dance, another flash catching the next part, and so on. The flash, however, has to be short enough so that the dancers don't move while the flash is on, and that is the challenge that professor Zewail overcame. What does professor Zewail want to do next? He would like to control the dancers' next moves. Beyond simply watching a chemical reaction in real time, a goal of Professor Zewail's work is to control the reaction. He wants to use ultrashort laser pulses to put energy into just the right part of a molecule at just the right time to enhance one reaction over another. It is just this kind of molecular control that chemists dream of having. Controlling the outcome of a chemical reaction has always been a major goal in chemistry because it allows chemists to produce useful substances while preventing harmful ones. Zewail is not the first to do this. Ever since humans learned of chemical reactions—probably primitive humans using fire—we have been trying to control them for our own purposes. The revolutionary nature of Zewail's work is the specificity with which he can observe, and perhaps one day control, a chemical reaction.

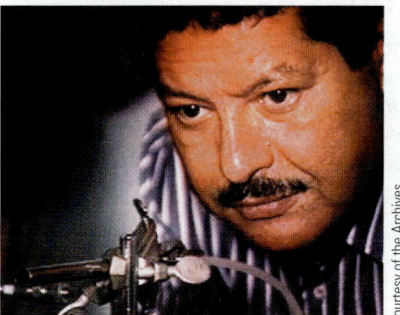

Ahmed H. Zewail

delivered through fiber-optic cable to places that are ordinarily difficult to reach; and its wavelength can be varied to produce certain desirable effects.

Eye surgeons routinely use lasers to cut away vision-impairing tissue in cataract patients or to drill microscopic holes in the cornea to relieve pressure buildup in glaucoma patients. Recent advances even allow laser reshaping of the cornea to correct both nearsightedness and farsightedness. In this process (called LASIK), a high-energy beam of UV light cleanly cuts into the cornea, changing its curvature and correcting the patient's vision.

The ability to deliver intense laser pulses through fiber-optic cable allows surgeons to operate on internal organs, often without breaking the skin. In this procedure, the surgeon uses an endoscope, a fiber-optic cable through which the surgeon sees, in combination with a second fiber-optic cable to deliver the laser light. The laser light interacts directly with internal organs removing tumors, clearing arteries, or destroying gallstones and kidney stones.

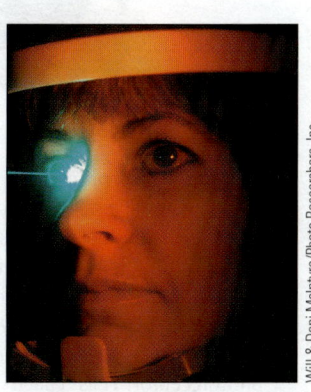

Laser beams can be used to reshape the cornea.

Unsightly port-wine stains, as shown in this picture, can be removed with laser light of the correct wavelength.

Dermatologists have employed lasers in the removal of skin cancer tumors and unsightly skin blemishes. To remove skin tumors, the patient receives a special dye that selectively attaches itself to tumorous tissue. A laser, producing light at the absorption wavelength of the dye, is then used to irradiate and destroy the tumor. The wavelength of the laser light can also be chosen to interact with unsightly blemishes beneath the skin. For example, port-wine stains—red birthmarks that often occur on the face—can be irradiated with laser light. The wavelength of the light is chosen to be transparent to the skin but to be absorbed by the blood vessels beneath it. The doctor burns out the birthmark while leaving the skin above intact. Such techniques have also been used to remove freckles, moles, dark spots, and even tattoos.

Molecular Focus

Retinal

Formula: $C_{20}H_{28}O$
Molar mass: 284.42 g/mol
Melting point: 62°C
Structure:

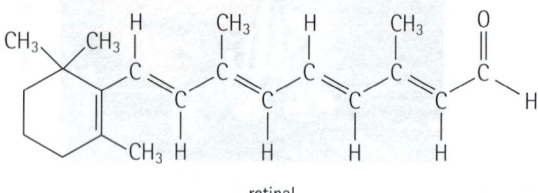

retinal

Three-dimensional structure:

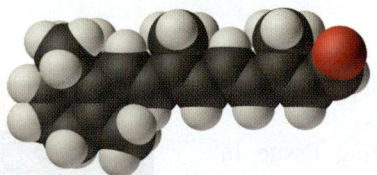

At room temperature, retinal forms orange-colored crystals that look similar to many other colored compounds. Yet this rather ordinary looking substance is responsible for our ability to see; retinal is the link between light and vision. Retinal has several isomers, two of which (11-*cis*-retinal and all-*trans*-retinal) act as a chemical switch that is activated by light.

The retina, or back portion of the eye, is coated with photoreceptor cells called rods and cones, each of which contains millions of 11-*cis*-retinal molecules. When a photon strikes an 11-*cis*-retinal molecule, one of the double bonds along the carbon backbone is broken. This causes the molecule to change its shape (Figure 21), which then—through several steps—produces an electrical signal that travels to the brain for processing. Under bright conditions, millions of photons strike the eye each second: The brain must combine many of these signals to form an entire image.

FIGURE 21 The first step in vision occurs when light converts 11-*cis*-retinal into its isomer, all-*trans*-retinal.

Chapter Summary

MOLECULAR CONCEPT

We have seen that light is a form of energy carried by electric and magnetic fields traveling at 186,000 miles per second (7.2). Light has both particle and wave characteristics. The wavelength of the light determines both its color and its energy; the shorter the wavelength, the higher the energy. Light, also called electromagnetic radiation, ranges in wavelength from gamma rays (10^{-15} m) through the visible region (500 nm) to the radio wave region (100 m). In the visible region, white light contains a spectrum of wavelengths from 400 nm (violet) to 780 nm (red); these can be seen in a rainbow or when light passes through a prism (7.3).

The color of substances depends on the color of light absorbed by the molecules or atoms that compose the substance. This, in turn, depends on the energy separations between electron orbits. When a molecule or atom absorbs light, electrons are excited from lower energy orbits to higher energy orbits (7.4). If the energy of the light is high enough, light can break chemical bonds and destroy or change molecules through photodecomposition; usually, however, the energy is simply given off again as heat or light through relaxation. The specific wavelengths at which molecules or atoms absorb or emit light serve as fingerprints for specific substances, making spectroscopy—the interaction of light and matter—a useful tool in identifying unknown substances (7.5). Magnetic resonance imaging and laser devices are two important applications of light and its interaction with matter (7.6, 7.7).

SOCIETAL IMPACT

Light is a fundamental part of our lives; by it, we see everything that we see. Sunlight is what keeps the Earth alive and is our ultimate energy source. Human eyes can see a narrow band of light called visible light, but humans use many other wavelengths for various purposes: X-rays and gamma rays are used in medicine, infrared light is used in night vision technology, microwaves are used in cooking, and radio waves are used in communication (7.3).

The human eye evolved to see not only different intensities of light (black-and-white vision) but different colors as well. The colors that we see depend on the interaction of light with the molecules or atoms within the thing we are looking at (7.3).

Just as we can identify some things by their color in the visible region of the spectrum, so scientists can identify substances by their "color" in other parts of the spectrum; this is called spectroscopy (7.5). Spectroscopy is used to make measurements important to society. For example, ozone levels in the upper atmosphere are monitored by spectroscopy, and magnetic resonance imaging (MRI) is a form of spectroscopy by which doctors image internal organs (7.6). The development of technology such as MRI is extremely beneficial to humans, but it also raises difficult questions—who should benefit from that technology? Does the high cost of technology leave some people unable to afford it? Is that fair?

The ability to make concentrated and pure forms of light with lasers has also impacted society (7.7). From CD players to supermarket scanners to laser-guided bombs, lasers have changed the way we live in the 40 years that have passed since their discovery.

Chemistry on the Web

For up-to-date URLs, visit the text website at www.chemistry.brookscole.com/tro3

- Color in Light
  http://science.howstuffworks.com/question41.htm
- Wave-Particle Duality
  http://nobelprize.org/physics/articles/ekspong/
- The Electromagnetic Spectrum
  http://imagine.gsfc.nasa.gov/docs/science/know_l1/emspectrum.html
- Ultraviolet Light
  http://www.srrb.noaa.gov/UV/
  http://www.epa.gov/sunwise/uvradiation.html
- X-rays
  http://health.howstuffworks.com/x-ray.htm
  http://nobelprize.org/physics/laureates/1901/rontgen-bio.html
- MRI
  http://electronics.howstuffworks.com/mri.htm
- Watching Molecules Dance
  http://nobelprize.org/chemistry/
  http://www.ca.sandia.gov/laserchemistry/research-interest.html
  http://chemistry.anl.gov/photochemistry/

Key Terms

absorption spectrum	fluorescence	magnetic field	radio waves
carotenes	frequency	magnetic resonance imaging (MRI)	relaxation time
chlorophylls	gamma rays		resonance frequency
electric field	Hertz, Heinrich	microwave radiation	Roentgen, Wilhelm
electromagnetic radiation	infrared (IR) light	nuclear magnetic resonance (NMR)	spectroscopy
electronic relaxation	laser	phosphorescence	ultraviolet (UV) light
electronic transition	laser cavity	photodecomposition	visible light
energy state	lasing medium	photons	wavelength
excited state			X-rays

Exercises

Assess your understanding of this chapter's topics with an online chapter quiz at **http://chemistry.brookscole.com/tro3**.

QUESTIONS

1. List the colors present in white light.
2. What kinds of molecules are responsible for the green color in spring leaves and the red color in fall leaves?
3. Describe electric and magnetic fields.
4. What is light? How fast does it travel? Make a sketch of its electric and magnetic fields.
5. What is the relationship between the wavelength of light and its color (its energy)?
6. List the different types of electromagnetic radiation and their approximate wavelength.
7. What is the function of sunscreen and of UV protection on sunglasses?
8. What prevents large amounts of high-energy UV and X-ray light from reaching the Earth's surface where it can adversely affect humans?
9. What is unique about X-rays?
10. What does the MHz dial on a radio signify?
11. Explain how night vision works. How do microwave ovens work?
12. What are phosphorescence and fluorescence? How do they differ?
13. What is spectroscopy?
14. What is nuclear magnetic resonance (NMR)?
15. How does magnetic resonance imaging (MRI) work?
16. Describe the difference between laser light and ordinary light.
17. Describe how a laser works.
18. Classify each of the following lasers as to type (solid-state, gas, dye, or semiconductor) and list its wavelength (include the region of the spectrum to which each wavelength corresponds):
    a. Nd:YAG
    b. helium–neon
    c. argon ion
    d. carbon dioxide
19. What is unique about a dye laser?
20. Why have semiconductor lasers become common in many consumer applications such as CD players?
21. What are the advantages of using laser light in medicine?
22. Describe the light-related discoveries of each of the following scientists and state their significance:
    a. Newton
    b. Roentgen
    c. Hertz
    d. Bloch and Purcell

PROBLEMS

23. The Sun is 1.5×10^8 km from the Earth. How long does it take light to travel from the sun to the Earth?
24. The speed of sound is 330 m/s. If fireworks are 1 km away, what is the length of the time delay between seeing the fireworks explode and hearing the sound?
25. Make a drawing, such as Figure 3, that shows how a red object interacts with white light to appear red.
26. Make a drawing, such as Figure 3, that shows how a yellow object interacts with white light to appear yellow.
27. Arrange the following three wavelengths of light in order of increasing energy per photon.
    a. 300 nm (n = nano = 10^{-9})
    b. 100 cm
    c. 10 nm
28. Which of the following types of light would be most likely to damage biological molecules? Why?
    a. infrared light
    b. visible light
    c. ultraviolet light

29. Calculate the wavelength of the radio waves used to produce the FM radio signal found at 93.6 megahertz (MHz) on the radio dial.
30. Calculate the wavelength of the radio waves used to produce the AM radio signal found at 1070 kilohertz (kHz) on the radio dial.
31. Cellular phone frequencies are approximately 850 megahertz. What is the wavelength of the radio waves used by cell phones?
32. If an FM radio wave signal has a wavelength of 2.75 m, what is its frequency?
33. Light from a yellow street lamp is analyzed and found to consist of 589 nm light. What element is present within the lamp? (Hint: Use Table 1.)
34. A lamp emits red light of 671 nm. What element is in the lamp?
35. Light from a distant star is analyzed and found to contain high intensities of light at the following wavelengths: 706, 656, 587, and 502 nm. What elements are present in the star?
36. Light from a fluorescent light source is analyzed and found to contain high intensities of light at the following wavelengths: 579 and 546 nm. What elements are present in the light tube?
37. A laser surgeon wants to remove a blue tattoo. What color laser would be appropriate? Why would the surgeon not want to use a blue laser?
38. A surgeon plans to remove a red port-wine stain. What color laser is appropriate?
39. Make a sketch representing the wave nature of light and label the wavelength. If the sketch you just made corresponds to visible light, how would UV light be different? How would infrared light be different?
40. The violet edge of the visible spectrum occurs at a wavelength of approximately 400 nm, and the red edge occurs at approximately 780 nm. By using the scale 1 nm = 0.1 mm, make sketches of the waves associated with these wavelengths. How long would X-ray waves and radio waves be on this scale? Comment on the range of wavelengths visible to our eyes.
41. Of the following types of light—microwave, infrared, ultraviolet, and X-ray—which is most energetic and which is least energetic?
42. Of the following types of light—radio waves, microwaves, gamma rays, and visible—which one has the longest wavelength? Which one has the highest frequency?

POINTS TO PONDER

43. Describe how white light interacts with a colored substance to make it look colored.
44. How would our lives change if our eyes could see infrared light as well as visible light?
45. A look at the entire spectrum of light reveals that we can see only a tiny fraction of wavelengths. What are the implications of this in terms of our picture of reality?
46. Explain why some clothes glow under black light and others do not.
47. MRI has been able to trace specific brain functions such as emotions and memory to specific locations on the brain where neurons are firing. How does this affect your impression of deep emotions such as love? Is it technically correct to say that you love someone with all your heart?
48. Read the What If . . . X-rays—Dangerous or Helpful? box in this chapter. Like X-rays, gamma rays are used in medicine in both diagnosis and treatment. One of its uses is in cancer therapy to irradiate and destroy tumors. However, the dose usually required is fairly high, so high that it increases the risk of future cancers by about 1%. Explain why this is justified.
49. Read the What If . . . The Mind–Body Problem box in this chapter. What is the mind–body problem? How does explaining free will become difficult under a purely physical description of human beings?

Feature Problems and Projects

50. The following drawing shows white light passing through a clear, colorless liquid. What is wrong with the picture? What color should the liquid be?

51. The following shows two pictures of the same scene, one taken in the light with ordinary film and one taken in the dark with infrared imaging. Explain the differences between the two pictures. Why do some objects appear brighter in one picture than in the other?

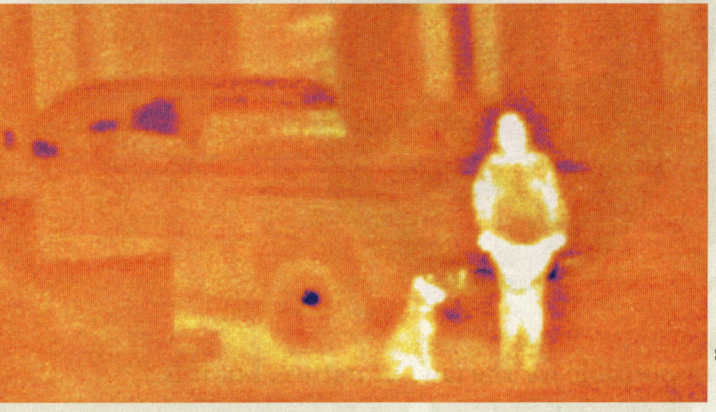

52. Read the What If . . . The Cost of Technology box in this chapter. The rich have always had better access to health care than the poor. Some argue that technology makes that gap wider. Write a short essay stating your position on this problem.

8 Nuclear Chemistry

The scientist is not responsible for the laws of nature, but it is a scientist's job to find out how these laws operate.... It is not the scientist's job to determine whether a hydrogen bomb should be used. This responsibility rests with the American people and their chosen representatives.

—J. R. Oppenheimer

In this chapter, we will see how changes in an atom's nucleus produce radiation, the emission of energetic particles. The study of radiation, and the processes that produce it, are called nuclear chemistry and have led to discoveries ranging from thermonuclear bombs to cancer therapy. As you read this chapter, think about the identity of elements—what defines an element's identity? What if, through nuclear processes, that identity could change? Unlike normal chemistry, where elements always retain their identity, nuclear chemistry often results in one element changing into another one.

The discovery of nuclear chemistry has had profound impacts on society. Since the testing of the first atomic bomb in 1945, humans have lived under the threat of possible nuclear annihilation. During the Cold War years, the United States and the Soviet Union built over 100,000 nuclear weapons, enough to destroy every major city in the world. Today, many of these weapons have been disarmed, and nuclear arsenals are much smaller.

Other major issues facing our society are related to power generation. Nuclear power still provides a way to generate electricity without many of the problems associated with the combustion of fossil fuels. In a world faced with fossil-fuel-related problems, including their inevitable depletion and ever increasing cost, nuclear electricity generation remains a viable alternative. Yet public perception of nuclear power has halted the growth of nuclear electricity generation in the United States. What are the benefits of nuclear power? What are the risks? Are the benefits worth the risks?

QUESTIONS FOR THOUGHT

- What is nuclear radiation?
- How was radiation discovered?
- What are the different kinds of radiation?
- What is nuclear decay? How long does it take?
- The splitting of atoms is called nuclear fission. How does it occur?
- What events led to the development of the first atomic bomb?
- How is nuclear power harnessed to generate electricity?
- What are the benefits of nuclear power? What are the risks?
- How does radiation affect humans?
- How are nuclear processes used to estimate the age of fossils and other artifacts?
- How are nuclear processes used to estimate the age of rocks and the age of the Earth?
- How is radiation used in medicine to heal people?

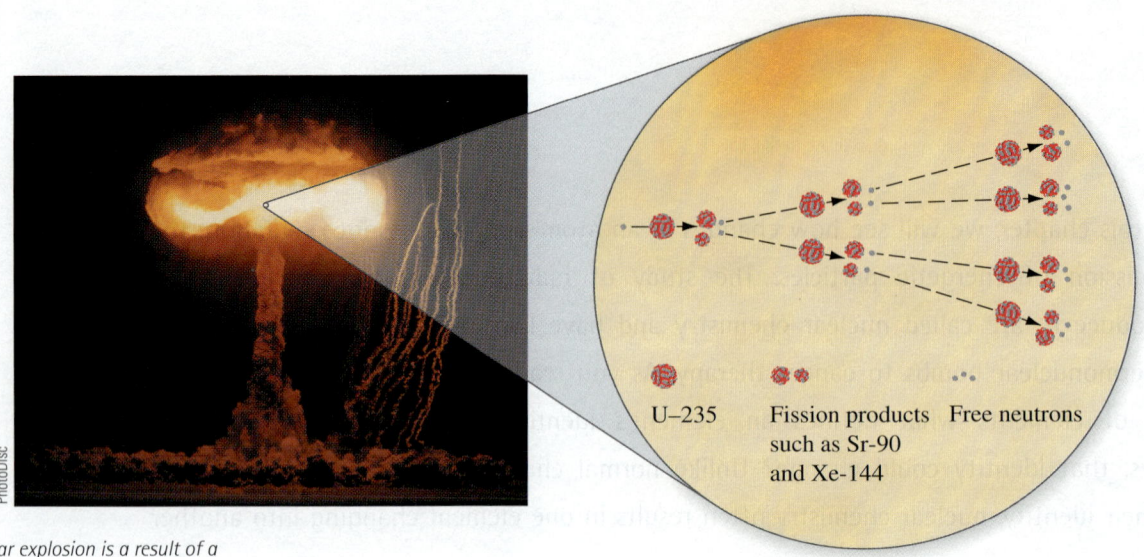

A nuclear explosion is a result of a chain reaction in which the splitting of one atomic nucleus causes the splitting of two more, which then causes the splitting of four more, and so on.

U–235   Fission products such as Sr-90 and Xe-144   Free neutrons

 Explore this topic on the **Chernobyl Health Effects Studies** website.

V. I. Lenin nuclear power plant in Chernobyl, USSR.

8.1 A Tragedy

At 1:24 A.M. on April 26, 1986, observers outside reactor 4 of the V. I. Lenin nuclear power plant in Chernobyl, USSR, heard two distinct explosions as the power in the reactor core spiraled to 120 times its normal levels. The explosions blew a hole through the roof of the reactor and sent hot pieces of the core flying throughout the facility. Hours later, a handful of rescue workers climbed onto a nearby structure and gazed through the hole in the roof, looking directly at the exposed reactor core; they died that night from acute radiation poisoning. The accident at Chernobyl was devastating. In the immediate aftermath, 31 people died, 230 were hospitalized, and countless others were exposed to high levels of radiation.

As those unfortunate rescue workers peered at the reactor core, invisible particles pelted them. The particles were emitted by uranium atoms whose nuclei are like popcorn kernels about to pop. When the pop occurs, the nucleus breaks apart, showering pieces of itself (radiation) in all directions. Radiation can pass through skin and destroy or damage important molecules within cells. The total amount of energy absorbed by these workers was less than would be absorbed by them if they had fallen off a chair; yet when delivered by radiation, it proved deadly.

The study of radiation is part of nuclear chemistry. In contrast to ordinary chemistry, which involves changes with the electrons of atoms and molecules, nuclear chemistry involves changes within the *nuclei* of atoms. When these changes occur, the nuclei emit energetic particles that we call radiation. We will see how the chance discovery of radiation eventually led to the development of the atomic and hydrogen bombs. We will also study several peacetime applications of radiation such as nuclear power and nuclear medicine.

8.2 An Accidental Discovery

The somewhat fortuitous discovery of nuclear radiation in 1896 was followed by developments in technology that culminated in the detonation of the first atomic bomb in Alamogordo, New Mexico, on July 16, 1945. In no other scientific discovery had mankind unleashed such enormous power. In no other scientific endeavor had scientists struggled so desperately with their creation. Yet their progress could not be reversed; the laws of nature, once learned, were not easily forgotten.

The story begins in Paris near the end of the 19th century with a physicist named **Antoine-Henri Becquerel** (1852–1908). Becquerel, a professor at the Paris Polytechnic Institute, had developed an interest in the newly discovered X-rays. He hypothesized that the production of X-rays was related to phosphorescence, the emission of visible light by some substances after exposure to ultraviolet light. To test his hypothesis, Becquerel chose a uranium salt crystal known to phosphoresce. He would irradiate the crystals with UV light from the Sun and look for the subsequent production of X-rays and phosphorescence.

Becquerel conducted his experiment by wrapping a photographic plate in black paper to prevent exposure to sunlight (Figure 1). He placed the uranium-containing crystals on the paper and took the whole assembly outside, allowing sunlight—which contains some UV light—to shine on the crystals, inducing phosphorescence. X-rays, if they were being produced with phosphorescence, would penetrate the black paper and expose the underlying photographic plate. On developing the plate, Becquerel was delighted to find that the film was exposed, and he wrongly concluded that phosphorescence and X-rays were related.

Like most scientists, Becquerel repeated his experiments several times to confirm the results. However, some bad weather interfered with his plans, and he was forced to delay his experiment. While waiting for sunshine, Becquerel stored his experimental materials, a freshly wrapped photographic plate and some uranium crystals, in his desk drawer. After several days with no sunshine, Becquerel developed the plate anyway. Perhaps a faint exposure would be seen due to ambient UV light. To his surprise, the film was just as exposed as in previous experiments. Becquerel wrote, "I particularly emphasize the following observation, which seems to me extremely important and quite unexpected: The same uranium crystals, placed on photographic plates as before, with the same paper wrappings, but shielded from incident radiations and kept in the dark, produced the same photographic impressions." Becquerel quickly concluded that the exposed plate had no connection with UV light or phosphorescence but came from the uranium salt crystals themselves. He knew he had discovered a new phenomenon and called these emissions uranic rays.

Becquerel's work was continued by a young Polish graduate student in Paris named **Marie Sklodowska Curie** (1867–1934). Curie chose the study of uranic rays as the topic for her doctoral dissertation. Rather than focus on the rays themselves, as Becquerel had, she began to search for other substances that might emit uranic rays. She was a brilliant chemist and, with the help of her husband, **Pierre Curie** (1859–1906), she found two other substances that emitted these rays, one of which was a previously undiscovered element. She writes, "We therefore think that the substance that we have extracted . . . contains a metal previously unknown. We propose to call it polonium, after the native land of one of us." Since the Curies discovered that uranic rays were not unique to uranium, they changed the name from uranic rays to ray activity or radioactivity. By further experimentation, they deduced that radioactivity was not a product of a chemical reaction, but instead, a product resulting from changes within the atom itself.

Phosphorescence is first discussed in Section 7.4.

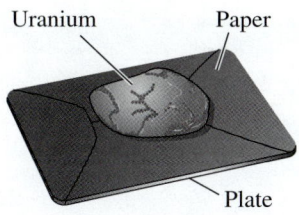

FIGURE 1 Becquerel wrapped a photographic plate in black paper and placed uranium-containing crystals on top of the black paper. If the uranium emitted X-rays, they would pass through paper and expose the plate.

Pierre and Marie Curie in their laboratory.

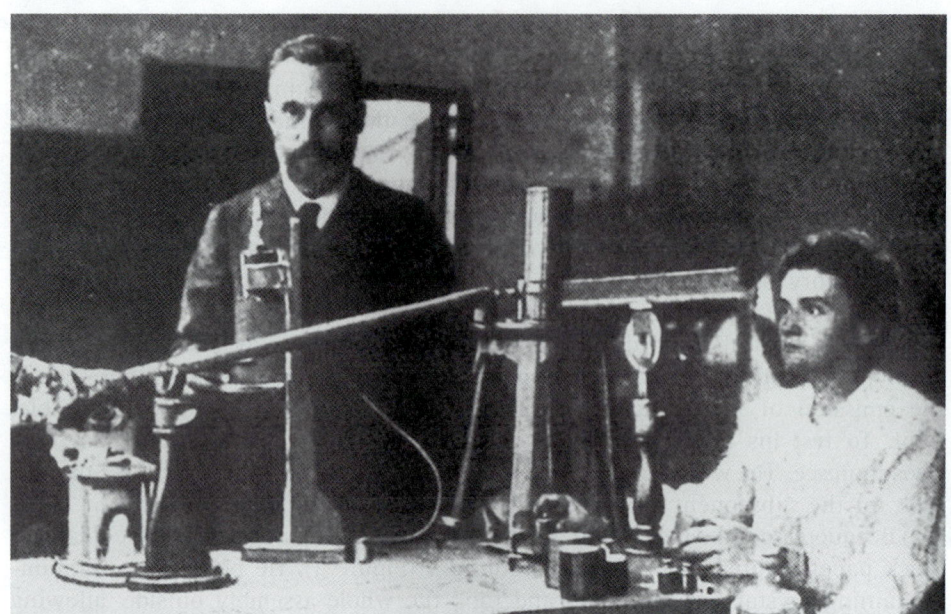

In 1898, the Curies discovered a second new element that they called radium due to its extreme radioactivity. They wrote, "The radioactivity of radium must be enormous . . . 900 times that of uranium." In fact, pure radium is so radioactive that it spontaneously glows. In 1903, Marie was awarded her Ph.D. and within a few months shared the Nobel Prize in physics with her husband Pierre and Henri Becquerel for the discovery of radioactivity. In 1911, Marie was awarded a second Nobel Prize, this time in chemistry, for her discovery of the two new elements radium and polonium. She was the first person to win two Nobel Prizes. As a further tribute to this amazing family, the Curies' daughter, Irene, shared a Nobel Prize in physics in 1935 with her husband, Frederick Joliot, for her work in radioactivity.

8.3 Radioactivity

Radioactivity itself was characterized by Ernest Rutherford around the turn of the century. Rutherford and others discovered that radioactivity was the result of nuclear instability; many nuclei, especially the heavy ones, are unstable and decay to attain stability, releasing parts of their nucleus in the process. These emitted particles were the rays that Becquerel and the Curies detected. There are three primary types of radiation emitted by decaying nuclei: alpha, beta, and gamma radiation.

ALPHA (α) RADIATION

An alpha particle consists of two protons and two neutrons—a helium nucleus—emitted by a decaying atomic nucleus. Using the notation developed in Chapter 3, we represent an alpha particle with the symbol $^4_2\text{He}^{2+}$. Because nuclear chemistry deals only with changes in the nucleus, we can ignore the electrons and represent an alpha particle more simply as ^4_2He. When alpha particles collide with matter, they generate a large number of ions and therefore can damage biological molecules. The ability of radiation to create ions in matter is called **ionizing power**. However, because of their relatively large size, alpha particles have low penetrating power; they do not get very far in air and are stopped with a sheet of ordinary paper. Alpha particles are the semi-trucks of radioactivity—they do a lot of damage in a collision but don't get very far in a traffic jam.

We represent radioactive decay with a **nuclear equation** that shows the symbol for the initial isotope on the left and the symbols for the products of the decay on the right. For example, U-238 decays via alpha emission to produce Th-234 as follows:

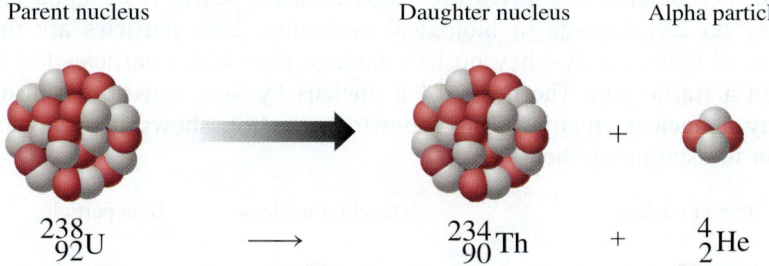

Parent nucleus   Daughter nucleus   Alpha particle

$$^{238}_{92}U \longrightarrow \, ^{234}_{90}Th \, + \, ^{4}_{2}He$$

The U-238 atom emits part of its nucleus, two protons and two neutrons, to form a thorium atom. Like chemical equations, nuclear equations must be balanced; the numbers of protons and neutrons on both sides of the equation must be equal. This equation is balanced because the sum of the atomic numbers on the right, 90 + 2, is equal to the atomic number on the left, 92. Likewise, the sum of the mass numbers on the right, 234 + 4, equals the mass number on the left, 238.

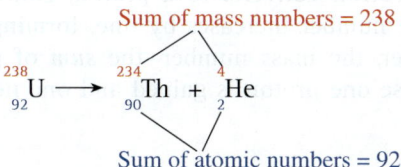

Sum of mass numbers = 238

$$^{238}_{92}U \longrightarrow \, ^{234}_{90}Th \, + \, ^{4}_{2}He$$

Sum of atomic numbers = 92

EXAMPLE 8.1

Writing Nuclear Equations for Alpha Decay

Write a nuclear equation to represent the alpha decay of Th-230.

SOLUTION

We write an equation showing the symbol for Th-230 ($^{230}_{90}$Th) on the left and the symbol for an alpha particle ($^{4}_{2}$He) on the right:

$$^{230}_{90}Th \longrightarrow ? + \, ^{4}_{2}He$$

The isotope that thorium decays *to* can be determined by calculating what atomic number and mass number will make both sides of the equation balanced. The atomic number of the product must be 88 and the mass number must be 226. The element with atomic number 88 is Ra; we write:

$$^{230}_{90}Th \longrightarrow \, ^{226}_{88}Ra \, + \, ^{4}_{2}He$$

Note that the sum of atomic numbers (90) is the same on both sides of the equation and that the sum of mass numbers (230) is the same on both sides.

YOUR TURN

Writing Nuclear Equations for Alpha Decay

Write a nuclear equation to represent the alpha decay of Ra-226.

BETA (β) RADIATION

A beta particle is an energetic electron emitted by an atomic nucleus and represented by the symbol $_{-1}^{0}e$. Because an electron is smaller than a helium nucleus, beta particles penetrate matter more easily than alpha particles, passing right through a sheet of paper. They require a sheet of metal or a book to stop them. However, their smaller size gives them less ionizing power than alpha particles; thus, they do less damage to biological molecules. Beta particles are the mid-sized cars of radioactivity—they do less damage than alpha particles but can get further in a traffic jam. The decay of a nucleus by beta emission is also represented by a nuclear equation. The following equation shows the conversion of polonium to astatine via beta decay:

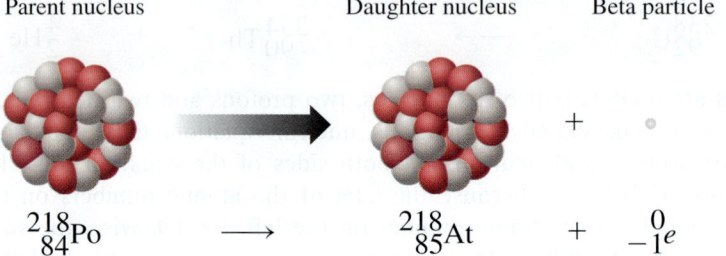

Parent nucleus		Daughter nucleus		Beta particle
$_{84}^{218}Po$	$\rightarrow$	$_{85}^{218}At$	$+$	$_{-1}^{0}e$

In beta decay, a neutron converts to a proton, emitting an electron in the process, and the atomic number increases by one, forming an element of higher atomic number. However, the mass number—the *sum* of protons and neutrons—remains constant because one proton is gained and one neutron is lost.

EXAMPLE 8.2

Writing Nuclear Equations for Beta Decay

Write an equation to represent the radioactive decay of Th-234 by beta emission.

SOLUTION

We write an equation showing the symbol for Th-234 ($_{90}^{234}Th$) on the left and the symbol for a beta particle ($_{-1}^{0}e$) on the right:

$$_{90}^{234}Th \rightarrow ? + _{-1}^{0}e$$

The isotope that thorium decays *to* can be determined by calculating what atomic number and mass number will make both sides of the equation balanced. Because a beta particle has a mass number of zero, the product will have the same mass number as Th-234. To determine the correct atomic number of the product, notice that beta particles have a *negative* atomic number. Therefore, the atomic number that makes both sides balanced is 91. Because the element with atomic number 91 is Pa, we write:

$$_{90}^{234}Th \rightarrow _{91}^{234}Pa + _{-1}^{0}e$$

Note that the sum of atomic numbers (90) is the same on both sides of the equation and that the sum of mass numbers (234) is the same on both sides.

YOUR TURN

Writing Nuclear Equations for Beta Decay

Write an equation to represent the radioactive decay of Pb-214 by beta emission.

TABLE 1				
Types of Radiation				
Name	Description	Symbol	Ionizing Power	Penetrating Power
Alpha (α)	He nucleus	^4_2He	High	Low
Beta (β)	Electron	$^0_{-1}e$	Moderate	Moderate
Gamma (γ)	High-energy photon	$^0_0\gamma$	Low	High

GAMMA (γ) RADIATION

A gamma ray is an energetic photon emitted by an atomic nucleus and represented by the symbol $^0_0\gamma$. Gamma rays are fundamentally different from alpha and beta rays because they are electromagnetic radiation, not matter (Table 1). Gamma rays are the motorcycles of radioactivity—they have the highest penetrating power, requiring up to several inches of lead to stop them, but the lowest ionizing power. Gamma rays are usually emitted in conjunction with other types of radiation.

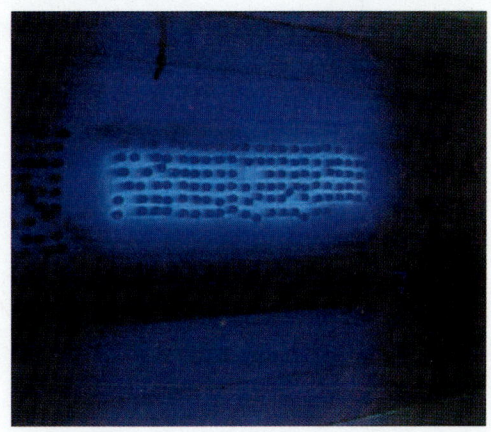

A sample of cesium-137, a gamma emitter used in cancer treatment.

APPLY YOUR KNOWLEDGE

A laboratory contains radioactive substances emitting alpha, beta, and gamma particles. If the lab were not well shielded, which of the three types of particles would you be most likely to detect in the building next door?

Answer: Gamma particles, because they have the highest penetrating power and are therefore most likely to pass through the walls of buildings.

8.4 Half-Life

Bismuth, atomic number 83, contains the heaviest stable nucleus. All elements heavier than bismuth have unstable nuclei and decay radioactively, some faster, some slower. In addition, some isotopes of elements with atomic number lower than 83 are also unstable and undergo radioactive decay. Beginning with U-238, the heaviest naturally occurring element, we can construct a radioactive decay

FIGURE 2 A natural radioactive decay series. Alpha decay is shown by arrows that go down and to the left. Beta decay is shown by arrows to the right. The half-life is shown for each decay process.

series that eventually ends in lead (Figure 2). One of the intermediates in this decay series is radon, a gas. Consequently, many regions with uranium deposits in the ground, including many parts of the United States, also have small amounts of radioactive radon gas in the soil and surrounding air. This radon can

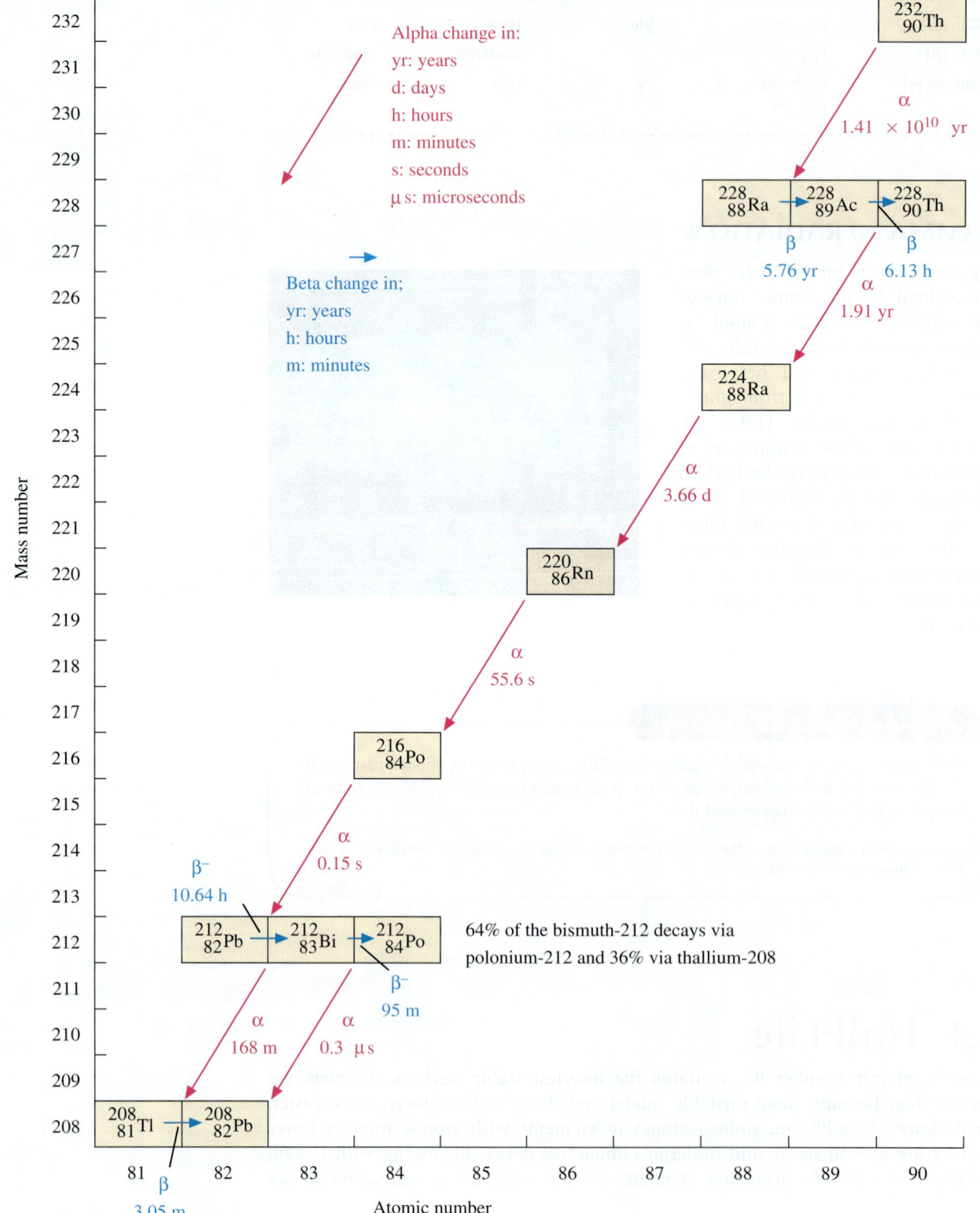

seep into homes and is a potential health hazard. We will examine this hazard in detail in Section 8.10.

The rate of decay for radioactive elements is specified by their half-life (Table 2). The **half-life** of a radioactive element is the time required for half the nuclei in a sample to decay. For example, Th-234, a beta emitter discussed earlier, has a half-life of 24.1 days, meaning that half of all nuclei in a particular sample decay to Pa-234 in 24.1 days. If we start with 1000 Th-234 atoms, there are 500 Th-234 atoms left after 24.1 days; the other 500 atoms have decayed to the daughter element Pa-234. After 24.1 *more* days (48.2 days total), 250 Th-234 atoms are left (Figure 3). U-238, in contrast, has a much longer half-life of 4.5 billion years. All the U-238 on the planet is undergoing radioactive decay; however, the slow rate means it will be around for many years.

TABLE 2
Representative Half-Lives

Isotope	Half-Life
C-14	5730 yr
P-32	14.3 day
K-40	1.25×10^9 yr
Po-214	1.64×10^{-4} s
Rn-222	3.82 day
Ra-226	1.60×10^3 yr
U-238	4.51×10^9 yr

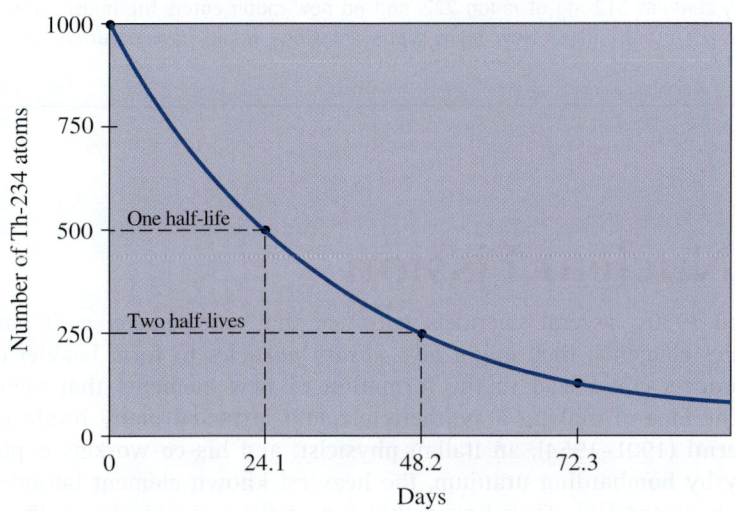

FIGURE 3 Radioactive decay of Th-234. The number of atoms decreases to one-half its original value with each half-life.

EXAMPLE 8.3

Half-Life

Radium-226 undergoes alpha decay to radon-222 with a half-life of 1.60×10^3 yr. A sample of radium ore initially contains 275 g of radium-226. (a) How many grams of Ra-226 are left in the sample after 4.80×10^3 yr? (b) How many alpha emissions would have occurred in this time? (Hint: The number of emissions equals the number of atoms that decayed.)

SOLUTION

a. To determine the number of half-lives that have passed, divide the total time by the time of one half-life as follows:

4.80×10^3 yr/1.60×10^3 yr = 3.00

The amount of Ra-226 falls to one-half its original value during each half-life:

initial amount:	275 g
after one half-life:	275 g/2 = 138 g
after two half-lives:	138 g/2 = 68.7 g
after three half-lives:	68.7 g/2 = 34.4 g

There will be 34.4 g of Ra-226 in the sample after 4.80×10^3 yr.

b. The total number of grams of Ra-226 that decayed is

275 g − 34.4 g = 241 g

The total number of Ra-226 atoms that decayed is

$$241 \text{ g} \times \frac{1 \text{ mol}}{226 \text{ g}} \times 6.022 \times 10^{23} \frac{\text{atoms}}{1 \text{ mol}} = 6.42 \times 10^{23} \text{ atoms}$$

Because the decay of each atom produces one alpha emission, there were 6.42×10^{23} alpha emissions.

YOUR TURN

Half-Life

Radon-222 decays via alpha emission to Po-218 with a half-life of 3.82 days. If a house initially contains 512 mg of radon 222, and no new radon enters the house, how much will be left in 19.1 days? How many alpha emissions would have occurred within the house?

8.5 Nuclear Fission

Enrico Fermi

In the mid-1930s, several scientists reasoned that since nuclei *emit particles* to form lighter elements, they might also *absorb particles* to form heavier elements. Such a process could lead to the formation of new elements that never before existed. The idea of making a synthetic element attracted many bright scientists. **Enrico Fermi** (1901–1954), an Italian physicist, and his co-workers explored this possibility by bombarding uranium, the heaviest known element (atomic number 92), with neutrons (^1_0n). They hoped that the nucleus would absorb the neutrons and then undergo a beta decay, transforming the neutron into a proton and forming a synthetic element with atomic number 93.

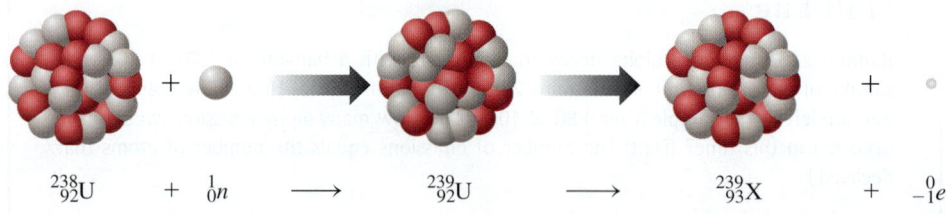

$$^{238}_{92}\text{U} + ^{1}_{0}n \longrightarrow ^{239}_{92}\text{U} \longrightarrow ^{239}_{93}\text{X} + ^{0}_{-1}e$$

To their delight, they detected beta emission following neutron bombardment of uranium and thought they had perhaps synthesized the first man-made element.

Experiments carried out in Germany during the following years by chemists **Otto Hahn** (1879–1968), **Lise Meitner** (1878–1968), and **Fritz Strassmann** (1902–1980) seemed to confirm Fermi's work. However, that changed in late 1938, one year before the outbreak of World War II.

In collaboration with Meitner, exiled from Germany because she was Jewish, Hahn and Strassmann reported that careful chemical examination of the products formed after bombardment of uranium with neutrons did not reveal a heavier element, but rather two lighter ones. Previously observed nuclear processes had

Otto Hahn and Lise Meitner, two of the discoverers of nuclear fission.

always been incremental, and Hahn puzzled over the results: Could they be witnessing the splitting of an atom in two? Hahn wrote, "As nuclear chemists . . . we have not yet been able to take this leap, which contradicts all previous experience in nuclear physics." Nevertheless, leap they did. Not only was the splitting—or **fission**—of uranium atoms occurring on absorption of neutrons, but large amounts of energy were being emitted in the process.

A nuclear equation describing one fission reaction is as follows:

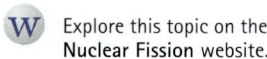
Explore this topic on the **Nuclear Fission** website.

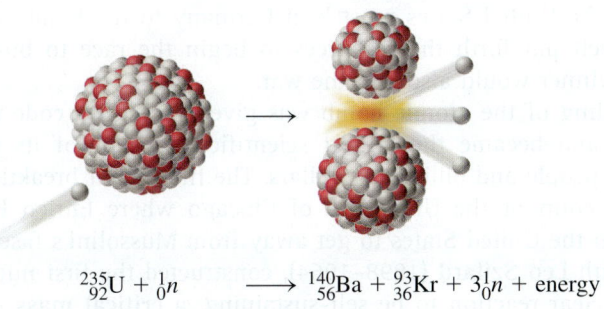

$$^{235}_{92}\text{U} + ^{1}_{0}n \longrightarrow ^{140}_{56}\text{Ba} + ^{93}_{36}\text{Kr} + 3^{1}_{0}n + \text{energy}$$

The fission reaction shown here produces three neutrons. Other fission reactions produce only two. In a sample of U-235, an average of about 2.5 neutrons are produced per fission event.

Other reactions with the same characteristics—the splitting of uranium producing lighter elements, neutrons, and energy—were also occurring. In the preceding reaction, the isotope undergoing fission is U-235. Most naturally occurring uranium, however, is U-238. Only the uranium isotope with mass number 235 (<1% of naturally occurring uranium) undergoes fission.

Within weeks, experiments demonstrated that fission could be the basis for a bomb of inconceivable power. The trick would be to get a large amount of uranium-235 atoms to undergo fission at the same time in a chain reaction. The fission of one uranium nucleus, producing neutrons, would induce fission in other uranium nuclei resulting in a continuous escalation (Figure 4). The amount of energy emitted by fission—1 million times more than a chemical reaction—would be violently released in an explosion never before imagined.

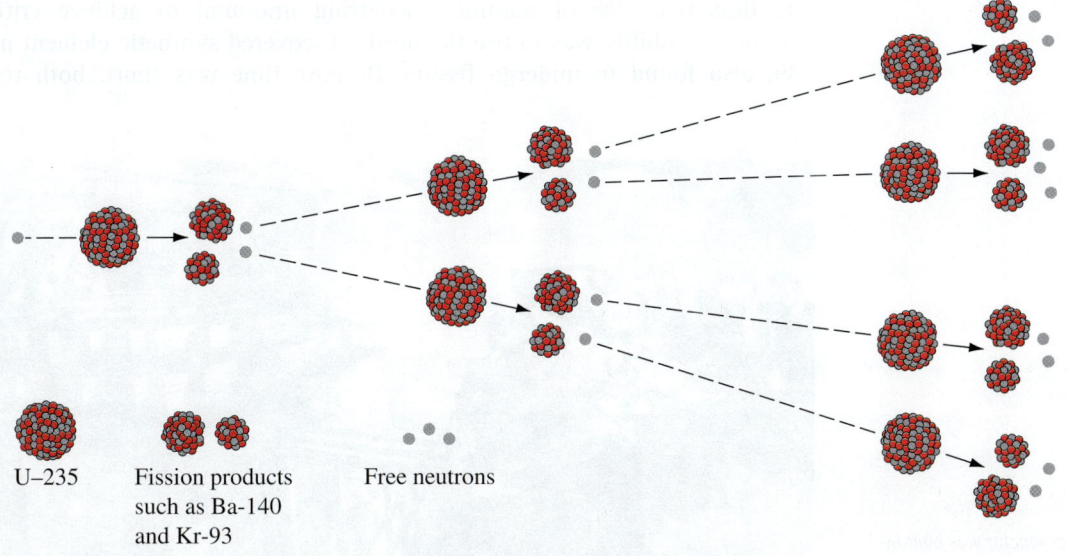

FIGURE 4 In a nuclear chain reaction, the fission from one nucleus produces fission in two or more others, which in turn produces fission in four others, resulting in an escalation of the reaction.

8.6 The Manhattan Project

Several American scientists feared that Nazi Germany, where fission was discovered, would develop a fission bomb. They persuaded **Albert Einstein** (1879–1955), the most famous scientist of the time, to write a letter to President Franklin Roosevelt warning him of this possibility. Einstein obliged and wrote, "This new phenomenon [fission] could also lead to the construction of bombs . . . extremely powerful bombs of a new type." Roosevelt was convinced by Einstein and decided that the United States must beat Germany to the bomb. On December 6, 1941, Roosevelt put forth the resources to begin the race to build the bomb—a race whose winner would also win the war.

The building of the atomic bomb was given the secret code name "Manhattan Project" and became the largest scientific endeavor of its time, involving thousands of people and billions of dollars. The first major breakthrough occurred on a squash court at the University of Chicago where Enrico Fermi, who had immigrated to the United States to get away from Mussolini's fascist Italy, in collaboration with **Leo Szilard** (1898–1964), constructed the first nuclear reactor.

For a nuclear reaction to be self-sustaining, a **critical mass** of uranium-235 is necessary. Lesser amounts of uranium will not undergo self-sustained fission because too many neutrons are lost to the surroundings instead of being absorbed by other U-235 nuclei. By using a pile of uranium and graphite, Fermi and Szilard achieved a critical mass, resulting in a self-sustaining fission reaction lasting 4.5 minutes. Uranium underwent *controlled fission*, the splitting of some atoms inducing the splitting of others, but never escalated out of control. Just in case their calculations were wrong, several junior researchers stood by with jugs of cadmium solution, a neutron absorber, to quench the reaction in case the first nuclear reactor turned into the first nuclear disaster. The cadmium solutions were not needed. The reaction worked as planned; the next step was to construct a device where the fission reaction would spiral upward exponentially—a bomb.

The main effort was in a top-secret research facility in the desert at Los Alamos, New Mexico, under the direction of **J. R. Oppenheimer** (1904–1967). A major problem in the construction of the bomb was obtaining enough uranium-235 (less than 1% of naturally occurring uranium) to achieve critical mass. Another possibility was to use the newly discovered synthetic element plutonium-239, also found to undergo fission. Because time was short, both routes were

The first nuclear reactor was built by nuclear physicists Enrico Fermi and Leo Szilard on a squash court at the University of Chicago.

The heart of the effort to build the bomb was at Los Alamos, New Mexico, under the direction of J. R. Oppenheimer.

simultaneously pursued. A plutonium manufacturing plant was constructed in Hanford, Washington, and a uranium processing plant was built in Oak Ridge, Tennessee, in hopes that one of the avenues would succeed.

At Los Alamos, America's best scientists gathered to design the bomb. The main difficulty was to achieve a critical mass of either U-235 or Pu-239 that would not blow itself apart and result in a fizzle. Experiments, called "tickling the dragon," were conducted in which subcritical masses of uranium or plutonium were momentarily combined, resulting in an instant of criticality. The results of these experiments were then used to determine the best design for the bomb. Two designs emerged: one using uranium and the other plutonium. The uranium design consisted of two subcritical masses of uranium at either end of a steel, cannonlike barrel. A conventional explosive accelerated one of these masses into the other one, resulting in criticality (Figure 5). The design for the plutonium bomb was more complicated. A subcritical mass of plutonium was to be squeezed into criticality by

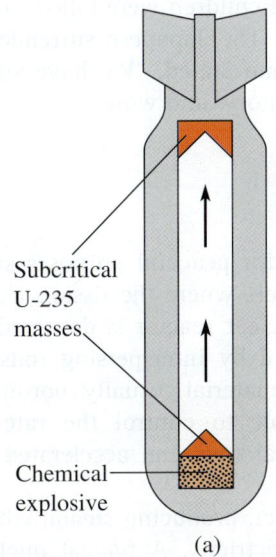
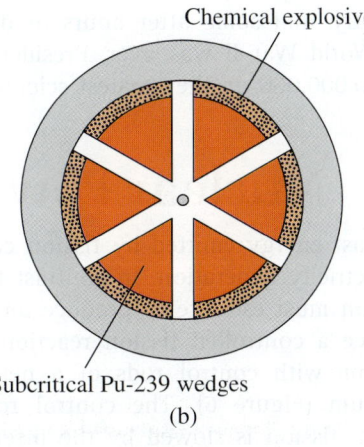

FIGURE 5 (a) Little Boy. Conventional explosives accelerate a subcritical slug of uranium into another subcritical slug. Once together, the two slugs exceed critical mass and fission is initiated. (b) Fat Man. Several conventional explosives compress a sphere of plutonium to criticality.

What if...

The Ethics of Science

Manhattan Project scientists struggled with the ethical implications of their work. They all knew that the result of their project had the potential to kill hundreds of thousands of people. However, many saw the creation of the bomb as inevitable and rationalized that it would be better in American hands than in Hitler's. C. P. Snow wrote, "If it is not made in America this year, it may be next year in Germany." The thought of Nazi Germany beating the United States to the atomic bomb kept many scientists at work. Even Oppenheimer himself struggled with the morality of the bomb, justifying his work by saying, "When you see something technically sweet, you go ahead and do it and you argue about what to do about it only after you have had your technical successes."

QUESTION: Do you agree with Oppenheimer? Should scientists pursue basic knowledge regardless of what the applications of that knowledge might be? Why or why not?

J. R. Oppenheimer, scientific director of the Manhattan Project.

an implosion; several conventional explosives were arrayed around a plutonium sphere and simultaneously detonated. The resulting shock wave was focused to compress the plutonium and achieve criticality.

Bombs using each of the designs were constructed, and on July 16, 1945, at the Trinity test site near Alamogordo, New Mexico, the world's first atomic bomb was tested. With the explosive force of 18,000 tons of dynamite, the test of a plutonium bomb ushered in a new age. Those who witnessed it were left awed. Oppenheimer said, "I am become death, the destroyer of worlds." Within weeks, two atomic bombs, Little Boy and Fat Man, were dropped on Hiroshima and Nagasaki, Japan. More than 100,000 men, women, and children were killed, some instantly, and some after hours or days of suffering. The Japanese surrendered and World War II was over. President Harry S. Truman stated, "We have spent $2,000,000,000 on the greatest scientific gamble in history, and won."

8.7 Nuclear Power

The vast energy emitted by fission can also be used for peaceful purposes such as electricity generation. In contrast to an atomic bomb, where the fission chain reaction must escalate to produce an explosion, a nuclear reactor is designed to produce a controlled fission reaction. This is achieved by interspersing rods of uranium with control rods of a neutron-absorbing material, usually boron or cadmium (Figure 6). The control rods are retractable to control the rate of fission; fission is slowed by the insertion of additional rods and accelerated by the retraction of rods.

The heat generated by controlled fission boils water, producing steam, which then turns the turbine on a generator to produce electricity. A typical nuclear power plant uses about 100 lb of fuel per day and produces enough electricity for

Loading of fuel rods into a reactor core.

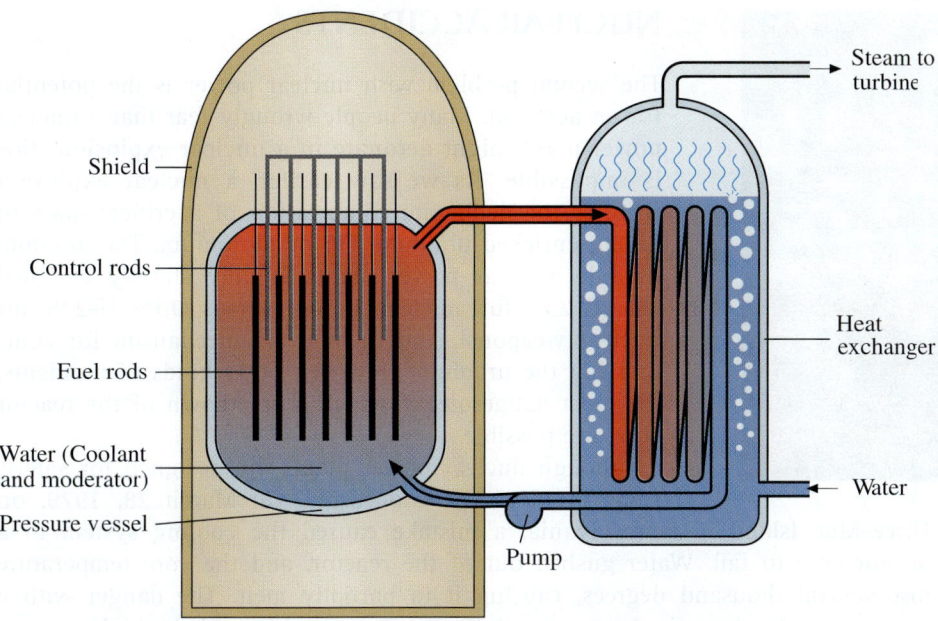

FIGURE 6 Schematic diagram of a nuclear reactor. The heat generated by fission is used to boil water and produce steam that turns the turbine on a generator to produce electricity.

a city of approximately 1 million people. In contrast, a coal-burning power plant requires 5 million lb of fuel to produce the equivalent amount of energy. Nuclear power plants do not have the problems associated with fossil fuels such as air pollution, greenhouse gases, or acid rain (discussed in detail in Chapter 9). However, they do have their own problems, the most significant of which are waste disposal and the potential for accidents.

WASTE DISPOSAL

A major problem with nuclear power is waste disposal. Nuclear fuel is composed of small uranium oxide pellets about the diameter of a pencil. These pellets are piled on top of each other to form long rods that are grouped together to form a fuel assembly. Approximately every 18 months, the fuel assembly must be replaced, leaving the spent fuel as a disposal problem. Most waste from nuclear power production in the United States is currently stored on the site where it is produced. The spent fuel rods are put into steel-lined concrete vaults filled with water; the vaults are kept in a temporary storage facility at the power plant.

In 1982, lawmakers passed the Nuclear Waste Policy Act, which established a program to build this country's first underground nuclear waste repository, a permanent disposal site for nuclear waste. In 1987, Yucca Mountain, Nevada, was chosen for study as a potential site. The stable rock formations deep underground combined with sparse population and little rainfall make it an ideal location for the site. Nuclear waste will be encased in several layers of containment material and placed in tunnels drilled out of the rock formations 1000 ft beneath the ground. The storage facility should keep these materials isolated from us and from the environment for the foreseeable future. However, as might be expected, the construction of the facility is controversial, with many opposing even the idea. Construction of the facility was scheduled to begin in 2005; however, political opposition has delayed the project, resulting in an uncertain future for Yucca Mountain.

 Explore this topic on the **Yucca Mountain** website.

Yucca Mountain in Nevada has been chosen as a potential site for a permanent nuclear waste facility. The stable rock formations 1000 ft beneath the surface should keep nuclear waste isolated from the environment for thousands of years.

The nuclear power plant at Three-Mile Island.

NUCLEAR ACCIDENTS

The second problem with nuclear power is the potential for an accident. Many people wrongly fear that a nuclear power plant might detonate in a nuclear explosion; this is impossible. As we saw earlier, a nuclear explosion requires the deliberate compression of a critical mass of highly enriched uranium-235 or plutonium. The uranium used in nuclear power plants is only slightly enriched (3% U-235 for nuclear power versus 90% U-235 for nuclear weapons), and there is no mechanism for compressing the uranium. However, other kinds of accidents, the most dangerous of which is meltdown of the reactor core, are possible.

Although nuclear power plants are designed for safety, some accidents have occurred. On March 28, 1979, on Three-Mile Island in Pennsylvania, a mistake caused the cooling system in a reactor core to fail. Water gushed out of the reactor, and the core temperature rose several thousand degrees, causing it to partially melt. The danger with a runaway nuclear core is that most substances cannot withstand the high temperatures that occur. The term "China syndrome" is often used in an exaggerative sense to describe the overheated core melting through the reactor floor and into the ground, presumably all the way to China. The reactor at Three-Mile Island was brought under control before it melted the reactor floor but not before an explosive and radioactive mixture of hydrogen, krypton, and xenon gases accumulated within the reactor. To avoid an explosion—not a nuclear detonation, but a conventional explosion—reactor operators chose to vent the gases into the atmosphere, releasing some radioactivity. Although the press emphasized the release of this radioactive material as dangerous, it was not significant in comparison to radiation from natural sources. For example, the amount of radiation released in the explosion of Mt. St. Helen's volcano was thousands of times greater.

A second, more devastating, nuclear accident occurred in Chernobyl, USSR, on April 26, 1986. In this incident, reactor operators were conducting an experiment to lower maintenance costs. Many of the reactor core safety features were turned off to conduct the experiment. The experiment failed, and the fission reaction spiraled out of control. The heat that evolved blew the 1000-ton lid off the reactor, and the graphite core began to burn, scattering radioactive debris into the atmosphere. As stated earlier, 31 people died in the immediate aftermath of the accident, 230 people were hospitalized, and countless others were exposed to high levels of radiation.

In contrast, there have been no deaths due to nuclear accidents in the United States, due primarily to superior power plant design. U.S. nuclear power plants have large containment structures made of steel-reinforced concrete that would prevent nuclear debris from escaping in the event of an accident. More importantly, reactor cores in the United States are not made of graphite, a flammable substance. However, accidents such as these, as well as problems with waste disposal, chilled public support for nuclear power in this country. In contrast to Europe and Japan, which are moving forward with nuclear power, there has been a plateau in nuclear power in the United States. However, recent increases in oil prices, as well as a growing realization of the seriousness of the problems associated with fossil fuels, have led to renewed consideration of nuclear power.

8.8 Mass Defect and Nuclear Binding Energy

We have seen that fission releases tremendous amounts of energy that can be used to generate electricity or, if precisely harnessed, to produce an explosion. Where does this energy come from? The source of the energy is related to the mass of atoms. We might expect the overall mass of an atom to simply be the sum of the masses of all its components, its protons, its neutrons, and its electrons. It is not, however. For example, helium-4, composed of two protons (mass = 1.0073 amu each), two neutrons (mass = 1.0087 amu each), and two electrons (mass = 0.0005 amu each), might be expected to have an overall mass of 4.0330 amu. However, the experimentally measured mass is 4.0015 amu. The difference between the experimentally measured mass and the sum of the masses of the protons, neutrons, and electrons is called the **mass defect**. For the helium-4 nucleus, this difference is 0.0315 amu.

The missing mass was converted to energy when helium formed from its constituent protons and neutrons. This energy, which is related to the missing mass by Einstein's equation $E = mc^2$ (E = energy, m = mass, and c = speed of light), is called the **nuclear binding energy** and represents the energy that holds a nucleus together. The nuclear binding energy per nucleon (proton or neutron) is not constant from one element to the next. The highest values are found for elements with a mass number of about 56 (Figure 7). This means that elements with a mass number near 56 have the most stable nuclei. Consequently, the conversion of nuclei with high atomic numbers, such as U, into nuclei with smaller atomic numbers, such as Ba and Kr, gives off large amounts of energy:

$$^{235}_{92}U + ^{1}_{0}n \rightarrow ^{140}_{56}Ba + ^{93}_{36}Kr + 3\,^{1}_{0}n + \text{energy}$$

The products of this reaction have a higher binding energy than the reactants; the difference in binding energy is the source of energy in fission. Because the products have a greater binding energy, they weigh less than the reactants. In nuclear fission, the missing mass is converted to energy according to Einstein's equation $E = mc^2$.

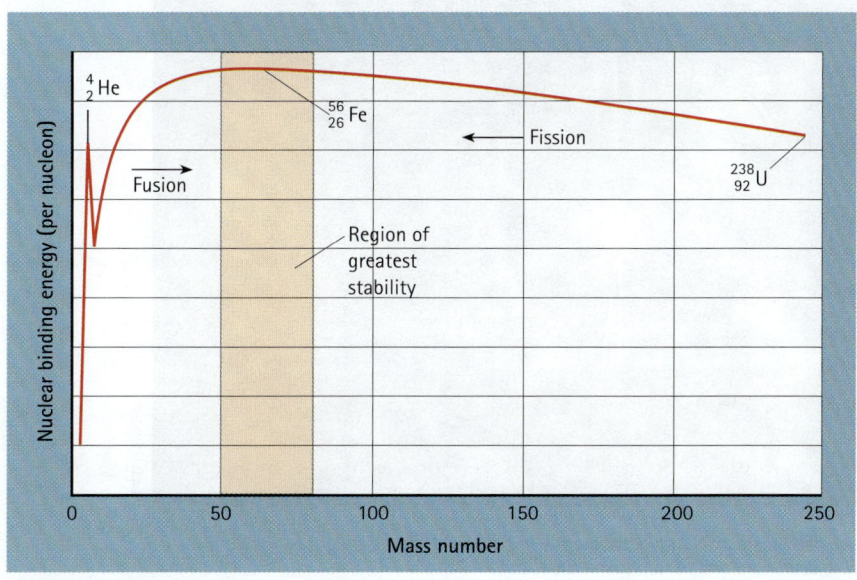

FIGURE 7 Nuclear binding energy (which reflects the stability of a nucleus) maximizes at a mass number of about 56. Consequently, the fission of a heavy nucleus (above mass 56) into lighter ones, or the fusion of lighter nuclei (below mass 56) into heavier ones, results in the emission of energy.

8.9 Fusion

In fission, heavy elements are split into lighter ones. In **fusion**, lighter elements are fused into heavier ones. In both cases, as you can see in Figure 7, the products have higher nuclear binding energies than the reactants and energy is released. Fusion is responsible for the Sun's energy. Fusion releases ten times more energy per gram than fission and is also the basis of modern nuclear weapons. These weapons are called hydrogen bombs and employ the following fusion reaction:

$$^2_1H + ^3_1H \rightarrow ^4_2He + ^1_0n + \text{energy}$$

^2_1H (deuterium) and ^3_1H (tritium) are isotopes of hydrogen. Because fusion requires two positively charged nuclei, which naturally repel each other, to come into contact, extremely high temperatures (millions of degrees) are required for the reaction to proceed. In a hydrogen bomb, a small fission bomb provides the necessary heat. By using this two-stage process, modern nuclear weapons are up to 1000 times more powerful than those dropped on Japan.

THE MOLECULAR REVOLUTION

Fusion Research

Fusion research has been steadily progressing for the past 30 years. The advances in the field have been significant, comparable in magnitude with the advancements in computer disk storage capacity. The first successful fusion experiments in the late 1970s barely produced enough power to run a 100 watt (W) light bulb. Recent experiments at the Joint European Torus (J.E.T.) project in England, however, produced 16 million W of fusion power. The main problem continues to be that all fusion experiments to date consume more power than they produce and are therefore not viable energy sources. The 16 million W produced to date came with the consumption of 24.6 million W. The ratio of power produced to power consumed, called Q, has increased from less than 0.01 in the early experiments to 0.65 today. The goal of fusion researchers is to reach the *break-even point*, where $Q = 1$, by the year 2010. If this happens, fusion power plants, which require $Q > 10$, could be a reality by about 2050. However, fusion research is very costly, and the United States has cut back substantially on funding for current and future fusion research.

The NOVA fusion laser at Lawrence Livermore National Laboratory in California. The laser light is fired at a 1-mm-diameter pellet containing deuterium and tritium to initiate fusion.

If fusion could be harnessed in a controlled fashion, it could provide an almost limitless source of energy for our society. However, the high temperatures required and the lack of materials to contain them—everything melts at 10 million degrees—have challenged our best efforts. There are currently two methods that hold promise. One involves the containment and acceleration of deuterium and tritium in a doughnut-shaped magnetic bottle, and the other uses lasers to heat a frozen deuterium–tritium pellet rapidly, causing it to collapse on itself and initiate fusion. Both methods have been successful in producing fusion; however, they currently consume more power than they produce. Fusion is advantageous over fission because there is enough naturally occurring deuterium in water to provide fuel for the foreseeable future. Furthermore, the waste products are less radioactive than fission products and have significantly shorter half-lives.

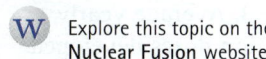

Explore this topic on the **Nuclear Fusion** website.

8.10 The Effect of Radiation on Human Life

The ability of radiation to destroy biological molecules presents a danger to human life. The primary danger lies in the uptake or ingestion of radioactive substances into our bodies. For example, a low-level alpha emitter *external* to our body presents little danger because alpha particles are stopped by clothes or skin and cannot damage internal organs. However, the *ingestion* of a low-level alpha emitter is hazardous because the alpha particles then have direct access to vital organs.

Extremely radioactive substances external to our bodies may also pose a threat. However, the danger is often overstated in the popular press, and most people are misinformed about the risks. Perhaps our inability to see radiation heightens our fears about it. The thought of an invisible ray of particles tearing away at our bodies without detection is frightful. However, barring catastrophic circumstances, such as detonation of a nuclear bomb or direct exposure to the core of a nuclear reactor, the dangers of low-level radioactive substances external to our bodies are minimal—unless they work their way into our bodies.

MEASURING EXPOSURE

The most common unit for measuring human exposure to radiation is called the **rem**. We are all exposed, on average, to a radiation exposure of approximately one third of a rem per year. The sources of this radiation are listed in Table 3. Naturally occurring radon is by far the largest contributor to this exposure. Man-made sources account for about 18% of the total. Although this background radiation may damage some of the body's cells, the cells can repair the damage, and no adverse effects are observed.

The term *rem* is an acronym for roentgen equivalent man.

If higher doses of radiation, up to 100 rem, are encountered in a short period of time, the cells may not be able to repair the damage and may die or change. The ones that die can usually be replaced by our bodies provided the number is not too large. The ones that change, however, may begin abnormal growth and lead to cancerous tumors after several years. Exposures greater than 100 rem may lead to radiation sickness, a condition in which the intestinal lining is damaged, hampering the body's ability to absorb nutrients and take in water. Exposures of several hundred rem damage the immune system, allowing infection to go unchecked. About 50% of people exposed to 400 rem die within 60 days, usually due to infection. It is important to remember that very few people have ever been

Explore this topic on the **National Council of Radiation Protection and Measurements** website.

Table 3. Average yearly radiation doses developed by the National Council on Radiation Protection and Measurement. These numbers are a breakdown of the sources of radiation for the population of the United States. They are averages and were obtained by estimating the total dose for the United States and dividing by the number of people in the country.

TABLE 3
Sources and Amounts of Average Radiation Exposure

Source	Dose (rem/yr)	Percent of Total (%)
Natural		
Radon	0.200	55
Cosmic	0.027	8
Terrestrial	0.028	8
Internal	0.039	11
Total natural	**0.294**	**82**
Artificial		
Medical X-ray	0.039	11
Nuclear medicine	0.014	4
Consumer products	0.010	3
Occupational	<0.001	<0.3
Nuclear fuel cycle	<0.001	<0.03
Fallout	<0.001	<0.03
Miscellaneous	<0.001	<0.03
Total artificial	**0.063**	**18**

TABLE 4
Results of Instantaneous Radiation Exposure

Dose (rem)	Probable Outcome
>500	Death
100–400	Lesions on skin, radiation sickness, hospitalization required but survival likely, increased risk of future cancers
20–100	Decrease in white blood cell count, possible risk of future cancers

exposed to doses exceeding 200 rem. The results of instantaneous exposure to a given amount of radiation are summarized in Table 4.

Another possible effect of radiation exposure is genetic defects in offspring. Genetic defects have occurred in the offspring of laboratory animals on exposure to high levels of radiation. In scientific studies of humans, however, the rate of genetic defects has not been increased by radiation exposure, even in studies of Hiroshima survivors. Few studies of this kind exist, however, due to the unacceptability of purposely exposing humans to large amounts of radiation.

RADON

Radon is by far the greatest single source of human radiation exposure. As we saw in Section 8.4, regions with underground uranium deposits also have small amounts of radon in the soil and surrounding air. Homes in these areas may have increased radon in indoor air. As air in a house is heated, it rises, producing a small suction at the bottom of the house that can pull radon out of the soil, through the floor, and into the house. When radon, or its daughter products, is inhaled, it lodges in lung tissue and can lead to increased lung cancer risk. Indeed, increased rates of lung cancer due to radon inhalation have been observed in uranium mine workers in the United States and Canada.

The concentration of radon in most homes is much lower than uranium mines, and therefore the risk is much lower. Although all scientists agree that high levels of radon exposure increase cancer risk, the level of radon that is considered safe is an area of controversy.

The Environmental Protection Agency (EPA) studied increases in lung cancer rates among uranium mine workers. The unit used to measure airborne radon is picocuries (pico = 10^{-12}) per liter (pCi/L). A curie is defined as 3.7×10^{10} radioactive disintegrations per second. The EPA examined mine workers exposed to radon levels of about 400 pCi/L. The EPA then used a linear risk-exposure model, which assumes that risk increases linearly with exposure, to estimate the risk associated with the much lower radon levels in homes. For example, by using this model, if 400 pCi/L increased cancer risk in uranium miners by 10%, then 40 pCi/L would increase the risk by 1.0%. The EPA, as shown in Table 5, has established that home radon levels above 4 pCi/L pose a significant threat to the occupants. Further studies demonstrated that the risk associated with radon exposure is greater for smokers.

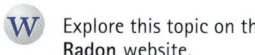

Explore this topic on the **Radon** website.

TABLE 5
EPA Radon Risk Chart

Radon Level (pCi/L)	Smokers[a]	Never Smokers[a]
20	14% or 135 people get lung cancer	0.8% or 8 people get lung cancer
10	7% or 71 people get lung cancer	0.4% or 4 people get lung cancer
4	3% or 29 people get lung cancer	0.2% or 2 people get lung cancer
2	2% or 15 people get lung cancer	0.1% or 1 person gets lung cancer

[a] Figures are based on 1000 people exposed to this level over a lifetime.

MOLECULAR THINKING

Radiation and Smoke Detectors

Many home smoke detectors contain small amounts of radioactive material. The radiation emitted by the material ionizes air within the detector, which in turn produces an electrical current. When smoke from the room enters the detector, ionized air attaches itself to the smoke particles, diminishing the electrical current and setting off the alarm.

QUESTION: Is there any risk associated with this kind of smoke detector? If so, why are they used? What are the benefits?

Some scientists believe that the EPA has overestimated the risk associated with radon exposure. All scientists agree that the radon levels of hundreds of picocuries per liter breathed by uranium workers pose a significant cancer risk; however, they have criticized the EPA's claim that much lower levels pose a risk. These criticisms are based partly on the linear risk-exposure model. It is not clear whether risk is linearly related to exposure or whether there might be some threshold level, below which the risk is negligible. In addition, a number of studies of lung cancer victims have failed to show a correlation between lung cancer rates and radon levels in homes. However, these studies, because of their small size, are not conclusive.

Relatively inexpensive kits are available to test for radon in your home. The EPA recommends action for levels of 4 pCi/L or above. Ways to reduce radon in homes range from the relatively simple, keeping a basement window open, to more costly measures such as installing a radon ventilation system.

8.11 Carbon Dating and the Shroud of Turin

Figure 8 shows a picture of a rectangular linen cloth bearing the image of a crucifixion victim. It is kept in the Chapel of the Holy Shroud in the Cathedral of St. John the Baptist in Turin, Italy. Curiously, the image on the cloth is a negative one and becomes clearer when photographed and viewed as a negative. (A negative of the negative image reveals the positive image.) Many have believed that the cloth was the original burial cloth of Jesus, miraculously imprinted with his

A home radon detector.

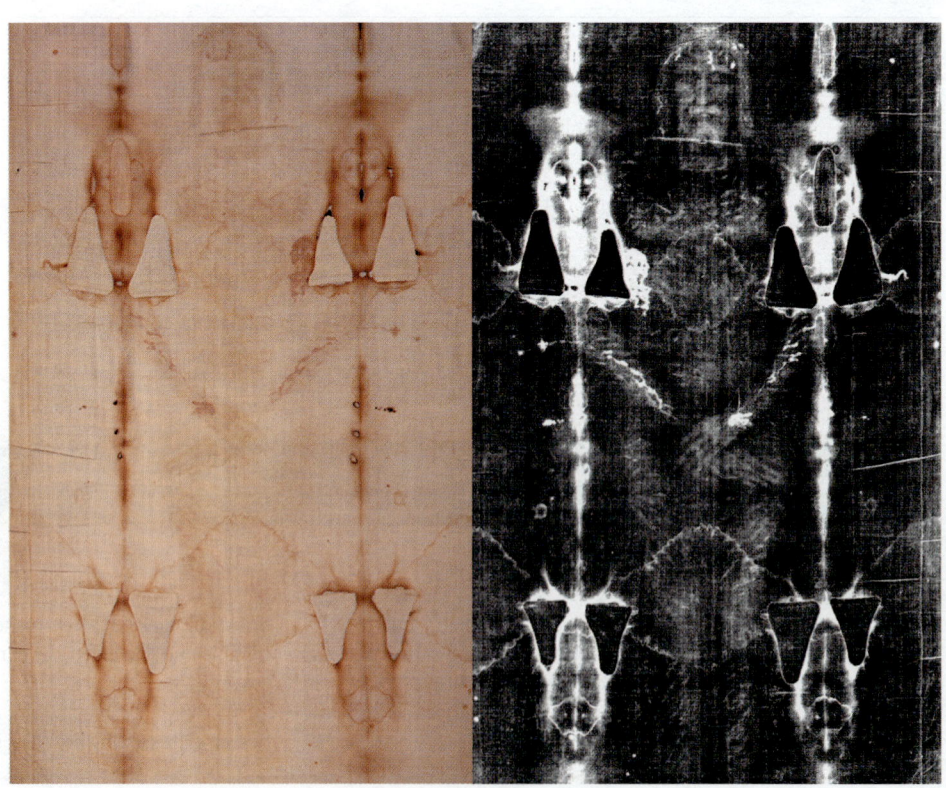

FIGURE 8 Image of the Shroud of Turin as seen by the naked eye (left) and in a photographic negative (right).

image. If the cloth was in fact Jesus' burial cloth, it would have to be close to 2000 years old. Nature has left a signature on the cloth that reveals its age; the ability to read that signature and determine its age would give insight into whether the cloth is authentic.

The age signature in the cloth is a result of radioactive elements in our environment. Their consistent decay provides a natural clock that can be used in dating artifacts such as the Shroud of Turin. One dating method involves the isotope carbon-14, which is constantly being formed from nitrogen by neutron bombardment in the upper atmosphere:

$$^{14}_{7}N + ^{1}_{0}n \rightarrow ^{14}_{6}C + ^{1}_{1}H$$

Carbon-14 subsequently reacts with O_2 to form CO_2, which is incorporated into plants by photosynthesis. Because animals and humans eat plants, carbon-14 is also incorporated into their tissues. As a result, all living organisms contain a residual amount of C-14. When the organism dies, it stops taking in C-14. The C-14 that was in the organism's body decays with a half-life of 5730 years. Because many ancient objects, such as the shroud, are made from organic matter that was alive in the past, the levels of C-14 in these objects can be used to determine their age.

In 1988, C-14 dating methods had advanced to the point that only a small sample of the object in question was required to estimate its age. The Vatican agreed to using this technique to test the authenticity of the Shroud of Turin, and

EXAMPLE 8.4

Carbon Dating

A fossil has a carbon-14 content that is 20% of that found in living organisms. Estimate the age of the fossil.

SOLUTION

It is easiest to construct a table showing the relative amount of C-14 in the fossil as a function of time:

Number of Half-Lives	% C-14 (relative to living organisms)	Age of Fossil (yr)
0	100%	0
1	50%	5,730
2	25%	11,460
3	12.5%	17,190

Because it would take two half-lives for the level of C-14 to reach 25% of that found in living organisms, the fossil must be slightly older than two half-lives, or slightly greater than 11,460 yr.

YOUR TURN

Carbon Dating

The carbon in an ancient linen had a carbon-14 content that was 50% of that found in living organisms. Estimate the age of the linen.

samples were sent to laboratories at Oxford University, University of Arizona, and the Swiss Federal Institute of Technology. Based on carbon-14 dating, all three laboratories estimated the age of the shroud to be no more than 750 years old. Although no experimental method is 100% foolproof, it appears that the shroud could not have been the original burial cloth of Jesus; it is much too young.

8.12 Uranium and the Age of the Earth

A second radiometric dating technique, used to measure much longer periods of time, involves U-238. U-238 decays with a half-life of 4.5×10^9 yr through a number of intermediates, and eventually ends as lead. As a result, all rocks on Earth that contain uranium also contain lead. If a rock is assumed to have been only uranium when it was formed (there are independent ways to test this that are beyond the scope of this book), the ratio of uranium to lead can be used to date the rock. For example, if one half-life has passed, the rock would contain 50% uranium atoms and 50% lead atoms. After two half-lives, the rock would be 25% uranium atoms and 75% lead atoms.

The oldest rocks on the Earth have a remarkably constant composition of approximately 50% uranium and 50% lead. The ratio of uranium to lead atoms in these rocks suggests that the Earth is about 4.5 billion years old. This is only one method among many to estimate the Earth's age; all have different assumptions, but all give approximately the same age for the Earth. In addition, the 4.5 billion-year age for the Earth is consistent with cosmological models for the formation of

EXAMPLE 8.5

Uranium to Lead Dating

A hypothetical meteorite is found to contain 25% uranium and 75% lead. How old is the meteorite?

SOLUTION

It is easiest to construct a table showing the percentages of each element as a function of time:

Number of Half-Lives	% Uranium	% Lead	Age of Rock
0	100	0	0
1	50	50	4.5×10^9 yr
2	25	75	9.0×10^9 yr
3	12.5	87.5	13.5×10^9 yr

The hypothetical meteorite would be 9×10^9 yr old.

YOUR TURN

Uranium to Lead Dating

A rock contains 60% uranium and 40% lead. How old is the rock?

WHAT IF...

Radiation—Killer or Healer?

People often wonder how radiation, which is known to induce cancer, can also be used to fight it. The answer to this question, like our previous one on X-rays, lies in the analysis of the risk involved. It is generally agreed that a radiation dose of approximately 100 rem will increase the risk of cancer by 1%. The doses of radiation required to kill tumors are in this range and are therefore known to increase the chance of future cancers. However, if your chance of dying from the cancer you already have is 100%, then the 1% increase in risk for future cancers is a relatively small one in comparison.

QUESTION: In some cases, cancer advances to the point where the chances of removing the cancer with radiation treatment become very small. In those cases, the treatment may prolong the life of the patient without offering any hope for long-term survival. This puts doctors and patients in a difficult position. Should the therapy be administered with little hope for long-term success even though it will certainly lower the quality of the patient's remaining life? What do you think?

---

the universe; these models suggest that the universe is approximately 13.7 billion years old. For example, we can see stars today that are billions of light years away. If the Earth and the universe were much younger, the light from these far-away stars would not have had enough time to reach the Earth. Because we can see these stars today, we know the Earth must be billions of years old.

8.13 Nuclear Medicine

Radioactivity also has important applications in medicine, both in the diagnosis and treatment of disease. For diagnosis, doctors often want to image a specific internal organ or tissue. One way to do this is to introduce a radioactive element, usually technetium-99m (m means metastable), into the patient that will concentrate in the area of interest. Technetium-99m is a gamma emitter with a half-life of approximately 6 hours. The gamma rays emitted by the technetium, which easily pass through the body, are allowed to expose a photographic film. The film then contains an image of the desired organ.

For example, appendicitis, the infection and inflammation of the appendix, is difficult to diagnose with certainty. Doctors can use radioactivity to determine if an appendix is indeed infected. Technetium-99m is incorporated into antibodies and administered to the patient. If the appendix is infected, the antibodies, and therefore the technetium-99m, concentrate there. A bright exposure on the developed film in the area of the appendix constitutes a positive test, allowing the surgeon to operate with confidence that he or she is removing an infected appendix.

Radiation is also used as a therapy to destroy cancerous tumors. Because radiation kills cells, and because it kills rapidly multiplying cells more efficiently than slowly multiplying ones, carefully focused gamma rays are used to target and obliterate internal tumors. To minimize the exposure of healthy cells, the gamma ray beam is usually focused sharply on the cancerous region and is moved in a circular path around it. This allows the tumor to receive maximum gamma ray exposure while keeping the damage to healthy cells at a minimum. In spite of this, healthy cells are damaged and patients undergoing this kind of treatment often have severe side effects such as skin burns, hair loss, nausea, and vomiting. Although these side effects are not pleasant, radiation therapy has been an effective tool in fighting many different types of cancer.

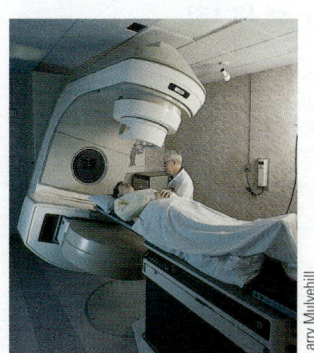

Radiation treatment for cancer.

Chapter Summary

MOLECULAR CONCEPT

Radioactivity, discovered by Becquerel and the Curies, consists of energetic particles emitted by unstable nuclei (8.1, 8.2). Alpha radiation consists of helium nuclei that have high ionizing power but low penetrating power. Beta radiation consists of electrons emitted when a neutron within an atomic nucleus converts into a proton. Beta particles have lower ionizing power than alpha particles but higher penetrating power. Gamma radiation is high-energy electromagnetic radiation with low ionizing power but high penetrating power (8.3). Unstable nuclei radioactively decay according to their half-life, the time it takes for one-half of the nuclei in a given sample to decay (8.4).

Some heavy elements such as U-235 and Pu-239 can become unstable and undergo fission when bombarded with neutrons (8.5). The atom splits to form lighter elements, neutrons, and energy. If fission is kept under control, the emitted energy can be used to generate electricity. If fission is forced to escalate, it results in an atomic bomb (8.6, 8.7). Hydrogen bombs, similar to the Sun, employ a different type of nuclear reaction called fusion in which the nuclei of lighter elements combine to form heavier ones. In all nuclear reactions that produce energy, some mass is converted to energy in the reaction (8.8, 8.9).

By measuring the levels of certain radioactive elements in fossils or rocks, radioactivity can be used to date objects. The age of Earth is estimated to be 4.5 billion years based on the ratio of uranium to lead in the oldest rocks (8.10–8.12). High levels of radioactivity can kill human life. Lower levels can be used in therapeutic fashion to either diagnose or treat disease (8.13).

SOCIETAL IMPACT

The discovery of radiation has had many impacts on our society. It ultimately led to the Manhattan Project, the construction and detonation of the first atomic bomb in 1945. For the first time, in a very tangible way, society could see the effects of the power that science had given to it (8.5, 8.6). Yet science itself did not drop the bomb on Japan, it was the people of the United States who did that, and the question remains—how do we use the power that technology can give? Since then, our society has struggled with the ethical implications of certain scientific discoveries. For the past decade, nuclear weapons have been disarmed at the rate of 2000 bombs per year. Today, we live in an age when the threat of nuclear annihilation is less severe.

Nuclear fission is used to generate electricity without the harmful side effects associated with fossil fuel combustion. Yet nuclear power has its own problems, namely the potential for accidents and waste disposal (8.7). Will the United States build a permanent site for nuclear waste disposal? Will we turn to nuclear power as the fossil fuel supply dwindles away? How many resources will we put into the development of fusion as a future energy source? These are all questions that our society faces as we begin this new millennium.

Nuclear processes have been able to tell us how old we are. Archaeological discoveries are fitted into a chronological puzzle that tells about human history from the very earliest times. We know that billions of years passed on the Earth before humans ever existed. We know how certain humans began to use tools and how they migrated and moved around on the Earth. We can date specific items such as the Shroud of Turin and determine if they are genuine (8.11, 8.12). What effect does this scientific viewpoint have on our society? On religion? What does it tell us about who we are?

Chemistry on the Web

For up-to-date URLs, visit the text website at **www.chemistry.brookscole.com/tro3**

- Chernobyl Health Effects Studies
  http://www.eh.doe.gov/health/hstudies/chern_hes.html
- Nuclear Fission
  http://www.atomicarchive.com/sciencemenu.shtml
  http://www.yale.edu/lawweb/avalon/abomb/mpmenu.htm
- Yucca Mountain
  http://www.ocrwm.doe.gov/ymp/
- Nuclear Fusion
  http://fusedweb.pppl.gov/CPEP/Chart.html
  http://www.ofes.fusion.doe.gov/
- National Council of Radiation Protection and Measurements
  http://www.ncrponline.org/
- Radon
  http://www.epa.gov/radon/

Key Terms

Becquerel, Antoine-Henri	Fermi, Enrico	ionizing power	Oppenheimer, J. R.
critical mass	fission	mass defect	radon
Curie, Marie Sklodowska	fusion	Meitner, Lise	rem
Curie, Pierre	Hahn, Otto	nuclear binding energy	Strassmann, Fritz
Einstein, Albert	half-life	nuclear equation	Szilard, Leo

Exercises

 Assess your understanding of this chapter's topics with an online chapter quiz at **http://chemistry.brookscole.com/tro3**.

QUESTIONS

1. On a molecular level, explain how radioactivity occurs and how it can be harmful to living organisms.
2. For each of the following scientists, explain their discoveries related to radioactivity and their significance:
   a. Becquerel
   b. Marie and Pierre Curie
   c. Rutherford
   d. Fermi
   e. Hahn, Meitner, and Strassmann
   f. Oppenheimer
3. Explain the difference between alpha, beta, and gamma radiation. Rank each in terms of ionizing power and penetrating power.
4. Explain what is meant by half-life. What is its relation to nuclear stability?
5. Where is radon found and why is it a potential hazard?
6. Describe nuclear fission and write a nuclear equation to describe the fission of U-235.
7. Who persuaded Roosevelt to put forth the resources to build the atomic bomb? Why?
8. What is a critical mass and how does it relate to the chain reaction that occurs in a fission reaction?
9. Explain the importance of each of the following locations with respect to the Manhattan Project:
   a. Los Alamos, New Mexico
   b. Hanford, Washington
   c. Oak Ridge, Tennessee
   d. Alamogordo, New Mexico
10. Describe experiments termed "tickling the dragon." Why were they important?
11. What is the difference between the fission reaction carried out by Fermi at the University of Chicago and the fission reaction required to produce an atomic bomb?

12. Describe the two designs, Fat Man and Little Boy, that were pursued in building the bomb.
13. What drove the scientists involved in the Manhattan Project to continue in spite of their concerns over the morality of the atomic bomb?
14. Sketch a nuclear reactor and explain how fission can be used to generate electricity. What is the role of the cadmium rods?
15. Define mass defect and nuclear binding energy.
16. Where does the energy associated with nuclear reactions come from?
17. What is the "China syndrome"?
18. What are the two main problems associated with nuclear power?
19. What is being done to deal with nuclear waste?
20. Describe the accidents at Three-Mile Island and Chernobyl. What was the outcome of each? Is it possible to have an *American* Chernobyl? Why or why not?
21. What is fusion? Write a nuclear equation to describe a fusion reaction.
22. What are the technical difficulties associated with a controlled fusion reaction?
23. Explain the difference between exposure to radioactivity from a source external to our bodies versus an internal source.
24. How does radiation affect humans at low exposures and at high exposures?
25. Can radiation induce genetic changes in humans or in their offspring? Have such changes been observed?
26. Explain how C-14 dating and U-238 dating work.
27. According to U-238 dating techniques, what is the age of the Earth? What assumptions are involved in the calculation of this age? Why are these assumptions valid?
28. Why can radiation, which is known to induce cancer, also be used as a treatment for cancer?
29. How is radiation used to obtain images of internal organs?

Problems

30. Write nuclear equations to represent each of the following:
    a. alpha decay of Th-230
    b. alpha decay of Po-210
    c. beta decay of Th-234
    d. beta decay of Po-218
31. Write nuclear equations to represent each of the following:
    a. alpha decay of Rn-220
    b. beta decay of Bi-212
    c. alpha decay of Ra-224
    d. beta decay of Tl-208
32. Write a set of nuclear equations to represent the series of decays that transforms Po-214 into Pb-206. The order is alpha, beta, beta, and alpha.
33. Write a set of nuclear equations to represent the series of decays that transforms Ra-228 into Rn-220. The order is beta, beta, alpha, alpha.
34. Francium-223 decays to polonium-215 via three decay steps. Propose a possible three-step sequence for the overall process.
35. Bismuth-215 decays to bismuth-211 via three decay steps. Propose a possible three-step sequence for the overall process.
36. I-131 has a half-life of 8.04 days. If a sample initially contains 30.0 g of I-131, approximately how much I-131 will be left after 32 days?
37. Tc-99 is often used in medicine and has a half-life of about 6 h. If a patient is given 20.0 micrograms of Tc-99, and assuming that none is lost due to other processes besides radioactive decay, how much Tc-99 is left after 24 h?
38. If the decay of each I-131 atom in problem 36 produces an alpha particle, how many alpha particles are emitted in 32 days?
39. If the decay of each Tc-99 atom in problem 37 produces a gamma ray, how many gamma rays are emitted in 24 hours?
40. Write a nuclear equation to represent the neutron-induced fission of uranium to produce Te-137, Zr-97, and two neutrons. (Hint: The symbol for a neutron is ^1_0n.)
41. Write a nuclear equation to represent the fusion of two deuterium atoms, H-2, to form He-3 and one neutron.
42. If a man is exposed to 250 mrem/yr of radiation because of radon and his total exposure is 380 mrem/yr, what percentage of his total exposure is due to radon?
43. If a woman is exposed to 78 mrem/yr of radiation because of medical X-rays and her total exposure is 360 mrem/yr, what percentage of her total exposure is due to medical X-rays?
44. The carbon in a fossil was found to contain 12.5% C-14 relative to living organisms. What is the age of the fossil?
45. The carbon in a human skull found in South America contained 25% C-14 relative to living organisms. How old is the skull?
46. The C-14 in an ancient scroll is found to be 50% of that found in living organisms. What is the age of the scroll?
47. The C-14 found in the skeleton of a wooly mammoth was 6.25% of that found in living organisms. How old is the skeleton?
48. The percentages of uranium and lead in a meteorite were 30 and 70%, respectively. What is the approximate age of the meteorite?
49. A rock that was originally all uranium contains 50% uranium and 50% lead. How old is the rock?

Points to Ponder

50. Even though his discovery was fortuitous, Becquerel shared the Nobel Prize with the Curies for the discovery of radiation. Why or why not was this fair?
51. Several medical applications require the ingestion of radioactive materials that concentrate in a particular organ. The emitted radiation is detected on a photographic plate to obtain an image of the organ. What type of radiation—alpha, beta, or gamma—would be best suited for this technique? Why?
52. Several radioactive isotopes are present in a sheet-metal storage cabinet. What type, if any, of radiation may be present outside of the cabinet?
53. If you were President Roosevelt, how would you have responded to Einstein's letter warning that Germany might be developing an atomic bomb? Why or why not would you have put forward the resources to initiate the Manhattan Project?
54. Why or why not is a scientist responsible for the ethical implications of his or her work?
55. Could terrorists build a nuclear bomb from uranium obtained from nuclear power plants? Explain.
56. In Chapter 1, we learned that Lavoisier formulated the law of conservation of mass. Now we learn that for nuclear reactions matter is not conserved. Was Lavoisier wrong? Is his law now useless?
57. Why do nations stockpile nuclear weapons instead of simply maintaining larger numbers of military troops?
58. Should we have dropped the bomb on unpopulated areas of Japan, instead of Nagasaki and Hiroshima, to show the Japanese the power of this technology without killing mostly civilians? Explain.
59. It is clearly unethical to study the effects of radiation on humans by intentionally exposing them to radiation. Is it ethical to study the effects on those who have been accidentally exposed? Why?
60. Is it ethical to intentionally expose animals to high levels of radiation to study its effects? If not, why not? If so, are there any species, such as dolphins, dogs, or monkeys, for which your answer would be different?

Feature Problems and Projects

61. The figure below represents the decay of N-16 by alpha emission. Count the protons (red) and neutrons (gray) and draw in the missing particle. Write a nuclear equation to represent this decay.

62. The figure below represents the decay of Na-20 by alpha emission. Count the protons (red) and neutrons (gray) and draw in the missing particle. Write a nuclear equation to represent this decay.

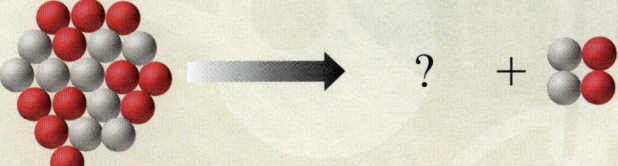

63. Read the What If . . . The Ethics of Science box in this chapter. What are some current scientific advances with ethical implications? Write a short essay explaining why scientists should or should not—depending on your opinion—be responsible for the implications of their work. Give at least two examples from current scientific advances.

W 64. Go to the U.S. Department of Energy's website (http://www.ocrwm.doe.gov/ymp/) on its current investigation of Yucca Mountain, Nevada, as a potential site for a permanent nuclear waste facility. Use information from this site to write a short essay about the proposed waste facility. In your essay, explain why this site has been chosen and describe how it is expected to operate. Also describe the current status of the site and when it is expected to be operational.

9 Energy for Today

There is a fact, or if you wish, a law, governing all natural phenomena that are known to date. There is no known exception to this law—it is exact as far as we know. The law is called the conservation of energy. It states that there is a certain quantity, which we call energy, that does not change in the manifold changes which nature undergoes.

—R. P. Feynman

In this chapter, we examine energy—one of the two fundamental components of the physical universe (the other fundamental component is matter). Energy is related to matter in a number of ways. For example, in Chapter 8 we learned that energy and matter are interchangeable. More importantly, the chemical and physical changes that we have discussed since the first page of this book are nearly always accompanied by energy changes (think of burning natural gas on a stove or burning gasoline in a car, for example). We now focus on energy itself and the relationship between chemical changes and energy. As you read these pages, think about your own energy use. What have you done today that used energy? Did you drive a car? Use electricity? Cook? Shower? We use energy every minute of every day, most of which comes (either directly or indirectly) from chemical reactions. Our lives would be drastically different without energy.

QUESTIONS FOR THOUGHT

- What is energy?
- What is heat? Work?
- How much energy do North Americans use? How much do people in other parts of the world use?
- Can energy be created or destroyed?
- What are the most efficient ways to use energy?
- How do we measure energy? How do we measure temperature?
- How is chemistry related to energy use?
- Where does our society get its energy today? In the past?
- What are fossil fuels? What are the environmental problems associated with their use?
- What are the main components of smog? How do they affect humans?
- What is acid rain? What causes it?
- Is the planet really getting warmer? Why?

The molecules and atoms that compose this hot skillet are moving and oscillating much faster than the molecules that compose your finger.

9.1 Molecules in Motion

A person working in the kitchen accidentally touches a hot iron skillet. She instinctively pulls her finger away but not soon enough to avoid a burn. After several minutes her finger swells and turns red. What happened? The atoms in

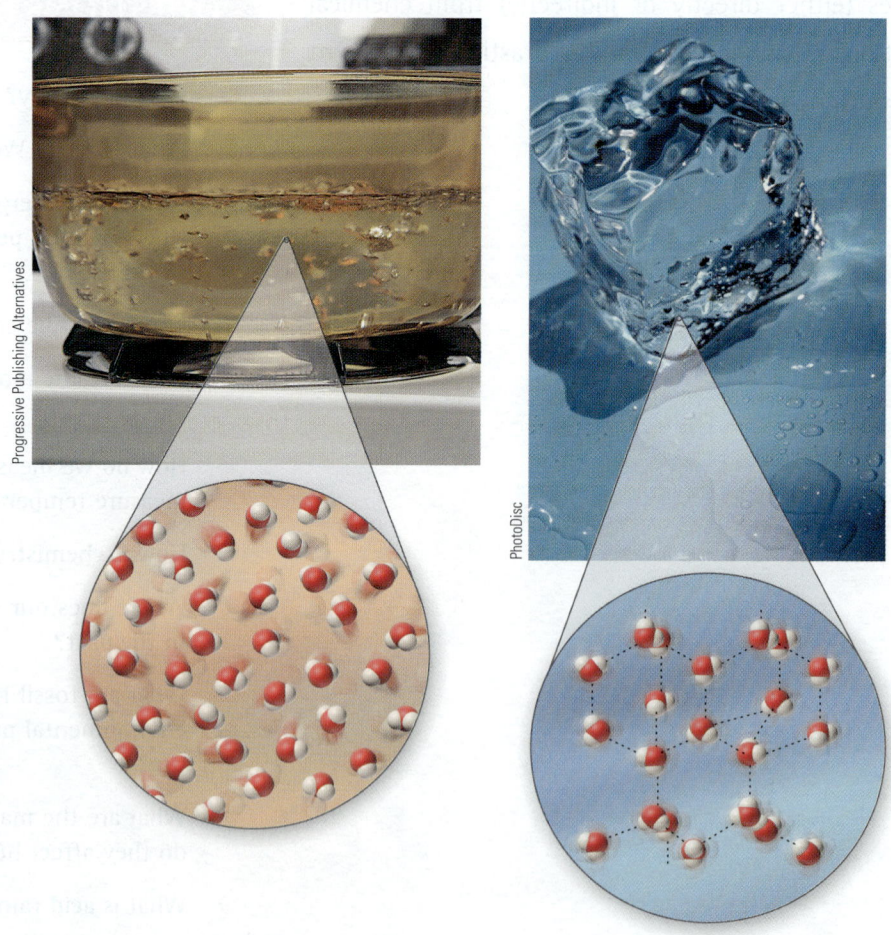

In a hot object, like boiling water, the molecules are moving relatively quickly. In a colder object the molecules are moving more slowly.

the skillet transferred some of their energy, in the form of heat, to the molecules in her skin, destroying them in the process. We call it a burn.

Heat is the flow of energy due to a temperature difference. Although invisible to our eyes, atoms and molecules are in constant motion, vibrating back and forth a trillion times per second. The hotter an object is, the faster its molecules move. A molecular view of the hot skillet would show iron atoms in a flurry of motion, vibrating, jostling, and occasionally even breaking off and shooting through the air. A human finger is composed of delicate organic molecules moving much slower. When the skillet touches the skin, a myriad of iron atoms slam into the skin molecules, destroying them instantly. The macroscopic result is a burn.

Energy, and our use of it, is ultimately tied to molecular motion. In the preceding example, we see that using energy for heating—as in cooking or heating—*increases* the random motion of molecules and atoms. Similarly, using energy for cooling, as in air-conditioning or refrigeration, *decreases* the random motion of molecules and atoms.

We often use energy to move atoms and molecules in a nonrandom or orderly fashion; using energy this way is called work. We use energy to do work when we drive a car or ride an elevator; we move the atoms and molecules that compose the car or the elevator in an orderly fashion—all in the same direction.

9.2 Our Absolute Reliance on Energy

We normally do not think about energy; however, without it, our most common tasks become impossible. In only a few hours without electricity, we realize how much we depend on energy. The world currently consumes approximately 412 quads (quad, a unit of energy that we will define later) of energy per year. The United States consumes approximately 24% of that total (98 quads) in the categories shown in Table 1.

The industrial sector—where energy is used to produce everything from jigsaw puzzles to jumbo jets—accounts for 33.1% of U.S. energy consumption, while the transportation sector—where energy is used to power cars, trucks, airplanes, and other forms of transportation—accounts for 27.4%. The residential sector—where

TABLE 1

U.S. Energy Use by Sector

Industrial	33.1%
Transportation	27.4%
Residential	21.6%
Commercial	17.9%

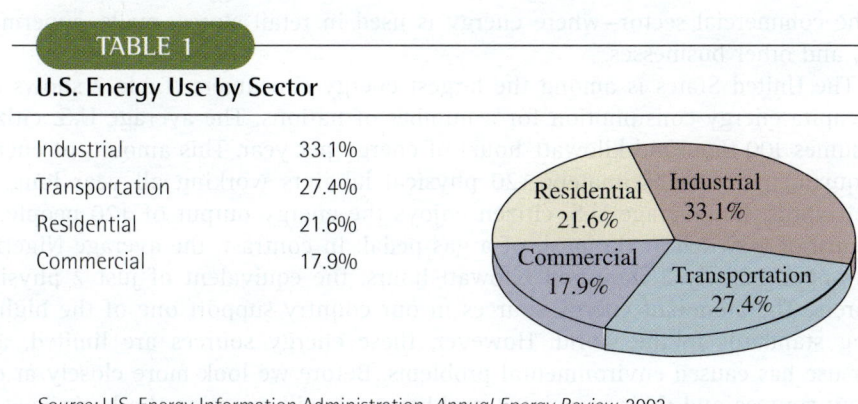

Source: U.S. Energy Information Administration, *Annual Energy Review*, 2003.

TABLE 2
Per Capita Yearly Energy Consumption

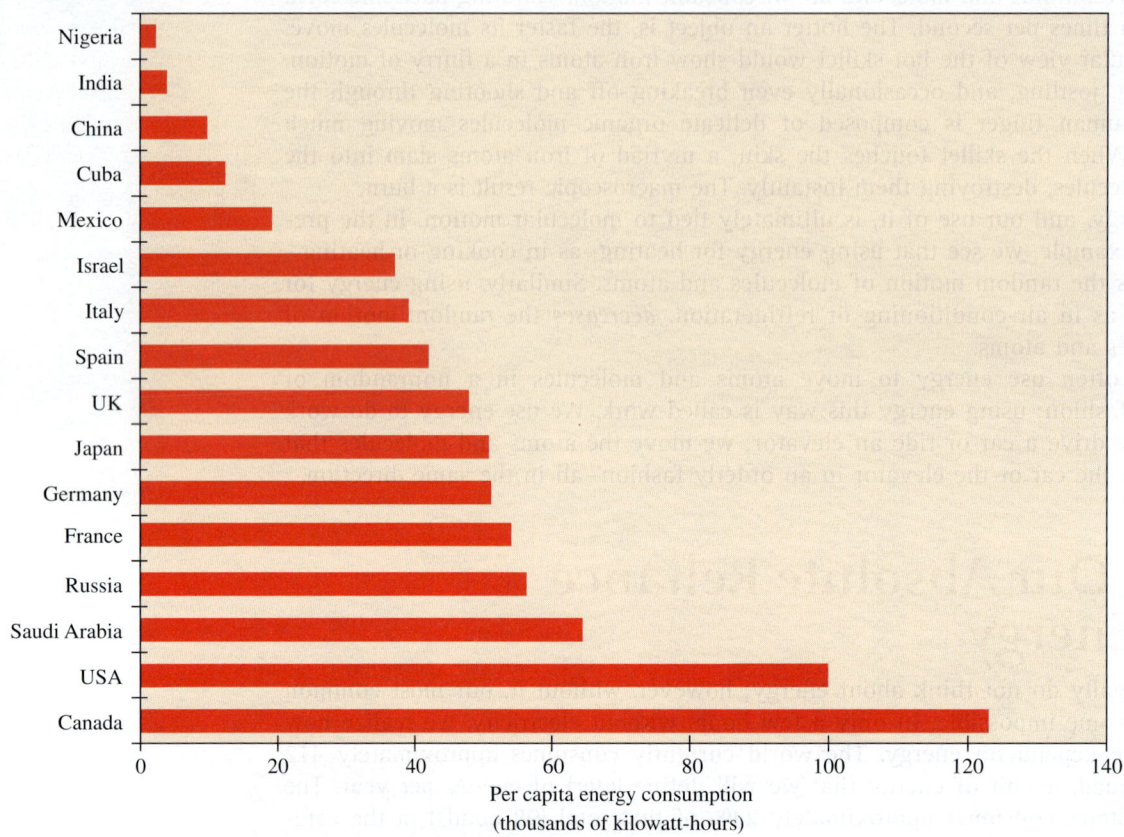

Source: Data compiled from the U.S. Energy Information Administration for the 2002 calendar year.

energy is used in homes for heating, cooling, cooking, lighting, and entertainment—uses 21.6% of total energy. The final 17.9% of energy consumption is due to the commercial sector—where energy is used in retail stores, malls, supermarkets, and other businesses.

The United States is among the largest energy consumers. Table 2 shows the per capita energy consumption for a number of nations. The average U.S. citizen consumes 100 thousand kilowatt-hours of energy per year. This amount of energy is equivalent to approximately 120 physical laborers working all year long. In other words, the average U.S. citizen enjoys the energy output of 120 people, at the turn of a switch or the push of a gas pedal. In contrast, the average Nigerian uses approximately 2 thousand kilowatt-hours, the equivalent of just 2 physical laborers. The abundant energy sources in our country support one of the highest living standards in the world. However, these energy sources are limited, and their use has caused environmental problems. Before we look more closely at our energy sources and their associated problems, we will examine the basic laws of nature related to energy and its use.

9.3 Energy and Its Transformations: You Cannot Get Something for Nothing

Chemical and physical changes are nearly always accompanied by energy changes. For example, when you burn wood in a fire (a chemical change), energy is released from the burning wood, warming the surrounding air. When ice melts in a glass of warm water (a physical change), energy is absorbed by the melting ice, cooling the surrounding water.

The study of energy and its transformation from one form to another is called **thermodynamics**. The formal definition of **energy** is *the capacity to do work*. An object possessing energy can perform work on another object. The formal definition of **work** is *a force acting through a distance*. If you push a chair across the room, you have done work; you have exerted a force over a distance.

The *total energy* of an object is a sum of its **kinetic energy**, the energy associated with its motion, and its **potential energy**, the energy associated with its position or composition. For example, if you hold a book at arm's length, it contains potential energy due to its position within Earth's gravitational field (Figure 1a). If you drop the book, the potential energy is converted to kinetic energy as the book accelerates toward the ground. When the book hits the ground, its kinetic energy is converted to **thermal energy**, the energy associated with the temperature of an object (Figure 1b). Although too small to be easily measurable, the temperature of the ground rises ever so slightly as it absorbs the kinetic energy of the rapidly falling book. Thermal energy is a kind of kinetic energy

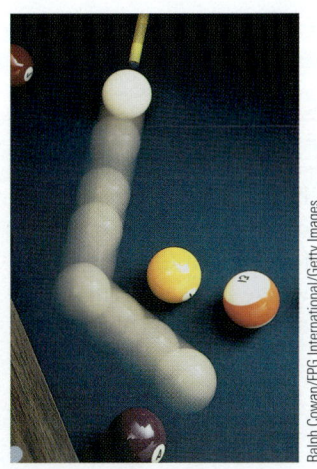

A rolling billiard ball contains energy associated with its motion. It can do work by colliding with another billiard ball. When it does work, some of its energy is transferred to the other ball.

Potential Energy

Kinetic Energy

FIGURE 1 Energy transformations.

because it is due to the motion of the individual atoms or molecules that make up an object.

In thermodynamics we distinguish between the subject we are studying, called the **thermodynamic system**, and the environment with which it is exchanging energy, called the **surroundings**. In the preceding example, we can think of the book as the system and the air and the ground as the surroundings. As the book falls and hits the ground, energy is transferred from the system (the book) to the surroundings (the air and the ground). However, the *total amount of energy remains constant*—the energy gained by the surroundings exactly equals the energy lost by the system. The universal observation that energy is always conserved is known as the **first law of thermodynamics** and can be stated as follows:

> Energy can neither be created nor destroyed, only transferred between the system and its surroundings.

An exception to the first law of thermodynamics occurs in nuclear processes (covered in Chapter 8), where mass and energy are interchangeable.

The total energy is constant and cannot change; it can be transferred but not created. If the system loses energy, the surroundings must gain energy and vice versa (Figure 2). When you "use" energy, therefore, you do not destroy it; you simply move it from the system to the surroundings or convert it from one form to another. For example, when you drive a car, you convert the **chemical energy**—a type of potential energy—stored in the chemical bonds of gasoline to the kinetic energy associated with the motion of the car and to thermal energy, which is dissipated as heat. In other words, the molecules that compose gasoline are a bit like the raised book—they contain potential energy, which can be converted into other kinds of energy. The problem with energy use, then, is not that we are using up energy—energy is constant—but instead that we are converting energy from easily usable and concentrated forms, such as petroleum, to forms that are less concentrated and more difficult to use (such as thermal energy).

The first law of thermodynamics has significant implications for energy use. According to the first law, we cannot create energy that was not there to begin with—a device that continuously produces energy without the need for energy input cannot exist. However, people have dreamed of such devices for centuries. Ideas like an electric car that recharges its batteries while driving, eliminating the need for fuel, or a machine that runs in the garage and powers an entire house without energy input have no basis in reality. Our society has a continual need for energy, and as our current energy resources dwindle, new energy sources will be required. Unfortunately, those sources must also follow the first law; we will not get something for nothing.

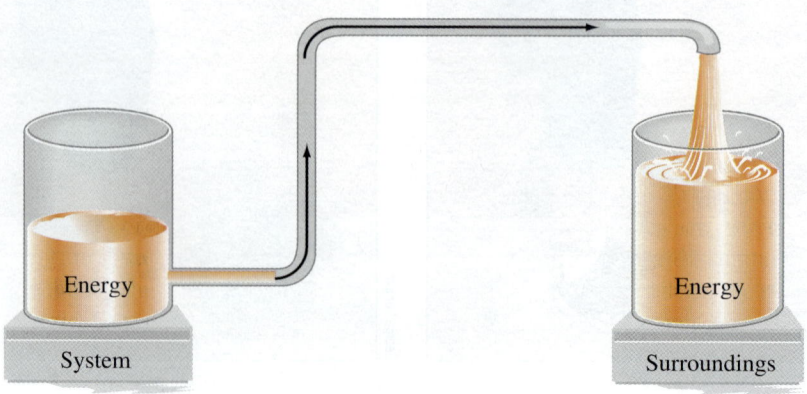

FIGURE 2 If the system loses energy, the surroundings must gain it and vice versa.

9.4 Nature's Heat Tax: Energy Must Be Dispersed

The first law of thermodynamics tells us that energy is conserved in any process. The **second law of thermodynamics** states that for any process to be *spontaneous*—to occur without outside intervention—energy must be dispersed (or become arranged in a more disorderly way). In other words, the second law of thermodynamics tells us that although energy is *conserved* in any process, it becomes *more spread out* in any spontaneous process. For example, consider a metal rod that is heated at one end (Figure 3). The heat deposited on one end of the rod will spontaneously disperse itself through the entire rod. Imagine a metal rod in which thermal energy did the opposite—one end of the rod would spontaneously get hot while the other end got cold. We know from experience that this does not happen. Thermal energy spontaneously flows from hot to cold (not the other way) as described by the second law of thermodynamics.

Another word for energy dispersal (or energy randomization) is **entropy**; thus, another way to state the second law is as follows:

For any spontaneous process, the entropy of the universe must increase.

When water spontaneously freezes below 0°C, the entropy of the water itself decreases. However, there is a correspondingly greater increase in the entropy of the surroundings because heat is evolved during freezing, dispersing energy into the surroundings.

We can probably think of spontaneous processes in which entropy *appears* to decrease (that is, in which energy *appears* to become less dispersed). For example, on a cold winter's day, a bucket of water left outdoors will spontaneously freeze; the water molecules go from a disorderly liquid (in which energy is more highly dispersed) to an orderly solid (in which energy is less dispersed). The water goes from a state of high entropy (the liquid) to a state of low entropy (the solid). However, on closer inspection, we find that it is only within the system, *the bucket of water*, that entropy decreases. In the surroundings, entropy increases because heat

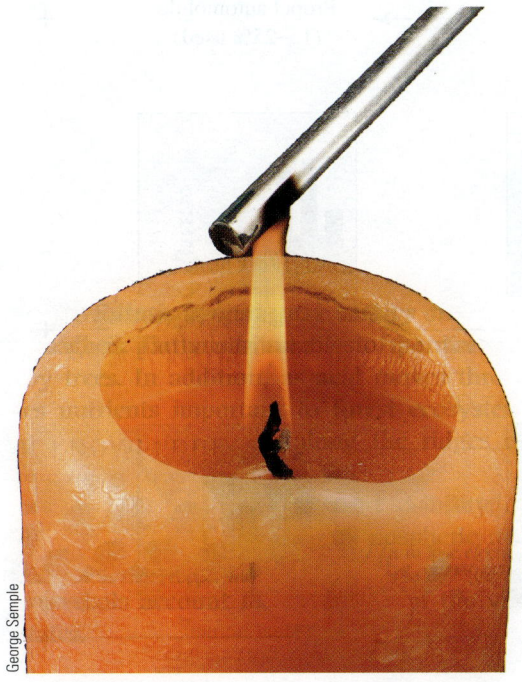

FIGURE 3 Energy has a pervasive tendency to disperse or spread itself out.

is evolved as water freezes. The heat generates more motion in the surrounding molecules, dispersing energy. According to the second law, the *increase* in entropy in the surroundings must be greater than the *decrease* in entropy achieved by the water molecules in the bucket; otherwise, the process would not be spontaneous.

How does the second law relate to energy use? Because the second law states that energy must be dispersed for a process to be spontaneous, the second law implies that no spontaneous process can be 100% efficient with respect to energy. For example, consider the use of a rechargeable battery. Suppose that you use the battery to turn an electric motor, converting the potential energy in the battery to the kinetic energy of the spinning motor. The amount of energy required to recharge the battery fully after it runs down will necessarily be more than the amount of kinetic energy that you obtained in the electric motor. Why? Because some energy must be dispersed (or lost to) the surroundings for the process to occur at all. Energy is not destroyed during the cycle of discharging and recharg-

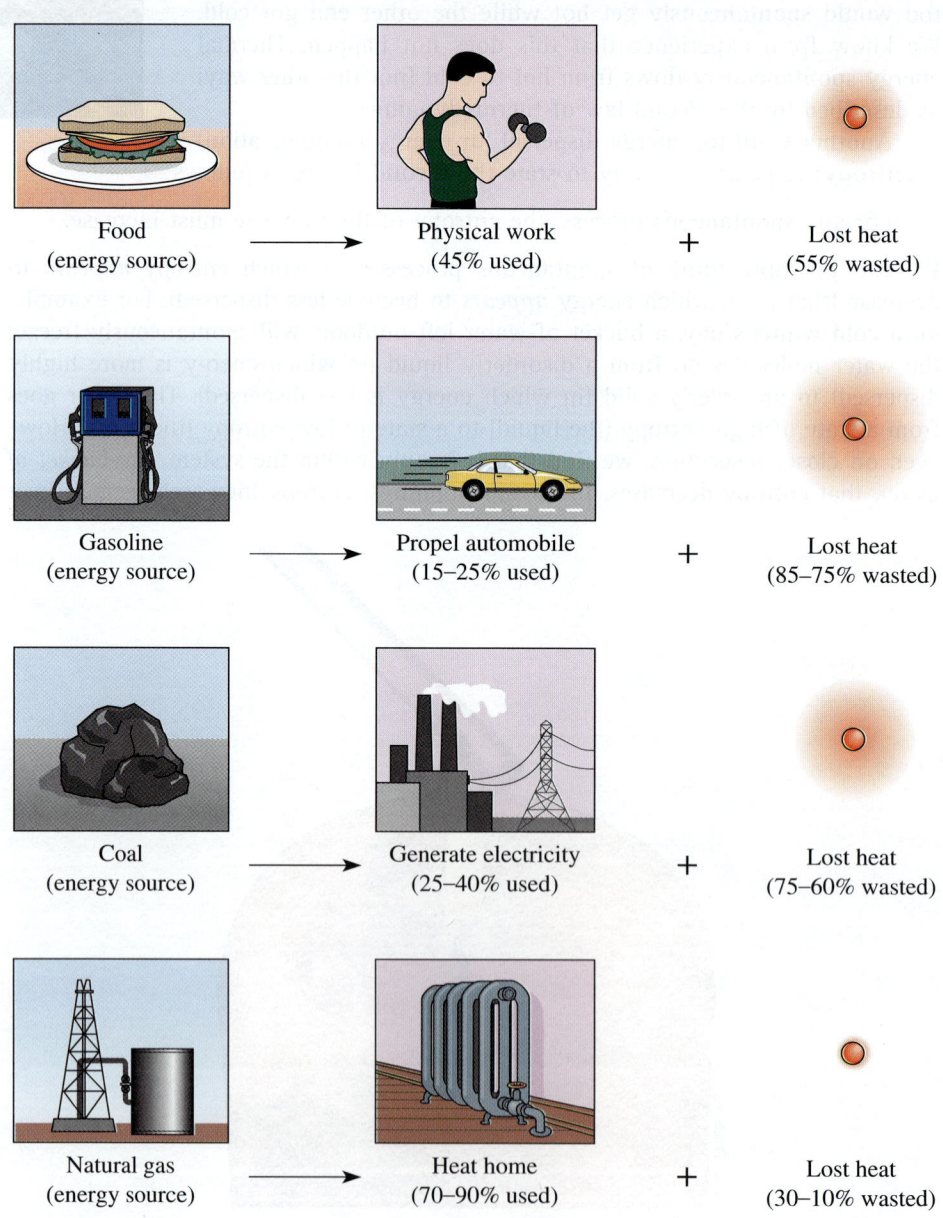

FIGURE 4 The efficiency of several energy conversions.

ing the battery, but some energy is lost to the surroundings in order for entropy to increase. In other words, nature takes a *heat tax*, an unavoidable cut of every energy transaction.

The implications of the second law for energy use are significant. First of all, according to the second law, not only are we unable to make a machine that creates energy out of nothing, but we cannot even make a machine that continues to move spontaneously without energy input—we cannot create a **perpetual motion machine.** If the machine is to be in motion, it must pay the heat tax with each cycle of its motion—over time, it will therefore run down and stop moving.

Second, in most energy transactions, not only is the *heat tax* lost to the surroundings, but additional energy is lost as well (due to inefficiencies). Most mechanical devices, like cars or electrical generators, convert energy with only 15–40% efficiency. In a car, for example, only 20% of the energy released in the combustion of gasoline goes into moving the car forward; the rest is lost as heat. Figure 4 shows the efficiency of several energy conversions. Because every step in an energy-consuming process results in a loss of energy to the surroundings, it is most efficient to minimize the number of energy conversions required to achieve a particular goal. For example, it is more efficient to use natural gas to heat a house directly than to use natural gas to produce electricity to then heat a house.

> **APPLY YOUR KNOWLEDGE**
>
> The first two laws of thermodynamics are sometimes paraphrased as (1) you can't win and (2) you can't break even. Explain.
>
> Answer: Because energy is conserved (first law), you can't win—in an energy transaction you will never have new energy that was not there before (in some other form). Because energy must be dispersed (second law), you can't break even—in an energy transaction you always to have pay the heat tax.

9.5 Units of Energy

ENERGY

The basic unit of energy is the **joule (J)**, named after the English scientist **James Joule** (1818–1889), who demonstrated that energy could be converted from one type to another as long as the total energy was conserved. A second unit of energy in common use is the **calorie (cal)**, the amount of energy required to heat 1 g of water by 1°C. A calorie is a larger unit than a joule (J), with the conversion 1 cal = 4.18 J. A related energy unit is the nutritional or capital "C" **Calorie (Cal)**, equivalent to 1000 little "c" calories. The amount of energy consumed by nations or by the world is usually reported in a very large energy unit called a **quad** (quad stands for quadrillion British thermal units and is equivalent to 1.06×10^{18} J). Table 3 shows various energy units and their conversion factors.

TABLE 3

Energy Conversion Factors

1 calorie (cal)	= 4.18 joules (J)
1 Calorie (Cal)	= 1000 calories (cal)
1 kilowatt-hour (kWh)	= 3.6×10^6 joules (J)
1 quad	= 1.06×10^{18} joules (J)

TABLE 4
Energy Uses in Various Units

Unit	Amount Required to Raise 1 g of Water by 1°C	Amount Required to Light 100-W Bulb for 1 h	Amount Used by Average U.S. Citizen in 1 day	Amount Used in United States in 1 yr
joule (J)	4.18	3.6×10^5	9.9×10^8	1.1×10^{20}
calorie (cal)	1	8.6×10^4	2.36×10^8	2.5×10^{19}
Calorie (Cal)	0.001	86	236,000	2.5×10^{16}
kWh	1.1×10^{-6}	0.10	274	2.9×10^{13}
quads	—	—	9.3×10^{-10}	98

Table 4 shows the amount of energy required for various processes in each of these units.

Another quantity related to energy is **power**, energy per unit of time. Power is the *rate* of energy output or input; *it is not energy*. The relationship between energy and power is similar to the relationship between distance and speed. Speed determines how long it will take to cover a given distance; power determines how long it will take to use a given amount of energy. The basic unit of power is the watt (W), equivalent to 1 J/s. Table 5 shows the power consumed, in watts, by several processes. The watt is familiar to most consumers because light bulbs are usually rated in watts. A 100-W light bulb uses 100 J/s. Power multiplied by time results in energy, just as speed multiplied by time results in distance:

$$\text{speed (mi/h)} \times \text{time (h)} = \text{distance (mi)}$$

$$\text{power (J/s)} \times \text{time (s)} = \text{energy (J)}$$

EXAMPLE 9.1

Conversion of Energy Units

A particular chocolate chip cookie contains 235 Calories of nutritional energy. How many joules does it contain?

SOLUTION

$$235 \text{ Cal} \times \frac{1000 \text{ cal}}{\text{Cal}} = 2.35 \times 10^5 \text{ cal}$$

$$2.35 \times 10^5 \text{ cal} \times \frac{4.18 \text{ J}}{\text{cal}} = 9.82 \times 10^5 \text{ J}$$

YOUR TURN

Conversion of Energy Units

The complete combustion of a small wooden match produces approximately 512 cal of heat. How many kilojoules (kJ) are produced?

9.5 Units of Energy

TABLE 5
Common Examples of Power Usage

Power use of 100-W light bulb	100 W
Average power use of human body	100 W
Average power use of average American	10,000 W
Average power use in a residential home	1000 W
Maximum power produced by typical coal-fired power plant	1000 MW (megawatts, mega = 10^6)

Electrical energy consumption is usually reported on electric bills in kilowatt-hours (kWh), a unit of energy because it has units of power multiplied by time. (Remember that kilo means 10^3 so that 1 kW is 1×10^3 W.) To determine how many kilowatt-hours a particular appliance uses, you simply multiply its electricity consumption in kilowatts by the number of hours that the appliance is in operation.

$$\text{energy use} = \text{power (in kW)} \times \text{hours of operation}$$

For example, a 100-W light bulb will consume 1 kWh in 10 h.

EXAMPLE 9.2

Calculating Energy Use in Kilowatt-Hours

A window air-conditioning unit is rated at 1566 W. (a) Calculate how many kilowatt-hours (kWh) the air conditioner consumes per month if it operates 6 hours per day. (b) If electricity costs 15 cents per kilowatt-hour, what is the monthly cost of operating the window air conditioner?

SOLUTION

a. The power rating of the air conditioner in kilowatts is

$$1566 \text{ W} \times \frac{1 \text{ kW}}{1000 \text{ W}} = 1.566 \text{ kW}$$

The total number of hours that it is in operation per month is

$$\frac{6 \text{ h}}{\text{day}} \times \frac{30 \text{ day}}{1 \text{ month}} = 180 \text{ h/month}$$

Therefore the number of kilowatt-hours used per month is

$$1.566 \text{ kW} \times 180 \text{ h} = 282 \text{ kWh}$$

b. Multiply the number of kilowatt-hours used by the cost per kilowatt-hour.

$$282 \text{ kWh} \times \frac{\$0.15}{\text{kWh}} = \$42.30$$

YOUR TURN

Calculating Energy Use in Kilowatt-Hours

What is the yearly cost of operating a 100-W television for 2 hours per day assuming the cost of electricity is 15 cents per kilowatt-hour?

9.6 Temperature and Heat Capacity

TEMPERATURE

The **temperature** of an object is a meaure of the kinetic energy associated with the motion of its composite atoms and molecules. Temperature is measured on three different scales, the most familiar being the **Fahrenheit scale.** In the Fahrenheit scale, water freezes at 32°F and boils at 212°F; room temperature is approximately 75°F. The scale most often used by scientists is the **Celsius scale,** in which water freezes at 0°C and boils at 100°C; room temperature is approximately 25°C. The Fahrenheit scale and the Celsius scale differ both in the size of their respective degrees and the temperature each calls zero (Figure 5). Both the Fahrenheit and Celsius scales contain negative temperatures. However, a third scale, called the **Kelvin scale,** avoids negative temperatures by assigning 0 K to the coldest temperature possible, absolute zero. Absolute zero (−273°C or −459°F) is the temperature at which all molecular motion stops; lower temperatures do not exist. The size of the Kelvin degree is identical to the Celsius degree; the only difference is the temperature that each calls zero.

Conversions between these temperature scales are relatively straightforward and are achieved using the following formulas:

$$K = °C + 273$$
$$°C = 5/9[°F - 32]$$
$$°F = 9/5[°C] + 32$$

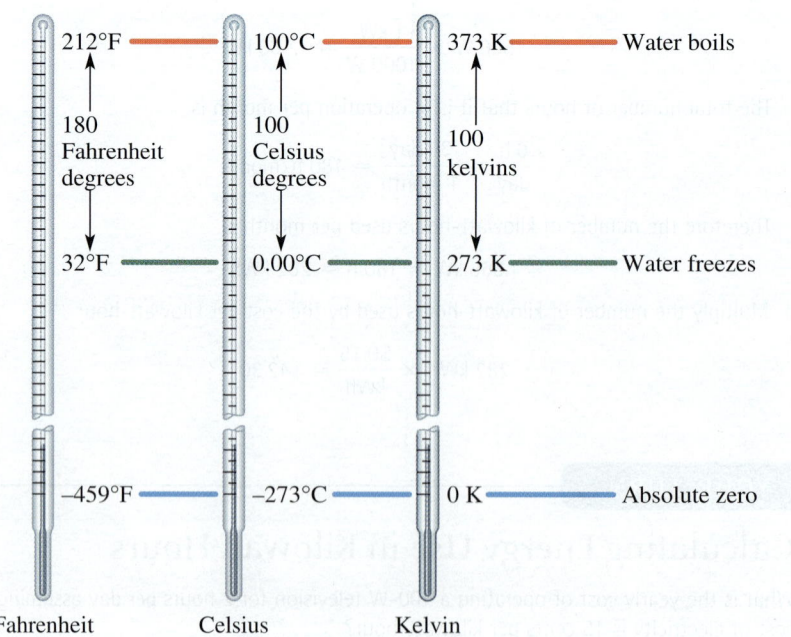

FIGURE 5 A comparison of the Fahrenheit, Celsius, and Kelvin temperature scales. The Fahrenheit scale uses graduations (degrees) that are roughly one-half the size of a Celsius degree. The only difference between the Celsius scale and the Kelvin scale is the temperature that each calls zero.

EXAMPLE 9.3

Temperature Conversion

Convert 80°F to Kelvin.

SOLUTION
We first convert to Celsius:

$$°C = 5/9[80 - 32] = 26.7°C$$

and then to Kelvin:

$$K = 26.7 + 273 = 299.7 \text{ K}$$

YOUR TURN

Temperature Conversion

Convert 400 K to degrees Fahrenheit

HEAT CAPACITY

Heat is the transfer of thermal energy due to a temperature difference. All substances change temperature when they are heated, but the magnitude of the change for a given amount of heat varies significantly from one substance to another. For example, if you put a steel skillet on a flame, its temperature rises rapidly. However, if you put some water in the skillet, the temperature increase is significantly slower. Why? The water resists changes in temperature more than steel because it has a higher **heat capacity**. The heat capacity of a substance is the quantity of heat energy required to change the temperature of a given amount of the substance by 1°C.

Water has one of the highest heat capacities known; it requires a lot of heat to change its temperature. If you have traveled from an inland region to the coast and felt the drop in temperature, you have experienced the effects of water's high-heat capacity. On a summer's day in California, for example, the temperature difference between Sacramento and San Francisco can be 30°F; San Francisco enjoys a cool 68°F, while Sacramento bakes near 100°F. Yet the intensity of sunlight falling on these two cities is the same. Why is there a large temperature difference? The difference between the two locations is the Pacific Ocean, which practically engulfs San Francisco. The high-heat-capacity water absorbs a lot of heat without a large increase in temperature, keeping San Francisco cool. The low-heat-capacity land surrounding Sacramento, on the other hand, cannot absorb a lot of heat without a large increase in temperature—it has a lower *capacity* to absorb heat for a given temperature change.

Water has a very high heat capacity, meaning that it takes a great deal of heat to change its temperature. Because of this, water will also release a great deal of heat as it cools, and therefore hot water can cause severe burns.

APPLY YOUR KNOWLEDGE

Which of the following would release more heat upon cooling from 50°C to 25°C: a 5-kg rock or a 5-kg jug of water?

Answer: The jug of water would release more heat because water has a higher heat capacity. It would take more energy to warm the water, and more energy is released as it cools.

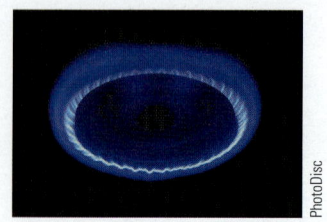

The burning of natural gas is an exothermic chemical reaction. Exothermic reactions have a negative ΔH; endothermic reactions have a positive ΔH.

9.7 Chemistry and Energy

As we have already seen, chemical reactions exchange energy with their surroundings. Those that give off energy to the surroundings are called **exothermic** reactions, whereas those that absorb energy from the surroundings are called **endothermic** reactions. The burning of natural gas is a good example of an exothermic chemical reaction. If you put your hand next to a natural gas flame, you feel heat; your hand becomes part of the surroundings, absorbing heat from the system. A chemical cold pack, often used to ice athletic injuries, is a good example of an endothermic reaction. The molecules absorb energy as they undergo a chemical reaction, and the energy comes from the surroundings. If the cold pack is put on your sore muscle, your muscle becomes part of the surroundings and loses energy to the molecules in the cold pack, cooling your muscle.

The amount of heat absorbed or emitted by a chemical reaction can be quantified using the **enthalpy of reaction** (ΔH_{rxn}). For example, the combustion of natural gas has $\Delta H_{rxn} = -11.8$ kcal/g CH_4, meaning that 11.8 kcal are emitted for each gram of CH_4 that is burned:

$$CH_4 + 2O_2 \longrightarrow CO_2 + 2H_2O \qquad \Delta H = -11.8 \text{ kcal/g } CH_4$$

By convention, the negative sign indicates an exothermic reaction, and a positive sign, an endothermic one because ΔH refers to enthalpy change from the point of view of the *system*. If the system loses energy (as in an exothermic reaction), the sign of ΔH is negative. (Think of money being withdrawn from a bank account, which also carries a negative sign.) If the system gains energy (as in an endothermic reaction), the sign of ΔH is positive. (Think of money being deposited into a bank account, which carries a positive sign.) The larger the magnitude of the enthalpy of reaction the greater the amount of heat absorbed or emitted by the reaction per gram of reactant.

For a substance to be a fuel, it must undergo an exothermic chemical reaction. Furthermore, it must release a significant amount of energy per gram of the substance. The enthalpy of reaction for *combustion* is often called the **enthalpy of combustion** (ΔH_{com}). The enthalpy of combustion allows the energy content of different fuels to be compared directly. Table 6 lists the heat of combustion of several fuels. Notice that the amount of energy released per gram varies from one

TABLE 6
Approximate Heat of Combustion of Some Fuels

Fuel	Heat of Combustion (kcal/g)	(kJ/g)
Pine wood	−5.1	−21
Methanol	−5.4	−23
Ethanol	−7.1	−30
Bituminous coal	−6.8	−28
Isooctane (a component of gasoline)	−8.7	−36
Natural gas	−11.8	−49.3
Hydrogen	−34.2	−143

fuel to another. For example, one of the advantages of burning isooctane—a component of gasoline—over wood or coal is its greater energy content.

EXAMPLE 9.4

Enthalpy of Reaction

How much energy in kilocalories is emitted by the complete combustion of 36.0 g of CH_4? (Note: $\Delta H_{rxn} = -11.8$ kcal/g CH_4.)

SOLUTION

$$36.0 \text{ g CH}_4 \times \frac{-11.8 \text{ kcal}}{\text{g CH}_4} = -425 \text{ kcal}$$

The negative sign indicates that the energy is emitted.

YOUR TURN

Enthalpy of Reaction

How much energy in kilocalories is emitted by the complete combustion of 386 g of isooctane (C_8H_{18})? (Note: $\Delta H_{rxn} = -8.7$ kcal/g C_8H_{18}.)

APPLY YOUR KNOWLEDGE

When two solutions are mixed in a beaker, a reaction occurs, and the temperature drops. Is the reaction exothermic or endothermic? Is the sign of ΔH positive or negative?

Answer: The reaction is endothermic, and the sign of ΔH is positive. The system absorbs energy from the surroundings, causing the drop in temperature.

9.8 Energy for Our Society

Prior to 1970, energy was taken for granted; most Americans believed there would always be more and that it would always be cheap. The 1970s ended this illusion when North America experienced an energy crisis. Gasoline became limited, long lines formed at filling stations, and the price of gasoline soared from 30 cents per gallon to $1 per gallon. Americans realized that current energy sources would not last forever. The 1990s, however, brought unprecedented economic expansion fueled by growth in technology and benign increases in energy costs. In inflation-adjusted dollars, even with recent increases in energy costs, the prices of fuels such as gasoline have only recently (at the time of this printing) reached the historical record high (which occurred in 1981). The prudency gained from our last energy crisis was largely forgotten until the recent spike in oil prices redirected our attention to the limited supply of oil.

Where do we get our energy? The historical pattern of energy use by source in the United States is shown in Figure 6. About 85% of our energy comes from fossil fuels (petroleum, natural gas, and coal). Our current major energy sources and

FIGURE 6 Historical energy use by source in the United States. (Source: Energy Information Administration Annual Energy Review for the 2003 calendar year.)

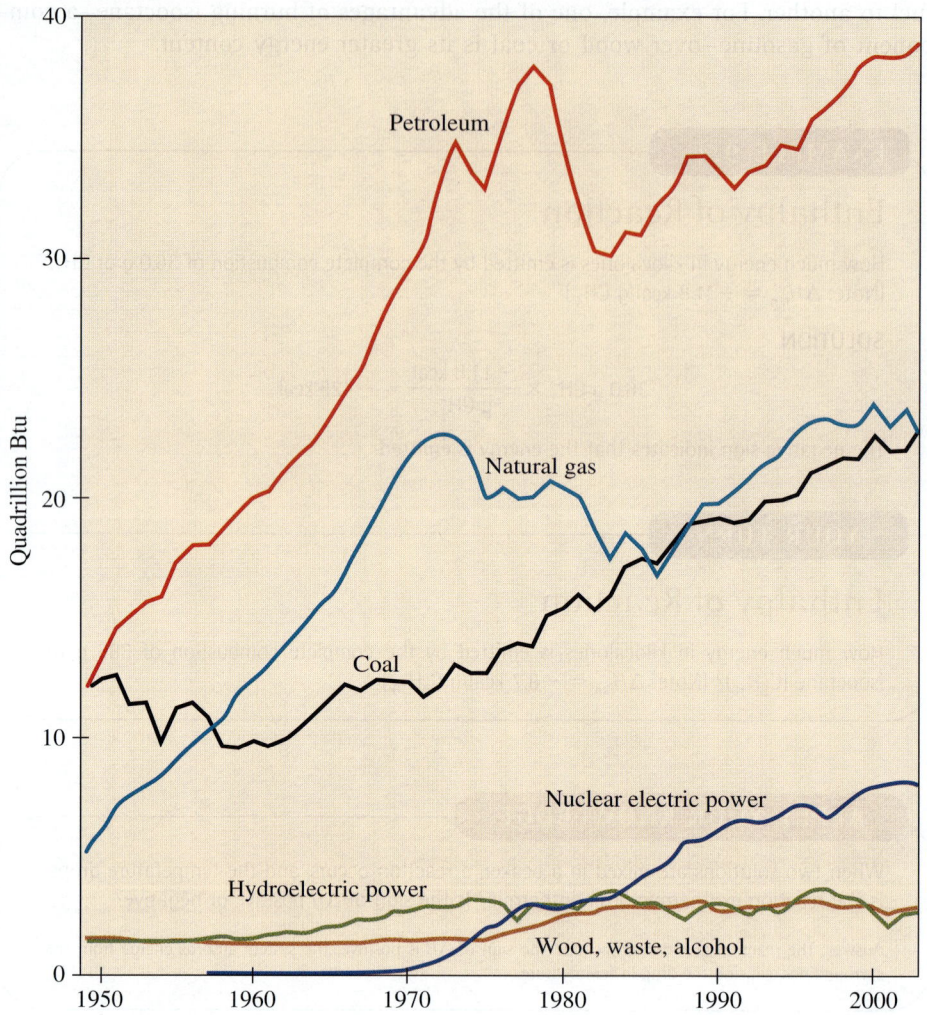

approximate times until their depletion, based on current rates of consumption, are shown in Table 7. Although the number of years until depletion is only an approximation, it is clear that fossil fuels will not last forever and that new sources must be developed.

Fossil fuels are primarily hydrocarbons. As we learned in Chapter 6, **natural gas** is a mixture of methane and ethane. **Petroleum** contains a wide variety of hydrocarbons, ranging from pentane with 5 carbon atoms to hydrocarbons with 18 carbon atoms or more. **Coal** is primarily long chains and rings of carbon with 200 or more atoms. The molecules that compose fossil fuels are highly energetic because they formed via endothermic reactions by plant and animal life from prehistoric times. The energy within these molecules ultimately came from sunlight as ancient plants used the sun's energy to synthesize energetic molecules in a reaction called **photosynthesis**:

$$\text{sunlight} + 6CO_2 + 6H_2O \longrightarrow 6O_2 + C_6H_{12}O_6$$

Plants capture the sun's energy in a process called photosynthesis. The energy is stored in the chemical bonds of glucose molecules.

Glucose, $C_6H_{12}O_6$, is an energy-containing molecule used by plants and animals. Glucose was converted into the fossil fuels we use today by a combination of chemical reactions within prehistoric living plants and a series of slow reactions

MOLECULAR THINKING

Campfire Smoke

Nothing feels better when camping than a hot, roaring fire on a cold night. However, if your fire-making skills are still in their infancy, your fires may be a smoky disappointment rather than a blazing delight. The smokiness of a fire depends on how hot the fire is; the hotter the fire, the less the smoke.

QUESTION: What causes a cool fire to be smoky and a hot one to be smoke free? Can you give a molecular reason for the smoke? Hint: If the temperature is too low, combustion will be incomplete.

TABLE 7

Total U.S. Energy Consumption by Source

Fuel	Percentage of Total	Years until Depletion[a]
Petroleum	39.8	50
Natural gas	22.9	110
Coal	23.1	1500
Nuclear	8.1	100[b]
Hydroelectric	2.8	indefinite
Other	3.3	

[a] At current rates of consumption. Includes approximations for undiscovered resources.

[b] This estimate is based on conventional nuclear power plants and excludes breeder reactors, which would increase this number 100 times.

Source: Data compiled from the U.S. Energy Information Administration, *Annual Energy Review* for the 2003 calendar year. Reserve estimates are from *Scientific American*, September 1990.

---

that occurred since their death. This process takes millions of years, so we have some time to wait before more fossil fuels naturally form.

The energy contained in the chemical bonds of fossil fuels is tapped by **combustion** reactions, the reverse of photosynthesis. For example, octane in gasoline is burned according to the reaction:

$$2C_8H_{18} + 25O_2 \longrightarrow 16\ CO_2 + 18H_2O + \text{energy}$$

Here, an energy-containing molecule, C_8H_{18}, reacts with O_2 to form CO_2 and H_2O, emitting energy. In cars, the energy expands the air in the engine's cylinders and drives it forward. The combustion of natural gas and coal proceeds in a similar fashion, consuming oxygen and producing CO_2, H_2O, and energy.

EXAMPLE 9.5

Writing and Balancing Combustion Reactions

Write a chemical equation for the combustion of propane (C_3H_8), the main component of LP gas.

SOLUTION

The combustion of any fossil fuel will have the fuel and oxygen as reactants and carbon dioxide and water as products.
The unbalanced equation is

$$C_3H_8 + O_2 \longrightarrow CO_2 + H_2O$$

We balance the equation, by first balancing carbon:

$$C_3H_8 + O_2 \longrightarrow 3CO_2 + H_2O$$

We balance hydrogen next:

$$C_3H_8 + O_2 \longrightarrow 3CO_2 + 4H_2O$$

We balance oxygen, which occurs as a pure element on the left-hand side of the equation, last:

$$C_3H_8 + 5O_2 \longrightarrow 3CO_2 + 4H_2O$$

YOUR TURN

Writing and Balancing Combustion Reactions

Write a chemical equation for the combustion of butane (C_4H_{10}), the fuel used in cigarette lighters.

9.9 Electricity from Fossil Fuels

About 70% of U.S. electricity (Figure 7) is generated by the burning of fossil fuels. Fossil-fuel-burning power plants (Figure 8) use the heat emitted in combustion reactions to boil water, creating steam that turns the turbine of an electric generator. The generator creates electricity that is transmitted to buildings through power lines. Power plants create electricity *on demand;* they do not store it for later

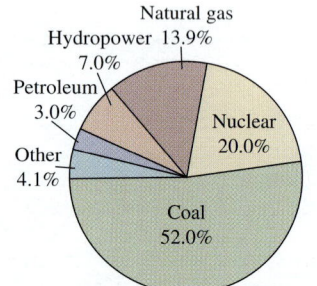

FIGURE 7 Sources of energy for electricity generation in the United States. (U.S. Energy Information Administration, *Annual Energy Review* for the 2003 calendar year.)

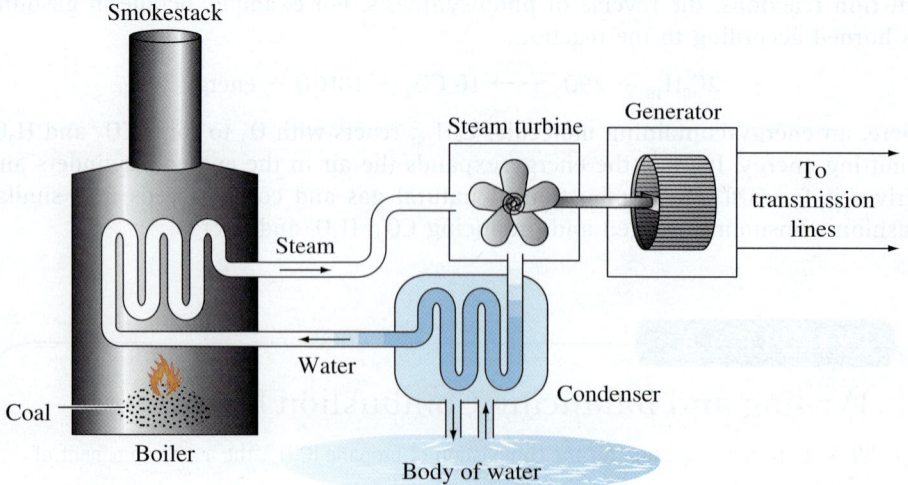

FIGURE 8 Fossil-fuel-burning power plants burn oil, coal, or natural gas to create steam. The steam turns a turbine on a generator to create electricity.

use. As the demand for electricity varies throughout the day, the amount of fuel is adjusted to meet the demand. When you turn on a light, your local plant uses a little more fuel. The energy output of power plants is measured in watts; a typical fossil-fuel power plant generates 1 gigawatt (GW, giga = 10^9) of power in the form of electricity, sufficient to power 1 million homes.

9.10 Smog

The combustion of hydrocarbons such as octane produces, in the ideal case, only carbon dioxide and water. However, because of other reactions that occur during combustion and because of impurities in fossil fuels, other substances are also produced, some of which have undesirable effects on the environment. We will look at these substances in more detail, including their levels in the air, in Chapter 11. For now, we introduce them and learn how they are related to fossil-fuel combustion.

W Explore this topic on the **Smog** website.

CARBON MONOXIDE: MAKING THE HEART WORK HARDER

In automobile engines, combustion is not always complete, producing carbon monoxide (CO) from some of the carbon that would, if completely burned, form carbon dioxide (CO_2). Carbon monoxide is a colorless, odorless, and tasteless gas that builds up in urban areas from automobile exhaust. It is harmful to humans because it binds to hemoglobin, the oxygen-carrying protein in blood, and diminishes the blood's capacity to carry oxygen. Carbon monoxide displaces oxygen on hemoglobin (Hb) according to the following reaction:

$$Hb\text{-}O_2 + CO \longrightarrow Hb\text{-}CO + O_2$$

Once Hb attaches carbon monoxide, it is no longer available to carry oxygen. In high concentrations, carbon monoxide is lethal; it is usually the cause of death in suicides involving automobile exhaust inhalation. At low concentrations, carbon monoxide causes the heart and respiratory system to work harder, possibly leading to increased risk of heart attack.

NITROGEN OXIDES: THE BROWN COLOR OF POLLUTED AIR

Because of nitrogen impurities in fossil fuels, and the presence of nitrogen in air, nitrogen monoxide (NO) forms as a byproduct of combustion according to the following reaction:

$$N_2 + O_2 \longrightarrow 2NO$$

Consequently, nitrogen monoxide is emitted in the exhaust of cars and fossil-fuel-burning power plants. Once in the air, nitrogen monoxide reacts with oxygen to produce nitrogen dioxide (NO_2):

$$2NO + O_2 \longrightarrow 2NO_2$$

Nitrogen dioxide is a brown gas that gives smog its characteristic color. It is also an eye and lung irritant; the burning feeling in your eyes on a smoggy day is partly because of nitrogen dioxide. We will see later that nitrogen dioxide also combines with moisture in the air to produce acid rain.

Photochemical smog over Mexico City. The brown color is due primarily to NO_2.

The cracking of rubber products such as this automobile tire are primarily caused by the formation of ozone and PAN.

OZONE AND PAN: STINGING EYES AND CRACKED RUBBER

Another consequence of incomplete combustion is the emission of partially burned hydrocarbons. These partially burned hydrocarbons combine with nitrogen dioxide (in polluted air) in reactions that require sunlight to form two particularly harmful components of smog, O_3 (ozone) and $CH_3CO_2NO_2$ (peroxyacetylnitrate, PAN):

$$\text{hydrocarbons} + NO_2 + \text{sunlight} \longrightarrow \underset{\text{ozone}}{O_3} + \underset{\text{PAN}}{CH_3CO_2NO_2}$$

Ozone and PAN are the main components of what is often called **photochemical smog**; they irritate the lungs, make breathing difficult, sting the eyes, damage rubber products, and damage crops. The dependence on sunlight for ozone and PAN to form is one reason that smog increases during the summer months.

CATALYTIC CONVERTERS

To reduce pollutant levels in urban areas, cars are equipped with catalytic converters (Figure 9). A catalytic converter employs **catalysts**, substances that promote chemical reactions without being consumed by them, to promote the *complete* combustion of hydrocarbons and to promote the decomposition of nitrogen monoxide. Partially burned hydrocarbons, carbon monoxide, and nitrogen monoxide in engine exhaust are channeled through the catalytic converter where they absorb or stick onto catalytic surfaces. The partially burned hydrocarbons react with oxygen to complete the combustion process in reactions such as

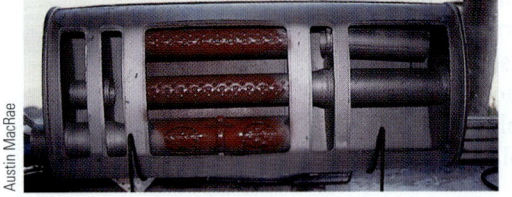

$$CH_3CH_2CH_3 + 5O_2 \xrightarrow{\text{catalyst}} 3CO_2 + 4H_2O$$

FIGURE 9 Cutaway of an automotive catalytic converter.

NO reacts with CO to form N_2 and CO_2 according to the following reaction:

$$2NO + 2CO \xrightarrow{\text{catalyst}} N_2 + 2CO_2$$

As a result of these reactions, catalytic converters lower the emissions of three harmful pollutants: partially burned hydrocarbons, carbon monoxide, and nitrogen monoxide (Figure 10). Because partially burned hydrocarbons and nitrogen monoxide are critical to photochemical smog formation, smog is also reduced by catalytic converters.

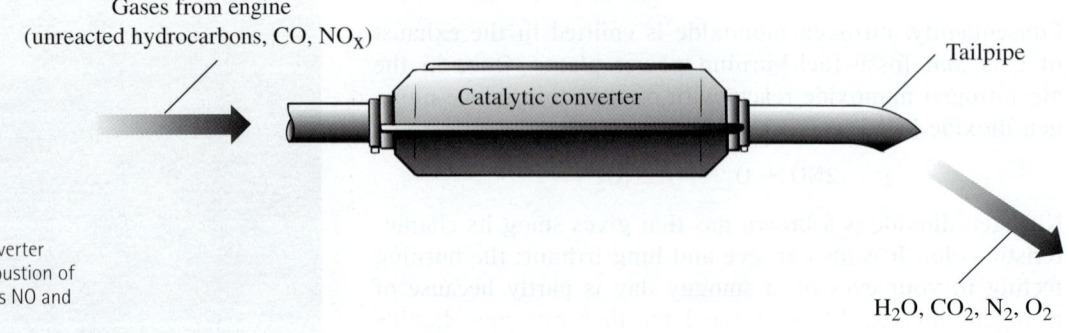

FIGURE 10 A catalytic converter promotes the complete combustion of hydrocarbons. It also converts NO and CO to N_2 and CO_2.

9.11 Acid Rain

The combustion of fossil fuels also produces an environmental problem called **acid rain**. We will look more closely at the details of acid chemistry and acid rain in Chapter 13. For now, we concentrate on identifying the emissions responsible for acid rain and its effects.

The primary cause of acid rain is the sulfur dioxide emission from fossil-fuel combustion. Sulfur dioxide forms because fossil fuels, especially coal, contain sulfur impurities; their combustion produces sulfur dioxide (SO_2):

$$S + O_2 \longrightarrow SO_2$$

The emissions of nitrogen monoxide (NO) and nitrogen dioxide (NO_2), together known as NO_x, also play a role in acid rain. In the United States, over 20 million tons of each of these pollutants—often called NO_x and SO_x—are emitted into the atmosphere each year. Electric utility plants account for 70% of annual SO_2 and 30% of annual NO_x emissions. When SO_2 and NO_x combine with water, they form acids that fall as acid rain (Figure 11). Acids are chemical compounds

 Explore this topic on the **Acid Rain** website.

Coal-burning power plants such as this one collectively emit millions of tons of SO_2 into the atmosphere each year.

FIGURE 11 Acid rain forms when SO_2, emitted primarily by coal-burning electric power plants, combines with water to form acids.

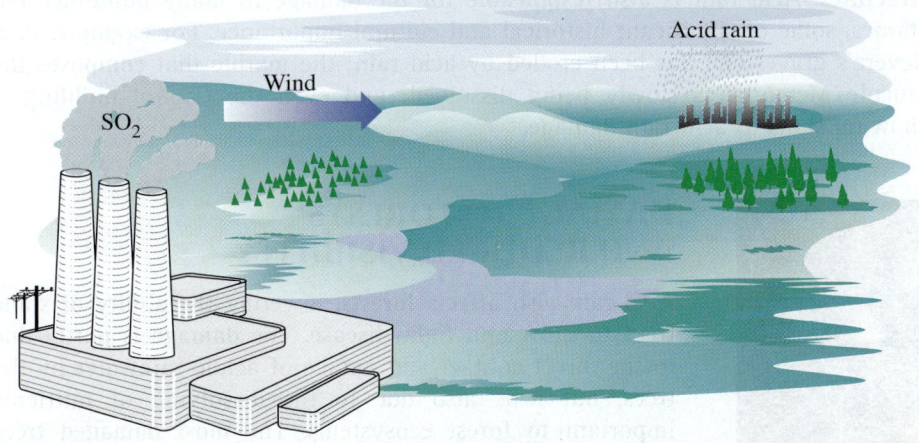

 Explore this topic on the **Sulfur Dioxide** website.

Molecular Focus

Sulfur Dioxide

Formula: SO_2
Molar mass: 64.07 g/mol
Melting point: $-72°C$
Boiling point: $-10°C$
Lewis structure:

$$:\ddot{O}-\ddot{S}=\ddot{O}:$$

Three-dimensional structure:

Sulfur dioxide is a colorless, nonflammable gas with a strong suffocating odor. It is formed by burning sulfur directly or by burning sulfur-containing compounds. Sulfur dioxide is emitted during fossil-fuel combustion, especially coal, and by industry in the extraction of iron and copper from their ores. Sulfur is also emitted naturally during volcanic eruptions.

Sulfur dioxide has some beneficial uses. It is a common preservative for fruits, vegetables, juices, and wine. It is used as a disinfectant in food factories and as a bleach for textiles and straw.

that produce H$^+$ ions. These H$^+$ ions are highly reactive, and in high enough concentration will kill fish, damage forests, and destroy building materials. The problem is most severe in the northeastern United States and eastern Canada because emissions from midwestern coal-burning U.S. power plants are carried east by prevailing winds.

DAMAGE TO LAKES AND STREAMS

Hundreds of lakes in the Adirondack Mountains now have acid levels that are unsuitable for the survival of certain fish species. Many are so acidic that nothing can survive in them; they are sterile. These lakes, along with thousands of others in the eastern part of our country, have suffered the effects of acid rain.

DAMAGE TO BUILDING MATERIALS

Acids dissolve metals and other building materials such as stone, marble, and paint. The inevitable rusting process of steel is accelerated by the presence of acids, resulting in damage to bridges, railroads, automobiles, and other steel structures. Acid rain is also responsible for the damage to many buildings and statues, some of significant historical and cultural importance. For example, Paul Revere's gravestone has been eroded by acid rain; the marble that composes the Lincoln Memorial is slowly being dissolved; and even the Capitol building is showing signs of acid rain damage.

A forest damaged by acid rain.

DAMAGE TO FORESTS AND REDUCED VISIBILITY

Acid rain also affects forests, lowering the ability of some trees to grow and fight disease. The damage is partly due to the direct contact and uptake of acidic rainwater by the trees, but it is also due to the dissolving of nutrients important to forest ecosystems. The most damaged trees are the red spruce trees that grow along the ridges of the Appalachian Mountains from Maine to Georgia.

Sulfur dioxide emissions also lead to decreased visibility, in some cases spoiling scenic vistas. The sulfur dioxide combines with atmospheric moisture to form tiny droplets called sulfate aerosols. According to the Environmental Protection Agency, these aerosols account for over 50% of the visibility problems in the eastern United States.

9.12 Environmental Problems Associated with Fossil-Fuel Use: Global Warming

A calculation of the Earth's temperature based on incoming light from the Sun and reradiated heat from Earth predicts an average global temperature about 60°F

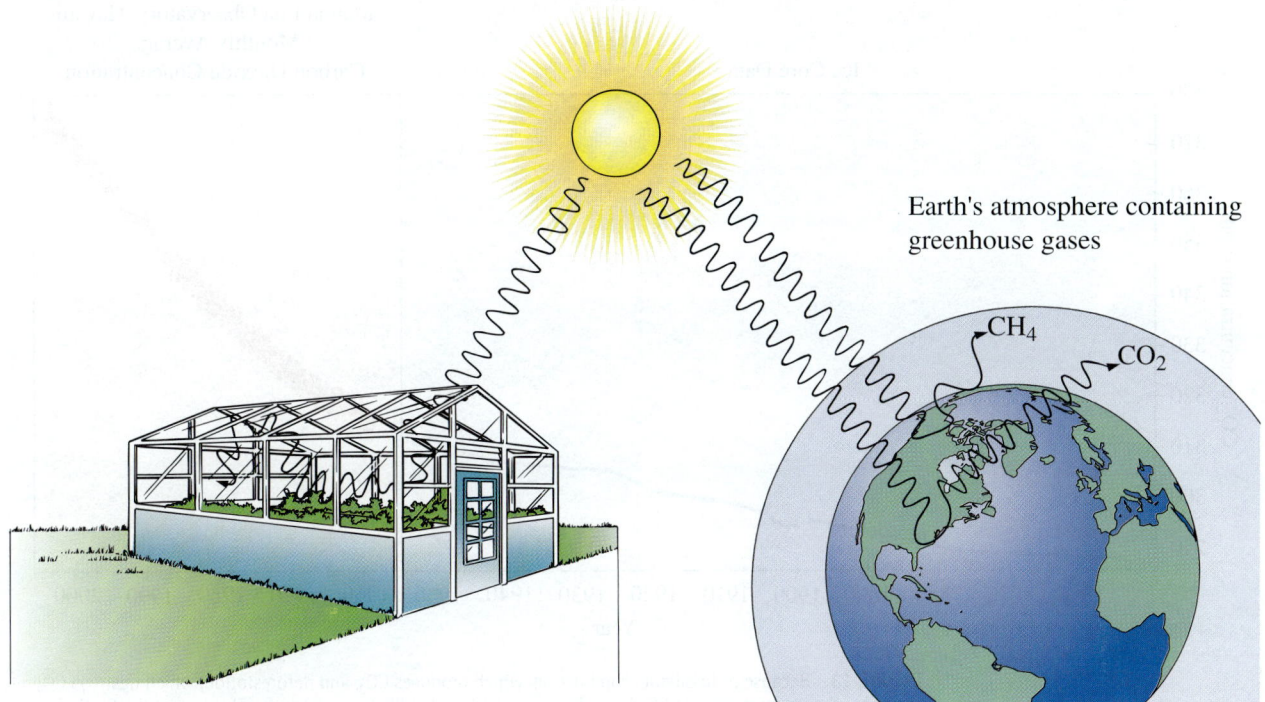

FIGURE 12 Visible sunlight passes through the windows in the greenhouse and is converted to infrared light when it is absorbed by objects inside. The IR light is then trapped inside because glass in opaque to IR light. Visible sunlight passes through Earth's atmosphere and is converted to IR light by objects on Earth. The IR light is then trapped inside Earth's atmosphere due to the presence of greenhouse gases such as CO_2 and CH_4. These gases are transparent to visible light but absorb IR light.

colder than we actually experience. The reason for this difference is the presence of our atmosphere, the thin layer of gases that surround the planet. Our atmosphere contains **greenhouse gases,** gases that allow visible light into the atmosphere but prevent heat in the form of infrared light from escaping (Figure 12). In a greenhouse, clear glass allows visible sunlight to enter. Inside the greenhouse, that visible sunlight is converted into heat as it is absorbed by the plants, soil, and other substances within the greenhouse. The heat energy is then reemitted as infrared light. Glass, however, absorbs infrared light, preventing its escape. The energy is therefore trapped in the greenhouse, causing warmer temperatures inside than outside. A similar situation occurs in a parked car on a sunny day; the glass windows allow visible light to enter but prevent IR light from escaping, warming the interior of the car to temperatures significantly above the outdoor temperature. Thanks to this natural greenhouse effect, the average temperature of the Earth is comfortable; without it, even tropical locations such as Hawaii would be below freezing throughout the year.

The most significant greenhouse gas is carbon dioxide, which naturally composes approximately 0.03% of our atmosphere. Carbon dioxide strongly absorbs IR light at 10 (μm), the peak wavelength of the Earth's IR emission. The combustion of fossil fuels on a global scale, in combination with deforestation,

 Explore this topic on the **Greenhouse Effect and Global Warming** website.

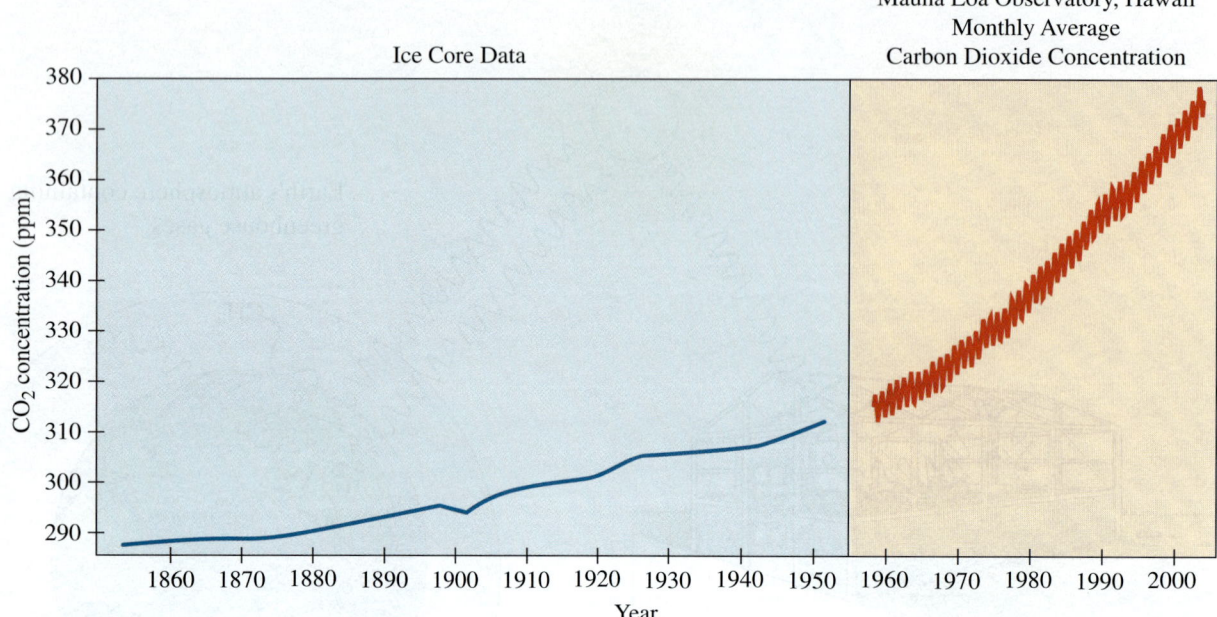

FIGURE 13 Because of fossil-fuel combustion, which produces CO_2, and deforestation, which destroys CO_2 absorbing forests, atmospheric CO_2 levels have been rising steadily for many years. The annual fluctuations are due to seasonal growth patterns in trees and plants. CO_2 levels are lowest in the summer when trees and plants undergo maximum growth, and highest in the winter when growth slows. Data were collected at the Mauna Loa Observatory in Hawaii and from ice cores. (Source: Carbon Dioxide Information Analysis Center, U.S. Department of Energy.)

has increased the level of CO_2 in the Earth's atmosphere (Figure 13) by about 20% over the last century. Computer-based climate models predict that this increase should produce an increase in global temperature. Has such an increase occurred? A continuing study has addressed that question, and the results are shown in Figure 14. As you can see from the graph, global temperatures vary over time. However, there appears to be a general increase over the last century, with the last ten years being among the warmest on record. Most scientists studying global warming agree that the observed increase in temperature is due to the increase in CO_2; it appears that global warming has begun.

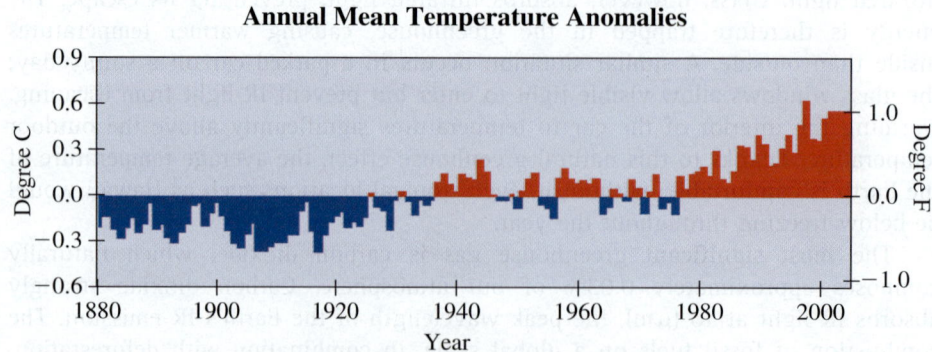

FIGURE 14 Global land and ocean temperatures from 1880–2004. Zero is 1880–2004 average temperature. (Source: National Oceanic and Atmospheric Administration, *Climate of 2004 Annual Review*.)

MOLECULAR THINKING

Are Some Fossil Fuels Better Than Others?

Not all fossil fuels are equal offenders when it comes to global warming. One way to categorize fossil fuels with relation to global warming is to look at their energy output per gram of CO_2 emitted. The higher the energy output per gram of CO_2, the better the fuel with respect to global warming because less CO_2 will be emitted in producing a given amount of energy. We can compare octane, as a representative of gasoline, with methane (natural gas) by looking at balanced chemical reactions for the combustion of each and their associated enthalpies of reaction. In this case, it is more convenient to express the enthalpy of reaction in units of kilocalories per mole (kcal/mol) rather than kilocalories per gram (kcal/gram).

$$C_8H_{18} + (25/2)\, O_2 \longrightarrow 8CO_2 + 9H_2O$$

$$\Delta H = \frac{-1309 \text{ kcal}}{\text{mole octane}}$$

$$CH_4 + 2O_2 \longrightarrow CO_2 + 2H_2O$$

$$\Delta H = \frac{-213 \text{ kcal}}{\text{mole methane}}$$

The combustion of 1 mol of octane produces 8 mol of CO_2 and 1309 kcal of energy, while the combustion of 1 mol of methane produces 1 mol of CO_2 and 213 kcal of energy. We calculate the number of grams of CO_2 produced by each reaction and then the amount of heat per gram of CO_2.

For octane:

$$8.00 \text{ mol } CO_2 \times \frac{44.0 \text{ g}}{1 \text{ mol}} = 352 \text{ g } CO_2$$

$$\frac{1309 \text{ kcal}}{352 \text{ g } CO_2} = \frac{3.72 \text{ kcal}}{\text{g } CO_2}$$

For methane:

$$1.00 \text{ mol } CO_2 \times \frac{44.0 \text{ g}}{1 \text{ mol}} = 44.0 \text{ g } CO_2$$

$$\frac{213 \text{ kcal}}{44.0 \text{ g } CO_2} = \frac{4.84 \text{ kcal}}{\text{g } CO_2}$$

We see that methane produces 4.84 kcal/gram of CO_2, 30% more than octane's 3.72 kcal/gram of CO_2; therefore, octane is worse than methane with respect to global warming. Coal is the worst offender, producing even less energy per gram of CO_2 than octane.

QUESTION: The enthalpy of reaction for the combustion of propane is 530 kcal/mol. With regard to global warming, how does it compare to octane and methane?

A continued increase in atmospheric CO_2 could lead to more global warming. If fossil-fuel combustion continues unabated, CO_2 levels should double from their preindustrial levels at about 2050. Computer-based climate models predict an average temperature increase of 3.5°F for a doubling of atmospheric CO_2. However, that temperature rise would not be uniform over the entire planet—some regions will warm more than others. North America would experience temperature increases of 5–10°F. This degree of warming would pose significant environmental problems. For example, precipitation regions would move northward, changing current forest and cropland geography. Weather is expected to become more extreme, producing more droughts and hurricanes. Polar ice caps could melt due to increasing temperatures, raising ocean levels and flooding low-lying urban areas.

However, many experts express a degree of caution in these predictions because the Earth's climate has changed in the past due to other factors. Fluctuations in the Sun's radiation, for example, have produced periods of colder climate such as the Ice Age. Also in question is the reliability of climate models. The observed temperature increases are at the lower end of what many models predict for the observed 20% increase in CO_2. The problem is that climate modeling is complicated, with many factors influencing the outcome. The Sun's radiation may change slightly from year to year, oceans absorb heat and CO_2, and air currents fluctuate. Add croplands, forests, and a multitude of other variables, and the situation becomes complex.

In spite of the complexity, most scientists agree that global warming is a serious concern. They differ, primarily, in their response. Many scientists believe

 Explore this topic on the **Carbon Sequestration** website.

THE MOLECULAR REVOLUTION

Taking Carbon Captive

The combustion of fossil fuels must produce carbon dioxide as a product, but does the carbon dioxide have to go into the atmosphere? The U.S. Department of Energy (DOE) does not think so; it has started a dialogue and research program that unites the government with academia and industry to develop technologies to "sequester" or trap carbon dioxide. Sorbents, substances that absorb carbon dioxide, are being developed that can significantly reduce and even eliminate carbon dioxide from exhaust with only about a 5% decrease in efficiency. These sorbents could be used in fossil-fuel-burning power plants and even in automobile tailpipes to capture carbon dioxide and prevent it from getting into the atmosphere.

The next problem is what to do with all that carbon dioxide—the world produces over 6 billion tons per year. Researchers are exploring ways to dispose of the carbon dioxide in environmentally benign ways. The top candidates are piping it deep into the ocean or injecting it into underground salt water aquifers or unmineable coal seams. There are currently several research projects in progress to assess the integrity and cost of these potential storage sites.

Alternatively, carbon dioxide could be sequestered directly out of the atmosphere. This could involve the enhancement of natural sinks, such as reforestation, or it could involve new technologies that mimic photosynthesis and use sunlight to convert carbon dioxide into other, perhaps useful, substances.

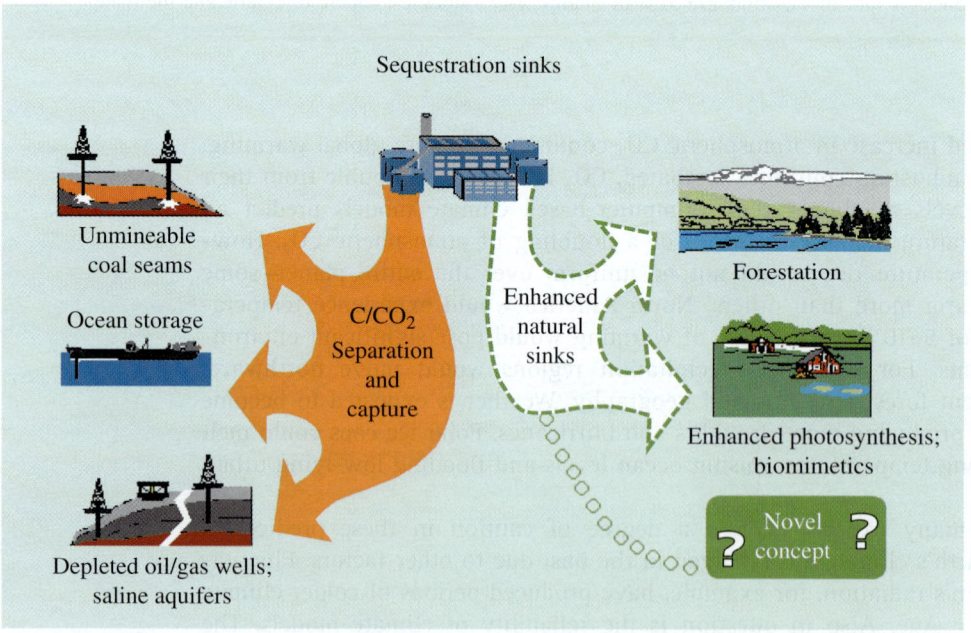

Scientists are researching ways to capture and store carbon dioxide, preventing it from entering the atmosphere where it enhances the greenhouse effect and causes global warming. (1999 Federal Energy Technology Center, U.S. Department of Energy.)

that governments should take immediate action to curb the emission of CO_2. Some believe that global changes in temperature will occur slowly and, in light of the uncertainty associated with climate models, the best response is to continue careful research until more certain answers are obtained.

THE KYOTO PROTOCOL: STEPS TO REDUCE GREENHOUSE GAS EMISSIONS

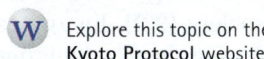
Explore this topic on the Kyoto Protocol website.

In 1997, an international meeting of the Climate Change Convention formalized a plan to begin the reduction of greenhouse gas emissions. Because the meeting was held in Kyoto, Japan, the plan is called the Kyoto Protocol. It commits industrialized nations to reducing their collective emissions of six greenhouse gases by at least 5% of 1990 levels by 2010. This reduction represents a 29% cut in the projected 2010 emission levels of greenhouse gases (because the emissions of these gases have increased since 1990). The Kyoto Protocol has been signed and ratified by 141 nations around the world and went into force in 2005. The only industrialized nations to not ratify the treaty were the United States and Australia.

Chapter Summary

MOLECULAR CONCEPT

We need energy, both individually to stay alive and as a society (9.2). Energy can be transferred between objects via heat, changing the temperature of an object, or it can be transferred via work, one object exerting a force on another object to make it move. In either case, these transfers involve changes in molecular motion. All energy transfers must obey the first law of thermodynamics, which states that energy can neither be created nor destroyed (9.3). They must also obey the second law of thermodynamics, which means that you always lose some energy in every transfer to pay nature's heat tax (9.4).

The majority of our society's energy needs are met by exothermic—heat-emitting—chemical reactions, specifically, the combustion of fossil fuels. The amount of heat produced or absorbed by a chemical reaction is called the enthalpy of reaction (9.5–9.7). Fossil fuels are good energy sources because their enthalpy of reaction for combustion is large and negative, meaning that they give off large amounts of heat when they are burned. Fossil fuels come from ancient plant and animal life and consist primarily of hydrocarbons such as methane, propane, and octane (9.8). The combustion of these hydrocarbons produces primarily carbon dioxide and water, but several other byproducts also form, some with environmental consequences (9.9–9.12).

SOCIETAL IMPACT

We use energy to transfer heat in heating and cooling (air conditioning or refrigeration) and to do work (moving cars, trucks, and airplanes). The first law of thermodynamics states that we cannot get energy out of nothing—there will never be a perpetual motion machine that can provide our energy needs without need for energy input (9.3). The second law of thermodynamics gives us a practical formula for energy use—the fewer the steps in any energy-consuming process the better, because energy will always be lost at each step (9.4). It is more efficient to heat your home by burning natural gas directly than by burning natural gas to generate electricity to then heat your home with an electric heater. Although fossil fuels—petroleum, natural gas, and coal—are good energy sources, they are not ideal. Their combustion produces smog—carbon monoxide, nitrogen oxides, and ozone—all of which pollute urban air and take their toll on human health (9.10). Their combustion also produces sulfur oxides, which together with nitrogen oxides, produce acid rain, which damages lakes, streams, building materials, and forests (9.11). A main product of combustion, carbon dioxide, enhances the natural greenhouse effect of the Earth and is causing small increases in average global temperatures (9.12). Climatic models predict that these temperature increases could get worse, especially if fossil-fuel combustion is not curbed.

Chemistry on the Web

For up-to-date URLs, visit the text website at www.chemistry.brookscole.com/tro3

- Smog
  http://www.smogcity.com/
  http://www.aqmd.gov/smog/inhealth.html

- Acid Rain
  http://www.epa.gov/airmarkets/acidrain/
  http://bqs.usgs.gov/acidrain/
  http://www.ec.gc.ca/acidrain/index.html

- Sulfur Dioxide
  http://www.epa.gov/airtrends/sulfur.html

- Greenhouse Effect and Global Warming
  http://www.pbs.org/wgbh/nova/ice/greenhouse.html
  http://yosemite.epa.gov/oar/globalwarming.nsf/content/index.html
  http://www.ncdc.noaa.gov/oa/climate/globalwarming.html

- Kyoto Protocol
  http://www.eia.doe.gov/oiaf/kyoto/kyotorpt.html
- Carbon Sequestration
  http://cdiac2.esd.ornl.gov/

Key Terms

acid rain	enthalpy of combustion	joule	power
calorie	enthalpy of reaction	Joule, James	quad
Calorie	entropy	Kelvin scale	second law of thermodynamics
catalysts	exothermic	kinetic energy	
Celsius scale	Fahrenheit scale	natural gas	surroundings
chemical energy	first law of thermodynamics	perpetual motion machine	temperature
coal	fossil fuels	petroleum	thermal energy
combustion	greenhouse gases	photochemical smog	thermodynamic system
endothermic	heat	photosynthesis	thermodynamics
energy	heat capacity	potential energy	work

Exercises

W Assess your understanding of this chapter's topics with an online chapter quiz at **http://chemistry.brookscole.com/tro3**.

QUESTIONS

1. What is the difference between a hot object and a cold one?
2. From a molecular standpoint, explain how thermal energy is transferred from a hot object to a cold one.
3. From a molecular standpoint, explain what heat and work are.
4. List the four sectors of U.S. energy consumption and their approximate percentages of total energy use.
5. How much energy does the United States consume each year?
6. Explain the first law of thermodynamics and its implications for energy use.
7. What is entropy? Why is entropy important?
8. Explain the second law of thermodynamics and its implications for energy use.
9. What is a perpetual motion machine? Why can such a device not exist?
10. Approximately how efficient are each of the following energy conversion processes?
    a. gasoline energy to forward motion of an automobile
    b. food energy to physical work
    c. fossil-fuel energy to electricity
    d. natural gas energy to heat
11. Define each of the following terms:
    a. heat
    b. energy
    c. work
    d. system
    e. surroundings
    f. exothermic reaction
    g. endothermic reaction
    h. enthalpy of reaction
    i. kinetic energy
    j. potential energy
12. Explain the difference between energy and power.
13. What happens to the temperature of the surroundings during an exothermic reaction? Endothermic reaction?
14. Which temperature scale(s)
    a. does not contain negative temperatures?
    b. sets the boiling point of water at 212°?
    c. has the same size of degree as the Kelvin scale?
    d. splits the difference between the boiling and freezing of water into 100 equally spaced degrees?
15. What is heat capacity? How is it related to changes in temperature?
16. What are our current energy sources and their approximate percentages of total energy use?

17. How long will each of the energy sources in the preceding question last at current rates of consumption?
18. Why are fossil fuels so named? Where do they come from?
19. Explain how a fossil-fuel-burning power plant works. How many watts are produced?
20. What are the environmental problems associated with fossil-fuel use?
21. List the four major components of smog and the effects of each.
22. What does a catalytic converter do?
23. What is the major cause of acid rain?
24. Explain how acid rain is formed and its effects on the environment and on building materials.
25. Explain the natural greenhouse effect.
26. What would happen to the Earth's temperature if there were no CO_2 in the atmosphere?
27. What is global warming? Is it occurring? What is the evidence?
28. What do computer-based climate models predict will happen to global temperatures if CO_2 levels double? What would be the effects of this?
29. How much have CO_2 levels increased in the last century? How much have temperatures changed?
30. Which fossil fuel is the worst offender when it comes to global warming? Which is the best?

PROBLEMS

31. Perform each of the following conversions:
    a. 1456 cal to Calories
    b. 450 cal to joules
    c. 20 kWh to calories
    d. 84 quad to joules
32. A person obtains approximately 2.5×10^3 Calories a day from his or her food. How much energy is that in
    a. calories
    b. joules
    c. kilowatt-hours
33. Perform each of the following conversions:
    a. 212°F to °C
    b. 77 K to °F (temperature of liquid nitrogen)
    c. 25°C to K
    d. 100°F to K
34. Perform each of the following conversions:
    a. 102°F to °C
    b. 0 K to °F
    c. 0°C to °F
    d. 273 K to °C
35. The coldest temperature ever measured in the United States is −80°F on January 23, 1971, in Prospect Creek, Alaska. Convert that temperature to Celsius and Kelvin.
36. The warmest temperature ever measured in the United States is 134°F on July 10, 1913, in Death Valley, California. Convert that temperature to Celsius and Kelvin.
37. A chocolate chip cookie contains about 200 kcal. How many kilowatt-hours of energy does it contain? How long could you light a 100-W light bulb with the energy from the cookie?
38. An average person consumes about 2.0×10^3 kcal of food energy per day. How many kilowatt-hours of energy are consumed? How long could you light a 40-W light bulb with that energy?
39. Assume that electricity costs 15 cents per kilowatt-hour. Calculate the monthly cost of operating each of the following:
    a. a 100-W light bulb, 5 h/day
    b. a 600-W refrigerator, 24 h/day
    c. a 12,000-W electric range, 1 h/day
    d. a 1000-W toaster, 10 min/day
40. Assume that electricity costs 15 cents per kilowatt-hour. Calculate the yearly cost of operating each of the following:
    a. a home computer that consumes 2.5 kWh per week
    b. a pool pump that consumes 300 kWh per week
    c. a hot tub that consumes 46 kWh per week
    d. a clothes dryer that consumes 20 kWh per week
41. Calculate the amount of energy in kilocalories produced by the combustion of each of the following:
    a. 50 kg of pine wood
    b. 2.0×10^3 kg of coal
    c. a tankful of gasoline (assume the tank holds 60 L of octane with a density of 0.7028 g/mL)
42. Calculate how many grams of each of the following would be required to produce 2.0×10^3 kcal of energy:
    a. pine wood
    b. octane
    c. ethanol
    d. coal
    e. natural gas
43. The useful energy that comes out of an energy transfer process is related to the efficiency of the process by the following equation:

    $$\frac{\text{total}}{\text{consumed}} \times \text{efficiency} = \frac{\text{useful}}{\text{energy}}$$

    where the efficiency is in decimal (not percent) form.
    a. If a process is 30% efficient, how much useful energy can be derived if 200 kcal are consumed?
    b. A person eats approximately 2200 kcal/day. How much of that energy is available to do physical work?
    c. If a car needs 1000 kcal to go a particular distance, how much energy will be consumed if the car is 20% efficient?
    d. If an electrical power plant produces 1000 kcal of electrical energy, how much energy will be consumed by the plant if it is 34% efficient?

44. A typical 1000 MW electrical generating power plant produces 1.0×10^3 megajoules (MJ) in 1 s.
    a. How many kilocalories does it produce in 1 s?
    b. If its efficiency is 34%, how much energy does it consume in 1 s?
    c. If the power plant were coal fired, how many kilograms of coal would it consume in 1 s?
    d. If the power plant were natural gas fired, how many kilograms of natural gas would it consume in 1 s?
45. Heating the air in a house by 10°C requires about 1.0×10^3 kcal of heat energy.
    a. By assuming the heat came from natural gas with 80% efficiency, how many grams of natural gas are required?
    b. By assuming the heat came from electricity with 80% efficiency, and the electricity was produced from the combustion of coal with 30% efficiency, how much coal would be required?
46. Heating the water in a 55-gallon water heater requires about 2.0×10^3 kJ.
    a. Assume the energy came from natural gas with 80% efficiency; how many grams of natural gas are required?
    b. Assume the energy came from electricity with 80% efficiency and that the electricity was produced from the combustion of coal with 30% efficiency; how many grams of coal are required?
47. For problem 45, calculate the number of grams of CO_2 emitted into the atmosphere.
    a. Use the balanced equation for the combustion of methane to determine how many grams of CO_2 would be produced from the amount of natural gas required.
    b. Assume that coal produces 1.25 kcal per gram of CO_2 and calculate how many grams of carbon dioxide would be produced.
48. For problem 46, calculate how much CO_2 in grams is emitted into the atmosphere.
    a. Use the balanced equation for the combustion of methane to determine how many grams of CO_2 are produced from the amount of natural gas required.
    b. Assume that coal produces 1.25 kcal of energy per gram of CO_2. Calculate how much CO_2 (in grams) is produced.
49. Calculate the amount of carbon dioxide (in kg) emitted into the atmosphere by the complete combustion of a 15.0-gallon tank of gasoline. Do this by following these steps:
    a. Assume that gasoline is composed of octane (C_8H_{18}). Write a balanced equation for the combustion of octane.
    b. Determine the number of moles of octane contained in a 15.0-gallon tank of gasoline (1 gallon = 3.78 L). Octane has a density of 0.79 g/mL.
    c. Use the balanced equation to convert from moles of octane to moles of carbon dioxide, then convert to grams of carbon dioxide, and finally to kg of carbon dioxide.
50. The amount of CO_2 in the atmosphere is 0.03% (0.03% = 0.0003 L CO_2/L atmosphere). The world uses the equivalent of approximately 7×10^{12} kg of petroleum per year to meet its energy needs. Determine how long it would take to double the amount of CO_2 in the atmosphere due to the combustion of petroleum. Follow each of the steps outlined to accomplish this:
    a. We need to know how much CO_2 is produced by the combustion of $.7 \times 10^{12}$ kg of petroleum. Assume that this petroleum is in the form of octane and is combusted according to the following balanced reaction:

    $$2C_8H_{18} \text{ (L)} + 25O_2 \text{ (g)} \longrightarrow 16CO_2 \text{ (g)} + 18H_2O \text{ (g)}$$

    By assuming that O_2 is in excess, determine how many moles of CO_2 are produced by the combustion of 7.0×10^{12} kg of C_8H_{18}. This will be the amount of CO_2 produced each year.
    b. By knowing that 1 mol of gas occupies 22.4 L, determine the volume occupied by the number of moles of CO_2 gas that you just calculated. This will be the volume of CO_2 produced per year.
    c. The volume of CO_2 presently in our atmosphere is 1.5×10^{18} L. By assuming that all CO_2 produced by the combustion of petroleum stays in our atmosphere, how many years will it take to double the amount of CO_2 currently present in the atmosphere?

POINTS TO PONDER

51. The second law of thermodynamics has been called "the arrow of time." Explain why this is so.
52. A bill has been proposed in Congress to lower our dependence on foreign oil by eliminating petroleum-burning power plants and replacing them with natural gas or coal. Which one would you argue for? Why? Which senators might argue for the other?
53. You are camping and contemplating placing some hot objects into your sleeping bag to warm it. You warm a rock and a canteen of water, of roughly equal mass, around the fire. Which would be more effective in warming your sleeping bag? Why?
54. List the hidden costs that go along with fossil-fuel use. Can you anticipate any that might come in the future?
55. A friend from high school calls you and says that he has solved the energy crisis. In his garage, he has created a machine that can constantly turn the turbine on a generator and create electricity, which he uses to power a second machine. The energy generated by the second machine is used to power the first machine and

to provide electrical power for his entire house. Your friend wants your financial backing to put his device into production. Would you invest in this device? Why?

56. Some people have criticized the idea that fossil-fuel combustion causes global warming. They say that carbon dioxide is a very small and natural component of the Earth's atmosphere and therefore does not pose a problem. Respond to this faulty argument.

Feature Problems and Projects

57. In the 1950s, the United States produced about the same amount of energy that it consumed. However, since then, a gap has grown between consumption and production. Today, the United States produces less energy than it consumes, and the difference is met by imported energy, mostly petroleum. Consider the following graph:

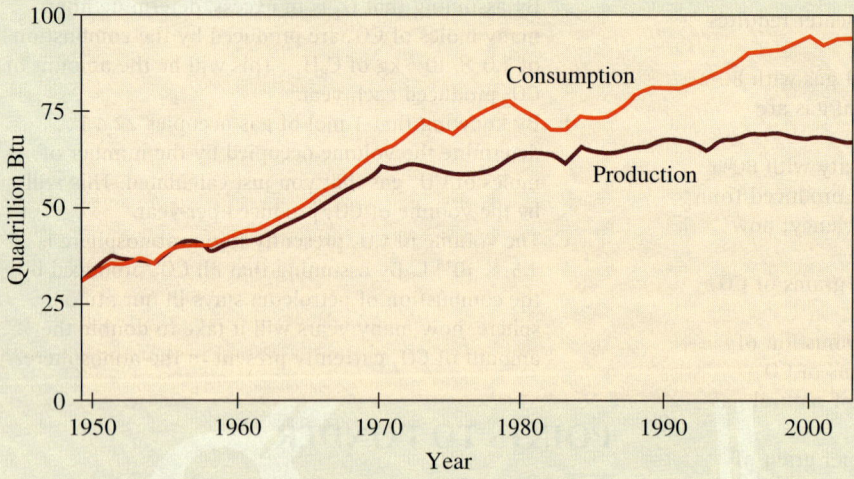

U.S. energy consumption and production from 1949 to 2003. (U.S. Energy Information Administration, *Annual Energy Review* for the 2003 calendar year.)

a. Calculate the increase in energy consumption (in quads) between 1959 and 2000. What is the average yearly increase during this period?
b. Calculate the increase in energy production (in quads) between 1959 and 2000. What is the average yearly increase during this period?
c. What is the gap (in quads) between production and consumption in 2000?
d. If average yearly increases that you calculated in parts (a) and (b) continue at the same rate, what will U.S. production be in 2025? Consumption in 2025? What will the gap (in quads) be between production and consumption?

58. The following graph shows U.S. carbon dioxide emissions between 1985 and 1997:

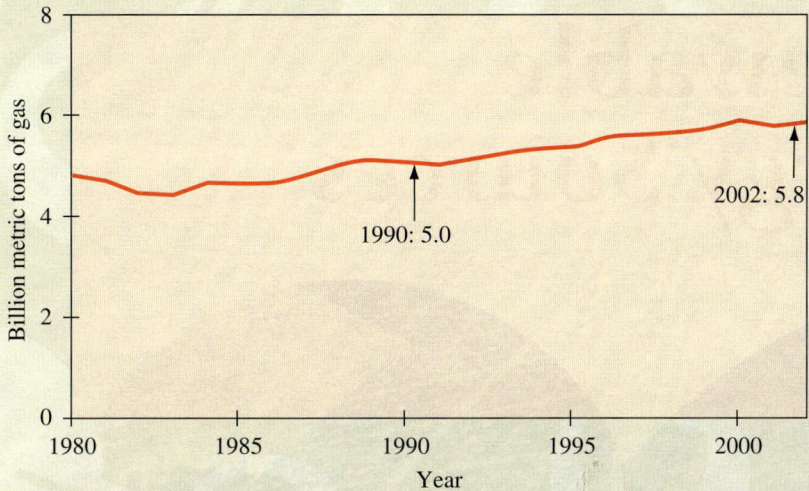

Carbon dioxide emissions 1980–2002. (U.S. Energy Information Administration, Annual Energy Review for the 2003 calendar year.)

a. Calculate the total increase in CO_2 emissions (in billion metric tons) between 1990 and 2002. Also calculate the percentage increase for this time period.
b. Calculate the average percentage increase in CO_2 emissions per year for this time period.
c. Assuming that the rate of increase remains constant, what will be the total CO_2 emissions (in billion metric tons) in 2025?

59. The Environmental Protection Agency maintains a website with environmental information. Go to their environmental atlas web site at (**http://www.epa.gov/ceisweb1/ceishome/atlas/index.html**). Choose the USA MAPS option, then the AIR option, and then the AIR QUALITY option. View the U.S. maps of pollution levels and find the area where you live. What are the approximate levels of CO and O_3?

60. Use the websites listed under Chemistry on the Web as well as the information in this chapter to write a two- to three-page essay on global warming. Within your essay; recommend whether or not the United States should reconsider its position on the Kyoto Protocol.

10 Energy for Tomorrow: Solar and Other Renewable Energy Sources

The mind, once expanded to the dimensions of larger ideas, never returns to its original size.

—Oliver Wendell Holmes

1 In this chapter, we examine renewable energy sources. It may not be obvious, but our energy sources and related technologies are slowly changing. In light of recent increases in the cost of fossil fuels, those changes will likely quicken in the coming years as our society moves from carbon-based, nonrenewable fuels to carbonless, renewable ones. As you read this chapter, think about where you get your energy. What does your car run on? Where do you purchase your electricity? How do you heat your home? The answers to these questions will be different in 50 years. The next half-century will bring changes in the fuels we use and the way we use them. Fossil fuels will not disappear from the overall energy equation—they are too convenient for certain kinds of energy production—but their role will be smaller as we supplement them with other forms of energy and new technologies. Wind power, solar thermal power, photovoltaic power, biomass, electric and hybrid vehicles—these technologies will slowly erode the supremacy of fossil fuels and ease the growing burden we have placed on our environment.

QUESTIONS FOR THOUGHT

- As our society runs low on fossil fuels, and as their effects on the environment continue, what energy sources will replace them?
- Is there enough sunlight on the Earth to meet our energy needs?
- What is hydroelectric power?
- What is the status of wind power?
- What is the status of solar power? What are the different kinds of direct solar power?
- What is biomass power?
- What is geothermal power?
- Will nuclear power play more or less of a role in the future?
- How can energy be conserved?
- What will our energy sources be in 2040?

The Sun, which has been burning for several billion years, is Earth's ultimate energy source.

 Explore this topic on the **Solar Power** website.

A Solarex MSX-120 module. The module is a polycrystalline module with an output of 120 watts.

 Explore this topic on the **Hydroelectric Power** website.

10.1 Earth's Ultimate Energy Source: The Sun

The Sun, a medium-sized, middle-aged star located in the Milky Way galaxy, is rather unremarkable when compared to the 100 billion or so other stars in the galaxy. The fusion reactions occurring in the Sun produce 10^{26} watts of power, constantly radiating into space. It has been that way for several billion years and will continue for several billion more. Of that 10^{26} watts, Earth intercepts approximately 1 part in a billion (10^{17} W). That would be like a billionaire spewing millions into the street, all going unnoticed except for $1.00 picked up by an occasional passerby. Yet, if we could capture and store even 0.01% of the energy represented by that dollar—1/100 of one penny—our energy problems would be solved.

The main obstacle to using solar energy is not its abundance—there is much more than we need—but its relatively low concentration. For example, even on a sunny summer day in the United States, the energy falling on 1 square meter (m^2) of land is approximately 1000 W. If we collected that energy with 100% efficiency (the efficiency of the typical solar cell is only about 10%), we would collect about 8000 kilocalories (kcal) of energy in 1 day. In contrast, a 15-gallon tank of gasoline contains about 400,000 kcal of energy. It is difficult to do better than fossil fuels for concentrated energy content. In spite of this problem, several technologies related to solar energy are emerging. As the cost of fossil fuels rises, some of these technologies are even becoming cost competitive with fossil fuels.

10.2 Hydroelectric Power: The World's Most Used Solar Energy Source

Hydroelectric power currently generates approximately 7% of U.S. electricity with no carbon dioxide emissions, no nitrogen oxide or sulfur oxide emissions, and no smog. In some countries, hydroelectric power provides over one-third of the electricity. Hydropower is the world's most utilized solar energy source. Hydroelectric power is an indirect form of solar energy because the energy ultimately comes from the Sun. Sunlight evaporates water from oceans, forming clouds that condense and fall as rain. The rain runs down mountains and into streams and rivers that are dammed to tap the energy in the moving water. As water flows through the dam, it turns a turbine on a generator that creates electricity (Figure 1). Because new rain always replaces the water that has run through a dam, hydropower is a renewable energy source; there is no foreseeable limit to its availability.

Hydropower seems, in many ways, like the ultimate energy source. It is clean, efficient, and renewable. However, it does have some disadvantages. The main disadvantage is the small number of dammable rivers. In the United States, 42% of the dammable rivers already have dams. Because of environmental concerns, new dams will probably not be constructed. Consequently, the main areas for growth in hydropower are developing nations. The World Energy Conference estimates that the hydroelectric potential in these countries is almost ten times current usage.

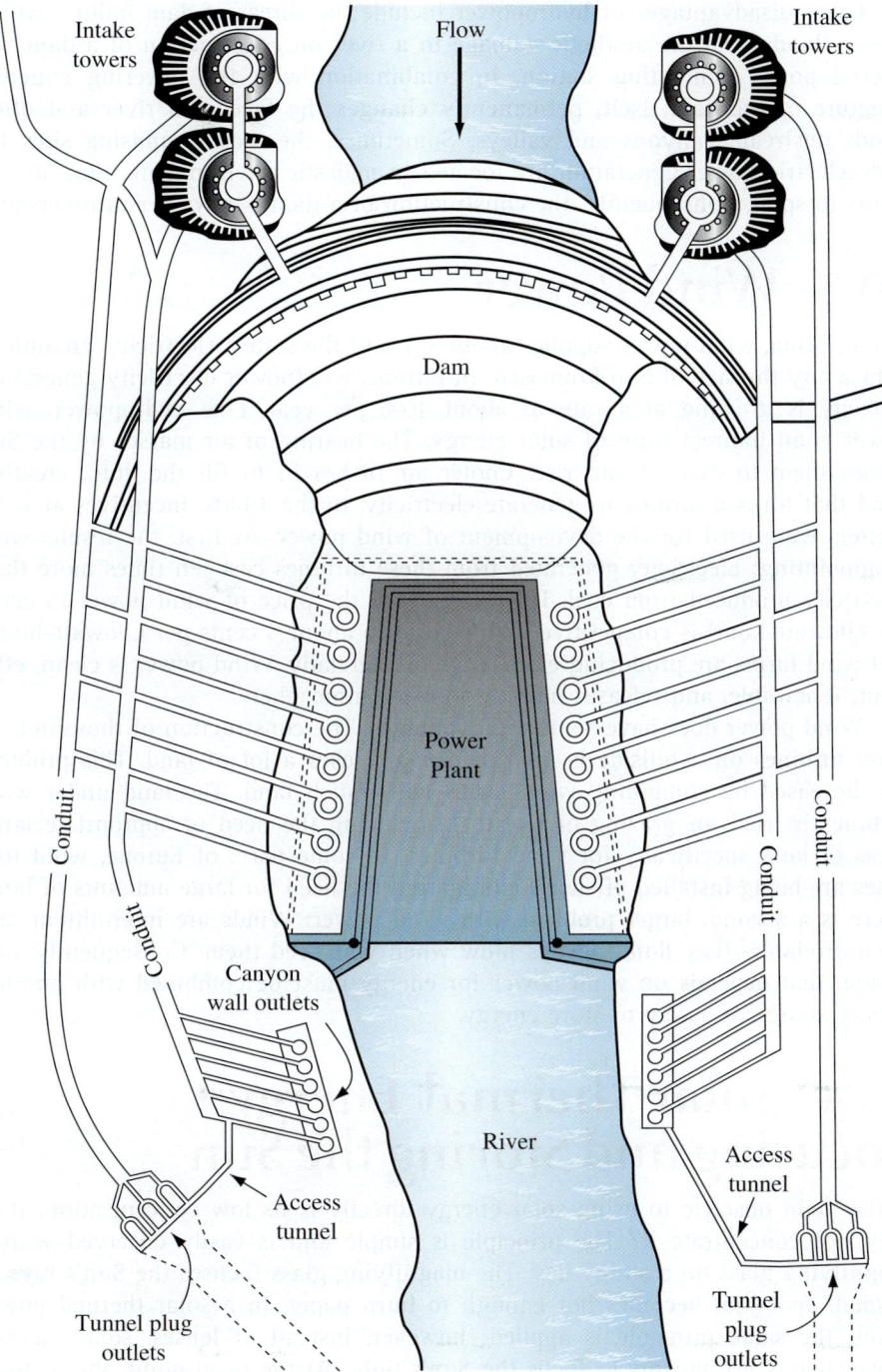

FIGURE 1 Water flowing through a dam turns the turbines on a generator to produce electricity. The potential energy stored in the water at the top of the dam is converted into electrical energy as it flows through the dam to the river below.

A second disadvantage to hydropower is the threat to marine life. Dams interfere with the spawning patterns of some fish such as salmon that must swim upstream to their spawning grounds. This problem is mitigated by the construction of fish ladders, a series of small pools that allow fish to get over the dam. Dams also stop the flow of freshwater and nutrients into downstream estuaries, disrupting ecosystems there.

Fish ladders, as shown here, allow fish swimming upstream to get through a dam.

Wind turbines near Palm Springs, California.

 Explore this topic on the **Windpower** website.

Other disadvantages of hydropower include the threat of dam failure, which causes floods, and the aesthetic damage to a river on construction of a dam. An electric power generating station, in combination with the towering concrete structure of the dam itself, permanently changes the face of a river and often floods upstream canyons and valleys. Sometimes, the most promising sites for hydroelectric power generation are located in majestic environments that no one wants to spoil. Consequently, the construction of a dam is always controversial.

10.3 Wind Power

In California, wind power supplies about 1.3% of the state's electricity, enough to light a city the size of San Francisco. In Europe, windpower electricity generation capacity is growing at a rate of about 10% per year. Like hydropower, wind power is an indirect form of solar energy. The heating of air masses by the Sun causes them to expand and rise. Cooler air rushes in to fill the void, creating wind that turns a turbine to generate electricity. In the 1980s, incentives and tax shelters were used for the development of wind power. At first, the results were disappointing: Electricity generated from these turbines cost ten times more than electricity produced from coal. However, today the price of wind power (5 cents per kilowatt-hour) is competitive with coal (also about 5 cents per kilowatt-hour), and wind farms are producing electricity commercially. Wind power is clean, efficient, renewable, and releases nothing into the atmosphere.

Wind power does have some disadvantages. The construction of thousands of wind turbines on a hillside is unattractive and uses a lot of land. This problem can be eased by combining wind farms with ranch land. The land under wind turbines is used for grazing of livestock, reducing the need to appropriate large areas of land specifically for wind turbines. In some parts of Europe, wind turbines are being installed offshore, eliminating the need for large amounts of land. There is a second, larger problem with wind power: Winds are intermittent and uncontrollable—they don't always blow when you need them. Consequently, any system that depends on wind power for energy must be combined with another energy source or a way to store energy.

10.4 Solar Thermal Energy: Focusing and Storing the Sun

If the main obstacle to using solar energy directly is its low concentration, then why not concentrate it? The principle is simple and is easily observed with a magnifying glass on a sunny day. The magnifying glass focuses the Sun's rays to a small spot that becomes hot enough to burn paper. In a solar thermal power plant, the same principle is applied; however, instead of lenses, solar thermal power plants use mirrors to focus the Sun's light. At the focal point, the temperature is high enough to produce steam and generate electricity. There are currently three different designs that utilize solar thermal energy: the solar power tower, the parabolic trough, and the dish/engine.

SOLAR POWER TOWERS

In 1982, a number of U.S. utilities, in cooperation with the U.S. Department of Energy, built Solar One, a 10-megawatt (MW) solar power tower in Dagget, California. In 1996, Solar One was upgraded to Solar Two, which has a storage mechanism that allows electricity generation for several hours in the dark. A

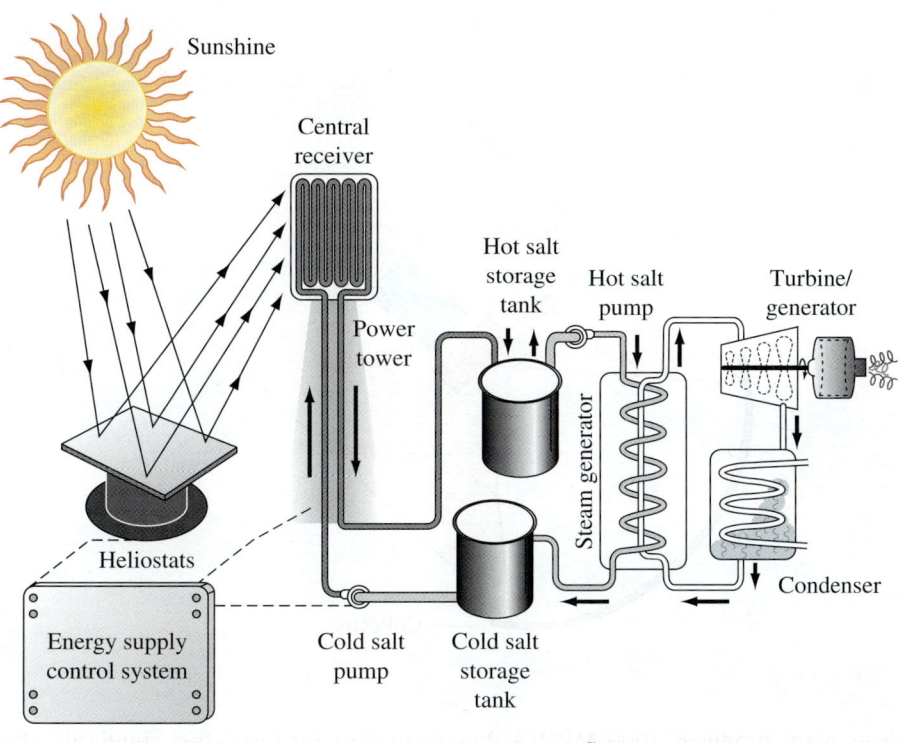

Energy supply system

Steam generator energy storage and balance of plant

FIGURE 2 Schematic diagram of Solar Two. The Sun's rays are focused by Sun-tracking mirrors, called heliostats, onto a central receiver. The high temperatures produced are used to heat a molten-salt liquid, which circulates into a storage tank. The molten salt is then pumped through a heat exchanger where it boils water to create steam. The steam turns a turbine on a generator to create electricity. Solar Two produces 10 MW of electricity, enough to light 10,000 homes. (Courtesy of Rocketdyne Propulsion and Power, a part of The Boeing Company.)

solar power tower consists of hundreds of Sun-tracking mirrors, called heliostats, that focus sunlight on a central receiver located on top of a tower (Figure 2). The focused sunlight heats a molten-salt liquid that circulates into an insulated storage tank. The molten salt is pumped out of the storage tank, as needed, to generate steam, which then turns a turbine on an electrical generator. The ability to store the high-temperature salt for a number of hours allows the plant to generate electricity in the absence of sunlight. The cost is currently 12 cents per kilowatt-hour (compared to 5 cents per kilowatt-hour for coal).

Solar Two in operation in the Mojave Desert at Dagget, California.

PARABOLIC TROUGHS

A second solar thermal technology uses large troughs to track and focus sunlight onto a receiver pipe that runs along the trough's length (Figure 3). The focused light heats synthetic oil flowing through the pipe. The hot oil then travels to a heat exchanger where it boils water to create steam that turns a turbine on an electrical generator. During cloudy days or at night, the oil in the pipes is heated with natural gas burners. Southern California has nine parabolic trough power plants in operation with a power-generating capacity of 350 MW. (A typical coal-fired

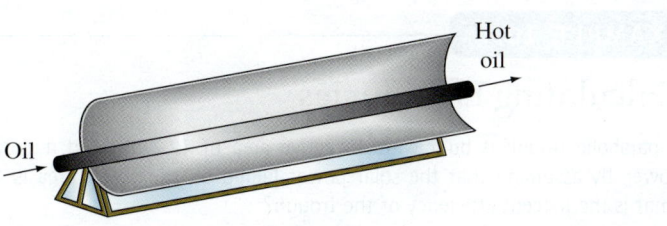

FIGURE 3 In a parabolic trough, the Sun's light is focused onto a receiver pipe that contains synthetic oil. The oil is circulated through a heat exchanger where it boils water to create steam.

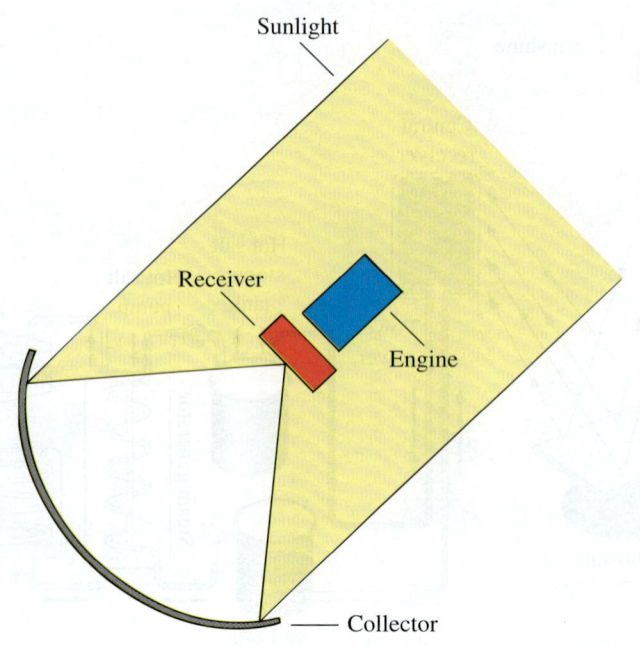

FIGURE 4 In a dish/engine solar power system, sunlight is focused onto a receiver by a reflective dish. The heat is then used to power an engine that generates electricity.

power plant produces 1000 MW.) Although modest by fossil-fuel standards, these power plants have been successfully providing electricity for approximately 350,000 homes for over 20 years. The owners of these plants have worked with the U.S. Department of Energy to reduce the operation and maintenance costs. Current cost is down to 12 cents per kilowatt-hour; however, continuing improvements in technology and new experimental plants are expected to reduce the cost in coming years.

DISH/ENGINE

The third solar thermal technology is the dish/engine. The dish/engine is composed of three parts: a collector, a receiver, and an engine (Figure 4). The collector is a dish-shaped reflector, or a collection of smaller reflectors shaped like a dish, that focus sunlight onto a central receiver. The receiver becomes the heat source for a conventional engine or turbine, which then generates electricity. Dish/engine systems can be hybridized to use other fuels, such as diesel or natural gas, when sunlight is not sufficient. The DOE is partnering with industry groups to develop this technology, and several dish/engine prototype systems are in operation in the southwestern United States. These systems generate electricity at a cost of approximately 15 cents per kilowatt-hour. However, that cost is expected to fall as the technology develops further.

A Solar thermal parabolic trough.

A 25-kilowatt dish/engine solar power system at the National Renewable Energy Laboratory.

EXAMPLE 10.1

Calculating Efficiencies

A parabolic trough is built with an active area of 12.2 m², and it produces 3.3 kW of power. By assuming that the solar power falling on the active area is 1.0×10^3 W/m², what is the percent efficiency of the trough?

SOLUTION

First, calculate the total power on the parabolic trough:

$$12.2 \text{ m}^2 \times 1.0 \times 10^3 \, \frac{W}{m^2} = 1.22 \times 10^4 \, W$$

$$\% \text{ efficiency} = \frac{\text{power out}}{\text{power in}} \times 100\%$$

$$= \frac{3.3 \times 10^3 \, W}{1.22 \times 10^4 \, W} \times 100\% = 27\%$$

> **YOUR TURN**
>
> **Calculating Efficiencies**
>
> A solar power tower has 50 heliostats each of which has an active area of 2.7 m². The tower produces 29.7 kW of power. By assuming that the solar power falling on the active area is 1000 W/m², what is the percent efficiency of the tower?

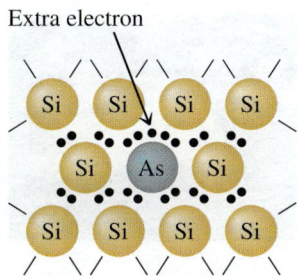

FIGURE 5 N-type silicon is doped with an element containing five valence electrons.

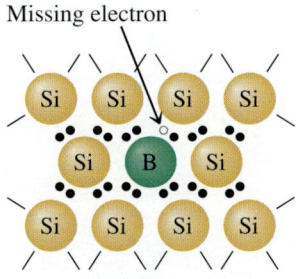

FIGURE 6 P-type silicon is doped with an element containing three valence electrons.

10.5 Photovoltaic Energy: From Light to Electricity with No Moving Parts

Perhaps the most familiar of all solar technologies are the photovoltaic (PV) cells found on solar watches and calculators. PV solar cells are, in some ways, the ultimate energy source; they convert light to electricity with no moving parts, no noise, and no pollution. PV cells are made of semiconductors, materials whose electrical conductivity is controllable.

A common semiconductor is silicon, which, like carbon, has four valence electrons. Silicon is useful because it can be mixed or **doped** with small amounts of other elements (called dopants) to change its electrical conductivity. Doping silicon with an element such as arsenic (Figure 5), which has five valence electrons, makes the silicon sample electron-rich, producing what is called **n-type** silicon (n stands for negative). Doping with an element like boron (Figure 6), which has three valence electrons, makes the silicon electron-poor, producing what is called **p-type** silicon (p stands for positive). In either case, the resulting mixture conducts electricity; the more dopant used, the greater the conductivity.

When an n-type silicon sample is brought in contact with a p-type sample, in what is called a **p-n junction**, there is a tendency for electrons to move from the n-type side to the p-type side; however, this requires energy, which can be provided by light. Light excites electrons to higher energy states and allows them to flow from the n-type side to the p-type side. If these mobile electrons are forced to travel through an external wire, an electrical current is produced (Figure 7). (Electricity is simply the flow of electrons through a wire.)

The main problem associated with PV cells is their expense. Semiconductors such as silicon are expensive to manufacture, and PV cells are expensive to produce. Furthermore, they are inefficient, converting only 10–20% of the incident energy into electricity. The cost of producing electricity from PV cells

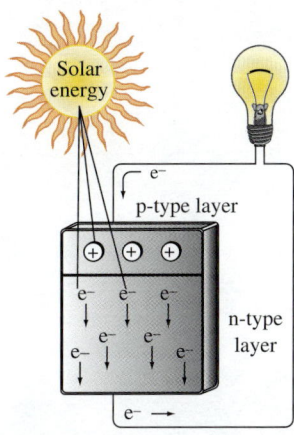

FIGURE 7 When light strikes the solar cell, electrons are excited to higher energy states, which allows them to move from the n-type side to the p-type side, producing an electrical current.

Explore this topic on the **Photovoltaic Power** website.

TABLE 1			
Price per Kilowatt-Hour of Electricity from PV Cells			
1970	1980	1990	2000
$5.00	$1.00	$0.25	$0.25

was $5.00 per kilowatt-hour in 1970. However, with dramatic advances in technology during the 1980s, that price has dropped to 25 cents per kilowatt-hour today (Table 1).

Because of their expense, PV cells are not extensively used to generate power grid electricity. They have a number of commercial applications, including watches, calculators, battery chargers, night-lights, and highway call boxes. They are also used in the space industry to provide electricity for satellites and other space-exploring vehicles. In addition, prototype cars and airplanes have been built that are fully solar powered. Although PV cells will not take over the energy market any time soon, their decreasing price, coupled with the limited supply and environmental problems of fossil fuels, makes them attractive future sources.

A one-passenger solar-powered car.

 Explore this topic on the **Solar Power** website.

EXAMPLE 10.2

Square Meters of Solar Cells

Suppose you are installing a 750-W water pump in a remote location and want to use PV cells to provide the power. By assuming that average solar power is 1.0×10^3 W/m² and that the PV cells are 15% efficient, how many square meters of PV cells are needed?

SOLUTION

Because the cell is 15% efficient, input power must be

$$750 \text{ W} \times \frac{1.0}{0.15} = 5.0 \times 10^3 \text{ W}$$

The total area of PV cells required is

$$5.0 \times 10^3 \text{ W} \times \frac{\text{m}^2}{1.0 \times 10^3 \text{ W}} = 5.0 \text{ m}^2$$

YOUR TURN

Square Meters of Solar Cells

Highway call boxes use PV cells for power. By assuming that each PV cell has an area of 0.16 m², calculate how much power it provides to the call box. As in the previous example, assume that solar power is 1.0×10^3 W/m² and that the cells are 15% efficient.

APPLY YOUR KNOWLEDGE

Is it feasible to design and build a full-size (multipassenger) solar-powered car? Why or why not?

Answer: Such a car is probably not feasible because sunlight is not sufficiently concentrated. Even if you could collect all of the solar energy falling on the surface of the car, it would not be enough to meet the energy requirements of the car. The solar-powered car shown on this page is much lighter than a conventional car and can hold only a single passenger.

10.6 Energy Storage: The Plague of Solar Sources

A significant problem with solar energy sources, besides low-energy concentration, is their intermittence. What happens at night or on cloudy days? Solar sources must be supplemented or hybridized with fossil-fuel sources or must have a mechanism for energy storage. The easiest way to store solar energy is through heat, such as the design in Solar Two. The main disadvantage is the short time duration of the storage period. Even the best-insulated systems could probably not store heat for longer than a day or two.

A more permanent storage option is a battery, described in detail in Chapter 14. However, storing energy in batteries is expensive and quickly increases the cost of energy. The expense is partly due to the high cost and large physical size of batteries. For example, a typical car battery stores approximately 700 kcal of energy. It would take 570 car batteries to store the energy equivalent (400,000 kcal) of a 15-gal tank of gasoline. A second problem with storing energy in a battery follows from the second law of thermodynamics. For every energy conversion, some energy is lost to the surroundings to pay nature's heat tax. As mentioned earlier, in most cases the energy lost to the surroundings exceeds the minimum required by thermodynamics. Consequently, it is always more efficient to use electricity as it is generated rather than storing it for later use.

MOLECULAR FOCUS

Hydrogen

Formula: H_2
Molar mass: 2.0 g/mol
Boiling point: $-252.8°C$ (20.4 K)
Lewis structure:

H : H

Three-dimensional structure:

Hydrogen, the primary constituent of all stars, is the most abundant element in the universe. However, elemental hydrogen, which occurs on Earth as H_2 gas, only composes 0.00005% of our atmosphere. Consequently, on Earth, hydrogen must be produced from hydrogen-containing compounds such as water. Hydrogen is a colorless, odorless, and tasteless gas that is flammable and explosive when mixed with air. On a per gram basis, its combustion produces almost four times more energy than gasoline, which explains its potential as an energy source. It is currently used as a fuel for rockets, in blowpipe welding, in the manufacture of ammonia and methanol, in the manufacture of hydrochloric acid, in the hydrogenation of fats and oils, and in nuclear fusion reactions. If cheaper ways to produce hydrogen become available, hydrogen, either by direct combustion or by electricity generation in fuel cells, could be the fuel of the future.

Hydrogen-powered Hummer.

Another way to store energy is directly in chemical bonds. For example, solar energy could be used to break bonds in water molecules, forming hydrogen and oxygen gas:

$$2H_2O + \text{energy} \longrightarrow 2H_2 + O_2$$

This could be accomplished through **electrolysis,** in which an electrical current splits water into hydrogen and oxygen. The hydrogen could then be stored in a gas cylinder, and the reverse reaction, the combustion of H_2, could be employed at a later time. Hydrogen burns cleanly and smoothly in air, producing only water as a product:

$$2H_2 + O_2 \longrightarrow 2H_2O + \text{energy}$$

The space program uses hydrogen as a fuel to propel rockets into space. There is danger, however, of the hydrogen exploding, as in the space shuttle *Challenger* disaster. A safer way to extract the energy of hydrogen gas is through a device called a fuel cell, described in detail in Chapter 14. In fuel cells, hydrogen and oxygen gas combine through an electrochemical process, never burning but producing electricity. The space shuttle uses fuel cells for onboard electricity generation. The product of the fuel cell reaction—water—is then used as drinking water for the astronauts. Again, the main disadvantage of this scenario is expense. In addition, as in the case of a battery, the process of storing the energy in chemical bonds and then retrieving it through a fuel cell results in the loss of some energy to surroundings because of nature's heat tax.

The space shuttle uses the combustion of hydrogen fuel for propulsion.

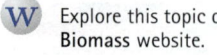

Explore this topic on the **Biomass** website.

10.7 Biomass: Energy from Plants

Nature efficiently captures and stores the sun's energy through **photosynthesis.** In photosynthesis, plants use energy from sunlight to convert CO_2 and H_2O into energetic glucose molecules ($C_6H_{12}O_6$) and O_2:

$$\text{sunlight} + 6CO_2 + 6H_2O \longrightarrow C_6H_{12}O_6 + 6O_2$$

Plants then use the energy in glucose to live and grow. They also link glucose units together to form starch or cellulose, which are conveniently stored by plants. Consequently, plants are storehouses of energetic molecules whose energy originally came from sunlight.

The energy contained in living plants can be harvested two ways. The most direct way is to burn the plant and use the heat to produce electricity. Indeed, the burning of wood to produce heat is a form of biomass energy. However, trees take too long to grow; thus, most biomass scenarios focus on rapidly growing plants such as corn or sugarcane. The second way is to produce liquid fuels such as ethanol from plants. Corn and sugarcane are easily fermented by yeasts, converting glucose into ethanol and carbon dioxide:

$$C_6H_{12}O_6 \xrightarrow{\text{yeast}} 2CH_3CH_2OH + 2CO_2$$

Ethanol is a clear liquid that is easily transported and burns cleanly to produce energy. Of course, the burning of plants, either directly or after fermentation, produces CO_2, the major contributor to global warming. However, plants consume CO_2 when they are grown. There is an exact balance between the amount of CO_2 absorbed while the plant is growing and the amount of CO_2 emitted when it is burned for energy. Consequently, there is little effect on global warming.

The major problem with biomass energy is the land required for it to become a significant energy source. Because most cropland is used to grow food, the

Ethanol is used as an additive to gasoline in some regions.

> ### WHAT IF...
> #### Politics and Energy
> Suppose you are a senator from a highly agricultural state. How would the use of biomass for energy affect your state's local economy? Would you argue for increasing or decreasing biomass energy? Why? Would a more global perspective change your thinking?

switch to growing fuel would be difficult. However, in a more limited scope, biomass may find a place among other alternative energy sources. Ethanol from corn is used as a gasoline additive in many states. Because ethanol contains oxygen, this mixture tends to burn more completely than pure gasoline, especially during cold weather. The result of more complete combustion is lower carbon monoxide and hydrocarbon emissions.

10.8 Geothermal Power

Geothermal energy is produced by the high temperatures present in the Earth's interior. In some seismically active regions, this heat leaks out through natural fissures or through holes drilled into the rock, sometimes producing temperatures over 300°C. In some locations, such as The Geysers in California, the steam emitted from the Earth's interior is used directly to turn turbines and produce electricity. Geothermal energy produces approximately 1900 MW of power in California, enough electricity for 1.9 million homes.

 Explore this topic on the **Geothermal Power** website.

While geothermal energy does not have most of the problems associated with fossil-fuels, it does have its own problems, the most significant of which is its limited availability. Only places of seismic or volcanic activity usually contain near-surface geothermal energy. In fact, the conditions found at The Geysers in California, where the emitted steam is used directly for electricity generation, are only known to exist in one other place in the world. A second problem associated with geothermal power plants is spent-steam disposal: It can be vented into the atmosphere, but it usually contains sulfurous gases and ammonia, both of which pose environmental dangers. Alternatively, the steam can be condensed and pumped back into the ground or disposed into the ocean. The high salinity of the water, however, is corrosive to the machinery and may pose environmental dangers.

Geothermal power plants, such as this one at Mammoth Lakes, California, generate electricity by using steam emitted from the Earth's interior.

10.9 Nuclear Power

As we saw in Chapter 8, the heat emitted in a controlled fission reaction can be used to generate electricity. Nuclear power plants do not produce smog, carbon dioxide, or other harmful emissions associated with fossil fuels. They do produce radioactive waste that must be disposed of. A second disadvantage to nuclear power is the limited supply of U-235, less than 1% of naturally occurring uranium; the other 99% is U-238, which does not undergo fission. Our best estimates indicate that our reserves of U-235 will last approximately 100 years at current rates of consumption. However, the incorporation of a new type of nuclear reactor, called a **breeder reactor**, would increase the supply 100 times. A

The growth of nuclear power in the United States has been halted in light of increasing public opposition.

breeder reactor converts nonfissile uranium-238 into fissile plutonium-239. In this process, U-238 is bombarded with neutrons to form U-239:

$$^{238}_{92}U + ^{1}_{0}n \rightarrow ^{239}_{92}U$$

U-239 is extremely unstable and undergoes two successive beta decays to form Pu-239:

$$^{239}_{92}U \longrightarrow ^{239}_{93}Np + ^{0}_{-1}e$$

$$^{239}_{93}Np \longrightarrow ^{239}_{94}Pu + ^{0}_{-1}e$$

Because a breeder reactor converts the other 99% of uranium into fissile plutonium, this process increases our reserves by 100 times, from 100 years to 10,000 years.

The growth of nuclear power in the United States has been halted by the public's fear of accidents and concerns over waste disposal. No new nuclear plants are currently being built in the United States, and plans for breeder reactors have been discontinued. However, recent increases in energy costs has resulted in a renewed interest in nuclear power. Some European countries now generate up to 50% of their electricity in nuclear power plants.

The development of controlled nuclear fusion could change this scenario. If the technical hurdles are overcome, the almost unlimited amount of fuel, as well as the lower radioactivity and shorter half-lives of waste products, would make fusion an attractive source of future energy.

10.10 Efficiency and Conservation

Although several technologies are developing that might together provide for future energy needs, energy will probably become more expensive. Consequently, the use of these new technologies must be coupled with energy conservation and efficiency if total energy costs are to remain stable. Efficiency and conservation also make sense in light of the limited supply and environmental problems associated with almost all energy sources. If all consumers became conservers— that is, if they did their part in conserving energy—the toll on the environment would be minimized.

Since about one-fourth of U.S. energy consumption is in transportation, significant energy savings can be realized by walking, riding a bike, sharing a ride, or using public transportation. Automobiles have become more efficient, with some models achieving gas mileages of about 40 miles per gallon (mpg). Most auto companies are investing in hybrid electric automobile technology; these cars, now available, boast mileages up to 70 mpg.

Household energy consumption can be divided as shown in Table 2. The largest single household energy use is space heating; thus, houses should be well insulated. Building codes have changed over the years so that newer projects require high efficiency standards. For example, the replacement of single-pane windows with double-pane insulated windows reduces heating costs by 10–20%. Substantial savings can also be realized by lowering thermostats and using an extra sweatshirt or blanket. Electric heaters should be avoided in favor of more efficient natural gas heaters. The savings here are substantial because, as we learned earlier, nature always takes a heat tax for every energy conversion. The electricity used in an electric heater must be generated by an electrical power plant, most likely from burning fossil fuels (Figure 8). It is three times more efficient to use the heat generated from burning the fossil fuel directly, as in natural gas heaters.

Double-pane insulated windows reduce heating energy consumption by 10-20% because heat is more efficiently retained.

TABLE 2		
Yearly Consumption by End Uses per U.S. Household (2001)		
Use	Consumption (quad)	Percent of Total
Space heating	4.61	47
Air conditioning	0.62	6
Water heating	1.69	17
Appliances (including refrigerator)	2.94	30

Source: Data adapted from the U.S. Energy Information Administration, 2001 Residential Energy Consumption Survey, conducted by the Office of Energy Markets and End Use.

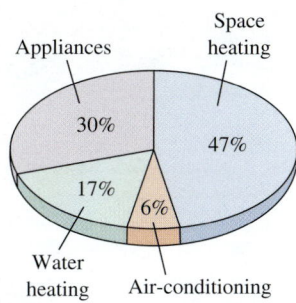

Yearly energy consumption by end uses per U.S. household.

The second-largest single consumer of energy in U.S. households is the water heater. Because water has a high-heat capacity, it requires a lot of energy to heat it. Lowering the thermostat on your water heater from 150 to 120°F reduces the energy used by about 30%. Additional savings are realized by insulating the water heater and replacing older models with newer, more efficient ones.

Appliances, including lighting and refrigeration, are responsible for almost one-third of household energy use. If possible, replace incandescent lights with fluorescent ones. Fluorescent lights are more efficient, as you can see by holding

FIGURE 8 Because energy is lost with each energy conversion, it is more efficient to burn natural gas directly to heat a home than to use electricity.

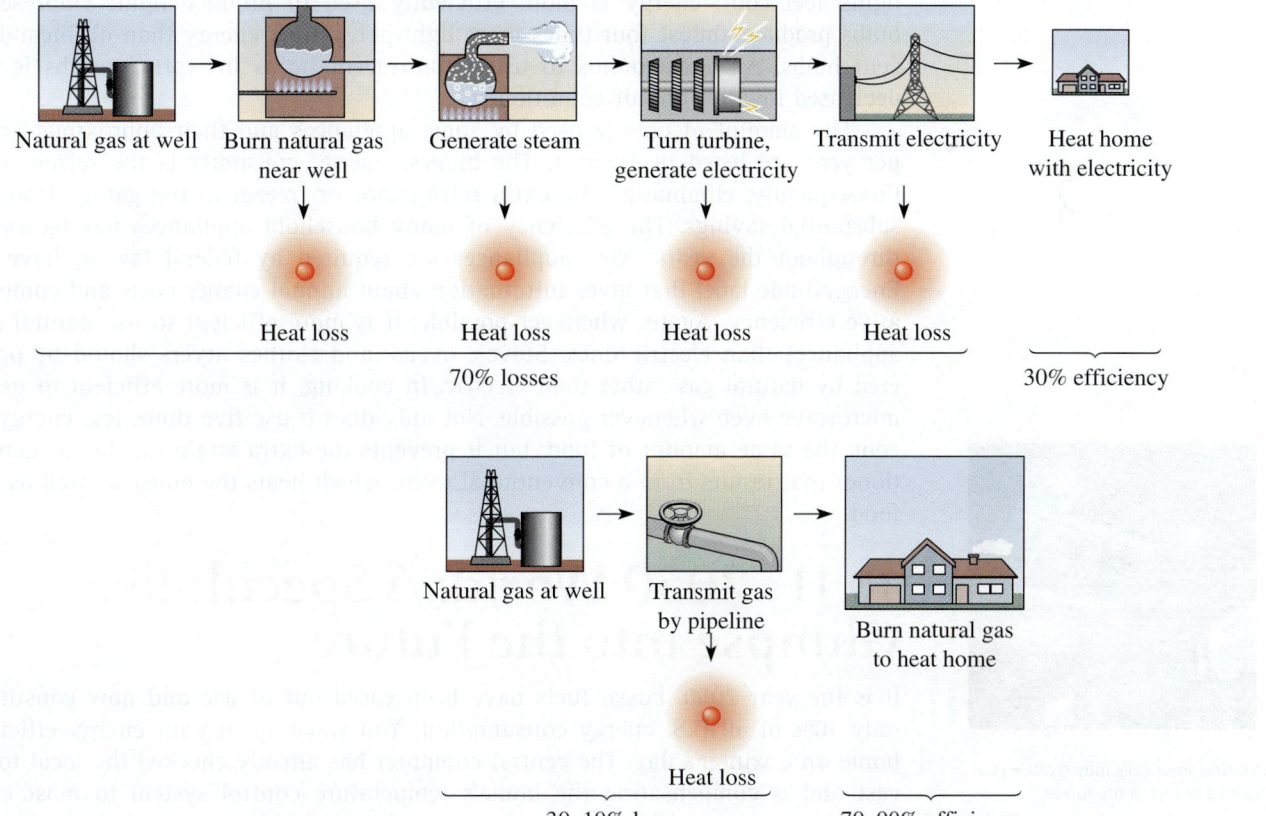

TABLE 3
Annual Electricity Use and Cost of Home Appliances

Appliance	Average Electricity Used per Yr (kWh)	Average Electricity Cost per Yr ($)[a]
Clothes dryer	1060	159
Clothes washer (including hot water)	1080	162
Coffeemaker	100	15
Dishwasher (including hot water)	935	141
Refrigerator-freezer (frostless)	650	98
Spa or hot tub	2300	345
Stereo	75	11
Television	197	30
Vacuum	25	4

[a]Based on typical usage at a rate of 15 cents per kilowatt-hour.

Source: Adapted from *Homemade Money: How to Save Energy and Dollars in Your Home*, by Richard Heede, Rocky Mountain Institute, Snow Mass, Co, 1995.

your hand near an incandescent light bulb and a fluorescent bulb. The incandescent bulb feels hot, meaning that energy is lost to heat. In contrast, fluorescent lights feel cool; energy is more efficiently used to produce light. Fluorescent bulbs produce almost four times more light per unit of energy than do incandescent bulbs. An extra bonus to using fluorescent lights in warm months is the decreased load on the air conditioner.

The amount of energy used by some appliances and their approximate cost per year are listed in Table 3. The biggest energy consumer is the refrigerator. Consequently, eliminating the extra refrigerator or freezer in the garage leads to substantial savings. The efficiency of many household appliances has increased throughout the years. New appliances are required by federal law to have an EnergyGuide label that gives information about annual energy costs and comparative efficiency. Again, whenever possible, it is more efficient to use natural gas appliances than electric ones. Stoves, ovens, and clothes dryers should be powered by natural gas rather than electric. In cooking, it is more efficient to use a microwave oven whenever possible. Not only does it use five times less energy to cook the same amount of food, but it prevents the extra strain on the air conditioner that results from a conventional oven, which heats the house as well as the food.

10.11 2050 World: A Speculative Glimpse into the Future

It is the year 2050. Fossil fuels have been eased out of use and now constitute only 10% of all U.S. energy consumption. You wake up in your energy-efficient home on a winter's day. The central computer has already checked the local forecast and is compensating the home's temperature control system to most effi-

Rooftop solar cells may replace centralized power in the future.

The Molecular Revolution

Fuel Cell and Hybrid Electric Vehicles

Most major automakers are developing fuel cell and hybrid electric vehicle technology. A fuel cell vehicle has no internal combustion engine. It runs on electricity generated by a fuel cell, a device that creates electrical current from a chemical reaction in which the reactants are constantly replenished. In 2003, General Motors (GM) demonstrated a vehicle called the *HydroGen3* to members of Congress. The vehicle seats five passengers, has a top speed of 100 miles per hour, and a range of 250 miles on one tank of fuel. Just two years later, General Motors demonstrated a vehicle called the *Sequel*, a fuel cell SUV with a 300-mile range and quick acceleration (0–60 mph in 10 seconds). According to GM, the Sequel is quicker, easier to handle, easier to build, and safer than gasoline-powered vehicles, and its only emission is water vapor. Other automakers have similar prototype models in development.

Most major automakers already have hybrid electric vehicles available. Hybrid electric vehicles run on both electricity (generated internally) and gasoline. The Honda *Insight*, first introduced in 2000, is a good example of a hybrid electric vehicle. The Insight has both a small gasoline-powered engine (1.0 liter, three cylinders) and an electric motor. The electric motor provides additional torque when needed, allowing the engine to be much smaller than normal. The Insight has an ultralight aluminum chassis, a teardrop shape with an ultralow drag coefficient, and a regenerative braking system that captures some of the kinetic energy normally lost in braking, storing it in the battery for later use. The Insight is the most fuel-efficient gasoline-powered automobile in the U.S. market, with an EPA mileage rating of 60 mpg city/66 mpg highway. Today, a number of vehicles—such as the Honda *Civic* and even the Chevrolet *Silverado*—are available as hybrids, increasing their gas mileage (compared to a conventional combustion engine) by as much as 40%.

The GM Sequel, a fuel cell vehicle with a 300-mile range and water vapor as its only emission.

The Honda Insight, a hybrid electric vehicle (HEV), has an EPA mileage rating of 60 mpg city/66 mpg highway.

W Explore this topic on the **Hybrid and Fuel Cell Vehicles** website.

ciently keep the home at a comfortable temperature throughout the day. The solar panels on your roof are tracking the Sun as it rises over the horizon. The hydrogen reserve in the fuel cell auxiliary system is down to half of its capacity due to last month's rains, but it is quickly building with the clear sky.

Centralized power, greatly reduced because most homes are solar powered, exists to provide electricity to urban centers where the concentration of people and corresponding energy need are too high for solar. Most of the fossil-fuel power plants of the past, however, have been replaced with nuclear fusion reactors. These reactors produce several thousand megawatts of power, about half of which is used for electricity production. The other half is used to produce hydrogen fuel by breaking O—H bonds in water. The hydrogen gas is piped to gas stations in the way that natural gas was piped to homes in the past. At the gas

> ## WHAT IF...
>
> ### Future Energy Scenarios
>
> The scenario given here may be overly optimistic. Can you think of some reasons why our energy future may not turn out to be as bright as described? What if the cost of energy doubled or tripled? In what ways would our lives change?

pump, the hydrogen fuel is transferred to your automobile's hydrogen gas storage system, a network of solid metal cylinders that absorb hydrogen gas. The hydrogen molecules are small enough to fit into the spaces between the metal atoms, giving your ultralight vehicle a hydrogen gas carrying capacity high enough to travel over 500 mi before refueling. The new airplanes, buses, trucks, and monorail systems are all powered by hydrogen gas.

Efficiency has kept energy costs only slightly higher than at the turn of the century. The new energy technologies—nuclear fusion, photovoltaic technology, and storage through hydrogen gas production—cost three times more than the old fossil-fuel energy sources. However, efficiency for most energy processes such as transportation and space heating has improved by over a factor of two, causing net energy costs to increase by only 15%.

The real winner from the new energy technologies is the environment. Global warming, which had continued for the first two decades of the 21st century, has now abated, leaving global temperatures at about 2000 levels. Because the combustion of hydrogen produces primarily water as an emission, most cities are pollution free. The only other emission, NO_2 (a byproduct of burning hydrogen gas in air), is captured by the tailpipe scrubbers required on all automobiles. Acid rain has also decreased. Because most coal-fired power plants of the midwestern United States have either been shut down or replaced by nuclear fusion reactors, man-made SO_2 emissions are very low. Many lakes and streams in the northwestern United States and in Canada have recovered and are alive again.

Chapter Summary

MOLECULAR CONCEPT

The Sun has always been the Earth's main energy source (10.1). Without it, our planet would be barren and devoid of life. The Sun's energy is sufficient to provide our society's energy needs, but there are several technical and economic hurdles to overcome. Our society has used indirect solar energy, in the form of hydroelectric power, for many years (10.2). Wind power, another form of indirect solar energy, has also been used recently with some success (10.3).

The fundamental problems with direct solar energy relate to its low concentration and its intermittence. Solar-thermal energy sources overcome the first of these problems by focusing the Sun's energy to reach temperatures high enough to boil water (10.4). The steam produced then turns the turbine on a generator to produce electricity. Several solar-thermal power plants generate electricity commercially in southern California.

Photovoltaic cells, in many ways the ultimate energy source, still suffer from high cost; however, their price has decreased and should continue to decrease in the future (10.5). Like other solar technologies, photovoltaic cells also have the problem of intermittence. A way to store energy cheaply and efficiently must be developed before solar energy can become a mainstay (10.6).

Other renewable energy sources include biomass, geothermal power, and nuclear power (10.7–10.9). Although not strictly renewable, nuclear fission with breeder reactors, or nuclear fusion, could provide energy for the foreseeable future.

Given the environmental problems associated with fossil fuels, their limited availability, and the high cost of alternatives, energy efficiency and conservation are important to the future viability of our current living standards (10.10).

SOCIETAL IMPACT

Our society has a constant and growing need for energy. The inevitable depletion of fossil fuels, and the environmental problems associated with their use, forces our society to seek alternatives. However, we have had convenient and relatively inexpensive energy sources for a long time; as a society, we are unwilling to accept inconvenient or more expensive sources. Consequently, the sources under development must become cheaper and more convenient.

In some areas, deregulation has given consumers the option of buying electricity from several different companies. Some of these companies provide electricity generated from only environmentally friendly, renewable energy sources. Many consumers now have a choice, but many lack the knowledge to choose wisely. In many cases, the electricity provided by these companies is cheaper than that from conventional sources.

Public opposition to nuclear power has halted the growth of nuclear power in the United States. Because public demand for electricity has not decreased, we must either depend more on fossil fuels or seek other alternatives.

Every person can take part in conservation while staying aware of alternatives as they develop. It is our society, as a whole, that determines our energy sources and patterns of use.

Chemistry on the Web

For up-to-date URLs, visit the text website at www.chemistry.brookscole.com/tro3

- Solar Power
  http://www.solarpower.com/
- Hydroelectric Power
  http://www.worldenergy.org/wec-geis/
  http://www.hydro.org/index.asp
- Windpower
  http://www.awea.org/
  http://telosnet.com/wind/
  http://windpower-monthly.com/
- Solar Power
  http://www.eere.energy.gov/RE/solar.html
  http://www.energylan.sandia.gov/sunlab/overview.htm
- Photovoltaic Power
  http://www.nrel.gov/ncpv/
- Biomass
  http://www.ems.org/biomass/intro.html
  http://www.amzenergy.com/
- Geothermal Power
  http://geothermal.id.doe.gov/
- Hybrid and Fuel Cell Vehicles
  http://www.hybridcars.com/
  http://www.fueleconomy.gov/feg/fcv_whatsnew.shtml
  http://www.energyquest.ca.gov/transportation/fuelcells.html

Key Terms

breeder reactor	electrolysis	photosynthesis	p-type
doping	n-type	p-n junction	

Exercises

Assess your understanding of this chapter's topics with an online chapter quiz at **http://chemistry.brookscole.com/tro3**.

QUESTIONS

1. What is the main obstacle to using solar power?
2. Explain why hydroelectric power and wind power are indirect forms of solar energy.
3. How does a dam produce electricity?
4. What are the advantages and disadvantages of hydropower?
5. How does a wind turbine produce electricity?
6. What are the advantages and disadvantages of wind power?
7. Explain how a solar power tower works.
8. Explain how a solar parabolic trough works.
9. What are the advantages and disadvantages of solar thermal technologies?
10. What solar technologies are currently used for on-grid electricity production? Are they a significant part of the electricity generation market?
11. What is a semiconductor? What do *n-type* and *p-type* mean?
12. Explain how a photovoltaic cell works.
13. What are the advantages and disadvantages of PV cells for electricity generation?
14. Why is intermittence a problem for solar energy sources? What are some possible solutions? What are the problems associated with those solutions?
15. What are the advantages and disadvantages of biomass energy?
16. How does a geothermal power plant generate electricity? What are the advantages and disadvantages of geothermal power?
17. What are the advantages and disadvantages of nuclear power? How may these be overcome? Why has the use of nuclear power in the United States not grown in recent years?
18. Why are efficiency and conservation of energy crucial to our society?
19. In what ways can energy be conserved in transportation and in home energy use?
20. Why should natural gas appliances always be favored over electric ones?
21. Explain why efficiency is important in keeping future energy costs down.

PROBLEMS

22. Suppose your monthly electric bill is $85.00 at a rate of 15 cents per kilowatt-hour. What would you pay if you converted to solar energy at a rate of 25 cents per kilowatt-hour?
23. Suppose your monthly electric bill is $195.00 at a rate of 15 cents per kilowatt-hour. What would you pay if you converted to solar thermal energy at a rate of 20 cents per kilowatt-hour?
24. Your monthly electric bill is $54 at a rate of 12 cents per kilowatt-hour. How many kilowatt-hours of electricity did you use? How many joules? (Use the tables in Chapter 9.)
25. Your monthly electric bill is $245 at a rate of 15 cents per kilowatt-hour. How many kilowatt-hours of electricity did you use? How many joules?
26. A PV cell intercepts 32.1 W of sunlight and produces 3.4 W of electrical power. What is its percent efficiency?
27. A PV cell intercepts 568 watts of sunlight and produces 71 watts of electrical power. What is its percent efficiency?
28. Several PV cells totaling 5.4 m^2 of active area are mounted on a rooftop. If you assume solar power to be 1000 W/m^2 and assume the cells are 17% efficient, how much power (in watts) can they produce?
29. A PV cell intercepts 1487 watts of sunlight and is 16% efficient. How much electrical power does it produce?
30. If a typical home requires an average power of 1 kW, calculate how many square meters of PV cells would be required to supply that power. Assume that solar power is 1.0×10^3 W/m^2 and that the cells are 15% efficient in converting sunlight to electrical energy. Could such an array meet the peak power demands? Could such an array fit on a rooftop?
31. One of the problems associated with electric vehicles is that most electricity is generated by fossil-fuel combustion; therefore, electric cars indirectly produce harmful emissions by using more electricity. This problem could be solved if electric cars were recharged using solar energy. Calculate how many square meters of solar PV cells are required to build a charging

station for an electric car. Assume that the station must produce 5.0 kW of power so that an electric car can be charged in 4–6 hours. Assume also that solar energy is 1.0×10^3 W/m² and that the PV cells are 18% efficient in converting solar energy to electrical power.

32. A rooftop in the southwestern United States receives an average solar power of 3.0×10^5 W (averaged over both day and night and seasons). Convert this to kilowatts and then multiply by the number of hours in a year to calculate the number of kilowatt-hours that fall on the rooftop in one year. If the home uses 1.0×10^3 kilowatt-hours per month, does its roof receive enough energy to meet its energy needs? What is the minimum efficiency required for solar cells to meet the entire energy need?

33. Use the energy conversion tables in the previous chapter to determine how many quads of solar energy fall on the earth each year, assuming that average solar power on earth is 10^{17} W. How does this number compare with world energy consumption—approximately 411 quad?

34. A window air conditioner requires approximately 1.5×10^3 W of power. How many square meters of PV cells would it require to provide this power? Assume that solar power is 1.0×10^3 W/m² and that the cells are 15% efficient in converting sunlight to electrical energy. Could you think of some good reasons to use solar power for air-conditioning units?

35. A hot tub requires 448 kWh of electricity per month to keep it hot. If solar energy could be converted directly to heat with 50% efficiency, would the surface of the hot tub (5.0 m²) receive enough solar energy to provide the energy needs of the hot tub? Assume that solar energy is 1.0×10^3 W/m² for 8 hours per day.

36. Does an average car have enough surface area to meet its power needs with solar power? Assume the car needs an average power of 100 kW, that average solar power is 1.0×10^3 W/m², that the car's surface area is 4.0 m², and that solar cells are 15% efficient.

37. How big a solar cell (in m²) is needed to provide enough power to light a 100-W light bulb? Assume that solar power is 1.0×10^3 W/m² and that solar cells are 15% efficient.

POINTS TO PONDER

38. List ways the government could encourage use of alternative energy sources today.

39. Suppose you are the president of a company that uses a large fleet of vehicles. You are contemplating converting a significant number of them to a nongasoline fuel. You have two options before you, natural gas power or electric power. Which would you choose and why?

40. Why might some members of congress be against legislation that might result in significant changes in current energy sources?

41. Write a letter to your congressperson asking him or her to develop renewable energy. Make sure to include your reasons for why renewable energy needs to be developed.

42. Some people choose renewable energy over conventional sources. For example, you can supplement home electricity with solar photovoltaic cells on your rooftop. In some areas, utility companies have programs that buy back unused electricity during peak sunshine and supply extra electricity at times of insufficient sunshine. The cost per kilowatt-hour, however, is still significantly more than if you just bought all your electricity directly from the utility company. Would you ever consider participating in such a program? Why or why not?

Feature Problems and Projects

43. Worldwide renewable energy consumption has been increasing over the last 30 years and is expected to continue increasing. Consider the following graph showing both historical and projected consumption of renewable energy.

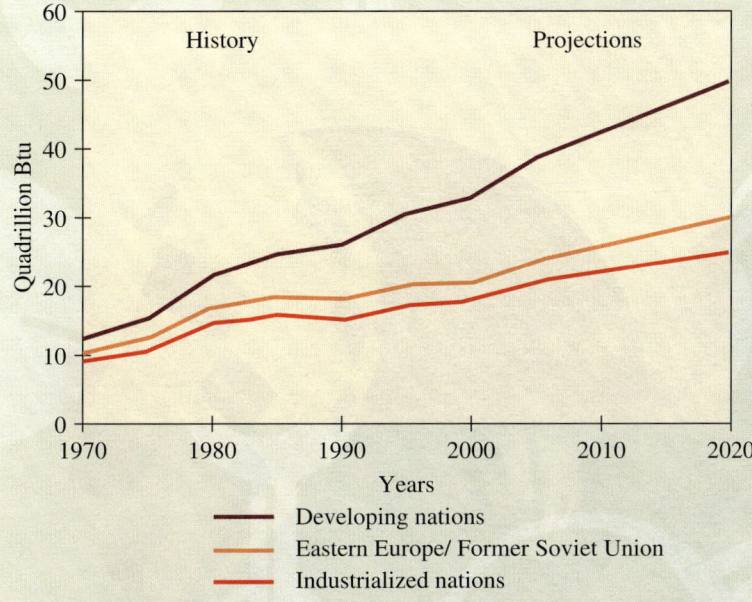

Worldwide renewable energy consumption. (Historical data from the U.S. Energy Information Administration (EIA) Office of Energy Markets and End Use, International Statistics Database and International Energy Annual, 1996, DOE/EIA-0219(96) (Washington, DC, February 1998). Projected data from EIA, World Energy Projection System (1999).)

   a. What is the total projected increase in world renewable energy consumption between 1970 and 2020? What is the yearly increase?
   b. What is the percent projected increase in world renewable energy consumption between 1970 and 2020? What is the yearly percent increase?

 44. The U.S. Department of Energy is the government agency that promotes and funds research related to renewable energy. Go to its website **(http://www.eren.doe.gov/power/)** and read the current hot topics. Write a two-page paper summarizing what is "hot" at the Department of Energy right now.

45. Look at your electric bill—if you are away from home, call home and ask a member of your family to look. How many kilowatt-hours of electricity do you (or your family) consume each month? How much do you pay per kilowatt-hour? How many square meters of solar cells would you need to provide your household with sufficient electricity? Assume that 1 m^2 of solar cells produces 25 kWh per month.

11 The Air Around Us

The generality of men are so accustomed to judge of things by their senses that, because air is invisible, they ascribe but little to it, and think it but one removed from nothing.

—Robert Boyle

In this chapter, we explore the atmosphere and the air that composes it. Air is a gas, which means that the molecules that compose it are not attached or held to each other as they are in a solid or liquid. As a result, there is a lot of empty space between gas molecules, allowing gases, unlike liquid or solids, to have a changeable volume. This property of gases is responsible for wind, storms, hot-air balloons, and our ability to drink out of a straw.

Air composes our atmosphere. What is in air? In what ways has air become polluted? Is air quality getting better or worse? We will examine these questions as we look at the effects of human activities on air quality. You will see that many, but not all, air pollutants are the result of the fossil-fuel combustion, described in Chapter 9. You will also see, however, that good legislation works. The Clean Air Act of 1970, together with the amendments that followed, has had positive effects on the air we breathe every day.

QUESTIONS FOR THOUGHT

- If we could see molecules, what would a gas look like?
- What is pressure? How is it measured?
- How does the volume of a gas change with changes in pressure and temperature?
- What gases compose our atmosphere?
- What is the structure of our atmosphere?
- What pollutants are in our air? What produces them?
- What is our society doing to reduce air pollution?
- What is ozone depletion? What causes it? What is being done to stop it?

11.1 Air Bags

The road is wet, the traffic heavy, and as you fumble with your cell phone the traffic light ahead of you turns red. You slam the brakes an instant too late and crash into the car in front of you. The next moment you have a nylon bag in your face and you are covered with talcum powder—the experience of a modern-day collision. Since their introduction in the 1980s, air bags have saved thousands of lives by providing—at just the right instant—a cushion of air between the driver and the rest of the car.

 Explore this topic on the **Air Bags** website.

Air bags work by converting atoms in a solid, which are relatively close together and occupy little volume, into gaseous atoms, which are relatively far apart and occupy a lot of volume (Figure 1). Any significant impact trips an electronic sensor that activates the following chemical reaction within the air bag:

$$2\ NaN_3(s) \rightarrow 2\ Na(s) + 3\ N_2(g)$$

This reaction releases 3 moles of nitrogen gas, which inflates a 45 liter bag for every 2 moles of solid NaN_3, which occupy less than 0.1 L. The sudden, 450-times increase in volume protects the driver from impact with the dashboard and windshield. Like all gas molecules, those released into the air bag are in constant motion, moving at velocities of about 600 miles per hour and constantly colliding with each other and the walls of the air bag. It is these collisions that give the air bag its shape. The sum of these collisions is called **pressure.**

Pressure is an inherent quantity associated with air and the gases that compose it. At sea level, the pressure due to the thin blanket of gas that covers Earth is about 14.7 pounds per square inch. That pressure is a result of the constant collisions between air molecules and everything else. The layer of air that surrounds the Earth is called the **atmosphere,** and without it, we could not exist. Earth's atmosphere provides the oxygen we need to breathe and protects us from short-wavelength radiation that would otherwise kill us. It is responsible for wind, rain, red-colored sunsets, and blue-colored midday skies. If the atmosphere were suddenly removed, we would die within minutes in the dark vacuum of space. The gas that fills the air bag, nitrogen, is the same one that composes 80% of the atmosphere. In this chapter, we learn about both the atmosphere and the gases that compose it.

The atmosphere is a thin blanket of gases held to planet Earth by gravity.

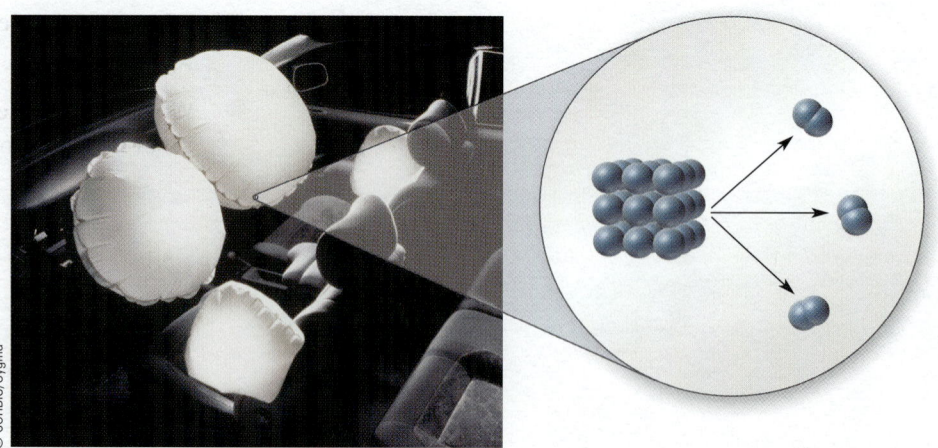

FIGURE 1 When an air bag system is triggered by a significant impact, a chemical reaction converts a solid to gas, which acts as a cushion to protect the driver.

11.2 A Gas Is a Swarm of Particles

Close as I ever came to seeing things
The way the physicists say things really are
Was out on Sudbury Marsh one summer eve
When a silhouetted tree against the sun
Seemed at my sudden glance to be afire:
A black and boiling smoke made all its shape.

Binoculars resolved the enciphered sigh
To make it clear the smoke was a cloud of gnats,
Their millions doing such a steady dance
As by the motion of the many made the one
Shape constant and kept it so in both the forms
I'd thought to see, the fire and the tree.

Strike through the mask? You find another mask,
Mirroring mirrors by analogy
Make visible. I watched till the greater smoke
Of night engulfed the other, standing out
On the marsh amid a hundred hidden streams
Meandering down from the Concord to the sea.
(Howard Nemerov, "Seeing Things")*

Short-wavelength light is scattered more efficiently by atmospheric molecules than long-wavelength light, resulting in the blue color of our sky. In the absence of the atmosphere, the sky would appear black.

This poet, in the hour of the sunset, glimpsed how nature might appear if we could see molecules. From this viewpoint, a gas looks much like the swarm of gnats that composed the poet's smoky tree. Each gnat represents a gas molecule, and like gas molecules, the gnats are very small in comparison to the space between them; yet, because of their constant motion, they collectively occupy a great deal of space. Imagine putting your hand in the midst of the gnat-filled swarm. You would feel a constant pelting of gnats against your hand; the same happens for molecules. As you sit reading this book, you are constantly pelted by gas molecules as they swarm around you. With each breath, you draw in a billion billion of them, use some, and force the rest out, along with some others that your lungs add to the mix. The air around us, a swarm of molecules not unlike a swarm of gnats, is tasteless, odorless, and invisible; yet we can feel its effects, and we depend on it at every moment of our existence.

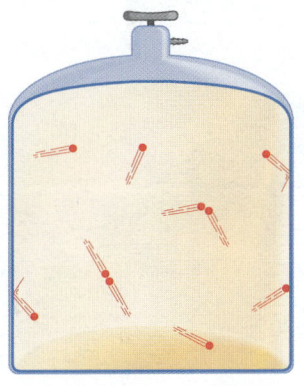

Gas molecules are very small compared to the space between them.

11.3 Pressure

Pressure is the direct result of constant collisions between gas molecules and the surfaces around them. Pressure keeps balloons and tires distended and allows us to drink from straws; it is constantly pushing on everything, including us. At constant temperature, pressure is directly proportional to the number of gas molecules per unit volume of air; the greater the number of molecules, the more collisions, and the greater the pressure. Pressure decreases with increasing altitude because the number of molecules in a given volume decreases with increasing altitude. As we climb a mountain or ascend in an airplane, we may feel ear pain because of the pressure decrease. Ears have small cavities within them that trap gas molecules. Under normal circumstances, the pressure within these cavities is equal to the external pressure;

---

* From *The Collected Poems of Howard Nemerov*, by Howard Nemerov, p. 768. Copyright © 1977 Little, Brown & Co. Reprinted by permission of the estate.

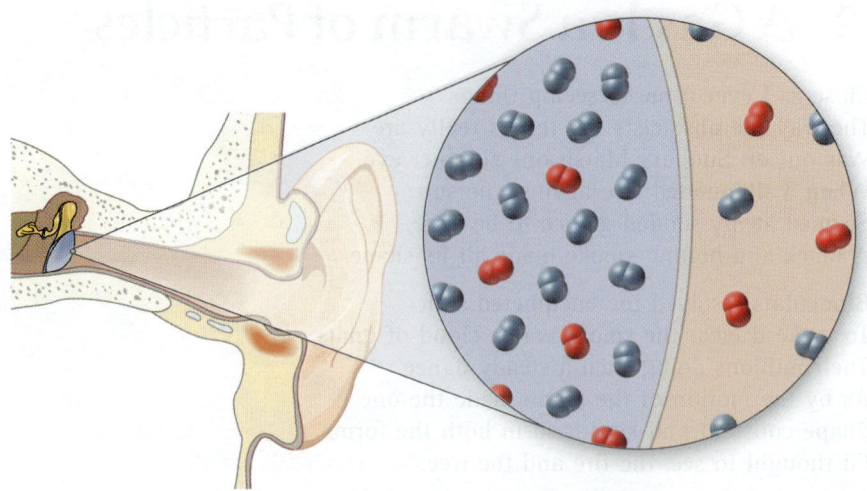

The pain we feel in our ears when changing altitude is due to a pressure imbalance between the cavities within our ears and the ambient air. The greater number of molecular collisions on one side versus the other stress the eardrum, resulting in pain. (Adapted from Human Information Processing: An Introduction to Psychology, Second Edition, *by Peter H. Lindsay and Donald A. Norman. Copyright © 1977 Harcourt Brace & Company.)*

therefore, the number of molecules colliding on either side of the eardrum is the same, and we do not feel any pain. However, when the external pressure decreases during an increase in altitude, the pressure imbalance causes more collisions on one side of the eardrum than the other. The eardrum is stressed, resulting in pain.

The units of pressure are those of force per unit area. In the English system, these are pounds per square inch (lb/in.^{2}), often abbreviated as psi; and in the metric system, these are newtons per square meter (N/m^2), also called a pascal (Pa). Several other units of pressure are also in common use (Table 1).

The first unit, mm Hg, is related to how pressure is measured using a barometer. Figure 2(a) shows a glass tube with one end immersed in a shallow bowl of water. Molecules push down on the surface of the water with equal pressure inside and outside of the tube. As a result, the water levels inside and outside the tube are equal. If we remove molecules from the top end of the tube by creating a vacuum, the pressures are no longer equal [Figure 2(b)]. The greater pressure over the surface of the water outside the tube pushes the water up into the tube. The same happens when we use a drinking straw. Our mouths create a partial vacuum, and the beverage is pushed up the straw by external pressure.

The height to which water rises in an evacuated tube depends on the external pressure pushing down on the surface of the water; the greater the pressure, the

TABLE 1

Several Units of Pressure and Their Values at Sea Level

Unit	Abbreviation	Sea Level Average
Millimeters of mercury	mm Hg	760
Torr	torr	760
Atmospheres	atm	1
Inches of mercury	in. Hg	29.92
Pounds per square inch	psi	14.7
Pascal	Pa	101,325
Newtons per square meter	N/m^2	101,325

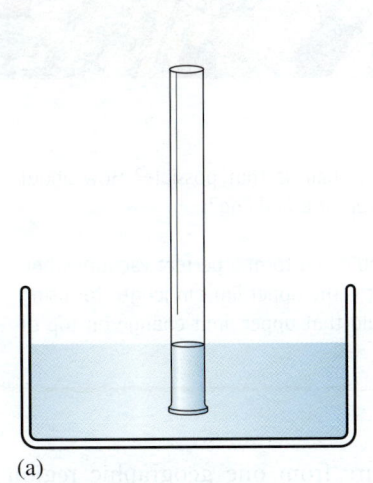
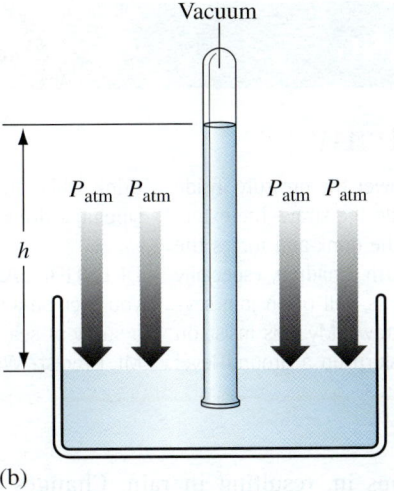

FIGURE 2 (a) A glass tube immersed in a bowl of water. The liquid level inside the tube is the same as the liquid level outside the tube. (b) If the top of the tube is evacuated, the liquid level within the tube rises because of atmospheric pressure pushing down on the surface of the liquid.

higher the water will rise. If we put a perfectly evacuated tube into a bowl of water at sea level, the pressure would push the water up 33 ft (10 m). At higher altitudes, where the pressure is less, the water would not be pushed up so far. On top of Mt. Everest, for instance, the water would rise only 10 ft.

The height of water in an evacuated tube can be used to measure pressure. However, a 33-ft tube is inconvenient. Consequently, a more dense liquid, usually mercury (Hg), is used in **barometers**, devices that measure pressure. At sea level on an average day, the height of mercury in a perfectly evacuated tube is 760 mm, about 2.5 ft. The barometer, therefore, gives us the unit of pressure, mm Hg. The pressure at sea level averages 760 mm Hg, while the pressure on top of Mt. Everest averages 240 mm Hg. Two other commonly used units of pressure are inches of mercury (in. Hg) and atmospheres (atm), both summarized in Table 1.

Weather forecasters use pressure, usually quoted in inches of mercury, to predict the likelihood of rain. High pressure over a region tends to direct storms away and is therefore a sign of fair weather. Low pressure, in contrast, tends to draw

An area of low pressure (L) will draw storms in and is therefore a sign of bad weather. An area of high pressure (H) tends to direct storms away.

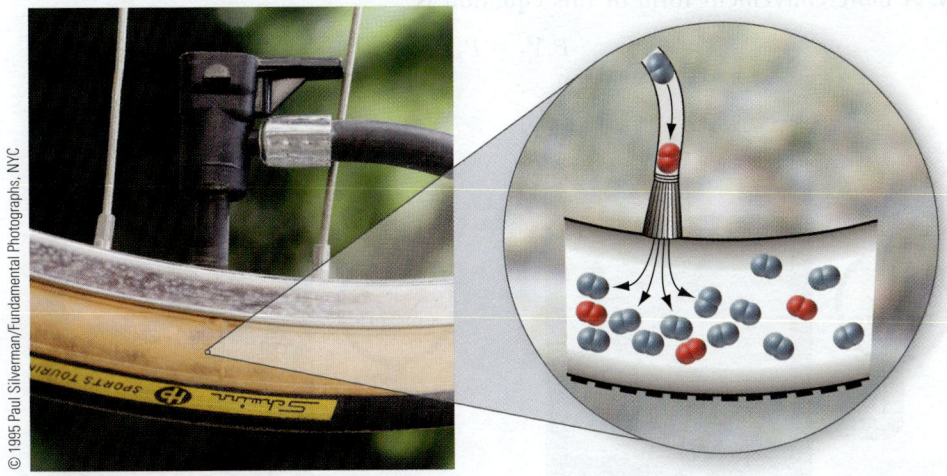

A pump forces air molecules into a tire. The more molecules within the tire, the more collisions occur between the molecules and the tire wall, resulting in a firmer tire.

MOLECULAR THINKING

Drinking from a Straw

Straws work because by sipping, you lower the pressure inside the straw relative to the pressure outside the straw. The external pressure pushes on the surface of the drink and forces the liquid up the straw and into your mouth. Children, especially those that frequent fast-food restaurants, will often join several straws together to make longer straws. My kids insist on constructing a straw long enough to slurp up a ground-level drink while standing on a chair. Is that possible? How about sipping a drink from the top of a building?

QUESTION: Assuming you could form a perfect vacuum when you suck on a straw, what is the upper limit in length for using a straw at sea level? Would that upper limit change on top of Mt. Everest? Why?

storms in, resulting in rain. Changes in pressure from one geographic region to another are responsible for wind.

In addition to pressure, three other fundamental properties are associated with a sample of gas. The first is the *amount of gas*, measured in moles and represented by the symbol n. The second is the *volume (V)* the gas occupies, usually expressed in liters; and the third is the *temperature (T)* of the gas, expressed in Kelvin.

11.4 The Relationships Between Gas Properties

The gas properties discussed above—moles, volume, temperature, and pressure—are interrelated. If we change one of these properties, others will also change. The relationships between changes in these properties are called the gas laws.

PRESSURE AND VOLUME: BOYLE'S LAW

The relationship between pressure and volume, called Boyle's law, states that if the pressure of a gas is increased at constant temperature, its volume will *decrease*. If we increase the pressure on a cylinder filled with gas by pushing down on a piston (Figure 3), the volume of the gas decreases. Pressure and volume are *inversely* related; the higher the pressure is, the lower the volume. Boyle's law can be expressed as $P = a/V$, where a is a constant of proportionality. A more convenient form of this equation is

$$P_1 V_1 = P_2 V_2$$

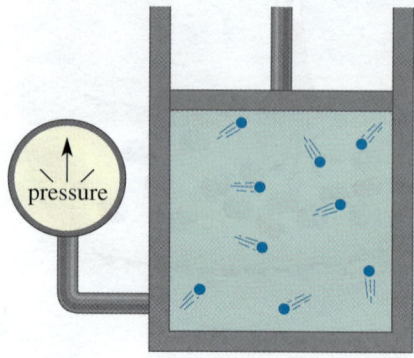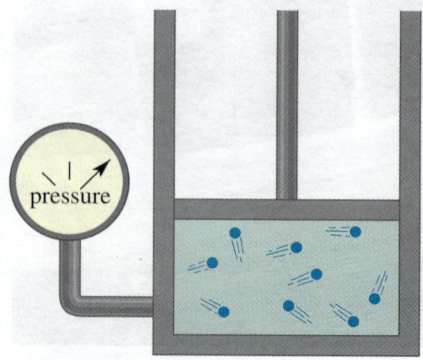

FIGURE 3 The volume of a sample of gas will decrease to one half its original volume if its pressure is increased by a factor of 2.

This equation allows us to calculate the changes in pressure or volume that will occur when the other is changed, as long as all other properties (temperature and moles of gas) are constant. P_1 and V_1 are the initial pressure and volume of a system, and P_2 and V_2 are the final pressure and volume. The molecular reason for the pressure increase on a volume decrease is that the molecules are now in a smaller space, colliding more frequently and producing greater pressure.

Boyle's law is important in scuba diving where divers experience pressure changes as they ascend and descend. For every 33 feet of depth, the pressure increases by 1 atmosphere. The pressure at sea level is 1 atm, at 33 ft it is 2 atm, at 66 ft it is 3 atm, and so on. In order to breathe normally under increased pressure, divers must breathe pressurized air that matches the external pressure. The regulator does the job of matching. For example, when a diver is at 33 ft, the regulator adjusts the air pressure to 2 atm; if the pressure were not adjusted, the diver could not breathe (Figure 4). However, breathing pressurized air carries some risk. For example, if a diver breathes 2 atm air and suddenly ascends to the surface where the pressure is 1 atm, what will happen to the volume of air in his lungs? According to Boyle's law, the air will expand, which is very dangerous and may cause severe injury. Consequently, scuba divers must ascend slowly, allowing time for normal breathing, which allows the regulator to slowly match the air pressure with the external pressure, preventing injury. For this reason, scuba divers should never hold their breath but should breathe normally, allowing the regulator to do its job.

FIGURE 4 (a) At 33 ft, a scuba diver experiences a pressure of 2 atm, and must breathe pressurized air at the same pressure. (b) If the diver ascends too quickly, the decrease in pressure will result in an increase of the volume occupied by the gas in his or her lungs, possibly leading to serious injury.

> ### EXAMPLE 11.1
>
> ### Boyle's Law
>
> A balloon is inflated to a volume of 1.00 L at sea level where the pressure is 760 mm Hg. Find the volume of the balloon on top of Mt. Everest where the pressure is 240 mm Hg. (Assume constant temperature.)
>
> **SOLUTION**
> First we must solve the equation for V_2:
>
> $$V_2 = \frac{P_1 V_1}{P_2}$$
>
> Now we substitute the values for P_1, V_1, and P_2.
>
> $$V_2 = \frac{760 \text{ mm Hg} \times 1.00 \text{ L}}{240 \text{ mm Hg}} = 3.17 \text{ L}$$
>
> ### YOUR TURN
>
> ### Boyle's Law
>
> An empty gasoline can has an initial volume of 2.0 L at sea level where the pressure is 1.0 atm. It is then submerged into water to a depth where the pressure is 4.0 atm. The walls fail, resulting in a crumpled can. What is the volume of the crumpled can? (Assume constant temperature.)

TEMPERATURE AND VOLUME: CHARLES'S LAW

The relationship between temperature and volume, Charles's law, states that if the temperature of a gas is increased at constant pressure, its volume will increase in direct proportion. Charles's law can be easily demonstrated by placing an inflated balloon above a warm toaster. As the temperature of the gas within the balloon increases, the balloon expands. Charles's law can be expressed as $V = bT$, where b is a constant of proportionality and T is the temperature in Kelvin. A more convenient form of this equation is

$$\frac{V_1}{T_1} = \frac{V_2}{T_2}$$

This equation allows us to calculate the changes in temperature or volume that occur when the other is changed. V_1 and T_1 represent the initial volume and temperature, and V_2 and T_2 represent their final values.

You can observe the effects of Charles's law in a heated room on a winter's day. The air near the ceiling is warmer than the air near the floor. When air is heated it expands according to Charles's law, lowering its density and causing it to rise. Hot-air balloons use the same effect to achieve buoyancy. When air in the balloon is heated, its volume increases, resulting in lower density and increased buoyancy. Afternoon winds in coastal regions are also a result of Charles's law. Inland air masses are warmed by the Sun and therefore rise; air from the coast, which remains cool because of the high heat capacity of oceans, rushes in to fill the void, resulting in wind.

The air in a syringe will increase in volume if the temperature increases.

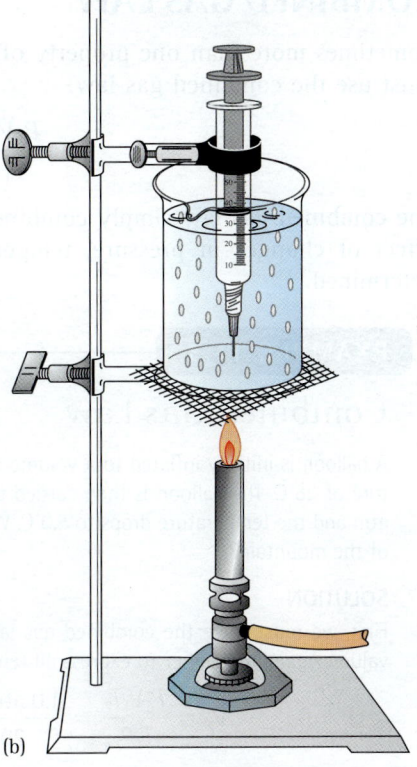
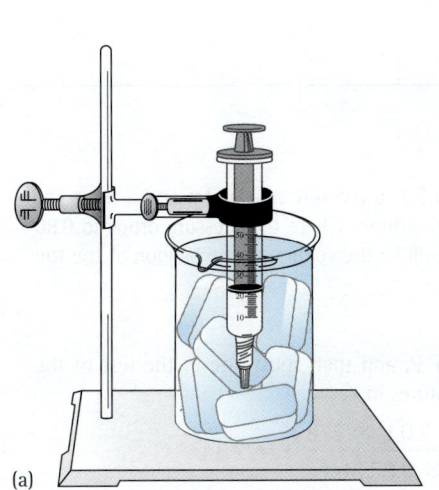

(a) Syringe in ice-water bath. Temperature = 0.0°C

(b) Syringe in boiling water bath. Temperature = 100.0°C

EXAMPLE 11.2

Charles's Law

A cylinder with a movable piston has an initial temperature of 25°C and an initial volume of 0.50 L. What is its volume at 75°C? (Assume constant pressure.)

SOLUTION
We are asked to find final volume so we solve the equation for V_2 and substitute the other values. Remember that all temperatures must be in Kelvin (°C + 273 = K).

$$V_2 = \frac{V_1}{T_1}T_2 = \frac{0.50 \text{ L}}{298 \text{ K}} \times 348 \text{ K} = 0.58 \text{ L}$$

YOUR TURN

Charles's Law

A balloon has an initial temperature of 30.0°C and an initial volume of 1.0 L. What would be its volume at 100.0°C? (Assume constant pressure.)

COMBINED GAS LAW

Sometimes more than one property of a gas changes at a time. In these cases, we must use the combined gas law:

$$\frac{P_1 V_1}{T_1} = \frac{P_2 V_2}{T_2}$$

The combined gas law simply combines Boyle's law and Charles's law so that the effect of changes in pressure, temperature, and volume can be simultaneously determined.

EXAMPLE 11.3

Combined Gas Law

A balloon is initially inflated to a volume of 2.0 L at a pressure of 1.0 atm and a temperature of 25°C. The balloon is then carried up a mountain where the pressure drops to 0.80 atm and the temperature drops to 5.0°C. What will be the volume of the balloon at the top of the mountain?

SOLUTION

First, we must solve the combined gas law for V_2 and then substitute in the rest of the values. Again, remember to express all temperatures in K.

$$V_2 = \frac{P_1 V_1 T_2}{T_1 P_2} = \frac{1.0 \text{ atm} \times 2.0 \text{ L} \times 278 \text{ K}}{298 \text{ K} \times 0.80 \text{ atm}} = 2.3 \text{ L}$$

YOUR TURN

Combined Gas Law

A bicycle tire has an initial volume of 0.30 L at 25°C and a pressure of 1.0 atm. What would be the volume of the tire on ascending a mountain where the temperature rises to 35°C and the pressure drops to 0.90 atm?

APPLY YOUR KNOWLEDGE

A gas is originally in a 3.0-L container at a given temperature and pressure. The gas is transferred into a 6.0-L container and its temperature (in Kelvins) is doubled. What is the net effect on the pressure?

a. The pressure increases by a factor of 2.
b. The pressure increases by a factor of 4.
c. The pressure decreases by a factor of 4.
d. The pressure remains constant.

Answer: (d) The effects of doubling the volume (which decreases the pressure by 2) and doubling the temperature (which increases the pressure by 2) cancel each other.

11.5 The Atmosphere: What Is in It?

The air that composes the atmosphere is a mixture of several different gases. The primary gases in our atmosphere and their relative percentages by volume are

TABLE 2		
Composition of Dry Air		
Gas	Percentage by Volume (%)	Number of Molecules per 1 Million (ppm) Molecules of Air
Nitrogen (N_2)	78	780,830
Oxygen (O_2)	21	209,450
Argon (Ar)	0.9	9,340
Carbon dioxide (CO_2)	0.03	350
Neon (Ne)	0.0018	18
Helium (He)	0.00052	5.2
All others	0.0004	4.0

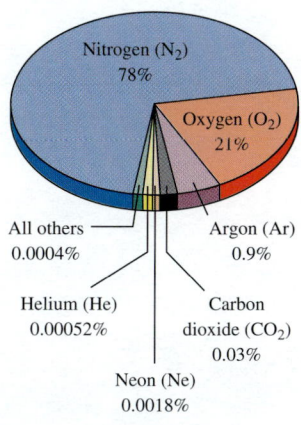

The primary gases present in our atmosphere (by volume).

shown in Table 2. The units on the right, parts per million (ppm), are often used in expressing fractional amounts that are very small. Just as percent means parts per hundred, ppm means parts per million. For example, neon is present in the atmosphere at a level of 18 ppm, meaning that for every 1 million molecules or atoms in the air there are 18 neon atoms.

N_2

Nitrogen is the most abundant gas in our atmosphere, composing nearly four-fifths of it. Nitrogen is tasteless, colorless, odorless, nonflammable, and relatively inert. The nitrogen that we breathe passes into and out of our lungs with little change. Plants need nitrogen, a vital element in protein synthesis, to survive. However, the strong triple bond in N_2 must be broken because plants need nitrogen atoms, not nitrogen molecules. Certain types of bacteria, present in soil, break nitrogen's triple bond in a process called **nitrogen fixation** to form nitrate compounds (contain NO_3^-) that plants use directly. Because nitrate compounds are one of the limiting factors in plant growth, fertilizers are also rich in them.

O_2

Oxygen, roughly one-fifth of our atmosphere, is more reactive than nitrogen. Oxygen is responsible for combustion, the rusting of iron, the dulling of paint, and the support of animal life. When we breathe air, we absorb about one-fourth of the oxygen into our bloodstream, where it is transported to cells. At the cell, oxygen reacts with glucose to provide energy, a process called respiration:

$$C_6H_{12}O_6 + 6O_2 \rightarrow 6CO_2 + 6H_2O + \text{energy (respiration)}$$

Carbon dioxide is transported back to our lungs, where it is exhaled.

CO_2

Carbon dioxide is central to plant growth through photosynthesis. Plants use carbon dioxide and water to form glucose in a reaction exactly opposite to respiration:

$$6CO_2 + 6H_2O + \text{energy} \rightarrow C_6H_{12}O_6 + 6O_2 \text{ (photosynthesis)}$$

In addition, large amounts of CO_2 are absorbed by the Earth's oceans and eventually form carbonates in rocks and soil. The carbon atoms do not stay in these places indefinitely, however. In a natural cycle, CO_2 is channeled back into the atmosphere through the decay of plants, the decomposition of rocks, the eruption of volcanoes, and respiration in animals. The Earth's carbon atoms thus travel a constant cycle, from air to plants and animals or rocks, and then back into the air. The average carbon atom in our own bodies has made this cycle approximately 20 times.

ARGON, NEON, AND HELIUM

Argon, neon, and helium are all inert gases, chemically unreactive. Neon is the gas used in neon signs. Argon is often used in the electronics industry when inert conditions are required. Helium is used in balloons and, in its liquid form, as a cryogenic to achieve temperatures near zero Kelvin.

11.6 The Atmosphere: A Layered Structure

The atmosphere changes in composition and pressure with increasing altitude. As altitude increases, pressure decreases (Figure 5). You may have experienced the shortness of breath associated with increased elevations. The lower pressure results in fewer air molecules. As a result, your lungs work harder to extract the same number of oxygen molecules as they did at sea level.

The atmosphere can be divided into four sections according to altitude (Figure 6). The lowest part of the atmosphere, that closest to the Earth, is called the **troposphere** and ranges from ground level to 10 km (6 mi). All Earth-bound life exists in the tropo-

FIGURE 5 Pressure decreases with increasing altitude.

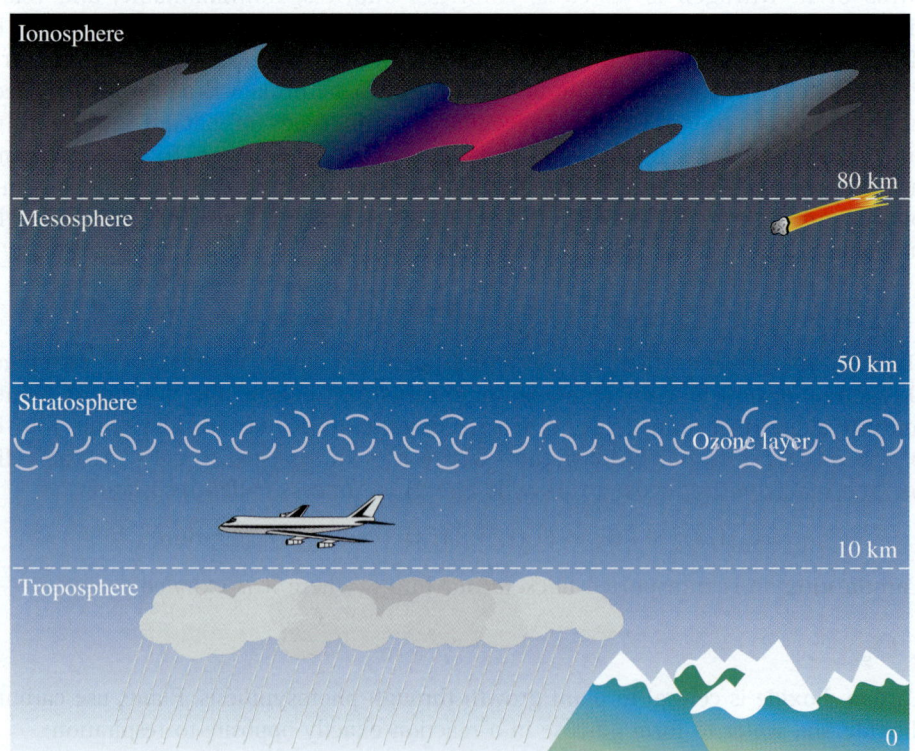

FIGURE 6 The structure of the atmosphere.

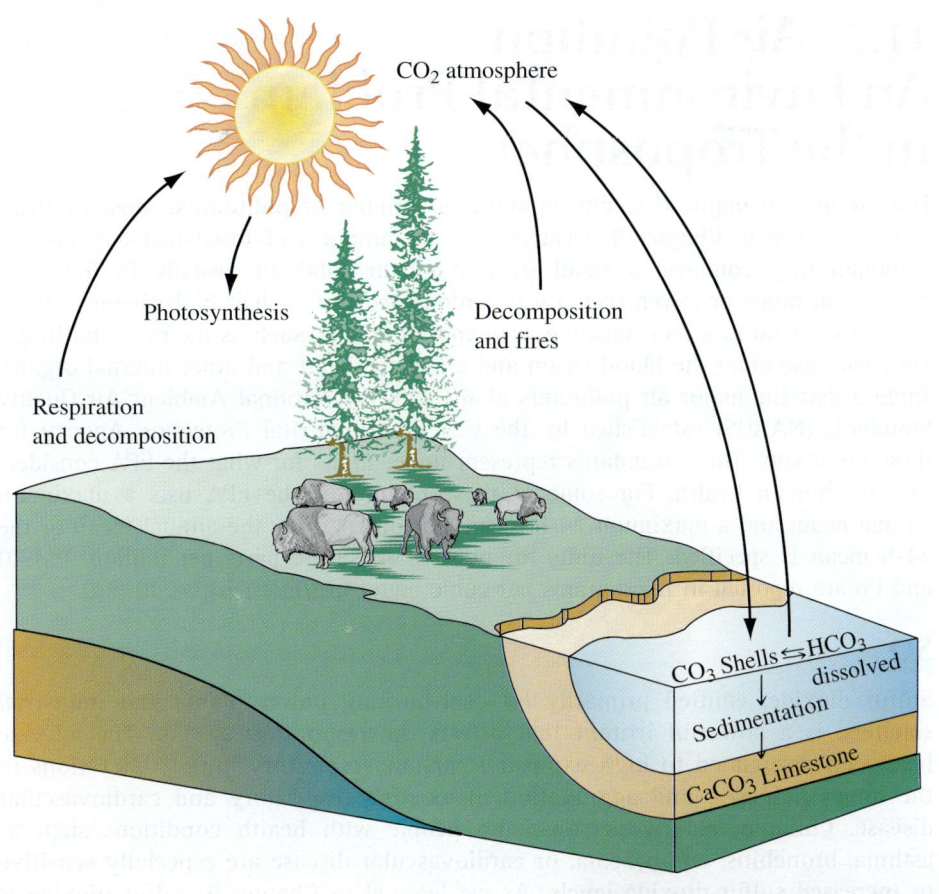

The carbon cycle. The average carbon atom has gone through the carbon cycle about 20 times. (Adapted from *Chemistry and Chemical Reactivity, Third Edition,* by John C. Kotz and Paul Treichel, Jr. Copyright © 1994, Saunders College Publishing. Reprinted by permission of the publisher.)

Explore this topic on **The Atmosphere** website.

The aurora borealis, or northern lights, occurs in the ionosphere.

sphere. Even the world's highest mountain, Mt. Everest at 9 km (5.5 mi), does not extend beyond the troposphere. All weather phenomena such as clouds and rainfall occur in the troposphere, making it turbulent. Consequently, most jet pilots choose to fly above the troposphere in the **stratosphere,** a region of the atmosphere ranging from 10 to 50 km. The stratosphere contains ozone (O_3). Even though ozone is a pollutant in the troposphere, it is a natural and necessary component of the stratosphere. Ozone absorbs harmful ultraviolet light, protecting life on the Earth from its ill effects.

Above the stratosphere, at >50 km, lie the **mesosphere** and the **ionosphere.** Meteors entering the Earth's atmosphere burn up in these regions and appear as "falling stars." The phenomenon called aurora borealis or northern lights is a result of ionic gas particles traveling at high speeds and emitting light in the ionosphere.

11.7 Air Pollution: An Environmental Problem in the Troposphere

> Explore this topic on the National Ambient Air Quality Standards (NAAQS) website.

The air in any major U.S. city contains a number of pollutants. Most of these were discussed in Chapter 9 because they are products of fossil-fuel combustion. Although they compose a small fraction of the total air, usually in the parts per million range or lower, they are hazardous because of their high chemical reactivity. They tend to affect sensitive, air-exposed organs, such as the eyes and lungs. They can also enter the bloodstream and affect the heart and other internal organs. Table 3 lists the major air pollutants along with the National Ambient Air Quality Standards (NAAQS) established by the U.S. Environmental Protection Agency for these pollutants. These standards represent upper limits for what the EPA considers safe for human health. For some of these pollutants, the EPA lists a maximum annual mean and a maximum 24-h mean. For others, only the annual mean or the 24-h mean is specified. The units for most of these are parts per million. PM-10 and Pb are reported in micrograms per cubic meter ($\mu g/m^3$, 1 $\mu g = 10^{-6}$ g).

SO_2

Sulfur dioxide, emitted primarily by coal-burning power plants and industrial smelters, is a powerful irritant that affects the respiratory system. The adverse health effects related to high exposures include respiratory illness, alterations in the lung's defenses, and aggravation of existing respiratory and cardiovascular disease. Children, elderly persons, and people with health conditions such as asthma, bronchitis, emphysema, or cardiovascular disease are especially sensitive to increased sulfur dioxide levels. As we learned in Chapter 9, sulfur dioxide is also the major precursor to acid rain.

PM-10

PM-10 is an abbreviation for particulate matter with a diameter less than or equal to 10 μm (about the size of a dust particle). The composition of these particles can vary, but their primary sources are agricultural tilling, construction sites, and

TABLE 3

National Primary Ambient Air Quality Standards (EPA)

	SO_2 (Annual Mean) (ppm)	PM-10 (Annual Mean) ($\mu g/m^3$)	CO (8-h Mean) (ppm)	O_3 (8-h Mean) (ppm)	NO_2 (Annual Mean) (ppm)	Pb (3-Month Mean) ($\mu g/m^3$)
Maximum amount	0.03	50	9	0.08	0.053	1.5
Primary sources	Electric utilities, industrial smelters	Agricultural tilling, construction, unpaved roads, fires	Motor vehicles	Motor vehicles (indirect)	Motor vehicles, electric utilities	Motor vehicles, smelters, battery plants

Source: EPA National Primary and Secondary Ambient Air Quality Standards.

unpaved roads. These sources produce airborne particles that stay suspended for long periods of time. Their adverse health effects include aggravation of existing respiratory and cardiovascular disease, alterations in the body's defense systems against foreign materials, damage to lung tissue, and cancer. Children, elderly persons, and people with pulmonary disease, cardiovascular disease, influenza, or asthma are especially sensitive. PM-10 can also lower visibility and soil-building materials.

The dust formed in agricultural tilling produces airborne particles that can stay suspended for long periods of time. PM-10 refers to those particles whose diameters are less than or equal to 10 μm. Because of their small size, these particles can get deep into the lungs when inhaled, creating adverse health effects.

CO

Carbon monoxide, emitted by motor vehicles, is toxic because it diminishes the blood's ability to carry oxygen. In high concentrations, carbon monoxide is lethal. In lower concentrations, it causes the heart and respiratory system to work harder. The adverse health effects related to carbon monoxide exposure include impairment of visual perception, decreased work capacity, decreased manual dexterity, and lower learning ability. Those who suffer from cardiovascular disease, especially those with angina or peripheral vascular disease, are most sensitive.

O_3

Ozone, as we learned in Chapter 9, is a component of photochemical smog and is formed from the action of sunlight on motor vehicle exhaust. It has a characteristic pungent odor often found around electrical equipment or during a lightning storm where ozone is formed in high concentrations. In the stratosphere, ozone acts as a UV shield, but tropospheric or ground-level ozone is a pollutant that irritates the eyes and lungs. Studies show that 6–7 h of exposure to ozone at relatively low concentrations can significantly reduce lung function—even in healthy adults—resulting in chest pain, coughing, nausea, and pulmonary congestion. Animal studies show that long-term exposure can produce permanent structural damage to the lungs. Ozone also damages rubber, agricultural crops, and many tree species.

Carbon monoxide is emitted by motor vehicles due to incomplete combustion. Some counties must now use oxygenated fuels. These fuels contain oxygen and promote the complete combustion of gasoline, reducing CO emissions.

NO_2

Nitrogen dioxide, which results from motor vehicles and (to a lesser extent) electric utility emissions, is also a lung and eye irritant. Nitrogen dioxide is partially responsible for the brown color of smog. It is a precursor to ozone formation and contributes to acid rain. Long-term exposure can result in acute respiratory disease in children.

Pb

Lead, emitted primarily from smelters and battery plants, enters the body in several different ways, including inhalation of lead in air, ingestion of lead in food, or consumption of lead in water. Lead can accumulate in the body over long periods of time and damage the kidneys, liver, and nervous system. Excessive exposure causes neurological damage, producing seizures, mental retardation, and other behavioral disorders. Fetuses, infants, and children are particularly sensitive.

11.8 Cleaning Up Air Pollution: The Clean Air Act

The severe state of air pollution in many of our nation's cities, in combination with the adverse health and environmental effects of these pollutants, prompted lawmakers to initiate legislation to control air pollution. In 1970, Congress passed

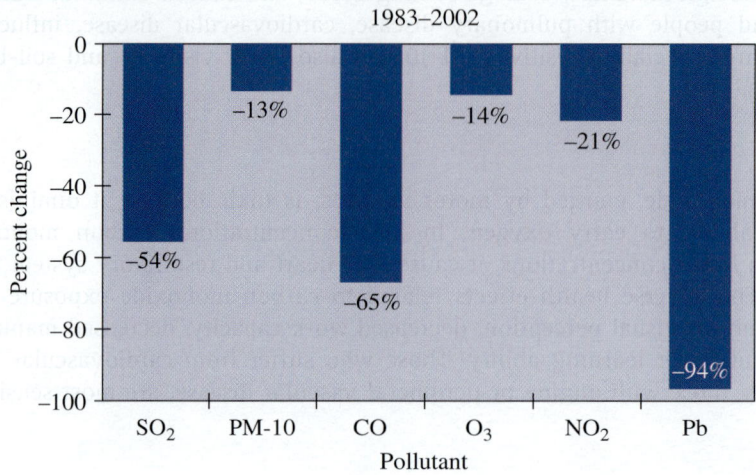

TABLE 4
Twenty-Year Air Pollution Trends[a]

1983–2002

Pollutant	Percent change
SO_2	−54%
PM-10	−13%
CO	−65%
O_3	−14%
NO_2	−21%
Pb	−94%

[a]Long-term trend not available. The 13% decrease is for the 10-yr period 1993–2002.

Source: The EPA *10-yr National Air Quality and Emissions Report*, 2003.

the Clean Air Act, a comprehensive federal law that regulates emissions and authorizes the EPA to establish the NAAQS shown in Table 3. The goal of the act was to achieve the NAAQS in every state by 1975. The goal was not met; thus in 1977 Congress passed amendments to the Clean Air Act to articulate new goals and make them more enforceable.

As a result of this legislation, pollution in urban areas today is significantly reduced. This progress has come in spite of large increases in the number of motor vehicles on the road and is attributed primarily to decreases in motor vehicle tailpipe emissions. Vehicles built today emit 60–80% fewer pollutants than those built in the 1960s. Table 4 shows the results of a continuing EPA study of air quality. The levels of all six major pollutants have significantly decreased over the past 20 years.

In spite of gains made by air pollution legislation, significant problems persist. Table 5 shows the pollution levels in several U.S. cities. As you can see, most cities meet the NAAQS for every pollutant except O_3. Table 6 shows the number of Americans living in counties that do not meet the NAAQS for each pollutant. One persistent problem is ground-level ozone. In response to this shortcoming, and to deal with other air pollution problems such as stratospheric ozone depletion and acid rain, the first President Bush in 1989 proposed further amendments to the Clean Air Act. These amendments were overwhelmingly approved by Congress and signed into law on November 15, 1990.

The Clean Air Act amendments of 1990 contain a variety of programs requiring those areas still above the NAAQS to come into compliance. The worse the air quality in a given area is, the more drastic the measures that must be taken. These measures include issuing control permits to businesses that must emit pollutants, using oxygenated fuels that promote complete combustion of gasoline, setting tighter pollution standards for emissions from cars and trucks, and establishing require-

This truck is powered by liquid propane, also known as LP gas, a clean fuel with very low emissions.

TABLE 5
Air Pollution Peak Statistics for Selected U.S. Cities (2002)

City	SO_2 (Annual Mean) (ppm)	PM-10 (Annual Mean) ($\mu g/m^3$)	CO (8-h Mean) (ppm)	O_3 (8-h Mean) (ppm)	NO_2 (Annual Mean) (ppm)	Pb (Quarter Max) ($\mu g/m^3$)
Atlanta, GA	0.003	26	4	0.1	0.019	0.04
Baltimore, MD	0.005	26	3	0.1	0.025	0.01
Chicago, IL	0.005	36	4	0.1	0.032	0.04
Dallas, TX	0.002	29	2	0.1	0.018	0.48
Detroit, MI	0.007	38	4	0.1	0.021	0.03
Houston, TX	0.003	34	3	0.1	0.019	0.01
Los Angeles, CA	0.003	46	9	0.13	0.040	0.04
New York, NY	0.012	NA	3	0.1	0.038	0.01
Philadelphia, PA	0.008	24	3	0.1	0.030	0.04
NAAQS	0.03	50	9	0.08	0.053	1.5

NA—not available.
Source: The EPA *10-yr National Air Quality and Emissions Report* (2003).

TABLE 6
Number of Americans Living in Counties with Pollutant Levels over the NAAQS

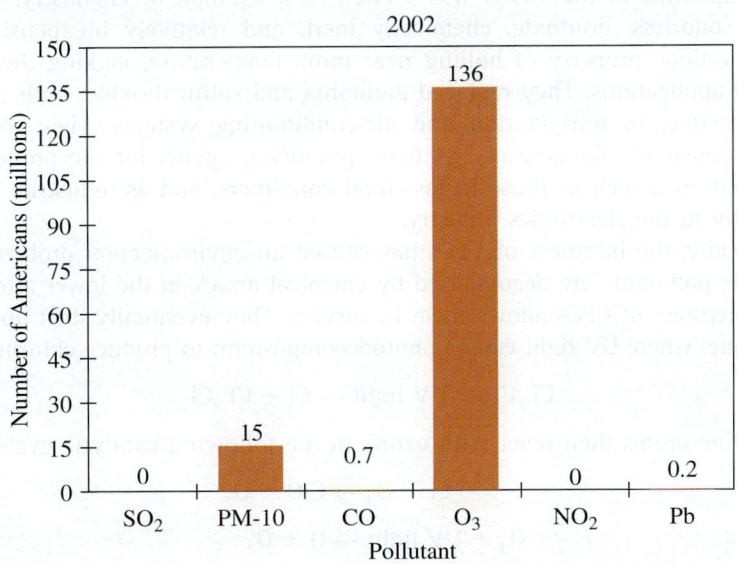

Source: The EPA *10-yr National Air Quality and Emissions Report* (2003).

ments for the conversion of fleet vehicles to alternative fuels. The amendments also establish provisions for how and when the federal government can impose sanctions on areas that do not comply. In spite of these measures, ground-level ozone continues to exceed the NAAQS in most major U.S. cities.

11.9 Ozone Depletion: An Environmental Problem in the Stratosphere

Ozone in the troposphere is a pollutant, but in the stratosphere, it is a natural and essential part of our atmosphere. Its most important function is the absorption of UV light emitted by the Sun. UV light is often divided into two categories, UV-A (320–400 nm) and UV-B (280–320 nm). UV-B light, shorter in wavelength and higher in energy, is most harmful to humans. Excessive exposure to UV-B increases the risk of skin cancer, promotes cataracts, weakens the immune system, and causes premature skin aging. UV-B also damages crops and marine ecosystems. Ozone absorbs UV-B, preventing it from reaching the Earth, via the following reaction:

$$O_3 + UV\ light \rightarrow O_2 + O$$

The ozone molecule absorbs UV light and breaks apart into O_2 and O. The O atom formed is very reactive and recombines with O_2 to re-form O_3 according to

$$O_2 + O \rightarrow O_3 + heat$$

This cycle repeats itself over and over again, each time converting UV light into heat and protecting life on the Earth from excessive UV exposure.

CHLOROFLUOROCARBONS: OZONE EATERS

Explore this topic on the Chlorofluorocarbons (CFCs) website.

In the 1970s, scientists began to suspect that a class of synthetic chemicals known as **chlorofluorocarbons (CFCs)** was destroying the stratospheric ozone layer. Two of the most common CFCs are Freon-11 and Freon-12 (Figure 7). The discovery of these compounds in the 1930s was viewed as a triumph of chemistry. CFCs are colorless, odorless, nontoxic, chemically inert, and relatively inexpensive. They have the unique property of boiling near room temperature, making them useful for many applications. They replaced ammonia and sulfur dioxide, both toxic and highly reactive, in refrigeration and air-conditioning systems. They have been used as propellants for aerosols, as foam-producing agents for the production of porous materials such as those in fast-food containers, and as industrial solvents, particularly in the electronics industry.

Ironically, the inertness of CFCs has caused an environmental problem. Many man-made pollutants are decomposed by chemical attack in the lower atmosphere, but the inertness of CFCs allows them to survive. They eventually drift up into the stratosphere, where UV light causes photodecomposition to produce chlorine atoms:

$$CF_2Cl_2 + UV\ light \rightarrow Cl + CF_2Cl$$

The chlorine atoms then react with ozone in the following catalytic cycle:

$$Cl + O_3 \rightarrow ClO + O_2$$
$$O_3 + UV\ light \rightarrow O + O_2$$
$$ClO + O \rightarrow Cl + O_2$$

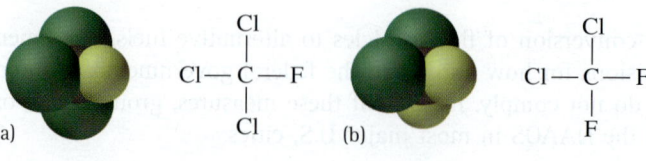

FIGURE 7 (a) Freon-11; (b) Freon-12.

In the first reaction, a chlorine atom reacts with an ozone molecule and destroys it, producing chlorine monoxide (ClO) and O_2. In the second reaction, UV light breaks ozone apart (a natural process in the stratosphere). However, in the third reaction, ClO, formed in the first step, reacts with O, preventing the re-formation of O_3. The net result is that one Cl atom destroys two ozone molecules and is regenerated in the process, allowing it to react over and over again.

OZONE DEPLETION OVER THE POLAR REGIONS

In 1985, scientists discovered a large hole in the ozone layer over Antarctica that has been attributed to CFCs. The amount of ozone depletion discovered over Antarctica was a startling 50%. The hole is transient, existing only in the Antarctic spring, late August to November. Examination of data from previous years showed that this ozone hole had formed each spring with growing intensity since 1977 (Figure 8). Furthermore, it correlated well with increased atmospheric chlorine from CFCs. If the Antarctic decrease in ozone occurred over North America or other populated areas, the results would be serious: The corresponding increase in UV light would increase the number of skin cancers and cataracts, and the effect on plant and animal life would be severe.

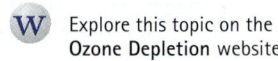

Explore this topic on the **Ozone Depletion** website.

For several years scientists wondered why such a large depletion of ozone occurred over Antarctica but not in other areas of the world. The presence of CFCs and other chlorine-containing compounds is roughly uniform throughout the stratosphere, so if these compounds were depleting ozone, why only over Antarctica? The answer to this question lies in the unique geography and climate of Antarctica. The stratosphere over Antarctica in winter is isolated by a swirling mass of air currents called the polar vortex. The temperatures within this vortex are cold enough that unique clouds, called polar stratospheric clouds (PSCs), form. The ice particles that compose these clouds act as catalysts to free chlorine from chemical reservoirs. These chemical reservoirs are compounds such as $ClONO_2$ that form from man-made stratospheric chlorine. The most important reaction in this process is

$$ClONO_2 + HCl \xrightarrow{PSCs} Cl_2 + HNO_3$$

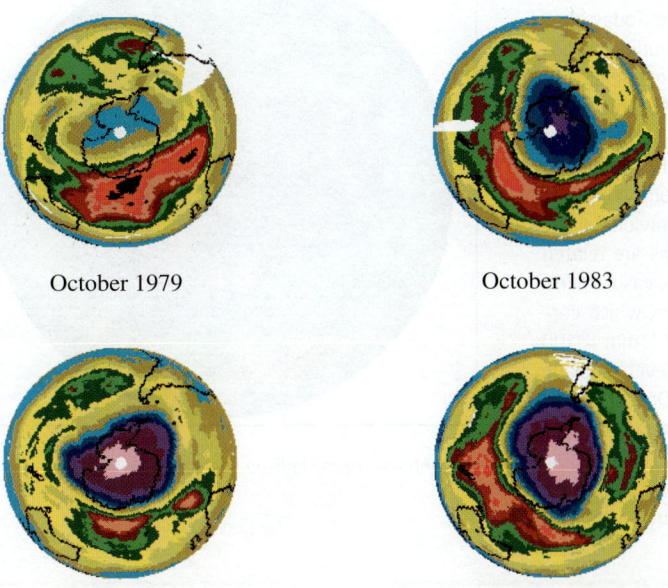

October 1979   October 1983

October 1987   October 1991

FIGURE 8 Antartic ozone levels each October from 1979 to 1991. The blue color indicates low ozone levels, which grow in intensity over this time period.

MOLECULAR FOCUS

Ozone

Formula: O_3
Molar Mass: 48.00 g/mol
Melting point: $-193°C$
Boiling point: $-111.9°C$
Three-dimensional structure:

Lewis structure:

$:\ddot{O}=\ddot{O}-\ddot{O}:$

Ozone is a bluish, extremely reactive gas. It has a pungent odor often associated with electrical storms that naturally form ozone. It is formed in the troposphere as a component of photochemical smog and is a natural component of the stratosphere, where it acts as a shield against UV radiation. Ozone is so effective in absorbing UV light that an alien with the ability to see only UV light (250 nm) would see a black sky, even at midday. Airplanes that fly in the stratosphere must filter cabin air to eliminate ozone; otherwise, passengers would experience chest pain, coughing, and nausea.

Ozone can be made in the laboratory by passing oxygen gas between high-voltage electrodes. It is used as a disinfectant for air and water and as bleach for textiles.

Chlorine gas is released into the gas phase. The Cl_2 bond is then broken by sunlight when the Sun rises in the Antarctic spring, producing chlorine atoms. These chlorine atoms then deplete ozone according to the preceding reactions. In the Antarctic summer, the polar vortex breaks down, along with the PSCs, and the

THE MOLECULAR REVOLUTION

Measuring Ozone

Since the early 1960s, scientists have launched satellites into space to view the Earth from above. At first, those satellites sent back only rough images of the Earth's surface. Today, they carefully monitor many of Earth's vital signs, including atmospheric ozone concentration. NASA has been using satellites to monitor atmospheric ozone since 1978. Today, NASA measures atmospheric ozone using a Total Ozone Mapping Spectrometer (TOMS) orbiting the Earth on the Earth Probe (EP) satellite at 740 km. This instrument measures ozone by comparing the intensities of two wavelengths of scattered ultraviolet light. Differences in the intensities of these wavelengths are related to the total ozone concentration at the point of measurement. Ozone concentrations are reported in Dobson Units, which correspond to the thickness of the entire ozone level (1mm = 100 Dobson Units) if it were brought down to sea level. Average worldwide ozone concentrations are about 300 Dobson Units. During the Antarctic ozone hole, ozone levels over Antarctica drop to about 100 Dobson Units. Data from TOMS is available to anyone on the World Wide Web. Just visit NASA's TOMS website at http://jwocky.gsfc.nasa.gov/ to get a bird's-eye view of global ozone concentrations.

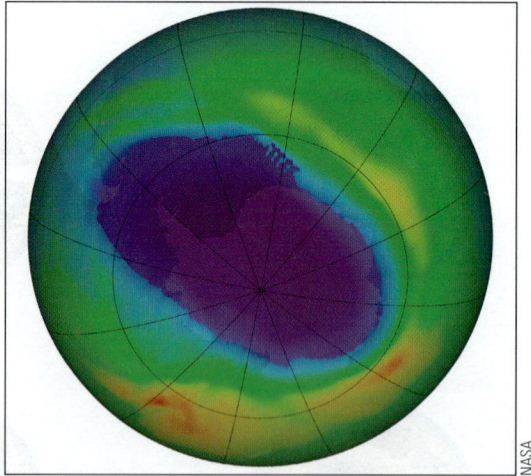

The Antarctic ozone hole on September 22, 2004.

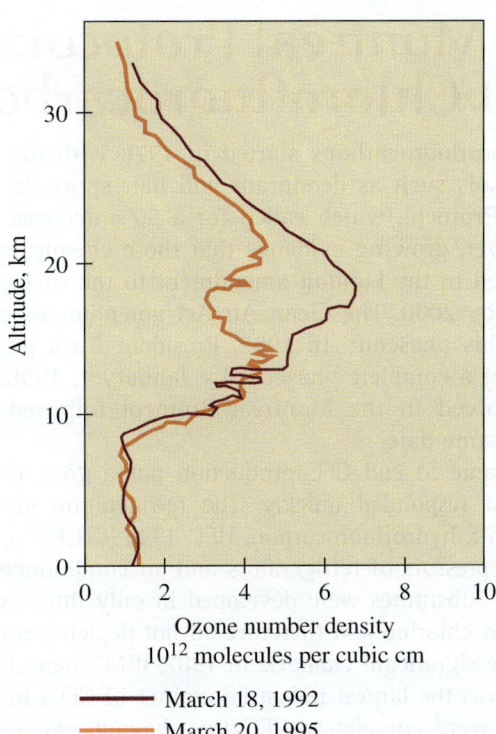

FIGURE 9 A comparison of Arctic ozone concentration versus altitude for 1992 and 1995. The cold winter of 1995 is believed to have allowed chlorine from CFCs to deplete the ozone by as much as 40%. (Courtesy of P. von der Gathen, Alfred Wegener Institute, in *Chemical and Engineering News*, May 24, 1993. Adapted by permission.)

ozone levels recover. The difference, then, between Antarctica and the rest of the world lies in the catalytic conversion—by PSCs—of reservoir chlorine to reactive chlorine atoms. The absence of the polar vortex and PSCs in other parts of the world prevents the same kind of devastating ozone destruction from occurring globally.

For many years, scientists suspected that significant ozone depletion might also occur over the Arctic circle. However, the Arctic polar vortex is much weaker, and the temperatures there are not as cold as in Antarctica, preventing the formation of PSCs. Scientists suspected that a particularly cold winter, like those in Antarctica, might result in significant ozone loss. In 1995, the Arctic circle experienced the coldest winter in recent history. Measurements of stratospheric ozone in the spring of 1995 showed record low levels (Figure 9). In some locations, ozone levels were as much as 40% below normal.

GLOBAL OZONE DEPLETION

Is the depletion of ozone over the poles a foreshadowing of what may happen worldwide? Although not as dramatic as the decreases seen in Antarctica, global stratospheric ozone levels have also fallen. A United Nations Environment Program Study called "Scientific Assessment of Ozone Depletion" concludes that ozone in the mid-northern latitudes has decreased about 6% since 1979. The closer to the equator, the smaller the observed decrease in ozone becomes. These decreases are more troublesome than polar ozone depletion because they occur over populated areas. The evidence for ozone depletion has brought worldwide cooperation for the phaseout of chlorofluorocarbons and other ozone-depleting compounds.

11.10 The Montreal Protocol: The End of Chlorofluorocarbons

CFCs have been banned from hair spray and other household aerosols since 1978.

The phaseout of chlorofluorocarbons started in 1978 with the banning of CFCs from household aerosols such as deodorant and hair spray. In 1987, 23 nations signed the Montreal Protocol, which called for a 50% decrease in CFC usage by the year 2000. However, growing evidence that these chemicals were responsible for ozone depletion led to the London amendment to the treaty, which called for a complete phaseout by 2000. The Clean Air Act amendments of 1990 contained legislation to enact this phaseout. In 1992, President Bush pushed the deadline even closer, calling for a complete phaseout by January 1, 1996. The other industrialized nations involved in the Montreal Protocol followed Bush's lead and banned CFCs on the same date.

The short time frame to end CFC production put a great deal of pressure on many industries. Most responded quickly. The refrigeration and air-conditioning industries converted to a hydrofluorocarbon, HFC-134a (CH_2FCF_3), to replace Freon-12 (CCl_2F_2) in the compressors of refrigerators and air conditioners. In a remarkable set of advances, these substitutes were developed in only three years. Hydrofluorocarbons do not contain chlorine and therefore do not deplete ozone. The electronics industry has also made significant changes. In 1987, IBM's manufacturing facility in San Jose, California, was the largest industrial emitter of CFCs in the United States. Ten years later, they were completely CFC free, having switched to water-based cleaning and drying technologies for electronic components.

Other industries, however, have been slower to switch over to ozone-safe compounds. For example, in foam production, HFC substitutes have not proved as useful. Many of these industries use transitional compounds called hydrochlorofluorocarbons or HCFCs (Figure 10). HCFCs contain chlorine and threaten the ozone layer. However, because they also contain hydrogen, HCFCs are more reactive and tend to decompose in the troposphere. The Montreal Protocol and the 1990 Clean Air Act amendments both have provisions that allow the use of these compounds as transitional substitutes. However, depending on their ozone destruction potential, they too will be banned. The ones with highest ozone depletion potential were banned in 2003; those with lower depletion potential will not be banned until 2030.

The average person has not noticed that their new refrigerator now uses HFC-134a instead of Freon-12, or that the foam in the walls of the refrigerator was formed with HCFCs. The most notable change to the consumer is at the time of replacing the freon in the air conditioner of an older-model automobile. The options are to buy stockpiled freon, which gets more expensive each year, or to convert the air-conditioning system to one using HFC-134a. To encourage people to choose the environmentally beneficial option, some automakers have kept the costs of the retrofit to reasonable levels, around $300.

Because of the Montreal Protocol, atmospheric concentrations of chlorine have leveled off and even begun to decrease as shown in Figure 11. The ozone hole has stabilized and is expected to disappear in about 2050 (Figure 12).

FIGURE 10 Hydrochlorofluorocarbons (HCFCs) like this one contain hydrogen, chlorine, and fluorine. Although the chlorine atoms have ozone depleting potential, the hydrogen atom makes the entire compound more reactive than CFCs and susceptible to decomposition in the troposphere.

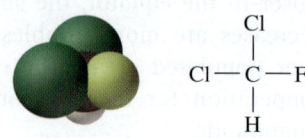

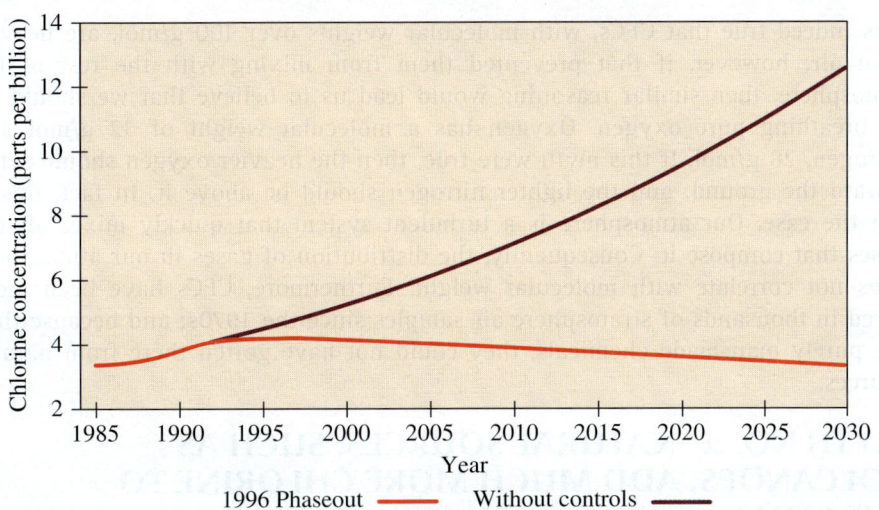

FIGURE 11 The Montreal Protocol has resulted in a decrease in the chlorine content of the stratosphere. The brown line shows how the chlorine content would have grown without the agreement. The red line shows how the chlorine content has decreased (and how it is extrapolated to continue to decrease) with the agreement. (http://www.epa.gov/ozone/science/indicat/)

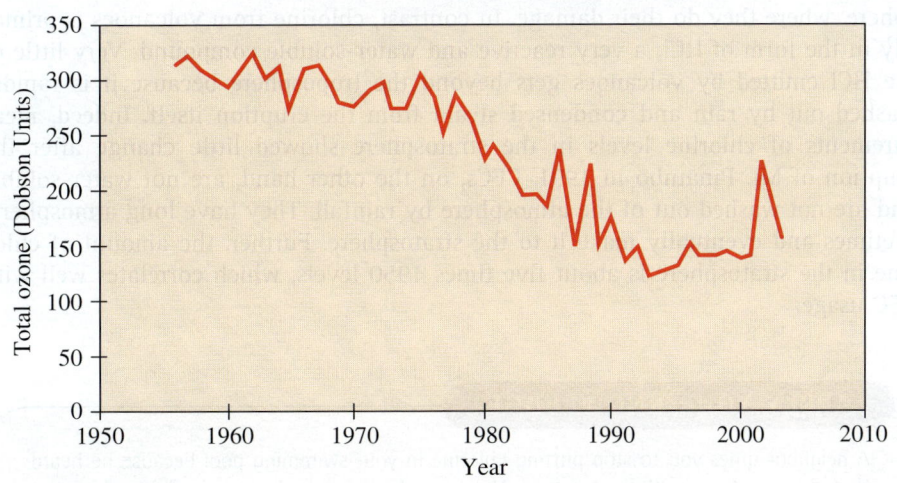

FIGURE 12 Mean October values of total Antarctic ozone. The hole has stabilized (and even begun to recover) as a result of the ban on ozone-depleting compounds. (http://www.epa.gov/ozone/science/indicat/)

11.11 Myths Concerning Ozone Depletion

In recent times, some public figures have spread several myths about the soundness of CFC-caused ozone depletion. Some have alleged that the CFC ban is a result of "environmental wackos" and that it is motivated by scientists trying to get more research funding. Although there may be people who will take a particular side of an issue for personal gain, the case for CFC-caused ozone depletion is a particularly strong one that is supported by the vast majority of the scientific community. Here are two myths concerning ozone depletion and rebuttals to each of them.

MYTH NO. 1—CFCs ARE HEAVIER THAN AIR AND DO NOT RISE INTO THE STRATOSPHERE

It is indeed true that CFCs, with molecular weights over 100 g/mol, are heavier than air; however, if that prevented them from mixing with the rest of the atmosphere, then similar reasoning would lead us to believe that we should all be breathing pure oxygen. Oxygen has a molecular weight of 32 g/mol and nitrogen, 28 g/mol. If this myth were true, then the heavier oxygen should settle toward the ground, and the lighter nitrogen should be above it. In fact, this is not the case. Our atmosphere is a turbulent system that quickly mixes all the gases that compose it. Consequently, the distribution of gases in our atmosphere does not correlate with molecular weight. Furthermore, CFCs have been measured in thousands of stratosphere air samples since the 1970s; and because they are purely man-made chemicals, they could not have gotten there from natural sources.

MYTH NO. 2—NATURAL SOURCES, SUCH AS VOLCANOES, ADD MUCH MORE CHLORINE TO THE STRATOSPHERE THAN CFCs

It is true that volcanoes emit vast amounts of chlorine, much more than chlorine in CFC emissions. However, only chlorine that makes it to the stratosphere depletes ozone. The chemical stability of CFCs allows them to drift to the stratosphere, where they do their damage. In contrast, chlorine from volcanoes is primarily in the form of HCl, a very reactive and water-soluble compound. Very little of the HCl emitted by volcanoes gets beyond the troposphere because it is rapidly washed out by rain and condensed steam from the eruption itself. Indeed, measurements of chlorine levels in the stratosphere showed little change after the eruption of Mt. Pinatubo in 1991. CFCs, on the other hand, are not water-soluble and are not washed out of the atmosphere by rainfall. They have long atmospheric lifetimes and eventually make it to the stratosphere. Further, the amount of chlorine in the stratosphere is about five times 1950 levels, which correlates well with CFC usage.

APPLY YOUR KNOWLEDGE

A neighbor urges you to stop putting chlorine in your swimming pool because he heard that chlorine depletes the ozone layer. How would you respond to your neighbor?

Answer: You should tell your neighbor that only chlorine that gets to the stratosphere depletes ozone. The chlorine you put in your pool is far too reactive to ever make it past the troposphere.

Chapter Summary

MOLECULAR CONCEPT

A sample of gas is composed of molecules, separated by large distances in comparison to their size, moving at velocities of about 600 mph and constantly colliding with each other and with their surroundings (11.2). The collective pelting of molecules on their surroundings is responsible for pressure. The pressure, volume, and temperature of a gas sample are all related by the gas laws (11.3). Boyle's law states that pressure and volume are inversely proportional—increasing one decreases the other—whereas Charles's law states that pressure and temperature are directly proportional—increasing one increases the other (11.4).

The air we breathe is a mixture of gases with nitrogen (4/5) and oxygen (1/5) being predominant (11.5–11.6). Humans have added pollutants to the air that, even at low concentrations, have adverse effects on human health and on the environment. (11.7–11.8).

CFCs, synthetic chemicals used in refrigeration, air conditioning, foam-blowing, and electronics, have been carrying chlorine atoms to the stratosphere, where they deplete stratospheric ozone. Unlike tropospheric ozone, a pollutant, stratospheric ozone is a vital and necessary part of our atmosphere because it absorbs harmful UV light. CFCs have caused ozone depletion over the poles and, to a lesser extent, over the entire globe (11.9–11.11).

SOCIETAL IMPACT

Air pressure, and changes in it, is responsible for wind, storms, hot-air balloon flight, and the pain in your ear when you change altitude. Humans have evolved in an environment with an average air pressure of 1.0 atm. Without that air, we could not survive (11.3).

The adverse effects of polluted air have prompted our society to pass legislation aimed at improving air quality. The Clean Air Act, passed by Congress in 1970 and amended several times since, has lowered pollutant levels over the past 30 years, but more remains to be done (11.8). Many Americans still live in cities where pollutant levels, especially ozone, exceed what the EPA considers safe.

The depletion of the ozone layer by CFCs affects society because increased amounts of UV light on Earth affects human health. Most notable is the increased skin cancer risk. The Montreal Protocol banned the production of these ozone-depleting chemicals on January 1, 1996 (11.10). Other, less harmful compounds like HCFCs face bans in the coming years. Like other good environmental legislation, the CFC ban has worked. Atmospheric CFC levels are expected to level out and then drop in the coming years. The ozone hole itself is expected to disappear by the middle of this century.

Chemistry on the Web

For up-to-date URLs, visit the text website at www.chemistry.brookscole.com/tro3.

- Air Bags
  http://www.nhtsa.dot.gov/portal/site/nhtsa/menuitem.dcee64704e76eeabbf30811060008a0c/

- The Atmosphere
  http://liftoff.msfc.nasa.gov/academy/space/atmosphere.html
  http://www.pbs.org/wgbh/nova/balloon/science/atmosphere.html

CHAPTER 11 — The Air Around Us

- National Ambient Air Quality Standards (NAAQS)
  http://www.epa.gov/air/criteria.html
  http://www.epa.gov/oar/oaq_caa.html/
  http://www.epa.gov/airtrends/ozone.html
- Chlorofluorocarbons (CFCs)
  http://www.cmdl.noaa.gov/noah/publictn/elkins/cfcs.html
  http://www.epa.gov/reg3artd/ozonelayer/ozonemain.htm
- Ozone Depletion
  http://www.cmdl.noaa.gov/ozone.html
  http://www.ucsusa.org/global_environment/archive/page.cfm?pageID=551
  http://aura.gsfc.nasa.gov/science/index.html

Key Terms

atmosphere
barometers
chlorofluorocarbons (CFCs)
ionosphere
mesosphere
nitrogen fixation
pressure
stratosphere
troposphere

Exercises

 Assess your understanding of this chapter's topics with an online chapter quiz at **http://chemistry.brookscole.com/tro3**.

QUESTIONS

1. What is pressure?
2. Describe a gas and its properties.
3. What causes the pain in our ears when we change altitude?
4. Explain how a barometer works.
5. How does pressure influence weather?
6. What is the relationship between the pressure and volume of a gas?
7. What is the relationship between the temperature and volume of a gas?
8. Why is it dangerous for a scuba diver to ascend rapidly?
9. How does a hot-air balloon function? How is windy coastal weather related to Charles's law?
10. What is the combined gas law? Why is it important?
11. What are the primary gases in the atmosphere and their approximate percentages?
12. What is nitrogen fixation?
13. Why is oxygen central to the existence of animal life? Use reactions to explain.
14. What component of the atmosphere causes iron to rust and paint to dull?
15. Explain the cycle that carbon atoms go through. How many times has the average carbon atom completed this cycle?
16. What are the chemical properties of argon, neon, and helium? What are they used for?
17. How would the Earth be different if it had no atmosphere?
18. What are the four major sections of the atmosphere? What happens in each of these sections?
19. What are the six major pollutants found in urban air? List the sources of each.
20. What are the adverse health and environmental effects associated with each of the following?
    a. SO_2
    b. PM-10
    c. CO
    d. O_3
    e. NO_2
    f. Pb
21. What is the Clean Air Act? When was it amended?
22. Explain the trends in levels of the six major pollutants in U.S. cities. Which pollutants are still above acceptable levels in many U.S. cities?
23. What type of urban pollutants are targeted by the Clean Air Act amendments of 1990?
24. Why is urban air pollution often termed "*photochemical smog*"?
25. With the help of chemical equations, explain how ozone absorbs UV light.
26. With the help of chemical equations, explain how CFCs deplete ozone in the stratosphere.
27. How were CFCs used before they were banned?

28. Why does ozone deplete severely over Antarctica each October but not over other areas of the planet?
29. Has ozone been depleted over other areas of the planet besides Antarctica? How severely?
30. Describe the steps taken to phase out CFCs.
31. What are HFCs? Do they deplete ozone?
32. What are HCFCs? Do they deplete ozone?
33. How has the average consumer felt the most direct impact of the ban on CFCs?
34. What has been the result of the CFC ban on the Antarctic ozone hole?
35. Explain the error in the argument that CFCs are heavier than air and therefore do not make it to the stratosphere.
36. Explain the error in the argument that natural sources such as volcanoes add much more chlorine to the stratosphere than CFCs.

PROBLEMS

37. The pressure on Mt. Everest is approximately 0.31 atm. What is this pressure in mm Hg?
38. The pressure at an ocean depth of 330 ft is approximately 11 atm. What is this pressure in torr? In psi?
39. Convert 7.0×10^2 mm Hg to
    a. torr
    b. atm
    c. in. Hg
    d. psi
    e. N/m^2
40. The pressure in Denver, Colorado, is approximately 0.95 atm. Convert that to
    a. torr
    b. mm Hg
    c. in. Hg
    d. psi
    e. N/m^2
41. A balloon is inflated to a volume of 1.5 L in an airplane where the pressure is 0.60 atm. What will be the volume of the balloon on returning to sea level, where the pressure is 1.0 atm?
42. A cylinder with a movable piston has a volume of 22.4 L at a pressure of 1.2 atm. What will be its volume if the pressure is increased to 1.7 atm?
43. Jet airplanes often fly in the lower stratosphere, where the pressure drops to approximately 245 mm Hg. In her suitcase, a passenger packed a sealed, crumpled plastic bag with an initial volume of 0.35 L. If the baggage compartment is not pressurized, what is the volume of the plastic bag at altitude?
44. A scuba diver carries a 1.5-L sealed plastic bottle from the surface, where the pressure is 1.0 atm, to a depth of 90 ft, where the pressure is 3.7 atm. The bottle compresses to what final volume?
45. A balloon is inflated at 25°C to a volume of 1.2 L and then immersed in liquid nitrogen at a temperature of 77 K. What will be the final volume of the balloon? (*Note:* If you actually try this experiment, the final volume of the balloon will probably be smaller because some of the gases within the balloon will condense. However, the experiment does work, and if undisturbed, the balloon will return to its original volume on warming.)
46. A hot-air balloon has a volume of 64,263 m^3 at 60°C. What is its volume at 20°C?
47. A cylinder and piston is left to warm in the Sun. If the initial volume is 1.30 L at 20°C, what is the final volume at 30°C?
48. A 2.8-L balloon is warmed over a toaster and its temperature rises from 25°C to 45°C. What is its final volume?
49. A cylinder with a movable piston has an initial volume of 0.4 L at a pressure of 7.0×10^2 torr and a temperature of 40°C. What will be its volume if the pressure changes to 2.0×10^2 torr and the temperature to 80°C?
50. A backpacker carries an empty, soft-plastic water container into the mountains. Its initial volume at 1.0 atm and 20°C is 0.50 L. The pressure at the mountaintop is 0.83 atm, and the midday temperature has reached 30°C. What is the new volume of the water container?
51. An automobile tire, with a maximum rating of 34.0 psi (gauge pressure), is inflated (while cold) to a volume of 9.8 L and a pressure of 33.0 psi at a temperature of 9.0°C. While driving on the highway on a hot day, the tire warms to 70.0°C and its volume expands to 10.1 L. Does the pressure in the tire exceed its maximum rating under these conditions? (*Note:* The *gauge pressure* is the *difference* between the total pressure and atmospheric pressure. Find the total pressure by assuming that atmospheric pressure is 14.7 psi.)
52. A weather balloon is inflated to a volume of 24.5 L at a pressure of 758 mm Hg and a temperature of 24.0°C. The balloon is released and rises to an altitude of approximately 25,000 ft at which the pressure is 382 mm Hg and the temperature is −14.0°C. Calculate the volume of the balloon at this altitude (assuming the balloon freely expands).

POINTS TO PONDER

53. Could an astronaut use a straw to drink a beverage on the moon? Why or why not?
54. Some scientists have suggested that the ozone layer be restored by chemically forming ozone and pumping it into the stratosphere. Is this a good solution? Why or why not?

55. Explain the difference between the effects of ground-level ozone and stratospheric ozone. Why is one beneficial to human life, while the other is not?
56. Environmental issues are extremely political with many people choosing sides simply because of their political ideology. Why is it important to avoid this pitfall? How can the average citizen avoid falling into this trap?
57. The governments of many developing nations opposed the CFC ban. Why? Should developed nations require these governments to ban CFCs? Why or why not?
58. Devise a method to actually determine the volume of your breath. How much air do you breathe in a 24-h period?

Feature Problems and Projects

59. Pressure is caused by the collisions of molecules with the walls of their container. Examine the following picture and explain, in molecular terms, why the pressure rises when the volume is decreased.

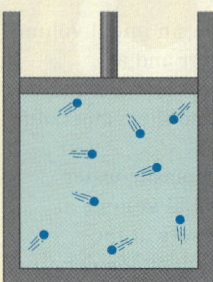 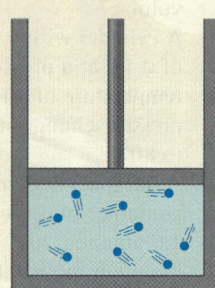

60. The following picture represents a sample of gas at 25°C and 1.0 atm. (Note that the volume is 1 L.) Use the combined gas law, and what you know about the molecular causes of temperature and pressure, to make a sketch of the same sample at 45°C and 1.4 atm.

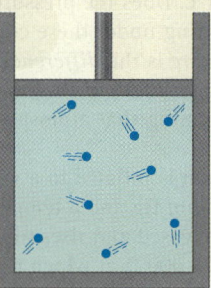

$V = 1.0$ L
$T = 25°C$
$P = 1.0$ atm

61. Read the Molecular Revolution box in this chapter entitled "Measuring Ozone." Prepare a graph of total ozone over the South Pole (in Dobson Units) as a function of time for the most recent entire calendar year. Get the data from NASA's TOMS website at http://jwocky.gsfc.nasa.gov/. Go to the page entitled *data product: ozone*. Select the *ozone over any location* option. You can get the ozone concentration on any date by entering it into the spaces provided on the website. For South Pole ozone levels, use -80 degrees as the latitude and 0 degrees as the longitude. Collect at least two readings per month (e.g., the 1st and the 15th of each month) and prepare a graph showing the yearly South Pole ozone cycle. *Note:* Some months will not be available because the TOMS does not measure South Pole ozone levels for every month.

62. In a manner similar to problem 61, use NASA's TOMS website (http://jwocky.gsfc.nasa.gov/) to prepare a graph of ozone levels over your hometown for an entire calendar year. Determine the latitude and longitude of your hometown by clicking on the *latitude and longitude* option on the *ozone over any location* page. Collect at least two data points per month and prepare a graph.

12 The Liquids and Solids Around Us: Especially Water

It will be found that everything depends on the composition of the forces with which the particles of matter act upon one another; and from these forces ... all phenomena of nature take their origin.

—Roger Joseph Boscovich

In the previous chapter, we explored the atmosphere and the gases that compose it; in this chapter, we explore liquids, solids, and water, with particular emphasis on water. We will see that liquids and solids exist because of attractive forces between molecules—the stronger these attractions are, the more likely that a substance will be a liquid or a solid at room temperature rather than a gas. The most important liquid on our planet is water. Water is unique in a number of ways. For example, water is one of the few substances whose solid form is less dense than its liquid form. Therefore, ice floats on water. In this chapter, we will see why this is so. As you read this, think about your own daily dependence on water. How long could you live without it? What if your water supply was contaminated? In the United States, The Safe Drinking Water Act of 1974 requires water providers to give you safe drinking water or let you know if it is not safe. Water is one of our most precious resources, and good legislation keeps it that way.

QUESTIONS FOR THOUGHT

- Why are raindrops spherical?
- What are the molecular differences between solids, liquids, and gases?
- What is melting and boiling?
- What are the forces that hold molecules together as liquids and solids?
- Can you tell by looking at a molecule's structure if it is likely to be a liquid, a solid, or a gas?
- What are mixtures of solids and liquids (salt water, for example)?
- Why is water so important? How is it unique?
- Is most water that we encounter pure? If not, what is in it?
- What substances contaminate drinking water?
- How can we further purify our water in the home?

12.1 No Gravity, No Spills

The properties of liquids on Earth are affected by gravity. However, one important property, the tendency of liquids to stick together, is somewhat masked by gravity. Have you ever seen a liquid in a gravity-free environment? It looks quite different. Space-bound beverages, for instance, are sealed in plastic pouches. To drink the beverage, an astronaut punctures the pouch and squeezes the liquid into his or her mouth. Space shuttle videos sometimes show astronauts playfully squeezing their drink directly into the air within the shuttle; since there is no gravity, there are no spills (Figure 1). The liquid hangs in space, wiggling like a suspended ball of Jell-O. The astronaut can then put his or her lips on the suspended ball and slurp it in.

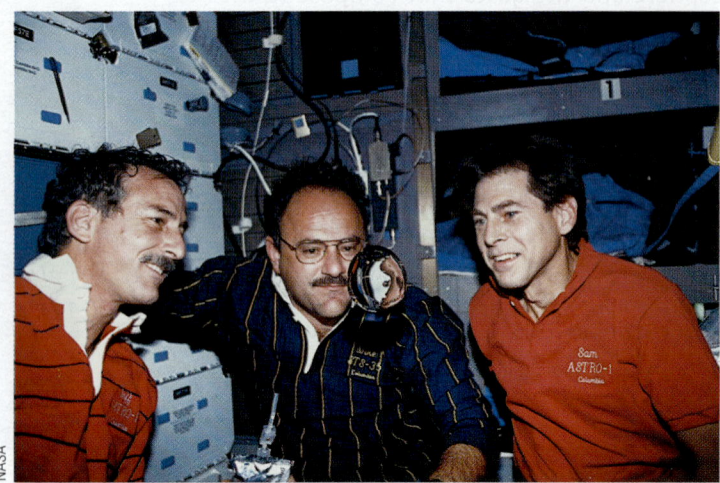

FIGURE 1 An astronaut on the space shuttle squeezes his liquid beverage into the air. The liquid hangs in space, held together by cohesive forces between its constituent molecules.

If not disturbed, a liquid in the absence of gravity assumes the shape of a sphere. Raindrops or water droplets on a waxed car would be perfectly spherical if not for the distorting influence of air and gravity. Why do water drops have spherical shapes? Why do liquids or solids even exist? The answer to both of these questions is the same: because of cohesive forces between molecules.

Liquids and solids are held together by **cohesive forces**—attractions between molecules. In the absence of these forces, liquids and solids would not form—all matter would be gaseous. These same cohesive forces also coax liquids into spherical shapes as molecules squeeze together to maximize their contact. The sphere is the geometrical shape with the lowest surface-area-to-volume ratio, allowing the maximum number of molecules to be completely surrounded by other molecules (Figure 2).

In this chapter, we explore liquids, solids, and solutions (homogeneous mixtures of liquids with solids or other liquids). We pay special attention to water, the most important liquid on Earth. We will also look at the substances that pollute our water, and how this pollution is regulated and controlled.

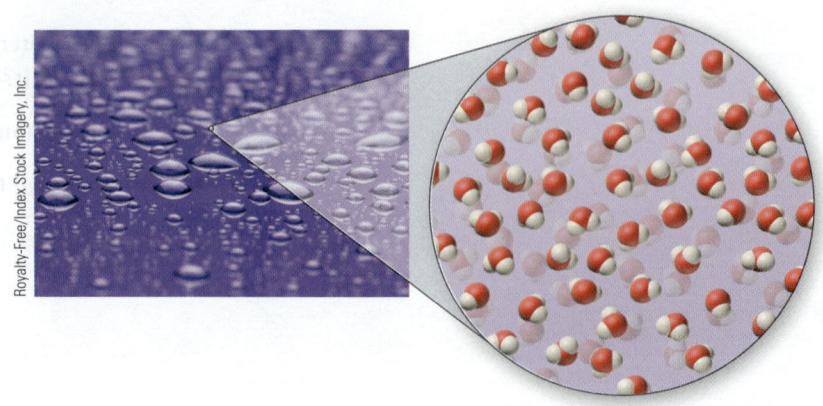

FIGURE 2 Water droplets form tight spherical beads on the surface of a freshly waxed car. The sphere has the lowest surface-area-to-volume ratio, allowing the maximum number of molecules to interact with each other.

12.2 Liquids and Solids

As we have seen, matter normally exists in three states: gas, liquid, and solid. The properties of these states are related to how the molecules that compose them interact. Molecules within gases are separated by large distances and move at high velocities, as shown in Figure 3(a). The attractions between them are weak relative to thermal energy, resulting in little interaction. Molecules in liquids, in contrast, are in intimate contact and interact strongly, as shown in Figure 3(b); however, they are still free to move around and past each other. Molecules in solids are locked in place and only vibrate about a fixed point, as shown in Figure 3(c).

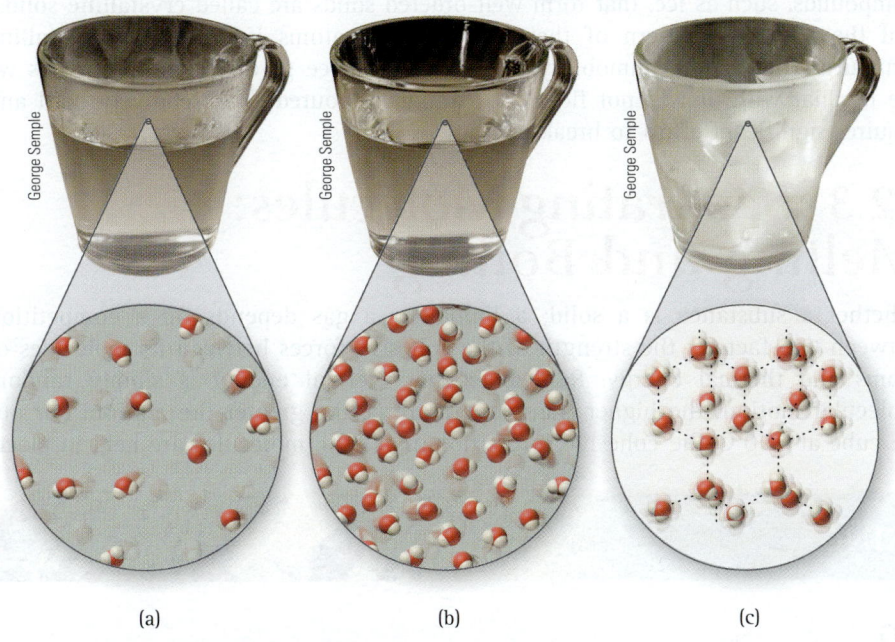

FIGURE 3 (a) Molecules in a gas, such as those in the steam pictured in this cup of hot water, are separated by large distances and do not interact with each other. (b) Molecules within a liquid are in intimate contact and interact strongly but are mobile. (c) Molecules in a solid are locked into a fixed structure and do not move except for small vibrations about a fixed point.

As an example of a liquid, consider a cup of water. Each molecule within the cup is attracted to its neighbors, much like a magnet is attracted to other magnets. However, unlike a static collection of magnets, thermal energy causes each molecule to be in constant motion. An individual water molecule moves around, vibrating, bumping into other water molecules, sliding around still others and traveling a random path. At each point in its path, any one molecule is attracted to its neighbors, but the attractions are not strong enough to hold the molecule in one place—it continually moves around.

In its random path, a water molecule might reach the surface of the liquid. This will not happen often because there is so little surface area compared with interior volume. The surface is not a stable place for a molecule because there are fewer other molecules with which to interact—think of a tiny magnet on the surface of a clump of magnets versus one in the interior. If the molecule happened to be moving particularly fast when it reached the surface, it could overcome the relatively fewer attractions at the surface and shoot off into the air (Figure 4). This is called **evaporation**.

The mobility of individual molecules gives liquids the properties we are so familiar with: Liquids do not have a fixed shape, but assume the shape of their container; they flow and can be poured; if left standing, they evaporate; and if heated, they boil. Heating increases thermal motion, giving a larger fraction of

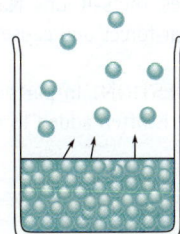

FIGURE 4 Molecules in a liquid, undergoing constant random motion, may acquire enough energy to overcome attractions with neighbors and fly into the gas phase.

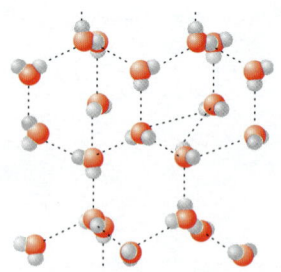

FIGURE 5 Like all crystalline solids, the molecules in crystalline ice are well ordered in a three-dimensional pattern called the crystalline structure of ice.

the moving molecules enough energy to break free and evaporate. The close contact between molecules in liquids is responsible for other properties such as their incompressibility and high density compared to gases.

Now, let us compare liquids with solids. As an example of a solid, consider a cube of ice. The water molecules that compose the ice are identical in every way to those composing liquid water; however, their movement and arrangement are different. In ice, the thermal energy is lower; molecules move less, allowing them to interact more. Imagine our collection of magnets again, only this time they are all fixed in an orderly arrangement, one next to another, in a three-dimensional array (Figure 5). Their positions and orientations are maintained by the attractive forces between them. Each molecule in ice vibrates, but its average position remains fixed. Compounds, such as ice, that form well-ordered solids are called **crystalline solids**; and the repeating pattern of their molecules or atoms is called the **crystalline structure**. The relative immobility of molecules in ice gives ice the properties we are familiar with: It will not flow and cannot be poured; it is relatively hard and requires significant effort to break apart.

12.3 Separating Molecules: Melting and Boiling

Whether a substance is a solid, a liquid, or a gas depends on a competition between two factors: the strength of the cohesive forces between its molecules or atoms and thermal energy. Remember that thermal energy is simply random molecular motion; the higher the temperature is, the greater the motion. For our ice cube at $-10°C$, the cohesive forces dominate. The molecules are held in place,

MOLECULAR THINKING

Making Ice Cream

If you have ever made ice cream, you have modified the melting point of water to get the liquid cream to solidify. In an ice cream maker, the outer section is filled with water and rock salt. Adding sufficient salt to water lowers its melting point—also called the freezing point because freezing and melting occur at the same temperature—to approximately $-10°C$. This phenomenon is called **freezing point depression** and is a result of interrupting the cohesive forces among water molecules. The salt ions, Na^+ and Cl^-, interfere with the intermolecular forces and cause the melting point to be lowered.

QUESTION: In parts of the country where snow is common, salt is often added to roads in winter months. Why?

In an ice cream maker, salt is used to lower the freezing point of water. The cold salt–ice–water slurry reaches temperatures of $-10°C$, causing the ice cream to solidify.

FIGURE 6 When water reaches its boiling point, molecules within the interior have enough energy to break free from their neighbors and become a gas. The gas appears as the familiar bubbles within boiling water.

and thermal energy only causes vibration about a fixed point. If we raise the ice cube's temperature, the vibration becomes greater, until at 0°C the molecules break free from their fixed points and become mobile; the ice melts. The temperature at which a solid melts is called the **melting point,** and it depends on the strength of the cohesive forces of the atoms or molecules that compose it. The stronger the force is, the higher the melting point.

If we continue to heat liquid water, molecules will move faster, until at 100°C the molecules have enough energy to completely overcome their cohesive forces and leave the liquid state; the water boils. At this point, molecules within the interior of the liquid, not just those near the surface, have enough thermal energy to overcome the cohesive forces among them. Molecules break away from their neighbors and appear as steam. Bubbles form as water molecules leap from the liquid to the gaseous state (Figure 6). The temperature at which a liquid boils is called the **boiling point.**

In melting and boiling water, thermal energy competes with cohesive forces, pulling molecules away from their neighbors and converting them into a liquid and then into a gas. Both the liquid and the gas, however, are still water—we have not broken any chemical bonds *within* molecules, only overcome the forces among them. The temperature required to break the chemical bonds *within* a water molecule is thousands of degrees higher.

The temperatures at which melting and boiling occur in a substance depend on the magnitude of the cohesive forces among the molecules or atoms that compose it. Those substances whose molecules or atoms have strong cohesive forces, such as diamond,* for example, have high melting and boiling points; it requires a great deal of energy to pry the atoms in diamond apart. In contrast, substances whose molecules or atoms have weak cohesive forces, such as those in nail polish remover (acetone) or gasoline, have low melting and boiling points.

Molecules with boiling points below room temperature (<25°C) are gases at room temperature. Butane (C_4H_{10}), for example, boils at about 0°C, making it a gas

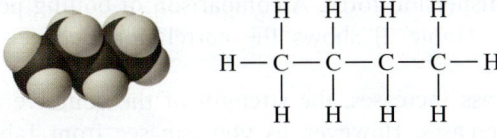

Butane

---

*Diamond is an example of an atomic solid in which the cohesive forces that hold the solid together are actually covalent chemical bonds. These forces are extremely strong and result in melting points of thousands of degrees.

at room temperature; by contrast, hexane (C_6H_{12}) boils at 69°C, making it a liquid. Some substances, such as diethyl ether, have their boiling point so close to room temperature that they are liquids on cold days but will boil into gas on hot days.

The strength of the cohesive forces among molecules is related to the molecule's structure and can often be estimated by examining the structure. We can therefore make a direct connection between the molecule and its likelihood of being a liquid or solid at room temperature. To make that connection, however, we must understand the types of cohesive forces that hold solids and liquids together.

> **APPLY YOUR KNOWLEDGE**
>
> Substance A is composed of molecules that have stronger intermolecular forces than the molecules that compose substance B. Which substance has a lower boiling point?
>
> Answer: Substance B. Weaker intermolecular forces result in lower melting and boiling points.

12.4 The Forces That Hold Us— and Everything Else—Together

All solids and liquids are held together by cohesive forces among molecules. Because our own bodies are a complex mixture of solids and liquids, even *we* are held together by these forces. If not for cohesive forces among molecules, all matter would be gaseous. The major cohesive forces in order of increasing strength are the dispersion force, the dipole force, and the hydrogen bond.

DISPERSION FORCE

The weakest cohesive force, present among all atoms and molecules, is the **dispersion** (or London) force. Dispersion forces are the result of small fluctuations in the electron clouds of atoms and molecules. These small fluctuations result in instances where electrons are not evenly distributed in the molecule or atom. This results in an **instantaneous dipole,** depicted in Figure 7(a). Instantaneous dipoles are similar to the polar bonds discussed in Chapter 5; the main difference is that polar bonds are permanent dipoles whereas instantaneous dipoles are only temporary. Nonetheless, instantaneous dipoles still cause attractions among molecules. An instantaneous dipole on one molecule or atom induces instantaneous dipoles in neighboring molecules or atoms, as shown in Figure 7(b). The molecules or atoms then attract each other; the positive end of one instantaneous dipole attracts the negative end of another. The magnitude of dispersion forces is proportional to the molar mass of the molecule or atom; the heavier a molecule or atom is, the stronger the dispersion force. A comparison of boiling points of noble gases (Table 1) or alkanes (Table 2) shows the correlation between boiling point and molar mass.

As the molar mass increases, the strength of the cohesive force increases and the boiling point increases. However, as you can see from Tables 1 and 2, molar mass alone does not determine boiling point. For example, krypton (with a molar mass of 83.8 g/mol) boils at 119.8 K, while pentane (with a lower molar mass) boils at the higher temperature of 309 K. Molar mass can only be used as a guide to the

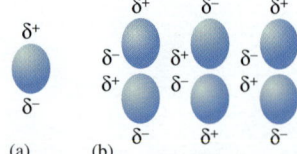

FIGURE 7 (a) Random fluctuations in the electron cloud of an atom or molecule produce an instantaneous dipole. (b) Instantaneous dipoles align so that the positive end of one dipole interacts with the negative end of the other.

TABLE 1
Noble Gas Boiling Points

Noble Gas	Molar Mass (g/mol)	Boiling Point (K)
He	4.0	4.2
Ne	20.2	27.1
Ar	39.9	87.3
Kr	83.8	119.8
Xe	131.3	165.0

Table 1. The dispersion force increases in magnitude with increasing atomic mass, resulting in higher boiling points with increasing atomic mass.

TABLE 2
Alkane Boiling Points

Alkane	Molar Mass (g/mol)	Boiling Point (K)
Pentane	72.2	309
Hexane	86.2	342
Heptane	100.2	371
Octane	114.2	399

Table 2. In the alkane family, boiling increases with increasing molar mass due to the increase in the dispersion force.

magnitude of dispersion forces when comparing a family of similar elements or compounds; it is not useful in comparing widely different elements or compounds.

DIPOLE FORCES

Dipole forces, stronger than dispersion forces, are present in **polar molecules**. Polar molecules have **permanent dipoles** that result in strong attractions between

---

EXAMPLE 12.1

Predicting Relative Boiling Points

Which has the higher boiling point, Cl_2 or Br_2?

SOLUTION
Because the molecules belong to the same family, the magnitude of the dispersion force should be proportional to molar mass. Because Br_2 has a higher molar mass than Cl_2, we would predict that it would have a higher boiling point.

YOUR TURN

Predicting Relative Boiling Points

Which has the higher boiling point, C_8H_{18} or $C_{10}H_{22}$?

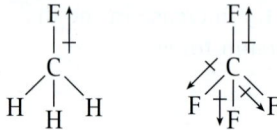

FIGURE 8 When atoms of different electronegativities form a bond, electrons are unevenly shared. This results in a partial positive charge on one of the bonding atoms and a partial negative charge on the other.

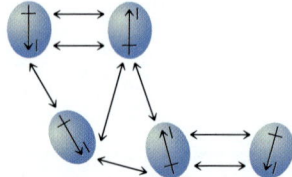

FIGURE 9 The two polar C=O bonds in carbon dioxide cancel each other, resulting in a nonpolar molecule.

FIGURE 10 (a) In methyl fluoride, the polar C—F bond is not canceled by the C—H bonds, making the molecule polar. (b) In carbon tetrafluoride, the four polar C—F bonds cancel each other, making the molecule nonpolar.

FIGURE 11 Polar molecules have strong cohesive interactions because the positive end of each molecule is attracted to the negative end of its neighbors.

neighboring molecules. As we learned in Chapter 5, atoms of different electronegativities form **polar bonds,** in which electrons are unevenly shared between the bonding atoms. This results in partial negative and partial positive charges on the two bonding atoms (Figure 8). Polar bonds within a molecule may or may not make the entire molecule polar; it depends on the molecular structure. Simple diatomic molecules with polar bonds, such as H—Cl or H—F, will always be polar. However, in polyatomic molecules, the polar bonds may cancel and result in a nonpolar molecule. For example, carbon dioxide (CO_2) has two polar bonds (each C=O bond is polar). However, the geometry is linear (Figure 9), which results in a cancellation of the polar bonds, making the molecule nonpolar.

Review Section 5.7 to refresh your memory on determining if molecules are polar. For organic compounds, recall that hydrocarbons—compounds that contain *only* carbon and hydrogen—are nonpolar. Therefore, compounds such as methane (CH_4), hexane (C_6H_{14}), and benzene (C_6H_6) are nonpolar. Organic compounds that contain polar groups, such as O—H, C—Cl, or C—F are polar unless their geometry is such that all polar bonds cancel (Figure 10).

Because polar molecules have *permanent* dipoles, the cohesive attractions between neighbors are stronger than in nonpolar molecules. The permanent dipoles align so that the positive end of one dipole interacts with the negative end of another (Figure 11). Consequently, a polar molecule has a higher boiling point than a nonpolar molecule of similar molecular weight. Consider ethane and formaldehyde:

Molecule	Molar Mass (g/mol)	Boiling Point (°C)
Ethane (CH_3CH_3)	30.0	−88.0
Formaldehyde (CH_2O)	30.0	−19.5

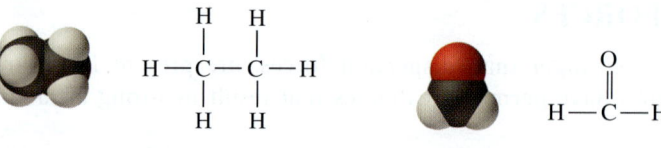

Ethane   Formaldehyde

Ethane is a hydrocarbon and therefore nonpolar. The only cohesive forces present in ethane are the relatively weak dispersion forces. Formaldehyde contains a polar CO bond. The cohesive forces present in formaldehyde are dipole forces, giving it a higher boiling point than ethane.

Although polar molecules are attracted to each other, they are not attracted to nonpolar molecules. The macroscopic result is dramatic. If you combine a highly polar substance with a nonpolar one—gasoline and water, for instance—they do not mix. Polar water molecules strongly attract other water molecules but do not attract the nonpolar hydrocarbon molecules that compose gasoline. The gasoline is squeezed out of the water by strong attractions among water molecules. The tendency for polar and nonpolar substances to separate is the molecular reason that salad dressing, composed of polar water and nonpolar vegetable oil, always separates into two layers (Figure 12). It also explains why plain water will not get your dishes or your clothes clean. Greasy hands will remain greasy if rinsed in nonsoapy water.

EXAMPLE 12.2

Determining If a Molecule Is Polar

Which of the following molecules are polar?

a. O_2
b. H—Cl
c. $CHCl_3$
d. CH_3CH_3
e. CH_3OH
f. CCl_4

SOLUTION

a. O_2 is not polar; it does not contain polar bonds.
b. H—Cl is polar; it contains a polar bond that is not canceled by others.
c. $CHCl_3$ is polar; it contains three polar C—Cl bonds and one slightly polar C—H bond. The bonds do not cancel.
d. CH_3CH_3 is a hydrocarbon; it is not polar.
e. CH_3OH contains a polar O—H bond; it is polar.
f. CCl_4 contains four polar C—Cl bonds, but the highly symmetrical tetrahedral geometry makes them cancel. It is not polar.

YOUR TURN

Determining If a Molecule Is Polar

Which of the following molecules are polar?

a. HF
b. N_2
c. CH_3Cl
d. CH_3CH_2OH
e. $CH_3CH_2CH_2CH_2CH_3$

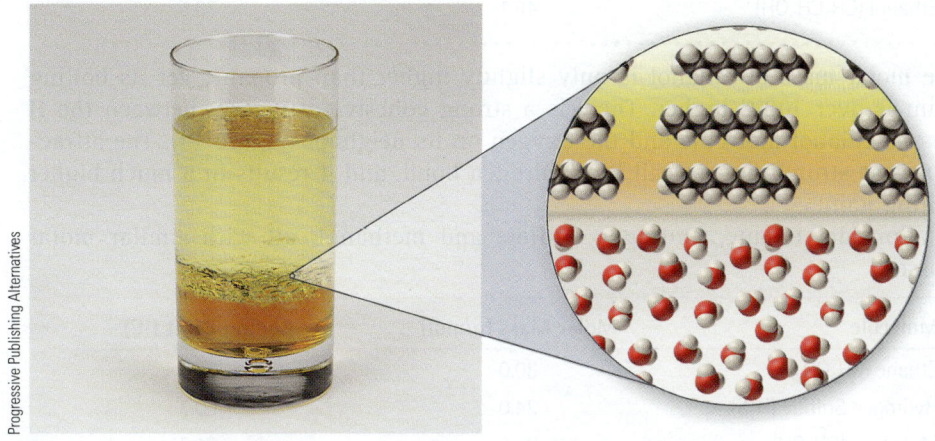

FIGURE 12 (a) Oil and vinegar separate into two layers because oil is composed of nonpolar molecules and vinegar is composed primarily of polar water. (b) Molecular view of interface. Water molecules are polar, but those that compose oil are not.

MOLECULAR THINKING

Soap—A Molecular Liaison

The inability of polar water to dissolve nonpolar grease and oil makes water an ineffective solvent for these substances. However, the addition of soap to water makes it dissolve oil easily. Greasy dirt is washed from clothes or skin with soapy water because of soap's unique ability to interact strongly with both the nonpolar grease and the polar water. One end of the molecule is polar, while the other end is nonpolar:

$$CH_3CH_2CH_2CH_2CH_2CH_2CH_2CH_2CH_2COO^-Na^+$$
Nonpolar · Polar

As a result, soap has cohesive attractions with both grease on one end and water on the other. As we pour the soap solution onto the greasy dish, soap molecules anchor themselves on one end to grease molecules:

$$\text{grease} \cdot CH_3(CH_2)_8CH_2COO^-Na^+ \cdot \text{water}$$

The other end of the soap molecule remains immersed in the water, holding firmly to water molecules and acting like a molecular link to allow the water to carry the grease down the drain.

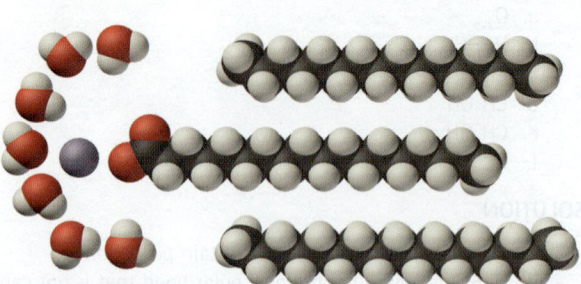

Soap molecules interact strongly with both water and grease.

QUESTION: Oil tankers that transport oil to refineries have sometimes run aground, resulting in the spilling of oil into our oceans. Does the oil mix with the water? Would the addition of soap be an effective way to clean up an oil spill? Why or why not?

HYDROGEN BONDING

Polar molecules containing H atoms bonded to either F, O, or N atoms contain a cohesive force called a **hydrogen bond**. A hydrogen bond is a cohesive attraction between molecules and should not be confused with a *chemical* bond that holds atoms together to form molecules. Substances composed of molecules with hydrogen bonding have high boiling points, much higher than expected from their molar mass. For example, consider propane and ethanol:

Molecule	Molar Mass (g/mol)	Boiling Point (°C)
Propane ($CH_3CH_2CH_3$)	44.1	−42.1
Ethanol (CH_3CH_2OH)	46.1	78.5

The molar mass of ethanol is only slightly higher than propane, yet its boiling point is over 100°C higher. There is a strong cohesive attraction between the H on one ethanol molecule and the oxygen on its neighbor (Figure 13). The attraction is so strong that we call it a hydrogen bond, and it results in a much higher boiling point.

Consider ethane, hydrogen sulfide, and methanol, all with similar molar masses:

Molecule	Molar Mass (g/mol)	Boiling Point (°C)
Ethane (CH_3CH_3)	30.0	−88.0
Hydrogen Sulfide (H_2S)	34.0	−60.3
Methanol (CH_3OH)	32.0	64.7

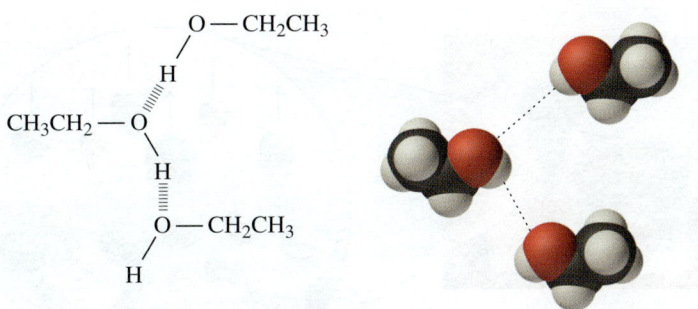

FIGURE 13 There is a strong cohesive attraction, called a hydrogen bond, between the H atom on one ethanol molecule and the O atom on its neighbor.

Ethane is nonpolar and has only dispersion interactions; it has the lowest boiling point. Hydrogen sulfide, a polar molecule, has dipole forces and a slightly higher boiling point. Methanol is polar, but because it contains a hydrogen atom bonded directly to an oxygen atom, it also has hydrogen bonds. Consequently, methanol has the highest boiling point of all and is a liquid at room temperature.

EXAMPLE 12.3

Predicting Relative Boiling Points

One of the following molecules is a liquid at room temperature. Which one and why?

$$CH_4 \quad CH_3OH \quad C_2H_6$$

SOLUTION
Although their molar masses differ somewhat, only one of these molecules, CH_3OH, displays hydrogen bonding and will therefore be most likely to be a liquid at room temperature. (The boiling points are CH_4: $-161.5°C$; CH_3OH: $64.7°C$; C_2H_6: $-88.6°C$.)

YOUR TURN

Predicting Relative Boiling Points

Which of the following molecules do you expect to have the highest boiling point?

$$NH_3, Ne, O_2, NO$$

12.5 Smelling Molecules: The Chemistry of Perfume

Imagine opening a new vial of perfume; you quickly notice a pleasant smell. Imagine now what must be happening with the molecules that compose the perfume. As in all liquids, the molecules within the vial are moving and jostling due to thermal energy. However, unlike a pure liquid, which contains only one kind of molecule, perfume is a mixture containing many different molecules with a range of cohesive forces. The molecules with the weakest cohesive forces are the ones that you smell first; they break free from their neighbors at room temperature, diffuse out of the vial and into the air. You inhale them, and they interact with sensors in your nose, which send a message to your brain that you interpret

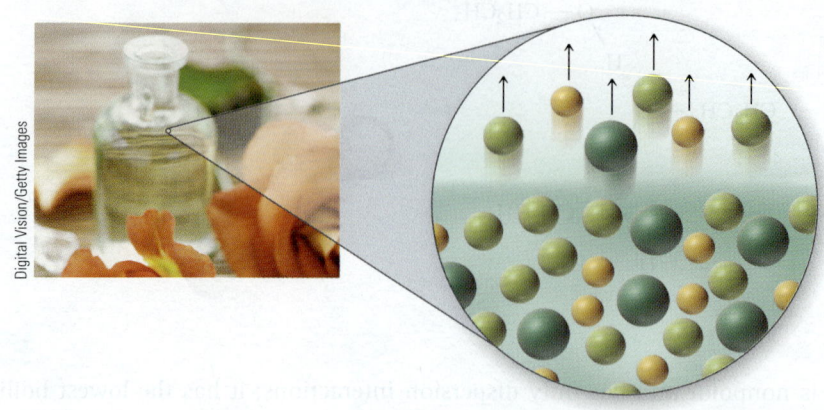

Molecules in perfume have relatively weak intermolecular forces, allowing many molecules to break free from the liquid and diffuse into the surrounding air.

Evaporation is an endothermic process, absorbing heat as it occurs. When overheated, the human body perspires; the moisture evaporates, cooling the body in the process.

as a pleasant smell. Of course, determining what most people find to be a pleasant smell is not always easy and is the main obsession of the perfume industry.

Liquids that vaporize easily have a high **vapor pressure** and are termed **volatile**. Liquids that do not vaporize easily have a low vapor pressure and are termed **nonvolatile**. In comparison with many of the components of perfume, water has a low vapor pressure. The same vaporization process happens in water, only slower. If we leave a glass of water out for several days, the level drops slowly as the water evaporates. As we saw earlier, water molecules at the surface occasionally acquire enough energy to leave the liquid and become a gas. Evaporation is slower in water than in perfume because water molecules form hydrogen bonds, a relatively strong cohesive force. The stronger the cohesive forces among the molecules that compose a liquid, the fewer molecules break free, and the lower the vapor pressure. Perfumes contain many different components, some with high vapor pressures that give an initial rush of fragrance, and others with lower vapor pressures to last the day. For this reason, perfumes smell differently several hours after application—a different molecule is responsible for the later fragrance.

Another example of a high-vapor-pressure liquid is nail polish remover, composed primarily of acetone (CH_3COCH_3), a polar molecule. Acetone does not hydrogen bond because the hydrogen atoms are not bonded to the oxygen directly. It is polar, however; thus its cohesive forces are dipole forces. These forces are strong enough to make it a liquid at room temperature but weak enough to result in a high vapor pressure. If you spill a few drops of acetone onto a table, they evaporate quickly. Acetone also demonstrates another property

> **APPLY YOUR KNOWLEDGE**
>
> A mixture initially contains equal masses of pentane, octane, and decane. The mixture is left in an open beaker overnight. Do you expect the mixture to contain equal masses of all three components in the morning? Explain.
>
> Answer: No. Because the three hydrocarbons only have dispersion forces, and because they are structurally similar, we can assume that the magnitude of the dispersion force increases with increasing molar mass. Therefore, pentane, with the lowest molar mass, will evaporate most rapidly (and will therefore have the least left in the beaker in the morning). Octane has the next lowest molar mass and will therefore evaporate more rapidly than decane, but less rapidly than pentane. There will be less octane than decane in the mixture, but more octane than pentane. Decane, with the highest molar mass, will evaporate most slowly. Consequently, the morning mixture will contain more decane than either of the other two components.

of vaporization—it is endothermic. If acetone spills on your skin, your skin feels cold as the acetone evaporates. The vaporization absorbs heat and lowers the temperature of your skin. The human body uses the vaporization of water to cool itself. When the body overheats, sweat glands produce water that evaporates from the skin, absorbing heat and cooling the body.

12.6 Chemists Have Solutions

A **solution** is defined as a homogeneous mixture (see Section 1.7) of two or more substances. Gasoline, seawater, coffee, and blood plasma are all examples of solution. The majority component of a solution is called the **solvent**, and the minority component is called the **solute**. For example, a sugar solution consists of sugar (the solute) and water (the solvent). In this book, we focus on aqueous solutions, those in which water is the solvent. Not all substances form solutions when mixed. For example, oil and water remain separate even when mixed vigorously. Whether or not a solution forms depends on the interactions between the components of the mixture. If the components attract each other, a solution forms. Sugar and water form a solution because sugar molecules and water molecules are both polar and therefore attract one another.

An important property of a solution is its **concentration**, the amount of solute relative to the amount of solvent. Two common ways to express concentration are **percent by mass** or **percent by volume**. For example, a 5% sugar solution by mass contains 5 g of sugar for every 100 g of solution. A 3% acetic acid solution by volume contains 3 mL of acetic acid for every 100 mL of solution.

Another common concentration unit is **molarity (M)**, the number of moles of solute divided by the number of liters of solution.

$$\text{molarity } (M) = \text{moles solute/liters solution}$$

EXAMPLE 12.4

Solution Concentrations in Molarity

20. g of NaCl is mixed with enough water to make 500. mL of solution. What is the molarity of the resulting sodium chloride solution?

SOLUTION
We first find the number of moles of NaCl and then divide by the total number of liters of solution:

$$20.\ g \times \frac{1\ mol}{58.4\ g} = 0.34\ mol$$

$$M = \frac{0.34\ mol}{0.500\ L} = 0.68\ M$$

YOUR TURN

Solution Concentrations in Molarity

40. g of sugar (molar mass = 342 g/mol) is mixed with enough water to make 200. mL of solution. What is the molarity of the resulting sugar solution?

Molecular Thinking

Flat Gasoline

We often store gasoline at home for use in lawn mowers, leaf blowers, or other garden tools. If the gasoline is stored in an open container, it will become flat over time. Flat gasoline does not ignite as easily as fresh gasoline and will make it harder to start a lawn mower.

QUESTION: What do you suppose is happening to gasoline when it becomes flat? Can you give a molecular reason? How can it be prevented? (Hint: Gasoline is a mixture of many different hydrocarbons.)

EXAMPLE 12.5

Finding Grams of Solute from Volume and Molarity

How many grams of NaCl are present in 2.8 L of a 0.53 M NaCl solution?

SOLUTION

Molarity has units of mol/L and is therefore a conversion factor between moles and liters. Use the volume and the molarity to find moles of NaCl and then use the molar mass of NaCl (58.5 g/mol) to find grams of NaCl:

$$2.8 \text{ L solution} \times \frac{0.53 \text{ mol NaCl}}{\text{L solution}} \times \frac{58.5 \text{ g}}{\text{mol}} = 87 \text{ g NaCl}$$

YOUR TURN

Finding Grams of Solute from Volume and Molarity

How many grams of NaF are present in 1.3 L of a 0.78 M solution?

Parts per million is similar to the unit percent. Percent or "per hundred" can be thought of as parts per hundred. A 1% solution means 1 g of solute per 100 g of solution. A 1-ppm solution means 1 g of solute per million grams of solution.

Another common unit of concentration, especially when reporting low concentrations, is parts per million. For solutions, parts per million is usually reported in terms of mass and refers to grams of solute per million grams of solution:

$$\text{ppm} = \frac{\text{grams solute}}{\text{grams solution}} \times 10^6$$

A fourth common unit of concentration is milligrams per liter (mg/L), defined as the number of milligrams of solute per liter of solution. This unit is often used in reporting impurity levels in drinking water. For solutions in water, milligrams per liter is numerically equivalent to parts per million because a liter of an aqueous solution weighs approximately 1000 g.

$$\text{mg/L} = \frac{\text{milligrams solute}}{\text{liters solution}}$$

EXAMPLE 12.6

Solution Concentrations in Parts per Million

A 100.0-g sample of hard water is found to contain 0.012 g of calcium carbonate ($CaCO_3$). What is the concentration of calcium carbonate in parts per million?

SOLUTION

$$\frac{0.012 \text{ g } CaCO_3}{100.0 \text{ g solution}} \times 10^6 = 120 \text{ ppm}$$

YOUR TURN

Solution Concentrations in Parts per Million

A 75-g sample of tap water is found to contain 8.1×10^{-5} g of sodium fluoride. What is the concentration of sodium fluoride in parts per million?

EXAMPLE 12.7

Solution Concentrations in Milligrams per Liter

A 250.-mL sample of tap water is found to contain 0.38 mg of lead. What is the concentration of lead in milligrams per liter?

SOLUTION

We first convert milliliters to liters:

$$250 \text{ mL} \times \frac{1 \text{ L}}{1000 \text{ mL}} = 0.250 \text{ L}$$

We then divide the amount of lead in milligrams by the amount of solution in liters:

$$\frac{0.38 \text{ mg}}{0.250 \text{ L}} = 1.5 \text{ mg/L}$$

YOUR TURN

Solution Concentrations in Milligrams per Liter

A 400.0-mL sample of tap water is found to contain 0.22 mg of copper. What is the concentration of copper in milligrams per liter?

12.7 Water: An Oddity Among Molecules

Water is the most common liquid on the Earth. It is the chief component of streams, lakes, oceans, clouds, ice, and snow and is an integral part of all living organisms. Water is also an unusual molecule. It has a molar mass of only 18.0 g/mol, yet it is a liquid at room temperature with a rather high boiling point of 100°C. No other compound of similar molar mass comes close to

 Explore this topic on the **Properties and Behavior of Water** website.

Water is ubiquitous in nature.

FIGURE 14 The structure of water.

Lettuce that has been frozen (right) will not return to its original state (left) because water within the cells expands on freezing, rupturing cells in the lettuce.

FIGURE 15 Hydrogen bonding in liquid water.

matching the boiling point of water. For example, nitrogen (MW = 28 g/mol), oxygen (MW = 32 g/mol), carbon dioxide (MW = 44 g/mol), and argon (MW = 40 g/mol) all have molar masses significantly above water, yet they are all gases at room temperature. Water is also central to life. On Earth, wherever there is water, there is life. Some scientists suspect that other planets in our solar system—especially Europa, a satellite of Jupiter—may also contain liquid water and as a result may also contain life.

The molecular reason for water's uniqueness is related to its structure (Figure 14). Water is highly polar and contains two O—H bonds that hydrogen bond to other water molecules (Figure 15). For water to become a gas, these hydrogen bonds must be broken; this requires a substantial amount of thermal energy, making the boiling point of water high.

Water has another unusual property: It expands when it freezes. Virtually every other liquid contracts on freezing. This property is related to the crystal structure of ice (Figure 16). The hexagonal open spaces in the crystal structure make water molecules in ice slightly more separated on average than they are in liquid water. Consequently, ice occupies a greater volume than the water from which it was formed. The consequences of this seemingly simple property are significant. The lower density of ice relative to liquid water makes ice float rather than sink. Consequently, we have icebergs and not ice sea floors. In winter, a relatively thin layer of ice on the surface of a lake insulates the water beneath from further freezing, allowing marine life to survive the winter. If ice sank, entire lakes could freeze solid, eradicating marine life.

The slow and constant erosion of mountains is also due in part to the expansion of water on freezing. Water flows into cracks within rocks during the day and then freezes slowly overnight. The expansion of the ice within the crack generates forces large enough to dislodge boulders. For this reason, it is dangerous to drive mountain roads in the early morning when temperatures are at their coldest and the potential for freezing water and falling boulders is at its highest.

The expansion of ice on freezing is also responsible for the damage to cells in biological tissue when frozen. The water in a cell expands as it freezes and ruptures the cell. If you ever put lettuce or broccoli in your freezer, you have seen the effects of cell rupture; the vegetable is limp and damaged when thawed. The frozen food industry minimizes the damage incurred on freezing by using a process called **flash freezing**. In this process, food is frozen very quickly. This prevents water molecules from ordering in their ideal crystalline structure and results in less expansion and therefore less cell rupture.

Another unique property of water is its ability to dissolve many organic and inorganic compounds. Water is the basic solvent for most of life's chemistry and is responsible for the flow of nutrients and other biologically important molecules throughout the body. The human body can survive over a month without food but only several days without water.

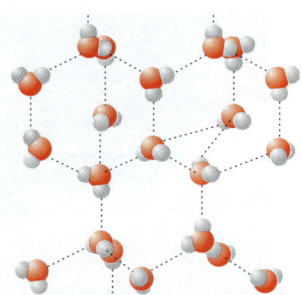

FIGURE 16 Hydrogen bonding in ice. The open structure of ice crystals results in a lower density for solid water than for liquid water.

12.8 Water: Where Is It and How Did It Get There?

The great majority of the Earth's water (97.4%) is in the oceans, which cover over two-thirds of the Earth's surface. The remaining 2.6% is primarily in ice caps, glaciers, and ground water. Only a tiny fraction (0.014%) of the Earth's total water is in readily usable forms such as lakes and streams.

Where did the Earth's water come from? Most geologists think water came from volcanic eruptions that spewed enormous amounts of water into the atmosphere. The surface of the primitive Earth was too hot for this atmospheric water to condense, so the water collected in a thick cloud layer that covered most of the planet. In time, the Earth's surface cooled, and the water held in clouds fell as torrential rains that filled low-lying areas to form oceans. Although much of the Earth's surface is covered with water, it is relatively shallow compared with the Earth's size. If the Earth were the size of a golf ball, its deepest oceans would be shallower than the dimples on the golf ball.

Land water comes from oceans through the **hydrologic cycle** (Figure 17). In this cycle, sunlight evaporates water from oceans, leaving the nonvolatile salts behind. The freshwater collects in the atmosphere as clouds, which eventually condense and fall as rain, most of it into the ocean. A small fraction of rain falls on land where it accumulates as snow, ice, lakes, streams, and ground water. In time, this water finds its way back to the ocean to complete the cycle.

If the Earth were the size of a golf ball, its deepest oceans would be shallower than the dimples on these golf balls.

Explore this topic on the **Water Conservation** website.

12.9 Water: Pure or Polluted?

The quality of water is important to human health. Water-borne diseases cause about 80% of the world's sickness. Because humans drink about 2 L of water per day, even small amounts of toxic compounds in drinking water have significant effects over time. But, what is pure water? What is polluted water? With numerous terms such as pure, purified, distilled, and disinfected to describe our drinking water, how do we know what we are getting?

First, we must understand that virtually no water is pure. Like most liquids we encounter, water is a solution containing a number of different elements and compounds. Pure water can be made only in the chemist's laboratory. Consequently, we will focus on what substances lurk in drinking water; some of these, such as calcium or magnesium ions, have positive health benefits, while others, such as bacteria or lead, are harmful.

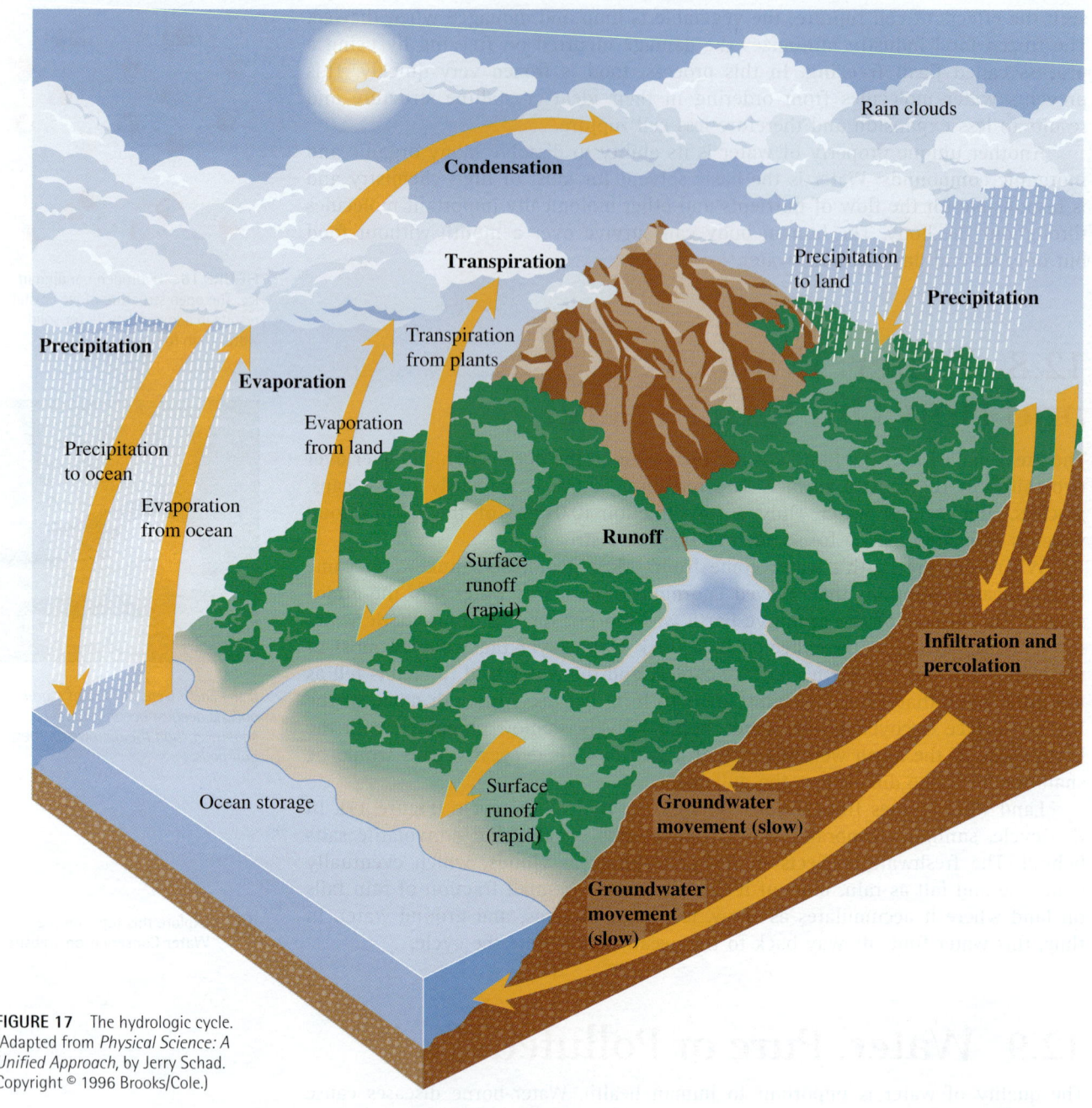

FIGURE 17 The hydrologic cycle. (Adapted from *Physical Science: A Unified Approach*, by Jerry Schad. Copyright © 1996 Brooks/Cole.)

12.10 Hard Water: Good for Our Health, Bad for Our Pipes

The most common impurities in tap water are calcium and magnesium ions. These ions dissolve into water as it runs through soils rich in limestone. Calcium and magnesium are both essential to proper nutrition, and many people take them as dietary supplements. Water rich in calcium and magnesium is called hard water and is characterized by the amount of calcium carbonate that would form if all the calcium ions in the water reacted to form calcium carbonate.

Classification	Hardness (ppm CaCO$_3$)
Very soft	<15
Soft	15–50
Medium	50–100
Hard	100–200
Very hard	>200

Hard water can leave white scaly deposits on plumbing fixtures.

Although hard water has no adverse health effects, it does have annoying properties. The calcium and magnesium in hard water leave a white scaly deposit on pipes, fixtures, cooking utensils, and dishes. Hard water also decreases the effectiveness of soap because calcium and magnesium ions react with soap, making it unavailable to dissolve grease and oil. The product of this reaction is a slimy, gray scum called curd that often deposits on skin or on the sides of the bathtub producing "bathtub ring."

To avoid these undesirable effects, some people install water softeners in their homes. These devices are charged with sodium ions, usually from sodium chloride, that exchange with the calcium and magnesium ions in hard water. Sodium ions do not form scaly deposits as calcium and magnesium do, and they do not react with soap. However, sodium does increase the risk of high blood pressure and must therefore be avoided by those who have high blood pressure or heart problems.

Water softeners remove calcium and magnesium ions from water.

EXAMPLE 12.8

Finding Grams of Impurity from ppm

A sample of hard water contains 55.0 ppm CaCO$_3$. If a person consumes 1.50 liters of this water, how many grams of CaCO$_3$ have they consumed? The density of water is 1.00 g/mL.

SOLUTION

The concentration of CaCO$_3$ (55 ppm) means 55.0 g CaCO$_3$/10^6 g water; it is a conversion factor between g water and g CaCO$_3$. We first convert the volume of water (1.50 L) into mass of water in grams:

$$1.50 \text{ L water} \times \frac{1000 \text{ mL}}{1 \text{ L}} \times \frac{1.0 \text{ g}}{\text{mL}} = 1.50 \times 10^3 \text{ g water}$$

We then convert from g water to g CaCO$_3$:

$$1.50 \times 10^3 \text{ g water} \times \frac{55.0 \text{ g CaCO}_3}{1 \times 10^6 \text{ g water}} = 8.25 \times 10^{-2} \text{ g CaCO}_3$$

YOUR TURN

Finding Grams of Impurity from ppm

A sample of water contains 2.0 ppm NaF. How many grams of NaF are contained in 25 L of water?

12.11 Biological Contaminants

The dumping of human and animal waste into the ground or nearest river has resulted in biological contamination of both surface and ground water. These contaminants are microorganisms that cause diseases such as hepatitis, cholera, typhoid, and dysentery. Many of these diseases have been eliminated in developed nations, but they are still a serious problem in the water supplies of developing or Third World countries.

The biological contaminants that occasionally contaminate North American water include bacteria such as *Giardia* or *Legionella* and a number of different types of viruses. *Giardia* and viruses come primarily from human and animal fecal waste and can lead to gastrointestinal disease if ingested. *Legionella* is indigenous to some natural waters and results in Legionnaire's disease (an acute respiratory infection similar to pneumonia) if ingested. Biological contaminants in drinking water pose immediate danger to human health, and water contaminated with these should be boiled before consumption. Boiling water vigorously for several minutes kills most microorganisms.

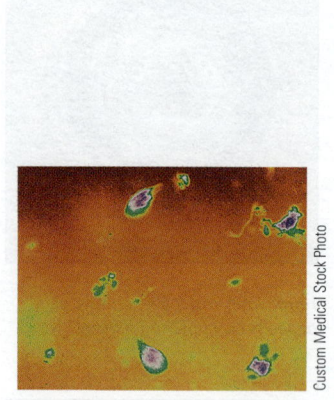

Bacteria such as Giardia *can contaminate water and result in gastrointestinal disease if ingested.*

12.12 Chemical Contaminants

Chemical contaminants get into ground water from a number of sources. The dumping of wastes by industry, either directly into streams and rivers or indirectly through atmospheric emissions, poses a significant problem. Pesticides and fertilizers from agriculture are also a problem; they diffuse into the soil and, with rain, eventually get into drinking water supplies. The dumping of household solvents, paints, and oils down the drain or into the ground is also a significant source of water pollution. We can broadly divide chemical water contaminants into three types: organic, inorganic, and radioactive.

ORGANIC CONTAMINANTS

The U.S. Environmental Protection Agency divides organic contaminants in water into two types, volatile organics and nonvolatile organics. Volatile organic contaminants include benzene, carbon tetrachloride, and a number of chlorohydrocarbons. Nonvolatile organic contaminants include ethylbenzene, chlorobenzene, and trichloroethylene (TCE). Both types of organic contaminants come from fertilizers, gasoline, pesticides, paints, and solvents. If ingested, they increase cancer risk, damage the liver and kidneys, and damage the central nervous system.

INORGANIC CONTAMINANTS

Inorganic contaminants include cancer-causing agents such as asbestos, which leaches into water from natural deposits or from asbestos cement in water-holding systems. They also include nitrates that get into water from animal waste, fertilizer, septic tanks, and sewage. Nitrates pose an immediate danger to humans, especially children under one year of age. The ingestion of nitrates by a baby can result in a condition known as **blue baby** in which nitrates react with hemoglobin in the baby's blood, diminishing its ability to carry oxygen. Because nitrates are nonvolatile, they are not removed from water by boiling. In fact, boiling nitrate-contaminated water actually increases the nitrate concentration. Boiling vaporizes some of the water, but the nonvolatile nitrates are left behind, resulting in water with a higher nitrate concentration (Figure 18).

Inorganic contaminants also include metals such as mercury and lead that damage the kidneys and central nervous system. Mercury comes from natural

Some organic water contaminants.

Benzene

Carbon tetrachloride

Ethylbenzene

Chlorobenzene

Trichloroethylene

deposits, batteries, and electrical switches. Lead comes primarily from plumbing in older homes. Since 1986, all plumbing fixtures must be made from lead-free pipe, solder, and flux. However, older plumbing systems may still contain significant amounts of lead.

If you live in a home or building with older plumbing, you can minimize lead intake by following two simple steps. When obtaining drinking water, flush the pipes with cold water until it becomes as cold as it will get. This may take anywhere from 5 seconds to several minutes but will flush out water that has

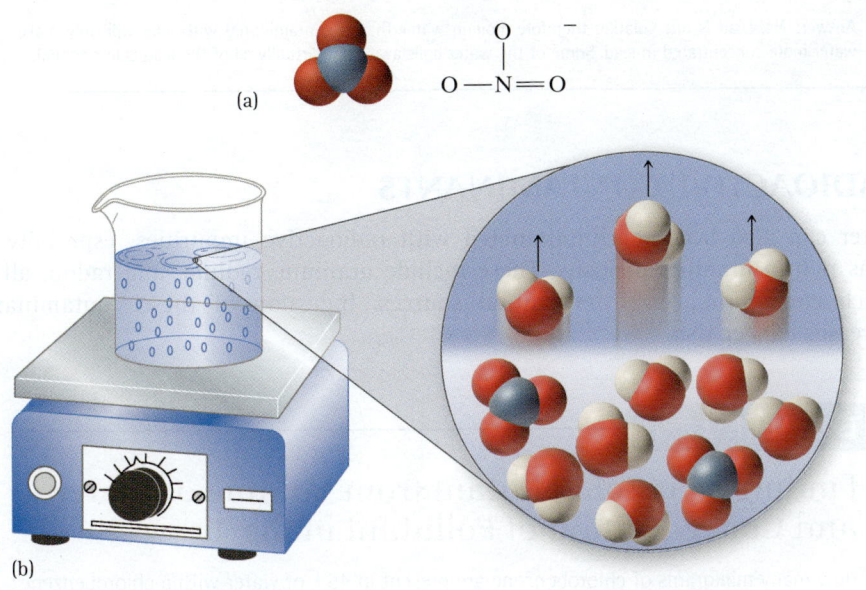

FIGURE 18 (a) Nitrates are substances that contain the NO_3^- ion. (b) When water containing nonvolatile substances such as nitrates is boiled, the water molecules evaporate, but the nitrates do not. The result is water with a higher nitrate concentration.

MOLECULAR FOCUS

Trichloroethylene (TCE)

Formula: $CCl_2=CHCl$
Molecular weight: 131.4 g/mol
Melting point: $-84.8°C$
Boiling point: $86.7°C$
Three-dimensional structure:

Lewis structure: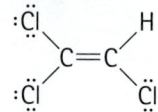

TCE is a colorless liquid. Because it is a chlorinated hydrocarbon, it is relatively inert and nonflammable. It is slightly polar and only dissolves in water in small amounts. It is used as a solvent for oils, paints, and varnishes, and as an industrial degreaser. TCE is also widely used in the dry-cleaning industry and in the manufacture of organic compounds and pharmaceuticals.

Moderate exposures to TCE cause effects similar to alcohol inebriation. Higher exposures result in death. Low exposures to TCE over a long period of time are expected to increase cancer risk.

TCE is 1 of the 100 or so compounds whose levels in tap water are controlled by the EPA. Current law requires that TCE levels in tap water be below 0.005 mg/L.

---

been sitting in the pipes—dissolving lead—for long periods of time. Second, because lead dissolves more easily in hot water than in cold, use only cold water for consumption, especially for making baby formula. If hot water is required, obtain cold water from the tap and heat it on the stove. These two steps should minimize your lead intake from drinking water and protect your family's health.

APPLY YOUR KNOWLEDGE

An older home has lead pipes. Should you boil the water that comes out of the pipes before drinking it? Why or why not?

Answer: No. Lead is not volatile; therefore, boiling water that is contaminated with lead will only make water more concentrated in lead. Some of the water boils away, but virtually all of the lead is left behind.

RADIOACTIVE CONTAMINANTS

Water can also become contaminated with radioactive impurities, especially in areas rich in uranium deposits. These include uranium, radium, and radon, all of which come primarily from natural sources. Ingestion of these contaminants increases cancer risk.

EXAMPLE 12.9

Finding Mass of Pollutant from Volume of Water and Concentration of Pollutant in mg/L

How many milligrams of chlorobenzene are present in 45 L of water with a chlorobenzene concentration of 0.12 mg/L?

SOLUTION

Use the concentration of chlorobenzene (in mg/L) as a conversion factor between L of water and mg of chlorobenzene:

$$45 \text{ L water} \times \frac{0.12 \text{ mg chlorobenzene}}{\text{L water}} = 5.4 \text{ mg chlorobenzene}$$

YOUR TURN

Finding Mass of Pollutant from Volume of Water and Concentration of Pollutant in mg/L

How many milligrams of nitrate are present in 185 L of water with a nitrate concentration of 15 mg/L?

12.13 Ensuring Good Water Quality: The Safe Drinking Water Act

To ensure high water quality for all Americans, Congress in 1974 enacted the Safe Drinking Water Act (SDWA). The SDWA, amended in 1986, authorized the EPA to establish maximum contaminant levels (MCLs) for 84 different contaminants likely to be found in public drinking water systems (Table 3). Every public water supply

TABLE 3

Some Selected Maximum Contaminant Levels (MCLs)

Contaminant	MCL (mg/L)
Biological	
Giardia lamblia	zero
Legionella	zero
Volatile organics	
benzene	0.005
carbon tetrachloride	0.005
1,1,1-trichloroethane	0.2
trichloroethylene	0.005
Nonvolatile organics	
chlorobenzene	0.1
dibromochloropropane	0.0002
dioxin	3.0×10^{-8}
Inorganics	
mercury	0.002
nitrate	10
lead	zero
copper	1.3
Radioactive	
uranium	0.02

in the country serving over 15 connections or 25 people must meet these minimum standards. Under the SDWA, public water suppliers must periodically sample and test the water supplied to your tap. If any of the contaminants controlled by the EPA are above the maximum contaminant level, the supplier must remove the contaminant as quickly as possible. They must also notify the appropriate state agency of the violation and must notify you, the consumer, either by a notice in the local media or by a direct letter. The notice must include any necessary precautions associated with using the tap water. Only biological contamination or nitrate contamination poses immediate health risks. Although a topic of debate, most of the other MCLs are set for safe consumption over a lifetime and usually have built-in safety margins. According to the EPA, consumption of water slightly above the MCL for a short period of time is harmless (with the exception of biological contaminants or nitrates).

For some of these contaminants, such as *Giardia* and lead, the EPA requires water suppliers to add substances to the water that prevent contamination during transport between the treatment facility and your tap.

12.14 Public Water Treatment

To meet the EPA requirements for drinking water, providers purify and treat water (Figure 19) before delivering it to the tap. Water from a reservoir is first screened

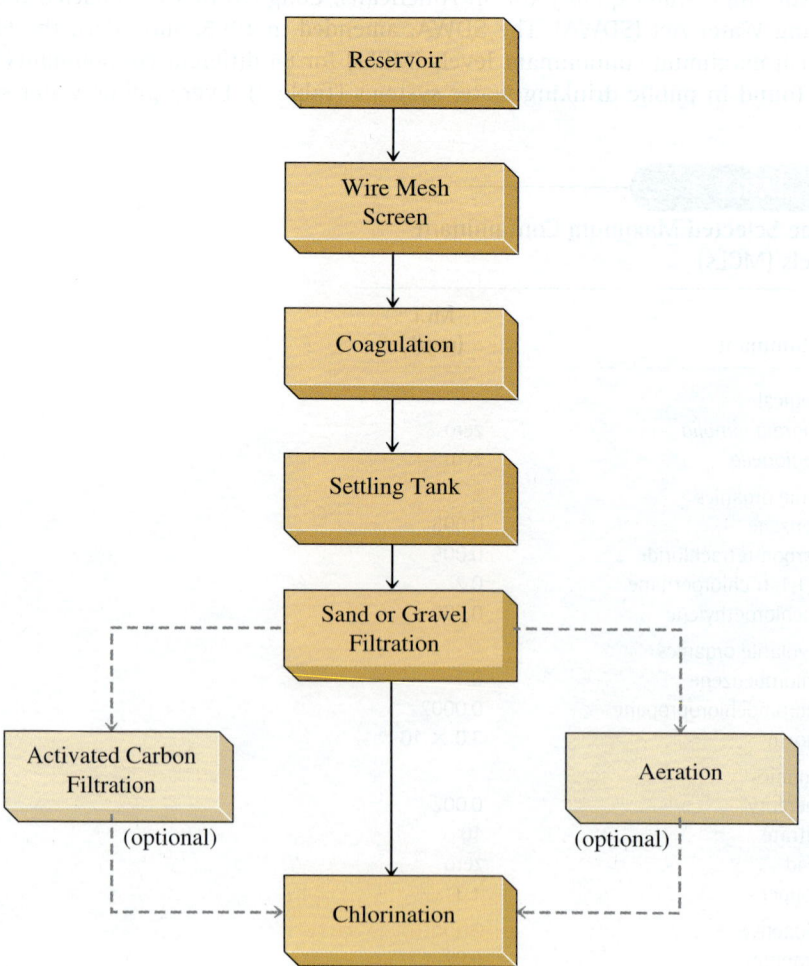

FIGURE 19 Water purification process.

WHAT IF...

Criticizing the EPA

The U.S. EPA, along with lawmakers that give the EPA its mandate, are often criticized on issues related to water quality. Industrial and business interests generally believe that water quality is fine and want less regulation. These groups want to minimize their costs and the cost of the service they provide to their customers. Environmental groups, on the other hand, generally believe that water quality is not as good as it could be and therefore push for more regulation of industry and business.

For example, in 1995 the Natural Resources Defense Council (NRDC) and the Environmental Workings Group (EWG) issued reports stating that over 53 million Americans are drinking tap water contaminated by lead, fecal bacteria, and toxic chemicals. The reports estimated that consumption of contaminated water caused 1200 deaths per year and more than 7 million cases of illness in the United States.

The June 2, 1995, issue of the *Washington Post* quotes the president of EWG as saying, "What our studies show is that our water laws need to be strengthened, not relaxed. This is America."

Critics, on the other hand, attacked the reports as alarmist. The *Washington Post* quoted the president of the American Water Works Association as saying, "The American people tell us that they are willing to pay for safe drinking water. But they are not willing to pay additional costs which provide them no additional protection."

Ultimately, the degree of regulation that municipal water suppliers and industry must adopt rests with the citizens who elect the politicians who make the laws. Scientists can provide good data, but almost all measurements carry uncertainty. Scientists are accustomed to living with a degree of uncertainty, but the public often wants answers that are either black or white. Such answers are often not possible.

One point to keep in mind, however, is that scientists can measure water impurities to fantastically small levels. For example, the U.S. EPA limits the amount of TCE in drinking water to 0.005 mg/L. At that level, you have to drink 200,000 L of water to consume 1 g of TCE; that is enough water to fill a backyard swimming pool. In contrast, the amount of TCE required to kill a human, based on extrapolations from experiments with rats, is about 500 g or 500 swimming pools worth of water.

Although it takes 500 g of TCE to cause immediate death, it could be that smaller amounts over long periods of time may have other effects. Experiments on mice show TCE to be a carcinogen, meaning that even small amounts may increase cancer risk. However, how much is too much? In many cases, we do not know. One of the most difficult problems facing politicians and regulatory agencies today is what to do in the face of uncertainty. Wisdom dictates that you err on the side of caution, but financial concerns limit how much unnecessary regulation can be implemented.

QUESTION: What is your point of view on this issue? Do we need more or less regulation? Why? Have you had any personal experiences where water quality has been below your expectations?

---

through wire meshes of successively smaller sizes to remove objects such as fish, other marine life, leaves, sticks, garbage, and debris. The water is then treated with coagulants, chemicals that surround impurities and coagulate into large particles that either float or sink. Particles that sink are removed in settling tanks, and floating particles are removed by skimming the surface of the water.

The water is then filtered through gravel and sand to remove remaining particulates. In some areas, the water is passed through high-surface-area activated carbon, which removes organic compounds from the water. Organic molecules adsorb onto the surface of the carbon, which is periodically treated to reactivate it. Another optional step is aeration, in which the water is sprayed into a fine mist allowing volatile compounds to evaporate. The water is then chlorinated to kill biological contaminants. The chlorine remains in the water all the way to the tap, preventing additional contamination from within pipes.

Other methods used to kill biological contaminants include bubbling ozone through the water or exposing the water to ultraviolet light. Both of these methods kill bacteria in the water. However, because ozone decomposes back to oxygen, and because the UV light is present only at the treatment plant, these methods do not provide protection from biological contamination in pipes or holding tanks beyond the treatment facility. Finally, if the water is acidic, it is neutralized. Acidic water must be neutralized because it is more likely to dissolve metals such as lead and copper within pipes. Many water districts also add small amounts of fluoride to the water to help prevent tooth decay.

12.15 Home Water Treatment

A number of companies offer home water treatment to improve the quality of tap water. You may wonder if home water treatment is necessary. Is your tap water safe to drink? If you receive your drinking water from a public water provider, the EPA believes that home treatment for the *protection of health* is not necessary. According to the EPA, you may want to treat your water for other reasons—such as hardness, poor clarity, or undesirable taste—but the EPA maintains that unless you get a notice from your water provider stating otherwise, the water from your tap is safe to drink over a lifetime.

A number of groups, however, disagree with the EPA on this point. They argue that home water treatment will at least give you clearer, better-tasting water and may remove unhealthful compounds overlooked by your local water provider.

CARBON FILTRATION

The most inexpensive home water treatment is filtration through activated carbon. Carbon filters attach to your tap and contain small carbon particles whose surfaces have been cleaned by chemical treatment; the clean surfaces adsorb impurities out of the water (Figure 20). These filters are particularly effective in removing organic contaminants and chlorine, often responsible for unpleasant taste or odor in drinking water. The filters have a limited lifetime because the carbon surfaces eventually fill with contaminants, making them ineffective. A number of systems contain a replaceable cartridge that houses the activated carbon. These cartridges are replaced periodically with new ones that contain fresh activated carbon.

WATER SOFTENING

Perhaps the most common home water treatment is water softening. Several companies will install these units in your home and maintain them on a regular

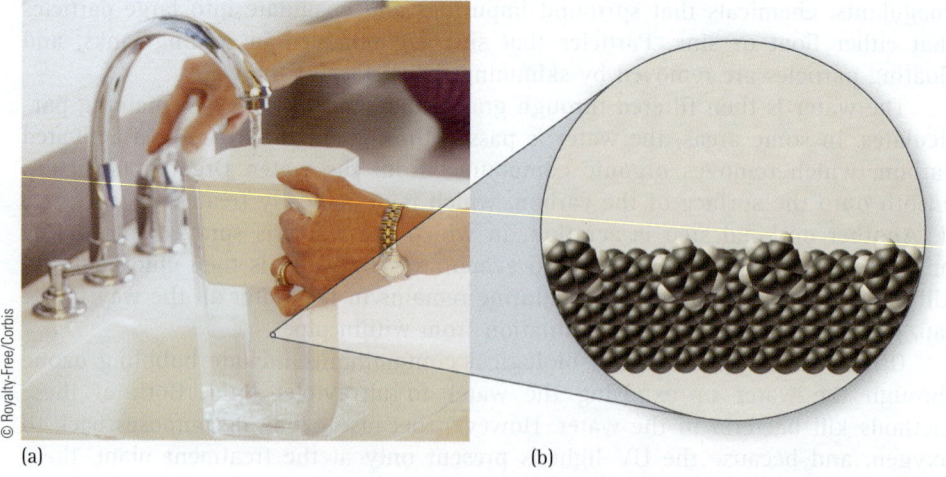

FIGURE 20 (a) Filters such as those shown here are often used in home taps to eliminate undesirable taste or odor from drinking water. (b) Activated carbon present in many home water filters will adsorb organic impurities from tap water. The organic molecules stick to the surface of the activated carbon.

basis. In water softening, hard water is passed through an ion-exchange material called a **zeolite**. Zeolites are solid materials with a rigid, cagelike structure that can hold metal ions. The zeolite initially contains sodium ions, which then exchange with calcium and magnesium ions as the hard water flows through the zeolite. Eventually, as water is softened, the zeolite becomes depleted in sodium content and full of calcium and magnesium. The zeolite is then recharged by back-flushing with concentrated sodium chloride, reversing the process.

REVERSE OSMOSIS

Osmosis is a naturally occurring process in which water flows from an area of low solute concentration to high solute concentration. It is most easily demonstrated in an osmosis cell, where two solutions are separated by a semipermeable membrane (Figure 21). A semipermeable membrane selectively allows water to pass through it but blocks other substances. If a salt water solution is placed on one side of an osmosis cell and fresh water on the other, osmosis will occur. The side of the membrane containing the salt water will rise as fresh water flows through the membrane and into the salt water solution.

In reverse osmosis (Figure 22), pressure is applied to the salt water side of the membrane, forcing water through the membrane in reverse. As the salt water solution flows through the membrane, water passes through, but the salt does not. Consequently, the salt water is purified. With different membranes, tap water can also be purified in this way. Reverse osmosis is also used on a much larger scale in desalination plants to produce freshwater from seawater.

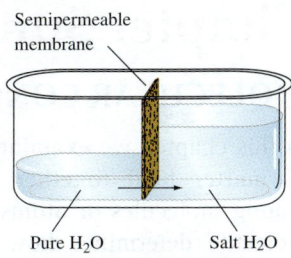

FIGURE 21 In an osmosis cell, water flows from the side containing pure water through a semipermeable membrane to the side containing a solution.

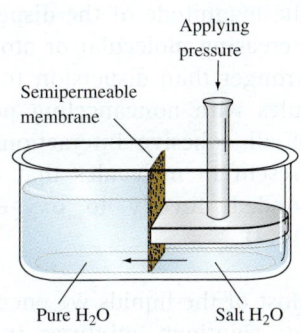

FIGURE 22 In reverse osmosis, pressure is applied to the solution side of the osmosis cell, forcing water to flow from the solution side to the pure water side. Because the membrane is semipermeable, water molecules are allowed to pass while the solute molecules are not.

Chapter Summary

MOLECULAR CONCEPT

In this chapter, we examined liquids and solids, states of matter held together by cohesive interactions among molecules or atoms (12.2). The structure of a molecule determines how strong these interactions will be, which in turn determines the macroscopic properties like melting point, boiling point, and volatility (12.3).

The weakest cohesive interaction is called the dispersion force and is present in all molecules and atoms. The magnitude of the dispersion force increases with increasing molecular or atomic weight. Dipole forces, stronger than dispersion forces, are present in molecules with noncanceling polar bonds. The strongest of all cohesive interactions, the hydrogen bond, is present in molecules that contain a hydrogen atom bonded directly to oxygen, nitrogen, or fluorine (12.4).

Most of the liquids we encounter are not pure liquids but solutions, mixtures in which a solid or liquid solute dissolves in a liquid solvent (12.6). Water solutions are the most common of all solutions. Water is a unique substance with unusually strong cohesive forces for its molecular weight (12.7). Water is constantly cycled through the oceans into clouds and back into the ocean in a process known as the hydrologic cycle (12.8).

The water we drink is not usually pure, but contains a number of other substances. Some of these substances are benign or even helpful to human health, whereas others are toxic (12.9–12.15).

SOCIETAL IMPACT

Cohesive forces hold together all solid and liquid matter. Solid, high-melting-point substances such as diamond have very strong cohesive forces, while gaseous, low-boiling-point substances such as helium have weak cohesive forces (12.3).

Hydrogen bonding is a particularly important cohesive force. It is responsible for the abnormally high boiling point of water and is important in the folding of human proteins (12.4). Hydrogen bonding is also the force that holds the two halves of human DNA together.

Water is one of society's most precious resources. Many countries have contaminated water supplies that cause disease and death to thousands. North America has particularly clean and reliable water supplies, partly because of environmental legislation such as the Safe Drinking Water Act (SDWA) (12.13). One question that continues to face our society is how much regulation is too much? Environmental groups often push for more regulation and business groups often push for less. Ultimately, the decision rests with citizens. How much regulation do you want?

Our current drinking water quality is controlled and monitored by the EPA, which has established maximum contaminant levels (MCLs) for a number of potentially dangerous contaminants (12.14–12.15). Because of this extensive monitoring, the EPA maintains that drinking water is safe to drink over a lifetime.

Chemistry on the Web

For up-to-date URLs, visit the text website at www.chemistry.brookscole.com/tro3.

- Properties and Behavior of Water
  http://www.visionlearning.com/library/module_viewer.php?mid=57

- Water Conservation
  http://www.nwf.org/getgreen/water.cfm
- Natural Resources Defense Council
  http://www.nrdc.org/water/drinking/default.asp
- World Health Organization
  http://www.who.int/topics/water/en/

Key Terms

blue baby	evaporation	melting point	polar molecules
boiling point	flash freezing	molarity (M)	solute
cohesive forces	freezing point depression	nonvolatile	solution
concentration	hard water	percent by mass	solvent
crystalline solids	hydrogen bond	percent by volume	vapor pressure
crystalline structure	hydrologic cycle	permanent dipoles	volatile
dispersion	instantaneous dipole	polar bonds	zeolite

Exercises

 Assess your understanding of this chapter's topics with an online chapter quiz at http://chemistry.brookscole.com/tro3.

QUESTIONS

1. Why are liquid drops spherical in shape?
2. Would there be solids and liquids if cohesive forces among molecules did not exist? Explain.
3. From a molecular viewpoint, explain the differences among gases, liquids, and solids.
4. Explain evaporation from a molecular point of view.
5. Define each of the following:
   a. boiling point
   b. melting point
   c. cohesive force
6. Describe the relationship between cohesive forces and
   a. boiling point
   b. melting point
7. Define each of the following cohesive forces and order them in terms of increasing strength:
   a. dispersion force
   b. dipole force
   c. hydrogen bonding
8. Why do oil and water not mix?
9. Explain how soap works.
10. Define volatile and nonvolatile. How are these related to cohesive forces?
11. Why does sweating cool the human body?
12. Define each of the following:
    a. solution
    b. solvent
    c. solute
    d. molarity
    e. ppm
    f. mg/L
13. What are the unique properties of water? Why are they important?
14. Where did the Earth's water come from?
15. Explain the hydrologic cycle.
16. What are the impurities present in hard water? What are the effects on health? Why are they bothersome?
17. What are the common classifications of water based on parts per million of $CaCO_3$?
18. How does a water softener work?
19. List the four types of contaminants commonly found in water, and give two examples of each.
20. Which water contaminants pose immediate health risks?
21. Which water contaminants can be eliminated by boiling?
22. What steps should be taken to minimize lead intake from tap water?
23. What is the SDWA?
24. Explain how drinking water is treated before being delivered to your tap.
25. Is home water treatment necessary for health reasons? Why would it be desirable?
26. Explain how each of the following home water treatments works and what impurities are eliminated:
    a. carbon filtration

b. water softening
c. reverse osmosis

27. Explain the concerns that groups like the EWG or NRDC have about drinking water quality.

PROBLEMS

28. Butane is a gas at room temperature, and pentane is a liquid. Why is this the case?
29. One of the following molecules is a liquid at room temperature. Which one? Why?
    a. Br_2
    b. O_2
    c. F_2
    d. Cl_2
30. Which of the following compounds would you expect to have the highest boiling point? Why?
    a. $CH_3CH_2OCH_3$
    b. $CH_3CH_2CH_2OH$
    c. $CH_3CH_2CH_3$
31. Which of the following compounds would you expect to have the highest boiling point? Why?
    a. H_2O
    b. H_2S
    c. HCl
32. All the following are liquids at room temperature. Which would you expect to be the most volatile?
    a. $CH_3CH_2CH_2OH$ (propanol)
    b. CH_3CCH_3 (acetone) with $\overset{O}{\underset{\parallel}{}}$
    c. CH_3CH_2OH (ethanol)
33. Which of the following liquids would you expect to be least volatile?
    a. pentane
    b. hexane
    c. octane
    d. decane
34. Classify each of the following as polar or nonpolar:
    a. H_2O
    b. $CH_3CH_2CH_3$
    c. CH_3CH_2OH
    d. H_2
    e. hexane
35. Classify each of the following as polar or nonpolar:
    a. CO_2
    b. CH_2Cl_2
    c. $CH_3CH_2CH_2CH_2CH_2CH_2CH_2CH_3$ (octane)
    d. CH_3CHF_2
36. What are the criteria used by the perfume industry when determining the substances to include in perfume?
37. Two nonpolar substances with similar structures but different scents are mixed in a perfume vial. One of the substances has a higher molar mass than the other. Which one is smelled when the perfume vial is first opened? Which one is smelled if the mixture is left on the skin for several hours?
38. What is the molarity of 300 mL of a KCl solution made using 10 g of KCl?
39. What is the molarity of 2.8 L of an $NaNO_3$ solution made with 34 grams of sodium nitrate?
40. How many grams of sucrose ($C_{12}H_{22}O_{11}$) are present in 2.8 L of a 2.2 M sucrose solution?
41. How many grams of glucose ($C_6H_{12}O_6$) are present in 7.8 L of a 1.3 M glucose solution?
42. How many grams of NaCl are present in 250 mL of a 2.3% solution by mass? Assume that the density of the solution is 1.0 g/mL.
43. How many grams of NaF are present in 3.8 L of a 1.1% NaF solution by mass? The density of the solution is 1.0 g/mL.
44. A 250-g sample of hard water contains 0.20 g of $Ca(CO_3)$. What is the hardness of the water in parts per million? How would the water be classified?
45. A 15-gram sample of water contains 1.1×10^{-3} g of $CaCO_3$. What is the hardness of the water in ppm? How would the water be classified?
46. Tap water is often fluoridated with a sodium fluoride (NaF) concentration of approximately 1 ppm. If you consume 0.8 kg of tap water per day, how many grams of NaF do you consume?
47. A softened water sample contains 245 ppm Na. If a person consumes 1.5 L of this water per day, how many mg of sodium are consumed?
48. A 450-mL sample of water is extracted from a well and analyzed for mercury. The mercury content is found to be 0.005 mg. What is the concentration of mercury in the well water in milligrams per liter? Is the water safe for human consumption?
49. A 225-mL sample of water is taken from a tap and analyzed for trichloroethylene (TCE, formula C_2HCl_3). The TCE content is 0.012 mg. What is the concentration of TCE in the water in mg/L? Is the water safe for human consumption?
50. A water sample is found to be high in lead content. Would boiling the water lower the lead concentration? Why or why not?
51. Your water provider posts an alert in the local newspaper stating that the nitrate concentration in the drinking water is temporarily above the maximum contaminant level recommended by the EPA. Would boiling the water make it safe to drink?

POINTS TO PONDER

52. The quote by Roger Joseph Boscovich at the beginning of this chapter states, "It will be found that everything depends on the composition of the forces with which the particles of matter act upon one another; and from these forces . . . all phenomena of nature take their origin." How is this true in regard to liquids and solids?

53. Explain the difference between a polar molecule and one that undergoes hydrogen bonding.
54. The EPA has become concerned over the use of lead in fishing weights. Many fishing lines break on snags, leaving the lead on the bottom of the lake. Why is this a problem? How would water temperature affect the scope of the problem?
55. When water is boiled, it is converted into a gas. When gasoline is burned, it, too is converted into a gas. Explain the fundamental difference between these two processes.
56. Suppose a perfume contained the following two compounds. What would the initial fragrance be? How would it change over time?

$$CH_3CH_2CH_2\overset{\overset{O}{\|}}{C}OCH_3$$
methyl butyrate (apple fragrance)

$$CH_3\overset{\overset{O}{\|}}{C}-O-CH_2-\text{(phenyl)}$$
Benzyl acetate (jasmine fragrance)

57. Explain what you think might be occurring in dry ice—frozen CO_2—that slowly disappears but never melts.
58. Over half of our nation's drinking water contains fluoride, an additive that helps prevent tooth decay. Some see the addition of fluoride to drinking water as an intrusion of government into the lives of individuals. After all, anyone who wants fluoride can easily get it from toothpaste or other sources. What do you think? Should municipal water supplies add fluoride to drinking water? Should they feel free to add other substances that they think would benefit society?

Feature Problems and Projects

59. The following drawing shows a molecular view of a water sample contaminated with lead. Suppose the water is boiled until half of the water has boiled away. Draw a picture of the water sample after boiling. Does boiling increase or decrease the lead concentration?

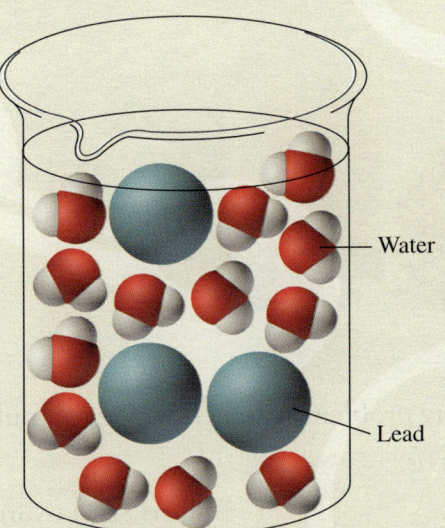

60. Visit the EPA's student website at **http://www.epa.gov/students/**. You may view several different aspects of the EPA's work—for now, choose *water*. Choose the *in your neighborhood* option and then the *surf your watershed* option. Find your hometown watershed either by clicking on the map or by entering your zip code. Where is your watershed?

61. The EPA maintains databases documenting violations of MCLs by water suppliers. These databases can be accessed at their enviro facts website at **http://www.epa.gov/enviro/**. Choose *water* and then enter your county and state to look up *public water systems violations*. Find your water system, and click on it. How many people are served by your water system? Are any violations reported? If so, what type?

13 Acids and Bases: The Molecules Responsible for Sour and Bitter

Knowledge is a sacred cow, and my problem will be how we can milk her while keeping clear of her horns.

—Albert Szent-Gyorgyi

In this chapter, we examine acids and bases. Acids and bases are found in many foods and in a number of consumer products such as toilet bowl cleaners and Drāno. Amino acids are the building blocks of proteins, and the genetic code is carried in a sequence involving four different bases. So acids and bases and their chemistry are vitally important. Did you drink orange juice this week? If you did, you consumed citric acid, a component of orange juice. Have you used soap? If you have, you experienced the slippery feel of bases. We will explore the properties of acids and bases and their molecular descriptions. We will also see several examples of common acids and bases that you encounter on a regular basis, and we will explore an environmental problem involving acids—acid rain. Acid rain—which threatens freshwater marine species, forests, and many building structures—is primarily caused by coal combustion and is a significant problem in the northeastern United States and eastern Canada.

QUESTIONS FOR THOUGHT

- What are the properties of acids?
- What are the properties of bases?
- How do you describe acids and bases from a molecular perspective?
- How do you measure acid strength?
- What are some common acids?
- What are some common bases?
- What are antacids?
- What is acid rain? What causes it?
- What are the effects of acid rain on the environment?
- What is being done about acid rain?

Most sour flavors are caused by the presence of acid.

Limes and lemons are high in acid content.

Acids dissolve some metals. Here nitric acid is dissolving the copper that composes a penny. The green-blue color of the solution is due to Cu^{2+} ions in solution. The Cu^{2+} ions are formed as the copper dissolves.

FIGURE 1 Acids turn blue litmus paper red.

13.1 If It Is Sour, It Is Probably an Acid

Children wince at it, chefs create with it, and wine makers make it an art form—the sour taste of acid. Acids cause the crisp tartness of an apple, the tang in lemonade, and the subtle bite of a good wine. Because they are present in so many foods, it would be impossible to list them. Sourness in foods is caused by acids, molecules that release protons. These protons, or hydrogen ions (H^+), react with protein molecules on the tongue. The reaction causes the protein molecule to change its shape, sending an impulse to the brain that we interpret as sour. The more acidic the food, the more sour the taste. Limes, very high in acid content, are extremely sour; tomatoes, having lower acid levels, have only a slight tang.

Acids, and their chemical opposite, **bases**, are all around us. We eat them, smell them, and use them in a number of household products and pharmaceuticals. In this chapter, we will examine their properties and their uses. We will also reexamine, this time with more chemical insight, a topic from a previous chapter, acid rain.

13.2 The Properties of Acids: Tasting Sour and Dissolving Metals

James Bond, the epic spy, is trapped in jail with little hope of escape. However, the scientific expertise within the British Secret Service has once again foreseen his unfortunate predicament. Bond pulls out his pen–gold, of course–filled with nitric acid. He squirts the liquid on the bars that imprison him, which quickly dissolve, and he escapes.

Although acids do not dissolve metals with the ease depicted in movies or television, they do dissolve many metals. A penny dropped in concentrated nitric acid or a small iron nail submerged in hydrochloric acid will dissolve away in several minutes.

A second property of acids, discussed earlier, is their characteristic sour taste. Although laboratory chemicals should never be tasted, many naturally occurring acids are present in foods, and we taste them daily. The sour taste of lemons, vinegar, plain yogurt, and vitamin C is due to their acid content.

A third property of acids is their ability to react with bases to form water and a **salt** through **neutralization** reactions. For example, hydrochloric acid (HCl) reacts with sodium hydroxide (NaOH) to form water and sodium chloride (NaCl).

$$\underset{\text{acid}}{HCl} + \underset{\text{base}}{NaOH} \longrightarrow \underset{\text{water}}{H_2O} + \underset{\text{salt}}{NaCl}$$

A fourth property of acids, particularly useful in laboratory tests, is their ability to turn **litmus paper** red (Figure 1). Litmus paper contains a dye that turns red in acid and blue in base. A litmus test simply involves dipping a piece of litmus paper into a solution to test its acidity. Red litmus indicates acid; blue indicates base.

In summary, the properties of acids are

- Acids dissolve many metals.
- Acids have a sour taste.

TABLE 1
Common Laboratory Acids

Name	Formula	Uses
Hydrochloric acid	HCl	Cleaning of metals, preparation of foods, refining of ores
Sulfuric acid	H_2SO_4	Manufacture of fertilizers, explosives, dyes, and glue
Nitric acid	HNO_3	Manufacture of fertilizers, explosives, and dyes
Phosphoric acid	H_3PO_4	Manufacture of fertilizers and detergents; flavor additive for food and drinks
Acetic acid	CH_3COOH	Present in vinegar; used in manufacture of plastics and rubber; used as preservative in foods and as solvent for resins and oils

- Acids react with bases to form water and a salt.
- Acids turn litmus paper red.

Some common laboratory acids are listed in Table 1. These include sulfuric acid (H_2SO_4) and nitric acid (HNO_3) used in the manufacture of fertilizers, explosives, dyes, paper, and glue. Hydrochloric acid (HCl) can be purchased at the hardware store under its trade name, muriatic acid. It is often used in cleaning metal products and in preparing various foods. Acetic acid (CH_3COOH), an organic acid, is present in vinegar in low concentrations and is often used as a preservative for foods.

In concentrated form, many acids are dangerous. If spilled on clothing, they dissolve the clothing material. If they contact the skin, they produce severe burns. If ingested, concentrated acids will damage the mouth, throat, stomach, and gastrointestinal tract. In larger amounts, they can kill.

Concentrated acids such as sulfuric acid shown here are dangerous. They may dissolve clothing and will cause severe burns on the skin.

13.3 The Properties of Bases: Tasting Bitter and Feeling Slippery

Like acids, bases can be identified by their properties. When washing your hands with soap, you experience one of the properties of bases, their slippery feel. Soap is basic and, like all bases, feels slippery on the skin. If you have ever tasted soap, you know another property of bases, their bitter taste. The bitter taste of coffee, milk of magnesia, and some medicines is due to their base content.

A third property of bases is their ability to react with acids to form water and a salt. For example, potassium hydroxide (KOH) reacts with sulfuric acid (H_2SO_4) to form water and the salt potassium sulfate (K_2SO_4).

$$\underset{\text{base}}{2KOH} + \underset{\text{acid}}{H_2SO_4} \rightarrow \underset{\text{water}}{2H_2O} + \underset{\text{salt}}{K_2SO_4}$$

TABLE 2
Common Laboratory Bases

Name	Formula	Uses
Sodium hydroxide	NaOH	Neutralization of acids, petroleum processing, and manufacture of soap and plastics
Potassium hydroxide	KOH	Manufacture of soap and paint remover; cotton processing; electroplating
Sodium bicarbonate	$NaHCO_3$	Antacid, source of CO_2, in fire extinguishers and cleaning products
Ammonia	NH_3	Detergent; removing stains; extracting plant colors; manufacture of fertilizers, explosives, and synthetic fibers

Through this reaction, bases can neutralize acids. Finally, bases turn litmus paper blue (Figure 2).

In summary, the properties of bases are

- Bases feel slippery.
- Bases taste bitter.
- Bases react with acids to form water and a salt.
- Bases turn litmus paper blue.

Some common laboratory bases are listed in Table 2. These include sodium hydroxide, also known as lye or caustic soda, used in the processing of petroleum and in the manufacture of soap and plastics. Sodium bicarbonate, also known as

FIGURE 2 Bases will turn red litmus paper blue.

EXAMPLE 13.1

Writing Acid/Base Neutralization Equations

Write a chemical equation to show the neutralization of nitric acid (HNO_3) by potassium hydroxide (KOH).

SOLUTION
All neutralization equations have the form

$$\text{acid} + \text{base} \rightarrow \text{water} + \text{salt}$$

where the salt is composed of the anion from the acid and the cation from the base. So in this case we write

$$\underset{\text{acid}}{HNO_3} + \underset{\text{base}}{KOH} \rightarrow \underset{\text{water}}{H_2O} + \underset{\text{salt}}{KNO_3}$$

YOUR TURN

Writing Acid/Base Neutralization Equations

Write a chemical equation to show the neutralization of sulfuric acid (H_2SO_4) by sodium hydroxide (NaOH).

MOLECULAR FOCUS

Cocaine

Formula: $C_{17}H_{21}NO_4$
Molecular weight: 303 g/mol
Melting point: 98°C
Structure:

Three-dimensional structure:

Cocaine is a white, volatile solid that dissolves in water and acts as a base. It belongs to a family of bases called **alkaloids**, organic compounds of nitrogen that act as bases much as ammonia does. Many alkaloids are poisonous. Cocaine is used in medicine as a topical anesthetic but is better known for its illicit use as a powerful and addictive stimulant. Cocaine acts by heightening the levels of neurotransmitters between nerve cells, producing feelings of euphoria and increased alertness. As the effect wears off, however, neurotransmitters return to normal levels, leaving the user depressed and craving more cocaine.

Cocaine is extracted from the leaves of the South American coca plant. Like other bases, cocaine reacts with acids to form salts. The reaction between cocaine and hydrochloric acid produces cocaine hydrochloride, a nonvolatile salt. In this form, cocaine is extracted from coca plants. Drug users get cocaine hydrochloride into their bloodstreams by snorting it into the nose. The white powder is readily absorbed through the mucous membranes of the nose, producing the desired, but dangerous, effect.

Drug users also abuse a more potent form of cocaine. If cocaine hydrochloride reacts with a base, the salt is converted to pure cocaine, often called *free-base* or *crack* cocaine. In this form, the volatile powder is smoked and reaches the brain in seconds, producing a sharp, intense high. Equally sharp and intense, however, is the depression that follows, leading crack users to crave increasingly more of the free-base form of cocaine.

Cocaine hydrochloride, a salt, is extracted from the leaves of the South American coca plant.

---

baking soda, is a common antacid and is found in many cleaning products. In concentrated forms, many bases are dangerous and will burn the skin on contact. If ingested, concentrated bases damage the mouth, throat, and gastrointestinal tract.

13.4 Acids and Bases: Molecular Definitions

We can define an *acid* as a substance that produces *hydrogen ions* (H^+) in solution and a *base* as a substance that produces *hydroxide ions* (OH^-) in solution. This definition of acids and bases is called the Arrhenius definition after its proponent, Swedish chemist **Svante Arrhenius** (1859–1927). Although it is a useful definition that describes a great deal of acid–base chemistry, it does not apply in

 Explore this topic on the **Acids and Bases** website.

all cases. For example, we said that ammonia (NH$_3$) is a base, yet it does not contain OH$^-$. In the Arrhenius definition, ammonia produces OH$^-$ when it combines with water:

$$NH_3 + H_2O \longrightarrow NH_4^+ + OH^-$$

A second and broader definition of acids and bases, the **Brønsted–Lowry** definition, applies more naturally to NH$_3$ and works in solutions that do not contain water.

The Brønsted–Lowry definition of acids and bases focuses on the *transfer* of H$^+$ ions (protons). When an acid, such as HCl, dissociates in water into a proton and a chlorine ion, the proton does not remain alone, but associates with a water molecule:

$$H:\!\ddot{Cl}: \longrightarrow H^+ + :\!\ddot{Cl}:^-$$

$$H^+ + H:\!\ddot{O}: \longrightarrow H:\!\ddot{O}:^{H^+}$$
$$HH$$

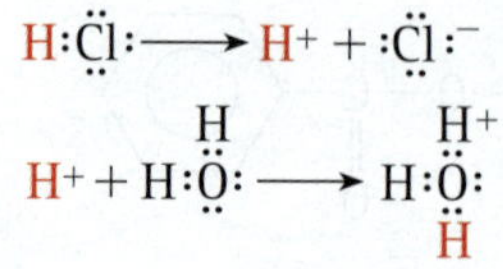

By combining the previous two equations into one, we can see that a proton is transferred from HCl to H$_2$O:

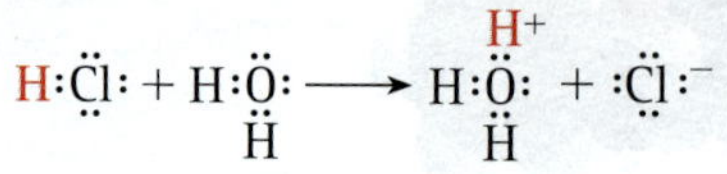

$$H:\!\ddot{Cl}: + H:\!\ddot{O}: \longrightarrow H:\!\ddot{O}:^{H^+} + :\!\ddot{Cl}:^-$$
$$\phantom{H:\ddot{Cl}: + }HH$$
$$\text{acid}\text{base}$$

> Notice that H$^+$ is simply a proton with no electrons. Its reactivity is caused by its attraction to electrons. A base will always have an electron pair that the proton can associate with and therefore can achieve a stable electron configuration.

In the Brønsted–Lowry definition, an *acid is a proton donor* and a *base is a proton acceptor*. In this reaction, HCl acts as an acid and H$_2$O acts as a base. The Brønsted–Lowry definition is broader and allows more substances to be considered acids and bases. For example, ammonia, with its lone electron pair, is a good proton acceptor:

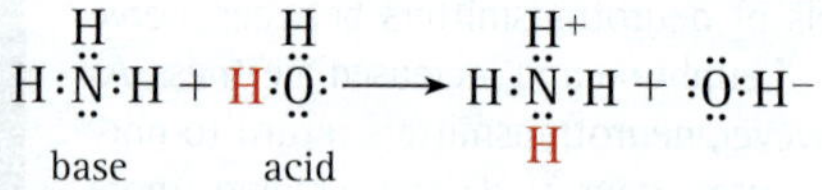

$$H:\!\ddot{N}:\!H + H:\!\ddot{O}: \longrightarrow H:\!\ddot{N}:\!H^{H^+} + :\!\ddot{O}:\!H^-$$
$$\phantom{H:\ddot{N}:H + }HH$$
$$\text{base}\text{acid}$$

Ammonia accepts a proton from water, acting as a base.

> ### EXAMPLE 13.2
> #### Identifying Brønsted–Lowry Acids and Bases
> Identify the Brønsted–Lowry acid and base in the following reaction:
>
> $$CH_3COOH + H_2O \longrightarrow CH_3COO^- + H_3O^+$$
>
> **SOLUTION**
> Because CH$_3$COOH (acetic acid) is the proton donor, it is the acid. H$_2$O is accepting the proton, making it the base.
>
> ### YOUR TURN
> #### Identifying Brønsted–Lowry Acids and Bases
> Identify the Brønsted–Lowry acid and base in the following reaction:
>
> $$C_5H_5N + H_2O \longrightarrow C_5H_5NH^+ + OH^-$$

13.5 Strong and Weak Acids and Bases

As we have seen, hydrochloric acid (HCl) dissociates in water according to the following reaction:

$$HCl + H_2O \longrightarrow H_3O^+ + Cl^-$$

An aqueous solution of HCl has virtually no HCl present; it all dissociates to H_3O^+ and Cl^- (Figure 3). Acids that completely dissociate, such as HCl, are called **strong acids**. Other strong acids include nitric acid (HNO_3) and sulfuric acid (H_2SO_4).

Contrast this behavior to that of acetic acid (CH_3COOH):

$$CH_3COOH + H_2O \rightleftharpoons H_3O^+ + CH_3COO^-$$

The equation looks similar to that of HCl, but there is an important difference: A solution of CH_3COOH contains significant amounts of CH_3COOH, H_3O^+, and CH_3COO^- (Figure 4). Unlike HCl, where all HCl molecules dissociate, a large fraction of CH_3COOH molecules does not dissociate. We indicate this by drawing a

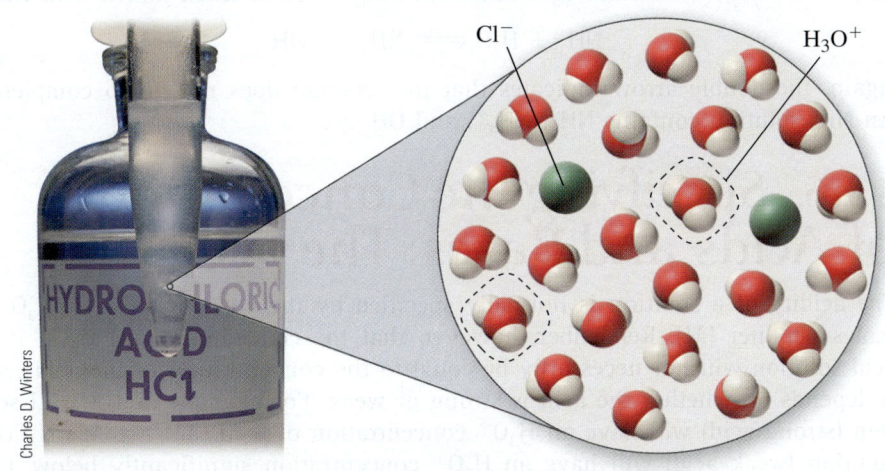

FIGURE 3 An HCl solution contains H_3O^+ and Cl^-, but no HCl. HCl is a strong acid, meaning that all the HCl dissociates into H_3O^+ and Cl^-.

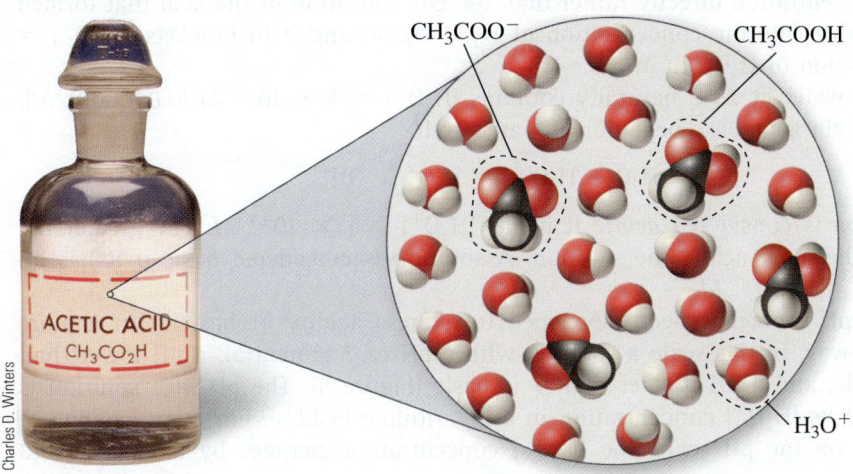

FIGURE 4 A CH_3COOH solution contains CH_3COOH, CH_3COO^-, and H_3O^+. CH_3COOH is a weak acid, meaning that only some of the CH_3COOH molecules dissociate.

TABLE 3
Some Common Weak Acids and Bases

Weak Acids		Weak Bases	
Acetic acid	CH_3COOH	Ammonia	NH_3
Formic acid	$HCOOH$	Ethylamine	$C_2H_5NH_2$
Benzoic acid	C_6H_5COOH	Pyridine	C_5H_5N
Citric acid	$C_3H_5O(COOH)_3$	Aniline	$C_6H_5NH_2$

double arrow in the reaction. Acids that do not completely dissociate, like CH_3COOH, are called **weak acids** (Table 3).

Strong and weak bases are similar, in this respect, to strong and weak acids. A **strong base**, such as NaOH, completely dissociates in solution:

$$NaOH \longrightarrow Na^+ + OH^-$$

A solution of NaOH contains Na^+ and OH^-, but no NaOH. A **weak base**, on the other hand, does not completely dissociate. A good example of a weak base is ammonia (NH_3), which undergoes the following reaction when mixed with water:

$$NH_3 + H_2O \rightleftharpoons NH_4^+ + OH^-$$

Again, the double arrow indicates that the reaction does not go to completion. An NH_3 solution contains NH_3, NH_4^+, and OH^-.

13.6 Specifying the Concentration of Acids and Bases: The pH Scale

The acidity of a solution is normally specified by the concentration of H_3O^+ in moles per liter (M). Remember, however, that the concentration of H_3O^+ in an acid solution will not necessarily be equal to the concentration of the acid itself; it depends on whether the acid is strong or weak. For example, a 1 M HCl solution (strong acid) will have an H_3O^+ concentration of 1 M, but a 1 M CH_3COOH solution (weak acid) will have an H_3O^+ concentration significantly below 1 M. Consequently, we usually specify the acidity of a solution by referring to the H_3O^+ concentration directly rather than the concentration of the acid that formed it. We abbreviate the concentration of H_3O^+ by putting it in brackets: $[H_3O^+]$ = concentration of H_3O^+ in M.

Pure water at 25°C naturally contains $[H_3O^+] = 1 \times 10^{-7}$ M (0.0000001 M), due to slight acid–base behavior in water itself:

$$H_2O + H_2O \rightleftharpoons H_3O^+ + OH^-$$

A solution is considered *acidic* if it has $[H_3O^+] > 1 \times 10^{-7}$ M. The greater the $[H_3O^+]$, the more acidic the solution. A solution is considered *basic* if it has an $[H_3O^+] < 1 \times 10^{-7}$ M.

The **pH scale** has been developed to express acidity or basicity in a more compact way. In this scale a solution with a pH of 7 is neutral, a pH lower than 7 is acidic, and a pH greater than 7 is basic (Figure 5). The pH of a solution is related to its $[H_3O^+]$ concentration in a logarithmic fashion. For every change of one unit on the pH scale, the $[H_3O^+]$ concentration changes by a factor of 10

W Explore this topic on the Interactive pH Monitoring of Acids, Bases, Salts, and Solutions website.

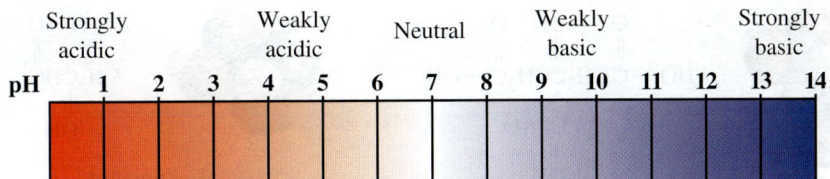

FIGURE 5 The pH scale.

TABLE 4
The Relationship Between pH and H_3^+O Concentration

	[H⁺]	pH	Example
	1.00	0	HCl (1 M)
	1.00×10	1	Stomach acid
	1.00×10^{-2}	2	Lemon juice
Acids	1.00×10^{-3}	3	Vinegar, apples
	1.00×10^{-4}	4	Soda, beer
	1.00×10^{-5}	5	Rainwater
	1.00×10^{-6}	6	Milk
Neutral	1.00×10^{-7}	7	Pure water
	1.00×10^{-8}	8	Egg whites
	1.00×10^{-9}	9	Baking soda solution
	1.00×10^{-10}	10	Milk of magnesia
Bases	1.00×10^{-11}	11	Ammonia
	1.00×10^{-12}	12	Mineral lime solution
	1.00×10^{-13}	13	Liquid Drāno
	1.00×10^{-14}	14	NaOH (1 M)

APPLY YOUR KNOWLEDGE

The ideal pH of a swimming pool is 7.2. You measure the pH of your pool to be 6.5. What should you add, acid or base, to restore your pool to the ideal pH?

Answer: The pH is too low (too acidic). You should therefore add base to neutralize some of the excess acid and bring the pH up.

(Table 4). Thus, a solution with a pH of 3 has $[H_3O^+] = 1 \times 10^{-3}$ M and a solution with a pH of 2 is 10 times more acidic with $[H_3O^+] = 1 \times 10^{-2}$ M.

The lower the pH of a solution, the greater its acidity.

13.7 Some Common Acids

Lemons, with a pH of about 2.0, are extremely sour. The acid responsible for this sourness, as well as that of limes, oranges, and grapefruit, is citric acid. Citrus fruits, like many acidic foods, are resistant to spoilage because many microorganisms cannot survive the low pH environment. Many low-acid foods can be

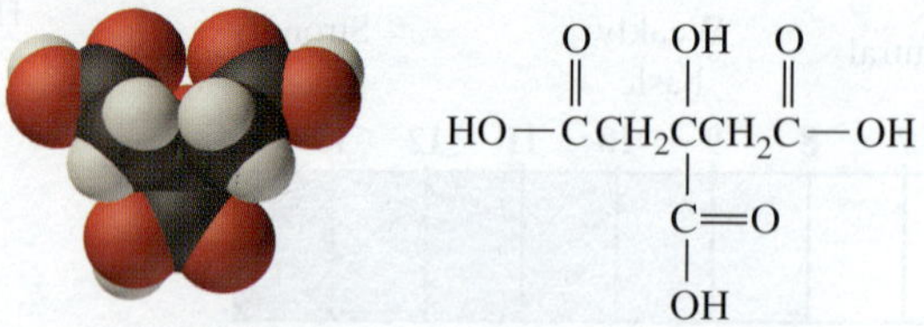

$$\begin{array}{c}
\overset{O}{\underset{\|}{}}OH\overset{O}{\underset{\|}{}}\\
HO-C\,CH_2\,C\,CH_2\,C-OH\\
\underset{\underset{OH}{|}}{\overset{|}{C=O}}
\end{array}$$

Citric acid

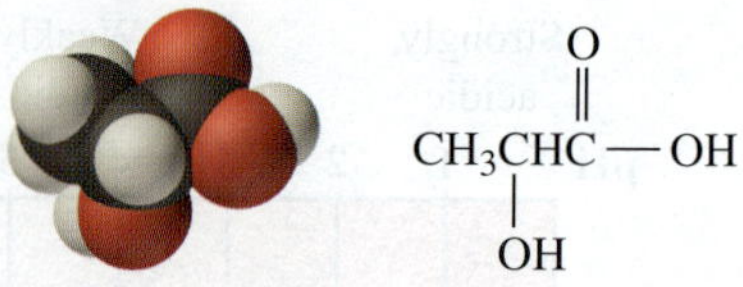

$$CH_3\underset{\underset{OH}{|}}{CH}\overset{O}{\underset{\|}{C}}-OH$$

Lactic acid

Citrus fruits contain citric acid.

made spoilage resistant by storing them in acid, or by fermentation with lactic-acid-forming bacteria. This technique is called pickling and is often used as a way to preserve cucumbers (pickles) and cabbage (sauerkraut).

The tartness of vinegar and therefore of salad dressing is due to acetic acid. Vinegar comes to us from the French words *vin aigre,* meaning sour wine. Acetic acid is formed in wine when the wine is exposed to oxygen for an extended period. The presence of acetic acid in a newly opened wine bottle indicates improper storage; oxygen has leaked into the wine and converted some ethanol into vinegar, making a chardonnay taste like salad dressing. Acetic acid is also produced by certain yeasts that metabolize sugars and produce acetic acid (as well as other acids) as a product. These yeasts are employed in bread making to turn the dough acidic or sour, producing sourdough bread.

$$CH_3\overset{O}{\underset{\|}{C}}-OH$$

Acetic acid

Vinegar is a solution of acetic acid and water.

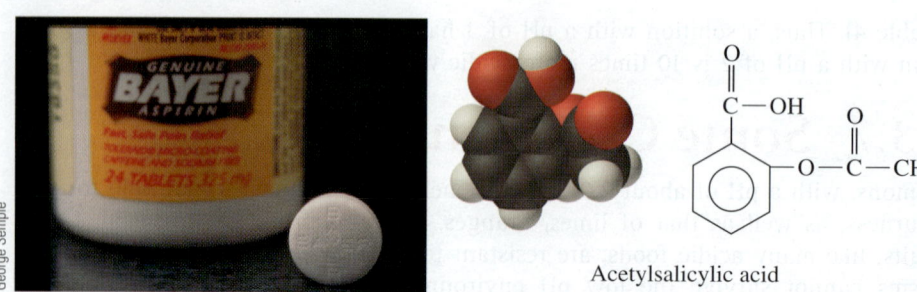

Acetylsalicylic acid is the active ingredient in aspirin.

Acetylsalicylic acid

Salicylic acid, originally obtained from the bark of the willow tree, is the parent molecule for acetylsalicylic acid, or aspirin, the most widely used of all drugs. The sour taste of aspirin is due to its acidic nature, but its medicinal qualities are not. Aspirin works because of its ability to hinder the formation of prostaglandins, compounds involved in the transmission of pain signals through the nervous system and in the production of fevers and inflammation. By hindering the body's ability to make prostaglandins, aspirin relieves all three of these conditions.

Other common acids include hydrochloric acid (HCl), found in relatively high concentrations in the stomach and responsible for the disagreeably sour taste in vomit; phosphoric acid (H_3PO_4), often added to soft drinks and beer to impart tartness; and carbonic acid, produced by the reaction between carbon dioxide and water and therefore present in all carbonated beverages.

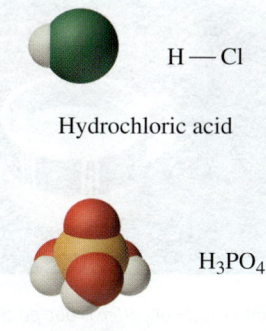

ACIDS IN WINES

All wines contain acid, with concentrations ranging between 0.60 and 0.80% of the total wine volume. Different acids, each with its own characteristic flavor, are responsible for wine's acidity. Part of the art of wine making is achieving a proper balance among these different acids. The acids come from two different sources: the grapes themselves, which naturally contain acids, and the fermentation process, in which bacteria convert sugars into ethyl alcohol and carbon dioxide, often producing or modifying acids in the process.

The citrus flavor in wine is due to citric acid, the same acid found in lemons, limes, and oranges. An apple flavor is found in wines that contain malic acid. The buttery flavor, often desired in chardonnays, is due to lactic acid. This flavor is often enhanced by using a strain of bacteria that converts the grape's malic acid into lactic acid, a process called malolactic fermentation. Wines also contain tartaric acid, a particularly sour acid, and acetic acid, the acid in vinegar. Too much acetic acid is considered a serious fault in most wines.

All wines contain acid.

Citric acid
(citrus flavor)

Malic acid
(apple flavor)

Lactic acid
(butter flavor)

Tartaric acid
(very sour)

Acetic acid
(vinegar flavor)

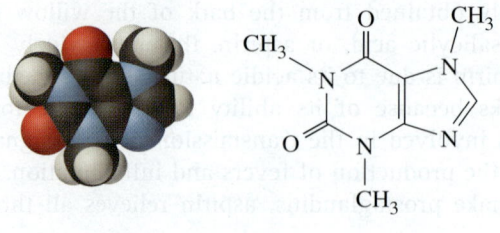
Caffeine

The bitter taste of coffee is due partly to the presence of caffeine, a base.

13.8 Some Common Bases

It is no accident that few of the foods and beverages listed in Table 4 are basic. The often disagreeably bitter taste associated with bases may be an evolutionary adaptation that warns against alkaloids, a family of organic, nitrogen-containing bases that are often poisonous. A few alkaloids, such as caffeine, are consumed for pleasure, but usually these are *acquired* tastes. Other alkaloids include morphine, nicotine, and cocaine.

Bases also constitute the active ingredient in antacids. There are hundreds of different brands of antacids, most of which contain one or more of the following bases: sodium bicarbonate ($NaHCO_3$), calcium carbonate ($CaCO_3$), magnesium carbonate ($MgCO_3$), magnesium hydroxide ($Mg(OH)_2$), and aluminum hydroxide ($Al(OH)_3$). These antacids dissociate in water to produce a metal ion and a base. Sodium bicarbonate, for example, produces sodium ions and basic bicarbonate ions (HCO_3^-) in solution:

Some common antacids.

$$NaHCO_3 \longrightarrow Na^+ + HCO_3^-$$

The bicarbonate ion is a good proton acceptor, making it basic.

MOLECULAR THINKING

Bee Stings and Baking Soda

A home treatment for bee stings is to remove the stinger, wash the wound with soap and water, and then apply a paste of baking soda ($NaHCO_3$) and water. When applied to the wound, this mixture soothes the pain and relieves the stinging sensation.

QUESTION: Since baking soda ($NaHCO_3$) eases the stinging of a bee sting, what do you suppose causes the stinging? Can you write a chemical reaction to show how the baking soda works?

Sodium bicarbonate, sold as baking soda, can be dissolved in water directly and taken as an antacid. It can also be purchased as Alka-Seltzer, a combination of sodium bicarbonate, citric acid, and aspirin. The citric acid reacts with some of the bicarbonate ions to produce gaseous CO_2, the familiar fizz:

$$\underset{\text{base}}{HCO_3^-} + \underset{\text{acid}}{H_3O^+} \longrightarrow CO_2(gas) + 2H_2O$$

This bubbling does not contribute to the antacid's ability to neutralize stomach acid, but it does make for great commercials and feels pleasant when consumed. The rest of the bicarbonate ions neutralize excess stomach acid via the same reaction. Because of its sodium content, people with hypertension (high blood pressure) should avoid this form of antacid.

Calcium carbonate ($CaCO_3$), the active ingredient in Tums, neutralizes stomach acid in a reaction similar to that of sodium bicarbonate:

$$\underset{\text{base}}{CO_3^{2-}} + \underset{\text{acid}}{2H_3O^+} \longrightarrow CO_2(gas) + 3H_2O$$

The calcium present in calcium carbonate is an essential part of good nutrition because calcium prevents osteoporosis, the degeneration of bone. One Tums tablet provides 20% of the recommended daily allowance of calcium. Too much calcium, however, should be avoided because it can cause constipation.

Magnesium hydroxide ($Mg(OH)_2$) is the active ingredient in milk of magnesia. The basic OH^- ions neutralize acid. The magnesium ions, however, have a laxative effect because they absorb water in the lower intestine. In higher doses, milk of magnesia works as a laxative. A number of antacids, such as Mylanta, combine $Mg(OH)_2$ with aluminum hydroxide ($Al(OH)_3$). The OH^- in aluminum hydroxide neutralizes stomach acid, but the aluminum ions have a constipating effect. The combination of magnesium ions (laxative) and aluminum ions (constipating) tends to cancel. Magnesium hydroxide is often used in combination with calcium carbonate to achieve the same result.

Bases are also found in several household cleaning products. Ammonia is a common floor and window cleaner. Sodium hydroxide or lye is the active ingredient in drain-opening products like Drāno. The sodium hydroxide dissolves hair and grease, but it does not dissolve copper or iron pipes the way an acid would.

BAKING WITH ACIDS AND BASES

Many baked goods are made with leavening agents that produce pockets of carbon dioxide gas in the dough. When the dough is baked, these carbon dioxide pockets produce light and fluffy cakes, biscuits, and breads. The ingredient often used to produce the carbon dioxide gas is *baking powder,* which contains sodium bicarbonate ($NaHCO_3$), sodium aluminum sulfate $NaAl(SO_4)_2$, and calcium acid phosphate ($CaHPO_4$). The $NaAl(SO_4)_2$ and $CaHPO_4$ act as acids when combined with water. The H_3O^+ then reacts with the HCO_3^- from the sodium bicarbonate to produce CO_2 and H_2O:

$$HCO_3^- + H_3O^+ \longrightarrow CO_2(gas) + 2H_2O$$

This reaction occurs as the food is baked, producing warm carbon dioxide gas that results in baked goods that can be two or three times larger than the original dough. You can achieve a similar effect outside of cooking dough by combining two other common kitchen products:

A number of household products contain bases.

Baked goods employ acid–base chemistry to produce pockets of CO_2 gas within the dough. These gas pockets produce light and fluffy cakes, biscuits, and breads.

W Explore this topic on the **Acid Rain** website.

baking soda (sodium bicarbonate) and vinegar (acetic acid). The solution will bubble as the acid and base neutralize each other and produce carbon dioxide gas.

Breads use a different, slower, mechanism to achieve leavening. In bread making, living organisms called yeasts are added to the dough. The dough and yeast are allowed to sit in a warm place while the bread rises. During rising, the yeast converts sugar into ethyl alcohol and carbon dioxide. The ethyl alcohol evaporates as the bread is baked, and the CO_2 expands to give light, fluffy bread.

13.9 Acid Rain: Extra Acidity from the Combustion of Fossil Fuels

As we learned in Chapter 9, sulfur and nitrogen impurities present in fossil fuels—especially coal—form gaseous SO_2 and NO_2 during combustion. These gases combine with atmospheric water and oxygen to form sulfuric acid and nitric acid, which then fall as **acid rain**:

$$2SO_2 + O_2 + 2H_2O \longrightarrow 2H_2SO_4$$

$$4NO_2 + O_2 + 2H_2O \longrightarrow 4HNO_3$$

How acidic is acid rain? We might expect pure, unpolluted rain to have a pH of 7.0, but it does not. Rain is naturally slightly acidic because of the presence of carbon dioxide in the atmosphere. Carbon dioxide combines with water to form carbonic acid, a weak acid:

$$CO_2 + H_2O \longrightarrow H_2CO_3$$

The pH of water saturated with CO_2 is about 5.6. The pH of rain in the United States varies from 4.1 to 6.1, more acidic than would be expected from CO_2 alone (Figure 6). The most acidic rainfall occurs in the northeastern United States where pollutants from midwestern coal-burning power plants tend to accumulate (due to the easterly flow of wind currents).

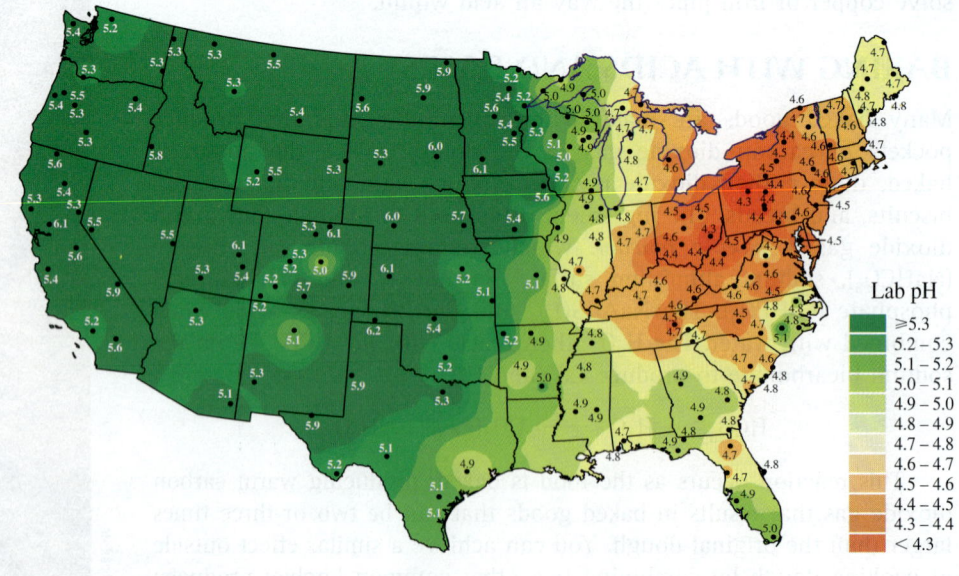

FIGURE 6 Average pH of the precipitation in the United States, 2003. (National Atmospheric Deposition Program (NRSP-3)/National Trends Network (2004). NADP Program Office, Illinois State Water Survey, 2204 Griffith Dr., Champaign, IL 61820.)

Water droplets at the base of clouds, where pollutants reach their highest concentrations, can be extremely acidic with measured pH values as low as 2.6 in the Northeast. Most of the acidity beyond what is expected from CO_2 is attributed to fossil-fuel pollutants; about 10% of the extra acidity is attributed to natural processes such as volcanic eruptions that also emit SO_2.

13.10 Acid Rain: The Effects

The fate of H_3O^+ that falls as acid rain depends on where it lands. Soils and natural waters often contain significant amounts of basic ions, such as bicarbonate (HCO_3^-), that come from rock weathering. If this is the case, the soil or lake can neutralize much of the incoming acid, and the danger is minimized. However, many soils and natural waters do not contain basic ions. These waters, often in otherwise pristine locations, are susceptible to rapid acidification.

The brook trout was formerly found in this Adirondack lake in New York, but, due to acid rain, it cannot survive.

DAMAGE TO LAKES AND STREAMS

A large study of acidity in U.S. lakes and streams, called the National Surface Water Survey (NSWS), concluded that over 2000 lakes and streams in the eastern United States have high acidity levels due to acid rain. In many sensitive lakes and streams, acid levels are high enough that fish species, such as the brook trout, have been eradicated. In the worst cases, acid levels are so high that the entire lake is effectively dead, supporting no marine life. Eastern Canada has also been affected by pollutants from U.S. power plants. The Canadian government estimates that 14,000 lakes in eastern Canada are acidic, at least partly due to U.S. emissions.

DAMAGE TO BUILDING MATERIALS

Acids dissolve metals and other building materials such as stone, marble, and paint. The rusting of steel is accelerated by the presence of acids, resulting in damage to bridges, railroads, automobiles, and other steel structures. Acid rain is also responsible for the damage of many buildings and statues, some of significant historical and cultural importance. As stated earlier, for example, Paul Revere's gravestone has been eroded by acid rain, the marble that composes the Lincoln Memorial is slowly being dissolved, and even the Capitol building is showing signs of acid rain damage.

DAMAGE TO FORESTS AND REDUCED VISIBILITY

Acid rain can also affect forests, lowering the ability of some trees to grow and fight disease. The damage is partly attributable to the direct contact and uptake of acidic rainwater by trees. In addition, as acid moves through the soil, it dissolves and eliminates nutrients important to forest ecosystems. The hardest hit trees are the red spruce trees that grow along the ridges of the Appalachian Mountains from Maine to Georgia.

Sulfur dioxide emissions also lead to decreased visibility, in some cases spoiling scenic vistas. Sulfur dioxide combines with atmospheric moisture to form tiny droplets called sulfate aerosols. According to the Environmental Protection Agency (EPA), these aerosols account for over 50% of the visibility problems in the eastern United States.

Acid rain damages many materials.

Explore this topic on the **Clean Air Act Acid Rain Program** website.

13.11 Cleaning Up Acid Rain: The Clean Air Act Amendments of 1990

In 1990, Congress—convinced by the evidence of acid rain damage—passed several amendments to the Clean Air Act that target acid rain. These amendments require electric utilities to cut their sulfur oxide emissions to half of 1980 levels by the year 2010 (Figure 7).

Utilities have a great deal of flexibility as to how they reduce their sulfur oxide emissions. One way to cut emissions is to use low-sulfur coal, or to remove sulfur from coal before burning it. Sulfur is removed from coal by crushing the coal to a powder and washing it with water, resulting in a 50% sulfur reduction. Another way to cut emissions is by removing the sulfur dioxide from the exhaust. **Flue gas scrubbers** can be installed in power plant smokestacks. These scrubbers spray a mixture of water and limestone into the exhaust, trapping the sulfur dioxide and preventing it from escaping. A third way for utilities to cut emissions is by encouraging and implementing conservation and efficiency programs for its customers.

To further increase flexibility among electric utilities, Congress arranged an "allowance trading system" for sulfur dioxide emissions. Under this system, each

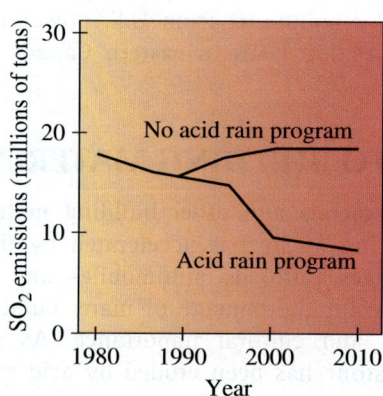

FIGURE 7 EPA projections of annual SO_2 emissions with and without the acid rain program.

WHAT IF...

Practical Environmental Protection

Certain environmental groups, such as the Clean Air Conservancy in Cleveland, Ohio, buy sulfur dioxide allowances and retire them, preventing them from ever being used. For approximately $700, an allowance equivalent to 1 ton of sulfur dioxide can be bought in the open market. Because there are a fixed number of allowances, the retirement of an allowance means that 1 ton less sulfur dioxide is emitted into the air.

QUESTION: What would happen to the price of each allowance if a substantial number of them are retired? How would that affect consumers?

The Molecular Revolution

Neutralizing the Effects of Acid Rain

Scientists have often used technology to undo the negative effects of other technology. In the case of acid rain, the burning of coal, which often contains sulfur impurities, forms sulfur dioxide. The sulfur dioxide reacts with atmospheric oxygen and water to form sulfuric acid, which then precipitates with rain and acidifies lakes and rivers. However, we know that acids can be neutralized with bases. Why not use base neutralization to undo the negative effects of electricity generation?

Scientists have done just that. In a process called liming, scientists add powdered limestone, a natural base, to waters acidified by acid rain. In Norway and Sweden, hundreds of lakes have been successfully restored with liming. In the United States, liming is less common. The problem is that liming is expensive and the effects are only temporary. If acid rain continues to fall, the water becomes acidic again. However, liming may be one way to keep lakes alive until acid rain levels are further reduced by the Clean Air Act.

---

power plant is allocated a number of allowances that represent how much sulfur dioxide it can emit each year. By the year 2010, the number of allowances will be equivalent to a 50% decrease in emissions from the year 1980. However, these allowances can be bought and sold among utilities. For example, a utility close to a mine with low-sulfur coal may decrease its emissions beyond what is required by switching to the cleaner coal. It may then sell its extra allowances to other utilities to pay for the difference in the cost of the coal.

Even though the effects of acid rain have been severe, most scientists believe they can be reversed. Indeed, recent studies showed that many lakes in North America have become less acidic due to SO_2 emission reductions. However, some lakes remain as acidic as ever, leaving some to wonder if the 1990 amendments go far enough in their reduction of SO_2 emissions.

 Explore this topic on the **SO_2 Allowances** website.

Chapter Summary

MOLECULAR CONCEPT

Acids are described by their properties (sour taste, ability to dissolve metals, ability to neutralize bases, ability to turn litmus red) **(13.1, 13.2)** or by their molecular definitions (H^+ producer, proton donor) **(13.4)**. Bases are described by their properties (bitter taste, slippery feel, ability to neutralize acids, ability to turn litmus blue) **(13.3)** or by their molecular definition (OH^- producer, proton acceptor) **(13.4)**. Acids and bases are either strong (complete dissociation in water) or weak (partial dissociation in water) **(13.5)**. The acidity or basicity of a solution is specified using the pH scale in which pH = 7 specifies a neutral solution; pH > 7, a basic solution; and pH < 7, an acidic solution **(13.6–13.8)**.

The combustion of fossil fuels produces nitrogen and sulfur oxides that then combine with oxygen and water to form acid rain. Acid rain causes lakes and streams to become acidic, harming marine life. It also damages building materials and can harm forests **(13.9–13.11)**.

SOCIETAL IMPACT

Acids and bases are integral to our daily lives, to many natural processes, and to the environment. We find acids in a number of foods, including limes, lemons, pickles, and wine. Bases are rarely found in foods except in a few acquired tastes such as coffee. Bases are commonly used as antacids and cleaning agents.

Acid rain is a significant problem in the northeastern United States and Canada. The Clean Air Act amendments of 1990 require coal-burning power plants to reduce their sulfur oxide emissions significantly. As a result, the pH of rain in the northeastern United States and Canada has improved; however, many lakes are still damaged by acid rain. Whether current legislation is enough to restore all of the damage remains to be seen.

Chemistry on the Web

For up-to-date URLs, visit the text website at **www.chemistry.brookscole.com/tro3**.

- Acids and Bases
  http://www.visionlearning.com/library/module_viewer.php?mid=58

- Interactive pH Monitoring of Acids, Bases, Salts, and Solutions
  http://www.chem.iastate.edu/group/Greenbowe/sections/projectfolder/flashfiles/acidbasepH/ph_meter.html

- Acid Rain
  http://www.epa.gov/acidrain/index.html

- Clean Air Act Acid Rain Program
  http://www.epa.gov/airmarkets/cmprpt/arp03/summary.html

- SO_2 Allowances
  http://www.evomarkets.com/emissions/index.php?xp1=so2#mdata

Key Terms

acid rain
acids
alkaloids
Arrhenius, Svante
bases
Brønsted-Lowry acids and bases
flue gas scrubbers
litmus paper
neutralization
pH scale
salt
strong acids
strong base
weak acids
weak base

Exercises

Assess your understanding of this chapter's topics with an online chapter quiz at http://chemistry.brookscole.com/tro3.

QUESTIONS

1. What is the flavor associated with acids in foods?
2. What are the properties of acids?
3. What are the properties of bases?
4. What is litmus paper? What is a litmus test?
5. List five common laboratory acids and their uses.
6. Why are bases not commonly found in foods?
7. List four common laboratory bases and their uses.
8. What are the Arrhenius definitions of acids and bases?
9. What are the Brønsted-Lowry definitions of acids and bases?
10. What is the difference between a strong acid and a weak acid?
11. The pH scale is a logarithmic scale. What is meant by this?
12. What pH range is considered acidic? basic? neutral?
13. What acid is responsible for the sour taste of lemons, limes, and oranges?
14. What is pickling? What acid is responsible for the sour taste of pickles?
15. Where can you find acetic acid?
16. What is aspirin? How does it work?
17. List several common acids and where they might be found.
18. What acids are present in wines? What kind of flavor does each impart to wine?
19. What is an alkaloid?
20. What causes acid indigestion? List some common antacid ingredients and their side effects.
21. How does an antacid work?
22. Explain how a leavening agent works.
23. Which pollutants are responsible for acid rain? Where do they come from?
24. Why is rain acidic even in the absence of pollutants? How acidic is it?
25. How acidic is rain in the United States? Can this acidity be attributed to natural causes?
26. Why can some lakes and soils tolerate acid rain better than others?
27. What are the effects of acid rain on the environment and on building materials?
28. What is being done to decrease the acidity of U.S. rainfall?

PROBLEMS

29. Write a chemical equation to show the neutralization of hydrochloric acid (HCl) by potassium hydroxide (KOH).
30. Write a chemical equation to show the neutralization of sulfuric acid (H_2SO_4) by calcium hydroxide ($Ca(OH)_2$).
31. Identify the Brønsted-Lowry acid and base in each of the following reactions:
    a. $HClO_2 + H_2O \longrightarrow H_3O^+ + ClO_2^-$
    b. $CH_3NH_2 + H_2O \rightleftharpoons OH^- + CH_3NH_3^+$
    c. $HF + NH_3 \rightleftharpoons NH_4^+ + F^-$
32. Identify the Brønsted-Lowry acid and base in each of the following reactions:
    a. $C_5H_5N + H_2O \rightleftharpoons OH^- + C_5H_5NH^+$
    b. $CH_3COOH + H_2O \rightleftharpoons H_3O^+ + C_2H_3O_2^-$
    c. $KOH + HBr \longrightarrow H_2O + KBr$
33. Write a chemical equation using Lewis structures for the reaction of water with ammonia. Identify the Brønsted-Lowry acid and base in the reaction.
34. Write a chemical equation using Lewis structures for the reaction of hydrochloric acid with ammonia. Identify the Brønsted-Lowry acid and base in the reaction.

35. A chemist makes a 0.01 M HCl solution and a 0.01 M CH_3COOH solution. The pH of the HCl solution is found to be 2. Would you expect the pH of the CH_3COOH solution to be greater than or less than 2? Explain.
36. A chemist makes a 0.001 M NaOH solution and a 0.001 M NH_3 solution. The pH of the NaOH solution is 11. Would you expect the pH of the NH_3 solution to be greater than or less than 11? Explain.
37. Give the pH that corresponds to each of the following solutions and classify them as acidic, basic, or neutral:
    a. $[H_3O^+] = 10^{-4}$
    b. $[H_3O^+] = 10^{-7}$
    c. $[H_3O^+] = 10^{-2}$
    d. $[H_3O^+] = 10^{-9}$
38. Give the pH that corresponds to each of the following solutions and classify them as acidic, basic, or neutral:
    a. $[H_3O^+] = 10^{-3}$
    b. $[H_3O^+] = 10^{-11}$
    c. $[H_3O^+] = 10^{-8}$
    d. $[H_3O^+] = 10^{-1}$
39. What is the $[H_3O^+]$ in a solution with a pH of 5?
40. What is the $[H_3O^+]$ in a solution with a pH of 12?
41. Write chemical reactions to show how each of the following antacid ingredients neutralize excess stomach acid:
    a. $NaHCO_3$
    b. $Mg(OH)_2$
42. Write chemical reactions to show how each of the following antacid ingredients neutralizes excess stomach acid.
    a. $CaCO_3$
    b. $Al(OH)_3$
43. Write a chemical reaction to show how SO_2 forms sulfuric acid in the atmosphere.
44. Write a chemical reaction to show how NO_2 forms nitric acid in the atmosphere.

POINTS TO PONDER

45. Write a short paragraph explaining why a person might want to consider purchasing a generic aspirin rather than a name brand. Try to be as persuasive as possible.
46. The phrase "litmus test," which originated in the chemistry laboratory, is often applied to a particular political issue like abortion or the death penalty. These issues become litmus test issues for particular groups in choosing political candidates to support. Their candidate must pass the litmus test if they are to gain their support.

    Suppose you are part of a proenvironment political watchdog group that examines how politicians vote on environmental issues. Use the information from this chapter and from Chapter 9 to determine where you would want your politician to stand on each of the following litmus test issues related to acid rain. Why?
    a. energy consumption taxes
    b. emissions controls for electrical utilities
    c. nuclear power
    d. fusion research
    e. solar energy research
    f. energy conservation measures
47. Suppose you are part of a political watchdog group that represents electrical utilities. Where would you want your politician to stand on the issues listed in problem 46? Why?
48. Examine the household chemicals shelf at your local supermarket or in your home. Identify five items that contain either acids or bases. List the product name and the acid or base that it contains.

Feature Problems and Projects

49. Determine from the following molecular view of a hydrofluoric acid (HF) solution if HF is a strong or a weak acid:

50. Determine from the following molecular view of a hydrobromic acid (HBr) solution if HBr is a strong or a weak acid:

51. Contact your local power company to find out the sources that it uses to generate electricity and the relative percentages of each source. Prepare a bar graph showing the percentages of each source. Which of those sources contribute to acid rain?

52. Find out if your local drinking water is treated for acidity. First, determine the source of your water. If it is a well, ask those in charge of the well if the water is treated for acidity. If it is a local water provider, call and ask them if the water is treated for acidity and if so, how. Ask them to tell you the pH of the water both before and after treatment. Write your results in a short report.

14 Oxidation and Reduction

The action of a beakerful of hot nitric acid on half a pound of sugar is one of the fantastic sights in elementary chemistry and to watch it is an astonishing experience: I have known it to interrupt a nearby game of lawn tennis.

—Theodore H. Savory

In this chapter, we explore an important class of chemical reactions, called oxidation-reduction reactions, which are prevalent in nature, in industry, and in many everyday processes. For example, you breathe so that cells in your body can carry out oxidation-reduction reactions that keep you alive. The rusting of iron, the bleaching of hair, and the disinfecting of a wound are all oxidation-reduction reactions. As you read these pages, try to visualize what happens with atoms and molecules as these familiar reactions happen. Remember, what you see with your eyes in a process such as hair bleaching has a molecular cause—in this case the cause is an oxidation/reduction reaction that permanently changes the color-causing molecules in hair. Oxidation-reduction reactions have significantly impacted society in a number of ways and may change a mainstay of American society—the internal combustion engine.

QUESTIONS FOR THOUGHT

- What is rusting?
- How do you define oxidation and reduction?
- What are some common oxidation and reduction reactions?
- How do batteries work?
- What are fuel cells and how do they work?
- How can you use chemistry to prevent rusting?
- How might oxidation be related to aging?

When a beaker of hot nitric acid is poured on ordinary sugar, a violent oxidation-reduction reaction occurs.

 Explore this topic on the **Rust** website.

Rust results from the oxidation of iron in which iron atoms react with oxygen atoms to form iron oxide (Fe_2O_3) commonly known as rust.

14.1 Rust

The rusting of iron destroys $10 billion worth of steel products each year. When iron is exposed to air and moisture, a chemical reaction occurs that changes the iron into iron oxide, the reddish-brown flakes also known as rust. This common chemical reaction damages pipes, cars, bikes, buildings, bridges, and other iron structures. You may have seen your bike develop reddish-brown spots that grow over the years. This macroscopic observation—the development of the reddish-brown flakes at the expense of the underlying metal—has a molecular explanation. The molecular explanation also accounts for the acceleration of rust under moist or salty conditions. Surprisingly, the principle involved in the molecular explanation of rust is the same one involved in the workings of the common flashlight battery.

The rusting of iron and the function of batteries both involve a particular type of chemical reaction called an **oxidation-reduction** or **redox** reaction. In a redox reaction, electrons are transferred from one substance to another. When iron rusts, electrons transfer from iron atoms, which hold electrons loosely, to oxygen atoms, which hold them tightly. The *oxidized* iron atoms *(iron atoms that lost electrons)* bond with the *reduced* oxygen atoms *(oxygen atoms that gained electrons)* to form iron oxide or rust.

As we will see, the overall chemical reaction that explains rusting also involves water. Consequently, rust is prevented if iron is kept dry. Rusting also requires the conduction of electrical charge. Consequently, rust is accelerated in the presence of salt water, a good electrical conductor.

Redox reactions are responsible for a number of common processes. The burning of hydrocarbons is a redox reaction in which hydrocarbons are oxidized. Respiration is a redox reaction in which our bodies oxidize sugars for energy. The refining of iron is a redox reaction in which iron ores are reduced to extract their iron content, only to oxidize again—in spite of our efforts to prevent it. Paints, metals, and some plastics lose their brilliance over time due to oxidation. Some theories about aging hypothesize that oxidation is the cause. The rusting that occurs in ordinary iron, transforming a shiny new car into an old wreck, may be the same process that changes us from youth to old age.

14.2 Oxidation and Reduction: Some Definitions

We have already encountered redox reactions in this book. Combustion reactions, such as those used for heat or energy production, are actually redox reactions. For example, the burning of coal is a redox reaction:

$$C + O_2 \longrightarrow CO_2$$

The combustion of natural gas is a redox reaction:

$$CH_4 + 2O_2 \longrightarrow CO_2 + 2H_2O$$

Other redox reactions include the rusting of iron:

$$4Fe + 3O_2 \longrightarrow 2Fe_2O_3$$

and the sometimes explosive reaction between hydrogen and oxygen to form water:

$$2H_2 + O_2 \longrightarrow 2H_2O$$

In each of these reactions, a substance reacts with oxygen: In the first reaction it is carbon, in the second reaction it is methane, in the third reaction it is iron, and in the fourth reaction it is hydrogen. Each of these substances reacts with oxygen and becomes oxidized. One definition of *oxidation* is simply the *gaining of oxygen*. If we were to reverse the arrows in these reactions, the elements named would all lose oxygen; they would be reduced. A definition of *reduction* is the *loss of oxygen*.

We can see another characteristic of oxidation and reduction by examining the second reaction in reverse:

$$CO_2 + 2H_2O \longrightarrow CH_4 + 2O_2$$

In this reaction, carbon loses oxygen (reduction) and gains hydrogen. The gaining of hydrogen is characteristic of reduction, and the loss of hydrogen is characteristic of oxidation.

The most general definitions of oxidation and reduction involve—not the transfer of oxygen or hydrogen—but the transfer of electrons. Consider the Lewis structures for the oxidation of sodium:

$$4 \text{Na} \cdot + \; :\ddot{\text{O}}=\ddot{\text{O}}: \; \longrightarrow 2[\text{Na}]^+[:\ddot{\text{O}}:]^{2-}[\text{Na}]^+$$

When sodium and oxygen react to form sodium oxide, each sodium atom transfers an electron to oxygen. The Lewis structures clearly show the loss of an electron (oxidation) by sodium and the gain of an electron (reduction) by oxygen. The broadest definition of oxidation is the *loss of electrons*, and of reduction, *the gain of electrons*. By this definition, reactions that do not involve oxygen may also be redox reactions. Consider the Lewis structures for the reaction between sodium and chlorine:

$$2\text{Na} \cdot + \; :\ddot{\text{Cl}}-\ddot{\text{Cl}}: \; \longrightarrow 2[\text{Na}]^+[:\ddot{\text{Cl}}:]^-$$

When sodium and chlorine react to form sodium chloride, each sodium atom transfers an electron to a chlorine atom. Sodium is oxidized and chlorine is reduced.

When ionic bonds form, as in Na_2O or $NaCl$, the transfer of electrons is complete. However, redox reactions need not involve ionic bonding. In covalent bonding the transfer is only partial, but the same definitions apply. Those atoms that lose electrons, even if only partially, are oxidized and those that gain electrons are reduced.

We will see that oxidation and reduction both have several definitions, all of which are useful.

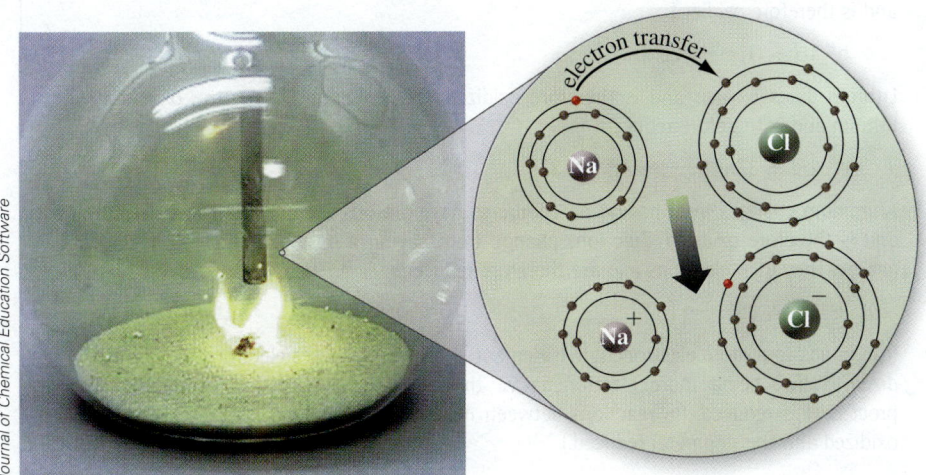

When sodium and chlorine react to form sodium chloride, each sodium atom transfers an electron to a chlorine atom. Sodium loses an electron and is therefore oxidized; chlorine gains an electron and is therefore reduced.

Because electron loss by one substance requires electron gain by another, *oxidation and reduction always occur together*. In a redox reaction, one substance is oxidized and another reduced. Substances such as elemental oxygen or chlorine, which tend to gain electrons easily, are called **oxidizing agents**; they cause the oxidation of other substances while they are themselves reduced. Similarly, substances like elemental sodium or potassium, which tend to lose electrons easily, are called **reducing agents**; they cause the reduction of other substances while they are themselves oxidized.

In summary:

Oxidation is defined as

- the gain of oxygen
- the loss of hydrogen
- the loss of electrons (the most fundamental definition)

The substance being oxidized is the reducing agent.

Reduction is defined as

- the loss of oxygen
- the gain of hydrogen
- the gain of electrons (the most fundamental definition)

The substance being reduced is the oxidizing agent.

Oxidation and reduction *must occur together*; one cannot occur without the other.

EXAMPLE 14.1

Identifying Substances Being Oxidized and Reduced

For the following redox reactions, identify the elements being oxidized and those being reduced:

    a. $2NO + 5H_2 \longrightarrow 2NH_3 + 2H_2O$

Nitrogen (in NO) gains hydrogen and is therefore reduced. Hydrogen (in H_2) gains oxygen and is therefore oxidized.

    b. $4Li + O_2 \longrightarrow 2Li_2O$

Lithium gains oxygen and is therefore oxidized. Oxygen brings about the oxidation of lithium and is therefore reduced.

    c. $2Al + 3Zn^{2+} \longrightarrow 2Al^{3+} + 3Zn$

Aluminum changes from having no charge to having a positive charge. It loses electrons and is therefore oxidized. Zinc ions change from having a positive charge to having no charge. They gain electrons and are therefore reduced.

    d. $2Fe + 3S \longrightarrow Fe_2S_3$

Iron bonds to a more electronegative element and therefore loses electrons to it. It is oxidized. Sulfur, being more electronegative than iron, gains electrons in the bonding process. It is reduced. (In reactions between metals and nonmetals, the metal is usually oxidized and the nonmetal reduced.)

> **YOUR TURN**
>
> ## Identifying Substances Being Oxidized and Reduced
>
> For the following redox reaction, identify the element being oxidized and that being reduced:
>
> $$CuO + H_2 \rightarrow Cu + H_2O$$

> **EXAMPLE 14.2**
>
> ## Identifying Oxidizing and Reducing Agents
>
> For the following redox reactions, identify the oxidizing and reducing agents:
>
> a. $2NO_2 + 7H_2 \rightarrow 2NH_3 + 4H_2O$
>
> H_2 is oxidized and is therefore the reducing agent. Nitrogen in NO_2 gains hydrogen and is therefore reduced. NO_2 is the oxidizing agent.
>
> b. $2K + Cl_2 \longrightarrow 2KCl$
>
> Chlorine (Cl_2), with its high affinity for electrons, is reduced in this reaction; therefore, it is the oxidizing agent. Potassium (K), with its tendency to lose electrons, is oxidized in this reaction; therefore, it is the reducing agent.
>
> **YOUR TURN**
>
> ## Identifying Oxidizing and Reducing Agents
>
> Identify the oxidizing and reducing agents in the following reaction:
>
> $$V_2O_5 + 2H_2 \longrightarrow V_2O_3 + 2H_2O$$

> ## MOLECULAR THINKING
>
> ### The Dulling of Automobile Paint
>
> New cars have brilliant, shiny paint. Over time, however, new paint often fades and loses its shine. The dulling of paint is caused, at least in part, by the oxidation of molecules within the paint by atmospheric oxygen. The dulling of paint can be prevented by the periodic application of wax to the paint. Automobiles that are regularly waxed may keep their new car shine for many years.
>
> **QUESTION:** Why does wax prevent oxidation of paint? Can you give a molecular reason for this?
>
>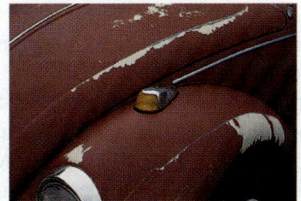
>
> *The dulling of paint over time is caused by a redox reaction.*

14.3 Some Common Oxidizing and Reducing Agents

Many antiseptics and disinfectants are oxidizing agents.

The most common oxidizing agent is oxygen, composing—as we learned in Chapter 11—approximately 20% of air. It is responsible for a number of everyday redox reactions such as the burning of wood in a fire or the combustion of gasoline in an automobile engine. The oxidation of carbon-containing materials always produces energy. Our bodies use oxygen to oxidize sugars for energy through respiration.

Oxidizing agents kill microorganisms and are often used as **antiseptics** and **disinfectants**. Antiseptics are applied to the skin or to minor cuts to prevent infection. Common antiseptics include hydrogen peroxide (H_2O_2), benzoyl peroxide [$(C_6H_5CO)_2O_2$], and iodine (I_2). Disinfectants are used to sterilize and sanitize. Common disinfectants include chlorine (Cl_2), used in the treatment of drinking water, and household bleach, an aqueous solution of sodium hypochlorite (NaOCl).

Because they can oxidize many color-causing molecules, oxidizing agents are often used for bleaching of fabrics, paper, and food. Chlorine (Cl_2) and hydrogen peroxide (H_2O_2) are both used as bleaches. Household bleach is a 5.25% aqueous solution of sodium hypochlorite (NaOCl), a strong oxidizing agent.

Reducing agents are less common in household products but are important in many industrial processes. The simplest and most common reducing agent is prob-

Molecular Focus

Hydrogen Peroxide

Formula: H_2O_2
Molecular weight: 34.02 g/mole
Melting point: $-0.43°C$

Three-dimensional structure:

Boiling point: 152°C
Structure:

$$H—\ddot{O}—\ddot{O}—H$$

Hydrogen peroxide is a colorless, rather unstable liquid, usually marketed as a 3% solution in water. It is a powerful oxidizing agent and oxidizes many organic molecules, especially pigments and dyes. Consequently, hydrogen peroxide is an excellent bleach for hair, food, flour, and silk and other textiles.

Hydrogen peroxide bleaches hair by oxidizing melanin, the molecule responsible for brown and black hair. Melanin, like most pigments, contains alternating single and double carbon-carbon bonds. Hydrogen peroxide severs the double bonds and inserts one of its own oxygen atoms between the carbon atoms. Once the double bonds are disrupted, the melanin loses its color.

Some toothpastes also contain small amounts of hydrogen peroxide with claims that they help whiten teeth. However, it is unclear whether the tiny amounts of hydrogen peroxide in toothpaste actually accomplish anything. It does produce tiny bubbles that makes it feel like something is happening, but whether any whitening actually occurs is yet to be demonstrated.

Hydrogen peroxide is also used as an antiseptic. When applied to the skin or a minor cut, hydrogen peroxide kills microorganisms that might otherwise cause infection.

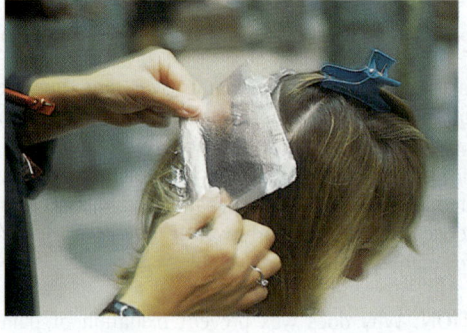

Hydrogen peroxide's ability to oxidize melanin, the color pigment in hair, makes it popular as a hair bleach.

ably hydrogen, used in the reduction of atmospheric nitrogen (nitrogen fixation) into ammonia by the **Haber process**:

$$3H_2 + N_2 \longrightarrow 2NH_3$$

Nitrogen gains hydrogen to form ammonia and is therefore reduced. The synthesis of ammonia from gaseous nitrogen is important because ammonia is a starting material in the manufacture of several important chemical products, including fertilizers and explosives.

Another important reducing agent is elemental carbon, also called coke, used in the reduction of metal ores. For example, carbon is used to reduce copper oxide to copper:

$$Cu_2O + C \longrightarrow 2Cu + CO$$

Carbon monoxide is also used in metal processing; it is used to reduce iron ore (iron oxide) to iron:

$$Fe_2O_3 + 3CO \longrightarrow 2Fe + 3CO_2$$

Elemental carbon, also called coke, is used in the reduction of metal ores.

14.4 Respiration and Photosynthesis

Animal and plant life survives through a set of redox reactions called respiration and photosynthesis. Animals, including humans, use **respiration** to oxidize food for energy. The complete digestion of food requires several steps, but energy comes from the oxidation of glucose ($C_6H_{12}O_6$) to form carbon dioxide and water:

$$C_6H_{12}O_6 + 6O_2 \longrightarrow 6CO_2 + 6H_2O \quad \text{(respiration)}$$

Notice that carbon *gains* oxygen (1 oxygen atom to every carbon atom on the left and 2 oxygen atoms to every carbon atom on the right) and *loses* hydrogen; carbon is *oxidized* in respiration. Respiration is exothermic and provides the energy that animals require to live.

Photosynthesis is the process by which plants form glucose; it is the reverse of respiration. In photosynthesis, plants reduce carbon to form glucose and oxygen:

$$6CO_2 + 6H_2O \xrightarrow{\text{light}} C_6H_{12}O_6 + 6O_2 \quad \text{(photosynthesis)}$$

Here carbon *loses* oxygen and *gains* hydrogen; carbon is *reduced*. This reaction is endothermic, absorbing energy provided by sunlight to produce glucose. In this way, nature captures the Sun's energy and stores it in chemical bonds. When animals eat plants, respiration releases that energy to be used by the animal.

Thus animals and plants depend on each other for life. Animals use oxygen to oxidize carbon and plants use water to reduce it.

14.5 Batteries: Making Electricity with Chemistry

When a redox reaction happens in a beaker or in air, the molecules or atoms are in direct contact with each other; electrons move directly from one molecule or atom to another. Batteries are based on the natural tendency for some substances to transfer electrons to others. By physically separating two substances that

FIGURE 1 (a) A copper nitrate solution that contains Cu^{2+} (blue), a zinc nitrate solution that contains Zn^{2+}, a strip of zinc metal, and a strip of copper metal. (b) If the zinc is placed into the copper solution, copper ions are immediately reduced to solid copper, which appears as a dark film on the zinc metal. Unseen is the dissolution of zinc metal to form zinc ions. (c) After a few hours a solid mass of copper becomes visible, and the blue color—originally due to copper ions—in the solution has faded. What has replaced the copper ions in solution? (d) If the copper strip is placed in a zinc solution, no reaction occurs.

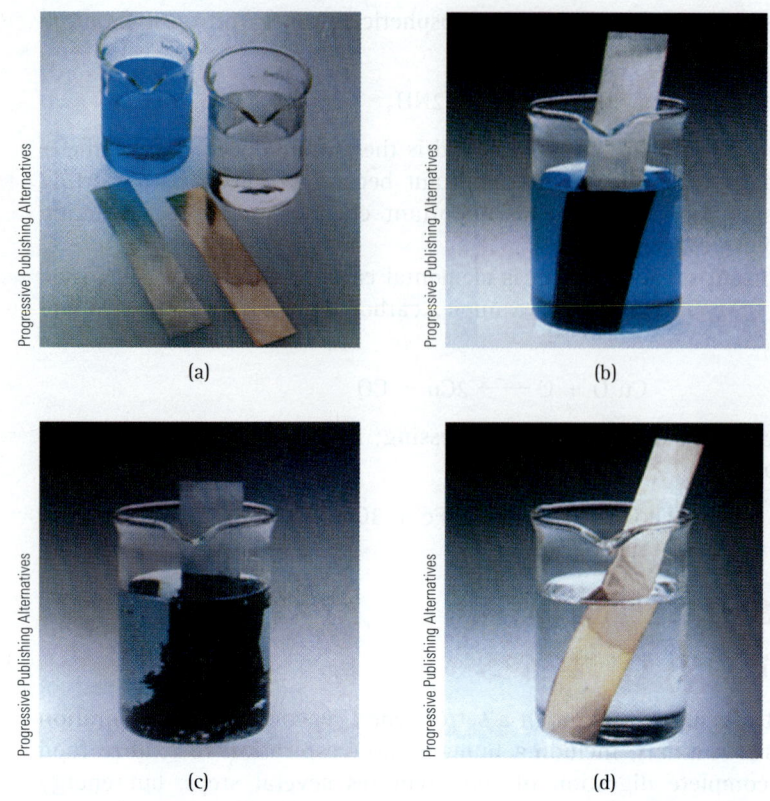

(a) (b) (c) (d)

W Explore this topic on the **Batteries** website.

undergo a redox reaction, the electrons can be forced to travel through an external circuit as an electric current.

For example, if zinc metal is immersed in a solution containing copper ions, a redox reaction occurs (Figure 1). The zinc loses electrons (oxidation) and the copper ions gain electrons (reduction):

$$Zn + Cu^{2+} \longrightarrow Zn^{2+} + Cu$$

The zinc dissolves into the solution as it changes from solid zinc metal to zinc ions, and the copper deposits as it changes from dissolved copper ions to solid copper.

A battery can be constructed from the preceding reaction by physically separating the zinc metal and the copper ions, so that the electrons' only path from the zinc to the copper is through a wire. This is done with an **electrochemical cell** (Figure 2). In this cell, a strip of zinc metal is immersed in a solution of zinc nitrate (Zn^{2+} ions and NO_3^- ions). A wire connects the zinc metal, through a light bulb or other electrical circuit, to a strip of copper metal immersed in a solution of copper nitrate (Cu^{2+} ions and NO_3^- ions). The two solutions are physically separated using a salt bridge.

Because Cu^{2+} ions have a greater affinity for electrons than zinc atoms do, electrons flow from the zinc, through the wire—forming an electrical current—and to the copper, where they combine with Cu^{2+} ions in solution. The Cu^{2+} ions deposit on the copper electrode as solid copper. As the reaction proceeds, the copper electrode gains mass, and the solution is depleted of copper ions. At the same time, zinc atoms lose electrons at the zinc electrode; they dissolve and

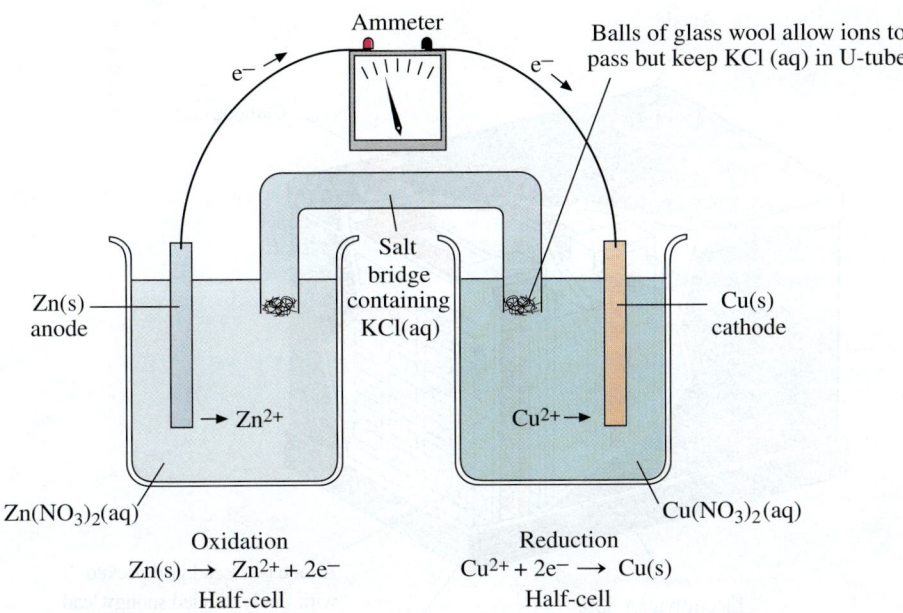

FIGURE 2 An electrochemical cell. In this cell, electrons flow from the zinc anode to the copper cathode. Zinc atoms are oxidized to Zn^{2+} ions and Cu^{2+} ions are reduced to copper atoms.

appear as Zn^{2+} ions in solution. As the reaction proceeds, the zinc electrode loses mass, and the solution develops more Zn^{2+} ions. Meanwhile, NO_3^- ions flow through the salt bridge, from the copper side to the zinc side, to correct the charge imbalance that develops as negatively charged electrons flow from the zinc to the copper.

By physically separating the reactants, we have directed the electrons to flow through a wire, producing an electrical current. By definition, the electrode where oxidation occurs is called the **anode** of the cell and is marked (−), while the electrode where reduction occurs is called the **cathode** and is marked (+). Electrons always flow from the anode (−) to the cathode (+).

We can now see why batteries have limited lifetimes. As the redox reaction proceeds in the cell just described, the reactants deplete; the zinc atoms oxidize and the copper ions reduce. Over time, the zinc electrode dissolves away, and the Cu^{2+} solution is depleted; the battery no longer produces electric current and must be discarded. In a rechargeable battery, the recharge cycle uses an external source to force electrons to travel in the opposite direction. The reaction goes in reverse and regenerates the reactants.

AUTOMOBILE BATTERIES

The battery in an automobile is a lead–acid battery that works much like the simple battery just described. The anode consists of porous lead plates in sulfuric

APPLY YOUR KNOWLEDGE

Closely examine Figure 1 and explain why no reaction occurs in part d.

Answer: The natural tendency, as you can see from parts a–c, is for Zn to *lose electrons* to Cu^{2+}. In part d, copper is present as Cu and zinc as Zn^{2+}; copper will not lose electrons to Zn^{2+} (the opposite of the natural tendency observed in parts a–c).

FIGURE 3 A lead–acid storage battery.

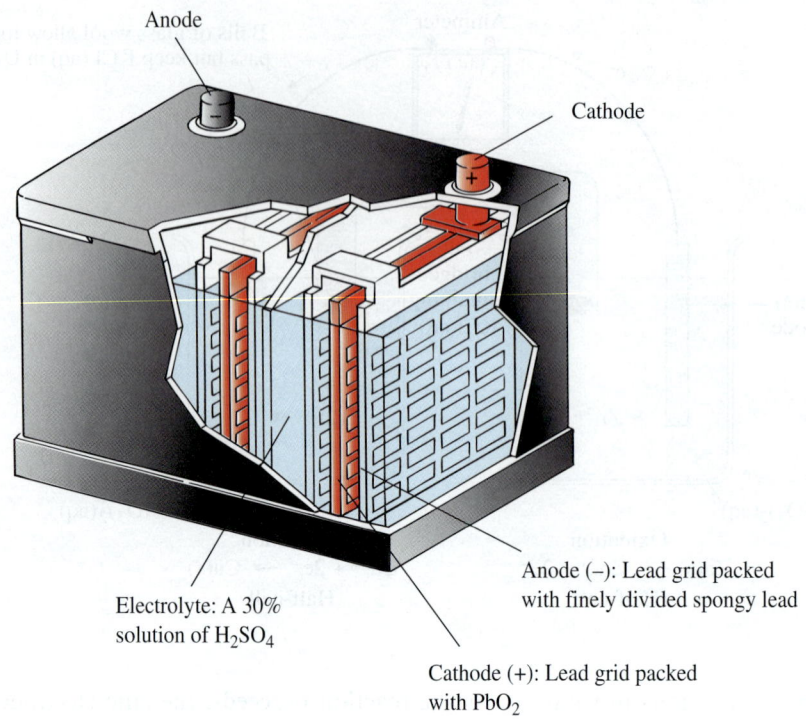

acid (H_2SO_4) solution (Figure 3). At the anode, lead is oxidized according to the following reaction:

$$Pb + SO_4^{2-} \longrightarrow PbSO_4 + 2e^- \quad \text{(oxidation)}$$

The cathode is composed of lead oxide also in sulfuric acid where reduction occurs according to the following reaction:

$$PbO_2 + 4H^+ + SO_4^{2-} + 2e^- \longrightarrow PbSO_4 + 2H_2O \quad \text{(reduction)}$$

The overall reaction is simply the sum of the oxidation and reduction reactions:

$$Pb + SO_4^{2-} \longrightarrow PbSO_4 + 2e^- \quad \text{(oxidation)}$$
$$PbO_2 + 4H^+ + SO_4^{2-} + 2e^- \longrightarrow PbSO_4 + 2H_2O \quad \text{(reduction)}$$
$$\overline{Pb + PbO_2 + 4H^+ + 2SO_4^{2-} \longrightarrow 2PbSO_4 + 2H_2O \quad \text{(overall reaction)}}$$

As the reaction proceeds, both electrodes become coated with lead sulfate, and the solution becomes depleted of sulfate ions; the battery becomes discharged. To charge it, an external charging source forces electrical current in the opposite direction. Electrons flow *to* the lead plates, converting lead sulfate back to lead and sulfate ions, and *away from* the lead oxide, converting that lead sulfate back into lead oxide and sulfate ions. In this way the battery can be charged and discharged thousands of times.

THE COMMON FLASHLIGHT BATTERY

The inexpensive flashlight batteries sold in retail stores use a design called a Leclanché dry cell, diagrammed in Figure 4. The body of the battery is made of zinc, which acts as the anode. The zinc is oxidized according to the following reaction:

$$Zn \longrightarrow Zn^{2+} + 2e^- \quad \text{(oxidation)}$$

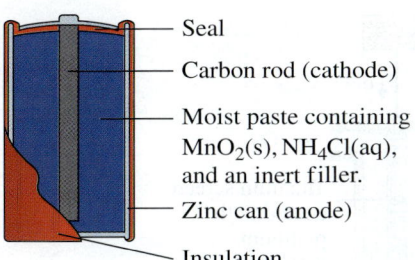

FIGURE 4 Cross section of an inexpensive flashlight battery.

The reduction reaction is more complex. A carbon rod acts as the cathode, but the substance being reduced is MnO_2, which is mixed into a moist paste in the interior of the battery. The MnO_2 is reduced to Mn_2O_3 in the following reaction:

$$2MnO_2 + H_2O + 2e^- \longrightarrow Mn_2O_3 + 2OH^- \text{ (reduction)}$$

Notice that Mn_2O_3 has fewer oxygen atoms per manganese atom than MnO_2, characteristic of reduction. The overall reaction can be obtained by summing the oxidation and reduction reactions:

$$\begin{aligned}
Zn &\longrightarrow Zn^{2+} + 2e^- &\text{(oxidation)}\\
2MnO_2 + H_2O + 2e^- &\longrightarrow Mn_2O_3 + 2OH^- &\text{(reduction)}\\
\hline
Zn + 2MnO_2 + H_2O &\longrightarrow Zn(OH)_2 + Mn_2O_3 &\text{(overall)}
\end{aligned}$$

The more expensive alkaline batteries use a slightly different reaction that employs a base, hence the name alkaline, in the oxidation of zinc. Alkaline batteries have a longer shelf life, longer battery life, and are less subject to corrosion than the standard Leclanché cell.

14.6 Fuel Cells

Fuel cells are batteries with continually renewed reactants. In a normal battery, the reactants deplete over time. In a fuel cell, the reactants flow into the cell as needed to produce electricity, and the products flow out. Fuel cells have great potential as future energy sources, both on a small scale to power electric cars or buses, and on a large scale to produce power grid electricity.

The efficiency of fuel cells is related to thermodynamics. Recall that for every energy conversion step, a certain amount of energy—the heat tax—is always lost to the surroundings. When electricity is produced from the combustion of fossil fuels, three steps are involved: Combustion converts chemical energy to heat energy; steam converts heat energy to mechanical energy; and a generator converts mechanical energy to electrical energy. In an electrochemical reaction, chemical energy is converted directly into electrical energy, eliminating the intermediate steps. Consequently, fuel cell efficiency is high, ranging from 50 to 80%; in contrast coal-burning power plants are about 40% efficient.

One of the simplest fuel cells is the hydrogen–oxygen fuel cell (Figure 5). Hydrogen and oxygen gases flow past porous electrodes that allow the gases to contact a potassium hydroxide (KOH) solution. At the anode, hydrogen gas is oxidized to water:

$$2H_2 + 4OH^- \longrightarrow 4H_2O + 4e^- \quad \text{(oxidation)}$$

At the cathode, oxygen gas is reduced to hydroxide ions:

$$O_2 + 2H_2O + 4e^- \longrightarrow 4OH^- \quad \text{(reduction)}$$

W Explore this topic on the **Fuel Cell** website.

FIGURE 5 The hydrogen–oxygen fuel cell used in the U.S. space program.

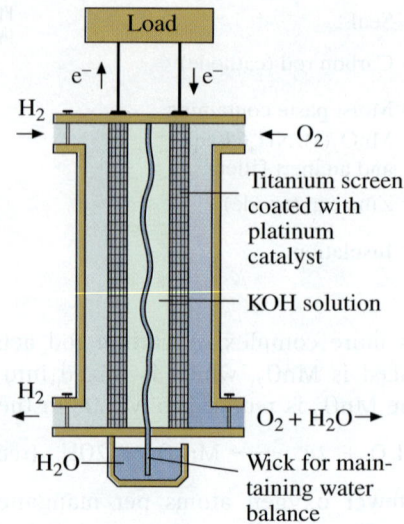

An MCFC fuel cell, like this one, can produce power grid electricity.

The amount of OH^- produced in the reduction is equal to the amount used in the oxidation. The overall reaction is simply the combination of hydrogen and oxygen to form water:

$$2H_2 + O_2 \longrightarrow 2H_2O \quad \text{(overall reaction)}$$

The space program uses the hydrogen-oxygen fuel cell for space vehicle electricity generation. In manned space flights, astronauts consume the only product of the reaction, water. Hydrogen–oxygen fuel cells are currently the fuel cell of choice for the automotive industry (see The Molecular Revolution box entitled "Fuel Cell Vehicles").

A number of fuel cells, currently in the development stage, may soon compete with fossil fuel electricity production. Several companies have developed fuel cell power plants that generate up to several megawatts of power. The most promising of these is called the molten carbonate fuel cell (MCFC), which uses potassium carbonate as the electron transfer medium and methane gas as the fuel. In an initial process called *reforming*, the methane gas reacts with water to form carbon dioxide and hydrogen gas:

$$CH_4 + 2H_2O \longrightarrow CO_2 + 4H_2$$

The hydrogen gas is then oxidized in the anode of the fuel cell:

$$4H_2 + 4CO_3^{2-} \longrightarrow 4CO_2 + 4H_2O + 8e^- \quad \text{(oxidation)}$$

Oxygen is reduced in the cathode:

$$4CO_2 + 2O_2 + 8e^- \longrightarrow 4CO_3^{2-} \quad \text{(reduction)}$$

The overall reaction is

$$CH_4 + 2O_2 \longrightarrow CO_2 + 2H_2O \quad \text{(overall reaction)}$$

The electrical generation efficiency of the MCFC ranges between 54 and 85% depending on how much secondary heat is recaptured and reused in power generation. One of the most attractive aspects of MCFCs is their relatively small size when compared with conventional hardware for electricity generation. A standard semitruck can easily transport a 1-MW MCFC. A disadvantage of MCFCs is their dependence on methane, a fossil fuel. Even though their efficiency is high and their emissions of sulfur and nitrogen oxides are low, the other problems

THE MOLECULAR REVOLUTION

Fuel Cell Vehicles

The internal combustion engine is noisy, dirty, and inefficient; like the coal-burning steam engine, it may be on its way to obsolescence. What was inconceivable just a few years ago is now becoming possible—a shift from the internal combustion engine. Exactly what will replace it remains to be seen, but fuel cells are a front-runner. They are quiet, clean, and highly efficient. Instead of explosively burning petroleum to produce energy, fuel cells quietly combine hydrogen and oxygen to produce energy—no sparks, no explosions, and no emissions except water. The electricity produced by the fuel cell powers a quiet electric motor that propels the car forward.

All major automobile manufacturers are either developing or have developed fuel cell prototype vehicles. In 2003, Daimler Chrysler introduced the Mercedes Benz A-class "F-cell" into practical field testing with 60 vehicles in small fleets within Europe, the United States, Japan, and Singapore. The F-cell runs on compressed hydrogen gas and emits only water. It has a range of 150 km, a top speed of 140 kilometers per hour (km/h), and accelerates from 0 to 100 km in 16 seconds.

Mercedes Benz A-class F-cell.

associated with fossil fuel use, such as their limited supply and carbon dioxide production, remain. The MCFC fuel cell, however, can be modified to accommodate other carbon- and hydrogen-containing fuels like ethanol or carbon dioxide and hydrogen. The challenge, of course, is to find cheap and renewable sources for these fuels.

14.7 Corrosion: The Chemistry of Rust

The rusting of iron, like the reaction occurring in batteries, is a redox reaction. In rusting, iron is oxidized according to the following reaction:

$$2Fe \longrightarrow 2Fe^{2+} + 4e^- \quad \text{(oxidation)}$$

Oxygen, from air, is reduced:

$$O_2 + 2H_2O + 4e^- \longrightarrow 4OH^- \quad \text{(reduction)}$$

Water is an important component in the reaction. We know from experience that iron rusts when wet or when exposed to moist air. One way to prevent rust is to keep iron dry. The overall reaction for the oxidation of iron is

$$2Fe + O_2 + 2H_2O \longrightarrow 2Fe(OH)_2 \quad \text{(overall reaction)}$$

$Fe(OH)_2$ then reacts via several steps to form iron oxide (Fe_2O_3), the familiar orange substance we normally call rust. Unlike iron, which has structural integrity, iron oxide is crumbly. It is a powder that flakes from the metal, exposing new metal, which is then susceptible to further rusting. If the conditions are right, entire pieces of iron will rust away.

Rusted iron.

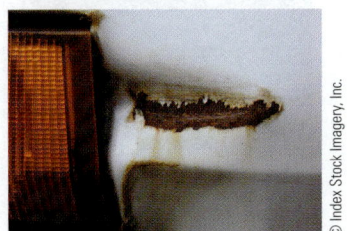

Paint will protect iron by forming a barrier to moisture and oxygen; however, a chip in the paint will result in rusting.

WHAT IF...

The Economics of New Technologies and Corporate Handouts

When start-up companies develop new products, they must often survive many years of negative profitability. Investors put money into a start-up company based on the company's future potential to generate profit. Companies developing new energy technologies, such as fuel cells or batteries, for instance, often find it difficult to attract enough investors to get through the development stage. There are many reasons for this, such as the immense technical difficulties that must be overcome to bring these products to market, or the lack of infrastructure for alternate energy technologies. For example, suppose a company was trying to develop a hydrogen fuel-cell automobile. It would never sell in the existing market because there are no roadside stations that sell hydrogen for refueling. Consequently, the federal government often funds companies developing new energy technologies. One publicly traded company developing fuel cell technology has operated at a profit for several years, even though it has no product to sell. Where does its income come from? The federal government provides it. We, as taxpayers, contribute to the profit of these companies.

Some believe that these types of corporate handouts are unjustified. They think that these companies should compete on the open market just like everyone else. If their product is good enough, they argue, it will sell and generate profit. Others, however, believe the hurdles to developing alternate energy sources are so high and the benefits of their development so great that additional help is justified. The government, they argue, currently subsidizes *fossil-fuel technology* by absorbing much of the costs associated with the environmental damage of fossil fuels or by fighting wars that guard oil interests. In their view, the government's indirect subsidies to fossil fuels far outweigh subsidies given to new technology companies.

QUESTION: What do you think? What if the federal government actually billed oil companies for maintaining stability in the Middle East or for acid rain damage? How would that influence oil company profitability? Should the government sustain the profits of new energy technology companies? What if these companies never generate a profit independently, but simply grow fat from government subsidies?

FIGURE 6 Zinc wire is attached to this underground pipe at intervals of 500–1000 ft. The zinc loses electrons more easily than the iron and is therefore oxidized instead of the iron.

Since one-fifth of the iron produced in the United States goes to replace rusted iron, the prevention of rust is a robust industry. Although no rust prevention system is completely effective, the rusting of iron can be temporarily halted or slowed. The simplest way to prevent rust is to cover iron with paint. The paint excludes water and acts as a barrier between air and the iron; because the redox reaction requires airborne oxygen and water, the metal is protected. The paint will only work, however, as long as it remains intact. Any break or nick in the paint will expose the iron and lead to rust.

Another way to protect iron from rusting, often used in underground pipes, is to attach a more active metal to it (Figure 6). The more active metal has a higher tendency to give up electrons than iron does. Zinc and magnesium are common choices because they are stable in air yet have a strong tendency to lose electrons. The active metal protects the iron because it oxidizes in place of the iron. Eventually, much of the active metal oxidizes and needs replacing. However, as long as the active metal remains, the iron is protected.

The rusting of iron can also be prevented by mixing or coating the iron with another metal whose oxide is structurally stable. Many metals—such as aluminum, for example—will oxidize in air much as iron does but form structurally stable oxides. The corrosion resistance of aluminum cans testifies to the structural stability of aluminum oxide (Al_2O_3). The aluminum oxide forms a tough film that protects the underlying metal from further oxidation. For iron, zinc is often used as a coating in a process called **galvanization**. Because zinc is more active than iron, it will oxidize instead of the underlying iron. The zinc oxide then forms a protective coat, preventing further oxidation. One advantage of this technique is that the zinc coating prevents oxidation of the underlying iron even if the coating becomes nicked or cracked. Zinc's greater tendency to lose electrons means that as long as some zinc is in contact with the iron, it will oxidize and the iron will remain intact.

14.8 Oxidation, Aging, and Antioxidants

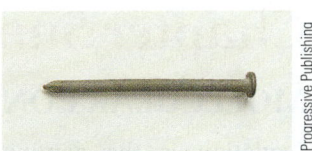

A galvanized nail has a thin layer of zinc on its surface. The zinc oxidizes instead of the underlying iron, forming a tough zinc oxide film that protects the iron.

Antioxidants have become popular vitamin supplements. Antioxidants include fat-soluble vitamin E, water-soluble vitamin C, and beta carotene. All these vitamins are reducing agents, which means they *promote* reduction and *retard* oxidation. For example, a freshly sliced apple will turn brown over time as the oxygen in air oxidizes it. However, a vitamin C solution spread on the surface of the apple prevents oxidation, and the apple remains fresh.

Studies suggest that a similar process may occur in human cells. The proposed oxidizing agents are **free radicals,** atoms or molecules with unpaired electrons. Researchers think that free radicals form from the combination of oxygen with pollutants or toxins present in food, water, and air. Free radicals are extremely reactive and will scavenge electrons from biological molecules, oxidizing them. The oxidized biomolecules can no longer serve their function, and disease develops.

One theory of aging postulates that free radicals extract electrons from large molecules within cell membranes. These molecules become reactive and link with one another, changing the cell wall properties. The body's immune system then mistakes the cell as foreign and destroys it. As cells are destroyed, the body cannot keep up with the demand for new cells, and the body becomes weaker and more susceptible to disease.

Free radicals can also attack other molecules within the cell such as DNA. Changes in DNA cause the cell to function improperly or divide abnormally, leading to diseases such as cancer. Antioxidants work by deactivating free radicals and preventing the oxidation of biological molecules in much the same way that magnesium metal prevents iron from rusting.

The best sources of antioxidants are fruits and vegetables. Many nutritionists and doctors suggest that a balanced diet provides as much of these vitamins as you need. Others, however, recommend supplements that increase the amounts of antioxidants in the diet. Medical science has not given us a definitive answer about the use of supplements, but research continues to probe the correlation between antioxidant intake and disease.

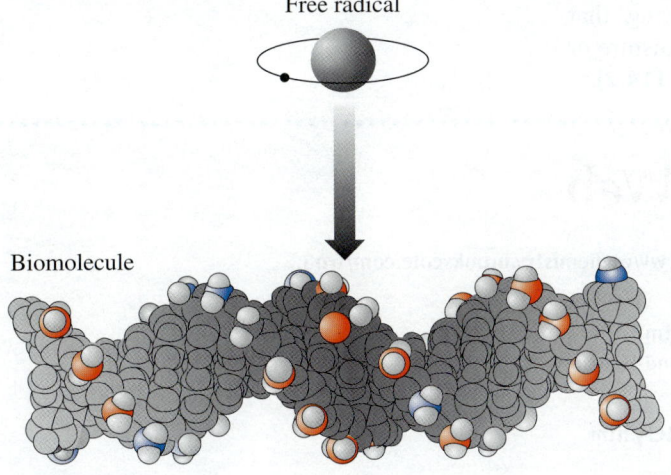

Free radicals, molecular or atomic species containing unpaired electrons, react with biomolecules to oxidize them.

Chapter Summary

MOLECULAR CONCEPT

Oxidation is the gaining of oxygen, the loss of hydrogen, or the loss of electrons. Reduction, the chemical opposite of oxidation, is the loss of oxygen, the gaining of hydrogen, or the gaining of electrons. The two processes occur together in redox reactions. An oxidizing agent promotes oxidation of other substances while it is reduced. A reducing agent promotes the reduction of other substances while it is oxidized (14.1–14.3).

Batteries work via redox reactions in which reactants are separated so that electrons must travel through an external circuit. The flowing electrons constitute an electric current that can power watches, flashlights, and even automobiles (14.5). Fuel cells are batteries in which the reactants are constantly replenished; they generate electricity more efficiently than combustion and may be a significant source of power in the future. Most promising for electrical-generating capacity is the molten carbonate fuel cell (MCFC). One-megawatt models are currently in operation at reasonable cost. This fuel cell must still overcome its dependence on fossil fuels, however, if it is to become a long-term solution to our energy needs (14.6).

A number of metals undergo redox reactions in air. For some metals, like aluminum or tin, the oxide formed is structurally stable and forms a protective layer over the underlying metal, preventing further oxidation. For other metals, most notably iron, the metal oxide formed is crumbly and flakes off to expose new metal to further oxidation. The oxidation of iron can be prevented by using a coating that keeps the iron out of contact with air and moisture or by attaching a more active metal to the iron (14.7).

SOCIETAL IMPACT

Oxidation-reduction reactions are responsible for many of the processes occurring in our everyday lives, including rusting, the function of batteries, and the utilization of food energy by our own bodies.

Common oxidizing agents include hydrogen peroxide (disinfectant and hair bleach), iodine (disinfectant), and household bleach. Common reducing agents include hydrogen (used in the Haber process) and elemental carbon (used in extracting metals from metal ores).

The development of efficient batteries and even more efficient fuel cells has led to the development of electric vehicles that, in some cases, are pollution-free. Fuel cells are also becoming cost-competitive as a way to generate power grid electricity. They are cleaner, more efficient, and environmentally friendlier than conventional electricity-generation technology. Will our society adopt the changes that come with this new, environmentally friendly technology, or will we continue as before?

Since a significant fraction of new iron goes to replace rusted iron, rusting is a major problem. However, it can be prevented, saving the cost of replacing the rusted iron.

Chemistry on the Web

For up-to-date URLs, visit the text website at www.chemistry.brookscole.com/tro3.

- Rust
  http://www.corrosion-doctors.org/index.htm
  http://science.howstuffworks.com/question445.htm
- Batteries
  http://electronics.howstuffworks.com/battery.htm

- Fuel Cells
  http://www.fuelcells.org/news/updates.html
  http://www.eere.energy.gov/hydrogenandfuelcells/
  http://www.ercc.com/index.html

Key Terms

anode	electrochemical cell	oxidation-reduction	reducing agents
antiseptics	free radicals	oxidizing agents	respiration
cathode	galvanization	photosynthesis	
disinfectants	Haber process	redox	

Exercises

Assess your understanding of this chapter's topics with an online chapter quiz at **http://chemistry.brookscole.com/tro3**.

QUESTIONS

1. Explain rusting from a molecular viewpoint.
2. Explain the basic principle involved in the common flashlight battery.
3. List some common processes that involve redox reactions.
4. Give three definitions for oxidation.
5. Give three definitions for reduction.
6. Can an oxidation reaction occur without a corresponding reduction reaction? Why or why not?
7. Why is carbon oxidized in the following reaction (i.e., in what sense is carbon losing electrons)?

    $$C + 2S \longrightarrow CS_2$$

8. Why is bromine reduced in the following reaction (i.e., in what sense is bromine gaining electrons)?

    $$H_2 + Br_2 \longrightarrow 2HBr$$

9. What is an oxidizing agent? a reducing agent?
10. Name some common oxidizing and reducing agents and explain their reactions.
11. Write equations for respiration and photosynthesis. For each reaction, label the substances that are oxidized and reduced. Why are these reactions important?
12. Explain how an electrochemical cell works. Include the definitions of anode and cathode in your explanation. Why can an electrochemical cell be used to generate electricity?
13. Use chemical equations to explain how an automobile battery works. What reaction is occurring at the anode, and at the cathode?
14. Use chemical equations to describe a Leclanché dry cell or inexpensive battery. How do alkaline batteries differ?
15. What is a fuel cell? How is it different from a battery?
16. Give a thermodynamic argument for why it is more efficient to generate electricity from methane oxidation in a fuel cell rather than from methane oxidation in a conventional combustion power plant.
17. Use chemical equations to explain how the hydrogen–oxygen fuel cell works.
18. Use chemical equations to explain how the molten carbonate fuel cell (MCFC) works. What are its advantages and its disadvantages?
19. Explain with the help of chemical equations why keeping iron dry prevents it from rusting.
20. Explain how each of the following helps prevent the rusting of iron:
    a. paint
    b. contact with an active metal
    c. galvanization
21. How might aging and oxidation be related?
22. What are free radicals and how might they affect human health?

PROBLEMS

23. What are the chemical similarities between the following processes:
    a. combustion of gasoline in an automobile engine
    b. metabolism of foods in our bodies
    c. rusting of iron
24. What are the chemical similarities among the following processes:
    a. the refining of iron ore to iron
    b. photosynthesis
    c. the Haber process
25. Draw Lewis structures for each of the following chemical reactions and identify what is oxidized and what is reduced:
    a. $2K + Cl_2 \longrightarrow 2KCl$
    b. $Ca + Br_2 \longrightarrow CaBr_2$
    c. $C + O \longrightarrow CO$

26. Draw Lewis structures for each of the following chemical reactions and identify what is oxidized and what is reduced:
    a. $4Al + 3O_2 \longrightarrow 2Al_2O_3$
    b. $2Li + F_2 \longrightarrow 2LiF$
    c. $2CH_3CH_3 + 7O_2 \longrightarrow 4CO_2 + 6H_2O$
27. For each of the following reactions, indicate which elements are oxidized and which are reduced:
    a. $4Fe + 3O_2 \longrightarrow 2Fe_2O_3$
    b. $NiO + C \longrightarrow Ni + CO$
    c. $2H_2 + O_2 \longrightarrow 2H_2O$
28. For each of the following reactions, indicate which elements are oxidized and which are reduced:
    a. $WO_3 + 3H_2 \longrightarrow W + 3H_2O$
    b. $2Na + Cl_2 \longrightarrow 2NaCl$
    c. $2Fe_2O_3 + 3C \longrightarrow 4Fe + 3CO_2$
29. For each of the following reactions, indicate which elements are oxidized and which are reduced:
    a. $2Fe^{3+} + Zn \longrightarrow 2Fe^{2+} + Zn^{2+}$
    b. $2Br^- + Cl_2 \longrightarrow Br_2 + 2Cl^-$
    c. $Zn + I_2 \longrightarrow Zn^{2+} + 2I^-$
30. For each of the following reactions, indicate which elements are oxidized and which are reduced:
    a. $2H^+ + Sn \longrightarrow H_2 + Sn^{2+}$
    b. $2I^- + Br_2 \longrightarrow I_2 + 2Br^-$
    c. $Cu^{2+} + Mg \longrightarrow Mg^{2+} + Cu$
31. For each of the following reactions, identify the oxidizing agent and the reducing agent:
    a. $C_2H_2 + H_2 \longrightarrow C_2H_4$
    b. $H_2CO + H_2O_2 \longrightarrow H_2CO_2 + H_2O$
    c. $Fe_2O_3 + 3CO \longrightarrow 2Fe + 3CO_2$
32. For each of the following reactions, identify the oxidizing agent and the reducing agent:
    a. $4NH_3 + 5O_2 \longrightarrow 4NO + 6H_2O$
    b. $C_6H_5OH + 14O_3 \longrightarrow 6CO_2 + 3H_2O + 14O_2$
33. Identify the oxidizing agent and the reducing agent in the photosynthesis reaction.
34. Identify the oxidizing agent and the reducing agent in the respiration reaction.
35. Sum each of the following oxidation and reduction reactions to obtain the overall reaction:
    a. $Zn \longrightarrow Zn^{2+} + 2e^-$ (oxidation); $Ca^{2+} + 2e^- \longrightarrow Ca$ (reduction)
    b. $2Al \longrightarrow 2Al^{3+} + 6e^-$ (oxidation); $6H^+ + 6e^- \longrightarrow 3H_2$ (reduction)
36. Sum each of the following oxidation and reduction reactions to obtain the overall reaction:
    a. $Ni \longrightarrow Ni^{2+} + 2e^-$ (oxidation); $2Ag^+ + 2e^- \longrightarrow Ag$ (reduction)
    b. $C_2H_5OH + 4OH^- \longrightarrow CH_3COOH + 3H_2O + 4e^-$ (oxidation); $O_2 + 2H_2O + 4e^- \longrightarrow 4OH^-$ (reduction)
37. A chemist attempts to protect two pieces of iron by placing them in contact with other substances. The chemist attaches a strip of strontium to one piece of iron and some sulfur to the other. After some time outdoors, one of the iron pieces has rusted while the other has not. Which one rusted? Why? (Hint: Use what you know about metals and nonmetals.)
38. A friend tells you that a health-related website recommended taking small amounts of iodine (I_2) as an antioxidant to prevent aging. What is wrong with this idea?
39. Which of the following factors might affect the lifetime of an automobile battery?
    a. concentration of the sulfuric acid solution
    b. size of the porous lead plates
    c. size of the battery casing
40. An engineer suggests using a plastic can instead of a zinc can as a way to reduce the cost of producing a common flashlight battery (Leclanché dry cell). What is wrong with this idea?

POINTS TO PONDER

41. A molten carbonate fuel cell (MCFC) costs approximately $1679 per kilowatt to build. By assuming the fuel cell has a serviceable lifetime of five years, what is the hardware cost per kilowatt-hour for this form of electricity generation? What other costs are involved in actually generating the electricity?
42. Fossil-fuel, nuclear, and hydroelectric power plants all require a transmission network to transport electricity from where it is generated to where it is used. MCFCs, with their smaller size and portability, can be placed much closer to where the power is needed, reducing the need for transmission lines. With this in mind, explain why MCFCs may have a greater advantage for providing power in developing countries versus in developed countries.
43. A new company called FC Power has spent the last two years developing MCFC prototypes. It has not made any sales yet but is projecting several in the near future. The company goes public, and you have the opportunity to purchase some of its stock at the initial offering. Would you buy the stock? Why or why not?
44. The U.S. Department of Energy (DOE) often funds research in new areas of energy production such as MCFCs. Recently, the DOE gave a company $104 million to produce a 1-MW MCFC power plant in Riverside, California. Is this expenditure justified given that this money comes from taxpayers like you and me? Should we pay to subsidize new companies like this one? Why or why not?
45. Michael Faraday (1791–1867) was an English physicist who, through basic research, discovered the electron. Mr. Gladstone, the chancellor of the exchequer, asked him if his discovery had any practical utility. Faraday replied, "One day, Sir, you may tax it." Has Faraday's prediction come true today? What can we learn about basic research (research that may not seem to have immediate practical utility) from the example of Faraday?

Feature Problems and Projects

46. The following diagram shows an electrochemical cell. Fill in a and b with the correct labels for the anode and the cathode. Fill in the blank spaces labeled c and d with the correct chemical species. Fill in the blank space labeled e with an arrow showing the correct direction of electron flow.

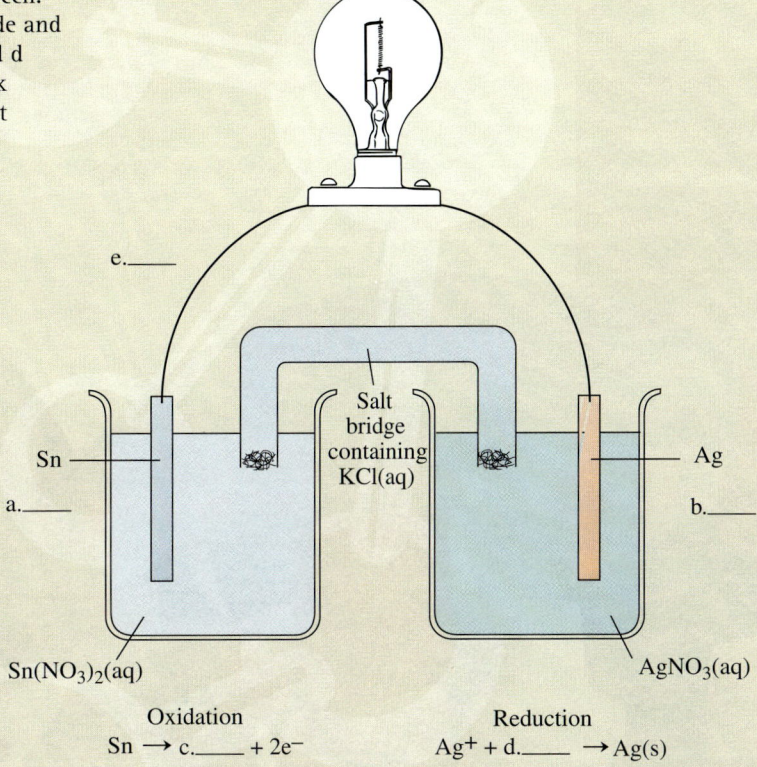

Oxidation
Sn → c.____ + 2e⁻

Reduction
Ag⁺ + d.____ → Ag(s)

47. The following diagram shows a molecular view of the electrochemical cell diagrammed in Figure 2 of this chapter. Draw a molecular view showing the changes that occur after the cell has produced electrical current for some extended period of time.

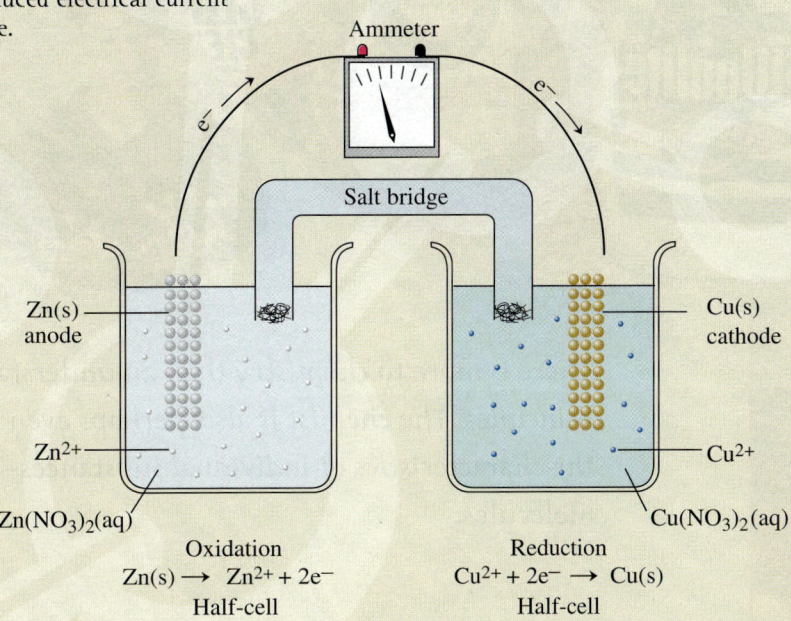

Oxidation
Zn(s) → Zn²⁺ + 2e⁻
Half-cell

Reduction
Cu²⁺ + 2e⁻ → Cu(s)
Half-cell

15 The Chemistry of Household Products

There is more to chemistry than an understanding of general principles. The chemist is also, perhaps even more, interested in the characteristics of individual substances—that is, of individual molecules.

—Linus Pauling

In this chapter, we explore the chemistry of household products. How do products like soap, shampoo, laundry detergent, and toilet bowl cleaner work? What kinds of compounds do they contain? As before, we will try to understand the properties of these various products in terms of the molecules that compose them. As you read this chapter, think about the household products you use; most of them rely on chemistry for their action. What are molecules doing when you wash your hair? When you unclog a drain? Many of the concepts in this chapter were developed in previous chapters. In explaining how household products work, we will draw on our knowledge of acids and bases, oxidation and reduction, polarity and volatility. Go back and review these concepts if you have forgotten them. Lastly, think about how chemical technology has affected your daily life. How would your life be different without these products?

QUESTIONS FOR THOUGHT

- What is soap and how does it work?
- What is the difference between soap and detergent?
- What is in a box of laundry detergent?
- How do toilet bowl and drain cleaners work?
- What happens when I "perm" my hair?
- What is in shampoo and conditioner?
- What happens when I color my hair?
- What is in hand lotions and creams?
- How does sunscreen work?
- What is in facial cosmetics?
- What is in perfumes and deodorants?
- What are plastics made of?
- What are polymers?
- What is rubber?

15.1 Cleaning Clothes with Molecules

When we do laundry, we take advantage of the properties of certain molecules to remove other (undesirable) molecules from clothing. For example, spilled pancake syrup is easy to clean. Just soak a rag in water and wipe up the mess. The sugar molecules that compose pancake syrup are polar, and they mix well with polar water molecules. The water dissolves the sugar and removes it. On the other hand, a greasy stain does not clean as easily. The molecules composing grease are nonpolar and do not mix with polar water; thus the greasy stain repels water. Water can remove grease from clothing but only with the help of molecules in soap. Soap molecules are polar on one end, attracting water molecules, and nonpolar on the other end, attracting grease. Once soapy water is applied to the greasy stain, the soap molecules interact firmly with both water and grease; the grease can then be washed away in the water.

Other stains require more severe treatment. Coffee, wine, or ink stains are particularly difficult to remove. In some cases, the molecules that compose these stains cling so tightly to fabrics that they can only be removed by chemically changing them. Bleaches do this quite nicely; they oxidize the stain-causing molecules, turning them into colorless products.

Most of the products we purchase and use around the house involve a significant amount of chemistry, either in their manufacture or in their actual use. In this chapter, we explore the chemistry of several of these household products. When we clean our hair, brush our teeth, or apply deodorant, we take advantage of the properties of certain molecules to achieve certain macroscopic results. In this chapter, we learn how these molecules do their jobs.

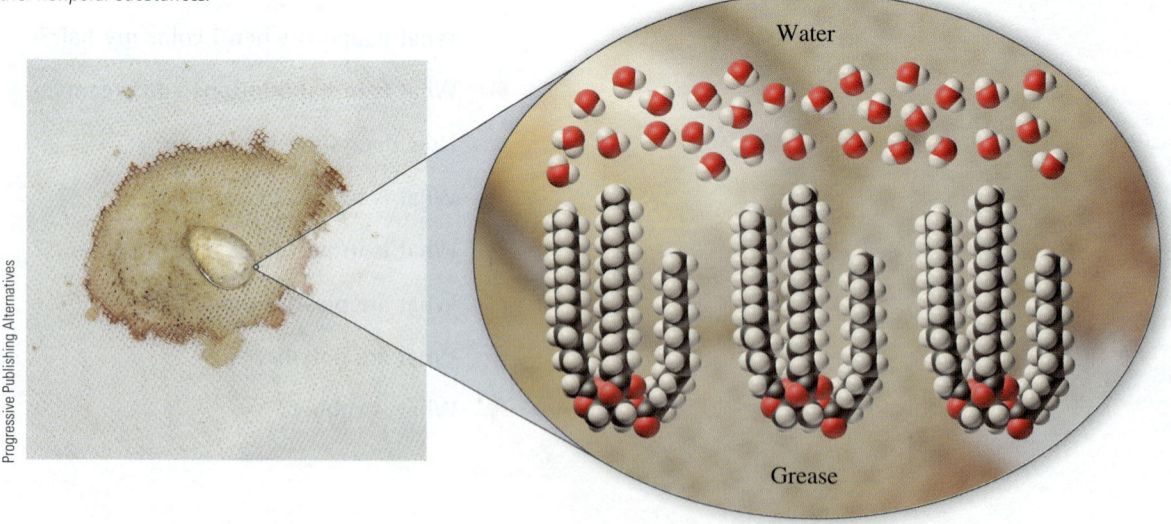

Grease and water do not mix because water is polar while grease is nonpolar. In general, polar substances mix well with other polar substances, and nonpolar substances mix well with other nonpolar substances.

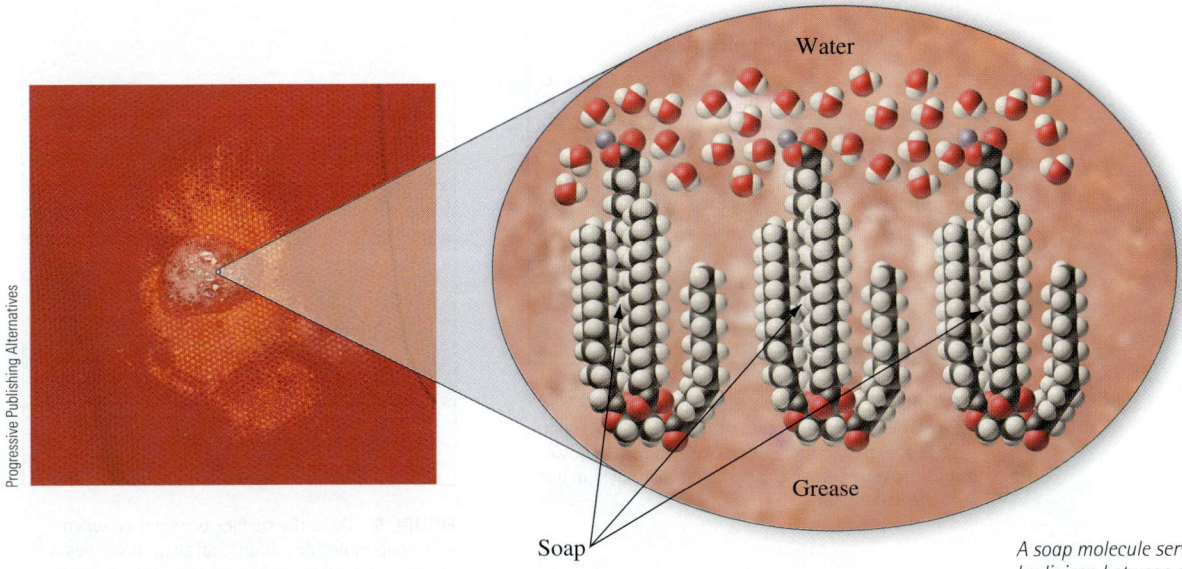

A soap molecule serves as a molecular liaison between polar water and nonpolar grease; it can serve this function because it has both a polar end and a nonpolar end.

15.2 Soap: A Surfactant

Recall from Chapter 5 that polar substances are composed of molecules with uneven electron distributions: One end of a polar molecule is more electron rich than the other. This distribution forms a dipole, a tiny separation of positive and negative charge. Polar molecules strongly attract other polar molecules because the positive end of one molecule aligns with the negative end of the other (Figure 1). The result is an attraction between molecules that is similar to the attraction between magnets. Nonpolar molecules, in contrast, have an even electron distribution and therefore are not attracted to polar molecules. When polar and nonpolar molecules are combined, the polar molecules clump together and squeeze the nonpolar molecules out, much like magnetic marbles clump together and squeeze out nonmagnetic ones (Figure 2).

The structure of soap (Figure 3) shows why it can interact with both nonpolar grease and polar water. One end of the soap molecule is ionic and polar, while the other end, a long hydrocarbon tail, is nonpolar. When soapy water is added to grease or oil, the hydrocarbon tail interacts with the nonpolar grease while the

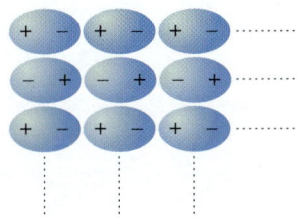

FIGURE 1 Dipoles will align so that the positive end of one dipole is next to the negative end of another. The result is an attraction similar to the attraction between magnets.

FIGURE 2 A mixture of magnetic marbles with nonmagnetic ones results in the separation of the two different kinds of marbles.

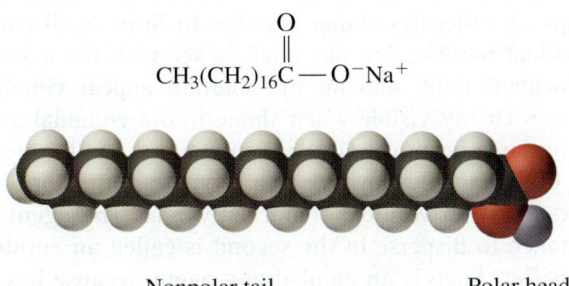

$$CH_3(CH_2)_{16}\overset{\overset{O}{\|}}{C}-O^-Na^+$$

Nonpolar tail   Polar head

FIGURE 3 Sodium stearate, a soap.

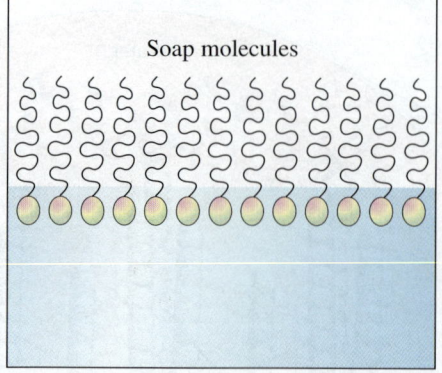

FIGURE 4 Soap molecules congregate at the surface of liquid water because they can accommodate their dual nature there. The polar head sticks down into the water, while the nonpolar tails stick up out of the water.

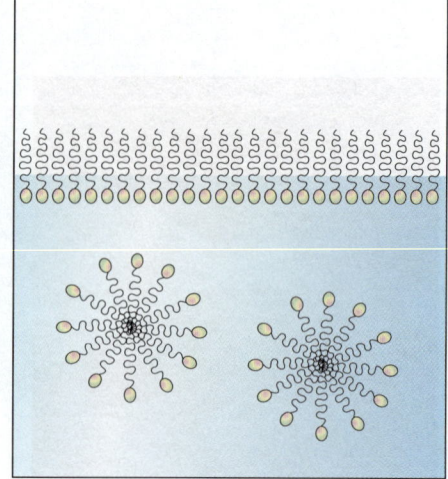

FIGURE 5 Once the surface becomes crowded with soap molecules, additional soap molecules will cluster in structures called micelles. Any grease in the solution will be held in the interior of these micelles.

ionic head interacts with water. The soap molecule holds the two dissimilar substances together and allows grease to be washed away.

Soap, and molecules like it, are called **surfactants** because they tend to act at surfaces. When soap is added to water, soap molecules aggregate near the surface where they can best accommodate their dual nature. At the surface, the ionic head submerges into the water while the nonpolar tail sticks up out of the water (Figure 4). If enough soap is added to water, however, the surface becomes crowded and soap molecules begin to aggregate in structures called **micelles** (Figure 5). In a micelle, the nonpolar tails crowd into a spherical ball as they maximize their interactions with each other and minimize their interaction with water molecules. The ionic heads form the surface of the sphere where they interact with water molecules. A mixture of soapy water and grease contains micelles with clumps of grease molecules trapped within the micelles. Because micelles have negatively charged surfaces they repel each other, preventing grease from coalescing (or aggregating) and allowing it to stay suspended in the water.

Micelle solutions, such as the one just described, are called colloidal suspensions. A **colloidal suspension** is a mixture of one substance dispersed in a finely divided state throughout another. Unlike normal solutions, where molecules mix on the molecular scale, colloidal suspensions are clumpy; groups of molecules clump together to form small particles. While the individual particles are too small to see with the naked eye, they scatter incident light, making the solution appear cloudy. A laser beam becomes clearly visible when shone into a colloidal suspension (Figure 6). The same beam cannot be seen when shone through pure water.

Colloidal solutions are also a type of **emulsion,** a mixture of two substances that would not normally mix. The agent that causes one substance to disperse in the second is called an **emulsifying agent** or **emulsifier.** Soap is an emulsifying agent because it allows grease and water to form an emulsion. Some food additives are emulsifiers because they keep foods that would ordinarily separate into layers, such as peanut butter, in just one smooth emulsion.

FIGURE 6 A colloidal suspension will scatter light, allowing a laser beam to be visible.

In this colloidal suspension of grease, soap, and water, we see how the soap micelles hold grease within their interior. The soap allows the water and grease to form a colloidal suspension, also called an emulsion. Because the soap allows the two to form an emulsion, it is called an emulsifying agent.

15.3 Synthetic Detergents: Surfactants for Hard Water

Certain impurities often found in tap water reduce the effectiveness of soap. Particularly troublesome are the calcium and magnesium ions in hard water. These ions react with soap molecules to form a slimy, gray scum called curd that deposits on skin when bathing or on the sides of the bathtub producing bathtub ring:

$$2CH_3(CH_2)_{16}COO^- + Ca^{2+} \longrightarrow (CH_3(CH_2)_{16}COO^-)_2\, Ca^{2+}$$
$$\text{soap} \qquad \text{hard water ions} \qquad \text{curd}$$

Because soap forms curd in hard water, it is a poor choice for washing hair, dishes, or clothes, all of which retain some of the curd, which appears as a gray film on otherwise clean surfaces.

In the 1950s, chemists developed synthetic detergents that, unlike soap, do not react with hard water ions. Among the first successful synthetic detergents were the **alkylbenzenesulfonates (ABS)** detergents (Figure 7). These detergents were

The ABS detergents would not biodegrade and consequently accumulated in some natural waters, producing suds and foam.

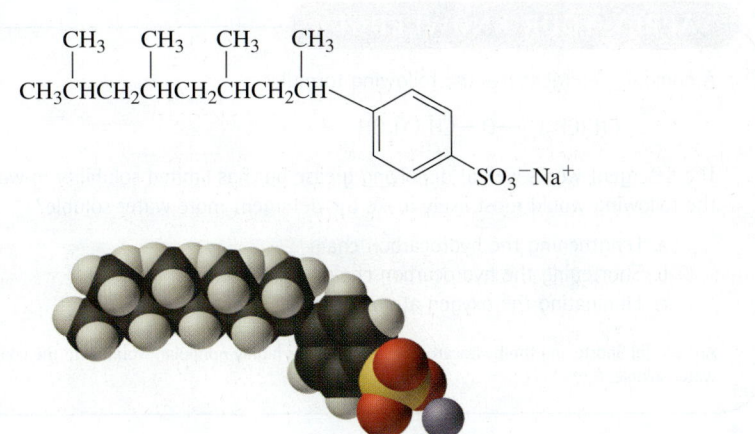

FIGURE 7 Sodium alkylbenzenesulfonate, an ABS detergent.

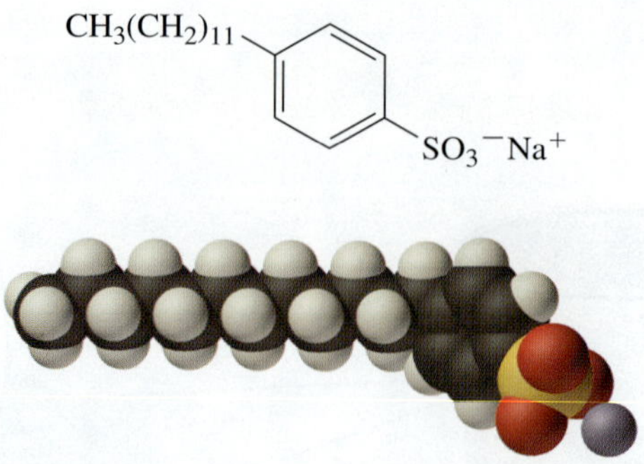

FIGURE 8 Sodium *para*-dodecylbenzene sulfonate, an LAS detergent. The linear alkyl chain allows microorganisms to decompose LAS detergents.

inexpensive to make and did not form curd. They found widespread use in laundry-cleaning formulations and other cleansers. It appeared that ABS detergents were the ideal substitute for soap; however, the ABS detergents were unusually stable, and unlike soap, they did not biodegrade. As wastewater containing ABS detergents was disposed, soapy suds began to appear in streams, rivers, lakes, and oceans. Foamy coastlines prompted researchers to investigate why the ABS detergents would not degrade. They found that it was due to their highly branched alkyl chains. Ordinary soap, in contrast, has straight alkyl chains. Here again we see the subtlety of the molecular world: The microorganisms that break down the straight alkyl chains in soap were not capable of breaking down the branched alkyl chains of ABS detergents. Consequently, ABS detergents accumulated in the environment.

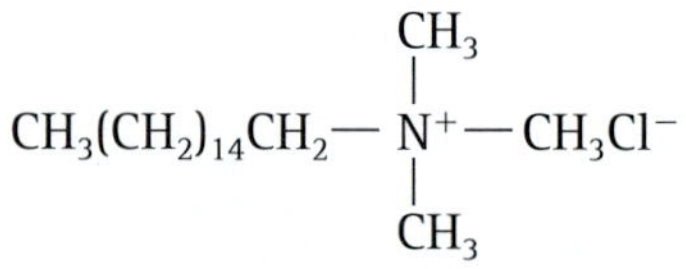

The detergent industry responded by designing and synthesizing modified versions of alkylbenzenesulfonate detergents. Chemists synthesized detergents in which the branched alkyl chain was replaced with a straight alkyl chain, more like soap. These **linear alkylsulfonate (LAS) detergents** (Figure 8) were biodegradable and alleviated the problem of detergent buildup. LAS detergents decompose over time forming CO_2, H_2O, and SO_4^{2-}, all common substances in the environment.

ABS and LAS detergents are referred to as **anionic detergents** because the polar part of the molecule is a negative ion. Chemists have also developed **cationic detergents** (Figure 9), where the polar part is a positive ion, and **nonionic detergents** (Figure 10), where the polar part is not ionic but maintains polarity because of OH or other polar groups.

FIGURE 9 Hexadecyltrimethyl-ammonium chloride, a cationic detergent.

Explore this topic on the Council for LAB/LAS Environmental Research (CLER) website.

APPLY YOUR KNOWLEDGE

A nonionic detergent has the following formula:

$$CH_3(CH_2)_{15}-O-CH_2CH_2OH$$

The detergent works well at dissolving grease but has limited solubility in water. Which of the following would most likely make the detergent more water soluble?

a. Lengthening the hydrocarbon chain
b. Shortening the hydrocarbon chain
c. Eliminating the oxygen atoms from the structure

Answer: (b) Shortening the hydrocarbon chain, which is highly nonpolar, would help the compound be more water soluble.

> ## MOLECULAR FOCUS
>
> ### Polyoxyethylene
>
> Formula: $C_{14}H_{30}O_2$
> Molar mass: 230 g/mol
> Structure: $CH_3(CH_2)_{11}$—O—CH_2CH_2—OH
> Three-dimensional structure:
>
>
>
> **FIGURE 10** Polyoxyethylene, a nonionic detergent.
>
> Polyoxyethylene (Figure 10) is an example of a nonionic detergent. One side of the molecule, the OCH_2CH_2OH group, is polar, while the other side, the R group, is nonpolar. Variations of polyoxyethylene have the general formula $R(OCH_2CH_2)_nOH$, where R is a long hydrocarbon chain and n represents the number of OCH_2CH_2 groups; common values of n range from 1 to 20. By varying n and by varying the number of CH_2 units in R, a broad range of properties can be achieved.
>
> Nonionic detergents foam less than their anionic counterparts, making them useful in applications where foam is not desired. They also are more effective at low temperatures, making them economical to use in laundry-cleaning formulations because clothes can be washed in colder water. In addition, nonionic detergents do not react with calcium and magnesium ions; therefore, these detergents work well in hard water.

15.4 Laundry-Cleaning Formulations

Laundry-cleaning formulations can have 40 or 50 different ingredients, all tailored to make clothes clean, bright, and sweet smelling. The most important ingredient, however, is the surfactant. LAS surfactants replaced ABS surfactants in the 1960s in most laundry formulations. Recent trends have favored a combination of anionic LAS surfactants with nonionic surfactants. This allows the use of fewer builders (described below) and therefore reduces the total amount of detergent per load of laundry. This in turn allows detergent manufacturers to use fewer packaging materials, which results in less waste.

Builders, substances that increase the effectiveness of the surfactant, are probably the next most important component in laundry formulations. Although synthetic surfactants are less affected by the presence of hard water ions than soaps are, they still work better in soft water. The primary function of builders is to soften water by removing Ca^{2+} and Mg^{2+} ions. Builders may also help suspend solid soil particles in solution once they are removed from fabrics. The best single builder for laundry detergents is sodium tripolyphosphate, $Na_5P_3O_{10}$. It eliminates Ca^{2+} and Mg^{2+} ions by binding them tightly (a process called sequestering) and preventing them from interfering with the action of the surfactant. Sodium tripolyphosphate is also relatively inexpensive and helps suspend solid dirt particles in solution.

The use of sodium tripolyphosphate and other phosphates in laundry formulations has been banned in many places because their presence in wastewater can—through a process called **eutrophication**—threaten marine life. Besides being excellent builders, phosphates are also excellent fertilizers. When phosphates are disposed in natural waters, they fertilize algae, producing abnormally large algae blooms. These algae eventually die and decompose via oxidation reactions. The

The decay of too much algae in a lake will deplete the lake of oxygen and harm marine life.

large amount of oxidation that occurs in a lake or bayou with large algae blooms depletes oxygen levels in the water, harming fish and other marine life. The role of laundry detergent phosphates in the eutrophication of natural waters has been greatly debated, and the magnitude of their contribution remains controversial. Nonetheless, phosphates are no longer used in laundry formulations in the United States and Canada.

There is no single replacement for phosphates in laundry detergents. Modern North American detergents contain a combination of builders including sodium carbonate ($NaCO_3$) and zeolites. Sodium carbonate softens water by forming solid precipitates with calcium and magnesium ions. Once the ions are bound in a solid, they no longer interfere with detergent action. **Zeolites** are small porous particles containing sodium ions. When a zeolite is added to hard water, calcium ions diffuse into the zeolite and exchange places with sodium ions. The sodium ions work their way back into the water but do not impede detergent action.

Other ingredients found in laundry-cleaning formulations include fillers, enzymes, brighteners, bleaching agents, and perfumes. **Fillers** serve no laundering purpose except to increase the bulk of the detergent and control consistency and density. Their use has decreased in recent years to reduce packaging materials. **Enzymes** are present to help degrade certain types of stains. One formulation contains an enzyme dubbed "Carezyme," which breaks down natural fibers and helps to reduce the "balling" that often occurs when laundering cotton fabrics. **Brighteners** are dye molecules that bind to clothes fibers and help make clothes appear "brighter than before." They absorb ultraviolet (UV) light and emit visible light, usually blue. The result is that some of the normally invisible UV light falling on clothes becomes visible, making the clothes appear brighter. *Bleaching agents* are oxidizing agents such as $NaBO_3 \cdot 3H_2O$ that help decompose tough stains like wine or fruit juice. They work by oxidizing the molecules that cause the stain, converting them into colorless substances. Perfumes and fragrances are added to most laundry formulations to make clothes smell fresh and clean.

15.5 Corrosive Cleaners

Some household products are very chemically reactive. Perhaps the most reactive of all products are those used to remove clogs from drains. These products usually consist of a strong base such as sodium hydroxide. When mixed with water, the solid sodium hydroxide dissolves, liberating significant amounts of heat and producing a hot NaOH solution. This solution melts the grease often associated with clogs and converts some of the grease molecules to soap molecules (soap is formed by the action of sodium hydroxide on grease); these in turn help dissolve more grease. The NaOH solution also dissolves protein strands in hair, another common component of clogged pipes. Some drain cleaners also incorporate small bits of aluminum. The aluminum dissolves in the NaOH solution and produces hydrogen gas. The bubbling gas helps to dislodge clogs physically. Care should be used with all NaOH solutions. They are corrosive and should not contact the skin.

Toilet bowl cleaners are often corrosive as well. Hard water leaves calcium carbonate deposits in the toilet bowl that accumulate dirt and grime. Acids react with carbonates, dissolving them. Consequently, many toilet bowl cleaners are acidic solutions such as hydrochloric acid or citric acid. As with drain cleaners, these products are corrosive and should not contact the skin. Drain cleaners contain strong bases, and toilet bowl cleaners contain strong acids; the two should not be mixed because the resulting neutralization reaction is vigorous and could splatter the caustic solution onto your skin.

Drain cleaners contain solid sodium hydroxide, a strong base.

15.6 Hair Products

CURLING HAIR

Hair is composed primarily of a protein called keratin. Keratin is composed of long, coiled protein chains held together by three types of interactions between the chains: hydrogen bonds, disulfide linkages, and salt bridges (Figure 11). The hydrogen bonds are the easiest of these interactions to modify. Simply wetting hair will interfere with hydrogen bonding between protein chains; the protein chains relax their coils to some degree, and consequently wet hair is longer and straighter than dry hair. Once the hair dries, the hydrogen bonds re-form and the hair returns to its normal state. If, however, curlers are put into hair when it is wet, the hydrogen bonds re-form in the positions maintained by the curlers. When the hair dries and the curlers are removed, the hydrogen bonds remain in their new configuration, resulting in curls.

Hair is sensitive to pH because of the salt bridges between keratin protein strands. These salt bridges are strongest under slightly acidic conditions. High pHs (very basic) will damage the salt bridges because H^+ ions are removed from $—NH_3^+$ groups resulting in neutral $—NH_2$. Because $—NH_2$ has no positive charge, it will not bond to $—COO^-$, and the salt bridge is broken. Consequently, the pH of hair products should be neutral to slightly acidic.

The third kind of interaction between keratin strands is the disulfide linkage, a covalent chemical bond that is not easily broken. With chemical treatment, however, disulfide linkages can be broken and re-formed to change permanently—at least until new hair grows—the shape of hair. To produce a permanent curl or "perm," a reducing agent such as thioglycolic acid is applied to hair. The acid attacks the disulfide bridges and breaks the sulfur–sulfur chemical bonds. The hair is then set in curlers in the desired shape. A mild oxidizing agent such as dilute hydrogen peroxide is then applied to the hair, reforming the disulfide linkages in the new shape. The result is permanently curled hair.

WASHING AND CONDITIONING HAIR

Hair is prevented from becoming overly dry by glands in the scalp that secrete an oily substance called sebum. Over time, the sebum accumulates, attracting dirt and making hair look oily. The primary purpose of shampoo is to remove dirty sebum from the hair and scalp. Like laundry detergents, shampoos contain synthetic surfactants such as sodium lauryl sulfate (Figure 12). These detergents remove sebum and do not form curd with hard water ions. Shampoos may also contain thickeners, suds boosters, and fragrances, most of which are added to make the shampoo more appealing.

Residual shampoo makes hair look dull. Consequently, many people use conditioners or rinses that rinse out residual shampoo while softening hair and making it more manageable. Conditioners usually contain cationic surfactants that bind with residual anionic surfactants and help rinse them out. These cationic surfactants are usually quaternary ammonium compounds (see Figure 9), so called because they contain four R groups attached to a nitrogen cation. Quaternary ammonium compounds also form thin coatings on individual hair shafts, reducing their tendency to tangle and producing softer, more manageable, hair.

COLORING HAIR

The natural color of hair depends primarily on two pigments: melanin, a dark pigment, and phaeomelanin, a red-orange pigment. Light blonde or white hair

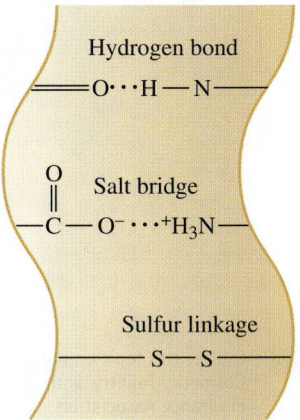

FIGURE 11 The shape of hair is determined by three kinds of interactions within keratin strands: hydrogen bonding, salt bridges, and sulfur linkages.

Wet hair can be curled because water interferes with hydrogen bonding. By shaping the wet hair with curlers, the hydrogen bonds can be made to reform in new configurations as the hair dries.

$$CH_3(CH_2)_{11}—O—SO_3^-Na^+$$

FIGURE 12 Sodium lauryl sulfate, a surfactant used in shampoos.

has very little of either pigment. Dark blonde to brown hair has varying amounts of melanin, and black hair has lots of melanin. Red hair has varying amounts of phaeomelanin, depending on its shade. The lightening of hair—whether it is red, brown, black, or blonde—requires the removal of these pigments, usually accomplished with bleaches such as hydrogen peroxide (H_2O_2). When applied to hair, hydrogen peroxide oxidizes the pigments to colorless products.

Darkening or adding color to hair can be accomplished by several different treatments. The simplest process involves applying a dye of the desired color to the hair. However, because dye molecules are fairly large, they only coat the surface of hair strands; the hair color is only temporary and will wash out after several shampoos. In more permanent coloring of hair, two or more components (usually smaller molecules) are added to hair. These molecules diffuse into hair strands and react to produce a larger, colored molecule. Because the molecule is formed within the hair strand, it tends to stay there, producing permanent color.

W Explore this topic on the **Cosmetic, Toiletry, and Fragrance Association** website.

15.7 Skin Products

The skin's outer layers are composed primarily of dead cells that have been pushed out from beneath. Under the dead cells are layers of live cells that are constantly dividing and pushing upward toward the skin's surface. As new cells migrate to the surface, old ones are lost. The dead cells that compose the skin's surface are lubricated by sebum (same as sebum in the scalp) excreted by glands just below the skin. The sebum regulates the moisture content of the skin, which is optimal at about 10% water. Too little moisture leads to excessive flaking of dead cells, while too much moisture leads to increased risk of infection by micro-organisms. The amount of moisture held in the skin varies with environmental conditions. Dry or windy weather dries the skin, while wet or humid weather moistens it. Products designed for use on the skin include creams, lotions, and sunscreen.

CREAMS AND LOTIONS

Creams and lotions are emulsions of water and oil designed to help the skin maintain enough moisture. The water adds moisture to the skin, while the oil holds the moisture in the skin. Lotions tend to contain more water than oil, and the oils tend to be liquids at room temperature. Creams tend to contain more oil than water, and the oils tend to be solids at room temperature. Both creams and

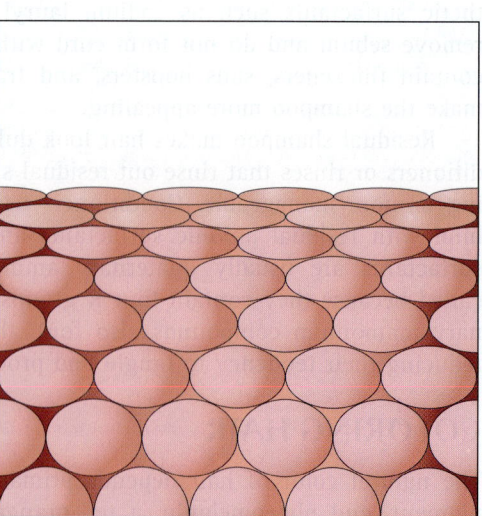

Skin cells showing large living cells at the bottom and flatter, smaller dead cells towards the surface.

MOLECULAR THINKING

Weather, Furnaces, and Dry Skin

Have you ever wondered why skin becomes dry in the winter months even though the weather seems wet and damp? The tendency for skin to dry is related to the **relative humidity**. Relative humidity is a measure of the actual water content of air relative to its maximum water content.

$$\text{relative humidity} = \frac{\text{actual water content}}{\text{maximum water content}} \times 100\%$$

The maximum water content of air increases with increasing temperature; thus relative humidity depends not only on how much water vapor is in the air but also on the air temperature. Suppose the relative humidity on a winter's day is 60% at an outdoor temperature of 40°F. The air is holding 60% of the water that it can hold at 40°F. Relative humidity in this range is fairly comfortable, feeling neither too dry nor too humid. However, when the outside air is brought into a house and warmed in a furnace, the increase in temperature increases the ability of air to hold water. Consequently the relative humidity—the amount of water the air holds relative to how much *it can hold*—decreases. The low relative humidity of indoor air during winter months causes water to evaporate more quickly from the skin, resulting in dry skin. It also can dry out mucous membranes in the nose and sinuses, increasing susceptibility to colds and other infections.

lotions also contain emulsifying agents, soaplike substances (as defined earlier) that keep the normally immiscible oil and water together in an emulsion.

Oils used in creams and lotions include olive oil, mineral oil, almond oil, and lanolin (oil from sheep's wool). Creams often include waxes such as beeswax or carnauba. In addition, most creams and lotions contain perfumes to add pleasant fragrances. Both creams and lotions can help skin retain its ideal moisture content of about 10%.

SUNSCREENS

Sunscreens are lotions with extra ingredients to absorb UV light. The absorption of UV light by skin triggers the production of melanin, the dark pigment also responsible for hair color. The skin produces melanin as a protection against further exposure to UV light. Because UV light is more energetic than visible light (see Chapter 7), it can damage the fragile biological molecules that compose skin, producing sunburn. Excessive exposure to UV light increases the risk of skin cancer.

UV light is divided into several categories, including UV-A and UV-B. UV-A extends from wavelengths of 320 to 400 nm, while UV-B extends from 280 to 320 nm (Figure 13). UV-B light, shorter in wavelength and higher in energy, is

UV light causes the skin to produce melanin, the dark pigment that is also responsible for the color of brown hair.

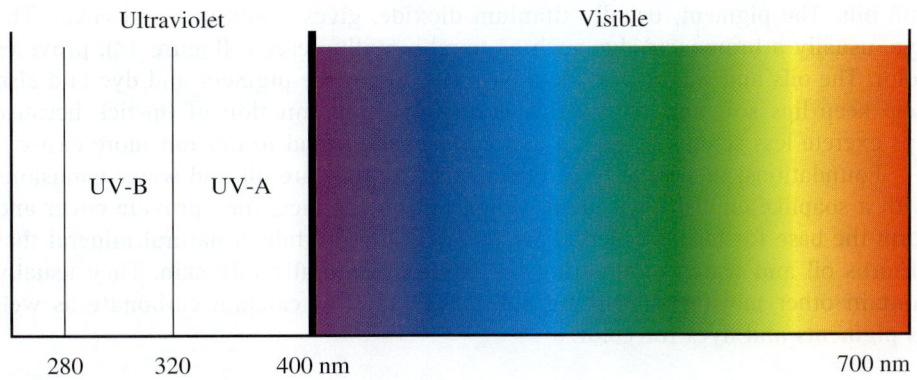

FIGURE 13 UV-B light is largely responsible for sunburns and skin cancers.

These compounds, common components of sunscreens, absorb UV light while letting visible light pass through.

PABA

benzophenone

oxybenzone

dihydroxyacetone

more harmful to humans, increasing skin cancer risk. Fortunately, most skin cancers are not the often fatal malignant melanoma, but rather a slower-spreading treatable form. In addition, excessive exposure to UV-B light causes premature skin aging and weakens the immune system. For these reasons, exposure to UV-B light should be minimized.

For those who spend time outdoors, the use of sunscreens and protective clothing are among the best ways to limit exposure. Sunscreens contain compounds that absorb light in the UV-B range. A thin layer of these compounds on top of the skin prevents UV-B from getting to the skin where it would otherwise do its damage. Common UV-absorbing compounds present in sunscreens include *p*-aminobenzoic acid (PABA), benzophenone, oxybenzone, and dihydroxy-acetone.

The ability of a sunscreen to absorb UV light is specified by its skin protection factor (SPF). These numbers, such as SPF 15 for example, indicate the number of hours you can spend in the sun while acquiring only 1 hour's worth of UV-B exposure. An SPF of 15 means that 15 hours in the sun with the sunscreen produces an exposure equivalent to 1 hour without it. The SPF number on a sunscreen simply reflects the concentration of the UV-B-absorbing compounds within them; higher SPF numbers indicate a higher concentration of UV-absorbing compounds.

15.8 Facial Cosmetics

Facial cosmetics are used to color lips, lashes, cheeks, and eyebrows. Most facial cosmetics consist of waxes, oils, pigments, dyes, and perfumes. Lipstick, for example, is composed of pigments and dyes suspended in a mixture of waxes and oils. The pigment, usually titanium dioxide, gives brightness and cover. The dye, usually a bromoacid dye such as tetrabromofluorescein (Figure 14), provides color. The oils and waxes provide a base to contain the pigment and dye and also help keep lips soft and moist. This is an important function of lipstick because lips excrete less sebum than skin and consequently tend to dry out more easily.

Foundations, in similarity to creams and lotions, are oil and water emulsions with a soaplike emulsifying agent. When put on the face, they provide cover and form the base for face powders. Face powders contain talc, a natural mineral that absorbs oil and water, eliminating the shine of naturally oily skin. They usually contain other moisture-absorbing substances such as calcium carbonate as well as pigments and dyes for color.

FIGURE 14 Tetrabromofluorescein, a common dye used in lipsticks.

Mascara, used to darken and thicken the eyelashes, is composed of soap, oils, waxes, and pigments. The possible pigments include iron oxide, which is brown, and carbon black, which is black.

15.9 Perfumes and Deodorants: Producing Pleasant Odors and Eliminating Unpleasant Ones

Our sense of smell is particularly sensitive. We smell a flower on a spring day because some of the molecules from the flower diffuse into the air, and we inhale them through the nose. In the nose lie about 50 million chemical receptors. The molecules from the flower precisely fit into some of these receptors, triggering a signal to the brain that we interpret as a sweet smell. Other molecules, such as those from rotting fish for instance, fit receptors that trigger a signal that we interpret as unpleasant. Still other molecules, like water, do not fit any receptors at all and therefore have no odor. The brain's processing centers for smell are linked to its processing centers for emotions; hence, smells often elicit strong emotions.

PERFUMES

The word *perfume* is derived from the Latin *per fumum*, which means "through smoke." Perfumes are composed of compounds with pleasant odors. Originally, perfumes were derived from plant and animal sources. Plant smells include floral

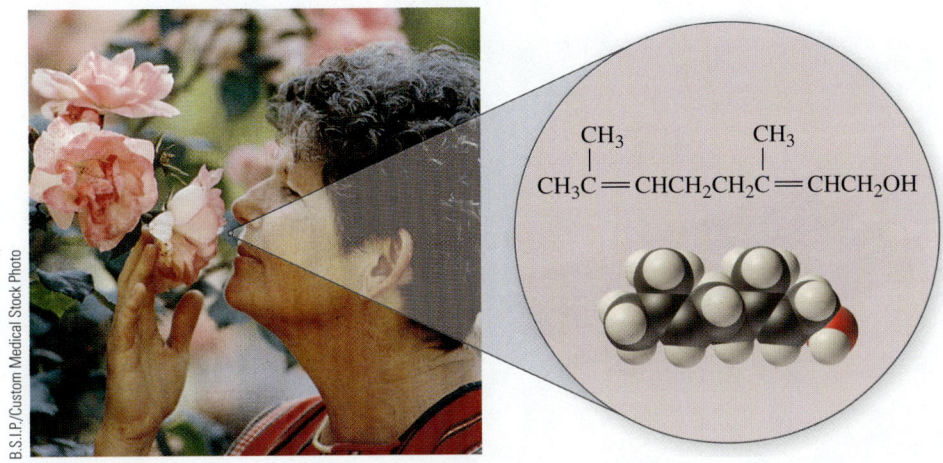

When people smell a rose, they inhale molecules of a compound called geraniol. These molecules trigger nerve signals that our brain interprets as a sweet smell.

fragrances such as jasmine (Figure 15), rose, or lily; spicy fragrances such as clove (Figure 16), cinnamon (Figure 17), or nutmeg; woody fragrances such as sandalwood or cedar; herbal fragrances such as clover or grass; and fruity fragrances such as orange, lemon, or apple (Figure 18). Animal fragrances include leather; castor from the beaver; civetone (Figure 19) from the civet cat; and musk (Figure 20), a sex attractant between musk deer.

FIGURE 15 Benzyl acetate, a component in the smell of jasmine flowers.

FIGURE 16 Eugenol, a component in the smell of cloves.

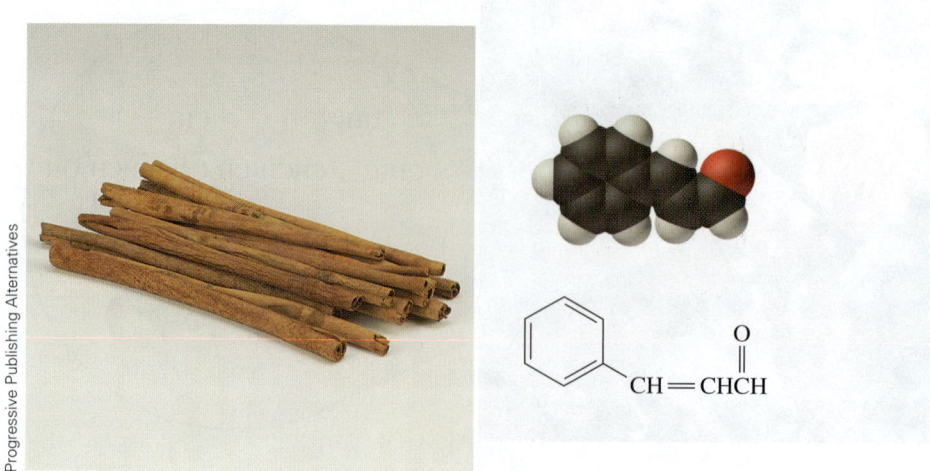

FIGURE 17 Cinnamaldehyde, a component in the smell of cinnamon.

15.9 Perfumes and Deodorants: Producing Pleasant Odors and Eliminating Unpleasant Ones

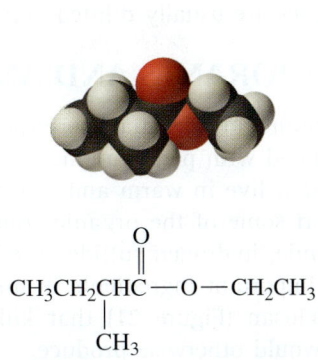

$$\text{CH}_3\text{CH}_2\text{CHC}\overset{\overset{\text{O}}{\|}}{}\text{—O—CH}_2\text{CH}_3$$
$$|$$
$$\text{CH}_3$$

FIGURE 18 Ethyl-2-methylbutanoate, a component in the smell of apples.

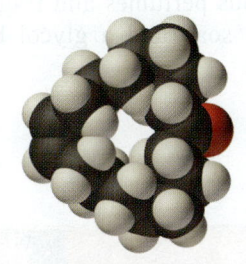

$$\begin{array}{c}\text{CHCH}_2\text{CH}_2\text{CH}_2\text{CH}_2\\\|\\\text{C}\qquad\qquad\text{CH}_2\\|\qquad\qquad|\\\text{CH}_2\qquad\qquad\text{CH}_2\\|\qquad\qquad|\\\text{CH}_2\qquad\qquad\text{C}=\text{O}\\\diagdown\qquad\diagup\\\text{CH}_2\text{CH}_2\text{CH}_2\text{CH}_2\end{array}$$

FIGURE 19 Civetone, a component in the smell of civet, which is an excretion of the civet cat.

$$\begin{array}{c}\qquad\qquad\qquad\overset{\text{O}}{\|}\\\text{CH}_2\text{CH}_2\text{CH}_2\text{CH}_2\text{CH}_2\text{CH}_2\text{CCH}_2\\|\qquad\qquad\qquad\qquad|\\\text{CH}_2\qquad\qquad\qquad\text{CHCH}_3\\\diagdown\qquad\qquad\diagup\\\text{CH}_2\text{CH}_2\text{CH}_2\text{CH}_2\text{CH}_2\end{array}$$

FIGURE 20 Muscone, a component of musk, which is a sex attractant between musk deer.

The difficulty of extracting these substances from plants and animals led chemists to synthesize many of the compounds responsible for characteristic plant and animal scents. Modern perfume manufacturers have thousands of ingredients, both natural and synthetic, to choose from. Blending them together to form a perfume is part art and part science.

Perfumes are usually designed to have three different stages, or notes, in which their fragrances are experienced. The top note is the first one encountered when opening a bottle of perfume. The scents in the top note are usually refreshing and are caused by small volatile molecules that quickly diffuse out of the bottle and into air. The middle note is detected only after a short time has passed and is usually full and solid in character. The molecules responsible for the scents in the middle note are larger and less volatile than those in the top note. They tend to persist on the skin. For this reason, perfume smells different on the skin than it does in the bottle; the warmth of the skin causes evaporation of the molecules in the top note, leaving a higher relative concentration of those in the middle and end notes. The end note usually contains more animal or earthy fragrances. It is composed of large molecules that evaporate slowly. The compounds in the end note also help hold the more volatile notes, preventing them from vaporizing too quickly.

Most perfumes are a 10–25% solution of the fragrant compounds in alcohol. Colognes are usually diluted versions with only 0.5–2% fragrant compounds.

DEODORANTS AND ANTIPERSPIRANTS

Deodorants are designed to either mask or reduce body odor. Body odor is often associated with perspiration; however, perspiration itself has little odor. The bacteria that live in warm and moist areas of the body, such as the armpit or groin, convert some of the organic compounds in perspiration into compounds such as ammonia, hydrogen sulfide, and others with unpleasant odors. Consequently, the most important ingredients in underarm deodorants are antibacterial agents such as triclosan (Figure 21) that kill bacteria and consequently eliminate the odor they would otherwise produce.

FIGURE 21 Triclosan, an antibacterial agent found in deodorants.

Antiperspirants actually reduce the amount of perspiration that sweat glands produce. Antiperspirants contain aluminum chlorohydrates such as $Al_2(OH)_4Cl_2$ that reduce sweating by constricting the opening of sweat glands. Both deodorants and antiperspirants also contain numerous perfumes and fragrances to mask foul odors that might develop, as well as a soap and a glycol base that holds everything together.

WHAT IF...

Consumer Chemistry and Consumerism

Consumerism is a preoccupation with buying consumer goods. Advances in science have led to an exponential rise in the number of products, gadgets, and formulations available to the consumer. Slick and polished advertising has been developed to convince us of our need for the latest and the best product.

Some argue that consumerism, as the root of economic growth, is necessary and good. The consumer, who constantly wants and demands more, drives the growth of the economic engine that has propelled American living standards to their present heights. Others, in contrast, have felt that the obsession with consumer products has led to a consumer culture in which all values are based on what you look like, what you have, and how you can get more. Such a culture forgets to ask questions about who we are and what is truly important in life.

QUESTION: What do you think? Has consumerism gone too far in America? Do you see any evidence that the consumer culture is diminishing, or growing? How are you influenced by advertising for consumer products?

Advertising appeals to our emotions to persuade us to buy.

> **APPLY YOUR KNOWLEDGE**
>
> Examine the structures ethyl-2-methylbutanoate (see Figure 18) and civetone (see Figure 19). Explain why a mixture of these two compounds would initially smell more like ethyl-2-methylbutanoate and later smell more like civetone.
>
> Answer: Civetone has a much higher molar mass than ethyl-2-methylbutanoate. Because the compounds are otherwise similar in structure, the higher molar mass results in greater intermolecular forces, which in turn results in lower volatility. Therefore, the ethyl-2-methylbutanoate evaporates more easily, resulting in the initial smell. The civetone evaporates more slowly, resulting in the later smell.

15.10 Polymers and Plastics

Many of the products we use in our homes either are composed of plastic or come wrapped in plastic. Plastic bottles, food containers, wrappings, straws, buckets, storage containers, toys, computers, television sets, floors, countertops, furniture, and even car parts are all made of plastic. The appeal of plastic lies in its high strength, low cost, variable properties, and ability to mold into almost any shape imaginable.

The molecules that compose plastics have long, chainlike structures composed of individual repeating units. The individual units are called *monomers*, and the chainlike molecules they compose are known as **polymers**. Perhaps the simplest polymer, as well as one of the most common, is **polyethylene**. Polyethylene is the plastic used in trash bags, soda bottles, plastic shopping bags, and wire insulation. It is formed from the monomer ethylene (Figure 22), also called ethene. Ethylene units string together by breaking their double bonds and forming single bonds with other ethylene units. The result is a long chain of repeating —CH_2— groups (Figure 23), sometimes thousands of monomers long.

Because polyethylene molecules resemble long-chain hydrocarbons, it is not surprising that the properties of polyethylene resemble those of long-chain hydrocarbon waxes; they both have smooth, water-repelling surfaces. However, because the hydrocarbon chains in polyethylene are so much longer than waxes, polyethylene is harder and melts at higher temperatures. Even though the nonpolar character of polyethylene expels water, it does not expel oils or other nonpolar materials;

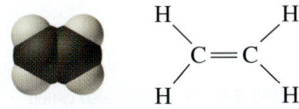 Explore this topic on the **Polymers** website.

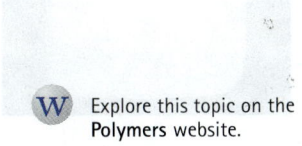

FIGURE 22 Ethylene, also known as ethene.

····$CH_2CH_2CH_2CH_2CH_2CH_2CH_2CH_2CH_2CH_2$····

FIGURE 23 Polyethylene.

Polymers are formed by the reaction of monomers to form long, chainlike structures.

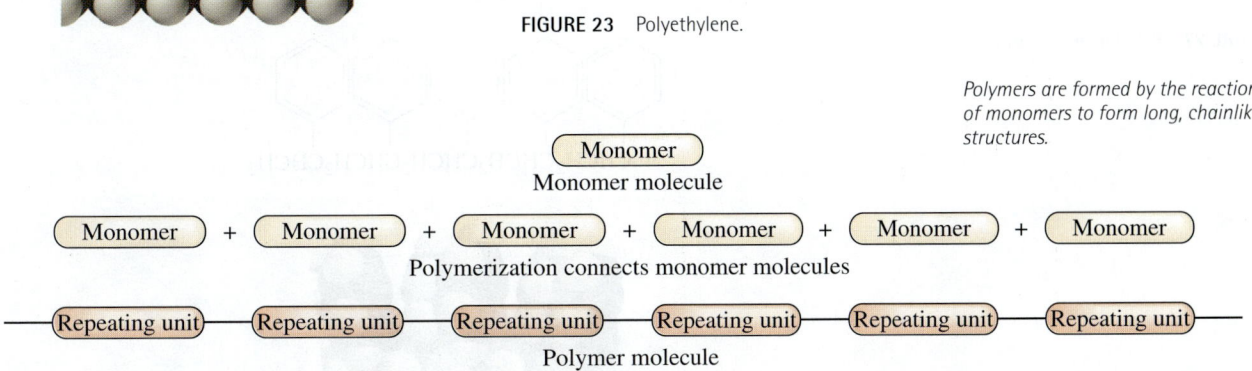

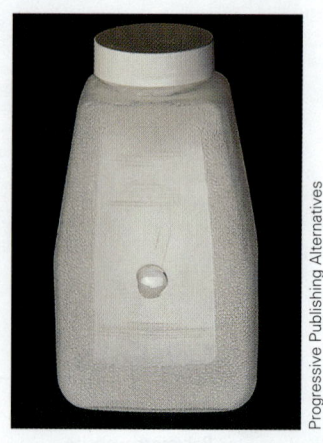

Because the structure of polyethylene resembles the structure of long-chain hydrocarbon waxes, its properties are similar to wax. Both wax and polyethylene repel water.

polyethylene can be stained by oils and greases. Polyethylene is an example of a **thermoplastic**; it softens when heated and hardens when cooled. Consequently, it can be heated and remolded many times. It is also an example of an **addition polymer**; it is formed by the addition of monomer units, one to the other without the elimination of any atoms.

Polyethylene is formed in two different ways. In one formulation, called low-density polyethylene (LDPE), the chains branch as they form. As we know from our study of fats and oils, branched hydrocarbon chains cannot pack as efficiently as straight ones. LDPE is the soft, flexible plastic used in garbage bags and shopping bags. In a second formulation, called high-density polyethylene (HDPE), the reaction conditions are controlled such that the ethylene units add to form long straight chains. These chains pack more efficiently, resulting in a denser, tougher plastic. HDPE is used in bottles that hold liquids such as fruit juices, bleach, and antifreeze.

Many other polymers can be thought of as substituted polyethylenes (Table 1). **Polypropylene**, for example, is formed from propylene, ethylene with a —CH_3 in place of one hydrogen atom (Figure 24). Propylene monomers can link together much like ethylene monomers to form long chains of carbon atoms that resemble polyethylene; the only difference is that in polypropylene every other carbon atom has a —CH_3 group attached (Figure 25). Polypropylene can be made under different conditions to give plastics with different characteristics. It can also be drawn into fibers and is often used in athletic clothing, fabrics, and carpets.

Other substituted polyethylenes include **polyvinyl chloride (PVC)** (Figure 26) and **polystyrene** (Figure 27). In PVC, a chlorine atom substitutes for one of the

FIGURE 24 Propylene, also called propene.

FIGURE 25 Polypropylene, a polymer that is often drawn into fibers for making athletic clothing.

FIGURE 26 Polyvinyl chloride (PVC).

FIGURE 27 Polystyrene structure.

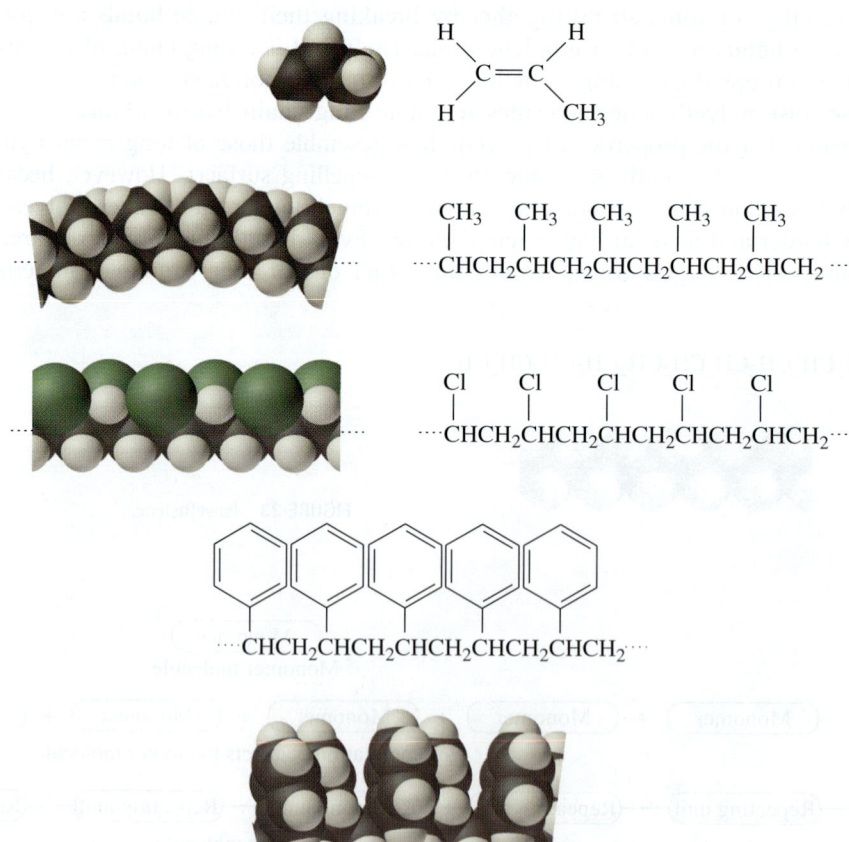

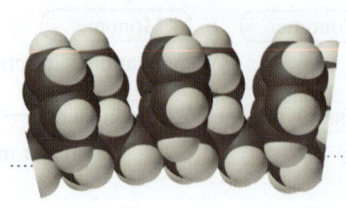

TABLE 1
Polymers Related to Polyethylene: Their Properties and Uses

Name	Formula — Monomer	Formula — Polymer	Recycling Symbol	Properties[a] and Uses
Polyethylene	$H_2C=CH_2$	$+CH_2-CH_2+_n$	2 HDPE, 4 LDPE	Unreactive, flexible, impermeable to water vapor; packaging films, containers, toys, housewares
Polypropylene	$CH_3CH=CH_2$	$+CH-CH_2+_n$ with CH_3	5 PP	Lowest density of any plastic; indoor-outdoor carpeting, upholstery, pipes, bottles
Polyisobutylene	$CH_2=C(CH_3)_2$	$+CH_2-C(CH_3)_2+_n$	7 OTHER	Elastomer; inner tubes, truck and bicycle tires
Polyvinyl chloride (PVC)	$H_2C=CHCl$	$+CH_2-CH(Cl)+_n$	3 V	Self-extinguishing to fire; pipe, siding, floor tile, raincoats, shower curtains, imitation leather upholstery, garden hoses
Polytetrafluoroethylene (Teflon)	$F_2C=CF_2$	$+CF_2-CF_2+_n$	7 OTHER	Very unreactive, nonstick, relatively high softening point; liners for pots and pans, greaseless bearings, artificial joints, heart valves, plumbers' tape, fabrics
Polystyrene	$CH=CH_2$ (with phenyl)	$+CH-CH_2+_n$ (with phenyl)	6 PS	Housings for large household appliances such as refrigerators, auto instruments and panels, clear cups and food containers, and foam cups and packing
Polyacrylonitrile (Orlon)	$H_2C=CHCN$	$+CH_2-CH(CN)+_n$	7 OTHER	Carpets and knitwear
Polymethyl methacrylate (Lucite, Plexiglas)	$H_2C=CCOOCH_3$ with CH_3	$+CH_2-C(COOCH_3)(CH_3)+_n$	7 OTHER	Substitute for glass, airplane windows, contact lenses, fiber optics, paint
Polyvinyl acetate	$H_2C=CH-O-C(=O)CH_3$	$+CH_2-CH(O-C(=O)CH_3)+_n$	7 OTHER	Adhesives, paint, chewing gum, safety glass

[a] Those that are outstanding or unusual.

Source: Adapted from *General Chemistry*, Third Edition, by Jean B. Umland and Jon M. Bellama. Copyright © 1999 Brooks/Cole.

Polytetrafluoroethylene (PTFE), a form of Teflon.

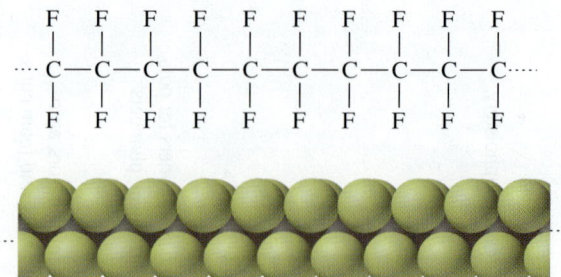

PVC is often used to make pipes.

hydrogen atoms on the ethylene monomer. The resulting polymer is a long chain of carbon atoms with a chlorine atom on every other carbon. PVC is the strong, tough plastic used in pipes, cable coverings, and gaskets. In polystyrene, a benzene ring is the substitute. Polystyrene can be formulated to make a tough, clear plastic called lucite, or it can be filled with tiny bubbles during its formation to make styrofoam.

A particularly tough and chemically resistant polymer is made by substituting all four of the hydrogen atoms on the ethylene monomer with fluorine atoms. The resulting monomer is called tetrafluoroethylene, and the polymer is polytetrafluoroethylene (PTFE), a form of Teflon. Like chlorinated hydrocarbons, fluorinated hydrocarbons are chemically inert; the C—F bond is particularly strong and resistant to chemical attack. Consequently, PTFE is a strong, inert plastic material that resists sticking or degradation by other compounds.

EXAMPLE 15.1

Polymers

Give the structure of the monomer from which the following addition polymer (saran, a substituted polyethylene) is formed:

$$-CH_2CCl_2CH_2CCl_2CH_2CCl_2CH_2CCl_2-$$

SOLUTION

The monomer is the simplest repeating unit with a double bond in place of the single bond, which in this case is

$$CH_2=CCl_2$$

Because substituted polyethylenes are addition polymers, every atom in the monomer appears in the polymer.

YOUR TURN

Polymers

Give the structure of the addition polymer that forms from the monomer $CH_2=CHF$.

The Molecular Revolution

Conducting Polymers

Plastics are usually not good electrical conductors. To conduct electricity, a material must have loose electrons that can move when a voltage is applied. The electrons in the polymers that compose plastics are tied up in strong covalent bonds, making plastics good insulators, not good conductors. However, in the 1970s a Japanese chemist made a mistake when following a recipe to make polyacetylene. The product he got was a thin, silvery film that resembled aluminum foil, but that could stretch and bend like a plastic. Even more importantly, the film conducted electricity. The field of conducting polymers was born, and the years that followed produced numerous advances, including the development of plastics that conduct electricity almost as well as copper.

The commercial applications of conducting polymers are still being developed, but the potential for plastics that conduct electricity are as broad as the uses of plastic itself. One potential application is in polymer circuits for flat panel displays. Because the properties of polymers are controllable, a display using them could be made flexible, allowing one to roll it up after use. The display could also be made crystal clear. Transparent plastic displays could be used to view images on a windshield or on the windows in an airplane cockpit. Imagine a car where the speedometer, gas gauge, and temperature gauge were all displayed right on the front windshield. Other applications of conducting polymers include plastic lights, plastic batteries, and antistatic coatings. The main barrier to these applications at this time is cost. However, as the technology develops, costs are likely to decrease, making conducting plastics a viable option.

15.11 Copolymers: Nylon, Polyethylene Terephthalate, and Polycarbonate

Many polymers are not composed of just one type of monomer, but two. These polymers are called **copolymers** and result in chains composed of alternating units, rather than a single repeating unit. One form of nylon, called nylon 66, is a copolymer of the monomers adipic acid and hexamethylenediamine (Figure 28). When nylon forms from these two monomers, a water molecule is expelled for each bond that forms (Figure 29). Polymers that expel atoms, usually water, during their formation are called **condensation polymers**. Nylon is usually drawn

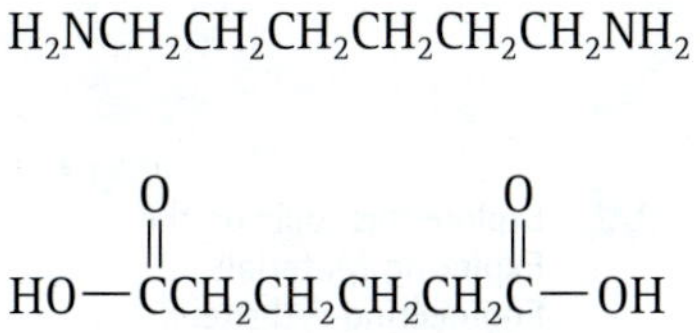

$H_2NCH_2CH_2CH_2CH_2CH_2CH_2NH_2$

$$HO-\overset{O}{\overset{\|}{C}}CH_2CH_2CH_2CH_2\overset{O}{\overset{\|}{C}}-OH$$

FIGURE 28 Hexamethylenediamine and adipic acid, the two monomers that form nylon.

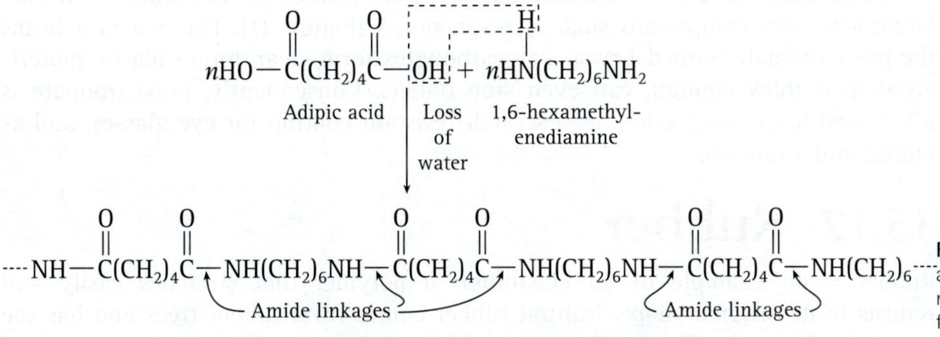

$$n\text{HO}-\overset{O}{\overset{\|}{C}}(CH_2)_4\overset{O}{\overset{\|}{C}}-OH + n\text{HN}(CH_2)_6NH_2$$

Adipic acid | Loss of water | 1,6-hexamethylenediamine

$$\cdots NH-\overset{O}{\overset{\|}{C}}(CH_2)_4\overset{O}{\overset{\|}{C}}-NH(CH_2)_6NH-\overset{O}{\overset{\|}{C}}(CH_2)_4\overset{O}{\overset{\|}{C}}-NH(CH_2)_6NH-\overset{O}{\overset{\|}{C}}(CH_2)_4\overset{O}{\overset{\|}{C}}-NH(CH_2)_6\cdots$$

Amide linkages — Amide linkages

A section of the polyamide chain in Nylon 66

FIGURE 29 When nylon forms from adipic acid and 1,6-hexamethylenediamine, a water molecule is expelled for each bond that forms. This kind of polymerization is called condensation polymerization.

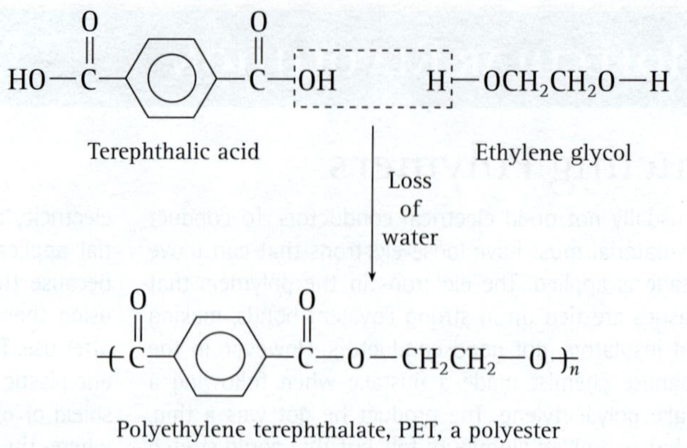

FIGURE 30 Terephthalic acid and ethylene glycol react to give the condensation polymer polyethylene terephthalate (PET), a polyester.

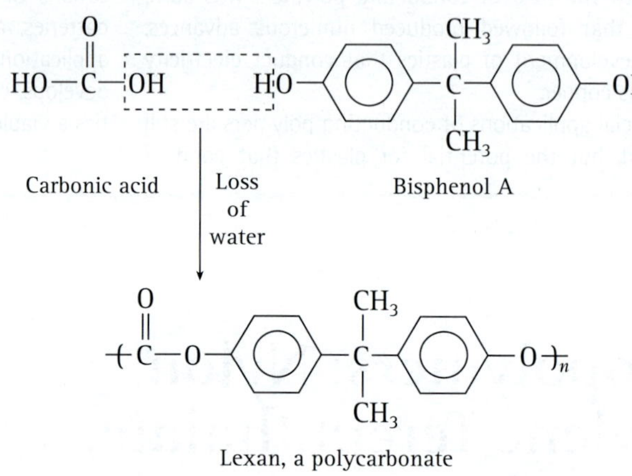

FIGURE 31 Carbonic acid and bisphenol A react to give the condensation polymer Lexan, a polycarbonate.

W Explore this topic on the Exploring Materials Engineering website.

into filaments and is best known for its use in nylon stockings, pantyhose, cords, and bristles.

Other copolymers include polyethylene terephthalate (PET) and polycarbonate. PET is a condensation copolymer formed by the reaction of terephthalic acid with ethylene glycol (Figure 30). Notice the repeating ester groups (—R—COO—R—) in the structure of PET; polymers with this structure are called *polyesters*. PET can be drawn into fibers to make Dacron, often used in clothing; or it can be made into thin sheets called Mylar, used in packaging, as a waterproof coating on sails, as a base for magnetic video and audio tapes, and as a base for photographic film.

Polycarbonates are condensation copolymers formed by the reaction of carbonic acid with compounds such as bisphenol A (Figure 31). This reaction forms the polycarbonate termed *Lexan*. Polycarbonates form clear, tough plastic materials that, if thick enough, can even stop bullets. Consequently, polycarbonate is often used in eye protection, as a scratch-resistant coating for eye glasses, and as bulletproof windows.

15.12 Rubber

Rubber is an example of an **elastomer,** a polymer that stretches easily and returns to its original shape. Natural rubber comes from rubber trees and has the structure shown in Figure 32; it is called polyisoprene because it consists of

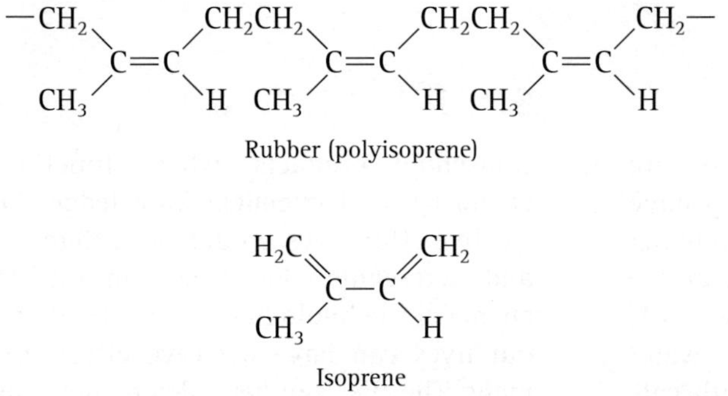

FIGURE 32 Natural rubber is polyisoprene, a polymer of the monomer isoprene.

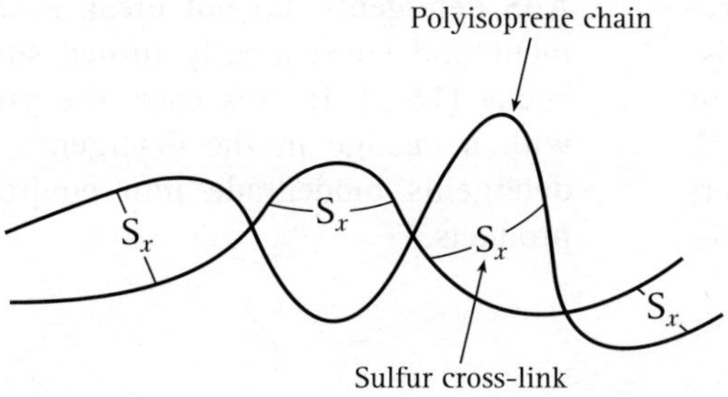

Rubber (polyisoprene)

Isoprene

FIGURE 33 Vulcanized rubber has sulfur cross-links between the polymer strands. These cross-links make the rubber harder, more elastic, and more heat resistant.

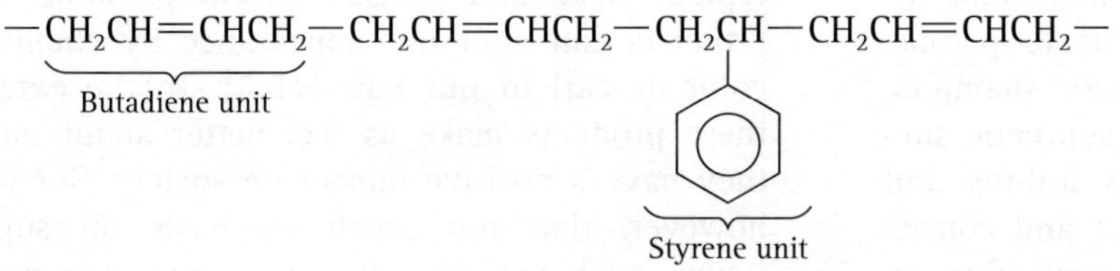

Polyisoprene chain

Sulfur cross-link

FIGURE 34 Styrene-butadiene rubber (SBR) is a copolymer of butadiene and styrene. The proportions of styrene and butadiene can be varied to achieve the desired properties.

$-CH_2CH=CHCH_2-CH_2CH=CHCH_2-CH_2CH-CH_2CH=CHCH_2-$

Butadiene unit

Styrene unit

repeating isoprene units. Polyisoprene polymers form coils that stretch under tension but recoil when tension is released, giving rubber its elasticity. However, when rubber is warm, tension causes the polymer chains to lose their original conformation and the rubber becomes soft and tacky. Rubber can be made harder and more elastic by a process called **vulcanization** invented in 1839 by **Charles Goodyear** (1800–1860). In this process, sulfur atoms form cross-links between the polyisoprene chains (Figure 33). These cross-links still allow the polymer chains to uncoil under stress, but they force them to snap back to their original position when the stress is released, even if the rubber is warm. Vulcanized rubber is commonly used in bicycle and automobile tires as well as in engine mounts, building foundations, and bridge bearings.

Chemists have succeeded in making several synthetic rubbers. A general purpose synthetic rubber is styrene–butadiene rubber (SBR). SBR is a copolymer formed from monomers styrene and butadiene (Figure 34). Unlike many copolymers that must be combined in a 1:1 ratio, SBR can be made with varying amounts of each of its constituents. A common formulation consists of 25% styrene and 75% butadiene. SBR, cheaper than natural rubber, is used extensively to replace natural rubber in automobile tires. SBR ages better than natural rubber, but it does not have as much strength or tack (stickiness). Other synthetic rubbers include polybutadiene, used in tires and footwear, and polychloropene (neoprene), used in hoses and protective clothing such as wet suits.

Chapter Summary

MOLECULAR CONCEPT

Soaps and detergents, also called surfactants, are composed of molecules with dual natures; they have a polar end and a nonpolar end. Surfactants interact with both polar water molecules and nonpolar grease molecules, allowing grease to be washed away (15.1, 15.2). Synthetic detergents work better in hard water than soap does because they do not react with calcium and magnesium ions to form curd (as soap does) (15.3). The most important ingredients in laundry-cleaning formulations are the surfactants, which dissolve grease and oil, and the builder that softens hard water and improves detergent action (15.4). Some household products—such as drain cleaners (strong bases) and toilet bowl cleaners (strong acids)—are strongly corrosive and should not contact the skin (15.5).

Hair products include chemicals such as thioglycolic acid and hydrogen peroxide, used to break and re-form disulfide linkages in hair proteins to permanently curl hair. Hair is kept clean with shampoo, whose most important ingredient is a synthetic surfactant (15.6). Skin products such as lotions and creams are designed to keep skin moist and consist of water and oil or wax along with emulsifiers to keep the two mixed. Sunscreens have similar compositions with an added ingredient to absorb UV-B light (15.7). Most facial cosmetics are mixtures of waxes, oils, pigments, dyes, and perfumes. Face powders also contain talc, a mineral that helps absorbs water and oil from the skin's surface (15.8). Perfumes are composed of several volatile substances with pleasant odors. Usually, the volatility of the ingredients varies so the smells are experienced in three different phases or notes. Deodorants contain antibacterial agents that kill odor-causing bacteria that live in the armpits. Antiperspirants contain aluminum chlorohydrates that actually reduce sweating by constricting sweat glands. Both deodorants and antiperspirants also contain fragrances, soap, and glycol (15.9).

Plastics are composed of polymers—long, chain-like molecules composed of repeating units called monomers. The plastic used in trash bags and soda bottles is polyethylene, an addition polymer. Many polymers, such as polypropylene (common in athletic

(continued on next page)

SOCIETAL IMPACT

Household products, whose functions depend on chemistry and chemical knowledge, impact the way we live. Our lives would be different without soaps and detergents. Sometimes, in the development of chemical technologies, products that help improve our lives can have negative effects on the environment. The first synthetic detergents, called ABS detergents, are a good example of this kind of problem. ABS detergents did not break down in the environment and consequently turned some natural waters sudsy (15.3). In this case, the problem was solved with a change in the detergent's structure. Today's detergents biodegrade into environmentally benign products.

Many personal care products have allowed us to remain clean and pleasant smelling. Some of these products can alter our appearance by changing the color or curl of our hair (15.6). To the extent that these products make us feel better about ourselves, they have a positive impact on society. Some worry, however, that too much emphasis on superficial issues, such as physical appearance, can make our society forget more important things such as who we are and what is truly important in life.

Plastics have had an impact on society by providing cheap, durable materials from which to build everything from sandwich wrappers to cars (15.10). These products improve the standard of living throughout the world, but they also present environmental problems related to their disposal. Plastics occupy many of our nation's landfills and their resistance to decay clothing), polyvinyl chloride (used to make plastic pipes), polystyrene (lucite and styrofoam), and polytetrafluoroethylene (Teflon), are simply substituted polyethylenes in which different functional groups replace one or more of the hydrogen atoms in the ethylene monomer (15.10). Copolymers consist of two or more different kinds of monomers linked to form a long chain. They are often also condensation polymers because they expel atoms, usually water, during their formation. Common copolymers include nylon 66 (used in hosiery), PET (used to make Dacron fabric), and polycarbonate (used in eye wear and bulletproof windows) (15.11). Rubber is an example of an elastomer, a polymer that stretches easily and returns to its original conformation. Natural rubber is usually vulcanized to increase its strength, heat tolerance, and elasticity. Synthetic rubbers include styrene–butadiene rubber (SBR), polybutadiene, and polychloropene (neoprene) (15.12).

makes the problem worse. Many communities have implemented extensive recycling efforts that allow plastics to be reused in other products, preventing them from landing in the nearest landfill.

Chemistry on the Web

For up-to-date URLs, visit the text website at www.chemistry.brookscole.com/tro3.

- Council for LAB/LAS Environmental Research (CLER)
  http://www.cler.com/lab.html
- Cosmetic, Toiletry, and Fragrance Association
  http://www.ctfa.org/Template.cfm?Section=Consumer_Information
- Polymers
  http://www.nationalgeographic.com/resources/ngo/education/plastics/
- Exploring Materials Engineering
  http://www.engr.sjsu.edu/WofMatE/

Key Terms

addition polymer
alkylbenzenesulfonate (ABS) detergents
anionic detergents
brighteners
builder
cationic detergents
colloidal suspension
condensation polymers
copolymers
elastomer
emulsifier
emulsifying agent
emulsion
enzymes
eutrophication
fillers
Goodyear, Charles
linear alkylsulfonate (LAS) detergents
micelles
nonionic detergents
polyethylene
polymers
polypropylene
polystyrene
polyvinyl chloride (PVC)
relative humidity
surfactants
thermoplastic
vulcanization
zeolites

Exercises

 Assess your understanding of this chapter's topics with an online chapter quiz at **http://chemistry.brookscole.com/tro3**.

QUESTIONS

1. Explain, in terms of molecules, why the stickiness of eating an orange is easily washed off hands with just plain water, while the greasiness of eating french fries requires soap.
2. Draw the structure of a soap molecule and explain how its structure is related to its function.
3. Why are soaps and detergents called surfactants?
4. What is a micelle? Why is it important in the process of washing?
5. What is a colloidal suspension? How can you tell if a solution is a normal solution or a colloidal suspension?
6. Define emulsion and emulsifying agent.
7. What are the problems associated with using soap in hard water?
8. What are the advantages of synthetic detergents over soap?
9. What is the difference between ABS synthetic detergents and LAS synthetic detergents?
10. Explain the difference between anionic detergents, cationic detergents, and nonionic detergents.
11. What kinds of surfactants are used in today's laundry-cleaning formulations?
12. What is a builder?
13. Why is sodium tripolyphosphate no longer used as a detergent additive in the United States?
14. Define eutrophication.
15. What substances have replaced tripolyphosphate in U.S. laundry formulations?
16. Explain the role of each of the following in laundry-cleaning formulations:
    a. enzymes
    b. brighteners
    c. bleaching agents
    d. fragrances
17. What are the active ingredients in drain cleaners, and how do they work?
18. What are the active ingredients in toilet bowl cleaners, and how do they work?
19. Why should drain cleaners and toilet bowl cleaners never be mixed?
20. What are the three types of interactions that occur within the protein keratin to give hair its shape?
21. Explain why allowing hair to dry in curlers results in curled hair.
22. Explain why high pH will damage hair.
23. Explain how permanent curling of hair is accomplished.
24. Why are shampoos more effective in cleaning hair than soap would be?
25. What is the purpose of hair rinses or conditioners? What kinds of compounds do they contain?
26. What pigments are responsible for the color of hair?
27. How can hair be lightened, and darkened?
28. Why is it important for skin to be moist but not too moist?
29. What are the main ingredients in creams and lotions? How do the two differ?
30. How do sunscreens protect your skin from the Sun's rays? What does SPF indicate?
31. What are the main ingredients in lipstick?
32. What are the main ingredients in facial foundations, facial powders, and mascara?
33. Explain what is happening with molecules in the process of smelling a meal that is cooking.
34. What kinds of fragrances are used in perfumes? Are they all natural?
35. Explain the three notes in a perfume and the kinds of molecules that you would expect in each note.
36. What is the cause of body odor?
37. How do deodorants help eliminate body odor?
38. How do antiperspirants work?
39. Define each of the following terms:
    a. monomer
    b. polymer
    c. copolymer
40. What are the main uses of high-density polyethylene and low-density polyethylene?
41. What are some of the uses of polypropylene, polyvinyl chloride, and polystyrene?
42. Explain why Teflon resists sticking and chemical attack.
43. Define each of the following:
    a. thermoplastic
    b. addition polymer
    c. condensation polymer
    d. copolymer
    e. elastomer
    f. vulcanization

44. What are each of the following polymers used for?
    a. nylon
    b. polyethylene terephthalate (PET)
    c. polycarbonate
    d. polyisoprene
    e. styrene–butadiene rubber (SBR)

PROBLEMS

45. Sodium stearate is a soap with the molecular formula $C_{18}H_{35}O_2Na$. Draw its structure.
46. Sodium palmitate is a soap with the molecular formula $C_{16}H_{31}O_2Na$. Draw its structure.
47. The salt bridges that hold hair protein (keratin) strands together rely on $-NH_3^+$ groups. Write an equation the reaction of this group with OH^- ions.
48. Draw the structure of a salt bridge between two keratin protein strands. What kinds of ions become common in high pH solutions? Would these damage the salt bridge?
49. The hydrochloric acid present in toilet bowl cleaners reacts with the calcium carbonate often deposited in toilet bowls to form calcium ions, carbon dioxide, water, and chlorine ions. Write a balanced chemical equation for the reaction.
50. Suppose you ran out of toilet cleaner in your home. What other household product might work to dissolve the calcium carbonate deposits in toilet bowls? Write a chemical equation to show how this product works.
51. When air cools at night, water often condenses out of the air to form dew. Explain dew in terms of relative humidity. (Hint: Water will condense out of air when the relative humidity nears 100%.)
52. When relative humidity is high, sweat does not easily evaporate from skin. Why?
53. A certain perfume tends to lack a top note. What kinds of molecules could you add to the perfume to improve this?
54. A certain perfume tends to lack a bottom note. What kinds of molecules could you add to the perfume to improve this?
55. Draw a section of the polyethylene polymer about five monomers long. Show how some hydrogen atoms can be substituted to form polypropylene and polystyrene.
56. Draw a section of the polyethylene polymer about five monomers long. Show how some of the hydrogen atoms can be substituted to form polyethylene and polyvinyl chloride (PVC).
57. Give the structure of the monomer from which the following addition polymer is formed:

    $-CH_2CHCNCH_2CHCNCH_2CHCN-$

58. Give the structure of the addition polymer that forms from the following monomer:

    $CH_2=C(CH_3)_2$

59. Give the structure of the condensation copolymer made from the following monomers:

    $HOOCCH_2CH_2COOH$
    and
    $H_2NCH_2CH_2CH_2CH_2NH_2$

60. Give the structure of the condensation copolymer made from the following monomers:

    $HOOCCH_2CH_2COOH$
    and
    $H_2NCH_2CH_2CH_2CH_2CH_2CH_2NH_2$

POINTS TO PONDER

61. You are planning a backpacking trip and notice soap marked "biodegradable" at the camping store. It costs four times as much as normal soap. Should you buy it? Why or why not?
62. One of the advantages of polymers is their inertness. In contrast to metals such as iron (which rusts over time) or paper and cardboard (which decompose), plastic resists chemical attack. Explain why this advantage can also be a problem. For what other kinds of compounds has this been a problem?
63. Many consumer products are used once and then disposed. Some will be used for only a few minutes and then will sit in a landfill for many years. Can you think of any products like this? Are there any environmentally friendlier alternatives to the products you thought of?

Feature Problems and Projects

64. Which of the following detergent molecules are biodegradable?

   a.

   b.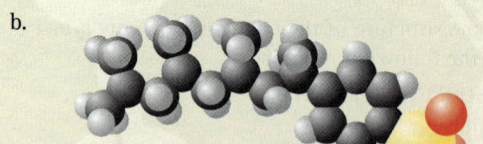

   c.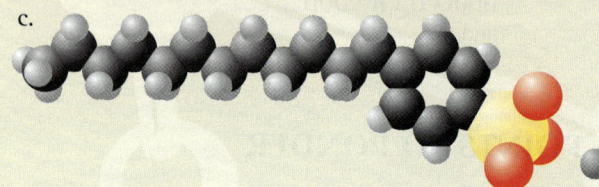

65. What is wrong with this molecular picture of a soap molecule in water?

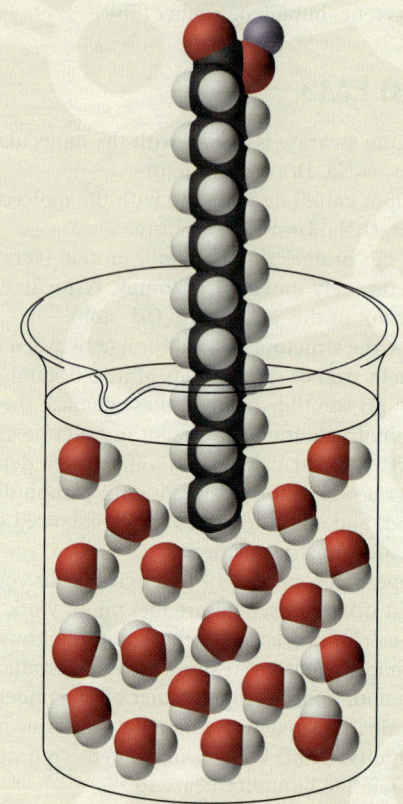

66. Get a bottle of shampoo and examine the ingredients label. List the ones that you recognize and the function they serve in the shampoo.
67. Get a box of laundry detergent and examine the ingredients label. List the ones that you recognize and the function they serve in the detergent.

16 Biochemistry and Biotechnology

Consider also the molecular structure of that stuff of the gene, the celebrated double helix of deoxyribonucleic acid. The structure's elucidation, in March of 1953, was an event of such surpassing explanatory power that it will reverberate through whatever time mankind has remaining. The structure is also perfectly economical and splendidly elegant. There is no sculpture made in this century that is so entrancing.

—Horace Freeland Judson

In this chapter, we explore biochemistry (the chemistry of life) and biotechnology (the application of biochemical knowledge to living organisms). Since 1828, when Friedrich Wöhler first synthesized an organic compound in the laboratory, to today, humans have pursued a chemical description of living organisms. As you read this chapter, reflect on the idea that you and all other living organisms are composed of molecules—complex molecules to be sure, but molecules nonetheless. Interactions and reactions between these molecules maintain life; we can explain a great deal about ourselves by understanding these molecules, their properties, and their reactions. The application of biochemical knowledge to living organisms is full of ethical questions. It is one thing to change the chemical composition of a polymer to create a plastic with different properties; it is quite another to change the chemical composition of an embryonic cell to create a living organism with different characteristics—and yet, this is exactly what we can do. Like all technology, biotechnology carries risks, but with careful consideration of the relevant issues, it can, with time, improve the quality of human life and also help us to better understand ourselves.

QUESTIONS FOR THOUGHT

- What are the basic molecules in living organisms?
- What are fats and how do they differ from one another? What function do they serve in living organisms?
- What are carbohydrates and how do they differ from one another? What functions do they serve in living organisms?
- What are proteins? What composes them?
- How is protein structure related to protein function and what kind of chemical interactions maintain that structure?
- What is DNA? How is it a molecular blueprint? What is it a molecular blueprint for? What are genes?
- How does DNA replicate and how do the blueprints within DNA get constructed?
- What is genetic engineering? How can the DNA from one organism be transferred to another?
- What are the main applications of biotechnology?

16.1 Brown Hair, Blue Eyes, and Big Mice

Eye color and hair color are both determined by molecular sequences called genes that are present in cells.

Have you ever wondered what makes you, you? Why do you have a short temper? Why do you look so much like your mother? Why do you have brown hair or blue eyes? We are finding that the answers to these questions are, at least in part, molecular. Your hair is brown because your cells contain molecular sequences, called genes, that specify how your cells build strands of hair; brown hair genes produce brown-colored hair. If you have blue eyes, the genes that control your eye color contain molecular codes for blue eyes. Almost all your physical appearance—the roundness of your nose, the dimple in your chin, or the color of your skin—is determined by the molecular sequences in your cells called genes. Genes are passed from parent to offspring, which is why you resemble your parents.

The more we study these molecular blueprints called genes, the more it appears that they not only control how we look but also are at least partially responsible for how we behave, how we think, and what diseases we might develop. Indeed, some diseases, such as sickle cell anemia, can be traced back to just a few dozen misplaced atoms in a gene. Other diseases and human traits are more complex, involving both genetic and environmental factors. Nonetheless, our genetic makeup has a lot to say about whether or not we will grow bald, develop cancer, or get Alzheimer's disease.

Our understanding of how genes work has grown exponentially in the last 30 years. With modern techniques, we can not only understand how a particular molecular sequence determines a particular trait, but we can take that molecular sequence out of one organism and implant it into another, which may then exhibit the same trait. For example, scientists have identified and isolated the gene in rats that codes for rat growth hormone, a substance that induces rats to grow to their normal adult size. The gene can be isolated, copied, and injected into fertilized mouse eggs. The gene incorporates into the mouse's cells and causes them to make the hormone that would ordinarily be found only in rats. That hormone makes the mouse grow abnormally large. Thus, we have the potential to design, or genetically engineer, organisms. How will our society respond to this kind of power? Should this kind of technology be applied to humans to treat disease?

In this chapter, we learn about the four major types of molecules present in living organisms: **lipids,** the primary means for long-term energy storage as well as the primary structural components of biomembranes; **carbohydrates,** the primary means for short-term energy storage; **proteins,** the molecules that make up much of our structure (such as muscle, skin, and hair) and control most of our body's chemistry; and **nucleic acids,** the molecules that contain the blueprints for proteins. We will look at biotechnology and particularly at recombinant DNA techniques that allow scientists to manipulate genes, the blueprints of life. This manipulation has led to new drugs and therapies to fight disease, better output and efficiency in agriculture, and the ability to predict and better address future health problems in people.

16.2 Lipids and Fats

Lipids are defined as those cellular components that are insoluble in water but extractable with nonpolar solvents. Lipids include fats, oils, fatty acids, steroids, and some vitamins (A, D, E, and K). Lipids form the structural components of biological membranes and serve as reservoirs for long-term energy storage. They contain more than twice as much energy per gram as any other class of biochemical compounds and are therefore an efficient way to store energy. One type of lipid is the **fatty acid**, an organic acid with a long hydrocarbon tail. Fatty acids have the general formula RCOOH, where R is a hydrocarbon containing between 7 and 21 carbon atoms:

fatty acid

Different fatty acids differ only in their R group (Table 1). For example, in stearic acid, R = $CH_3(CH_2)_{16}$:

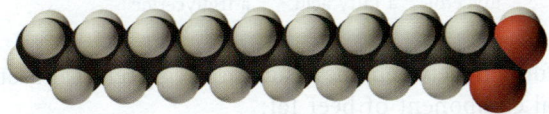

$$CH_3(CH_2)_{16}COH$$

Stearic acid
(melting point 71°C)

TABLE 1
Some Important Fatty Acids

Compound Type and Number of Carbons	Name	Formula	Melting Point (°C)	Common Sources
Saturated				
14	Myristic acid	$CH_3(CH_2)_{12}$—COOH	54	Butterfat, coconut oil, nutmeg oil
16	Palmitic acid	$CH_3(CH_2)_{14}$—COOH	63	Lard, beef fat, butterfat, cottonseed oil
18	Stearic acid	$CH_3(CH_2)_{16}$—COOH	70	Lard, beef fat, butterfat, cottonseed oil
20	Arachidic acid	$CH_3(CH_2)_{18}$—COOH	76	Peanut oil
Monounsaturated				
16	Palmitoleic acid	$CH_3(CH_2)_5CH=CH(CH_2)_7$—COOH	−1	Cod liver oil, butterfat
18	Oleic acid	$CH_3(CH_2)_7CH=CH(CH_2)_7$—COOH	13	Lard, beef fat, olive oil, peanut oil
Polyunsaturated				
18	Linoleic acid	$CH_3(CH_2)_4(CH=CHCH_2)_2(CH_2)_6$—COOH	−5	Cottonseed oil, soybean oil, corn oil, linseed oil
18	Linolenic acid	$CH_3CH_2(CH=CHCH_2)_3(CH_2)_6$—COOH	−11	Linseed oil, corn oil
20	Arachidonic acid	$CH_3(CH_2)_4(CH=CHCH_2)_4(CH_2)_2$—COOH	−50	Corn oil, linseed oil, animal tissues
20	Eicosapentaenoic acid	$CH_3CH_2(CH=CHCH_2)_5(CH_2)_2$—COOH		Fish oil, seafoods
22	Docosahexaenoic acid	$CH_3CH_2(CH=CHCH_2)_6CH_2$—COOH		Fish oil, seafoods

Fats and oils are *triglycerides*, the combination of glycerol (a trihydroxy alcohol) with three fatty acids:

$$\begin{array}{c}CH_2OH\\|\\CHOH\\|\\CH_2OH\end{array} + 3RCOH \longrightarrow \begin{array}{c}CH_2OCR\\|\\CHOCR\\|\\CH_2OCR\end{array} + 3H_2O$$

glycerol   a fatty acid   a triglyceride

The triglyceride formed between the reaction of glycerol and stearic acid is tristearin, a principal component of beef fat:

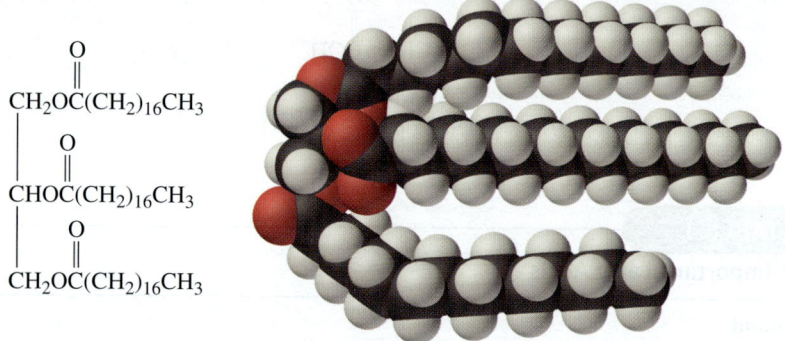

Tristearin

By now, we know enough about the relation between structure and properties to predict some of the properties of tristearin. Like all fats, tristearin has three long hydrocarbon chains, making it nonpolar and imiscible with water.

The long hydrocarbon chains on fats resemble the hydrocarbons such as octane or decane that compose gasoline. The energy stored in fats, much like the energy stored in gasoline, is extracted via the oxidation of these long chains. The main differences between fat and gasoline are the length of the hydrocarbon chains, and the oxygen atoms present in fat but absent in octane. Octane (C_8H_{18}) is in a completely reduced state, while fats, containing some oxygen, are slightly oxidized.

For tristearin, the R groups are composed of **saturated** hydrocarbons; they are *saturated* with hydrogen and do not contain double bonds. This allows neighboring molecules to pack efficiently, making the lipid a solid at room temperature. Fat is conveniently stored in the body and also provides thermal insulation.

$$CH_3CH_2CH_2CH_2CH_2CH_2CH_2CH_2CH_2CH_3$$

Decane, a component of gasoline.

Not all triglycerides are solids at room temperature, however, and this too relates to their molecular structure. Contrast the structure of tristearin, a saturated fat, to triolein, an unsaturated fat that is the main component of olive oil:

$$\begin{array}{l}\text{CH}_2\text{OC}(\text{CH}_2)_7\text{CH}=\text{CH}(\text{CH}_2)_7\text{CH}_3 \\ \quad | \quad \overset{\text{O}}{\underset{\|}{}} \\ \text{CHOC}(\text{CH}_2)_7\text{CH}=\text{CH}(\text{CH}_2)_7\text{CH}_3 \\ \quad | \quad \overset{\text{O}}{\underset{\|}{}} \\ \text{CH}_2\text{OC}(\text{CH}_2)_7\text{CH}=\text{CH}(\text{CH}_2)_7\text{CH}_3 \end{array}$$

Triolein

The R groups of triolein each contain a double bond. The double bond interferes with the efficient packing of neighboring molecules and makes triolein a liquid at room temperature. The differences between tristearin and triolein are general: Saturated fats have no double bonds in their hydrocarbon chains and tend to be solids at room temperature; unsaturated fats contain one or more double bonds and tend to be liquids (oils) at room temperature.

Trilinolenin, which occurs in linseed oil, contains three double bonds in each of its hydrocarbon chains:

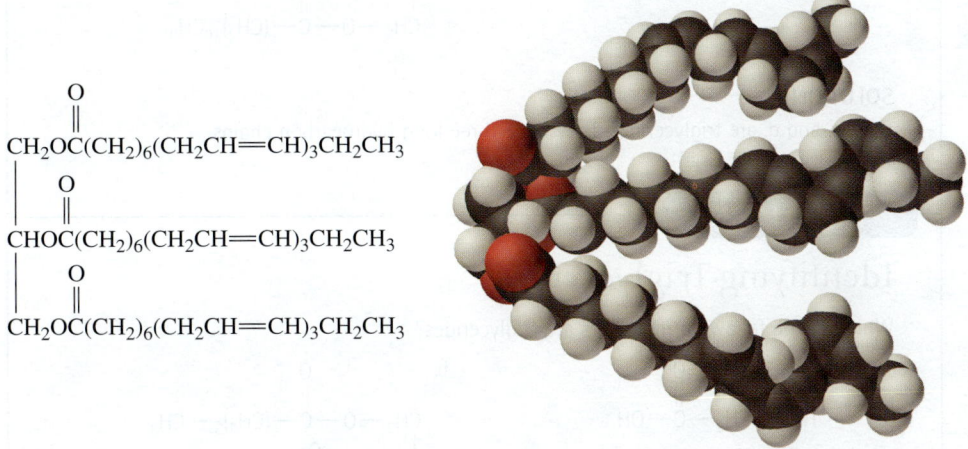

$$\begin{array}{l}\text{CH}_2\text{OC}(\text{CH}_2)_6(\text{CH}_2\text{CH}=\text{CH})_3\text{CH}_2\text{CH}_3 \\ \quad | \quad \overset{\text{O}}{\underset{\|}{}} \\ \text{CHOC}(\text{CH}_2)_6(\text{CH}_2\text{CH}=\text{CH})_3\text{CH}_2\text{CH}_3 \\ \quad | \quad \overset{\text{O}}{\underset{\|}{}} \\ \text{CH}_2\text{OC}(\text{CH}_2)_6(\text{CH}_2\text{CH}=\text{CH})_3\text{CH}_2\text{CH}_3 \end{array}$$

Trilinolenin

It is an example of a polyunsaturated fat. Like its monounsaturated cousin triolein, trilinolenin is a liquid at room temperature. Studies in humans have correlated saturated-fat intake to an increased risk of developing heart disease. Consequently, many diets call for reducing total fat intake and substituting unsaturated fats for saturated fats. Although there are a number of exceptions, fats derived from animals tend to be saturated, while those derived from plants tend to be unsaturated.

We have seen how variations in the structures of triglycerides result in fats and oils with different properties. This variation allows fats and oils to serve a number of purposes in the human body. The same basic structures can act as reservoirs for energy, insulators for heat, barriers to the outside environment, and moisturizers to prevent dryness. We will see this principle at work in many of the biological molecules we examine. Like a composer playing on slight variations in a general theme to create a symphony, nature uses slight variations in a general structure to achieve the many functions that molecules must serve to sustain life.

EXAMPLE 16.1

Identifying Triglycerides

Which of the following molecules are triglycerides?

a.
$$CH_2-O-\overset{\overset{O}{\|}}{C}-(CH_2)_{14}CH_3$$
$$CH-O-\overset{\overset{O}{\|}}{C}-(CH_2)_{14}CH_3$$
$$CH_2-O-\overset{\overset{O}{\|}}{C}-(CH_2)_{14}CH_3$$

b.
$$CH_3-O-\overset{\overset{O}{\|}}{C}-(CH_2)_{14}CH_3$$

c.
$$CH_3CH_2\overset{\overset{O}{\|}}{C}-CH_3$$

d.
$$CH_2-O-\overset{\overset{O}{\|}}{C}-(CH_2)_{16}CH_3$$
$$CH-O-\overset{\overset{O}{\|}}{C}-(CH_2)_{16}CH_3$$
$$CH_2-O-\overset{\overset{O}{\|}}{C}-(CH_2)_{16}CH_3$$

SOLUTION
Both a. and d. are triglycerides. They have three long hydrocarbon chains.

YOUR TURN

Identifying Triglycerides

Which of the following molecules are triglycerides?

a.
$$H_2N-CH_2-\overset{\overset{O}{\|}}{C}-OH$$

b.
$$CH_2-O-\overset{\overset{O}{\|}}{C}-(CH_2)_3-CH_3$$
$$CH-O-\overset{\overset{O}{\|}}{C}-(CH_2)_3-CH_3$$
$$CH_2-O-\overset{\overset{O}{\|}}{C}-(CH_2)_3-CH_3$$

c.
$$CH_2-O-\overset{\overset{O}{\|}}{C}-(CH_2)_7CH=CHCH_3$$
$$CH-O-\overset{\overset{O}{\|}}{C}-(CH_2)_7CH=CHCH_3$$
$$CH_2-O-\overset{\overset{O}{\|}}{C}-(CH_2)_7CH=CHCH_3$$

d.
$$H_2N-\underset{\underset{CH_3}{|}}{CH}-\overset{\overset{O}{\|}}{C}-OH$$

EXAMPLE 16.2

Saturated and Unsaturated Fats

Which of the following triglycerides would you expect to be a solid at room temperature?

a.
$$CH_2-O-\overset{O}{\underset{\|}{C}}-(CH_2)_{14}-CH_3$$
$$CH-O-\overset{O}{\underset{\|}{C}}-(CH_2)_{14}-CH_3$$
$$CH_2-O-\overset{O}{\underset{\|}{C}}-(CH_2)_{14}-CH_3$$

b.
$$CH_2-O-\overset{O}{\underset{\|}{C}}-CH_2CH=CHCH_3$$
$$CH-O-\overset{O}{\underset{\|}{C}}-CH_2CH=CHCH_3$$
$$CH_2-O-\overset{O}{\underset{\|}{C}}-CH_2CH=CHCH_3$$

c.
$$CH_2-O-\overset{O}{\underset{\|}{C}}-(CH_2)_6-(CH=CH)_2-CH_3$$
$$CH-O-\overset{O}{\underset{\|}{C}}-(CH_2)_6-(CH=CH)_2-CH_3$$
$$CH_2-O-\overset{O}{\underset{\|}{C}}-(CH_2)_6-(CH=CH)_2-CH_3$$

d.
$$CH_2-O-\overset{O}{\underset{\|}{C}}-CH=CH-CH_3$$
$$CH-O-\overset{O}{\underset{\|}{C}}-CH=CH-CH_3$$
$$CH-O-\overset{O}{\underset{\|}{C}}-CH=CH-CH_3$$

SOLUTION
a. will be a solid at room temperature because it is a saturated fat. The others are unsaturated oils and will therefore be liquids.

YOUR TURN

Saturated and Unsaturated Fats

Which of the following triglycerides will be a liquid at room temperature?

a.
$$CH_2-O-\overset{O}{\underset{\|}{C}}-(CH_2)_{12}-CH_3$$
$$CH-O-\overset{O}{\underset{\|}{C}}-(CH_2)_{12}-CH_3$$
$$CH_2-O-\overset{O}{\underset{\|}{C}}-(CH_2)_{12}-CH_3$$

b.
$$CH_2-O-\overset{O}{\underset{\|}{C}}-CH_2CH=CHCH_3$$
$$CH-O-\overset{O}{\underset{\|}{C}}-CH_2CH=CHCH_3$$
$$CH_2-O-\overset{O}{\underset{\|}{C}}-CH_2CH=CHCH_3$$

c.

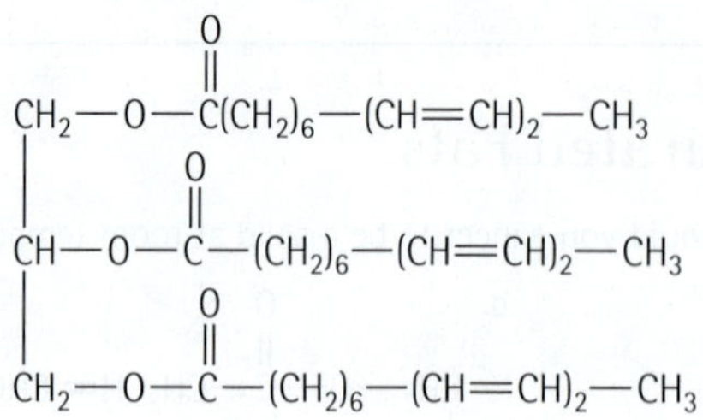

$$CH_2-O-\overset{O}{\underset{\|}{C}}(CH_2)_6-(CH=CH)_2-CH_3$$
$$CH-O-\overset{O}{\underset{\|}{C}}-(CH_2)_6-(CH=CH)_2-CH_3$$
$$CH_2-O-\overset{O}{\underset{\|}{C}}-(CH_2)_6-(CH=CH)_2-CH_3$$

d.

$$CH_2-O-\overset{O}{\underset{\|}{C}}-(CH_2)_{18}-CH_3$$
$$CH-O-\overset{O}{\underset{\|}{C}}-(CH_2)_{18}-CH_3$$
$$CH_2-O-\overset{O}{\underset{\|}{C}}-(CH_2)_{18}-CH_3$$

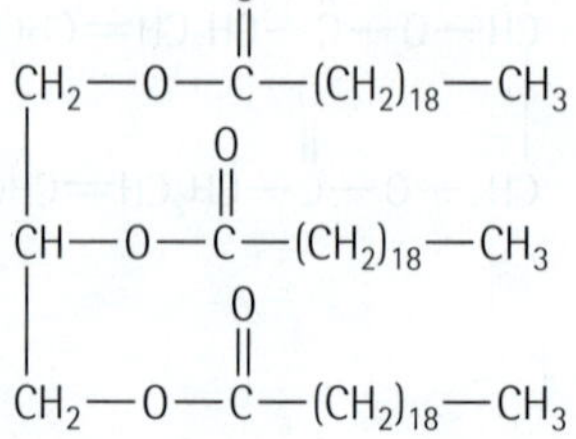

16.3 Carbohydrates: Sugar, Starch, and Sawdust

Examples of carbohydrates include the sugars and starches found in our bodies and in foods. A major function of carbohydrates is short-term energy storage. Carbohydrates are transported throughout the body to cells that oxidize them for energy.

The name *carbohydrate* comes from *carbo* meaning carbon, and *hydrate*, meaning water. Carbohydrates often have chemical formulas that are a multiple of CH_2O, carbon and water. Their chemical structure is related to three classes of organic compounds covered in Chapter 6, aldehydes, ketones, and alcohols. Recall that alcohols contain the —OH group, and aldehydes and ketones contain the C=O group:

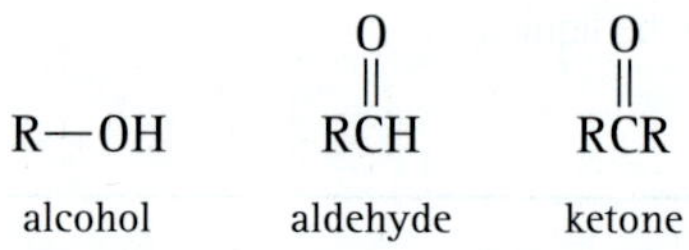

$$R-OH \qquad R\overset{O}{\underset{\|}{C}}H \qquad R\overset{O}{\underset{\|}{C}}R$$

alcohol   aldehyde   ketone

Carbohydrates are polyhydroxy aldehydes or ketones, or their derivatives. A common carbohydrate is glucose ($C_6H_{12}O_6$):

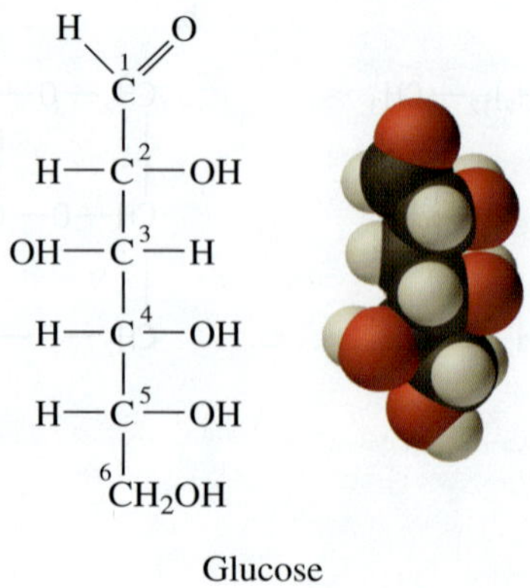

$$\begin{array}{c}
H\diagdownO\\
\overset{1}{C}\diagup\\
|\\
H-\overset{2}{C}-OH\\
|\\
OH-\overset{3}{C}-H\\
|\\
H-\overset{4}{C}-OH\\
|\\
H-\overset{5}{C}-OH\\
|\\
\overset{6}{C}H_2OH
\end{array}$$

Glucose

Carbohydrates undergo an intermolecular cyclization reaction in which one of the —OH groups reacts with the CO group to form a ring:

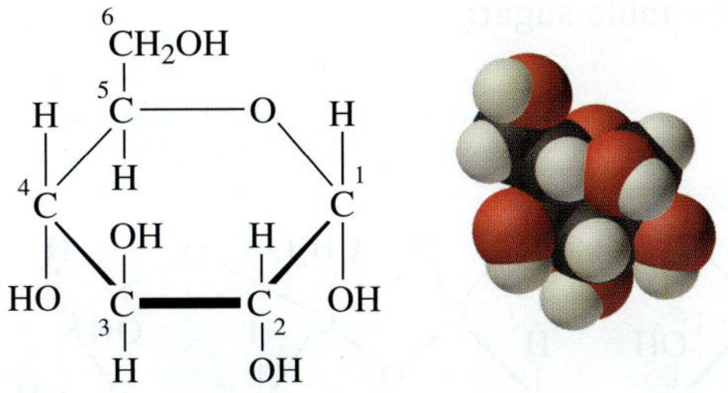

Glucose (ring form)

For glucose, the ring itself contains five carbon atoms and one oxygen atom. Most of the carbon atoms in the ring have an —OH group attached and one has a —CH_2OH group. You should think of a glucose solution at room temperature as a dynamic system in which the interconversion between the two forms constantly occurs. However, at any one instant, the majority of the molecules is in the ring form.

The structure of glucose should tell you something about its properties. The large number of hydroxy (—OH) groups on the ring form strong hydrogen bonds with each other and with water. Consequently, glucose and water mix readily. This is important to the function of glucose as a quick energy source; it must be soluble in bodily fluids. Compared with octane or fats, glucose contains more oxygen atoms per carbon; it is partially oxidized. Consequently, the oxidation of glucose, the reaction that releases its chemical energy in cells, evolves less energy per gram than the oxidation of octane or lipids. Thus, even though glucose does not store energy as efficiently as gasoline or fat, it stores it in a form that is readily transported. Its structure can be viewed as a compromise between the need to store the energy efficiently, and the need to transport it to cells where it is needed.

Fructose, another common carbohydrate, occurs widely in fruits and vegetables and is an isomer of glucose. It has the same chemical formula ($C_6H_{12}O_6$) but a slightly different structure. In its open-chain form, its CO group is located on the second carbon atom rather than on the first. In its ring form, the ring consists of five atoms rather than six:

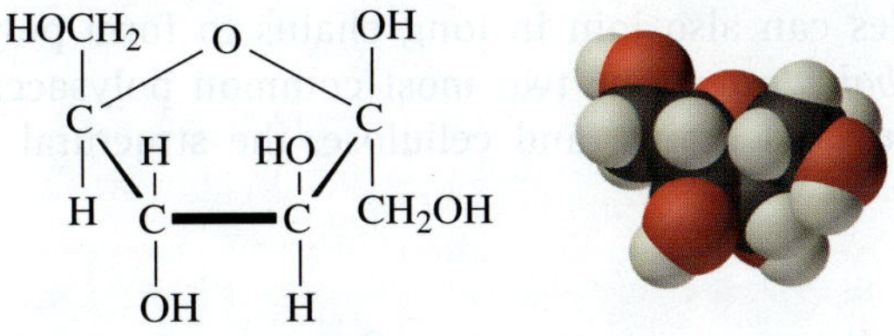

Fructose (ring form)

Consequently, *two* of the carbon atoms in the ring have —CH_2OH groups attached. This difference makes fructose sweeter than glucose and results in an even greater ability to mix with water. Consequently, fructose is sometimes used in low-calorie diets; it takes less carbohydrate to achieve the same level of sweetness.

Table sugar is sucrose, a disaccharide.

Carbohydrates, such as glucose or fructose, composed of a single ring are called **monosaccharides**. Monosaccharides can join together to form double-ringed structures called **disaccharides**. Like their single-ringed cousins, disaccharides mix with water and have a sweet flavor. The most well-known disaccharide is sucrose ($C_{12}H_{22}O_{11}$), or common table sugar:

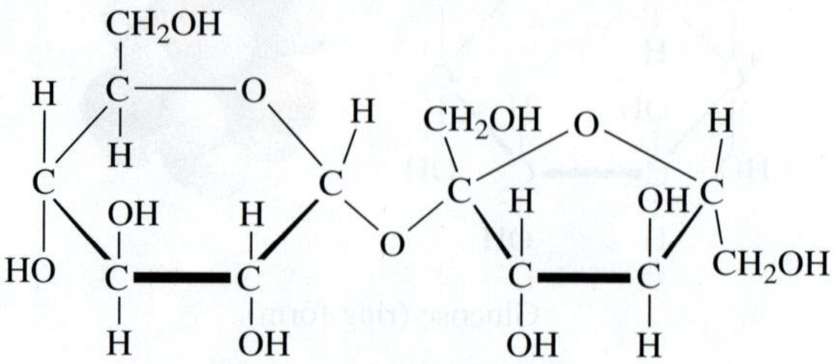

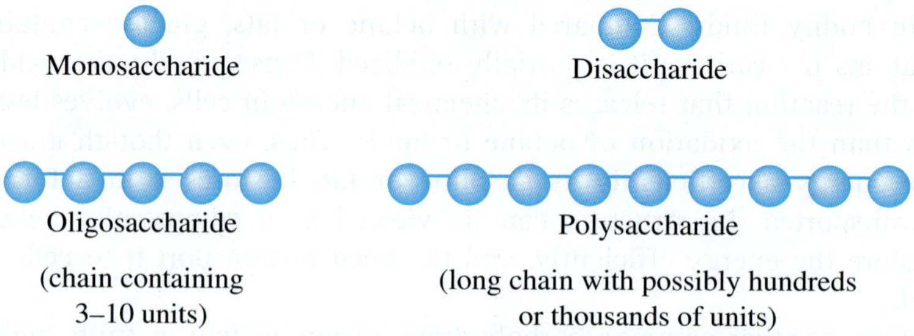

Monosaccharide | Disaccharide

Oligosaccharide | Polysaccharide

(chain containing 3–10 units) | (long chain with possibly hundreds or thousands of units)

Carbohydrate classification. (Source: Reprinted with permission of H. R. Muinos, Muinos Medical Visuals.)

In sucrose, a glucose unit and a fructose unit are linked together to form the two-ringed structure. Sucrose occurs in high concentrations in sugarcane, sugar beet, and maple tree sap (maple syrup).

Monosaccharides can also join in long chains to form **polysaccharides**, also called *complex carbohydrates*. The two most common polysaccharides are starch, common in potatoes and grains, and cellulose, the structural material of plants and trees:

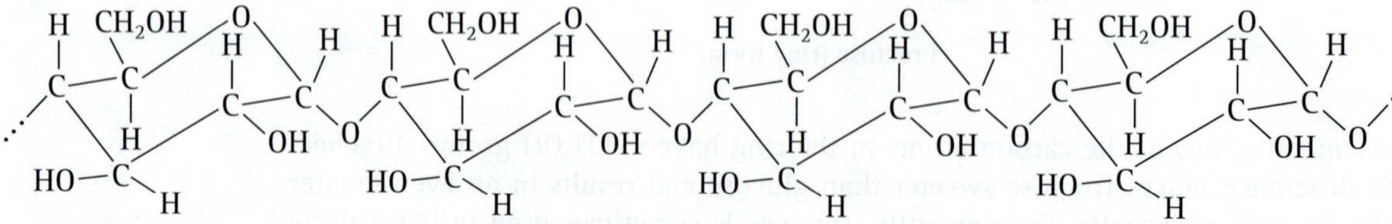

starch

cellulose

The monosaccharide unit of both starch and cellulose is glucose; the difference is in how the units bond together. In starch, the oxygen atom joining the glucose units points down relative to the planes of the rings *(alpha linkage)*, while in cellulose, the oxygen atoms are parallel to the planes of the rings *(beta linkage)*. Here again we see nature capitalize on subtle molecular differences; the twist of a bond has a dramatic macroscopic result. In the first case, oxygen pointing down, you have potatoes (mostly starch); in the second, oxygen parallel, you have sawdust (mostly cellulose).

In the cellulose structure, neighboring units form hydrogen bonds between the —OH group on one unit and the ring oxygen atom on the other. This gives each cellulose molecule a rigid structure that packs efficiently with its neighbors, resulting in the structural strength of cellulose. Starch molecules, however, do not form hydrogen bonds between neighboring glucose units. The molecules are floppy, and consequently starch is softer.

When we eat starches, specialized molecules in our bodies called enzymes (more on these later) have the ability to cut the long starch chains into their individual glucose units; these units are then oxidized in our cells for energy. The process of snipping the chains begins in the mouth, which is why a mouthful of potato or other starch begins to taste sweet if savored for several minutes. Our bodily enzymes

EXAMPLE 16.3

Identifying Carbohydrates

Which of the following are carbohydrates?

a. [structure: aldehyde with H–C=O at top, then H–C–OH, H–C–OH, H–C–OH, CH$_2$OH]

b. [ring structure: HOCH$_2$, O, OH, with H, HO, CH$_2$OH, OH, H substituents]

c. [benzene ring with C(=O)–O–CH$_3$ and OH substituents]

d. CH$_3$CHCH$_2$CH$_2$CH$_3$ with CH$_3$ branch

444 CHAPTER 16 Biochemistry and Biotechnology

SOLUTION
Both a. and b. are carbohydrates. The carbohydrate in a. is in the straight-chain form and can be identified by the presence of C=O and OH groups along its chain. The carbohydrate in b. is in the ring form and can be identified by its ring that contains carbon atoms and one oxygen atom. The ring also has —OH groups attached to it as well as a —CH$_2$OH group, characteristic of carbohydrates.

> ### YOUR TURN
>
> ## Identifying Carbohydrates
>
> Which of the following is a carbohydrate?
>
> a.
> $$CH_2-O-\overset{O}{\underset{\|}{C}}-(CH_2)-CH=CH(CH_2)-CH_3$$
> $$CH-O-\overset{O}{\underset{\|}{C}}-(CH_2)-CH=CH(CH_2)-CH_3$$
> $$CH_2-O-\overset{O}{\underset{\|}{C}}-(CH_2)_{16}CH_3$$
>
> b. phenol (OH on benzene ring)
>
> c. glucose ring structure (CH$_2$OH, OH groups on six-membered ring with O)
>
> d. $CH_2-CH-CH_2$ with OH, OH, OH

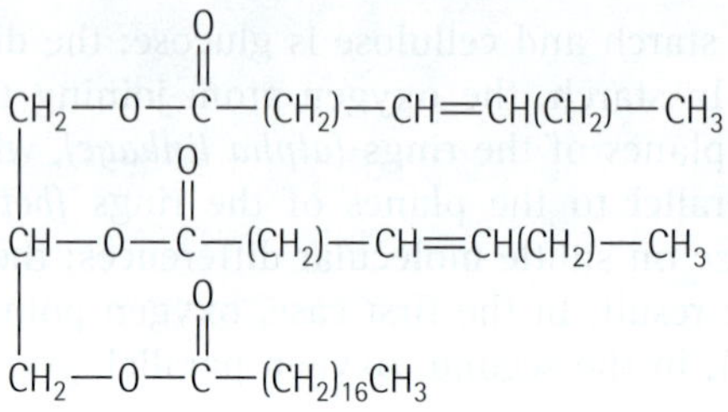

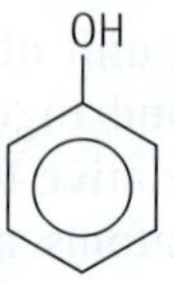

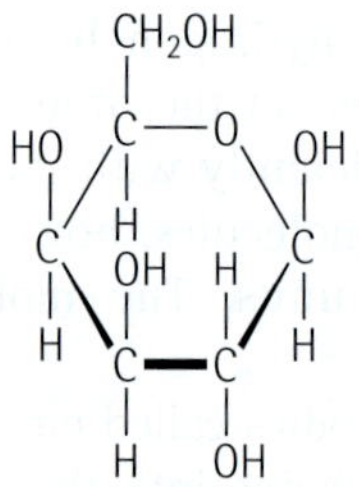

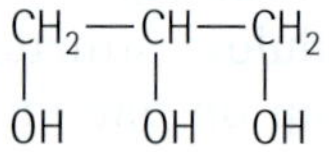

cannot, however, cut the long chains of cellulose. When consumed, cellulose passes through the body without being digested. In the lower intestine, cellulose, also called fiber, gives bulk to the remnants of the digestive process, allowing them to move through the intestine smoothly, preventing constipation.

> ### EXAMPLE 16.4
>
> ## Classifying Saccharides
>
> Classify each of the following carbohydrates as monosaccharides, disaccharides, or polysaccharides.
>
> a. five-membered ring with HOCH$_2$, CH$_2$OH, OCH$_2$CH$_3$, OH substituents
>
> b. two six-membered rings linked by O (disaccharide structure) with CH$_2$OH groups and OH groups
>
> c. six-membered ring with CH$_2$OH and OH groups

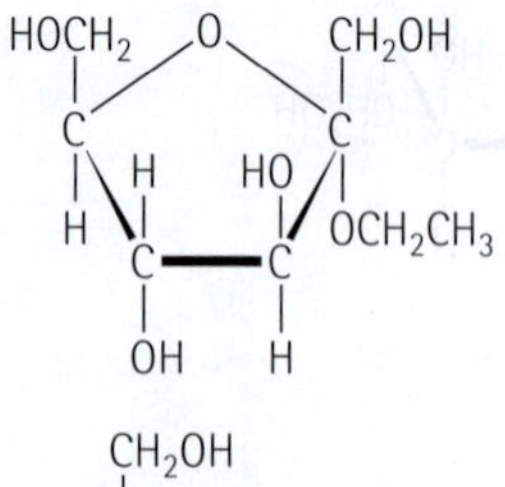

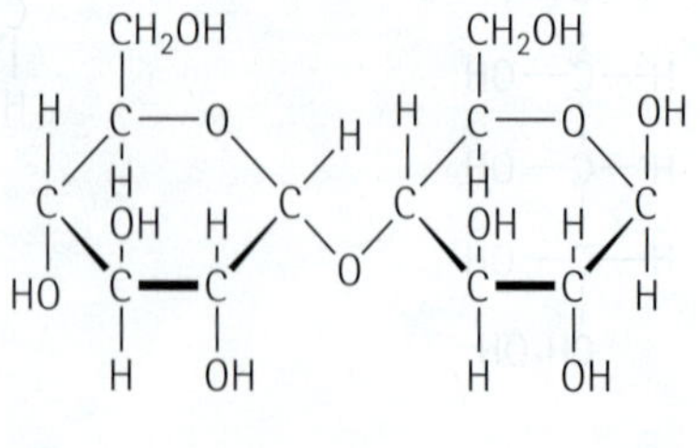

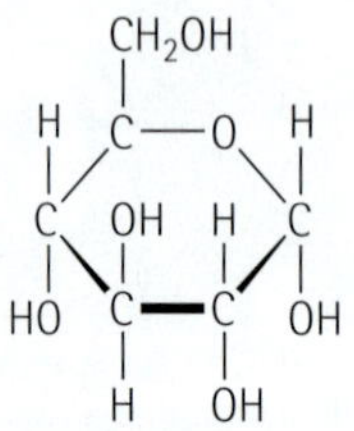

16.3 Carbohydrates: Sugar, Starch, and Sawdust

d. 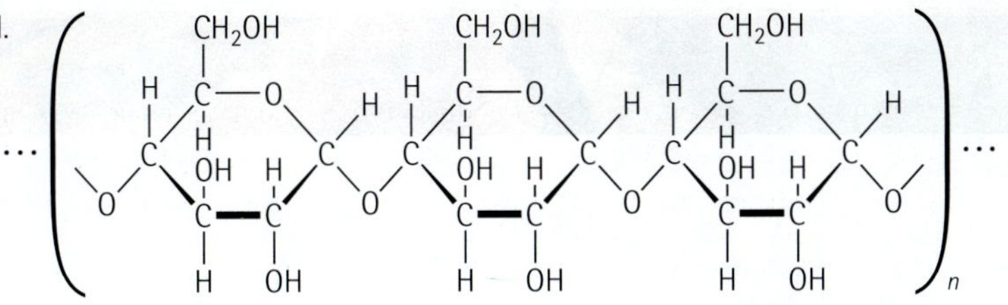 [polysaccharide structure with repeating glucose units]

SOLUTION

Both a. and c. are monosaccharides; b. is a disaccharide; and d. is a polysaccharide.

YOUR TURN

Classifying Saccharides

Classify each of the following carbohydrates as monosaccharides, disaccharides, or polysaccharides.

a. 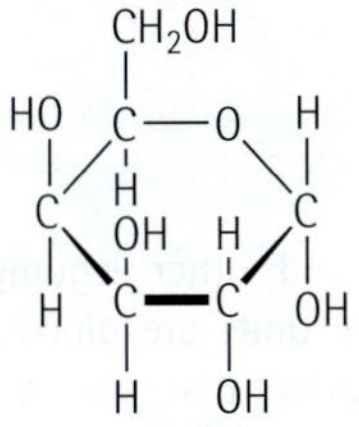 [monosaccharide ring structure]

b. 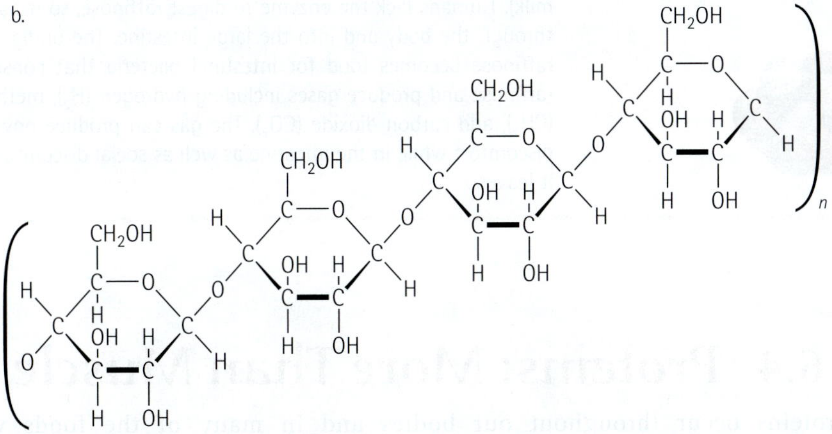 [disaccharide/oligosaccharide structure with repeating units, subscript n]

c. 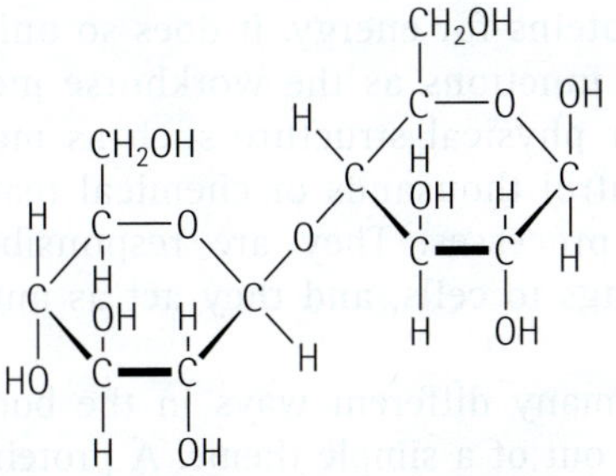 [disaccharide structure]

d. 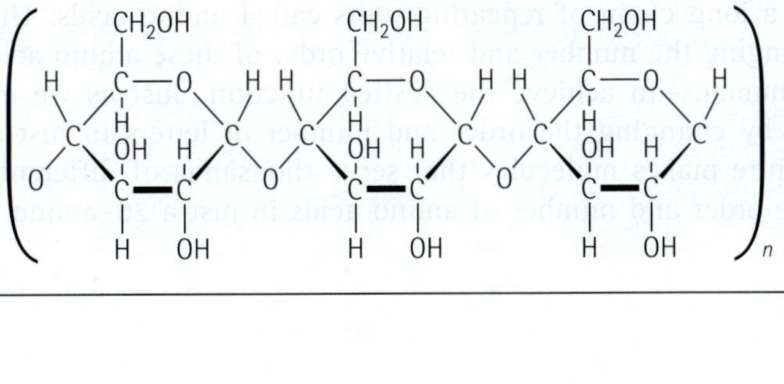 [polysaccharide structure with repeating units, subscript n]

Molecular Focus

Raffinose

Formula: $C_{18}H_{32}O_{16}$
Molecular weight: 504.46 g/mol
Melting point: 80°C
Structure:

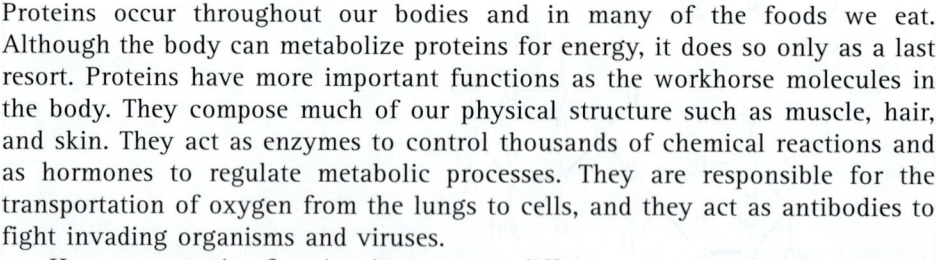

Three-dimensional structure:

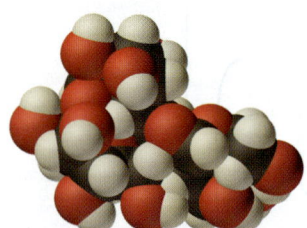

Raffinose, which occurs in beans and other legumes, is a trisaccharide in which the individual units are glucose, fructose, and galactose (a monosaccharide that also occurs in milk). Humans lack the enzyme to digest raffinose, so it passes through the body and into the large intestine. The undigested raffinose becomes food for intestinal bacteria that consume raffinose and produce gases including hydrogen (H_2), methane (CH_4), and carbon dioxide (CO_2). The gas can produce physical discomfort while in the intestine as well as social discomfort as it leaves.

Meats are good sources of protein.

16.4 Proteins: More Than Muscle

Proteins occur throughout our bodies and in many of the foods we eat. Although the body can metabolize proteins for energy, it does so only as a last resort. Proteins have more important functions as the workhorse molecules in the body. They compose much of our physical structure such as muscle, hair, and skin. They act as enzymes to control thousands of chemical reactions and as hormones to regulate metabolic processes. They are responsible for the transportation of oxygen from the lungs to cells, and they act as antibodies to fight invading organisms and viruses.

How can proteins function in so many different ways in the body? Nature again derives complexity and diversity out of a simple theme. A protein molecule is composed of a long chain of repeating units called amino acids. The diversity comes from changing the number and relative order of these amino acids—20 different ones in humans—to achieve the desired function. Just as we make thousands of words by changing the order and number of letters in just a 26-letter alphabet, so nature makes molecules that serve thousands of different functions by changing the order and number of amino acids in just a 20–amino acid set.

The structure of amino acids is related to two classes of organic compounds discussed in Chapter 6, amines and carboxylic acids:

$$\underset{\text{amine}}{\text{RNH}_{\text{H}}} \qquad \underset{\text{carboxylic acid}}{\text{RCOH}}^{\text{O}}$$

Amino acids are molecules that contain both the amine group and the carboxylic acid group bonded to a central carbon atom:

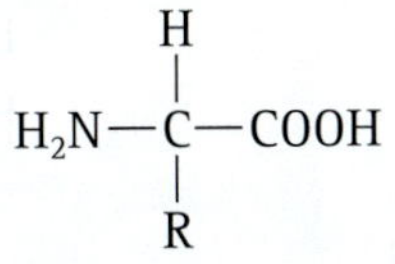

$$\underset{\text{an amino acid}}{\text{H}_2\text{N}-\underset{\underset{R}{|}}{\overset{\overset{H}{|}}{C}}-\text{COOH}}$$

The central carbon atom also has an R group attached. The differences among the 20 different amino acids are different R groups. For example, in the simplest amino acid, glycine, the R group is just a hydrogen atom:

$$\text{H}_2\text{N}-\underset{\underset{H}{|}}{\overset{\overset{H}{|}}{C}}-\text{COOH}$$

In valine, the R group is a hydrocarbon (R = $(CH_3)_2CH-$), and in serine it is an alcohol (R = $HOCH_2-$):

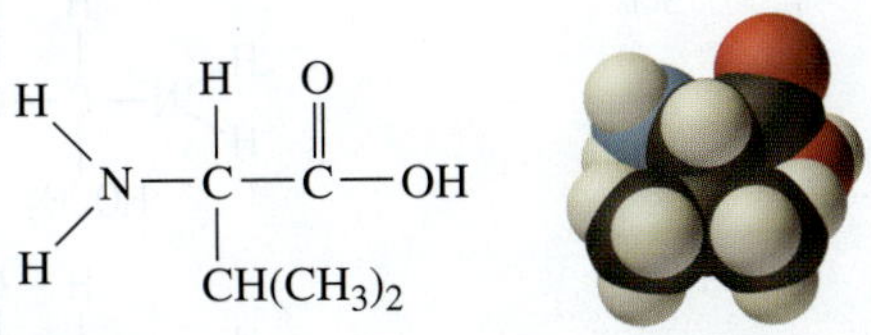

Valine

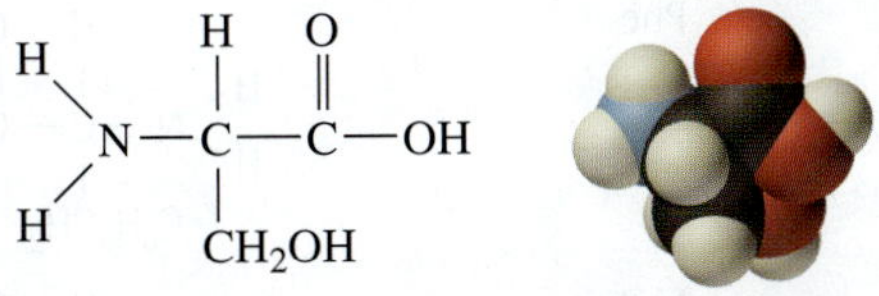

Serine

The 20 different amino acids differ only in their R groups (Table 2). Each R group has a different molecular structure and therefore a different property. For example, valine with its nonpolar hydrocarbon R group gives its section of the protein a hydrophobic characteristic; it will interact well with fats and oils but

TABLE 2
Some Common Amino Acids

Name	Symbol	Formula
Neutral Amino Acids		
Glycine	Gly	H₂N–CH₂–COOH (H,H)N–C(H,H)–C(=O)–OH
Alanine	Ala	H₂N–CH(CH₃)–COOH
Valine	Val	H₂N–CH(CH(CH₃)₂)–COOH [side chain: CH₃CH, CH₃]
Leucine	Leu	H₂N–CH(CH₂CH(CH₃)₂)–COOH [side chain: CH₃CHCH₂, CH₃]
Isoleucine	Ileu or Ile	H₂N–CH(CH(CH₃)CH₂CH₃)–COOH [side chain: CH₃CH₂CHCH₃]
Serine	Ser	H₂N–CH(CH₂OH)–COOH [side chain: HOCH₂]
Threonine	Thr	H₂N–CH(CH(OH)CH₃)–COOH [side chain: CH₃CHOH]
Phenylalanine	Phe	H₂N–CH(CH₂C₆H₅)–COOH [side chain: C₆H₅CH₂]
Methionine	Met	H₂N–CH(CH₂CH₂SCH₃)–COOH [side chain: CH₃SCH₂CH₂]

448 CHAPTER 16 Biochemistry and Biotechnology

TABLE 2 (continued)

Some Common Amino Acids

Name	Symbol	Formula
Neutral Amino Acids		
Cysteine	Cys	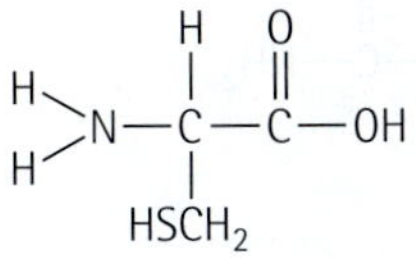
Tryptophan	Trp	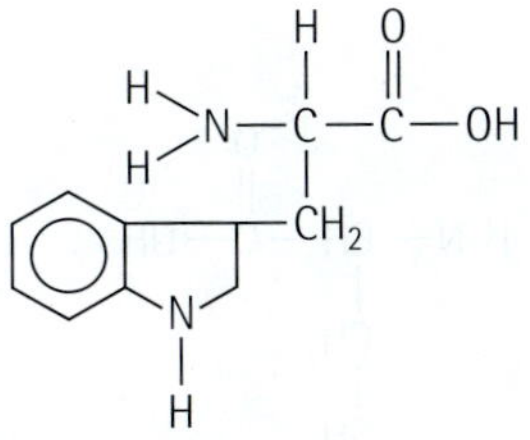
Proline	Pro	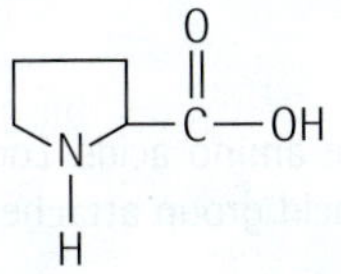
Acidic Amino Acids		
Aspartic acid	Asp	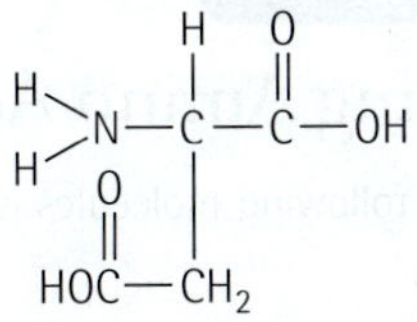
Glutamic acid	Glu	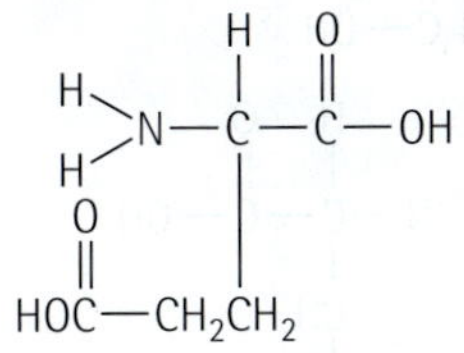
Basic Amino Acids		
Lysine	Lys	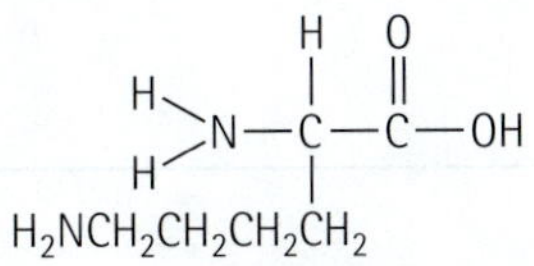
Histidine	His	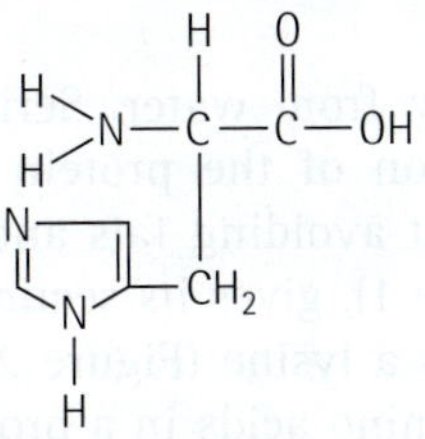

EXAMPLE 16.5

Identifying Amino Acids

Which of the following molecules are amino acids?

a.
$$CH_3-\underset{\underset{O}{\|}}{C}-NH_2$$

b. (cyclic sugar structure with CH₂OH, HO, OH, OH groups)

c.
$$H_2N-\underset{\underset{\underset{SH}{|}}{\underset{CH_2}{|}}}{CH}-\underset{\underset{O}{\|}}{C}-OH$$

d.
$$H_2N-\underset{\underset{CH_3}{|}}{CH}-\underset{\underset{O}{\|}}{C}-OH$$

SOLUTION

Both c. and d. are amino acids. Look for the central carbon atom with an amine group and a carboxylic acid group attached to it.

YOUR TURN

Identifying Amino Acids

Which of the following molecules is an amino acid?

a.
$$CH_3\underset{\underset{O}{\|}}{C}-OH$$

b.
$$CH_3-\underset{\underset{H}{|}}{\overset{\overset{CH_2CH_3}{|}}{N}}-H$$

c.
$$\underset{H}{\overset{H}{\diagdown}}N-\underset{\underset{\underset{C_6H_5}{|}}{\underset{CH_2}{|}}}{\overset{\overset{H}{|}}{C}}-\underset{\underset{O}{\|}}{C}-OH$$

d. (cyclic sugar structure with CH₂OH, OH groups)

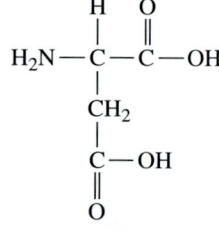

FIGURE 1 Aspartic acid.

FIGURE 2 Lysine.

will stay away from water. Serine, on the other hand, with its polar R group, gives its section of the protein a hydrophilic character, attracting and mixing with water, but avoiding fats and oils. An acidic R group, such as that on aspartic acid (Figure 1), gives its section of the protein acidic character, while a basic R group, such as a lysine (Figure 2), gives its section a basic character. By varying the order of amino acids in a protein, an almost infinite number of properties can be achieved. In this way nature has fine-tuned proteins to perform very specific functions with almost endless variety.

Amino acids link together via peptide bonds to form proteins. The acidic end of one amino acid reacts with the amine side of another to form this bond:

$$\text{H}_2\text{N}-\underset{\underset{R}{|}}{\text{CH}}-\overset{\overset{O}{\|}}{\text{C}}-\text{OH} + \text{H}_2\text{N}-\underset{\underset{R'}{|}}{\text{CH}}-\overset{\overset{O}{\|}}{\text{C}}-\text{OH} \longrightarrow \text{H}_2\text{N}-\underset{\underset{R}{|}}{\text{CH}}-\overset{\overset{O}{\|}}{\text{C}}-\underset{\underset{\text{peptide bond}}{\uparrow}}{\overset{\overset{H}{|}}{\text{N}}}-\underset{\underset{R'}{|}}{\text{CH}}-\overset{\overset{O}{\|}}{\text{C}}-\text{OH} + \text{H}_2\text{O}$$

The result is a dipeptide, two amino acids linked together. The dipeptide can link with a third amino acid to form a tripeptide, which can link with a fourth, and so on. By continuing this pattern, amino acids form long chains where each link is an amino acid bonded to its neighbors by peptide bonds. Chains of amino acids with 50 units or less are called polypeptides; chains with over 50 units are proteins.

EXAMPLE 16.6

Drawing Peptide Structures

Draw the dipeptide that would result from two glycine units forming a peptide bond.

SOLUTION

$$\text{H}_2\text{N}-\underset{\underset{H}{|}}{\overset{\overset{H}{|}}{\text{C}}}-\overset{\overset{O}{\|}}{\text{C}}-\underset{\underset{H}{|}}{\overset{\overset{H}{|}}{\text{N}}}-\underset{\underset{H}{|}}{\overset{\overset{H}{|}}{\text{C}}}-\overset{\overset{O}{\|}}{\text{C}}-\text{OH}$$

YOUR TURN

Drawing Peptide Structures

Draw the tripeptide that would result from cysteine attaching to the previous dipeptide on the acidic side.

Functional proteins often contain hundreds of amino acids in a specific order to carry out a specific function. For example, hemoglobin (Hb), a small to medium protein that transports oxygen in the blood, consists of four protein subchains, each with 146 amino acid units for a total of 584 amino acids. It is a huge molecule with a molecular weight exceeding 60,000 amu and a molecular formula that contains close to 10,000 atoms, $C_{2952}H_{4664}O_{832}N_{812}S_8Fe_4$. The order and number of amino acids in the protein chains are crucial to its function. For example, replacing polar glutamate with nonpolar valine at one position on two of these chains results in a disease called *sickle cell anemia*. In this disease, the solubility of Hb is lowered, resulting in the deformation of red blood cells into a crescent or sickle shape. The deformed cells block the flow of blood (and therefore oxygen) into small capillaries. The cells in the vicinity die, causing pain. Over time, sickle cell anemia can be fatal.

16.5 Protein Structure

 Explore this topic on the **Ras Mol Gallery (Images of DNA, RNA, and Proteins)** website.

Just as the structure of a guitar is finely tuned to achieve a specific sound, the structure of a protein is finely tuned to achieve a specific function. We can characterize protein structure in four different categories: primary, secondary, tertiary, and quaternary.

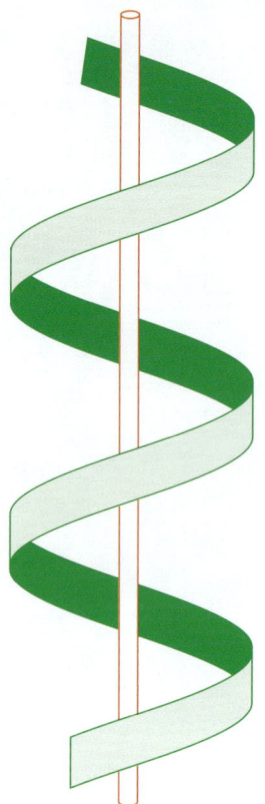
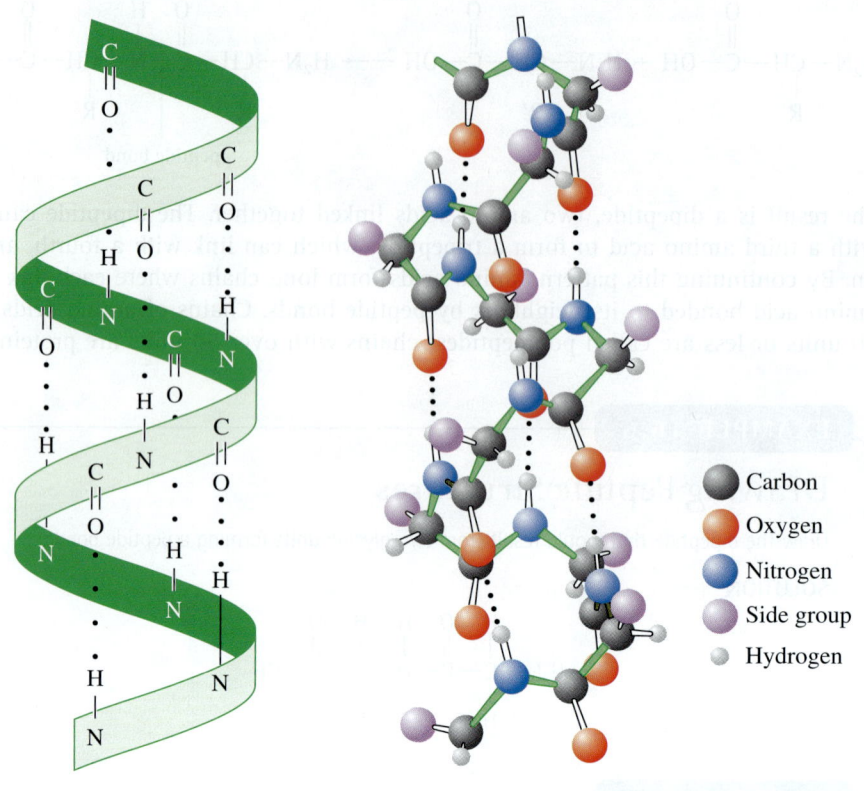

FIGURE 3 The α-helix structure of proteins. The coil is held in shape by hydrogen bonding between different amino acids along the chain.

PRIMARY STRUCTURE

The **primary structure** of a protein is simply the amino acid sequence. Biochemists often use a shorthand notation to specify primary structure. For example, gly-val-ala-asp is the abbreviation for a short polypeptide consisting of the four amino acids, glycine, valine, alanine, and aspartic acid. As we have seen, the primary structure of a protein is held together by peptide bonds between the acid side of one amino acid and the amine side of its neighbor.

SECONDARY STRUCTURE

The **secondary structure** of a protein is the way a chain of amino acids orients itself along its axis. A common secondary structure of proteins is the *alpha-helix (α-helix)* (Figure 3). The helical shape is maintained by hydrogen bonds between different amino acids along the protein chain. α-Keratin, the protein that composes hair and wool, contains chains of glycine, leucine, and several other amino acids in an α-helical structure. The slight elasticity of hair and wool fibers is due to the helical structure of the α-keratin protein. Just as a spring can stretch and return to its original configuration, the coils within the α-helix of hair or wool protein can unwind under tension and snap back when the tension is released.

Another secondary structure found in some proteins is the *pleated sheet* (Figure 4). Here the protein forms zigzag chains that stack neatly to form sheets held together by hydrogen bonds. The smooth feeling of silk is related to the pleated sheet structure of its protein. The inelasticity of silk is related to the near full extension of the protein chains in the pleated-sheet arrangement. The zigzag structure in the pleated sheet is not due to compression of the chain, but rather to the bond angles associated with the polypeptide backbone; consequently the chain

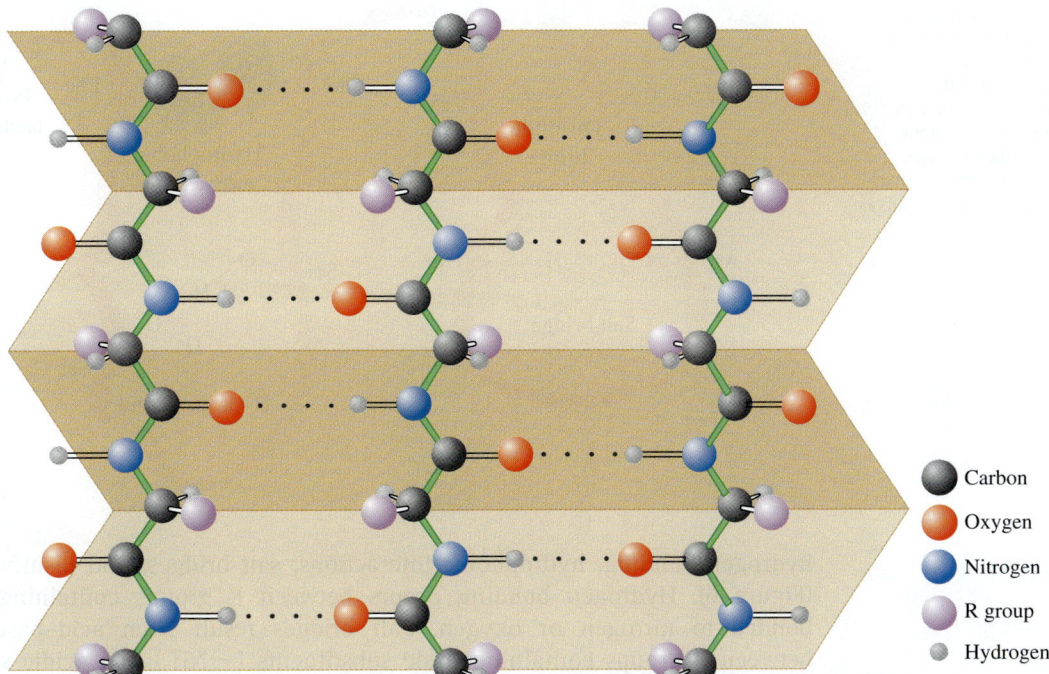

FIGURE 4 The pleated-sheet structure of some proteins. Individual protein chains are held to others via hydrogen bonding. (*Source:* Reprinted with permission of H. R. Muinos, Muinos Medical Visuals.)

is close to maximum extension. The flexibility and softness of silk are related to the ease with which neighboring sheets slide past and around each other.

TERTIARY AND QUATERNARY STRUCTURE

The **tertiary structure** of a protein is the bending and folding due to interactions between amino acids separated by significant distances on the chain (Figure 5). In some proteins, the entire chain is extended, while in others it folds into globular or ball-like structures. Even though the overall shape of a particular protein strand may seem random, it is not; the shape is critical to its function and is reliably reproduced from one molecule to another. For example, as long as the body's cells reliably reproduce the correct sequence of amino acids when synthesizing Hb, Hb molecules will all fold in exactly the same manner to carry out their function.

Many proteins consist of two or more subchains that are themselves held together by interactions between the chains. The arrangement of these subunits in space is called the **quaternary structure** of the protein.

The tertiary and quaternary structures of proteins are maintained by four kinds of interactions between the R groups on different parts of a protein strand:

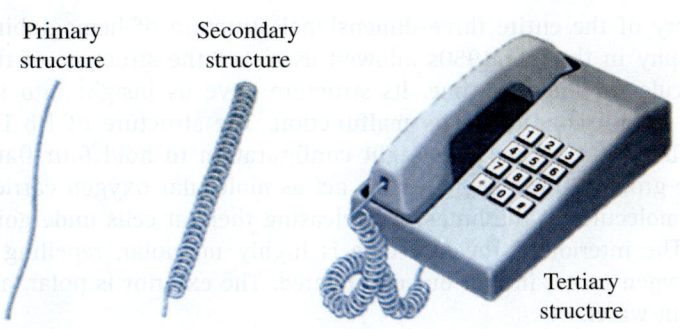

FIGURE 5 In proteins, primary structure simply refers to the linear order of amino acids in the chain. The secondary structure refers to the way the chain might be coiled due to interactions between closely spaced amino acids on the chain. The tertiary structure refers to how the coil orients due to interactions between amino acids that are separated by long distances on the chain.

FIGURE 6 The tertiary and quaternary structures of proteins are maintained by four kinds of interaction between the R groups on different parts of a protein strand: hydrogen bonding, hydrophobic interactions, salt bridges, and disulfide linkages. (*Source:* Reprinted with permission of H. R. Muinos, Muinos Medical Visuals.)

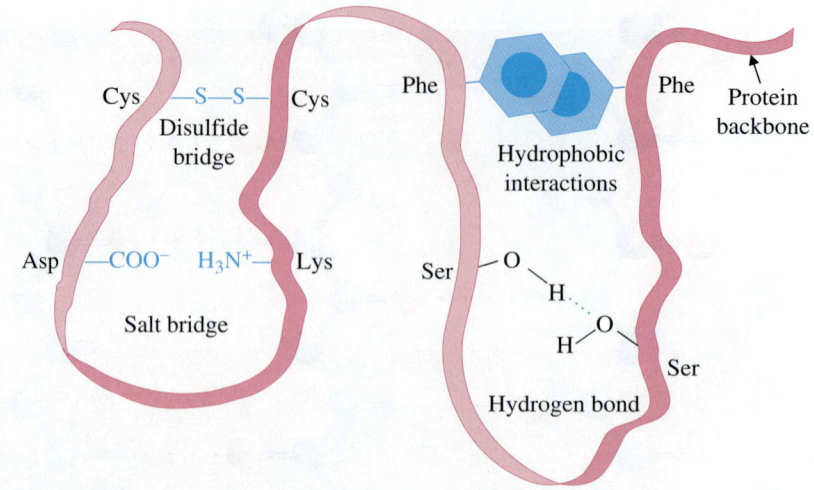

hydrogen bonding, hydrophobic interactions, salt bridges, and disulfide linkages (Figure 6). Hydrogen bonding occurs between R groups containing hydrogen bonded to nitrogen or oxygen. Salt bridges result from acid–base reactions between R groups containing basic substituents ($-NH_2$) and acidic substituents ($-COOH$). Disulfide bridges result from the covalent bonding that occurs between R groups containing sulfur, and hydrophobic interactions occur between nonpolar R groups.

> **APPLY YOUR KNOWLEDGE**
>
> Proteins normally fold so that nonpolar R groups orient toward the interior of the protein. Why?
>
> a. Because nonpolar R groups avoid the aqueous environment in which proteins are normally found.
> b. Because nonpolar R groups are bulky.
> c. Because nonpolar R groups form hydrogen bonds with other nonpolar R groups.
>
> Answer: a. Nonpolar R groups orient toward the interior of proteins to avoid the highly polar aqueous environment in which proteins are normally found.

16.6 Some Common Proteins

HEMOGLOBIN

The discovery of the entire three-dimensional structure of hemoglobin by X-ray crystallography in the late 1950s allowed us to see the structure of this complicated molecule for the first time. Its structure gave us insight into its function and, under certain conditions, its malfunction. The structure of Hb is shown in Figure 7. Hb folds in exactly the right configuration to hold four flat molecules called heme groups. These heme groups act as molecular oxygen carriers, picking up oxygen molecules at the lungs and releasing them at cells undergoing glucose oxidation. The interior of the molecule is highly nonpolar, repelling water and allowing oxygen to get in and out unhindered. The exterior is polar, allowing Hb to dissolve in water.

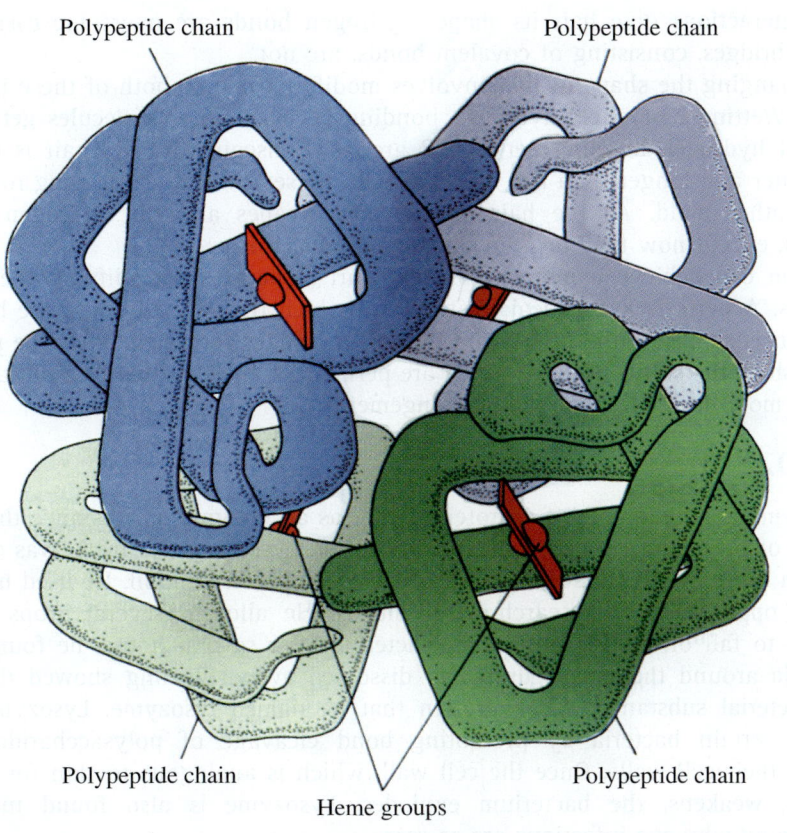

FIGURE 7 The structure of hemoglobin (Hb).

α-KERATIN

α-Keratin is the protein that composes hair and wool. Its α-helix structure is maintained by hydrogen bonding between amino acids along the protein chain. In a hair fiber, three of these α-helices wrap around each other in a coil held together by two kinds of interactions, hydrogen bonds and sulfur bridges. These

MOLECULAR THINKING

Wool

A wool sweater has a tendency to either shrink or stretch when washed. If a wool sweater is washed in hot water and then dried on a hanger, it will stretch beyond its original size. Why do you think this is so? Hint: Consider the effect of washing the sweater in hot water on the hydrogen bonds that maintain the α-helix structure.

QUESTION: If the sweater that was stretched was washed again, but this time laid flat to dry, what would happen? Why?

A wool sweater will stretch beyond its original size if dried on a hanger but will maintain its original size if dried flat.

two interactions give hair its shape. Hydrogen bonds are overcome easily, but sulfur bridges, consisting of covalent bonds, are not.

Changing the shape of hair involves modifying one or both of these interactions. Wetting hair alters hydrogen bonding because water molecules get in the way of hydrogen bonding between R groups. Consequently, wet hair is usually straighter and longer than dry hair. Hair can be set while wet by using rollers or some other mold. As the hair dries, water escapes and the hydrogen bonds reform, except now they take on their imposed shape.

You can achieve a more permanent curl (a perm) by modifying the sulfur bridges. A perm is a chemical treatment in which the sulfur bridges are broken; the hair is shaped with curlers, and the sulfur bridges reformed. After the process is finished, the sulfur bridges, which are permanent chemical bonds, hold the hair in the more desirable (hopefully) arrangement.

LYSOZYME

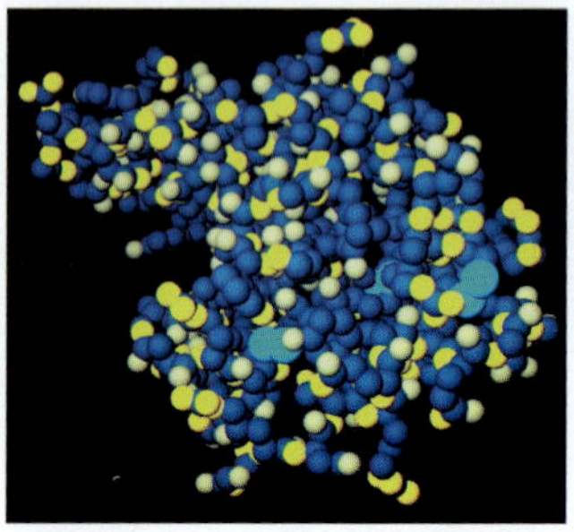

Lysozyme is an example of a protein acting as an **enzyme**, a substance that catalyzes or promotes a specific chemical reaction. Lysozyme (Figure 8) was discovered in 1922 by Alexander Fleming, a bacteriologist in London. He used his cold as an opportunity to research nasal mucus. He allowed several drops of his mucus to fall on a dish containing bacteria. After several hours, he found that bacteria around the mucus drop had dissolved away. Fleming showed that the antibacterial substance was a protein that he named lysozyme. Lysozyme dissolves certain bacteria by promoting bond cleavage of polysaccharide units within their cell walls. Once the cell wall, which is analogous to skin for a bacterium, weakens, the bacterium explodes. Lysozyme is also found in tears, explaining why eye infections are so rare.

FIGURE 8 Structure of lysozyme, an antibacterial protein found in tears and mucus.

INSULIN

Insulin, synthesized in the human pancreas, is an example of a protein acting as a **hormone**, a substance that regulates specific processes within the body. Insulin is relatively small for a protein, composed of only 51 amino acids (Figure 9); yet its function, regulating glucose levels in the blood, is crucial to the correct function of cells. Insulin does its job by promoting the entry of glucose into muscle and fat cells, lowering the glucose level in blood. Diabetes is a disease in which the body fails to make enough insulin, resulting in excessively high blood glucose levels. Many diabetics inject insulin to supply what their bodies fail to make.

FIGURE 9 Structure of insulin.

16.7 Nucleic Acids: The Blueprint for Proteins

The sequence of amino acids in a protein is so important to its function that it must be reproduced with each protein-molecule synthesis. Further, for the offspring of an organism to function properly, it must have the correct sequence of amino acids in its proteins. Nature's blueprints for making proteins are found in nucleic acids, the templates from which all proteins are made. We classify nucleic acids into two categories, deoxyribonucleic acid (**DNA**) and ribonucleic acid (**RNA**). DNA occurs primarily in the information center of the cell called the nucleus. RNA occurs throughout the entire interior of cells.

Nucleic acids are long, chainlike molecules composed of thousands of repeating units called **nucleotides**. Each nucleotide consists of a phosphate group, a sugar group, and a base (Figure 10). The phosphate group and sugar group are

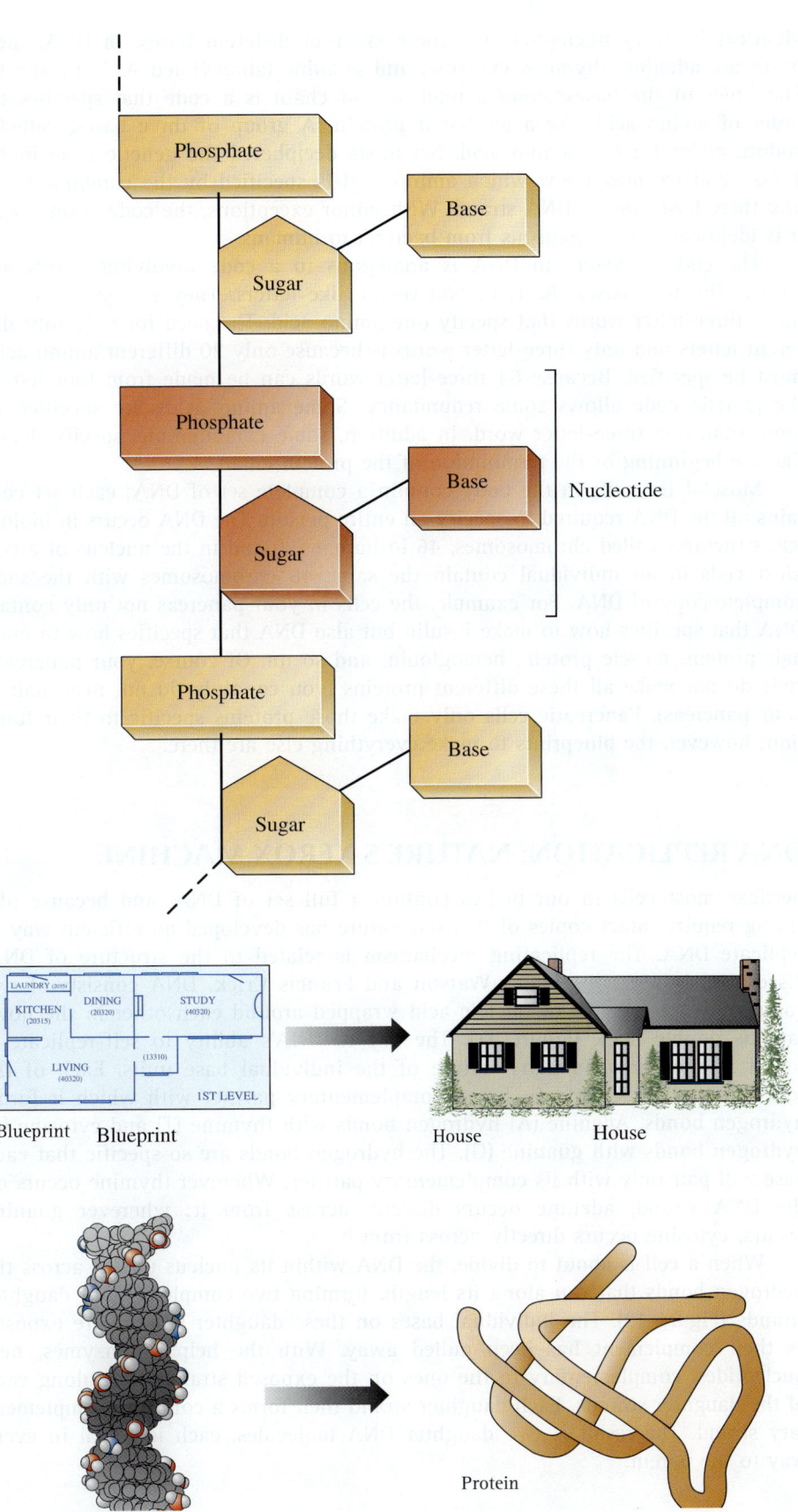

FIGURE 10 Nucleic acids are long, chainlike molecules whose individual units are called nucleotides. A nucleotide consists of three parts: a phosphate group, a sugar group, and a base.

Just as a blueprint specifies all the structure of a house, DNA specifies the structure of a protein molecule.

Three bases constitute a codon that codes for one amino acid. For example, GCA codes for alanine; AGC codes for serine.

 Explore this topic on the **Human Genome Project DOE** website.

identical in every nucleotide, but there are four different bases. In DNA, these bases are adenine, thymine, cytosine, and guanine (abbreviated A, T, C, and G). The order of the bases along a nucleic acid chain is a code that specifies the order of amino acids for a particular protein. A group of three bases, called a **codon,** codes for one amino acid. Scientists deciphered the genetic code in the 1960s, and we now know which amino acid is specified by the combination of any three bases in the DNA strand. With minor exceptions, the code is universal; it is identical in all organisms from bacteria to humans.

The coding system in DNA is analogous to a code involving letters and words. The four bases—A, T, C, and G—are like letters. They group together to make three-letter words that specify one amino acid. The need for only four different letters and only three-letter words is because only 20 different amino acids must be specified. Because 64 three-letter words can be made from four letters, the genetic code allows some redundancy. Some amino acids are specified by more than one three-letter word. In addition, some combinations specify details like the beginning or the termination of the protein chain.

Most of the cells in the body contain a complete set of DNA; each set contains all the DNA required to specify an entire person. The DNA occurs in biological structures called chromosomes, 46 in humans, found in the nucleus of a cell. Most cells in an individual contain the same 46 chromosomes with the same complete copy of DNA. For example, the cells in your pancreas not only contain DNA that specifies how to make insulin but also DNA that specifies how to make hair protein, muscle protein, hemoglobin, and so on. Of course, your pancreatic cells do not make all these different proteins (you certainly do not need hair in your pancreas). Pancreatic cells only make those proteins specific to their function; however, the blueprints to make everything else are there.

DNA REPLICATION: NATURE'S XEROX MACHINE

Watson and Crick were aided in their discovery by the X-ray diffraction pattern of DNA obtained by Rosalind Franklin (1920–1958).

Because most cells in our bodies contain a full set of DNA, and because offspring require intact copies of this set, nature has developed an efficient way to replicate DNA. The replicating mechanism is related to the structure of DNA, discovered in 1953 by **James Watson** and **Francis Crick.** DNA consists of two complementary strands of nucleic acid wrapped around each other in the now-famous double helix (Figure 11). The key to DNA's ability to self-replicate is found in the complementary nature of the individual base units. Each of the four bases, A, T, C, and G, has a complementary partner with which it forms hydrogen bonds. Adenine (A) hydrogen bonds with thymine (T) and cytosine (C) hydrogen bonds with guanine (G). The hydrogen bonds are so specific that each base will pair only with its complementary partner: Wherever thymine occurs on the DNA strand, adenine occurs directly across from it; wherever guanine occurs, cytosine occurs directly across from it.

When a cell is about to divide, the DNA within its nucleus unzips across the hydrogen bonds that run along its length, forming two complementary daughter strands (Figure 12). The individual bases on these daughter strands are exposed as their complement has been pulled away. With the help of enzymes, new nucleotides, complementary to the ones on the exposed strands, add along each of the daughter strands. Each daughter strand then forms a complete complementary strand. The result is two daughter DNA molecules, each identical in every way to the parent.

J. D. Watson (left) and F. H. C. Crick (right) with a model of the DNA structure they discovered.

16.7 Nucleic Acids: The Blueprint for Proteins 459

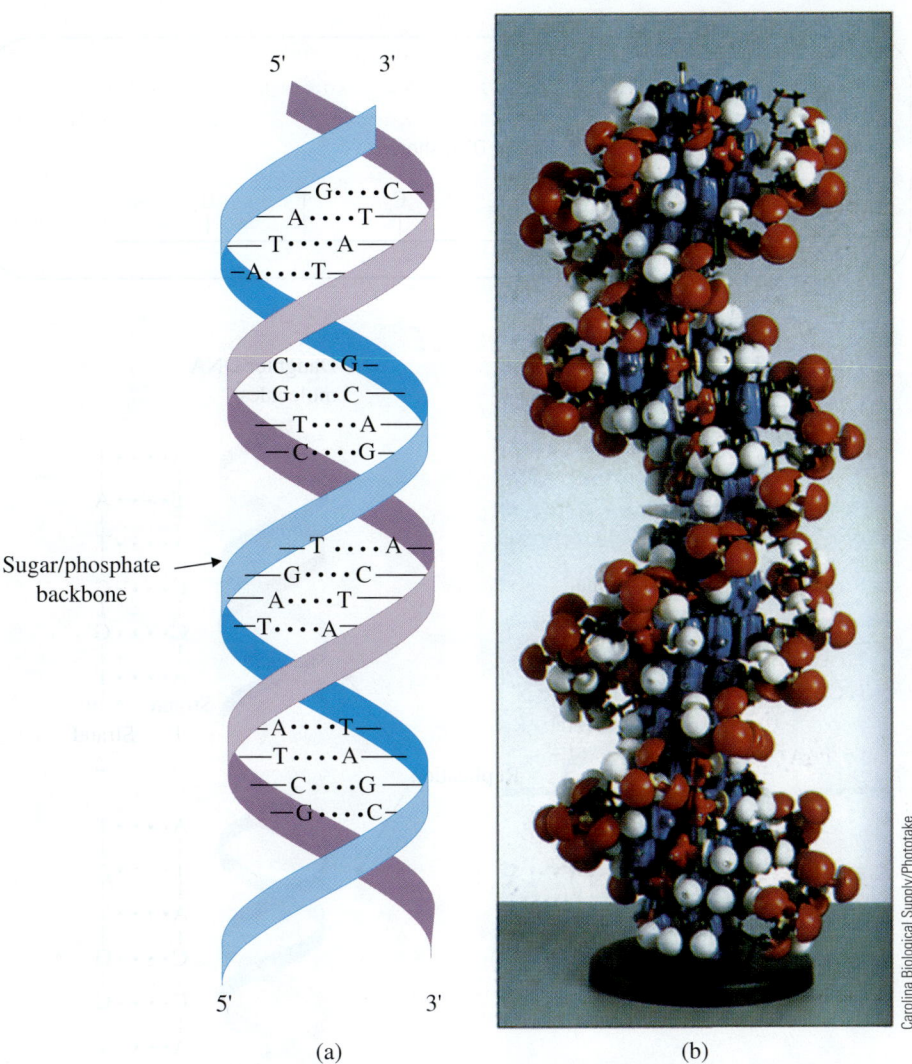

FIGURE 11 The structure of DNA. (a) Schematic diagram. (b) Space-filling model. (*Source:* Reprinted with permission of H. R. Muinos, Muinos Medical Visuals.)

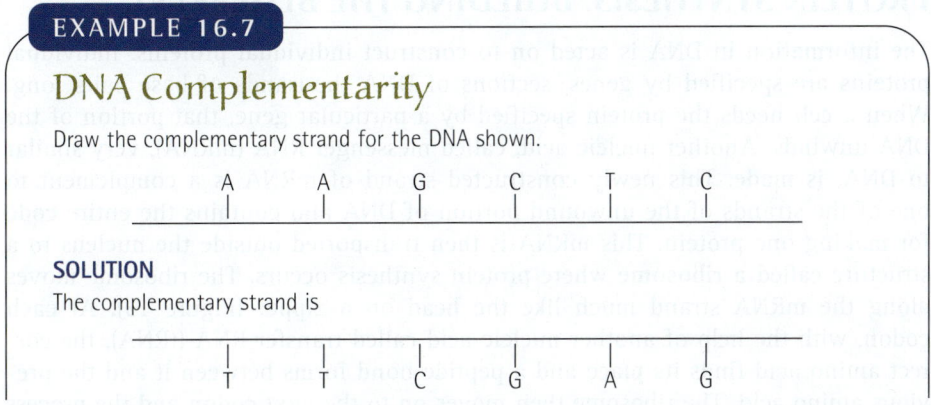

EXAMPLE 16.7

DNA Complementarity

Draw the complementary strand for the DNA shown.

A A G C T C

SOLUTION
The complementary strand is

T T C G A G

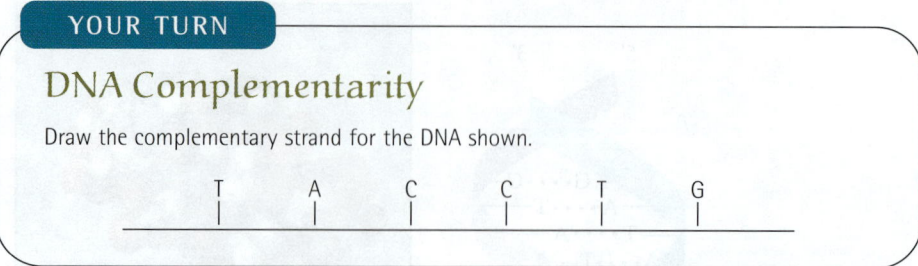

YOUR TURN

DNA Complementarity

Draw the complementary strand for the DNA shown.

```
T   A   C   C   T   G
|   |   |   |   |   |
```

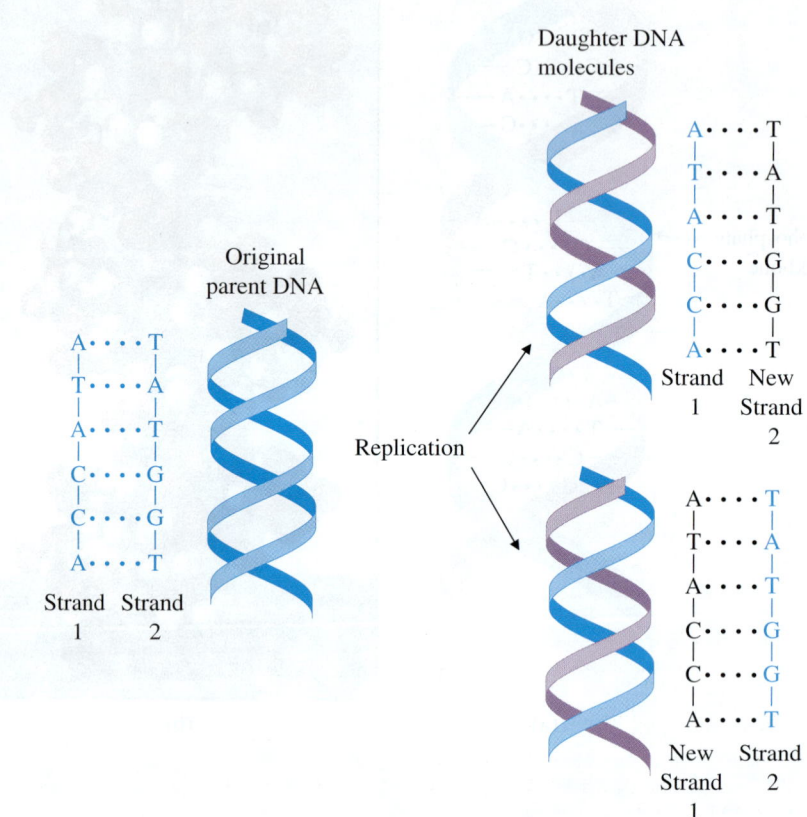

FIGURE 12 DNA replication. The original DNA replicates by forming copies of each of its complementary strands to produce two daughter DNA molecules. (*Source:* Reprinted with permission of H. R. Muinos, Muinos Medical Visuals.)

PROTEIN SYNTHESIS: BUILDING THE BLUEPRINT

The information in DNA is acted on to construct individual proteins. Individual proteins are specified by genes, sections of DNA thousands of base pairs long. When a cell needs the protein specified by a particular gene, that portion of the DNA unwinds. Another nucleic acid, called messenger RNA (mRNA), very similar to DNA, is made. This newly constructed strand of mRNA is a complement to one of the strands of the unwound portion of DNA and contains the entire code for making one protein. This mRNA is then transported outside the nucleus to a structure called a **ribosome** where protein synthesis occurs. The ribosome moves along the mRNA strand much like the head on a zipper (Figure 13). At each codon, with the help of another nucleic acid called transfer RNA (tRNA), the correct amino acid finds its place and a peptide bond forms between it and the previous amino acid. The ribosome then moves on to the next codon and the process

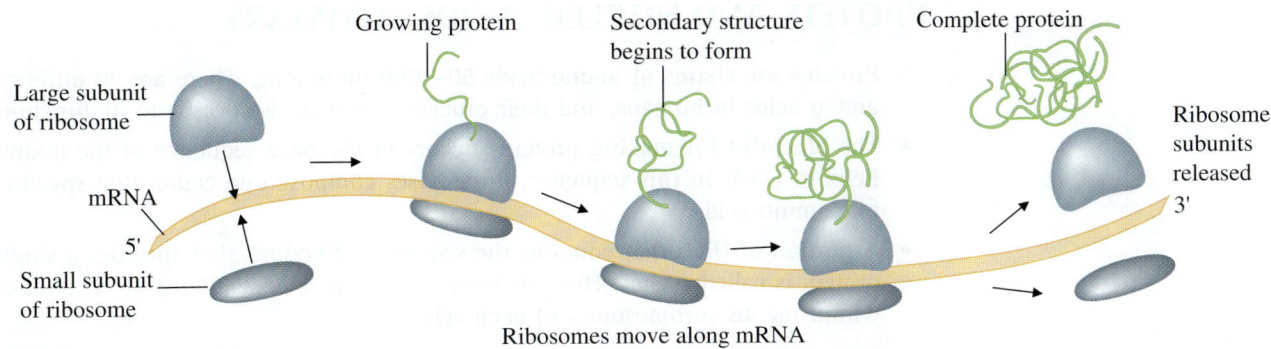

FIGURE 13 Protein synthesis. The ribosome moves along a chain of mRNA, linking together the amino acids that are specified by the RNA to form the complete protein.

repeats itself. When the ribosome reaches the end of the mRNA strand, the protein is complete and moves off to carry out its function.

VIRUSES AND AIDS

In the classification of life and nonlife, viruses fall somewhere in the middle. They consist primarily of DNA or RNA and a protein coat. Because they are not really alive, they are difficult to kill. A viral infection does not respond to antibiotics the way that a bacterial infection does.

Viruses require the machinery of a host cell to reproduce. A virus takes over a host cell and inserts its own DNA into the chromosomes of the host. The host then expresses (or synthesizes) the viral protein specified by the viral DNA. The host cell also copies the viral DNA, which combines with viral proteins to produce more viruses. When the host cell dies, the daughter viruses are released to infect again.

A number of diseases are caused by viruses including the common cold, flu, measles, mumps, polio, smallpox, and ebola. In the past, viral epidemics have killed millions. The smallpox virus, for example, was one of the world's most dreaded diseases, killing over a million people per year in the 1960s. It has since been completely eradicated through vaccination (see Section 17.4). The most widespread and serious viral epidemic of our day is caused by the human immunodeficiency virus (HIV), which causes acquired immune deficiency syndrome (AIDS). It is estimated that 60 million people worldwide have been infected with HIV since it appeared in the early 1980s.

The HIV consists of a bar-shaped core containing two short strands of RNA about 9200 nucleotides in length. The virus also stores a number of proteins, including reverse transcriptase, an enzyme that causes the reverse transcription of RNA into DNA. The HIV attacks cells associated with the human immune system. Once inside the cell, the virus uncoats and releases its RNA into the cell. The reverse transcriptase then uses the cell's resources to form viral DNA from viral RNA. Another enzyme inserts the viral DNA into the chromosomes of the host cell, which then replicates the HIV. When the cell dies, daughter HIVs are released and the cycle is repeated.

As the cells of the immune system are slowly destroyed, a process that may take many years, the body can no longer keep up, and clinical AIDS develops. Once this stage is reached, the body becomes subject to a number of infections and cancers that eventually cause death.

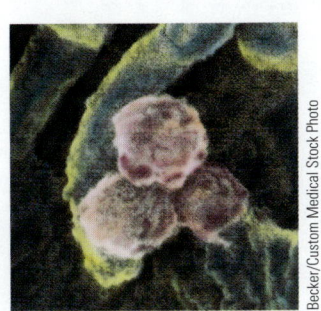

Human immunodeficiency virus (HIV).

PROTEIN AND NUCLEIC ACIDS SUMMARY

- Proteins are chains of amino acids 50–1000 units long. There are 20 different amino acids in humans, and their order in a protein is specific to its function.
- The blueprint for making proteins occurs in the base sequence of the double helix of DNA. In this sequence, three bases compose one codon that specifies one amino acid.
- A portion of DNA that contains the sequence of codons that specifies a single protein is called a gene. Humans have on the order of 25,000 genes contained within the 46 chromosomes of each cell.

Although every cell has a complete copy of the DNA required to make an entire person, it is not all expressed. Expression of DNA (making the specified protein) requires the formation of mRNA followed by protein synthesis at the ribosome. Cells express only the proteins specific to their function. For example, pancreatic cells express the DNA to make insulin but do not express the DNA to make hemoglobin.

16.8 Recombinant DNA Technology

Technology that allows scientists to transfer DNA from one organism to another is now available. The new DNA, including the foreign component, is called **recombinant DNA**. If the foreign DNA is human, and the organism into which it is introduced is a bacteria, the bacteria may synthesize the protein specified by the human DNA and therefore become tiny factories for the human protein. Alternatively, if the DNA of a fertilized egg is modified, the organism that results will have characteristics determined by the recombinant DNA. By carefully choosing what kind of DNA is introduced into a fertilized egg, the characteristics of the resulting organism can be engineered.

Recombinant DNA techniques employ **restriction enzymes**, enzymes that cut DNA at specific places along the DNA strand. This allows scientists to snip long sections of DNA into smaller portions. By choosing among over 100 different restrictive enzymes, scientists can control where the DNA is cut and the lengths of the resulting fragments. After treatment of a DNA sample with restriction enzymes, the scientist is left with a mixture of DNA fragments of different lengths and composition (Figure 14). The different fragments are then separated by **gel electrophoresis**, in which the mixture of DNA fragments is placed on a thin gel. An electric current causes the mixture to separate into distinct bands, each representing a different DNA fragment. The DNA within these bands can then be isolated. By using restrictive enzymes and gel electrophoresis, as well as other techniques, scientists have isolated short DNA strands that contain several genes, or in some cases, even a single gene. This is a major accomplishment considering that many animals have tens of thousands of different genes.

A DNA strand obtained from one organism, such as a human, can be introduced into another, such as a bacterium. The bacterium containing the recombinant DNA can then be grown in culture. As the bacteria multiply, they replicate the foreign DNA along with their own, making millions of copies. In some cases, the bacteria also express the foreign DNA, providing a source for the protein coded for by the DNA.

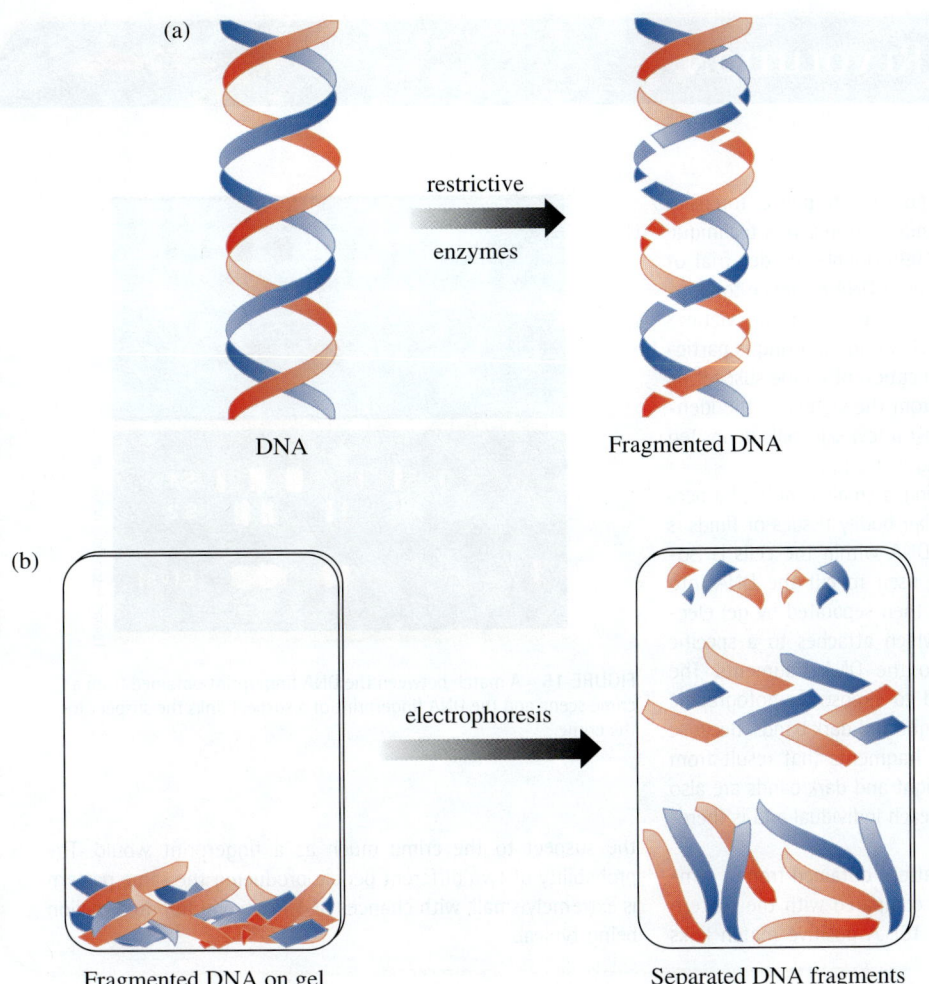

FIGURE 14 (a) Showing long DNA strands cut into shorter pieces with restrictive enzymes. (b) Showing shorter pieces mixed on gel at the bottom and separation of fragments according to size as they move up the gel.

PHARMACEUTICALS

By using recombinant DNA techniques, scientists have made pharmaceuticals not available by other means. For example, for many years diabetics injected themselves with insulin derived from pigs or cattle. The animal insulin was similar enough to human insulin that it worked to lower blood glucose levels. However, some patients could not tolerate this insulin. One of the early successes of genetic engineering was the isolation of the gene that codes for the production of human insulin. This gene was introduced into bacteria, which not only copied the gene but also expressed it. These bacteria, which normally have no need for human insulin, became factories of human insulin, allowing pharmaceutical companies to produce it in large enough quantities to market. Today, most insulin-dependent diabetics take genetically engineered human insulin to control their blood glucose levels.

Scientists have also used recombinant DNA techniques to produce human growth hormone. Some children do not make sufficient amounts of this protein and consequently fail to grow to normal adult size. The drug was previously available in only very small quantities from human sources. However, through isolation of the human gene and introduction of it into bacteria that express it,

The Molecular Revolution

Forensics

DNA technology has become important in police forensics through a technique called DNA fingerprinting. This technique gained national attention in the 1995 double-murder trial of O. J. Simpson in which prosecutors used DNA evidence to identify Simpson's blood at the murder scene and the victim's blood in Simpson's car and home. DNA fingerprinting is particularly sensitive, allowing the identification of a rape suspect by analysis of just a few sperm cells from the victim, or the identification of a murder suspect by just a few skin cells deposited under the victim's fingernails during a struggle.

In one form of DNA fingerprinting, a small sample of a person's blood, hair, skin, semen, or other bodily tissues or fluids is collected at the crime scene. The DNA within the cells is isolated, and restriction enzymes are used to cut the DNA into fragments. The DNA fragments are then separated by gel electrophoresis. A radioactive probe, which attaches to a specific base sequence, is then applied to the DNA fragments. The radioactive sample is then allowed to expose a photographic film, which produces a pattern of light and dark bands. Because every person's DNA is unique, the fragments that result from this treatment and the pattern of light and dark bands are also unique. The pattern is different for each individual and is therefore called a DNA fingerprint.

For example, in forensics, the pattern obtained from a sample collected at the crime scene is compared with the pattern obtained from the suspect (Figure 15). A positive match links the suspect to the crime much as a fingerprint would. The probability of two different people producing the same pattern is extremely small, with chances of 1 in several hundred million being typical.

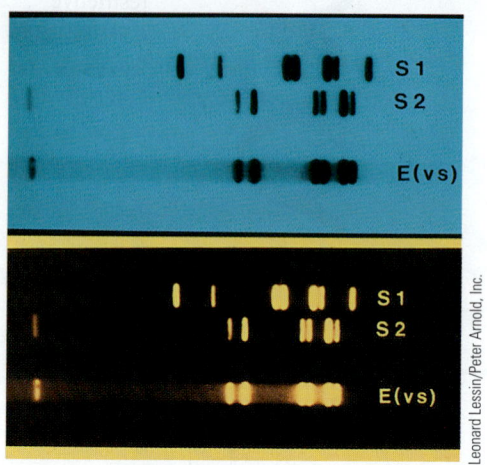

FIGURE 15 A match between the DNA fingerprint obtained from a crime scene and the DNA fingerprint of a suspect links the suspect to the crime.

 Explore this topic on the **DNA Fingerprinting** website.

large amounts of human growth hormone are now available as a therapy for dwarfism. Two forms of another genetically engineered drug called interferon, also produced from the human gene that codes for it, are used as treatments for multiple sclerosis. As genetic engineering continues to advance, it will no doubt produce more new pharmaceuticals, and in this way prove to be an important part of human health care.

AGRICULTURE

The application of genetic engineering to agriculture has produced novelties such as rot-resistant tomatoes and extra-beefy cattle. However, its wholesale application has met with some resistance. The impacts of introducing genetically engineered organisms—organisms that never existed before—into the environment are not always clear. For example, a group of geneticists has engineered bacteria that should help prevent frost damage in crops. The natural version of these bacteria, often found on crop leaves, produces a protein that accelerates ice crystal production on crops during cold weather. The formation of these ice crystals leads to crop damage.

Geneticists have succeeded in engineering these same bacteria without the DNA sequence that codes for the frost-promoting protein. The engineered bacteria do not produce the protein and therefore do not accelerate the freezing of

Genetic engineering has helped create agricultural products such as the rot-resistant tomato. These tomatoes have a longer shelf life than ordinary tomatoes.

crops. Researchers hope to release the modified bacteria on crops in large enough numbers to compete with and hopefully replace the natural bacteria. This is likely to lessen crop damage in cold weather, but what other impacts might it have on the environment? Are there other important functions that these bacteria serve? Might they be responsible for the initiation of frost in other systems where it is beneficial? These are the kinds of questions that need addressing before we release modified life forms into an environment where we may no longer have control of them.

GENETIC SCREENING AND DISEASE THERAPY

A small sample of blood, skin, semen, or almost any other tissue or bodily fluid contains cells with complete copies of DNA within them. That DNA can be isolated and screened for genes that may increase a person's chances of developing certain diseases such as cancer, heart disease, and Alzheimer's. The knowledge of a predisposition to a certain kind of disease may help the person take preventive steps to avoid it. Again, genetic engineering raises some important questions. How well do we want to know our future? Should such information be available to employers and insurance companies?

Although still in its infancy, and meeting with limited success to date, genetic engineering techniques may one day treat genetic diseases directly. For example, gene therapy is being researched as a way to treat cystic fibrosis (CF), a condition in which patients develop a thick, infection-prone mucus in their lungs. CF patients usually do not live past the age of 30. The faulty gene responsible for CF was discovered in 1989. Researchers have since conducted several experiments that attempt to substitute the correct gene for the faulty one within the lungs of CF patients. So far, the research has met with limited success. However, researchers in the field are optimistic that the problem is one of delivery and may be overcome soon. Other genetic diseases such as Huntington's disease, inheritable breast cancer, muscular dystrophy, and immune deficiency may one day be treatable via gene therapy.

16.9 Cloning

Genetic engineering becomes even more intriguing and perhaps more questionable when scientists modify the DNA in eggs. Unlike bacteria or blood cells, which simply produce more bacteria or more blood cells when grown in culture, eggs develop into whole organisms. If the genetic material in several fertilized eggs is made to be genetically identical, then those eggs will develop into genetically identical whole organisms. Even stranger, if the genetic material in an egg is made to be genetically identical to that of an existing adult, then the egg will develop into the genetic twin of that adult. Such genetic feats once were only possible in science fiction, but today they are possible in reality.

The cloning of *embryos*, cells in the early stages of development, has been achieved in animals. The process works by allowing a fertilized egg to divide several times to produce a total of up to eight cells. The DNA from these divided cells is then transferred into eight other fertilized eggs, producing eight embryos with identical DNA. The eight embryos are then implanted into the wombs of surrogate mothers and eight identical offspring result. This technique has been used successfully in cattle to increase the offspring of a good milk- or beef-producing cow.

The cloning of adult organisms has recently been achieved in mammals. In 1997, several British scientists shocked the world by announcing that they had successfully cloned an adult sheep by a technique known as *nuclear transfer.* In

Explore this topic on the **Genetically Modified Foods** website.

Explore this topic on **The First Human Cloned Embryo—Scientific American Article** website.

This lamb, Dolly, was cloned from the DNA of a 6-year-old adult ewe.

FIGURE 16 Cloning. The DNA from the donor cell is transferred into an unfertilized egg that has had its own DNA removed. The newly formed cell divides and becomes a blastocyst, which can then be implanted into a surrogate mother (for reproductive cloning). In therapeutic cloning, the blastocyst becomes a source for embryonic stem cells.

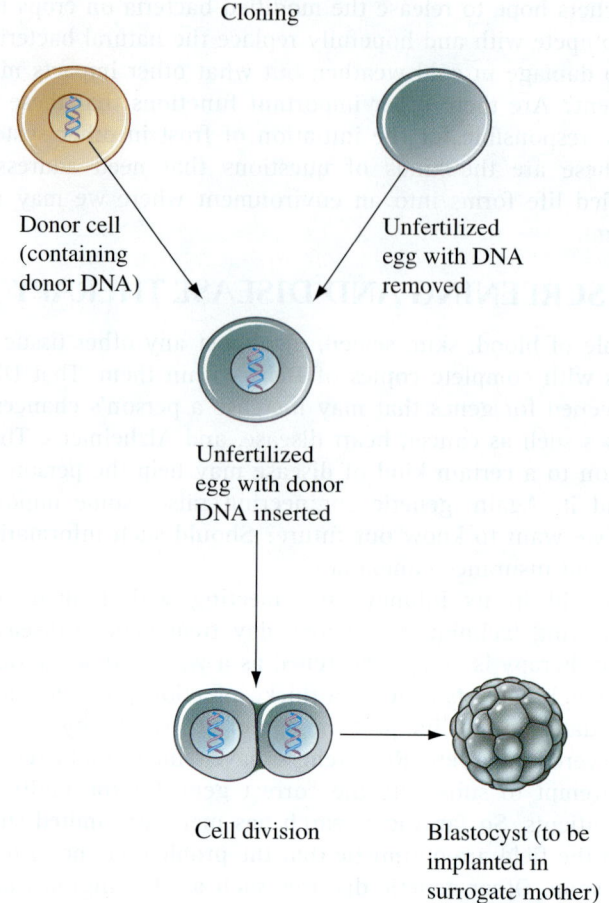

this experiment, the DNA of a 6-year-old adult ewe was transferred into an unfertilized egg that had its own DNA removed. The egg was then coaxed to divide, forming an embryo that was implanted into a surrogate mother (Figure 16). The embryo developed normally and resulted in the birth of a lamb named Dolly. Dolly was the genetic twin of the 6-year-old ewe whose DNA she shared. Since Dolly, many other types of mammals, including cattle, pigs, and mice, have been cloned in this way.

THERAPEUTIC CLONING AND STEM CELLS

The cloning of an adult organism for the purpose of producing genetically identical offspring (as occurred with Dolly) is called *reproductive cloning*. For a number of reasons, including the large number of trials that it takes to successfully produce a cloned mammal, most scientists view the application of reproductive cloning to humans as unethical. However, another kind of cloning—called *therapeutic cloning*—is viewed by many scientists as acceptable. The goal of therapeutic cloning is not to produce *offspring* that are genetically identical to an adult donor but to produce *embryonic stem cells* that are genetically identical to an adult donor.

Embryonic stem cells are the master cells normally present in embryos days after the fertilization of an egg. Stem cells have the inherent capability of becoming any of the cells within the human body. For example, some of the stem cells within a days-old embryo will become liver cells, others will differentiate into brain cells, and still others will differentiate into heart or muscle cells. For years, scientists have been trying to understand stem cells and the chemical signals required to coax them into one or another type of specialized cell in the hopes of being able to grow any kind of human tissue from the stem cells. This tissue could then be used to treat a number of conditions such as diabetes, heart disease, liver disease, and even spinal cord injuries.

Therapeutic cloning offers the potential to make stem cells that are a perfect genetic match to the donor of the DNA from whom the cells are cloned. Therefore, any tissue grown from those stem cells would not be rejected by the immune system of the donor. For example, the DNA from the cell of a liver patient could be transferred into an unfertilized egg. The egg would then be coaxed to divide into a days-old embryo. The stem cells would then be removed from the embryo and used to grow liver tissue. The resulting liver would be transplanted into the patient with assurance that it would not be rejected (because the DNA is a perfect match). The potential of therapeutic cloning and stem cell research to treat a seemingly endless number of diseases has led to the pursuit of this research by scientists all over the world. However, because the research involves human embryos and potential life, it has also been surrounded with controversy (see the following What If . . . box).

What if. . .

The Ethics of Therapeutic Cloning and Stem Cell Research

In October of 2001, scientists at Advanced Cell Technologies created the first cloned early-stage human embryos. Those scientists had hoped to get those embryos to divide into blastocysts—hollow spheres containing about 100 stem cells, but the cells stopped dividing at the 6-cell stage. Nevertheless, scientists at Advanced Cell Technologies are at the center of the controversy over therapeutic cloning and stem cell research. Those opposing this research argue that the blastocyst has the potential to become a human being, and that harvesting stem cells from a blastocyst is the equivalent of taking a human life. To their minds, no amount of benefit justifies what they see as murder. Those who support the research argue that the blastocyst is not a human life, but a clump of cells with no brain, no liver, no muscle, or anything else that you would classify as human. Therefore, they argue, the use of blastocysts to obtain stem cells that could cure a host of diseases is perfectly justified. In the United States, therapeutic cloning and stem cell research are legal at this time. However, federal funding for stem cell research is severely limited, and legislation to make therapeutic cloning illegal has been proposed by some.

QUESTION: What do you think about therapeutic cloning and stem cell research? Should therapeutic cloning remain legal? Should the federal government have fewer restrictions and more funding for stem cell research?

Chapter Summary

MOLECULAR CONCEPT

Biochemical compounds are divided into four major types: lipids, whose primary function is long-term energy storage; carbohydrates, whose primary function is short-term energy storage and transport; proteins, whose functions vary from composing nonskeletal structure to controlling bodily reactions; and nucleic acids, whose function is storing and replicating the blueprints by which proteins are made (16.1).

A subclass of lipids, fats, are triesters containing three long hydrocarbon chains. The primary difference among fats lies in the degree of saturation of their hydrocarbon chains. Saturated fats have no double bonds within their chains and tend to be solid at room temperature (16.2). Unsaturated fats (oils) have one or more double bonds and tend to be liquids at room temperature. Carbohydrates are ketones or aldehydes (contain C=O) that also have OH groups attached to their carbon atoms. They exist primarily in their ring form. Carbohydrates that contain just one ring, such as glucose or fructose, are called monosaccharides; those that contain two linked rings, such as sucrose, are called disaccharides; and those that contain many rings, such as starch, are called polysaccharides (16.3).

Proteins consist of long amino acid chains held together by peptide bonds. Amino acids contain an amine group ($-NH_2$), a carboxylic acid group ($-COOH$), and an R group. The difference between amino acids—20 different ones in humans—is the R group (16.4). The order of amino acids within a protein determines its structure and function (16.5). The blueprint for stringing amino acids together in the correct order is provided by nucleic acids. Nucleic acids consist of long chains of nucleotides that each contain one of four different bases. A series of three nucleotides is called a codon and specifies one amino acid. The order of codons along the DNA molecule determines the order of amino acids in the protein that it codes for. A section of DNA that codes for one protein is called a gene. Humans are estimated to have 25,000 genes located on 46 biological structures called chromosomes (16.6–16.9).

SOCIETAL IMPACT

The chemical study of the molecules that compose life has led to incredible advances in medicine, agriculture, and other related technologies. Yet, this field also raises important societal questions and ethical issues related to the chemical modification of living organisms. Is life sacred? Does a chemical description make it less sacred? In what cases is it proper to change the molecular blueprint for living organisms and in what cases is it not? These are questions that society, together with the scientific community, must answer as our technology continues to advance.

The description of living things on the molecular level has also had an impact on how humans view themselves. Are the chemical reactions occurring in human bodies and brains no different than those that occur outside of life? It appears they are not. If that is so, then are humans just complicated chemical reactions? In this century, humans have struggled with the advance of science into the realm of what was once held sacred. To some, a detailed knowledge of how it all works somehow makes life less sacred. To others, however, it simply makes life more fascinating.

Proteins are the workhorse molecules of living organisms (16.4–16.7). When humans lack the gene to make a particular protein, or when the gene to make a particular protein is faulty, disease results. In some cases, the pharmaceutical industry can synthesize particular proteins by transferring the human gene that codes for the protein into bacteria. Because bacteria, like all living organisms, share the same genetic code, they can construct the specified protein, and a previously unavailable drug becomes available.

Scientists have also modified the DNA in fertilized eggs to produce genetically engineered organisms (16.8). They have produced several fertilized eggs with identical DNA which then develop into offspring that are genetically identical clones, and they have produced eggs that are genetically identical to an adult, resulting in offspring that are clones of the adult (16.9). Even though these technologies have met with success in animals, their application to humans has met with caution. It is up to our society to decide just how far to go with the genetic modification of humans.

Chemistry on the Web

For up-to-date URLs, visit the text website at **www.chemistry.brookscole.com/tro3**.

- RasMol Gallery (Images of DNA, RNA, and Proteins)
  http://www.umass.edu/microbio/rasmol/galmz.htm#dna
- Human Genome Project DOE
  http://www.doegenomes.org/
- DNA Fingerprinting
  http://protist.biology.washington.edu/fingerprint/dnaintro.html
- Genetically Modified Foods
  http://www.fda.gov/fdac/features/2003/603_food.html
- Human Cloned Embryo
  http://www.sciam.com/article.cfm?articleID=0008B8F9-AC62-1C75-9B81809EC588EF21&pageNumber=1&catID=4
  http://news.bbc.co.uk/1/hi/health/4563607.stm

Key Terms

carbohydrates
codon
Crick, Francis
disaccharides
DNA
enzyme
fatty acid
gel electrophoresis
hormone
lipids
monosaccharides
nucleic acids
nucleotides
polysaccharides
primary structure
proteins
quaternary structure
recombinant DNA
restriction enzymes
ribosome
RNA
saturated
secondary structure
tertiary structure
Watson, James

Exercises

Assess your understanding of this chapter's topics with an online chapter quiz at **http://chemistry.brookscole.com/tro3**.

QUESTIONS

1. List the four major classes of biochemical compounds and their functions within the body.
2. Why are fats more efficient than carbohydrates for long-term energy storage?
3. What is the definition of a lipid? What is the general structure of a triglyceride?
4. What is the difference between a saturated and an unsaturated fat, both in terms of molecular structure and of properties?
5. What does the name *carbohydrate* mean?
6. Why do carbohydrates contain less energy per gram than fats?
7. What are the general structural features of carbohydrates?
8. What are the two different forms that a carbohydrate can take? Which form is dominant under normal conditions?
9. Why is it important for carbohydrates to contain —OH groups?
10. Draw the structure of glucose, both in its straight chain and in its ring form.
11. What is the difference between a monosaccharide, a disaccharide, and a polysaccharide? Give one example of each.
12. What is the difference in structure between starch and cellulose? What are the results of this difference?
13. What are proteins? What functions do they serve in the body?
14. What are the general structural features of amino acids?

15. How many different amino acids exist in human proteins? How are these amino acids different from one another?
16. What is primary structure in proteins? What kinds of interactions or bonds hold a protein's primary structure together?
17. Define secondary structure in proteins. What kinds of interactions or bonds hold a protein's secondary structure together?
18. Define tertiary and quaternary structure in proteins. What kinds of interactions or bonds hold a protein's tertiary and quaternary structure together?
19. What is hemoglobin (Hb)? What is its function?
20. What is α-keratin? What is its function? How can it be modified?
21. What is lysozyme? How was it discovered?
22. What is insulin? What is its function?
23. Draw a block diagram for the structure of a nucleic acid.
24. What is the difference between DNA and RNA?
25. What are the four bases found in DNA? List the complement of each one.
26. What is a codon? What is its function? How many different codons exist?
27. What are chromosomes? How many exist in humans?
28. What are genes? How many exist in humans?
29. Does each cell make every protein specified by the DNA in its nucleus?
30. Explain how DNA is replicated within cells.
31. Draw a schematic diagram of DNA. Show the double-helix structure, the molecular backbone, and the complementary nature of the base pairs.
32. Explain how proteins are synthesized within cells.
33. What is recombinant DNA?
34. Describe how a particular gene or set of genes might be isolated.
35. Explain how recombinant DNA technology has made available pharmaceuticals not available by other means.
36. What are the dangers of applying genetic engineering to organisms that may find their way into the environment?
37. How can genetic engineering be used to prevent certain diseases?
38. How can genetic engineering be used to treat certain diseases?
39. Explain how DNA fingerprinting works. How can it be used in a trial?
40. What is cloning? What kind of cloning has been accomplished?
41. Has cloning been accomplished in humans? In what sense?
42. What are the dangers inherent in applying genetic engineering to human embryos? What are the benefits?
43. Explain the difference between therapeutic cloning and reproductive cloning.
44. What are stem cells? Why are scientists interested in them?

PROBLEMS

45. Which of the following is a triglyceride?
    a. [benzene ring with OCH$_3$, C(CH$_3$)$_3$, and OH substituents]
    b. CH$_2$OH—CHOH—CH$_2$OH
    c. [disaccharide structure]
    d. CH$_2$OC(CH$_2$)$_{16}$CH$_3$ / CHOC(CH$_2$)$_{16}$CH$_3$ / CH$_2$OC(CH$_2$)$_{16}$CH$_3$

46. Which of the following is a triglyceride?
    a. CH$_2$OH—CHOH—CH$_2$OH
    b. HO—C(=O)(CH$_2$)$_{14}$CH$_3$
    c. H$_2$N—CH(CH(CH$_3$)CH$_3$)—C(=O)—OH
    d. CH$_2$OC(=O)(CH$_2$)$_7$CH=CH(CH$_2$)$_7$CH$_3$ / CHOC(=O)(CH$_2$)$_7$CH=CH(CH$_2$)$_7$CH$_3$ / CH$_2$OC(=O)(CH$_2$)$_7$CH=CH(CH$_2$)$_7$CH$_3$

47. Which of the following is a saturated fat?

a.
$$\begin{array}{l} CH_2OC(CH_2)_7CH{=}CH(CH_2)_7CH_3 \\ \quad\;\; \| \\ \quad\;\; O \\ CHOC(CH_2)_7CH{=}CH(CH_2)_7CH_3 \\ \quad\;\; \| \\ \quad\;\; O \\ CH_2OC(CH_2)_7CH{=}CH(CH_2)_7CH_3 \end{array}$$

b.
$$\begin{array}{l} CH_2OC(CH_2)_{14}CH_3 \\ \quad\;\; \| \\ \quad\;\; O \\ CHOC(CH_2)_{14}CH_3 \\ \quad\;\; \| \\ \quad\;\; O \\ CH_2OC(CH_2)_{14}CH_3 \end{array}$$

c.
$$\begin{array}{l} CH_2OH \\ CHOH \\ CH_2OH \end{array}$$

d.
$$CH_3-\underset{OH}{\underset{|}{\overset{O}{\overset{\|}{C}}-CHCH_3}}$$

48. Classify each of the following fatty acids' R groups as saturated or unsaturated. Which would form fats that are solid at room temperature?

a. $-\overset{O}{\overset{\|}{C}}(CH_2)_{14}CH_3$

b. $-\overset{O}{\overset{\|}{C}}(CH_2)_7CH{=}CH(CH_2)_7CH_3$

c. $-\overset{O}{\overset{\|}{C}}(CH_2)_{16}CH_3$

49. Which of the following is a carbohydrate?

a. [glucose ring structure with CH₂OH]

b.
$$\begin{array}{l} CH_2OH \\ CHOH \\ CH_2OH \end{array}$$

c. $H_2N-\underset{CH_3}{\underset{|}{\overset{H}{\overset{|}{C}}}}-COOH$

d.
$$\begin{array}{l} CH_2OC(CH_2)_2CH_3 \\ \quad\;\; \| \\ \quad\;\; O \\ CHOC(CH_2)_{14}CH_3 \\ \quad\;\; \| \\ \quad\;\; O \\ CH_2OC(CH_2)_7CH{=}CH(CH_2)_7CH_3 \end{array}$$

50. Which of the following is a carbohydrate?

a. $\underset{CH_3SCH_2CH_2}{\underset{|}{\overset{H}{\overset{|}{\underset{H}{\overset{|}{N}}}}-\overset{H}{\overset{|}{C}}-\overset{O}{\overset{\|}{C}}-OH}}$

b. [three-ring polysaccharide structure]

c.
$$\begin{array}{l} CH_2OC(CH_2)_{16}CH_3 \\ \quad\;\; \| \\ \quad\;\; O \\ CHOC(CH_2)_{16}CH_3 \\ \quad\;\; \| \\ \quad\;\; O \\ CH_2OC(CH_2)_{16}CH_3 \end{array}$$

d. $CH_3CH_2CH_2CH_2CH_2CH_2CH_2CH_2CH_2CH_3$

51. Draw the structure of sucrose.
52. Draw the structure of fructose in both its open- and closed-chain forms.

53. Classify each of the following carbohydrates as a monosaccharide, disaccharide, or polysaccharide:

a. 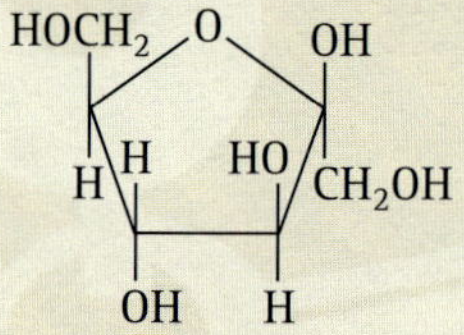

b. 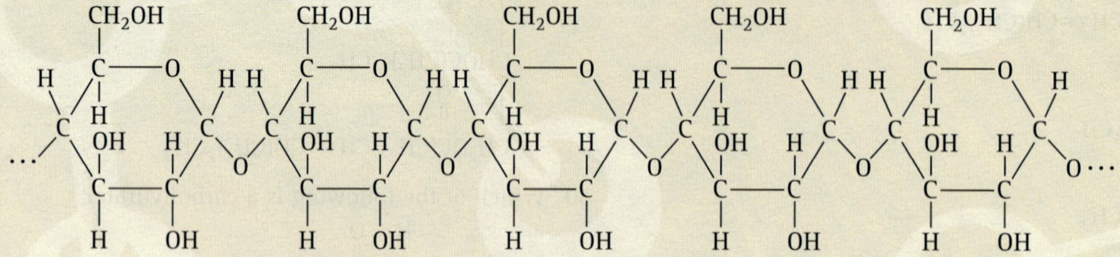

c. 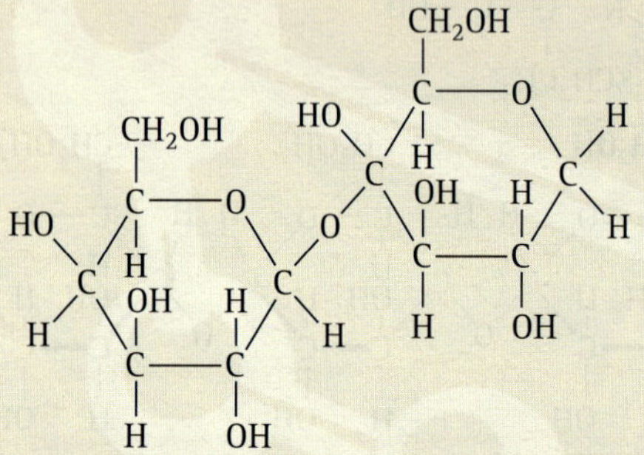

d. 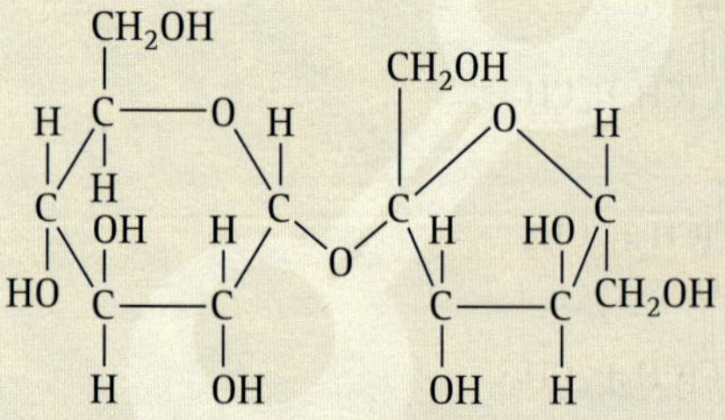

54. Classify each of the following carbohydrates as a monosaccharide, dissaccharide, or polysaccharide:

a. 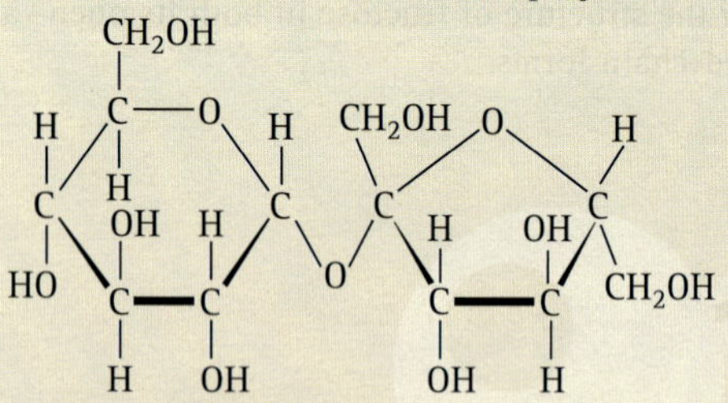

b. 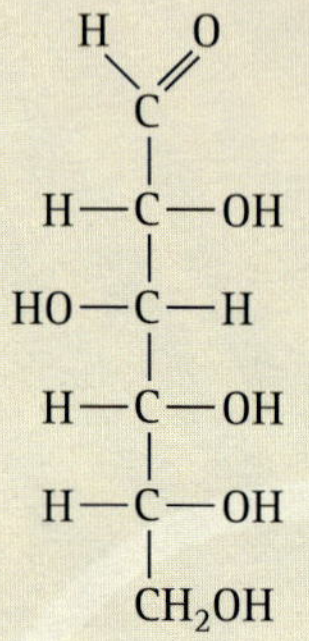

c. 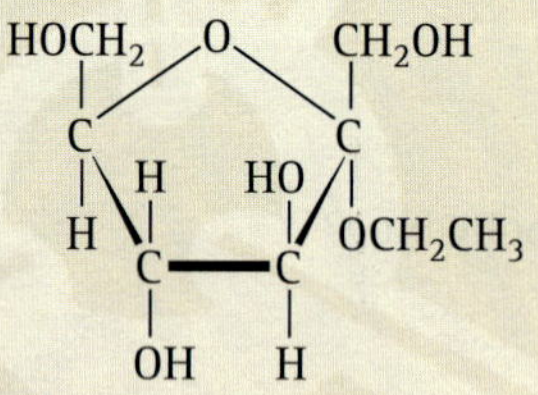

d. 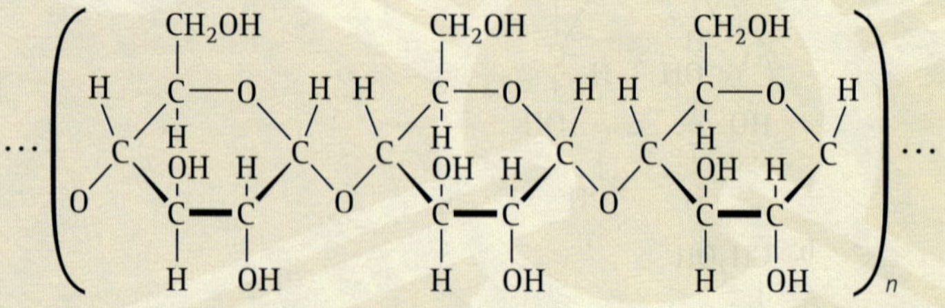

55. Which of the following is an amino acid?

a. [cyclic sugar structure with CH₂OH, OH groups — glucose]

b.
$$H_2N-\underset{CH_2OH}{\underset{|}{\overset{H}{\overset{|}{C}}}}-\overset{O}{\overset{\|}{C}}-OH$$

c.
$$\begin{array}{l} CH_2OC(CH_2)_{14}CH_3 \\ | \\ CHOC(CH_2)_{14}CH_3 \\ | \\ CH_2OC(CH_2)_{14}CH_3 \end{array}$$
(each C=O)

d. $CH_3CH_2CH_2NH_2$

56. Which of the following molecules is an amino acid?

a.
$$CH_3-\overset{O}{\overset{\|}{C}}-\underset{OH}{\underset{|}{CHCH_3}}$$

b.
$$\begin{array}{l} CH_2OC(CH_2)_7CH=CH(CH_2)_7CH_3 \\ | \\ CHOC(CH_2)_7CH=CH(CH_2)_7CH_3 \\ | \\ CH_2OC(CH_2)_7CH=CH(CH_2)_7CH_3 \end{array}$$

c. [cyclic sugar structure with CH₂OH, OH groups — glucose]

d.
$$\underset{H}{\overset{H}{\underset{|}{N}}}-\underset{C_6H_5CH_2}{\underset{|}{\overset{H}{\overset{|}{C}}}}-\overset{O}{\overset{\|}{C}}-OH$$

57. Classify each of the following molecules as a lipid, carbohydrate, or amino acid:

a. [cyclic sugar structure — glucose]

b.
$$\begin{array}{l} CH_2OC(CH_2)_{16}CH_3 \\ | \\ CHOC(CH_2)_{16}CH_3 \\ | \\ CH_2OC(CH_2)_{16}CH_3 \end{array}$$

c.
$$\begin{array}{l} CH_2OC(CH_2)_7CH=CH(CH_2)_7CH_3 \\ | \\ CHOC(CH_2)_7CH=CH(CH_2)_7CH_3 \\ | \\ CH_2OC(CH_2)_7CH=CH(CH_2)_7CH_3 \end{array}$$

d.
$$H_2N-\underset{CH_2}{\underset{|}{\overset{H}{\overset{|}{C}}}}-\overset{O}{\overset{\|}{C}}-OH$$
with CH_2 attached to a phenyl ring

58. Classify each of the following molecules as a lipid, carbohydrate, or amino acid:

a.
$$\begin{array}{l} CH_2OC(CH_2)_{12}CH_3 \\ | \\ CHOC(CH_2)_{14}CH_3 \\ | \\ CH_2OC(CH_2)_7CH=CH(CH_2)_7CH_3 \end{array}$$

b. [disaccharide/polysaccharide ring structures connected]

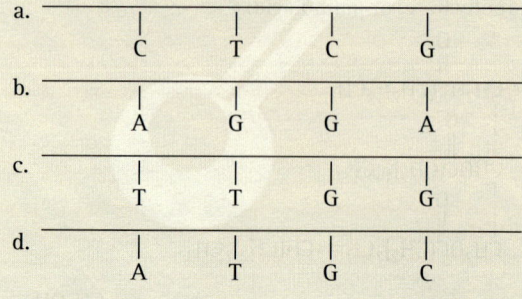

c. [structure of disaccharide with CH₂OH groups]

d. [structure of amino acid with CH₂SH side chain]

59. Draw the structure for the dipeptide that would result from valine and serine, Val-Ser.
60. Draw the structure for the dipeptide Ser-Val. How is this dipeptide different from the one in problem 59?
61. Draw the structure for the tripeptide, Leu-Leu-Leu. Would you expect this tripeptide to mix readily with water?
62. Draw the structure for the tripeptide Ser-Ser-Ser. Would you expect this tripeptide to mix readily with water?
63. Give the complement of each of the following bases:
    a. adenine
    b. cytosine
    c. guanine
    d. thymine
64. Draw the complementary strand for each of the following DNA fragments:
    a. C T C G
    b. A G G A
    c. T T G G
    d. A T G C

POINTS TO PONDER

65. Comment on the relationship between the structure of lipids and their function as vehicles of long-term energy storage. Describe a straight chain hydrocarbon that might accomplish the same thing.
66. Compare ethanol (CH_3CH_2OH), the alcohol in alcoholic beverages, with glucose. How are the two molecules similar? How are they different? Is ethanol transportable in the blood? What do you think its energy content is (on a per gram basis) relative to glucose?
67. Suppose you were a legislator looking at a bill that proposed to severely curtail genetic engineering experiments in humans. Would you argue for or against the bill? Why?
68. Suppose that scientists produce a genetically modified bee that is incapable of stinging. They believe that this bee could compete with the natural honeybee and replace it in the environment. Proponents argue that the bee should be released to save the lives that are lost each year due to bee sting allergies. How would you respond to the proposed release of this bee? Why?
69. What are the ethical issues involved in the genetic screening of human embryos? Would you favor or oppose the screening of embryos? Are there some kinds of genetic screens that would be acceptable while others would not?
70. Suppose a scientist succeeded in taking one human fertilized egg and producing two genetically identical embryos (clones). The embryos are implanted into two surrogate mothers and both embryos survive to birth. How would the two children be alike? How might they be different?

Feature Problems and Projects

71. Classify each of the following as a saccharide, disaccharide, or amino acid:

    a.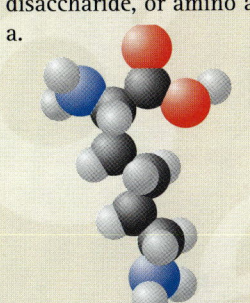

    b.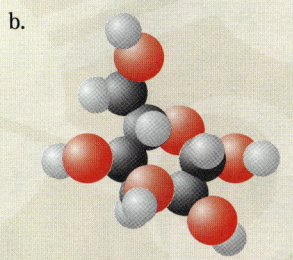

    c.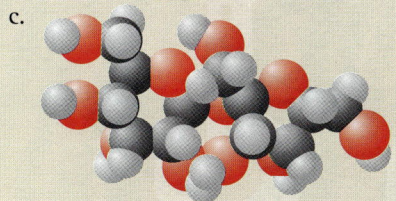

    d.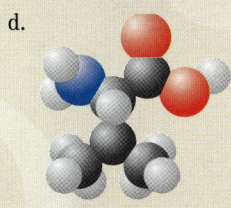

    c.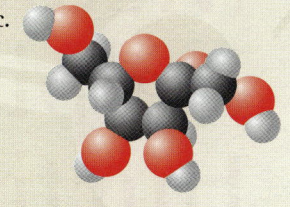

    d.

72. Classify each of the following as a triglyceride, saccharide, or amino acid:

    a.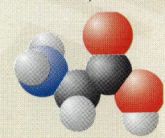

    b.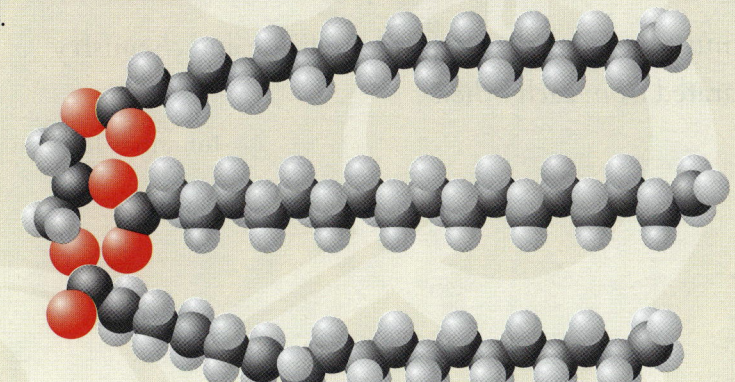

W 73. Genetically engineered foods have met with some public opposition in recent times. Read an FDA published article that originally appeared in the December 2003 *FDA Consumer* entitled, "Genetic Engineering: The Future of Foods?" It is available on the World Wide Web at: **http://www.fda.gov/fdac/features/2003/603_food.html**. What do you think about genetically engineered foods? Write a one-paragraph response.

W 74. Read the *Scientific American* article entitled, "The First Human Cloned Embryo," which can be found at the following website: **http://www.sciam.com/article.cfm?articleID=0008B8F9-AC62-1C75-9B81809EC588EF21&pageNumber=1&catID=4**. What do you think about therapeutic cloning? How is therapeutic cloning different from reproductive cloning? In your opinion, is therapeutic cloning ethical? Write a one-page essay addressing these questions.

17 Drugs and Medicine: Healing, Helping, and Hurting

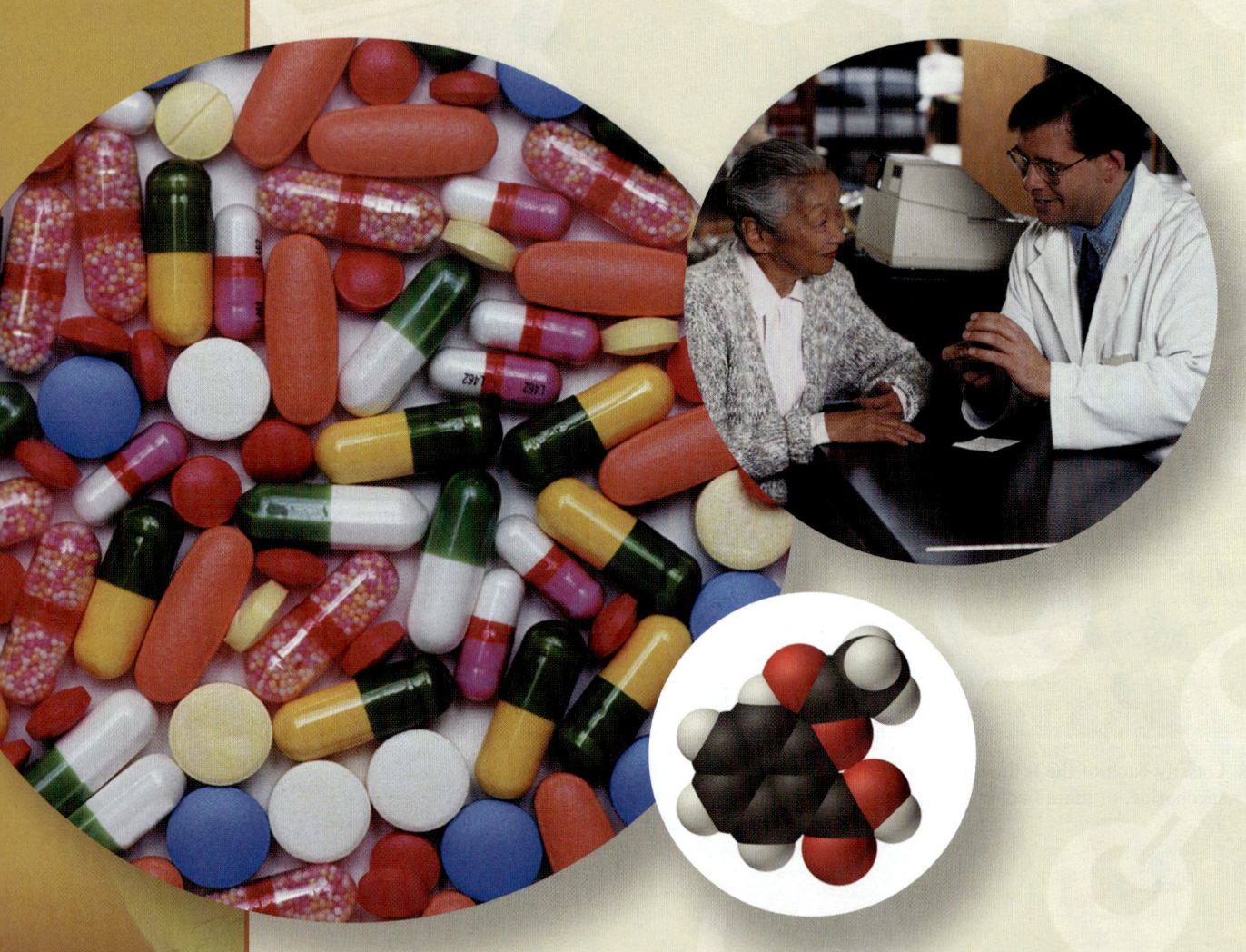

...the doctor must be a chemist also, and medicine and chemistry cannot be separated from each other.

—Johannes Janssen

In this chapter, we will examine drugs—molecules that alter physical and mental conditions. These molecules have been used to treat disease, increase life expectancy, and improve the quality of life. Some drugs, however, can be abused, and we will discuss those as well. Morphine, for example, is an indispensable drug for treating severe pain; however, it can also be abused by those seeking a temporary high. A number of over-the-counter drugs, such as aspirin and its substitutes, are used frequently to relieve minor aches and pains or cold and flu symptoms. Antibiotics are the miracle drugs of the 20th century, and their discovery has all but eradicated a number of once-deadly diseases. Antiviral and anticancer drugs have been developed in recent years with some success. As you read this chapter, think about the drugs that you have used. Where do they come from? What are their effects? What are their side effects?

QUESTIONS FOR THOUGHT

- Are emotions associated with molecules?
- How does aspirin work? What are aspirin substitutes?
- What are antibiotics and how do they work?
- What are antiviral drugs?
- What drugs are used to treat AIDS?
- How do birth control pills work?
- What are steroids?
- What is chemotherapy and how does it kill cancer cells?
- What are depressants?
- What are narcotics?
- What are stimulants?
- What are hallucinogenic drugs?
- What are the effects of smoking marijuana?
- How do antidepressants such as Prozac work?

FIGURE 1 Serotonin, a molecule that helps transmit nerve signals. Depression is associated with decreased levels of brain serotonin.

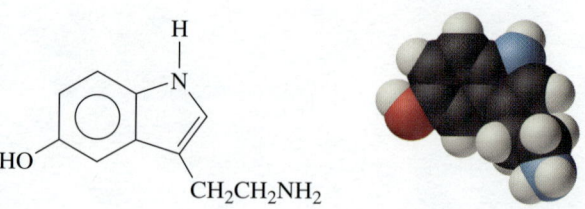

The feelings of being "in love" are associated with elevated levels of phenylethylamine in the brain.

17.1 Love and Depression

You have probably experienced the feelings of attraction to a particular person: excitement, nervousness, happiness, and inability to concentrate on other matters—the signs of "being in love." You may have also experienced opposite feelings at low points in your life: sadness, despair, helplessness, and pessimism—the signs of being depressed. In our search for molecular reasons, we have seen how many everyday observations are linked to interactions at the atomic and molecular level. Could even feelings of love or depression have a molecular origin?

While the answer is not simple, we know that the feelings of love and depression are associated with certain molecules in the human brain. Feelings of love are associated with elevated levels of a molecule called phenylethylamine in the brain. When this compound is administered to a test patient, they report feelings identical to those of "being in love." Similarly, feelings of depression are associated with the absence of a molecule called serotonin (Figure 1) in the brain. Serotonin is a **neurotransmitter**, a substance that helps transmit messages among nerve cells. When a test patient is given reserpine, a drug that depletes serotonin and other neurotransmitters, the patient suffers severe emotional depression just like a person undergoing a major life crisis.

As our understanding of the chemical basis for biology has grown, we have learned much about the molecules that constitute life. We are beginning to understand the connections between physical and psychological conditions and the molecules that cause them. Consequently, we can now engineer molecules to combat dangerous or unpleasant conditions. Depression, for example, can be treated with drugs that restore brain serotonin to normal levels.

Many medicines and drugs alter physical and mental conditions, improving the quality and quantity of life. The average person today lives about 20 years longer than the average person at the beginning of the century. On the other hand, some drugs have severe addictive and destructive power. The heroin junkie destroys his or her life through a compound that did not exist before humans synthesized it about 100 years ago.

Explore this topic on the U.S. Food and Drug Administration—Home Page website.

17.2 Relieving Pain, Reducing Fever, and Lowering Inflammation

The most popular drug in the world is aspirin (Figure 2), a nonsteroidal anti-inflammatory drug (NSAID) that relieves minor pain (analgesic), lowers fever (antipyretic), and reduces swelling (anti-inflammatory). In some cases, aspirin is also used to reduce the risk of stroke and heart attack. Aspirin (acetylsalicylic acid) was first marketed around the beginning of 20th century by the German company Bayer. It is a synthetic modification of salicylic acid, a compound that occurs naturally in the bark of the willow tree.

17.2 Relieving Pain, Reducing Fever, and Lowering Inflammation

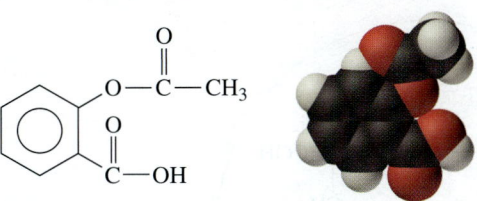

FIGURE 2 Aspirin, the most common drug in the world.

Aspirin reduces pain and lowers fever by preventing the formation of **prostaglandins**, fatty acid derivatives involved in a number of physiological processes. Prostaglandins sensitize pain receptors in nerve cells. By inhibiting prostaglandin formation, aspirin reduces pain.

Prostaglandins also mediate fever. When foreign organisms invade the body and spread into the bloodstream, specialized cells release *pyrogens,* which in turn stimulate the release of prostaglandins. Prostaglandins act as chemical messengers that turn up the body's thermostat, producing fever. By inhibiting prostaglandin formation, aspirin reduces fever. The role of fever in the immune response, however, remains unclear. Because fever is such a common result of infection, it may play a beneficial role in the body's fight against infection. Extremely high fevers, however, are dangerous and should be treated.

Inflammation, a physiological response to tissue damage or foreign invasion, is mediated by powerful chemical substances called **histamines**. Aspirin reduces inflammation by decreasing histamine release.

While aspirin has relatively minor side effects, they are worth noting. The acidity of aspirin can irritate the stomach. Although unnoticeable, aspirin usually causes a small amount of stomach bleeding. For this reason, people with ulcers should avoid aspirin. Also, aspirin reduces the formation of blood platelets that initiate blood clotting. This causes the blood to be thinner and is probably responsible for aspirin's ability to lower stroke and heart attack risk. However, the same thinning effect also increases susceptibility to bruising and blood loss during injury.

There is evidence that ingestion of aspirin by children or teenagers with the chicken pox or influenza is associated with Reye's syndrome, a rare but serious illness that may cause death if not treated early. For this reason, children or teenagers with the chicken pox or flulike symptoms should not take aspirin. Finally, aspirin is toxic in large doses. More children are accidentally poisoned by aspirin than any other drug. Like all drugs, aspirin should be kept out of the reach of children.

 Explore this topic on the **RX List: The Internet Drug Index** website.

ASPIRIN SUBSTITUTES

Because of aspirin's minor side effects, and because of the allergic responses some people have to it, several aspirin substitutes have been developed (Figure 3). The most common aspirin substitute is acetaminophen (Tylenol, Anacin-3), which reduces fever and relieves pain much like aspirin but does not reduce inflammation. Overuse of acetaminophen can lead to kidney and liver damage. A second common aspirin substitute is ibuprofen (Advil, Nuprin, and Motrin IB), which acts just like aspirin to reduce pain, fever, and inflammation. Another addition to the family of over-the-counter aspirin substitutes is naproxen (Aleve). Its main advantage is its long-lasting effect. While most pain relievers require doses every 4–6 hours, a single naproxen dose lasts 8–12 hours.

Progressive Publishing Alternatives with permission from BayerRoche L.L.C. and McNeil Consumer Products Company

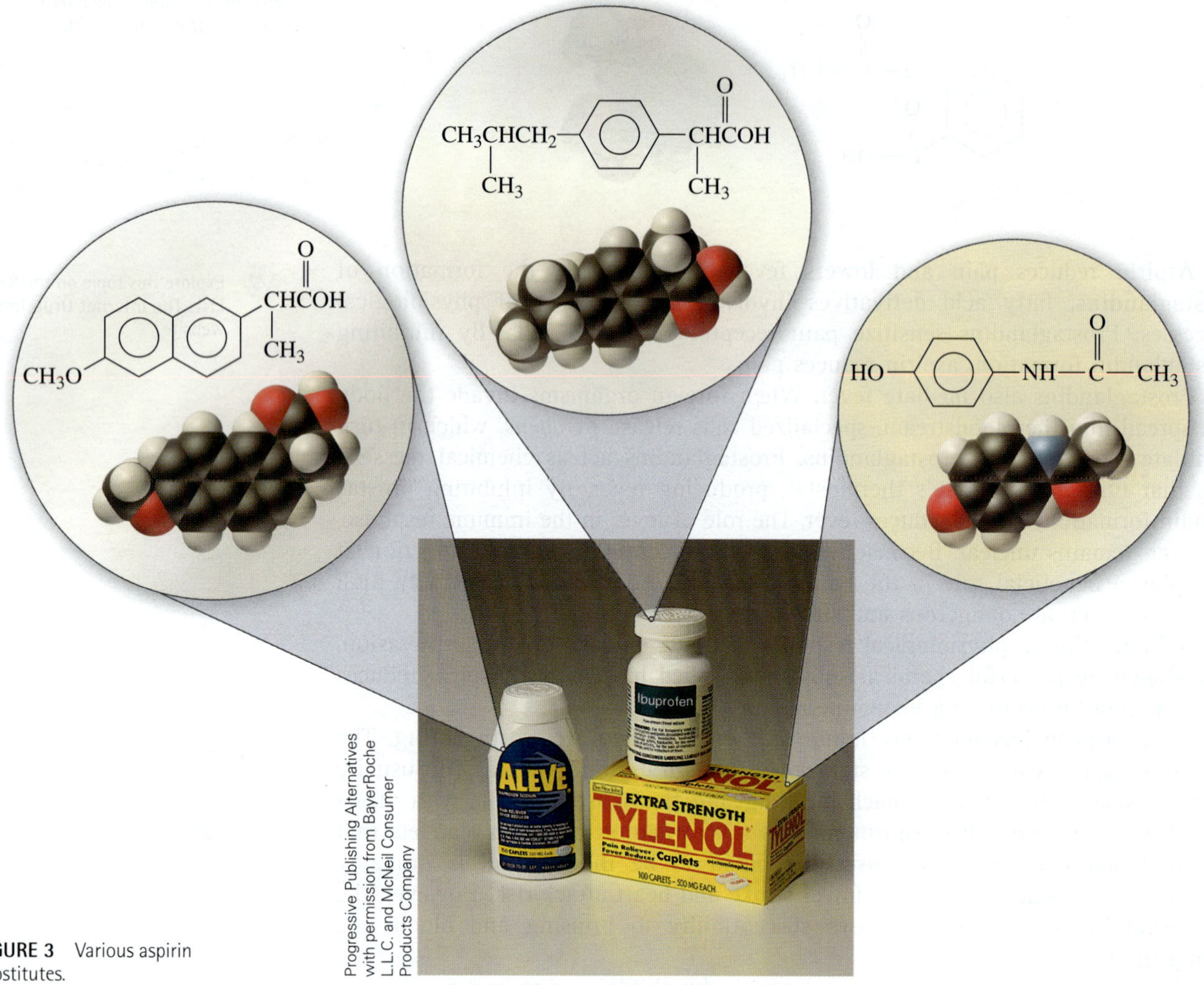

FIGURE 3 Various aspirin substitutes.

17.3 Killing Microscopic Bugs: Antibiotics

Explore this topic on the **World Health Organization** website.

Hossler/Custom Medical Stock Photo

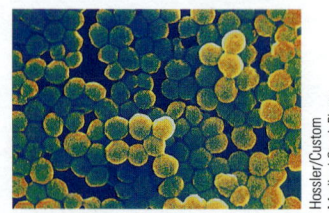

Bacteria, like those shown here, are responsible for a number of infectious diseases, including diphtheria, tuberculosis, cholera, and pneumonia.

The most prescribed medicines in the United States are antibiotics, drugs that fight infectious diseases like diphtheria, tuberculosis, cholera, and pneumonia. These diseases are caused by **bacteria**, microorganisms that reproduce within the human body. Antibiotics work by targeting the unique physiology of bacteria, selectively killing them. Before antibiotics were available, bacterial infections and their associated diseases were among the top causes of death in the United States. Today, many of these diseases have been virtually eradicated from developed nations. Unfortunately, bacterial diseases still claim many lives in underdeveloped countries where antibiotics are less available and sanitary conditions remain poor.

Antibiotics are classified into several different categories. The most common categories are penicillins, cephalosporins, and tetracyclines. The penicillins were discovered in 1928 by Sir Alexander Fleming, a bacteriologist at the University of London. One of his bacterial cultures, which he planned to use for his research, became contaminated with a fungus. While he was on vacation, a period of cool weather slowed the bacterial growth, allowing the fungus to flourish. When Flem-

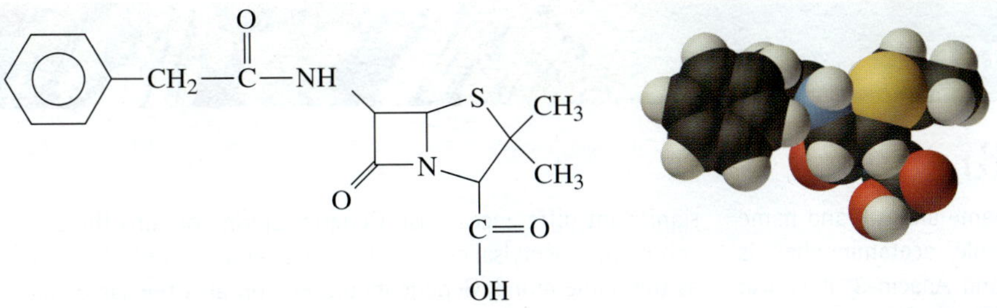

FIGURE 4 Penicillin G.

ing returned to his laboratory, he noticed that the bacteria immediately surrounding the fungus colonies had been destroyed. Fleming went on to show that the fungus contained an antibacterial agent that he called **penicillin** (Figure 4).

Fleming did not realize the magnitude of his discovery. He never imagined that penicillin could be introduced into the human body to cure infection. It was not until 1941, 13 years later, that penicillin was tried on humans by other researchers. These trials demonstrated that penicillin was highly effective in killing a number of pathogenic bacteria, while showing almost no toxicity to humans. One of the researchers, Professor Howard Florey at Oxford University, remarked about the results, "[it is] a truth so gratifying as to be at times almost unbelievable." That previously fatal diseases could be cured with the simple administration of a single drug with few or no side effects is truly a medical miracle.

The use of penicillin to cure infectious diseases began a new era in the chemical treatment of illness. However, even in the early years of penicillin treatment, researchers discovered that some bacteria were resistant to penicillin, and, even worse, bacteria that initially responded to penicillin could over time develop resistance to it.

Today, we have characterized many penicillins and penicillin-like compounds, many with clinical value. The early penicillins were followed by a number of modified penicillins that proved even more effective. Some could be taken orally; others had to be injected. Bacteria that did not respond to one kind of penicillin could be treated with another. Also discovered were the chemically related compounds called cephalosporins (Figure 5), found to be active against a number of penicillin-resistant bacteria. The cephalosporins could also be chemically modified to achieve specific desirable properties. The penicillins and the cephalosporins both work by preventing the normal development of bacterial cell walls. Once weakened, the cell wall ruptures, and the bacteria die. Because human cells are different, they are not affected. Other antibiotics function in different ways. The tetracyclines (e.g., Achromycin, Terramycin), for example, bind to bacterial ribosomes and inhibit protein synthesis. Without the necessary proteins, the bacteria cannot survive.

In spite of our arsenal of antibiotics, some resistant bacteria have evolved. For example, tuberculosis (TB), which had all but disappeared until the mid-1980s, has begun to spread among people with inadequate health care and among those with an AIDS-weakened immune system. Although most strains of

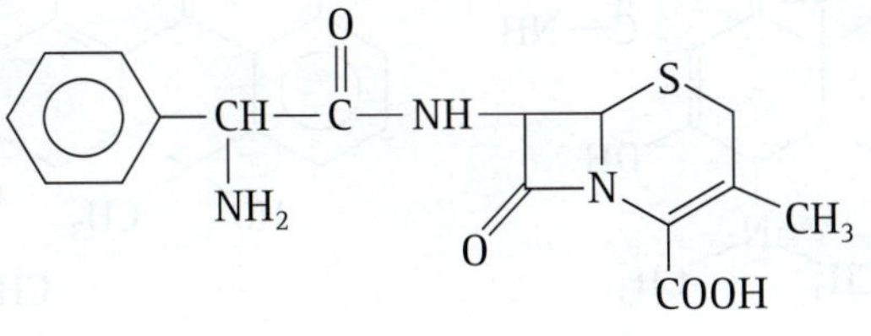

Alexander Fleming discovered penicillin.

FIGURE 5 Cephalexin, a cephalosporin.

MOLECULAR THINKING

Generic or Name Brands?

All drugs have a generic or chemical name and a brand name given by the manufacturer. For example, acetaminophen is available under brand names Tylenol and Anacin-3; it is also available as generic acetaminophen. When a drug company discovers a new drug and brings it to market, the firm usually enjoys several years of patent protection in which it has exclusive rights to the drug. However, once the patent expires, other manufacturers can market the same drug, either as generic or under a different brand name.

The common pain relievers available today have been in production for many years, so no single drug company has exclusive rights to them. Consequently, today's pain-relief shopper is bombarded by choices at the local drugstore. Each pain reliever is available under several different brand names as well as under a generic label. Is there any difference? While some may argue that doses are more carefully controlled by the name brands, there is little evidence to suggest that any significant differences exist. Generic aspirin contains the same compound (acetylsalicylic acid) in the same quantity (325 mg) as the name brand. Generic ibuprofen contains the same compound (ibuprofen) in the same quantity (200 mg) as the name brand. The situation is similar with the other pain relievers. Consequently, it is not worth paying extra for the name brands. The generic products are identical without the added expense.

Most pain relievers are also available in regular and extra strength. What is the difference? The extra-strength pain reliever contains more of the compound per tablet. For example, extra-strength acetaminophen contains 500 mg per tablet while regular strength contains 325 mg per tablet. The same effect can be obtained by taking more of the regular-strength tablets.

QUESTION: How many regular-strength acetaminophen tablets should you take to approximate the two-tablet, extra-strength dose?

TB bacteria respond to some form of antibiotic, one strain has evolved that is completely drug resistant. Patients with this form of TB suffer a 50% death rate. Scientists are continually working to produce new and better antibodies.

17.4 Antiviral Drugs and Acquired Immune Deficiency Syndrome

In contrast to bacteria, which are single-celled living organisms that reproduce on their own, viruses require the machinery of a host cell to reproduce. A **virus**, which usually consists of only DNA and protein, takes over a host cell and inserts its own DNA into the chromosomes of the host. The virus then uses the machinery of the host cell to synthesize viral protein and copy viral DNA, producing more viruses. These new viruses infect other host cells, and the cycle is repeated. Because viruses use host cells to reproduce, they are difficult to attack without also attacking the healthy host cells. Viruses do not respond to antibiotics.

The common cold, flu, measles, polio, and, most recently, acquired immune deficiency syndrome (AIDS) are all caused by viruses. The human immunodeficiency virus (HIV) that causes AIDS is a **retrovirus**. Unlike most viruses that consist of DNA and protein, retroviruses consist of RNA and protein. Once in the

W Explore this topic on the HIV/AIDS Treatment, Prevention, and Research website.

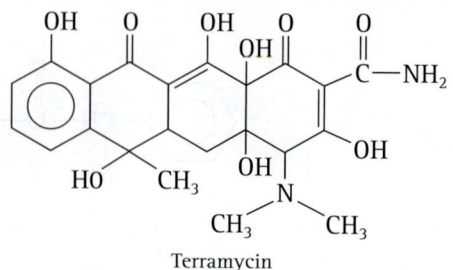

Tetracyclines are so named because their structures contain four rings. The structures of Aureomycin and Terramycin are shown here.

Aureomycin   Terramycin

17.4 Antiviral Drugs and Acquired Immune Deficiency Syndrome

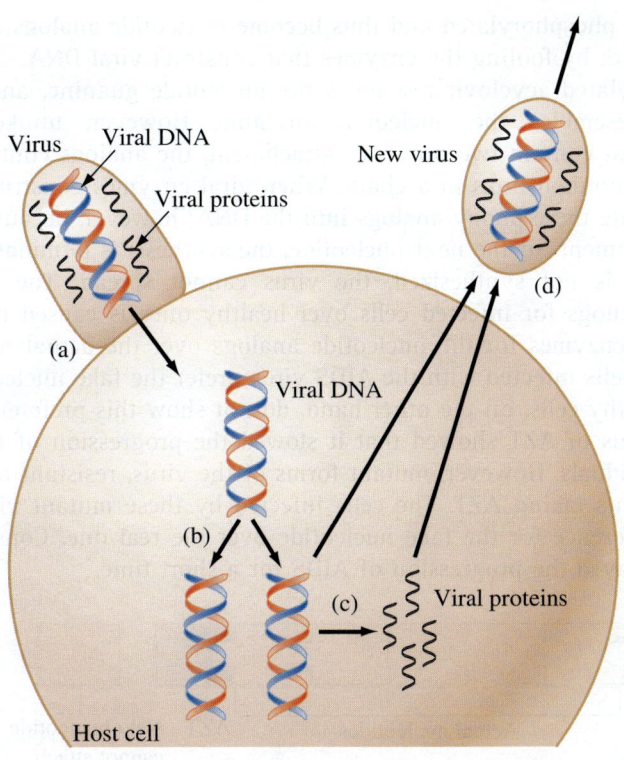

Reproductive cycle of a virus. (a) A virus, usually composed of DNA and proteins, inserts itself into a host cell. (b) The virus then uses the machinery of the host cell to copy viral DNA. (c) The virus also uses the machinery of the host cell to synthesize viral protein. (d) New viruses form that then go on to infect other host cells.

host cell, the viral RNA is *reverse transcribed* into DNA, which then goes on to direct the synthesis of viral protein. Because reverse transcription is unique to retroviruses and does not occur in normal cells, it has been a target of attack for many researchers combating HIV.

Until recently, the only way to fight viruses was to prevent them through vaccination, a procedure that uses the body's own defenses to combat disease. A vaccination consists of introducing a small number of inactivated viruses into the body, usually through injection. The body then develops antibodies to the virus, which remain in the body and are available to fight off the disease should infection with the real virus occur. Vaccinations have helped to control many, but not all, viral diseases. The common cold, for example, is caused by so many different strains of the cold virus that it is almost impossible to vaccinate against it. The AIDS virus has also resisted our attempt to create a vaccination. However, with the increased understanding of how viruses work, some antiviral drugs, including *nucleoside analogs* and *protease inhibitors,* are now becoming available.

NUCLEOSIDE ANALOGS

One class of antiviral drugs is **nucleoside analogs**. Nucleosides are nucleotides (the individual links in DNA) without the phosphate groups. Acyclovir (Figure 6), used against the herpes virus, and azidothymidine (AZT, Figure 7), used against HIV, are both examples of nucleoside analogs. When these drugs enter infected

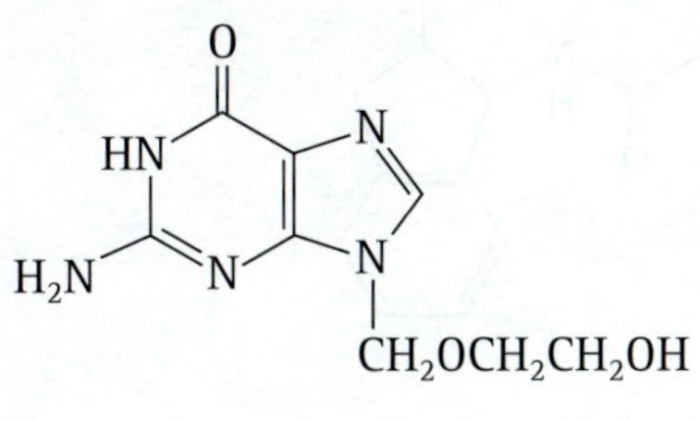

FIGURE 6 Acyclovir, an antiviral drug used against the herpes virus.

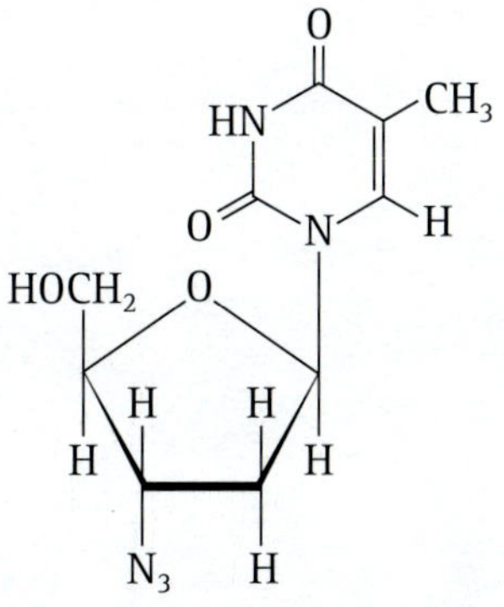

FIGURE 7 AZT, antiviral nucleoside analog used against the AIDS virus.

cells, they are phosphorylated and thus become nucle*otide* analogs. In this active form, they work by fooling the enzymes that construct viral DNA.

Phosphorylated acyclovir resembles the nucleotide guanine, and phosphorylated AZT resembles the nucleotide thymine. However, unlike the actual nucleotides that contain two points of attachment, the analogs contain only one; they are like terminal links in a chain. When viral enzymes construct viral DNA, they incorporate these phony analogs into the DNA; however, because they lack a point of attachment for the next nucleotide, the synthesis is terminated (Figure 8). If viral DNA is not synthesized, the virus cannot spread. The selectivity of nucleoside analogs for infected cells over healthy ones is caused by the preference of viral enzymes for the nucleotide analogs over the actual nucleotides. In other words, cells infected with the AIDS virus prefer the fake nucleotide over the real one. Healthy cells, on the other hand, do not show this preference.

Initial trials of AZT showed that it slowed the progression of the disease in infected individuals. However, mutant forms of the virus, resistant to AZT, developed in patients taking AZT. The cells infected by these mutant viruses did not show the preference for the fake nucleotide over the real one. Consequently, the drug only delayed the progression of AIDS for a short time.

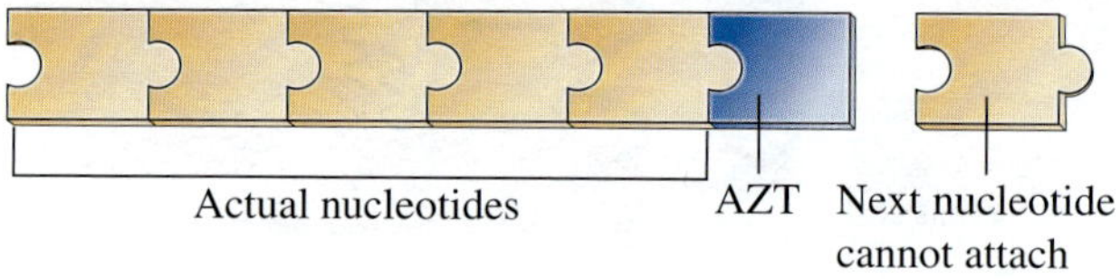

FIGURE 8 AZT acts as a phony nucleotide that terminates the synthesis of viral DNA.

PROTEASE INHIBITORS

More recently, a new class of antiviral drugs, called **protease inhibitors** (e.g., Indinavir, Ritonavir), have been developed against HIV. They attack the replication of the virus at a different stage, the construction of viral protein. Once viral DNA is incorporated into host cell DNA, it directs the construction of viral proteins. In the construction, viral enzymes, called **proteases**, act like molecular scissors to cut freshly made viral proteins to the correct size. In 1989, researchers discovered the three-dimensional structure of HIV's protease (Figure 9). The molecule looks like two butterfly wings joined at the active site, the place where the actual cutting occurs.

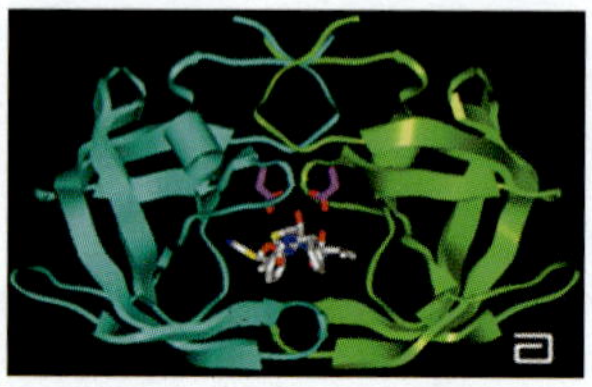

FIGURE 9 HIV's protease. The center of this protein molecule is the active site where freshly minted AIDS proteins are cut to the right size. Ritonavir (green) jams the active site.

After the discovery of this structure, researchers set out to design a molecule that would jam the active site, disabling the protease. In the early 1990s, after many failed attempts, a number of drug companies succeeded in developing drugs that seemed to work. Because these drugs inhibit the action of protease, they are called protease inhibitors. By jamming the active site, they prevent protease from cutting viral proteins to the correct size; without these proteins, the AIDS virus cannot reproduce.

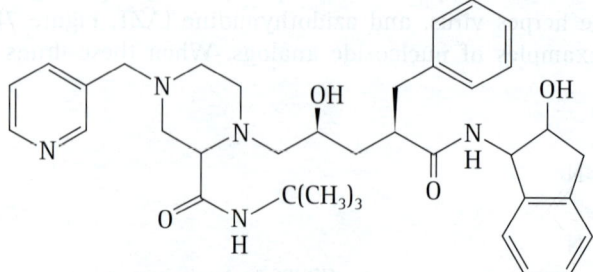

Indinavir, a protease inhibitor used against the AIDS virus.

Molecular Focus

Azidothymidine (AZT)

Formula: $C_{10}H_{13}N_5O_4$
Molecular mass: 267 g/mol
Structure:

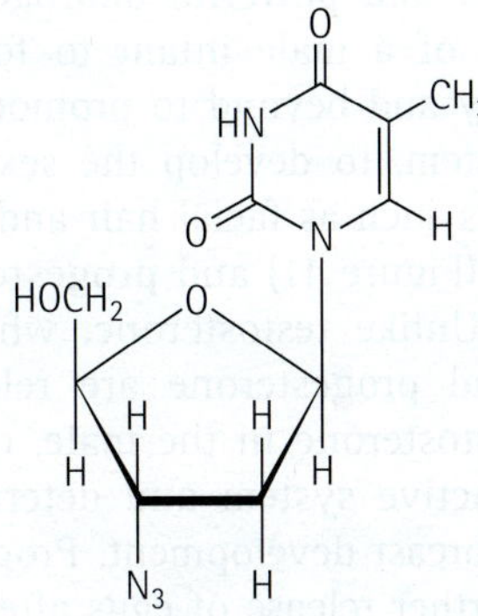

AZT is a replica of the nucleoside thymidine with one important difference: The hydroxy group on the thymidine is replaced by three nitrogen atoms. This difference terminates the growth of the DNA chain during the reverse transcription of viral RNA. AZT was first synthesized in 1964 as a potential anticancer drug, but it failed. In 1985, researchers realized its potential as an inhibitor of reverse transcription and demonstrated its potency *in vitro* (in a test tube) against the AIDS virus. Although initial clinical trials showed promising results, extended trials were disappointing because the AIDS virus quickly developed resistance to AZT. Today, AZT is a crucial component of the triple-drug therapy used to treat AIDS.

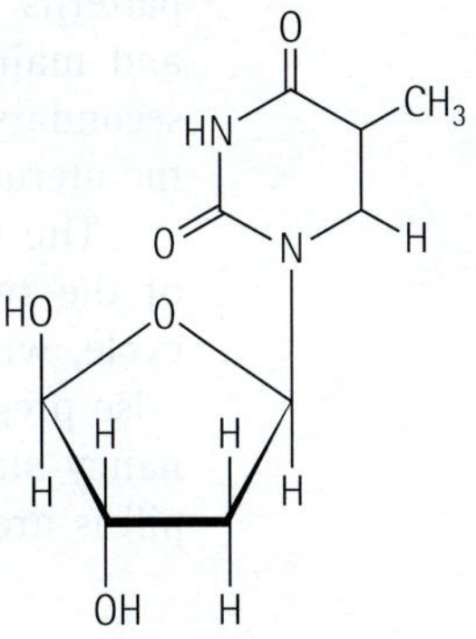

Three-dimensional structure:

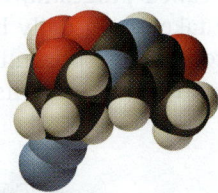

thymidine

Protease inhibitors, when given in combination with AZT and other nucleoside analogs (3TC or ddI), have shown promising results. Clinical trials showed that a three-drug protease cocktail (AZT/ddI/Indinavir) reduced the AIDS virus to undetectable levels in 90% of infected participants. The powerful combination of three drugs kills the virus at two different points in its replication cycle. For resistant viruses to survive, they would have to develop resistance to all three drugs simultaneously, an unlikely event. The triple-drug cocktail is not without its drawbacks, however. Patients must maintain a strict regimen of taking the

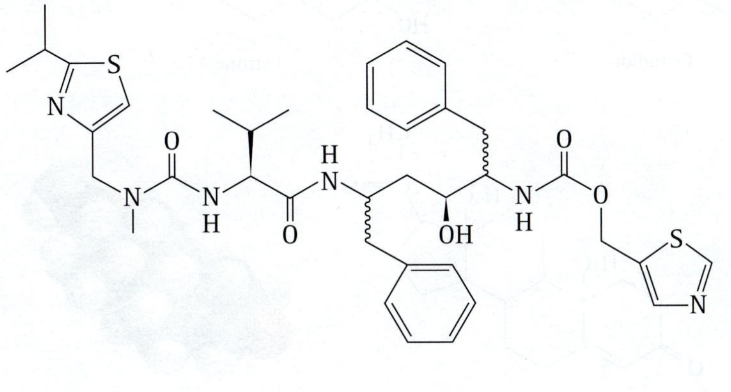

Ritonavir, a protease inhibitor used against the AIDS virus.

pills on a daily basis, and the drugs have unpleasant side effects. The cost is also high, about $10,000 per year for the triple drug therapy, which may be needed over a lifetime.

17.5 Sex Hormones and the Pill

Sexual characteristics are governed by molecules called **hormones**, chemical messengers that act on target cells usually located a long distance from where the hormone is secreted. The male sexual hormones, called **androgens**, are secreted primarily by the testes. The most important and powerful androgen is **testosterone** (Figure 10), which acts before the birth of a male infant to form the male sex organs. Testosterone acts again at puberty and beyond to promote the growth and maturation of the male reproductive system, to develop the sexual drive, and to determine secondary sexual characteristics such as facial hair and voice deepening.

The female sex hormones, **estrogen** (Figure 11) and **progesterone** (Figure 12), are secreted primarily by the ovaries. Unlike testosterone, which is constantly secreted by mature males, estrogen and progesterone are released in cyclical patterns. Estrogen in the female, like testosterone in the male, causes maturation and maintenance of the female reproductive system and determines the female secondary sexual characteristics such as breast development. Progesterone prepares the uterus for pregnancy and prevents further release of eggs after pregnancy.

The birth control pill, developed in the 1950s, consists of synthetic analogs of the two female sex hormones. The estrogen analog regulates the menstrual cycle, while the progesterone analog (also called progestin) establishes a state of false pregnancy. When these hormones are taken on a daily basis, the false pregnancy state is maintained, and no eggs are released, preventing conception. The pill is greater than 98% effective when taken regularly, but its effectiveness drops

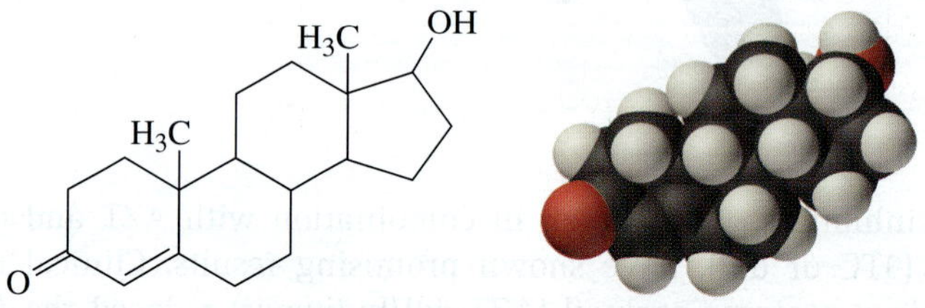

FIGURE 10 Testosterone, the primary male sexual hormone.

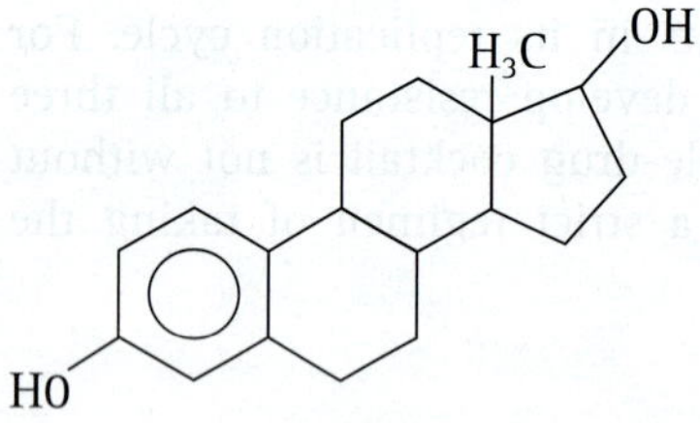

FIGURE 11 The estrogens, estradiol and estrone.

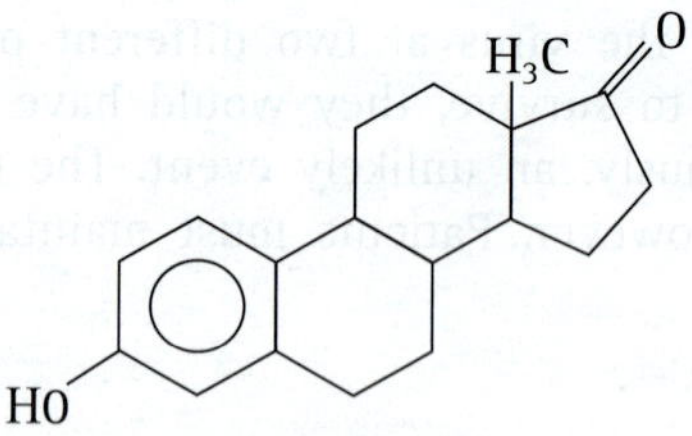

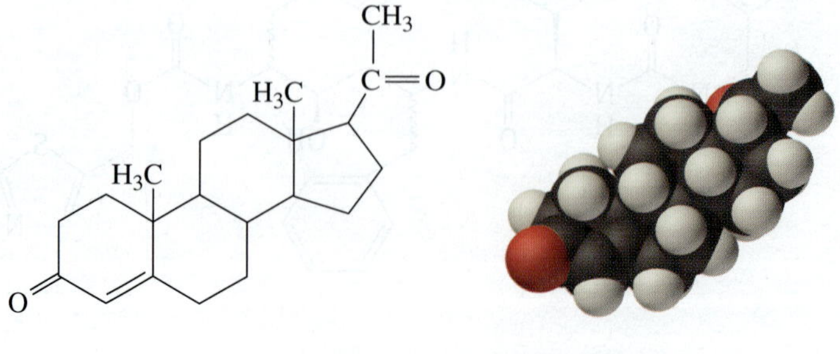

Estradiol

Estrone

FIGURE 12 Progesterone, one of the primary female sexual hormones.

WHAT IF...

The Controversy of Abortion

The ethics of abortion have been argued for centuries, and the ease of abortions with a pill such as RU-486 has inflamed the passions of both sides. Those arguing against abortion see little reason to distinguish the developing fetus from a newborn infant. By the end of the second month, they argue, all major organ systems are developed; and although the fetus is only about 1 in. long, it is recognizably human in form. To pro-life activists, taking the life of this developing fetus is murder, and should be treated as such.

On the other side are those who see few human characteristics in the developing fetus. The fetus cannot survive apart from the mother, and therefore it is the mother's right—not the state's—to decide whether or not to continue the pregnancy. To pro-choice activists, restrictions on abortions are an impingement of women's rights. Abortions were illegal in many U.S. states until the landmark *Roe versus Wade* decision of 1973, in which the Supreme Court declared state regulation of abortion unconstitutional before *viability*. (A fetus is *viable* when it can survive outside of the mother. Viability occurs at the beginning of the third trimester.) The Supreme Court has upheld certain state restrictions on abortion after viability but has never overturned *Roe versus Wade*.

QUESTION: What do you think? Should restrictions on abortion become more or less severe? Why? What if you or a close friend became pregnant? Would you choose or advise abortion as an option? Would the availability of an abortion pill influence your decision?

---

off when doses are missed. The pill does have some side effects including nausea, weight gain, fluid retention, breast tenderness, and acne.

In 1990, the U.S. FDA approved Norplant, a form of progestin that does not require the daily ingestion of pills. Norplant consists of six surgically implanted capsules of progestin, each about the size of a matchstick. The capsules are implanted under the skin in the upper arm and slowly release progestin to provide contraceptive action for up to 5 years. The efficacy of Norplant is greater than 98%. Because there are no daily pills to take, there is no decrease in efficacy due to missed doses. The side effects are similar to those of the oral birth control pill, with the added possible side effect of an irregular menstrual cycle.

Pregnancy can be terminated within five weeks of conception with a controversial drug called RU-486 (mifepristone). RU-486 works by blocking the action of progesterone, the hormone required to sustain pregnancy. When a pregnant woman takes RU-486, pregnancy cannot be maintained and a menstrual cycle is triggered, flushing the developing fetus out of the uterus.

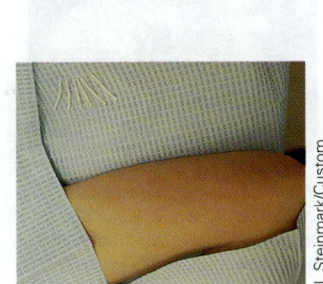

Norplant consists of five surgically implanted capsules of progestin, a synthetic progesterone analog.

17.6 Steroids

The sexual hormones discussed previously are part of a larger class of compounds called **steroids**, characterized by their 17-carbon-atom, four-ring skeletal structure. Progestin is an example of a steroid used as a drug. Other steroids used as drugs include the adrenocortical steroids and the anabolic steroids.

ADRENOCORTICAL STEROIDS

The **adrenocortical steroids**—drugs such as cortisone, hydrocortisone, prednisone, and prednisolone—are used primarily for their potent anti-inflammatory effects. Inflammation is part of the body's defense systems; when a part of the body is invaded by a foreign organism or when tissue is damaged, the body responds by transporting blood proteins and other substances to the affected area, resulting in inflammation. The adrenocortical steroids work by suppressing the body's inflammatory and immune responses and therefore reducing inflammation. This is useful when the inflammation causes pain and discomfort as in allergic reactions

(a) Steroids all have a 17-carbon-atom, four-ring skeletal structure. (b) Cortisone and prednisone are both steroids used as drugs for their potent anti-inflammatory effect.

(rashes), rheumatoid arthritis, or colitis (inflammation of the colon). However, the suppression of the immune system leaves the patient subject to infection, and adrenocortical steroids have numerous other side effects such as liquid retention, muscle loss, and osteoporosis (loss of bone mass).

ANABOLIC STEROIDS

The **anabolic steroids** are usually synthetic analogs of testosterone, the male sex hormone. When taken in high doses and coupled with exercise or weight lifting, these drugs increase muscle mass. Although illegal, some athletes and bodybuilders take anabolic steroids to improve their performance. They do so at great risk, however, because the use of anabolic steroids has several serious side effects. In males, anabolic steroids cause lower sperm production and testicular atrophy. In females, they promote male secondary sexual characteristics such as facial hair growth, voice deepening, and breast diminution as well as menstrual irregularities and ovulation failure. In both sexes, the use of anabolic steroids damages the liver and increases the risk of stroke and heart attack.

When taken in high doses and coupled with exercise, anabolic steroids help build muscle mass; however, they have serious side effects.

17.7 Chemicals to Fight Cancer

About 40% of Americans develop cancer in their lifetimes, and about one in five die from it. Cancer kills about 6 million people worldwide each year. Cancer is a particularly difficult disease to combat because cancer cells are a modified form of human cells. Cancer begins when DNA within healthy cells is changed (mutated), probably by carcinogenic chemicals, ionizing radiation, viruses, or a combination of these. Mutated cells may just die, as often happens, or they may become cancerous, in which case they divide at an uncontrolled rate, often much faster than normal cells. The resulting mass of tissue is called a tumor and may invade vital organs, destroying their function and often resulting in death.

Chemicals that fight cancer must destroy cancerous cells. Unfortunately, cancerous cells are so similar to normal cells, that these chemicals destroy normal cells as well. The main exploitable difference between cancerous cells and healthy ones is their faster division rate. Consequently, most anticancer chemicals target rapidly dividing cells, often by disrupting DNA synthesis, a crucial step in cell division. Some of the body's cells—such as those within the bone marrow, the

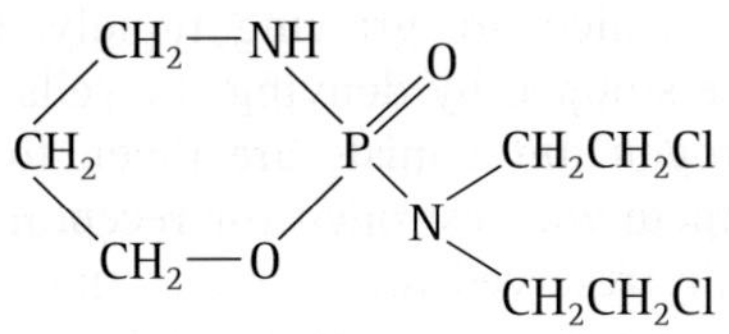

Cyclophosphamide, an alkylating agent used in cancer treatment.

digestive tract, the reproductive system, and the hair follicles—are naturally rapidly dividing and are therefore most adversely affected by anticancer agents. Damage to these cells causes severe side effects such as anemia, inability to fight infection, nausea, vomiting, infertility, hair loss, fatigue, and internal bleeding. Most of these side effects diminish once treatment ends. The chemicals that fight cancer are divided into several groups, including alkylating agents, antimetabolites, topoisomerase inhibitors, and hormones.

Alkylating agents are highly reactive compounds that add alkyl groups (such as CH_3CH_2-) to other organic or biological molecules. When these compounds enter a cell, they add alkyl groups to DNA, producing defects in the DNA that result in cell death. The most commonly used alkylating agent is cyclophosphamide, called a nitrogen mustard because its structure is similar to the mustard gas used in chemical warfare during World War I.

Like AZT, **antimetabolites** are chemical impostors; they work by mimicking compounds normally found in the cell but have subtle differences that usually stop DNA synthesis. For example, methotrexate is a chemical impostor for folic acid. It binds to the enzyme that normally converts folic acid into two building blocks (adenine and guanine) required for DNA synthesis. Without these building blocks, DNA cannot be synthesized and the cell cannot replicate.

Topoisomerase inhibitors work by modifying the action of a topoisomerase, an enzyme that pulls the individual strands of DNA apart in preparation for DNA synthesis. The topoisomerase inhibitor results in DNA damage and therefore in cell death during the action of topoisomerase. Effectively, these compounds cause cells to die whenever they initiate replication. The most common anticancer drug is a topoisomerase inhibitor called cisplatin, a platinum-containing compound.

Hormone treatment is available primarily for those cancers involving the breasts or sexual organs. These tissues need hormones to grow, and cancerous

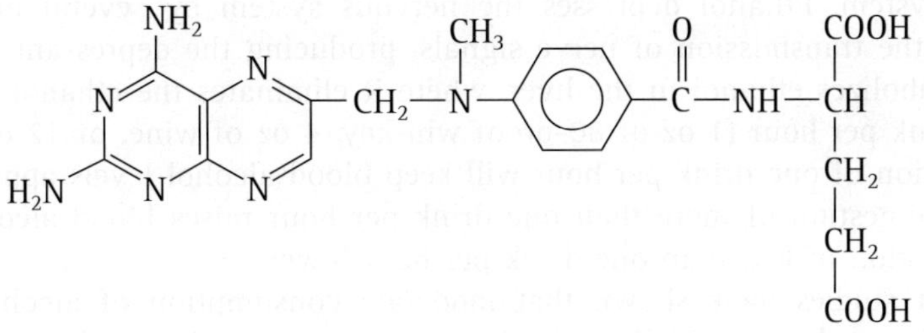

methotrexate

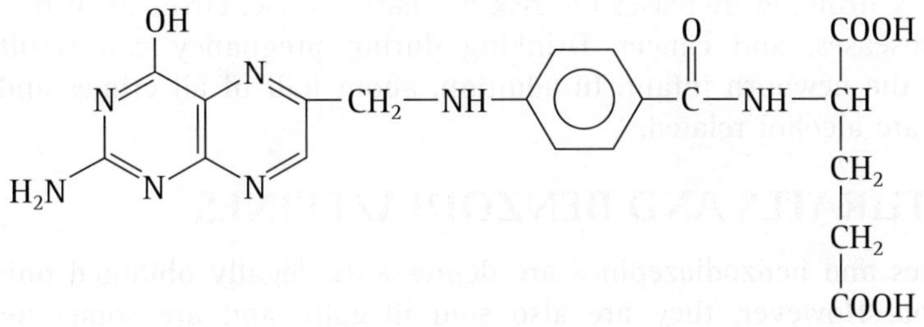

folic acid

Methotrexate, an antimetabolite used in cancer treatment, is a chemical impostor. It mimics folic acid but blocks DNA synthesis.

Cl   Cl
 \\ //
  Pt
 / \\
NH₃   NH₃

Cisplatin, a topoisomerase inhibitor used in cancer therapy. Cisplatin causes cells to die on initiation of cell division.

cells within these tissues, which are growing rapidly, need a particularly large supply. The growth can be stopped by denying the cells the needed hormone. For example, in women, estrogen antagonists are given to treat breast cancer. The estrogen antagonists compete with estrogen for receptor sites in breast tissue. By occupying the site normally reserved for estrogen, they halt the growth of breast tissue and therefore stop tumor growth. Because hormone therapy is specific to those tissues requiring the hormone, it has fewer side effects than any of the other cancer therapies.

Chemotherapy is the administration of these drugs, either singly or in combination, over an extended period of time. It is often combined with other treatments, such as surgery or radiation treatment (see Section 8.13), to achieve eradication of the cancerous tissue. For example, breast cancer is often treated by removing the cancerous tissue surgically and then treating with chemotherapy to remove smaller clumps of cancerous cells that may have spread to other parts of the body. Through these treatments many types of cancer, especially Hodgkin's disease, testicular cancer, some bone and muscle cancers, and many children's cancers, have significantly reduced death rates.

17.8 Depressants: Drugs That Dull the Mind

Depressants, often called tranquilizers or downers, are those drugs that depress or dull the central nervous system. The amount of depression usually depends on the dose of the drug; low doses produce a relaxed sensation and anxiety release, while increasingly higher doses produce loss of inhibition, sedation, sleep, unconsciousness, general anesthesia, coma, and death. Depressants are used medically to treat anxiety and insomnia and to anesthetize patients undergoing surgery; however, depressants are also widely abused as recreational drugs.

ALCOHOL

The alcohol found in alcoholic drinks is **ethanol** (Figure 13), or grain alcohol. When ingested, ethanol is absorbed out of the stomach and upper intestine and distributes itself equally in all bodily fluids and tissues, including the central nervous system. Ethanol depresses the nervous system by several mechanisms that slow the transmission of nerve signals, producing the depressant effect. The body metabolizes ethanol in the liver, where it eliminates the ethanol equivalent of one drink per hour (1 oz of 80-proof whiskey, 4 oz of wine, or 12 oz of beer). The ingestion of one drink per hour will keep blood alcohol levels approximately constant; ingestion of more than one drink per hour raises blood alcohol levels, while ingestion of less than one drink per hour lowers them.

CH_3CH_2OH

FIGURE 13 Ethanol, the alcohol found in alcoholic drinks.

Recent studies have shown that moderate consumption of alcohol (one to two drinks per day, especially red wine) may increase lifespan by lowering the risk of coronary artery disease, but it is clear that heavy drinking decreases lifespan. Heavy drinking increases the risk of heart disease, cirrhosis of the liver, respiratory diseases, and cancer. Drinking during pregnancy can result in birth defects in the newborn infant. In addition, about half of all crimes and highway accidents are alcohol related.

BARBITURATES AND BENZODIAZEPINES

Barbiturates and benzodiazepines are depressants, legally obtained only through prescription. However, they are also sold illegally and are sometimes abused.

WHAT IF...

Alcoholism

Alcoholism is a disease estimated to affect over 5 million Americans. Although the line between the moderate use of alcohol and its abuse is blurred, certain signs separate the alcoholic from the moderate user. Alcoholism is marked by a loss of control. The alcoholic keeps drinking even when it begins to have a negative impact on his or her work, family, and relationships. Alcoholics often drink in the morning, become depressed when they do not drink, drink alone or in secret, and may suffer blackouts or memory loss after heavy drinking.

If you think you might have an alcohol problem, keep careful track of the number of drinks you have over a given period of time. Medical experts recommend no more than one drink per day for women and two drinks per day for men. A drink is defined as one 12-oz beer, one 4-oz glass of wine, or 1 oz of 80-proof liquor. Persons who suspect alcoholism in themselves should try to stop drinking on their own. Failed attempts to stop drinking are a sure sign of alcoholism and require medical attention.

QUESTION: Have you ever been affected by alcoholism, either directly or indirectly? Do students at your institution indulge in alcohol abuse? If so, how does this abuse affect academic performance?

They depress the nervous system by binding to specific receptor sites in nerve cells. Once attached to these sites, the barbiturates and benzodiazepines enhance the action of γ-aminobutyric acid (GABA), a neurotransmitter that makes nerve endings less excitable by other neurotransmitters. In other words, when GABA is present at nerve junctions, the action of the chemical messengers that carry nerve signals is suppressed (Figure 14). The barbiturates and the benzodiazepines make GABA even more efficient at suppressing nerve signals.

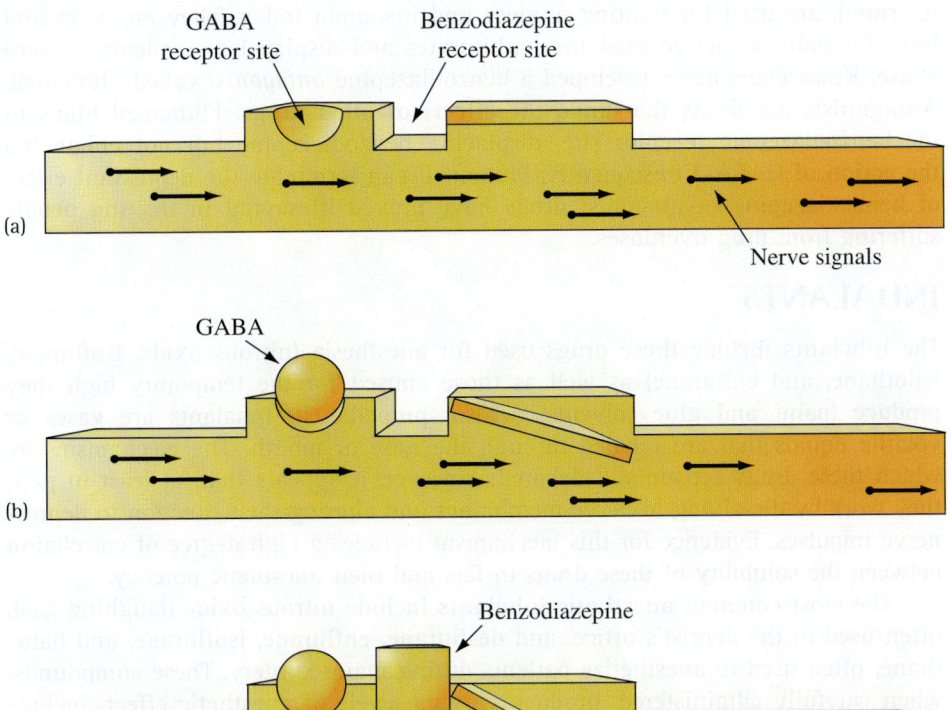

FIGURE 14 (a) Nerves transmit signals to the brain much like particles flow through a pipeline. (b) GABA restricts the flow of nerve signals. (c) Benzodiazepines enhance the action of GABA to further restrict flow.

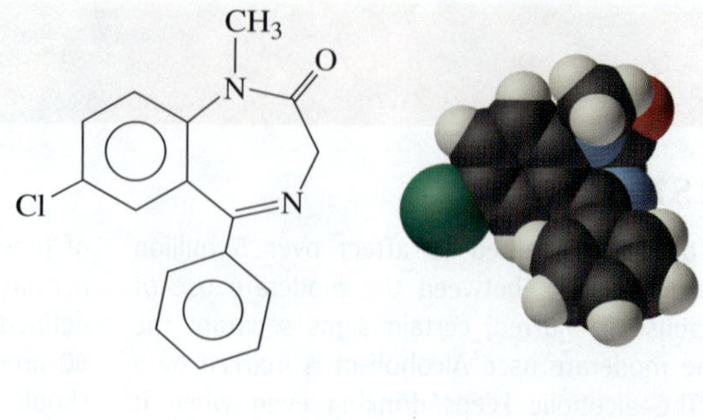

Diazepam (Valium), a benzodiazepine depressant.

FIGURE 15 Barbital and phenobarbital.

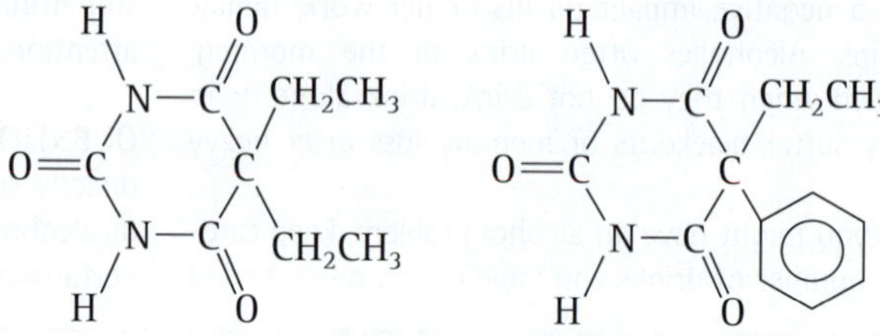

barbital

phenobarbital

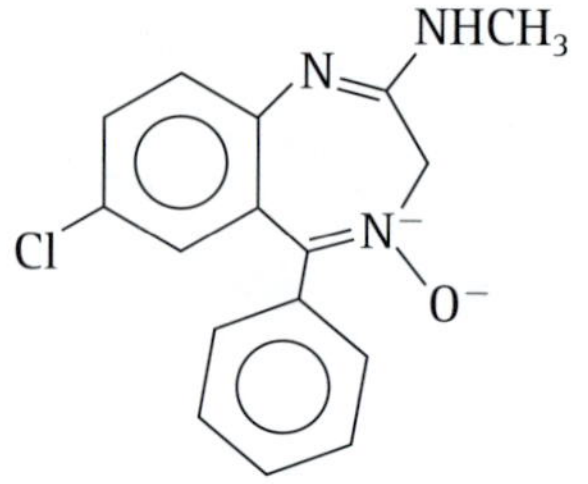

Chlordiazepoxide (Librium), a benzodiazepine depressant.

Barbiturates, drugs like barbital and phenobarbital (Figure 15), were the drugs of choice for treating anxiety and insomnia between 1912 and 1960. They are taken orally and, in low doses, produce anxiety relief and loss of inhibition. In higher doses or when taken with alcohol, barbiturates are dangerous, resulting in coma or death. Barbiturates induce both physical and psychological dependence with withdrawal symptoms such as hallucinations, restlessness, disorientation, and convulsions.

The **benzodiazepines,** drugs like diazepam (Valium) and chlordiazepoxide (Librium), are used for treating anxiety and insomnia today. They are safer and less physically addictive than the barbiturates and display less tendency toward abuse. Researchers have developed a benzodiazepine *antagonist* called Flumanzil. **Antagonists** are drugs that undo the effects of other drugs. Flumanzil binds to the benzodiazepine receptor site, displacing benzodiazepine but not enhancing the action of GABA. Consequently, Flumanzil can terminate the depressant effect of benzodiazepines. Antagonist drugs have proved lifesaving in treating people suffering from drug overdoses.

INHALANTS

The **inhalants** include those drugs used for anesthesia (nitrous oxide, isoflurane, halothane, and enflurane) as well as those abused for the temporary high they produce (paint and glue solvents, aerosol propellants). Inhalants are gases or volatile liquids that are inhaled through the nose or mouth. The mechanisms by which these drugs act are still debated; however, it appears that, at least in part, they work by dissolving in nerve membranes and altering their function to depress nerve impulses. Evidence for this mechanism includes a high degree of correlation between the solubility of these drugs in fats and their anesthetic potency.

The most common **anesthetic inhalants** include nitrous oxide (laughing gas), often used in the dentist's office, and desflurane, enflurane, isoflurane, and halothane, often used to anesthetize patients during major surgery. These compounds, when carefully administered, produce varying levels of anesthetic effect, including complete loss of sensation and unconsciousness.

Occasionally, some anesthetic inhalants are abused, with nitrous oxide being the most common. The recreational use of nitrous oxide is particularly dangerous, however, because the effective dose requires that 50% of the inhaled gas be nitrous oxide. When administered in the doctor's office, the nitrous oxide is carefully mixed with pure oxygen to ensure an adequate oxygen supply to the patient. If nitrous oxide were mixed with room air, however, the patient would suffer from **hypoxia**, oxygen deprivation. The effects of hypoxia include irreversible brain damage and suffocation. Indeed, the press occasionally reports on nitrous abusers who have suffocated themselves by opening a bottle of nitrous oxide in their car. The nitrous oxide displaces the air in the car, and the victims—in their state of euphoria—do not realize that they are suffocating.

Abused inhalants, besides nitrous oxide, include household chemicals such as paint thinners, degreasers, glue solvents, marker pen solvents, butane, propane, and aerosol propellants. These compounds have no medical value, but are often inhaled by those seeking a temporary high. Like other depressants, they produce varying states of sedation varying from relaxation, to alcohol-like intoxication, to loss of consciousness, to death. Because they are often inhaled by directly spraying the gas into the mouth or by inhaling vapors from a rag or paper bag, the dangers of hypoxia and suffocation are ever present. The majority of deaths involving these inhalants occur among male teenagers. Long-term problems, including brain damage, kidney damage, and liver damage, are common side effects of inhaling these uncontrolled mixtures of chemicals.

Nitrous oxide, also known as laughing gas.

$$\text{F}-\underset{\underset{\text{F}}{|}}{\overset{\overset{\text{H}}{|}}{\text{C}}}-\text{O}-\underset{\underset{\text{F}}{|}}{\overset{\overset{\text{F}}{|}}{\text{C}}}-\underset{\underset{\text{F}}{|}}{\overset{\overset{\text{H}}{|}}{\text{C}}}-\text{Cl}$$

Enflurane, an anesthetic inhalant.

17.9 Narcotics: Drugs That Diminish Pain

The **narcotics** are those drugs that act on the central nervous system to produce an analgesic and sedative effect. They include morphine, codeine, heroin, meperidine (Demerol), and fentanyl. They differ from depressants in their greater ability to reduce pain and their lower tendency to produce generalized sedation. The narcotics act by binding to **opioid receptors**, specific sites on nerve cells in the spinal cord and brain. The presence of a narcotic molecule on an opioid receptor inhibits certain neurotransmitters involved in the transmission of pain impulses (Figure 16).

 Explore this topic on the **Drug Enforcement Administration—Home Page** website.

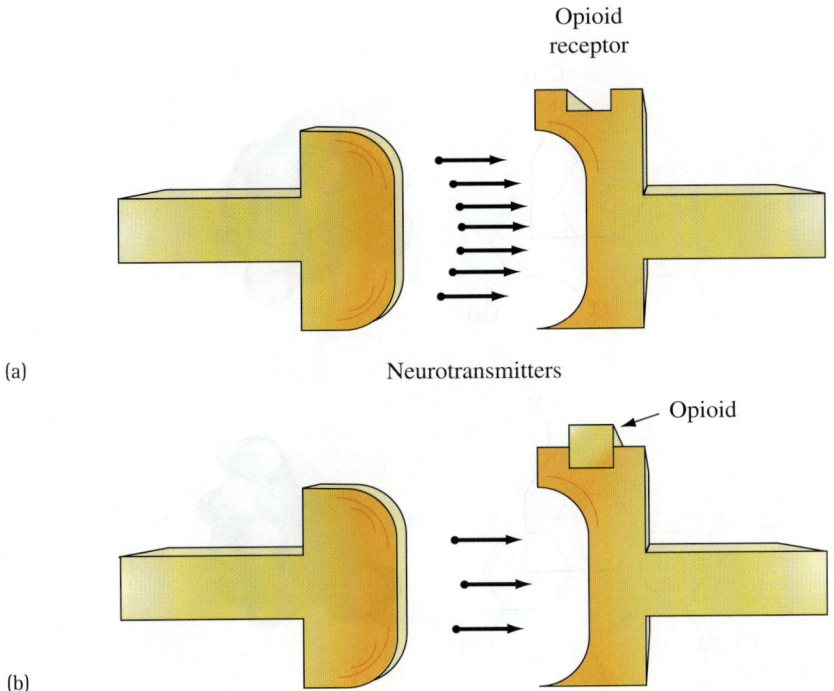

FIGURE 16 (a) Nerve signals are transmitted between nerve cells via chemical messengers called neurotransmitters. (b) When an opioid receptor is occupied by an opioid agonist, neurotransmitter flow is inhibited.

494 CHAPTER 17 Drugs and Medicine: Healing, Helping, and Hurting

Opium poppy.

Opium, extracted from the opium poppy, is a naturally occurring narcotic that has been used for thousands of years to produce euphoria, relieve pain, and induce sleep. In the early 1800s, the primary component of opium, **morphine** (Figure 17), was isolated from the opium poppy sap. Since that time, morphine has been the drug of choice to treat severe pain. When injected into the bloodstream, morphine quickly binds to opioid receptors in the central nervous system to produce its potent analgesic effect. It provides relief from and indifference to pain. For those who are dying from excruciatingly painful ailments, the administration of morphine is often the only relief available from an otherwise tormented existence.

Morphine also produces euphoria and feelings of contentment and well-being. This euphoria is the reason for the illicit use of morphine and other narcotics. Once the euphoria wears off, after 4 to 5 hours, the drug user is left craving more. As morphine use continues over an extended period, achieving the same euphoria requires a greater and greater dose. Soon, the user becomes addicted. The desire for the drug becomes uncontrollable, and the user must constantly self-administer the drug to avoid the vicious effects of physical withdrawal.

In the late 19th and early 20th centuries, the use of opium and morphine was widespread and uncontrolled. In 1914, amid growing concern about abuse and addiction, the U.S. Congress passed the Harrison Narcotic Act, which controlled the use of opium and its derivatives. Nonmedical uses of opium were banned.

Codeine (Figure 18) is another **opioid** drug found in opium. It is present in much lower amounts than morphine and is about one tenth as potent. Codeine is taken orally and often combined with aspirin or ibuprofen to relieve minor to moderate pain. **Heroin** (Figure 19), produced from morphine through a slight chemical modification, is less polar than morphine and consequently reaches the brain faster, producing a sharper, more intense effect. It is also three times more potent than morphine and extremely addictive. Heroin addicts will do almost anything to obtain their next fix.

An opioid is any drug that binds to opioid receptors in the central nervous system.

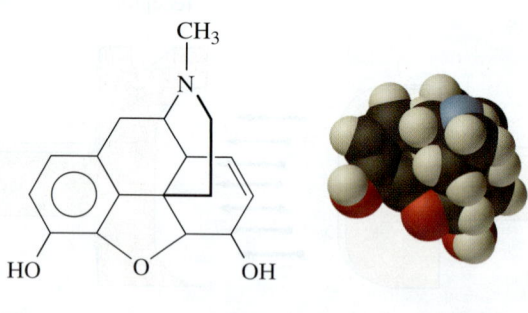

FIGURE 17 Morphine, the primary narcotic found in opium.

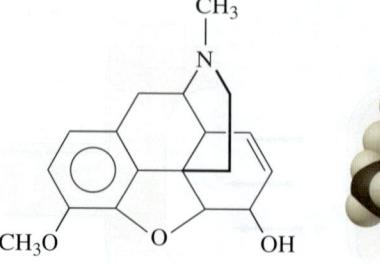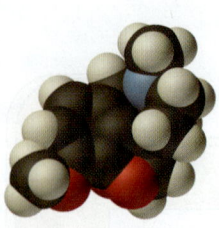

FIGURE 18 Codeine, a narcotic often used to treat moderate pain.

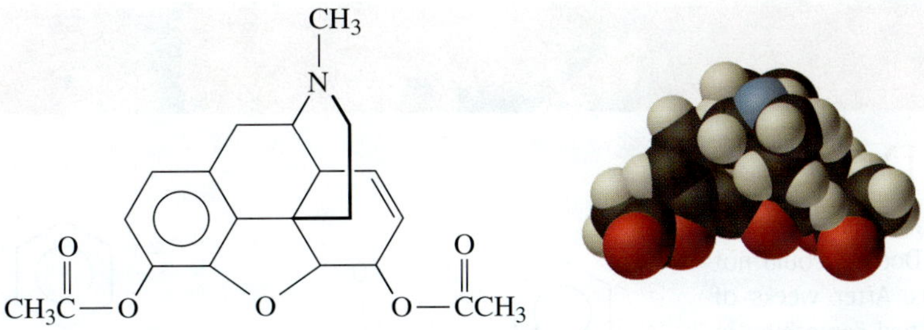

FIGURE 19 Heroin, a synthetic modification of morphine.

Other narcotics include meperidine (Demerol) and fentanyl (Sublimaze). Meperidine is a synthetic opioid with widespread medical use; it is about one-tenth as potent as morphine. Fentanyl, also synthetic, is more potent than morphine and is often used after surgery to relieve pain. It also is a drug of significant abuse under its street name, "China White."

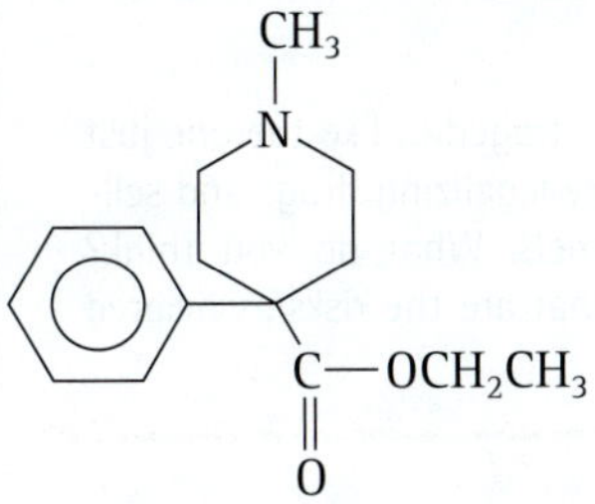

Meperidine (Demerol), a synthetic narcotic with widespread medical use.

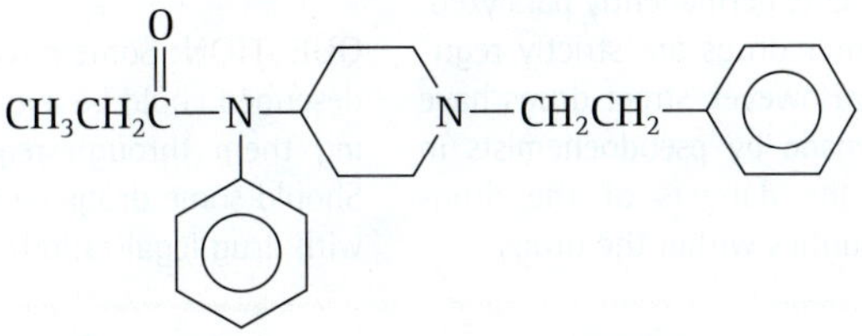

Fentanyl (Sublimaze), a synthetic narcotic often used to relieve postsurgical pain. Fentanyl is also a drug of abuse under its street name, "China White."

DRUGS THAT FIGHT NARCOTIC OVERDOSE AND ADDICTION

Breaking a narcotic addiction is difficult but not impossible. Some drugs have been developed to help the addict recover. One of those drugs, naloxone (Narcan), is used to treat narcotic overdoses. In contrast to morphine or heroin, which are *opioid agonists*, naloxone is an **opioid antagonist**. Opioid antagonists bind to opioid receptor sites, *displacing opioids* but *not producing* the narcotic effects. The injection of naloxone cancels the effect of the narcotic. Naloxone is used in emergency rooms to treat patients who have overdosed on heroin or other narcotics. A similar drug, naltrexone (Trexan), has the same effect but is taken orally. When naltrexone is taken daily, any injection of an opioid narcotic is ineffective. Addicts who really want to quit can take supervised doses of naltrexone as part of their recovery program. The craving for the narcotic is partly overcome by the knowledge that the drug will not produce the desired high.

A second drug used by recovering addicts is **methadone** (Figure 20), a synthetic opioid drug that also binds to opioid receptor sites. Unlike other narcotics, methadone does not produce stupor; it does, however, prevent the physical symptoms associated with withdrawal, which include restlessness, sweating, anxiety, depression, fever, chills, vomiting, cramping, panting, diarrhea, and insomnia. The addict taking methadone can quit using heroin or morphine without suffering these effects. Unfortunately, the symptoms of withdrawal return if methadone treatment is terminated. Some addicts remain on methadone for an extended period of time, while others try to taper off slowly.

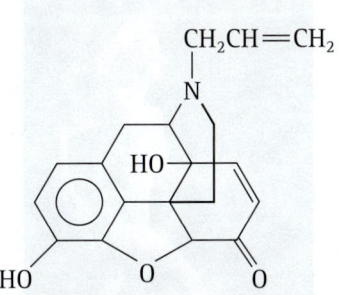

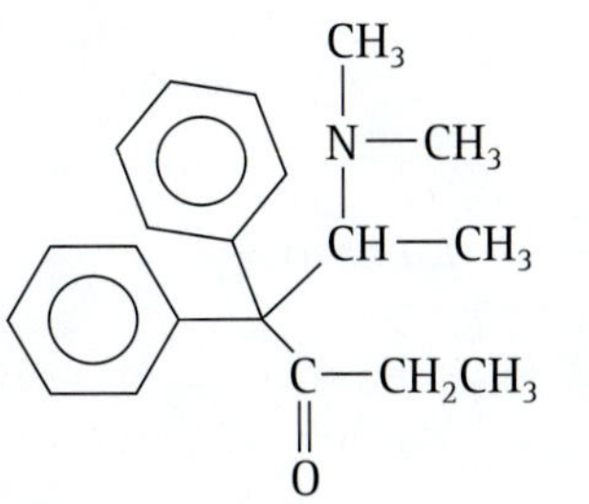

Naloxone (Narcan), an opioid antagonist. Opioid antagonists fit opioid receptor sites but do not produce the narcotic effect.

FIGURE 20 Methadone, a synthetic opioid that prevents the symptoms associated with narcotic withdrawal.

What if...

The Danger of Street Drugs

In the summer of 1982, several young California men and women came to hospitals unable to move. Doctors could not explain their immediate and total paralysis. After weeks of inquiry, doctors found that the patients all had something in common; they had recently taken heroin bought off the street. However, the heroin they injected into their bloodstream was not heroin at all, but a botched batch of the designer drug methylphenylpiperidylpropionate (MPPP). The basement chemist who had done the synthesis rushed it, and the product was tainted with a nerve toxin called N-methylphenyltetrahydropyridine (MPTP). This toxin crossed the blood–brain barrier and killed critical, dopamine-producing nerve cells in the patients' brains. Without dopamine, nerve signals could not be transmitted, leaving the patients permanently paralyzed.

Prescription and over-the-counter drugs are strictly regulated to ensure quality and purity. However, street drugs have no regulatory process; they are made by pseudochemists in clandestine laboratories. Besides the dangers of the drugs themselves, there is danger of impurities within the drugs.

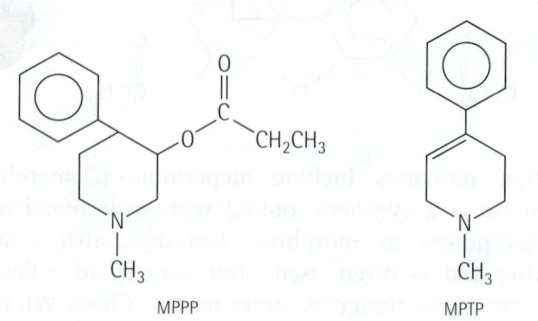

MPPP   MPTP

QUESTION: Some have argued that tragedies like the one just described could be avoided by simply legalizing drugs and selling them through regulated channels. What do you think? Should some drugs be legalized? What are the risks associated with drug legalization?

ENDORPHINS: THE BODY'S OWN PAINKILLERS

The presence of opioid receptors in the brain and throughout the nervous system led researchers to ask why these receptors existed. They certainly did not exist for the sake of interacting with plant-based compounds to relieve pain and produce euphoria. Researchers found that the human body produces its own opioid compounds called **endorphins**. Endorphins bind to opioid receptors and suppress pain. It appears that the body releases endorphins as a way to diminish high levels of pain. Birthing mothers have high blood–endorphin levels, as do marathon runners whose blood–endorphin levels have been measured at four times greater than normal. These compounds are thought to cause the common effect of a "runner's high."

17.10 Stimulants: Cocaine and Amphetamine

Stimulants, often called "uppers," are those drugs—such as cocaine or amphetamine—that stimulate the central nervous system. They produce an increased level of alertness, increased energy, decreased fatigue, decreased appetite, and a sense of euphoria. These effects are similar to the body's fight-or-flight response to potentially dangerous situations. The therapeutic uses of stimulants are limited and often controversial.

Cocaine (Figure 21) occurs naturally in the leaves of *Erythroxylum coca*, a tree indigenous to Peru, Bolivia, and Ecuador. Natives in these countries chewed on the leaves of *E. coca* for mystical and religious reasons as well as for the stimulant and anesthetic effects. In 1884, Sigmund Freud advocated the use of cocaine to treat depression and chronic fatigue and even used it himself. Until 1903, Coca-Cola contained approximately 60 mg of cocaine per serving. It was

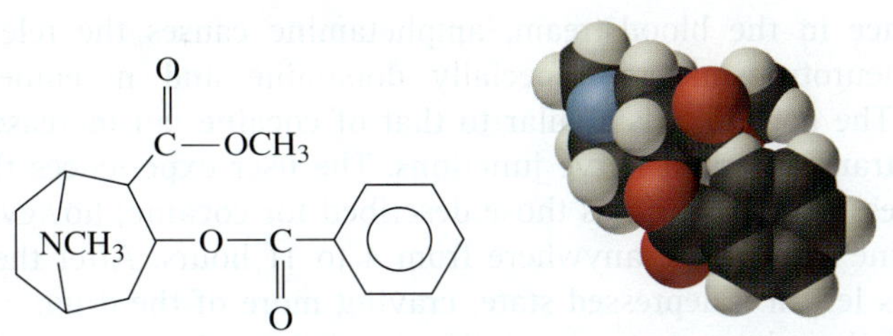

17.10 Stimulants: Cocaine and Amphetamine 497

FIGURE 21 Cocaine.

not until the Harrison Act of 1914 that recreational cocaine use was banned and cocaine became a controlled substance.

Cocaine is either inhaled (snorting), injected (mainlining), or smoked in its freebase form (freebasing; see Section 13.3). Once in the bloodstream, cocaine interacts with nerve cells to block the re-uptake of neurotransmitters, especially **dopamine** and **norepinephrine** (Figure 22). The net effect is that more of these neurotransmitters are available to transmit nerve impulses in the brain and spinal chord. The user experiences an increased sense of alertness, a feeling of excitement and euphoria, and a sensation of immense power. These effects wear off quickly, usually in 30–40 minutes, leaving the user depressed and craving another high. Side effects include anxiety, sleep deprivation, paranoia, and aggression. Although the cocaine user does not develop the physical signs of addiction associated with narcotics, the psychological dependence can be just as intense, making a cocaine addiction difficult to overcome.

Amphetamine (Figure 23) was first synthesized in the late 1880s. Its only medical use is in the treatment of narcolepsy, a condition that produces constant

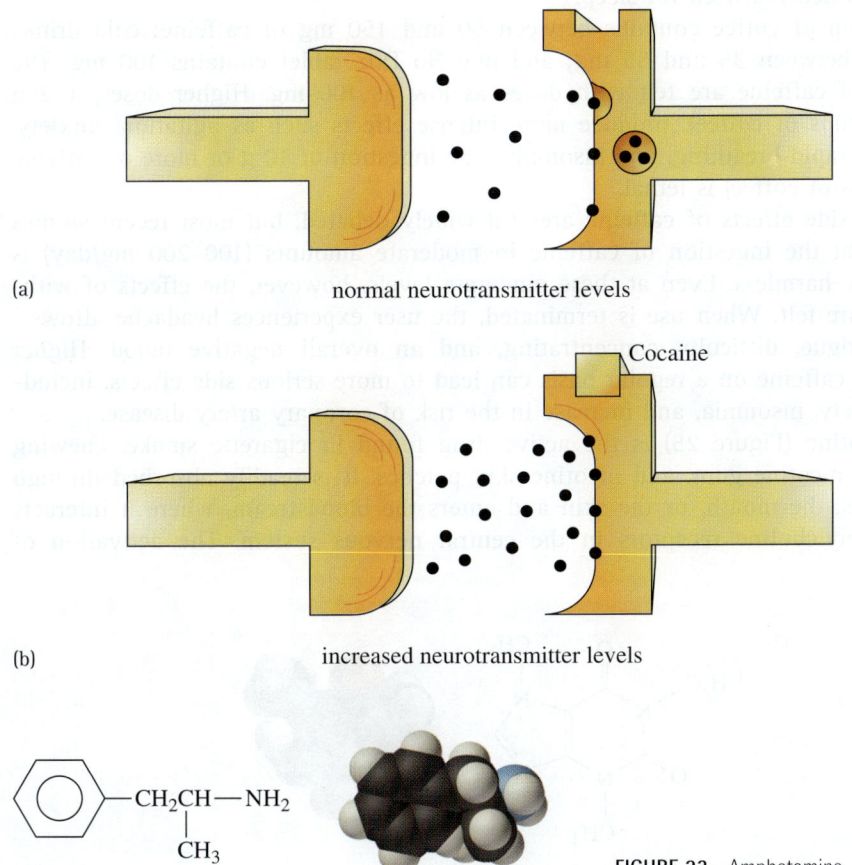

(a) normal neurotransmitter levels

(b) increased neurotransmitter levels

FIGURE 22 (a) Nerve junction showing how the re-uptake of neurotransmitters occurs. (b) Same diagram showing how cocaine blocks the re-uptake, resulting in a net increase of neurotransmitters.

FIGURE 23 Amphetamine.

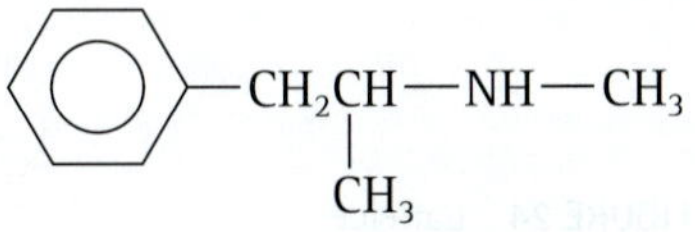

methamphetamine

sleepiness. Once in the bloodstream, amphetamine causes the release of newly synthesized neurotransmitters, especially dopamine and norepinephrine, from storage sites. The net effect is similar to that of cocaine: an increased concentration of neurotransmitters at nerve junctions. The user experiences the same feelings of alertness and euphoria as those described for cocaine; however, the effects of amphetamine last longer, anywhere from 4 to 11 hours. After the effect wears off, the user is left in a depressed state, craving more of the drug.

The most abused amphetamine today is **methamphetamine**, also known as speed. Like amphetamine, methamphetamine causes the release of stored neurotransmitters to produce the stimulant effect. Methamphetamine is easily and cheaply made in basement laboratories from readily available chemicals. Although its primary entry route is injection, it is also available in a freebase, smokable form known as "ice" or "crystal meth."

17.11 Legal Stimulants: Caffeine and Nicotine

Caffeine (Figure 24) is the stimulant found in coffee, tea, and soft drinks. It is also available as an over-the-counter medicine to prevent sleep (No Doz, Vivarin). When ingested, caffeine is absorbed into the bloodstream where it interacts with nerve cells to block adenosine receptors. Under normal conditions, adenosine (a neurotransmitter suppressant) binds to these receptors and suppresses the excretion of neurotransmitters, especially dopamine and norepinephrine. By blocking the adenosine receptor, caffeine causes an increase in dopamine and norepinephrine. The result is a feeling of alertness, competence, and wakefulness, as well as a delayed need for sleep.

A cup of coffee contains between 50 and 150 mg of caffeine; cola drinks contain between 35 and 55 mg; and one No Doz tablet contains 100 mg. The effects of caffeine are felt with doses as low as 100 mg. Higher doses, 1–2 g (10–20 cups of coffee), produce more intense effects such as agitation, anxiety, tremors, rapid breathing, and insomnia. The ingestion of 10 g or more of caffeine (100 cups of coffee) is lethal.

The side effects of caffeine are still widely debated, but most recent studies show that the ingestion of caffeine in moderate amounts (100–200 mg/day) is relatively harmless. Even at these moderate levels, however, the effects of withdrawal are felt. When use is terminated, the user experiences headache, drowsiness, fatigue, difficulty concentrating, and an overall negative mood. Higher doses of caffeine on a regular basis can lead to more serious side effects, including anxiety, insomnia, and increase in the risk of coronary artery disease.

Nicotine (Figure 25) is the active drug found in cigarette smoke, chewing tobacco, nicotine gum, and nicotine skin patches. It is readily absorbed through the lungs, the mouth, or the skin and enters the bloodstream, where it interacts with acetylcholine receptors in the central nervous system. The activation of

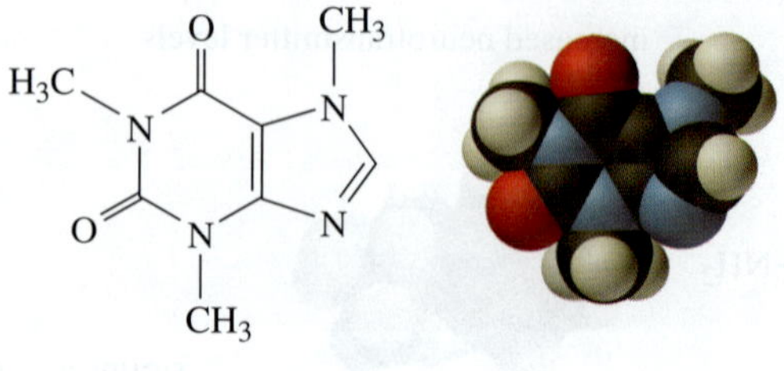

FIGURE 24 Caffeine.

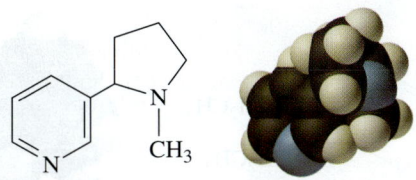

FIGURE 25 Nicotine.

these receptors increases the heart rate and blood pressure and causes the release of adrenaline, promoting a fight–flight response. The smoker experiences feelings of alertness and attentiveness. Overdosage can result in tremors and convulsions.

Chronic use of nicotine produces a powerful psychological and physiological addiction. The veteran smoker must continue to smoke to avoid the effects of withdrawal, which include craving for nicotine, irritability, anxiety, anger, difficulty in concentration, restlessness, and insomnia. For those trying to quit, these effects can last for months. All smokers experience these symptoms in the morning because they have gone six hours or more without nicotine. The first cigarette after waking is often the most craved, and because it alleviates the symptoms of withdrawal, it provides strong positive reinforcement to continue smoking.

The negative effects of chronic smoking—including lung cancer, emphysema, bronchitis, cardiovascular disease, and cancers of the mouth and throat—are well documented. Each cigarette is estimated to shorten the smoker's life by about 14 min. A two-pack-a-day smoker who smokes for 20 years loses about 8 years of life.

17.12 Hallucinogenic Drugs: Mescaline and Lysergic Acid Diethylamide

The **hallucinogenic drugs,** also called the psychedelics or "mindbenders," are those drugs—such as mescaline and lysergic acid diethylamide (LSD)—that distort and disturb cognition and perception. In high doses, they produce vivid hallucinations including dancing colors and lights. They gained popularity in the 1960s and 1970s as drugs that supposedly allowed people to get in touch with their inner selves.

Mescaline (Figure 26) occurs naturally in the crown of the peyote cactus, common in the southwestern United States and Mexico. When taken orally, mescaline is absorbed quickly into the blood where it interacts with nerve cells via serotonin receptors. While the exact mechanism by which mescaline and other hallucinogenic drugs act is still unclear, it is postulated that they inhibit the filtering of sensory stimuli in the nerves. Nerve cells are bombarded by a great deal of stimulation; under normal circumstances, the nerves filter out meaningless or unimportant signals. The hallucinogens seem to suppress this filtering ability, resulting in the distortion of perception.

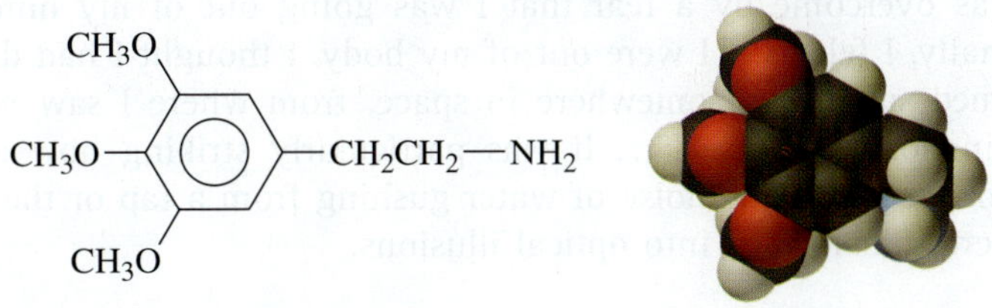

FIGURE 26 Mescaline, a hallucinogen found in the peyote cactus.

FIGURE 27 Lysergic acid diethylamide (LSD).

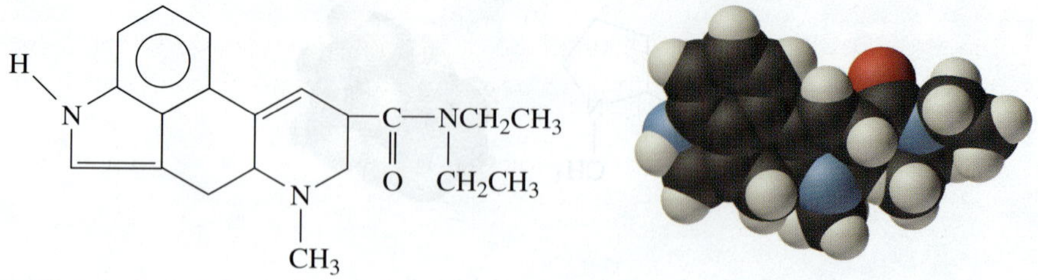

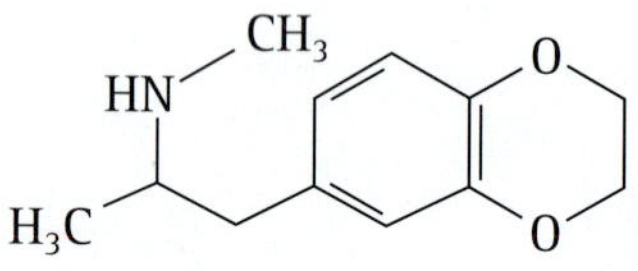

MDMA, a designer drug sold on the street as "ecstasy."

In low doses, mescaline has a stimulant effect similar to amphetamine and causes mild distortions of perception. In large doses, it produces time distortion, altered perception of color and sound, hallucinations, and feelings of separation from the physical body. Mescaline has been used in the sacramental religious rites of Aztec and other Indian tribes for centuries. It is legally available in some states for religious use by the Native American Church.

Other drugs with mescaline-type action include the so-called "designer drugs" such as MDA (methylenedioxyamphetamine), MMDA (methoxymethylenedioxyamphetamine), and MDMA (methylenedioxymethamphetamine) (ecstasy). All these drugs, like mescaline, show a stimulant effect at low doses with more hallucinogenic effects at higher doses. Most of these drugs do not develop physical addictions in the user, although psychological addiction is a constant concern. Also of concern is potential brain damage. For example, MDMA (ecstasy) has been shown to produce irreversible destruction of serotonin neurons in laboratory animals.

LSD (Figure 27) is a synthetic drug first made in 1938 by Dr. Albert Hoffman, a Swiss chemist developing therapeutic drugs. In 1943, Dr. Hoffman had a strange experience after working with LSD. He wrote of his experience,

> I was seized by a peculiar sensation of vertigo and restlessness. Objects, as well as the shape of my associates in the laboratory, appeared to undergo optical changes. I was unable to concentrate on my work. In a dreamlike state I left for home, where an irresistible urge to lie down came over me. I drew the curtains and immediately fell into a peculiar state similar to drunkenness, characterized by an exaggerated imagination. With my eyes closed, fantastic pictures of extraordinary plasticity and intensive color seemed to surge toward me.

Later, Dr. Hoffman, seeking to confirm his hypothesis that LSD had caused his experience, took an intentional dose of LSD. Although this second dose was minuscule by common drug standards, it was significantly larger than the first, and many times more than usually required for the hallucinogenic effect. He further wrote,

> I noticed with dismay that my environment was undergoing progressive changes. My visual field wavered and everything appeared deformed as in a faulty mirror. Space and time became more and more disorganized and I was overcome by a fear that I was going out of my mind. . . . Occasionally, I felt as if I were out of my body. I thought I had died. My ego seemed suspended somewhere in space, from where I saw my dead body lying on the sofa. . . . It was particularly striking how acoustic perceptions, such as the noise of water gushing from a tap or the spoken word, were transformed into optical illusions.

LSD is a more powerful hallucinogen and a less powerful stimulant than mescaline. The active dose is very small, so small that it is usually added to other substances like sugar cubes or mushrooms that are ingested orally. (Although some mushrooms naturally contain hallucinogenic compounds, many mushrooms are laced with LSD and sold in the street as "magic mushrooms.") When LSD enters the bloodstream, it interacts with serotonin receptors to produce the psychedelic effect. The exact response to LSD is unpredictable and depends largely on the user's state of mind and his or her environment. Bad or extremely unpleasant trips are common. LSD can produce rapid alterations in mood and emotion as well as severe perceptual distortions and hallucinations. Users report "seeing sounds" or "hearing colors" as well as significant slowing in the passage of time.

Although LSD does not produce a physical addiction, it can be psychologically addictive. Adverse reactions include bad trips in which the user enters an almost psychotic and paranoid state. Also, flashbacks can occur, even weeks or months after drug use is terminated. It is suspected that repeated, high-dose use of LSD results in a lowered capacity for abstract thinking as well as permanent brain damage.

17.13 Marijuana

Marijuana comes from the plant *Cannabis sativa*, which grows untended in most moderate climates. The active ingredient in marijuana is Δ-9-tetrahydrocannabinol (THC). The most potent form of marijuana, called hashish, comes from the dried resinous material obtained from the female flowers and contains about 7–14% THC. A preparation composed of the dried tops of the plants is called ganja or sinsemilla (without seeds) and contains approximately 4–5% THC. The dried stems, leaves, and seeds of the plant are called bhang or simply marijuana and contain between 1 and 3% THC.

Marijuana is usually smoked, and THC is absorbed out of the lungs and into the bloodstream. Although the exact way in which THC acts is still debated, it appears that nerve cells have special receptors, originally called cannabinoid receptors, that bind THC. THC changes nerve signals to produce both a sedative and a mild hallucinogenic effect. Low doses produce relaxation, relief from anxiety, increased sense of well-being, alteration in sense of time, and mild euphoria. Moderate to high doses produce sensory distortions and mild hallucinations. Very high doses result in depression, panic reactions, and paranoia. The behavioral effects occur within minutes of smoking THC and last 3–4 hours.

The side effects of marijuana are still under investigation and are controversial. It appears that marijuana produces only mild physical dependence with few withdrawal symptoms. However, because the body metabolizes THC very slowly—THC remains in the body for several days or weeks after it is ingested—the

Marijuana comes from the plant Cannabis sativa.

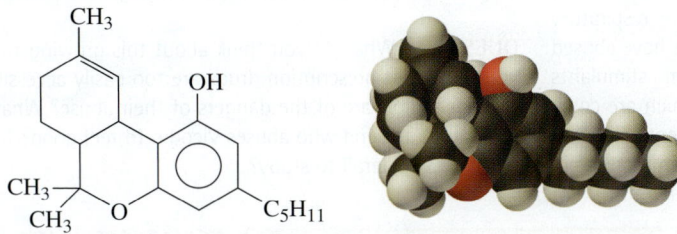

Δ-9-Tetrahydrocannabinol (THC), the active ingredient in marijuana.

chronic, long-term marijuana user has a constant background THC in his or her bloodstream. This residual amount can interfere with cognitive function, hindering learning, reasoning, and academic achievement. Other long-term effects include lung and throat irritation, possible increase in lung cancer risk, and partial immune system suppression. THC also appears to interact with the sex hormones inhibiting sperm formation in males and disturbing the menstrual cycle in females. On the other hand, THC is surprisingly nontoxic; no human has ever been known to die of a THC overdose.

17.14 Prozac and Zoloft: SSRIs

In the 1960s and early 1970s, Valium gained immense popularity as the drug of choice to treat anxiety. It became a household name and patients began asking their doctors for Valium prescriptions. Newscasters reported on it, authors wrote about it, and musicians sang about it. The Rolling Stones wrote a song called "Mother's Little Helper" that described how Valium helped Mother "get through her day."

Today "mother's little helper" is arguably Prozac, a serotonin-specific antidepressant approved for U.S. distribution in 1988. Prozac and similar drugs such as sertraline (Zoloft) and Paxil have been prescribed to millions of patients worldwide to treat depression. Unlike Valium, which has significant sedative effects, Prozac treats depression without sedation. In fact, these new wonder drugs have surprisingly few side effects.

While feelings of sadness and depression are normal human emotions, extended or unremitting depression that interferes with normal activities can be a sign of an underlying mood disorder. These mood disorders are often caused by biochemical imbalances in the brain that can be corrected with drugs. Clinical depression is diagnosed if four of the following eight symptoms are present in

Sertraline (Zoloft), an antidepressant.

WHAT IF...

Prescription Drug Abuse

Prescription drugs can be dangerous without the supervision of a doctor, yet these are increasingly becoming the drugs of choice among some of America's youth. Prescription drugs are easy to access, either from a parent's medicine cabinet or from the Internet. Many teens believe that they are safer because they are doctor prescribed and FDA approved. The main types of abused prescription drugs include pain relievers, stimulants, sedatives, and tranquilizers.

Popular pain relievers include Codeine, OxyContin, Percocet, and Vicodin. All of these drugs are opioids, which can become addictive and have the potential to cause severe respiratory depression that can be fatal. One in five teenagers have abused prescription painkillers to get high. Commom stimulants include Adderall, Dexedrine, and Ritalin, all of which are commonly prescribed to treat conditions such as attention deficit hyperactive disorder (ADHD). Abusers experience an increase in alertness and a sense of euphoria, but they are at risk of cardiovascular failure (heart attack) and potientially lethal seizures. About one in ten teens have abused stimulants. College students are particularly at risk for addiction to stimulants as a sort of "study drug" to pull "all-nighters" and get through finals. Used less frequently are prescription sedatives and tranquilizers, which include Mebaral, Quaaludes, Xanax, and Valium. These drugs are prescribed to treat anxiety, tension, stress, and sleep disorders. A physical dependence can develop with extended use, and higher doses can lead to memory and judgment impairment, paranoia, and thoughts of suicide.

QUESTION: What do you think about this growing trend? Do you think that prescription drugs are too easily accessible? Are teenagers unaware of the dangers of their abuse? What would you say to a friend who abuses Vicodin to feel good? To a peer who takes Adderall to study?

Fluoxetine (Prozac), an antidepressant.

the patient on a daily basis for a two-week period: change in appetite, change in sleep, psychomotor agitation or retardation, loss of interest in usual activities or decrease in sexual drive, increased fatigue, feelings of guilt or worthlessness, slowed thinking or impaired concentration, and suicide attempt or suicidal ideation.

Clinical depression is at least partly caused by a deficit of certain neurotransmitters in the brain, especially serotonin. It appears that serotonin deficits induce adaptive changes in nerve cell receptors that produce the depressed state. If serotonin levels are brought back to normal, the adaptive changes are reversed, and the depression is relieved.

First-generation antidepressant agents, called *tricyclic antidepressants*, affected the brain levels of several neurotransmitters including norepinephrine, serotonin, and dopamine. Although they did relieve depression, they also had a number of

THE MOLECULAR REVOLUTION

Consciousness

In this chapter, we have seen how certain molecules in the brain can alter emotions and perceptions. Our emotions and perceptions are susceptible to molecules because they are mediated by molecules. In recent years, scientists have made remarkable progress in understanding just what those molecules are and how their levels can be modified. The development of Prozac and other antidepressants is just one example of how this understanding has benefited society.

Magnetic resonance imaging (discussed in Chapter 7) and other technologies have been able to reveal the brain at work by monitoring oxygen consumption and blood flow to different parts of the brain while a patient performs specific mental tasks. For example, scientists can watch the firing of neurons in a specific part of the brain as patients view a particular image or as they reconstruct a particular memory. This kind of unprecedented understanding led President Bush (Sr.) to call the 1990s the decade of the brain. However, President Bush may have done well to extend his definition far into the 21st century because much remains to be understood.

The most important question remains controversial: What is consciousness and how does it arise? The debate on consciousness is not new; Plato and Aristotle wondered about it over 2000 years ago. Whatever consciousness is, it is central to being human. Rene Descartes's famous 17th-century phrase "Cogito, ergo sum" ("I think, therefore I am") equated consciousness with existence, and we constantly differentiate ourselves from the rest of the universe based on our concept of self, a central part of consciousness. But scientists struggle with explaining how the physical brain creates consciousness. For example, a scientist may explain the processes associated with seeing the color blue. Light of approximately 450 nm strikes the retina, which causes the isomerization of a molecule called *cis*-retinal, which then causes an electrical signal to be transmitted to a certain part of the brain. All of this is known. Yet this description cannot describe what it is like to *experience* the color blue. A scientist with perfect knowledge of vision might be able to describe every step of the vision process, but if that scientist had never seen the color blue, she would not know what it looked like. This is the gulf that confronts neuroscience: How do electrical signals in the brain form conscious experience? How does the physical brain create the mind? Some think this question can never be answered. According to them, the mind will never be understood because it must rely on itself for an explanation. Others, however, are more optimistic. They believe that with continued research and a greater understanding of the brain, the secret of the mind will emerge.

side effects including sedation. The new antidepressants, called *serotonin-specific re-uptake inhibitors (SSRIs)*, only affect the levels of serotonin. Because the deficit of serotonin appears to be the cause of depression, these drugs target only the depression and show minimal side effects.

The first SSRI to be available in the United States was **fluoxetine** (Prozac) in 1988. It is the most widely prescribed antidepressant today. The usual dose of 20 mg/day produces a slow increase in the blood levels of fluoxetine. Fluoxetine works in nerve cells to inhibit the re-uptake of serotonin, producing a net increase in brain serotonin levels. Because the half-life of fluoxetine in the blood is one to four days, the antidepressant action builds with repeated daily doses. Unlike Valium, whose effects are felt almost immediately on ingestion, the antidepressant effects of fluoxetine take weeks to develop. The side effects, which are experienced by only a small fraction of patients, include nervousness, dizziness, anxiety, sexual dysfunction, insomnia, nausea, and loss of appetite.

Sertraline (Zoloft) was introduced in the United States shortly after fluoxetine and has become almost as popular. The usual dose of 50 mg/day produces antidepressant action in most patients. Sertraline is slightly more selective for serotonin than fluoxetine and has a shorter half-life in blood. Consequently, steady levels in the blood are reached in a shorter time period (4 to 5 days). The side effects of sertraline include dizziness, sexual dysfunction, insomnia, nausea, diarrhea, and dry mouth.

Chapter Summary

MOLECULAR CONCEPT

Aspirin, acetaminophen, ibuprofen, and naproxen are common drugs used to relieve minor pain, fever, and inflammation (17.2). Antibiotics, including penicillins, cephalosporins, and tetracyclines, are used to treat bacterial infections, many of which were fatal before the discovery of antibiotics (17.3). Although successful treatment of viral diseases is less common, several new antiviral drugs have been developed. Perhaps most stunning is the recent advances in treating AIDS with protease cocktails (17.4).

Sexual hormones—testosterone in males and estrogen and progesterone in females—govern sexual characteristics, and synthetic analogs are often used as drugs. Birth control pills and Norplant are both synthetic analogs of the female sexual hormones (17.5). Sexual hormones are also examples of steroids, compounds characterized by their 17-carbon-atom, four-ring skeletal structure. Other steroids used as drugs include the adrenocortical steroids used to relieve inflammation and the anabolic steroids, illegally used to help build muscle mass in athletes (17.6). Cancer treatment often involves the administration of certain compounds—classified as alkylating agents, antimetabolites, topoisomerase inhibitors, and hormones—to kill cancerous cells. Most of these work by targeting the DNA synthesis that must occur for cells to divide (17.7).

A number of drugs affect mental and cognitive function by interfering or modifying the transmission of nerve signals in the central nervous system. Depressants—such as alcohol, barbiturates, benzodiazepines, and inhalants—all dull the nervous system, producing sedation, sleep, unconsciousness, and even death (17.8). Narcotics—such as morphine, codeine, and heroin—have an analgesic (pain-relieving) as well as a sedative effect. They are among the most physically addictive drugs (17.9). Stimulants—such as cocaine, amphetamine, caffeine, and nicotine—all stimulate the central nervous system, producing increased levels of alertness, increased energy, and decreased fatigue (17.10, 17.11). Hallucinogenic drugs, such as mescaline and LSD, alter the central nervous system's ability to filter information, resulting in distortion of cognition and perception. THC, the active drug in marijuana, appears to act as both a mild depressant and a mild hallucinogen (17.12, 17.13).

SOCIETAL IMPACT

The idea that physical ailments are treatable with drugs is not new to our society; however, the idea that emotions and feelings have molecular origins and that emotional problems are treatable with drugs is more recent. The positive impact of drug therapy on society is obvious—the average person today lives 20 years longer than the average person in 1900. However, the easy access of drugs to our society also raises many issues regarding their use and abuse.

The advent of the birth control pill in the 1950s (17.5) set the stage for the sexual revolution in the 1960s. With the risk of pregnancy lowered to almost zero, society's sexual practices changed. However, the threat of AIDS in the 1980s has caused society to reconsider and the sexual freedom felt in the 1960s, while still felt, has diminished.

The use of steroids by athletes to build strength has been questioned by our society (17.6). To what degree should athletes be allowed to use chemicals to improve their performance?

Legal and illegal drug use is a constant problem in our society that carries with it a number of issues (17.8–17.12). What drugs should be legal? What drugs should not? Why is alcohol legal while marijuana is not? Science can tell us what the effects of certain drugs are, but it is up to society to determine which should be allowed and which should be forbidden.

Recently, highly specific drugs with limited side effects have been developed to treat depression. The so-called SSRIs work by increasing the levels of the neurotransmitter serotonin in the brain (17.1, 17.14).

Antidepressants have found widespread use in our society (17.14), but some think their use has gone too far. They think that a society that copes with problems by taking a pill is doomed to fail. Others, however, feel that antidepressants have helped many clinically depressed people to cope with their depression, and since they have relatively few side effects, there is no problem.

Chemistry on the Web

For up-to-date URLs, visit the text website at www.chemistry.brookscole.com/tro3.

- U.S. Food and Drug Administration—Home Page
  http://www.fda.gov/
- RX List: The Internet Drug Indcx
  http://www.rxlist.com
- World Health Organization—Home Page
  http://www.who.int/en/
- HIV/AIDS Treatment, Prevention, and Research
  http://www.aidsinfo.nih.gov/
- Drug Enforcement Administration—Home Page
  http://www.usdoj.gov/dea/index.htm

Key Terms

abused inhalants
adrenocortical steroids
alkylating agents
amphetamine
anabolic steroids
androgens
anesthetic inhalants
antagonists
antimetabolites
bacteria
barbiturates
benzodiazepines
caffeine
cocaine

codeine
depressants
dopamine
endorphins
estrogen
ethanol
fluoxetine (Prozac)
hallucinogenic drugs
heroin
histamines
hormones
hypoxia
inhalants

lysergic acid diethylamide (LSD)
mescaline
methadone
methamphetamine
morphine
narcotics
neurotransmitter
nicotine
norepinephrine
nucleoside analogs
opioid
opioid antagonists
opioid receptors

opium
penicillin
progesterone
prostaglandins
protease inhibitors
proteases
retrovirus
sertraline (Zoloft)
steroids
stimulants
testosterone
topoisomerase inhibitors
virus

Exercises

 Assess your understanding of this chapter's topics with an online chapter quiz at **http://chemistry.brookscole.com/tro3**.

QUESTIONS

1. What molecule is associated with feelings of being in love?
2. What are the three main effects of aspirin?
3. Explain how aspirin produces the effects in question 2.
4. What are pyrogens and histamines?
5. What are the side effects of aspirin?
6. Name the common aspirin substitutes and how they differ from aspirin.
7. What is the difference between the generic brand of a drug and its name brand?
8. What is the difference between extra-strength and regular-strength aspirin?
9. What are antibiotics?
10. Explain how antibiotics were discovered.
11. How do penicillins and cephalosporins kill bacteria?
12. How do tetracyclines kill bacteria?
13. Why can antibiotics lose their effectiveness over time?
14. Explain the difference between bacteria and viruses.
15. What is a retrovirus?
16. How do nucleoside analogs kill viruses?
17. Explain the role of AZT in the battle against AIDS.
18. How do protease inhibitors work? What is their role in the battle against AIDS?
19. Why is the multidrug approach important in attacking AIDS?
20. What is a hormone?
21. What is the role of testosterone in males?
22. What is the role of estrogen and progesterone in females?
23. How does the birth control pill work? What are its side effects?
24. What is Norplant? How does it work?
25. What are steroids? Give three examples of steroids used as drugs.
26. What do cortisone and prednisone do? What are the side effects?
27. What do the anabolic steroids do? What are the side effects?
28. What is the difference between cancerous cells and normal ones?
29. What are the side effects of chemotherapy?
30. How do cancer-fighting chemicals target cancerous cells over normal ones? How specific can these chemicals be?
31. Explain how each of the following cancer drugs work:
    a. alkylating agents
    b. antimetabolites
    c. topoisomerase inhibitors
    d. hormones
32. What is a depressant? What are the effects of depressants?
33. Where and how fast is ethanol metabolized?
34. What are the risks associated with heavy drinking?
35. What are the symptoms of alcoholism?
36. How do barbiturates and benzodiazepines work? Give examples of each.
37. List some common depressant inhalants used for anesthetic purposes.
38. What are the dangers associated with self-administering inhalants?
39. What are the side effects of inhaling the vapors of household chemicals?
40. What is a narcotic? Give some examples.
41. How do narcotics work?
42. Where does morphine come from and what are its primary effects?
43. What are the effects of codeine and heroin? How do they compare with morphine?
44. What are agonists and antagonists?
45. How do drugs like naltrexone and methadone help a narcotic addict overcome his or her addiction?
46. What are endorphins?
47. What is a stimulant? Give some examples.
48. What are the effects of cocaine and amphetamine? How do the drugs cause these effects?
49. How does caffeine produce its stimulant effect?
50. What are the side effects of caffeine?
51. How does nicotine produce its stimulant effect?
52. What are the side effects of smoking?
53. What is a hallucinogen? Give some examples.
54. How do hallucinogens produce their perception-distorting effects?
55. Who discovered LSD? How?
56. What are the primary effects of LSD and of mescaline?
57. What are the side effects of LSD?
58. What is the active ingredient in marijuana?
59. What are the primary effects of marijuana? What are the side effects?
60. What qualifies a depressed state as one requiring medication?
61. How do Prozac and Zoloft produce their antidepressant effect? What are their side effects?

PROBLEMS

62. Draw the chemical structure of aspirin and identify the functional groups that are present. You may want to refer to Chapter 6 to review functional groups.
63. Draw the chemical structure of ibuprofen and identify the functional groups that are present.
64. Children's Tylenol is a solution containing 32 mg/mL. The recommended dose for a two-year-old child is 120 mg. How many milliliters should be given to the child?

65. Children's ibuprofen is a solution containing 20 mg/mL. The recommended dose for a two-year-old child is 75 mg. How many milliliters should be given to the child?
66. The toxicity of many drugs is quantified by a standard called the LD_{50}, the amount of drug per kilogram of body weight, required to kill 50% of the animals in a laboratory study. The LD_{50} of morphine in mice is 500 mg/kg. By assuming that toxicities for humans are similar to those for mice, what would be the lethal dose of morphine for an 80-kg person?
67. The LD_{50} (see problem 66) of cocaine in rats is 17.5 mg/kg. What would be the approximate lethal dose for an 80-kg person?

POINTS TO PONDER

68. The growth in our knowledge of the chemical origin of feelings and emotions is sometimes viewed by society with skepticism or fear. Does it bother you that feelings such as love can be correlated to specific molecules in the brain? Why or why not?
69. Our senses produce a chemically mediated signal that results in our perception of the world. It is clear that drugs that alter these chemical signals also alter our perceptions. Is our current chemical makeup such that we get the "most accurate" perception of reality? How would our perceptions change if our chemical makeup changed? How do these findings affect our notion of "reality"?
70. A continuing debate in our society is over the legalization of certain drugs. Proponents argue that some drugs will always attract users and consequently can never be stopped. They argue that their legalization would at least remove the criminal element from the distribution process and result in safer, more controllable forms of the drugs. Opponents of drug legalization argue that it would only increase drug use and the subsequent destruction of lives. What do you think? Should some drugs be legalized? Which ones should be legalized and why?
71. Although our society has condoned the recreational use of certain drugs, most notably alcohol and nicotine, it has had mixed feelings about it. Alcohol was banned during Prohibition and smoking is becoming more and more controlled. Why do you think society is more accepting of these drugs than others? Should more controls be issued on their use? Why or why not?
72. A controversial book called *Listening to Prozac* discusses how the use of Prozac causes changes in a person's personality so that he or she becomes "someone else." Do you think it is beneficial for people to be able to change their personality by taking a pill? Why or why not? Would further advances in this area produce a society of "nice" people, with little to differentiate one chemically modified nice person from the next?

Feature Problems and Projects

73. The life expectancy in the United States has risen steadily throughout the twentieth century due to the discovery and improvements of drugs as well as increasing access to quality medical care. The graph below shows the life expectancy at birth by race and sex as a function of time from 1970 to 2002.

    a. What is the total increase (in years) in life expectancy for a white female over this period?
    b. What is the average yearly increase in life expectancy for a white female over this period?
    c. If life expectancy continues to rise at the average rate shown in this graph, what would be the life expectancy of a white female baby born in 2030?

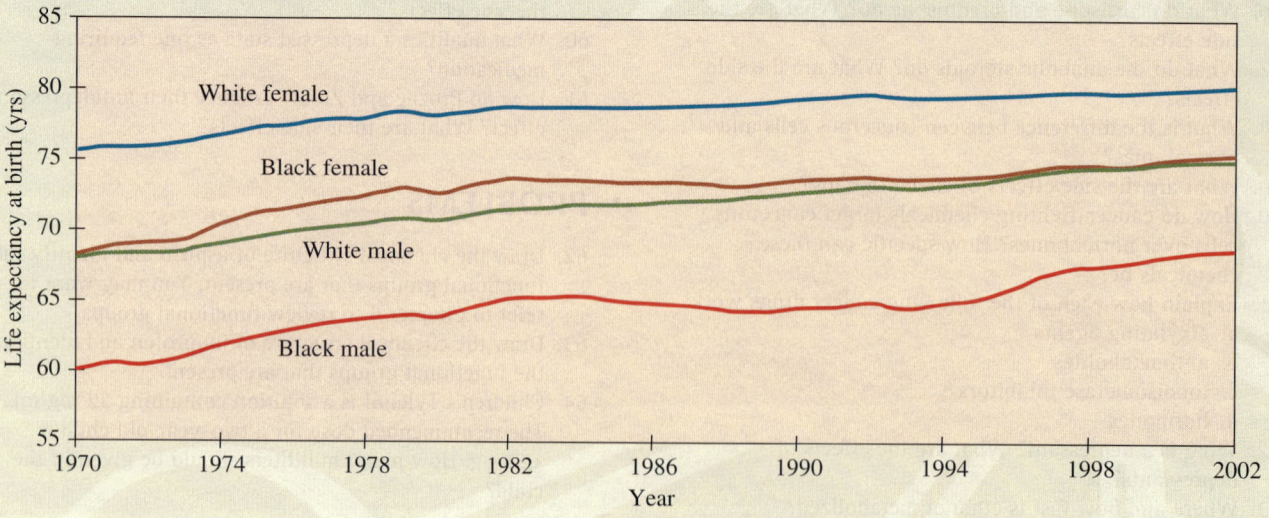

Life expectancy at birth by race and sex: 1970–2002. (National Vital Statistics Report, Vol. 53, No. 6, November 10, 2004.)

74. The cost of prescription drugs has significantly increased in the last decade. The graph below shows the total cost of prescription drugs in the United States as a function of time from 1993 to 2003.
    a. What is the total increase (in billions of dollars) over this time?
    b. What is the average yearly increase (in billions of dollars) over this time?
    c. What is the total percentage increase over this time?

75. Write down the last five prescription or over-the-counter drugs that you have taken. How many can you categorize according to the categories discussed in this chapter?

76. Survey 30 classmates, asking them to recall the last two prescription or over-the-counter drugs that they have taken. Prepare a table listing these drugs in order of decreasing use. Which drug was most used by college students?

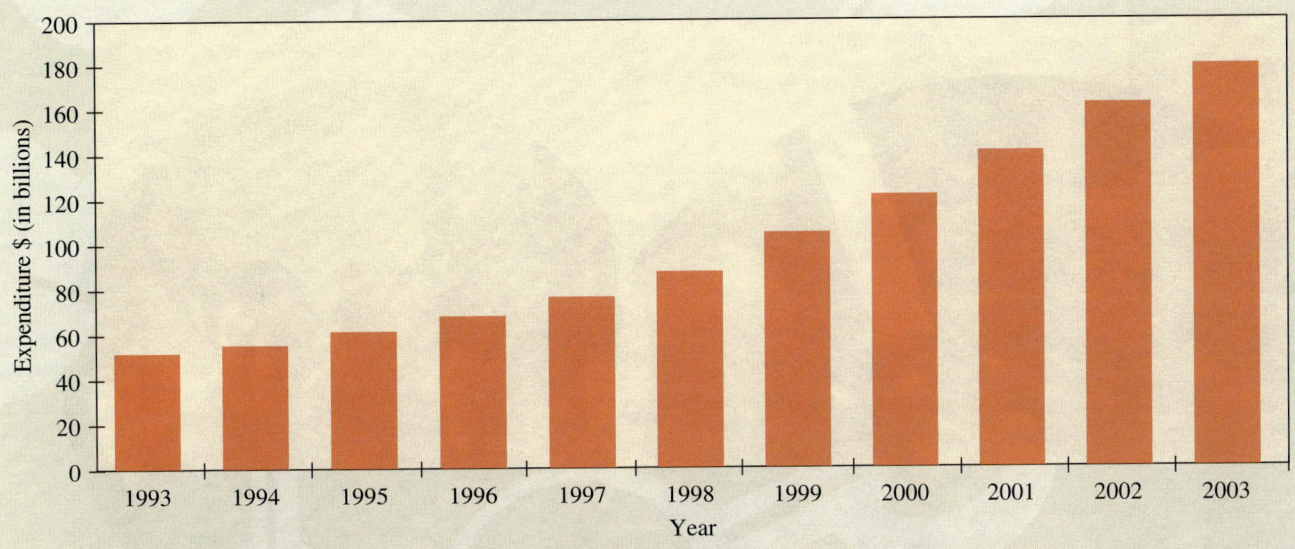

U.S. national health expenditures for prescription drugs and percentage of change 1993–2003. (Centers for Medicare and Medicaid Services.)

18 The Chemistry of Food

The discovery of a new dish does more for human happiness than the discovery of a star.

—Anthelme Brillat-Savarin

In this chapter, we explore the chemistry of food. Like everything else, food is composed of molecules. Virtually every atom and molecule in your body got there from the food you have eaten. Good nutrition involves getting the right combination of molecules in the proportions that your body needs. If your diet lacks a particular kind of necessary molecule, disease can develop. As you read this chapter, think about the foods you eat. Are you getting the right molecules in the right proportions?

The amount of food you eat, along with your activity level, determines whether you gain or lose weight. How much food will make you gain weight? How much should you cut out of your diet to lose weight? We can answer these questions by looking at the energy content of the molecules within food. The principles of energy consumption, covered in Chapter 9, apply here.

We will also discuss vitamins and minerals, substances your body needs in much smaller amounts than food. Lastly, we examine food additives, pesticides, and fertilizers. Pesticides and fertilizers help farmers grow food, and additives help to preserve or improve its appearance and flavor.

QUESTIONS FOR THOUGHT

- Where do the molecules that compose our bodies come from?
- What are carbohydrates and in what foods are they found?
- What is fiber?
- What are proteins and in what foods are they found?
- What are fats and in what foods are they found?
- What is the difference between saturated and unsaturated fat?
- What is cholesterol? What are its effects on health?
- How does caloric intake relate to weight gain or weight loss?
- How many calories do you burn per day?
- What are vitamins? Are vitamin supplements necessary?
- What are minerals? Are mineral supplements necessary?
- What are food additives and what do they do? Are they harmful?
- What are fertilizers and why do farmers use them?
- What are insecticides and herbicides? What are their effects on the environment?

18.1 You Are What You Eat, Literally

Our eyes, skin, bones, and muscle, along with the rest of our bodies, are all composed of molecules and atoms that we borrowed from food.

We all begin life as tiny, single-celled eggs fertilized by sperm. From this combination, we develop into embryos and after a few months are born as living, breathing babies. We then grow over the years, gaining weight and height, and eventually become mature adults. Where did all the atoms and molecules that compose our adult bodies—roughly 70 kg or 3.5×10^{27} atoms—come from? They came from food. The steak we ate yesterday becomes part of our muscle tomorrow. Our eyes, skin, bones, and brains are ultimately composed of atoms borrowed from food. We use these atoms for three-fourths of a century and then return them to the environment, where they are recycled and used by other living organisms. The average carbon atom in our own bodies has been through the life cycle many times before we got it.

The process that converts a McDonald's hamburger into energy, muscle, and various other bodily components begins in the mouth where food is crushed into smaller bits. The bits mix with saliva and then travel a twisting journey through the continuous tube we call the digestive tract (Figure 1). The tube originates at the mouth and terminates at the rectum. Along the way, the hamburger is plundered of its nutritionally valuable molecules. These molecules are chemically broken down into simpler units that pass through the walls of the digestive tract and into the bloodstream. The remaining, nutritionally void components of the hamburger continue through the digestive tract to the rectum, where they are excreted. In this way, the body extracts the required atoms and molecules from food. Some of these molecules are oxidized to meet the body's energy needs, while others are rearranged to make the necessary structures and components of the body. So, the old saying, *you are what you eat*, is literally true; virtually every atom or molecule found in our bodies originated from the food we ate at some point in our lives.

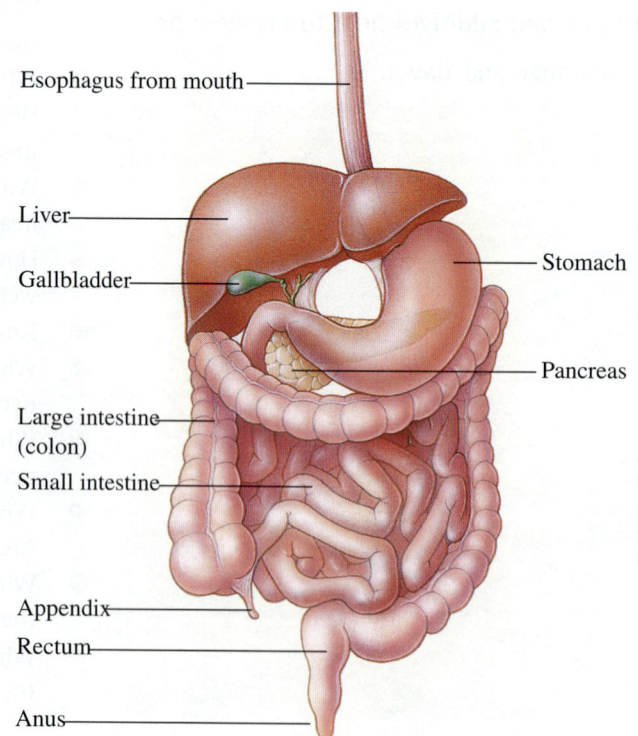

FIGURE 1 As food travels through the digestive tract, nutritionally important molecules are extracted. (Adapted from *Biology: Concepts and Applications, Fourth Edition*, by Cecie Starr. Copyright © 2000 Brooks/Cole.)

The chemistry of food is related to biochemistry (see Chapter 16) because the main nutritional substances in food are the biomolecules that include carbohydrates, proteins, and fats. Our bodies also need other substances (vitamins and minerals) in much smaller amounts. In addition, foods often contain additives, substances that enhance flavor and appearance and retard spoilage.

18.2 Carbohydrates: Sugars, Starches, and Fibers

The carbohydrates are the sugars, starches, and fibers in food. Foods such as candy, fruits, and sweetened desserts are rich in sugars; foods such as pasta, rice, beans, and potatoes are rich in starches; and the hulls of wheat, rice, and corn, as well as the structural material of fruits and vegetables, are rich in fiber. Recall from Chapter 16 that carbohydrates often have the general formula $(CH_2O)_n$ and are polyhydroxy aldehydes or ketones, or their derivatives. The OH groups (polyhydroxy) make carbohydrates very soluble in water. Consequently, simple carbohydrates are easily and efficiently transported in the bloodstream to all areas of the body.

Simple carbohydrates include the monosaccharides, such as glucose or fructose (Figure 2), and the disaccharides, such as sucrose (table sugar) and lactose (milk sugar) (Figure 3). **Glucose** is the body's primary fuel and the exclusive fuel of the

Potatoes, rice, and pasta are all rich in starch.

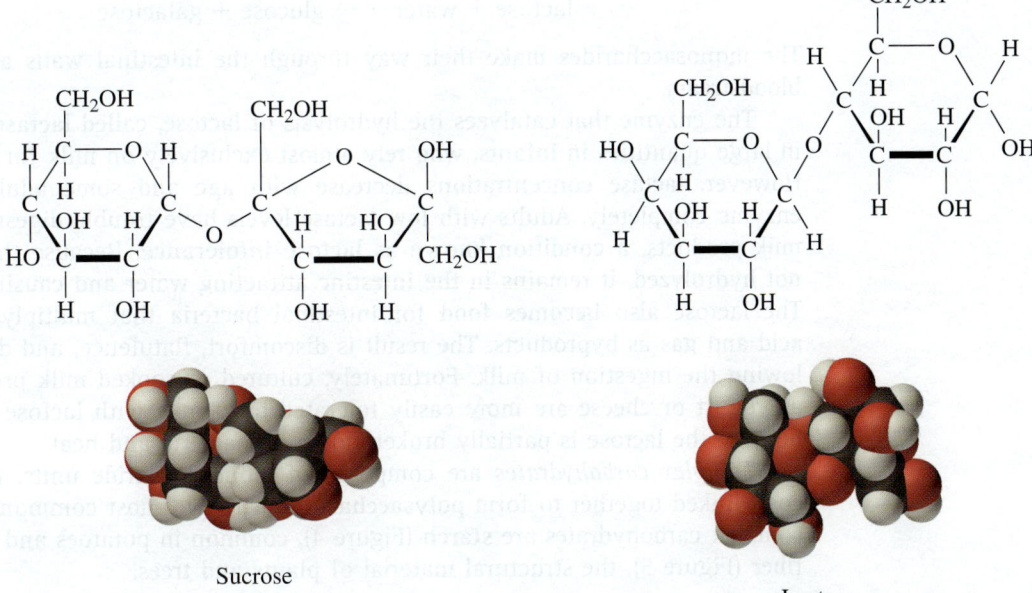

FIGURE 2 Glucose and fructose are both monosaccharides.

FIGURE 3 Sucrose and lactose are both disaccharides.

MOLECULAR THINKING

Sugar Versus Honey

Honey is formed by bees who harvest sugars, especially sucrose, from plants. The bees break down much of the disaccharides in plant nectars into their component monosaccharides; consequently, honey is a mixture of sucrose, glucose, and fructose.

QUESTION: In considering the digestive process, is there any significant difference between eating honey versus table sugar (sucrose)? Why or why not?

brain. It is either eaten directly from fruits or plants or formed by breaking down disaccharides or polysaccharides. Glucose is easily transported in the blood to cells where it is oxidized in a number of steps to form H_2O, CO_2, and energy. Its oxidation yields approximately 4 Calories (Cal) of energy per gram of carbohydrate.

Fructose, like glucose, consists of a single saccharide group and is found in fruits, plants, honey, and products sweetened by high fructose corn sweetener (HFCS). It is an isomer of glucose, having the same chemical formula ($C_6H_{12}O_6$) but a slightly different structure. It is the sweetest of all sugars, about 50% sweeter than table sugar.

Sucrose, or table sugar, is a disaccharide, two monosaccharide units—glucose and fructose—linked together. Sucrose occurs naturally in sugarcane, sugar beets, and many plant nectars. During digestion, the link between the monosaccharide units in sucrose is broken with water in a process called hydrolysis:

$$\text{sucrose} + \text{water} \longrightarrow \text{glucose} + \text{fructose}$$

The component monosaccharide units pass through the walls of the digestive tract and into the bloodstream.

Lactose is a disaccharide consisting of the monosaccharides glucose and galactose. It is found primarily in milk and milk products and is therefore called milk sugar. Like sucrose, lactose is hydrolyzed during digestion into its component monosaccharides, glucose and galactose:

$$\text{lactose} + \text{water} \longrightarrow \text{glucose} + \text{galactose}$$

The monosaccharides make their way through the intestinal walls and into the bloodstream.

The enzyme that catalyzes the hydrolysis of lactose, called **lactase**, is present in large quantities in infants, who rely almost exclusively on milk for sustenance. However, lactase concentrations decrease with age and some adults lack the enzyme completely. Adults with low lactase levels have trouble digesting milk or milk products, a condition known as **lactose intolerance**. Because the lactose is not hydrolyzed, it remains in the intestine attracting water and causing bloating. The lactose also becomes food for intestinal bacteria that multiply, producing acid and gas as byproducts. The result is discomfort, flatulence, and diarrhea following the ingestion of milk. Fortunately, cultured or cooked milk products such as yogurt or cheese are more easily tolerated by people with lactose intolerance because the lactose is partially broken down by bacteria and heat.

Complex carbohydrates are composed of monosaccharide units, mostly glucose, linked together to form **polysaccharides**. The two most common nutritional complex carbohydrates are **starch** (Figure 4), common in potatoes and grains, and **fiber** (Figure 5), the structural material of plants and trees.

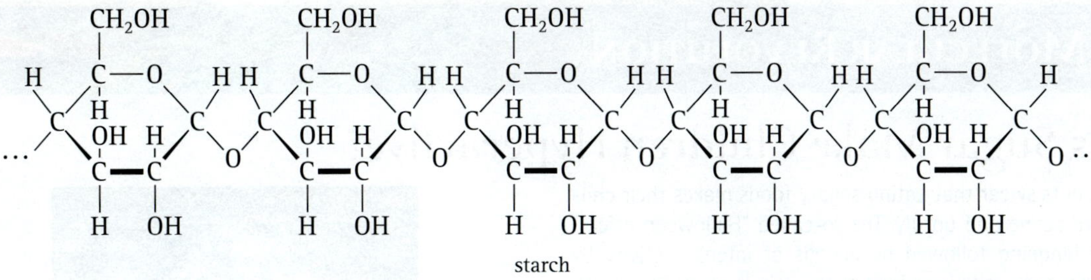

FIGURE 4 Starch is a polysaccharide composed of glucose units linked together.

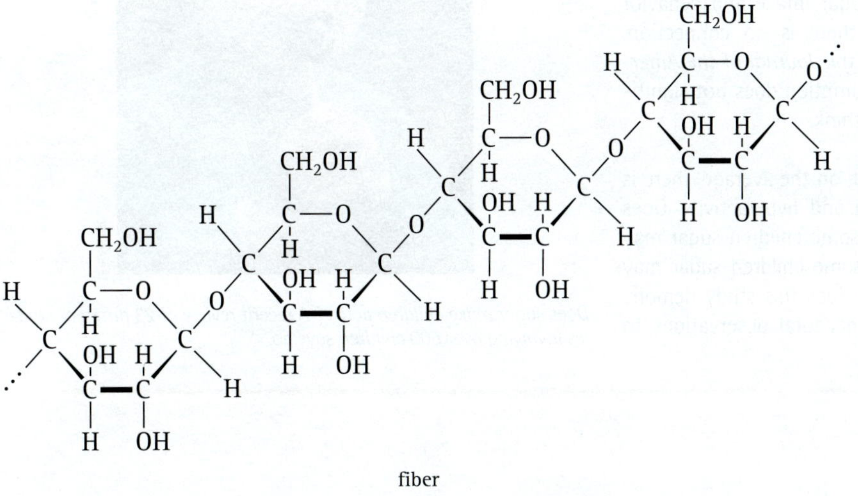

FIGURE 5 Fiber is a polysaccharide composed of glucose units linked together; however, the link is slightly different than that in starch, making fiber indigestible.

Starch is composed of glucose units linked together via alpha linkages (see Chapter 16). In plants such as potatoes, wheat, rice, or corn, starch stores energy for growth. When we eat the plant, our bodies digest the starch (Figure 6) by breaking it into its composite glucose units, which pass through the intestinal wall and into the bloodstream. Other important sources of starch include rye, barley, oats, and legumes.

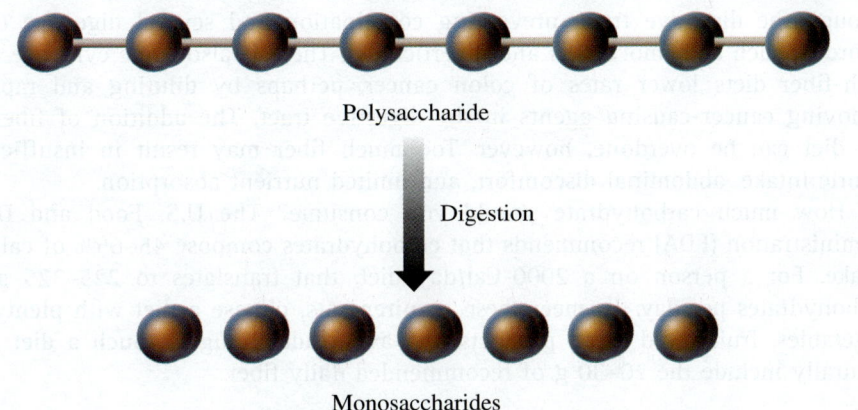

FIGURE 6 In the digestive tract, starch is cleaved into individual glucose units.

The Molecular Revolution

Does Sugar Make Children Hyperactive?

Many parents swear that eating sugary foods makes their children more active and unruly. The so-called "Halloween effect" of sugar bingeing followed by periods of intense activity by youngsters is virtually ingrained in the minds of most parents. Indeed, two studies from 1980 and 1986 linked sugar intake to hyperactivity. However, scrutiny of those studies indicates that the experimental methods did not differentiate between sugar causing the hyperactivity or the children's hyperactivity causing the sugar eating. A 1995 review of 23 different studies involving the relationship between sugar intake and behavior in 500 youngsters concludes that there is no connection. According to this review, published in the *Journal of the American Medical Association,* sugar consumption does not significantly affect the way children act or think.

QUESTION: This study concludes that, on the average, there is no connection between sugar intake and hyperactivity. Does that rule out the possibility that for some children sugar may make them hyperactive, or that for some children sugar may make them more sedate? Also, what does this study demonstrate about the wisdom of using anecdotal observations to make generalized statements?

Does sugar make children active? A recent review of 23 different studies involving over 500 children says no.

Fiber is also composed of long chains of glucose units, but the links between them, called beta links, are slightly different. This molecular difference prevents intestinal enzymes from breaking down fiber into its composite glucose units. Consequently, fiber cannot pass through the intestinal wall and has little caloric value; it passes through the digestive tract unabsorbed. It does serve a useful purpose, however. Fiber in foods increases their bulk without adding significantly to their caloric content. Consequently, a meal rich in fiber will lead to feelings of fullness and satiety with lower caloric content, resulting in natural weight control. Also, fiber adds bulk to stools, enlarging them and speeding up transit time through the digestive tract, preventing constipation and several digestive tract disorders such as hemorrhoids and diverticulitis. There is also some evidence that high-fiber diets lower rates of colon cancer, perhaps by diluting and rapidly removing cancer-causing agents in the digestive tract. The addition of fiber in the diet can be overdone, however. Too much fiber may result in insufficient caloric intake, abdominal discomfort, and limited nutrient absorption.

How much carbohydrate should you consume? The U.S. Food and Drug Administration (FDA) recommends that carbohydrates compose 45–65% of caloric intake. For a person on a 2000-Cal/day diet, that translates to 225–325 g of carbohydrates per day. To meet these requirements, choose a diet with plenty of vegetables, fruits, and grain products and avoid added sugars. Such a diet will naturally include the 20–30 g of recommended daily fiber.

Explore this topic on **The Food Pyramid** website.

18.3 Proteins

Proteins, as we learned in Chapter 16, are polymers of amino acids (Figure 7). They serve a host of different functions in the body, many more than carbohydrates. Human proteins are composed of 20 different kinds of amino acids, about half of which can be synthesized by the body itself. The other half, called the **essential amino acids** (Table 1), must come from food. Foods that are high in proteins include meat, cheese, eggs, milk, grains, legumes, and nuts.

In the digestive tract, proteins are chemically cut by enzymes into their amino acid components (Figure 8), which are absorbed through the intestinal wall and into the bloodstream. Intact proteins do not pass into the bloodstream. Eating muscle protein from an animal does not give you muscle protein; it gives you the amino acids that muscle protein contains. Once in the bloodstream, amino acids are transported throughout the body to cells where protein synthesis is taking place. The cells use the amino acids to construct their own proteins.

Although the body can metabolize proteins for energy, providing about 4 Cal/g, it does so only as a last resort. Proteins have more important functions as the workhorse molecules in the body. They compose much of our physical structure such as muscle, hair, and skin. They act as enzymes to control thousands of chemical reactions and as hormones to regulate metabolic processes. They are responsible for the transportation of oxygen from the lungs to cells and act as antibodies to fight invading organisms and viruses.

Eggs, meats, legumes, and nuts are all high in protein.

TABLE 1
Essential Amino Acids

Histidine
Isoleucine
Leucine
Lysine
Methionine
Phenylalanine
Threonine
Tryptophan
Valine

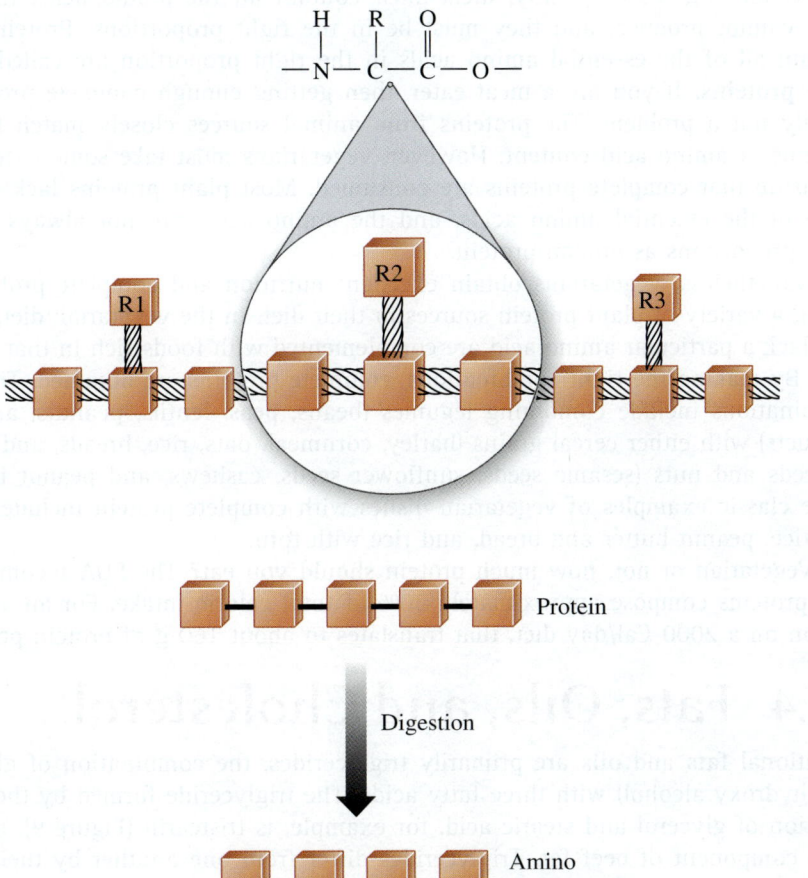

FIGURE 7 Proteins are long, chainlike molecules in which the individual links are amino acids.

FIGURE 8 When protein is digested, it is broken into its individual amino acid units.

WHAT IF...

The Second Law and Food Energy

The laws of thermodynamics that we learned in Chapter 9 apply to the energy transformations that occur in the growing and eating of food. Recall that a result of the second law of thermodynamics is that nature takes a "heat tax" in each step of an energy conversion or utilization process; therefore, it is more efficient to burn natural gas directly for heat than to burn it to generate electricity and then use the electricity to produce heat. If we apply the same reasoning to food energy, which diet is most energy efficient, the vegetarian diet or the meat-eating diet? In your thinking, consider each step that occurs in producing the final food and remember that each time that energy is converted from one form to another, nature takes a heat tax.

QUESTION: Construct an argument either for or against a vegetarian or meat-eating diet. Feel free to argue that the most efficient way may not be the best, if that is your opinion. Keep in mind, however, that in light of world hunger, some have argued that the most efficient use of food energy should be utilized. Do you agree or disagree? Why?

The Hispanic combination of beans and rice provides complementary proteins so that all the essential amino acids are consumed.

Proteins can function in so many different ways because of the great variety in structure that is achieved by changing the number and relative order of amino acids in the protein chain. Cells must have all 20 different kinds of amino acids in the right proportions to build functional proteins. A protein, composed of hundreds or thousands of amino acids, will probably not function if even one amino acid is missing. Consequently, diets must contain all the amino acids that the body cannot produce, and they must be in the right proportions. Proteins that contain all of the essential amino acids in the right proportion are called **complete proteins**. If you are a meat eater, then getting enough complete protein is usually not a problem. The proteins from animal sources closely match human proteins in amino acid content. However, vegetarians must take some extra care to ensure that complete proteins are consumed. Most plant proteins lack one or more of the essential amino acids, and the amino acids are not always in the same proportions as human protein.

Nonetheless, vegetarians obtain excellent nutrition and complete protein by eating a variety of plant protein sources in their diet. In the vegetarian diet, foods that lack a particular amino acid are complemented with foods rich in that amino acid. By making the right combinations, complete proteins are obtained. The best combinations include combining legumes (beans, peas, lentils, peanuts, and soy products) with either cereal grains (barley, cornmeal, oats, rice, breads, and pasta) or seeds and nuts (sesame seeds, sunflower seeds, cashews, and peanut butter). Some classic examples of vegetarian dishes with complete protein include beans and rice, peanut butter and bread, and rice with tofu.

Vegetarian or not, how much protein should you eat? The FDA recommends that proteins compose approximately 30% of total caloric intake. For an average person on a 2000-Cal/day diet, that translates to about 160 g of protein per day.

18.4 Fats, Oils, and Cholesterol

Nutritional **fats** and **oils** are primarily triglycerides, the combination of glycerol (a trihydroxy alcohol) with three fatty acids. The triglyceride formed by the combination of glycerol and stearic acid, for example, is tristearin (Figure 9), a principal component of beef fat. Triglycerides differ from one another by their fatty acid side chains. If the side chains contain no double bonds, the triglyceride is called a saturated fat and tends to be solid at room temperature. If the side

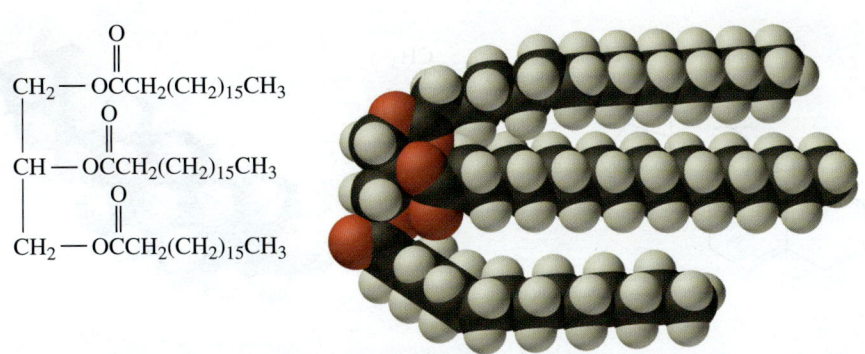

FIGURE 9 Tristearin, the principal component of beef fat.

chains contain one or many double bonds, the triglyceride is called monounsaturated or polyunsaturated, respectively, and tends to be liquid at room temperature. Fats occur in high levels in egg yolks, animal meat, vegetable oils, butter, cheese, cream, and ice cream. Fats from animal sources tend to be more saturated, while those from plant sources tend to be more unsaturated.

In the digestive tract, fats are broken down into fatty acids, glycerol, monoglycerides, and diglycerides (Figure 10); however, the process is slower than the breakdown of carbohydrates. The slow digestion of fats provides a lingering sense of fullness after a meal. The products of fat digestion pass through the intestinal wall and are largely reassembled into triglycerides before absorption into the blood. Because triglycerides are nonpolar, they cannot dissolve in the blood without the help of **lipoproteins**, specialized proteins that carry fat molecules and their derivatives in the bloodstream. Fats are transported to cells throughout the body and to the liver for dismantling and reassembly as other fats. Fat can be used for energy directly, providing about 9 Cal/g of fat, or it can be stored in fat cells for later use.

Oils, like corn oil or safflower oil, are composed primarily of unsaturated triglycerides.

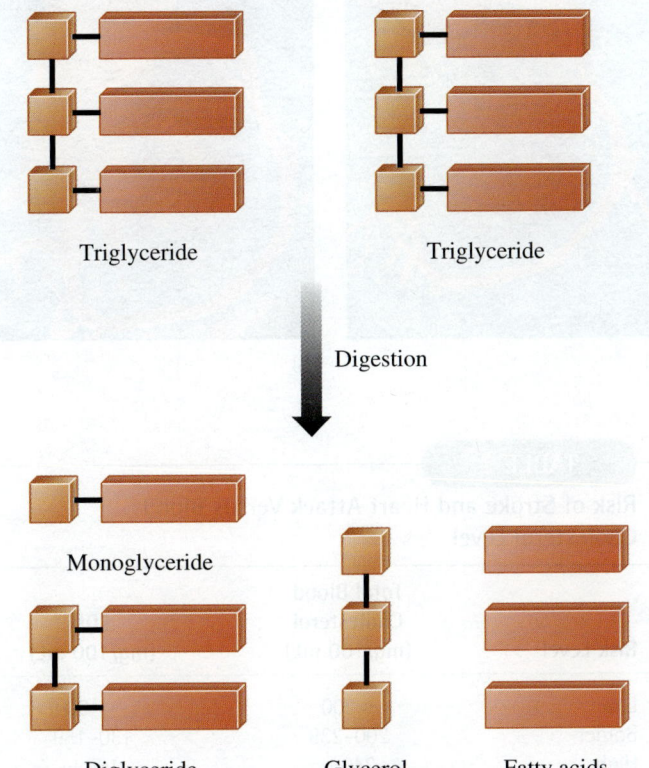

FIGURE 10 During digestion, fats and oils are broken into fatty acids, glycerol, monoglyceride, and diglyceride; however, these products are largely reassembled into fats and oils once they pass through the intestinal wall.

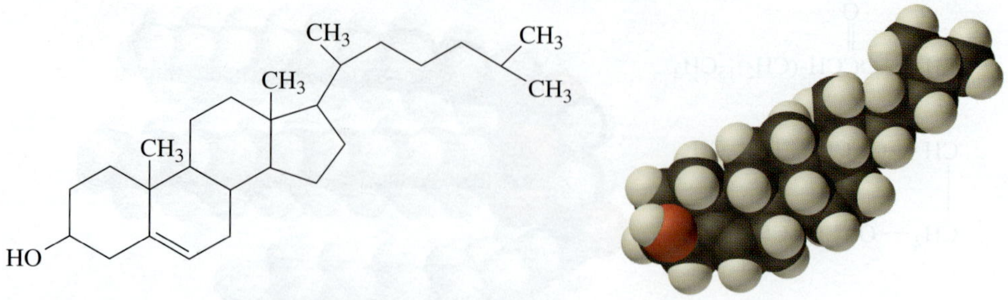

FIGURE 11 Cholesterol, the main cause of arteriosclerosis or blocking of the arteries.

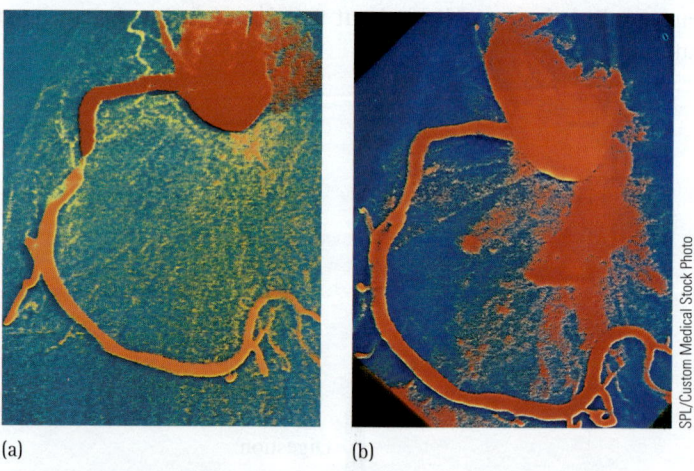

Cholesterol is a nonpolar compound with the four-ring structure (Figure 11) characteristic of steroids. It is found in animal foods such as beef, eggs, fish, poultry, and milk products. Cholesterol is an integral component of cell membrane structure and a starting material for the synthesis of several hormones. It is so important in the body that the liver constantly synthesizes it. However, excessive amounts of cholesterol in the blood—controlled by both genetic factors and diet—may result in the excessive deposition of cholesterol on arterial walls, leading to a condition called arteriosclerosis, or blocking of the arteries (Figure 12). Such blockages are dangerous because they inhibit blood flow to important organs. If blood flow to the heart is blocked, a heart attack results; if blood flow to the brain is blocked, a stroke results. The risk of stroke and heart attack increases with increasing blood cholesterol levels (Table 2).

Like all nonpolar substances, cholesterol is carried in the blood by lipoproteins that are often classified by their density. The main carriers of blood cholesterol are

FIGURE 12 Blocked artery before (a) and after (b) being cleared with a procedure known as angioplasty.

TABLE 2

Risk of Stroke and Heart Attack Versus Blood Cholesterol Level

Risk Level	Total Blood Cholesterol (mg/100 mL)	LDL (mg/100 mL)
Low	<200	<130
Border	200–239	130–159
High	240+	160+

low-density lipoproteins (LDLs). They transport cholesterol from the liver to areas throughout the body where cholesterol is needed. However, they also tend to deposit cholesterol on arterial walls; consequently, LDLs are often called "bad" cholesterol. The risk of stroke and heart attack increases with increasing levels of LDL. Cholesterol is also carried by high-density lipoproteins (HDLs). These proteins transport cholesterol to the liver for processing and excretion and therefore have the tendency to reduce cholesterol deposition on arterial walls. Consequently, HDLs are often called "good" cholesterol. Exercise along with a diet low in saturated fats is believed to raise HDL levels in the blood while lowering LDL levels.

How much fat and cholesterol should you consume? The FDA recommends that fats compose less than 20% of total caloric intake. For a 2000-Cal/day diet, that translates to about 40 g of fat. The consumption of saturated fat, trans-fat, and cholesterol should be minimized.

18.5 Caloric Intake and the First Law: Extra Calories Lead to Fat

Food provides energy for our bodies much like gasoline provides energy for a car. Each of the food types discussed earlier has a caloric value that the body gains when those foods are consumed. For example, 10 g of fat gives the body 90 Cal of energy; fat contains 9 Cal/g. That energy must be either used or stored. In the body, just as everywhere else in nature, the first law of thermodynamics holds: Energy can neither be created nor destroyed. As a result, the following equation applies:

$$\text{energy intake} = \text{energy expended} + \text{energy stored}$$

The *energy intake* corresponds to total caloric intake, the amount of food we eat. The *energy expended* is energy used by the body, and the *energy stored* is energy kept as fat. In spite of our best attempts, we cannot fool nature and get around the first law. If our energy intake exceeds our energy expenditure, our waistline is sure to increase.

Because of the difference in fat content, adding a pat of butter to bread increases its caloric content by 50%.

ENERGY INTAKE

We have seen that different food types have different caloric contents, as summarized in Table 3. Fats and oils contain over twice the number of calories per gram as that of carbohydrates or proteins. To follow the nutritional recommendation that a healthy diet consists of less than 20% of total calories from fat, the total mass of fat should be less than about 8% of the diet.

Explore this topic on the **Counting Calories** website.

TABLE 3
Caloric Content of Several Food Types

Food Type	Caloric Content[a] (Cal/g)
Fats and oils	9
Carbohydrates	4
Proteins	4
Fiber	0

[a]Remember that the nutritional, or big C, Calorie is equivalent to 1000 scientific, or little c, calories.

EXAMPLE 18.1

Calculating the Percentage of Calories from Fat

A particular snack mix lists the following nutritional information:

fat	7 g
protein	4 g
carbohydrate	16 g

What percentage of the total calories is obtained from fat?

SOLUTION

We first calculate the caloric contribution of each:

fat	7 g × 9 Cal/g =	63 Cal
protein	4 g × 4 Cal/g =	16 Cal
carbohydrate	16 g × 4 Cal/g =	64 Cal
total		143 Cal

The percentage of calories from fat:

$$\frac{63 \text{ Cal}}{143 \text{ Cal}} \times 100\% = 44\%$$

YOUR TURN

Calculating the Percentage of Calories from Fat

Extra-light milk lists the following nutritional information:

fat	2 g
protein	11 g
carbohydrate	15 g

What percentage of total calories is obtained from fat?

ENERGY EXPENDED

The body expends energy to breathe, circulate blood, move, think, walk, or run. These energy expenditures are of two types: energy required to stay alive and energy associated with exercise in the broadest sense of the word. The energy expended to stay alive, or more correctly to maintain basal metabolism, includes the energy used to pump the heart, breathe, and keep body temperature constant. Basal metabolism requires approximately 0.5 Cal/hour per pound of body weight. A 120-lb person expends approximately 60 Cal/h in basal metabolism, while a 180-lb person expends 90 Cal/h. This represents the minimum amount of energy required to stay alive even when sleeping.

Energy expended during waking hours varies according to activity level. Sedentary activities such as sitting and reading use 70–100 Cal/h, while strenuous activities such as running use 500–700 Cal/h. An average-size male who exercises moderately for 1 h/day expends approximately 2870 Cal, while a female with similar exercise expends approximately 2130 Cal. The energy expenditures are tabulated as:

Caloric Expenditures for 180-lb Person

Activity	No. of Hours	Energy per Hour	Total
Awake	15	120 Cal	1800
Asleep	8	90 Cal	720
Exercise	1	350 Cal	350
Total	24		2870

Caloric Expenditures for 120-lb Person

Activity	No. of Hours	Energy per Hour	Total
Awake	15	90 Cal	1350
Asleep	8	60 Cal	480
Exercise	1	300 Cal	300
Total	24		2130

ENERGY STORED

If the energy intake exceeds the energy output, the remainder is stored as fat. For every 3500 Cal of excess energy intake, the body stores 1 lb of fat. Fat is the most efficient way to store excess energy because it has the highest caloric content. In contrast, if the body stored excess energy in the form of carbohydrate—with less than half the caloric content—we would gain weight twice as fast.

If the body's energy intake is less than its energy output, weight is lost. For every 3500 Cal of energy expenditure over energy intake, the body burns 1 lb of fat (as long as fat is available).

EXAMPLE 18.2

Relating Caloric Intake and Expenditure to Changes in Weight

If a male who expends 2870 Cal/day wants to lose 5 lb in 1 month (30 days), what should his daily caloric intake be?

SOLUTION

To lose 5 lb in 1 month, the male must have a caloric deficit each day. This deficit can be calculated in the following way:

$$\frac{5 \text{ lb}}{1 \text{ month}} \times \frac{1 \text{ month}}{30 \text{ day}} \times \frac{-3500 \text{ Cal}}{1 \text{ lb}} = -583 \text{ Cal/day}$$

Because the person expends 2870 Cal, the intake that would provide a 583 Cal deficit is

$$2870 \text{ Cal} - 583 \text{ Cal} = 2287 \text{ Cal}$$

YOUR TURN

Relating Caloric Intake and Expenditure to Changes in Weight

If an average female who expends 2130 Cal/day eats an average of 2500 Cal/day, how many pounds will she gain in 1 month?

> **APPLY YOUR KNOWLEDGE**
>
> Some diets claim that as long as you avoid a certain kind of food—such as carbohydrate or fat—you can eat as much as you want and still lose weight. Explain why this cannot be true.
>
> Answer: The first law of thermodynamics tells us that energy can be neither created nor destroyed. If you eat more calories than you expend, the extra energy cannot just be destroyed—it must be stored as fat.

Brightly colored fruits and vegetables are excellent sources of vitamin A.

18.6 Vitamins

Vitamins are those organic compounds essential in the diet in small amounts but that have little or no caloric value. Their chemical composition is varied, and they serve a number of important roles as helpers in cell functions. Vitamins are normally classified according to their polarity. The nonpolar, or *fat-soluble*, vitamins include vitamins A, D, E, and K. Because these vitamins are soluble in fatty tissues, they are stored in the body for long periods of time; they are not easily excreted and can therefore be overconsumed, especially by overzealous users of vitamin supplements. The polar, or *water-soluble*, vitamins include the eight B vitamins and vitamin C. Because they are soluble in water, these vitamins are excreted in the urine and pose little threat of overdose. They must, however, be consumed in sufficient amounts on a daily basis to maintain good health.

FAT-SOLUBLE VITAMINS: A, D, E, AND K

Vitamin A (Figure 13) is important in vision, immune defenses, and maintenance of body lining and skin. Its most familiar role is probably in the eye, where it is part of the retinal pigments that absorb visible light. Vitamin A deficiency results in night blindness, a condition in which night vision recovers only slowly from bright flashes. Vitamin A is present in a number of animal-based foods such as liver, fish liver oils, milk, butter, and eggs. It is also available in a slightly different form, called beta carotene, in plant pigments. Once in the body, beta carotene is converted into active vitamin A. Foods rich in beta carotene include the brightly colored vegetables and fruits such as carrots, squash, tomatoes, and cantaloupe. The Recommended Daily Allowance (RDA) for vitamin A is 1000 retinol equivalents (RE) for men and 800 RE for women. One-half cup of cooked carrots

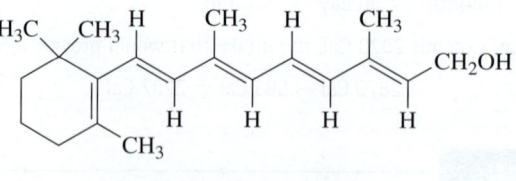

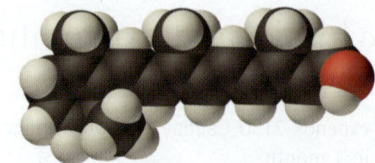

FIGURE 13 Vitamin A, important in vision, immune defenses, and maintenance of skin.

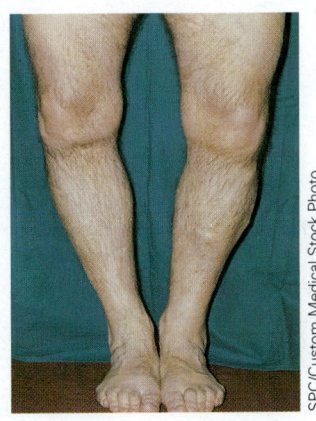

FIGURE 14 Vitamin D is important in the development and maintenance of bone structure.

provides about twice the RDA. Consuming too much vitamin A, more than 10 to 15 times the RDA, can have detrimental effects such as blurred vision, hair loss, muscle soreness, and nausea.

Vitamin D (Figure 14) promotes the absorption of calcium through the intestinal wall and into the blood; thus vitamin D is important in the development and maintenance of bone structure. Vitamin D deficiency, especially in children, causes rickets, a bone disease resulting in bowed legs, knock knees, and a protruding chest. Because infants and children, whose bones are developing, need more vitamin D than mature adults, formula and milk are fortified with vitamin D. Vitamin D fortification of milk has virtually eliminated rickets in developed countries. Vitamin D is also present in egg yolks, liver, and butter. It is unique among the vitamins in that the body can synthesize its own under the skin in the presence of sunlight. However, unless a person spends a lot of time in the sun, the body cannot make enough vitamin D to meet its needs. The RDA for vitamin D is 10 micrograms (μg)/day for young adults (19-24). One cup of vitamin D fortified milk provides about 2.5 μg. Too much vitamin D, 4-5 times the RDA, consumed on a consistent basis causes excess absorption of calcium into the bloodstream and can result in the deposition of calcium in soft body tissues, including the major organs. In extreme cases, this can result in death.

Vitamin D deficiency causes rickets, which results in bone deformities.

Vitamin E (Figure 15) is an antioxidant in the body, preventing oxidative damage to cellular components, especially cell membranes. Its presence in food is so widespread that deficiencies are extremely rare. Likewise, it has a low toxicity, and overdoses are just as rare as deficiencies. The RDA for vitamin E is 10 mg/day for men and 8 mg/day for women.

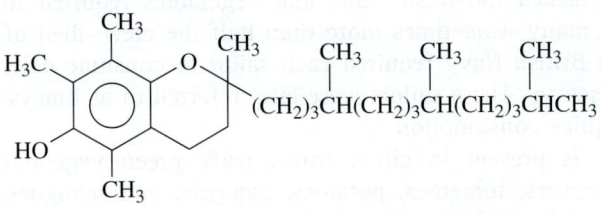

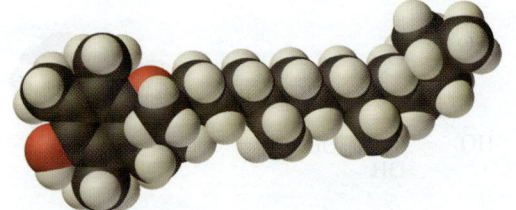

FIGURE 15 Vitamin E is an antioxidant and protects cell components from oxidation.

FIGURE 16 Vitamin K is important in the clotting of blood.

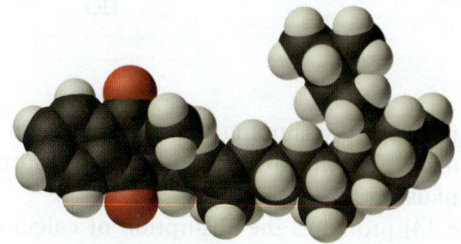

Vitamin K (Figure 16) is necessary for the synthesis of four proteins involved in blood clotting. Because blood clotting stops bleeding, Vitamin K deficiency, although rare, causes excess or uncontrollable bleeding. Vitamin K is present in leafy green vegetables and in milk and is also synthesized by intestinal bacteria. Because newborn infants lack intestinal bacteria, they are usually given a vitamin K supplement at birth. The RDA for vitamin K is 80 μg/day for men (19–24) and 60 μg/day for women (19–24). Persons over 25 need slightly more. Vitamin K overuse can lead to excessive blood clotting and result in brain damage.

WATER-SOLUBLE VITAMINS: C AND B COMPLEX

Vitamin C (Figure 17) plays several roles in the body, including synthesis of connective tissue called **collagen**, protection against infection, and absorption of iron through the intestinal wall. Like vitamin E, vitamin C is an antioxidant, protecting against oxidation in water-soluble cellular components. Vitamin C deficiency results in a condition called **scurvy**, in which bodily tissues and blood vessels become weakened. The first sign of scurvy is bleeding gums, followed by the slow degeneration of skin and muscle. Because collagen, a main component of scar tissue, cannot form in people with scurvy, wounds do not heal and massive bleeding often results. If untreated, scurvy ends in death. Early sailors on long voyages often lacked the fresh fruits and vegetables required to get sufficient vitamin C and many—sometimes more than half the crew—died of scurvy. In the late 1700s, the British Navy required each sailor to consume daily doses of lime juice to avoid scurvy. These sailors were later referred to as limeys in jest of their constant lime juice consumption.

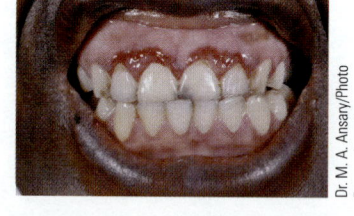

Vitamin C deficiency results in a condition called scurvy, which is first manifested as bleeding gums.

Vitamin C is present in citrus fruits, leafy green vegetables, cantaloupe, strawberries, peppers, tomatoes, potatoes, papayas, and mangoes. The RDA for vitamin C for adults is 60 mg/day. A three-quarter-cup serving of orange juice

FIGURE 17 Vitamin C is important in the synthesis of the body's connective tissue called collagen.

provides 1.5 times the RDA for vitamin C. Because vitamin C is water soluble, excess amounts are easily excreted in the urine and therefore pose little threat to health. Megadoses of vitamin C (2–3 g/day), however, can lead to nausea, cramps, and diarrhea.

B complex vitamins are a family that includes thiamin, riboflavin, niacin, vitamin B_6, folate, and vitamin B_{12}. As a whole, B complex vitamins play central roles in metabolism (the extraction of energy from food), protein synthesis, and cell multiplication. B complex vitamins are found in enriched flour and, depending on the specific vitamin, spread through all food groups. Many B complex vitamins are concentrated in leafy green vegetables and in the hulls of grain that are often removed during processing. Deficiencies in B complex vitamins result in the inability of cells to extract energy from foods. The symptoms of deficiency include exhaustion, irritability, depression, forgetfulness, partial paralysis, abnormal heart action, and severe skin problems. RDAs for several B vitamins are shown in Table 4.

18.7 Minerals

The nutritional minerals are those elements, other than carbon, hydrogen, nitrogen, and oxygen, needed for good health. Many of these are present in the body as ions rather than as neutral atoms. The **major minerals** compose about 4% of the body's weight and include Ca, P, Mg, Na, K, Cl, and S. The **minor minerals** are present in trace amounts and include Fe, Cu, Zn, I, Se, Mn, F, Cr, and Mo.

THE MAJOR MINERALS

As the main structural material for bones and teeth, **calcium** is the most abundant mineral in the body. Calcium also plays an important role in nerve signal transmission and in blood clotting. If blood calcium levels get too low, calcium stored in bones will release into the blood causing loss of bone mass. Persistently low blood calcium levels result in osteoporosis, a condition in which bones become weakened by calcium loss. The adult RDA for calcium is 800 mg/day. Growing children and young adults need significantly more. One cup of milk contains approximately one-third of the adult RDA. Other calcium-rich foods include cheese, sardines, oysters, and broccoli.

Phosphorus is the second most abundant mineral in the body and is largely bound with calcium in bones and teeth. Phosphorus also assists in energy metabolism and is part of DNA. The adult RDA for phosphorus is 800 mg/day and is easily met with almost any diet, making deficiencies unknown.

Sodium, usually eaten as salt (NaCl), is the main element involved in bodily fluid-level regulation. Blood sodium levels and bodily fluid levels move in unison. Therefore, if a person eats a salty meal, thirst will follow to ensure that water levels increase to meet the increased sodium levels. Both the excess sodium and water are later eliminated in the urine. Similarly, if blood sodium levels drop, as often happens after heavy vomiting or diarrhea, body water is eliminated in the urine and both sodium and water must be consumed to restore normal levels. Consistently high blood sodium levels can result in high bodily fluid levels and consequently high blood pressure (hypertension). People with a tendency toward hypertension should lower their salt intake by reducing the amount of salt they put on foods and by reducing their intake of highly processed foods. Because eating enough salt is usually not a problem, there is no RDA for sodium.

TABLE 4 RDAs for Vitamins and Minerals

Age (years)	Weight (kg)	Weight (lb)	Height (cm)	Height (inches)	Protein (g)	Vitamin A (RE)	Vitamin D (μg)	Vitamin E (mg)	Vitamin K (μg)	Vitamin C (mg)	Thiamin (mg)	Riboflavin (mg)	Niacin (mg equiv.)	Vitamin B$_6$ (mg)	Folate (μg)	Vitamin B$_{12}$ (μg)	Calcium (mg)	Phosphorus (mg)	Magnesium (mg)	Iron (mg)	Zinc (mg)	Iodine (μg)	Selenium (μg)
Infants																							
0.0–0.5	6	13	60	24	13	375	7.5	3	5	30	0.3	0.4	5	0.3	25	0.3	400	300	40	6	5	40	10
0.5–1.0	9	20	71	28	14	375	10	4	10	35	0.4	0.5	6	0.6	35	0.5	600	500	60	10	5	50	15
Children																							
1–3	13	29	90	35	16	400	10	6	15	40	0.7	0.8	9	1.0	50	0.7	800	800	80	10	10	70	20
4–6	20	44	112	44	24	500	10	7	20	45	0.9	1.1	12	1.1	75	1.0	800	800	120	10	10	90	20
7–10	28	62	132	52	28	700	10	7	30	45	1.0	1.2	13	1.4	100	1.4	800	800	170	10	10	120	30
Males																							
11–14	45	99	157	62	45	1000	10	10	45	50	1.3	1.5	17	1.7	150	2.0	1200	1200	270	12	15	150	40
15–18	66	145	176	69	59	1000	10	10	65	60	1.5	1.8	20	2.0	200	2.0	1200	1200	400	12	15	150	50
19–24	72	160	177	70	58	1000	10	10	70	60	1.5	1.7	19	2.0	200	2.0	1200	1200	350	10	15	150	70
25–50	79	174	176	70	63	1000	5	10	80	60	1.5	1.7	19	2.0	200	2.0	800	800	350	10	15	150	70
51+	77	170	173	68	63	1000	5	10	80	60	1.2	1.4	15	2.0	200	2.0	800	800	350	10	15	150	70
Females																							
11–14	46	101	157	62	46	800	10	8	45	50	1.1	1.3	15	1.4	150	2.0	1200	1200	280	15	12	150	45
15–18	55	120	163	64	44	800	10	8	55	60	1.1	1.3	15	1.5	180	2.0	1200	1200	300	15	12	150	50
19–24	58	128	164	65	46	800	10	8	60	60	1.1	1.3	15	1.6	180	2.0	1200	1200	280	15	12	150	55
25–50	63	138	163	64	50	800	5	8	65	60	1.1	1.3	15	1.6	180	2.0	800	800	280	15	12	150	55
51+	65	143	160	63	50	800	5	8	65	60	1.0	1.2	13	1.6	180	2.0	800	800	280	10	12	150	55
Pregnant					60	800	10	10	65	70	1.5	1.6	17	2.2	400	2.2	1200	1200	320	30	15	175	65
Lactating																							
1st 6 months					65	1300	10	12	65	95	1.6	1.8	20	2.1	280	2.6	1200	1200	355	15	19	200	75
2nd 6 months					62	1200	10	11	65	90	1.6	1.7	20	2.1	260	2.6	1200	1200	340	15	16	200	75

[a]The allowances are intended to provide for individual variations among most normal, healthy people in the United States under usual environmental stresses. Diets should be based on a variety of common foods to provide other nutrients for which human requirements have been less well defined. See the text for a more detailed discussion of the RDA and of nutrients not tabulated.

Source: Recommended Dietary Allowances. Copyright © 1989 National Academy of Sciences, *National Academy Press*, Washington, D.C.

Potassium and magnesium are both involved in maintaining electrolyte balance in and around cells. Magnesium also plays an important role in bone formation and in the operations of many enzymes, including those involved in the release of energy. Potassium is found in significant amounts in most fresh fruits and vegetables, including bananas, cantaloupe, and lima beans. Magnesium occurs in oysters, sunflower seeds, spinach, and in "hard" drinking water.

THE MINOR MINERALS

Iodine is a trace mineral needed by the body in only very small quantities. It is used in the synthesis of a hormone that regulates basal metabolic rate. Its deficiency results in goiter, a condition in which the thyroid gland swells as it tries to capture as many iodine particles as possible. Iodine deficiency during pregnancy can result in severe retardation of the developing baby. Iodine is readily available in iodized salt, seafood, and milk. The addition of iodine to salt has all but eliminated the effects of its deficiency in developed countries. The RDA for iodine is 150 μg/day, provided by about one-half teaspoon of iodized salt.

Iron composes a critical part of hemoglobin (Hb), the protein that carries oxygen in the blood. All cells need oxygen for the oxidation of food for energy. Consequently, the body hoards iron and uses it efficiently. Its only significant loss is through bleeding. Nonetheless, iron deficiency can occur and results in the body's inability to make enough Hb, which in turn causes anemia—a condition in which the patient feels tired, apathetic, and susceptible to cold temperatures. Infants, young children, and pregnant women are particularly vulnerable. Women need about 50% more iron than men due to the loss of blood in menstrual bleeding. The RDA for iron is 10 mg/day for men and 15 mg/day for women of childbearing age. Foods that are rich in iron include meat, fish, poultry, clams, legumes, and leafy green vegetables. Cooking foods in an iron skillet also increases their iron content.

Zinc, although it occurs only in very small amounts throughout the body, is essential to the function of more than 100 enzymes. It is important in growth and development, immune function, learning, wound healing, and sperm production. The zinc RDA is 15 mg/day for men and 12 mg/day for women. The best sources of zinc are animal products like meat, shellfish, and poultry. Because plant sources contain only small amounts of zinc, vegetarians are most at risk for zinc deficiency. Too much zinc, however, is toxic and should be avoided.

Other important minor minerals include selenium, another antioxidant; chromium, important in the control of blood glucose levels; and copper, needed to form Hb and collagen. Selenium and copper are present in many foods and meeting the RDA is usually not a problem. Chromium deficiencies, however, are a growing concern because food processing often depletes their chromium content. Insufficient dietary chromium results in diabetes-like symptoms of high blood sugar. The RDA for chromium is 50 mg/day and can be obtained from foods such as liver, whole grains, nuts, and cheeses.

18.8 Food Additives

People have added substances to food since the earliest times. Ancient civilizations used salt to preserve meats, added herbs and spices to improve flavor, and preserved fruit with sugar and vegetables with vinegar. In recent times, the move from rural life to urban life has required that food be preserved, packaged, and shipped—sometimes thousands of miles—before it is consumed. Consequently, the number of

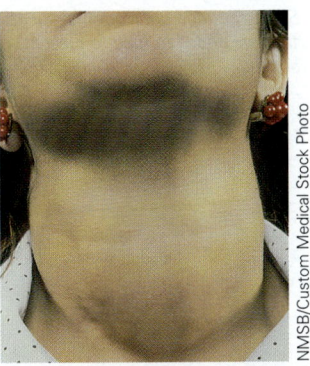

An iodine deficiency leads to a condition called goiter in which the thyroid gland swells as it tries to capture as much iodine as possible.

Cooking food in an iron skillet significantly increases its iron content because iron atoms from the skillet incorporate into the food.

 Explore this topic on the U.S. Food and Drug Administration (FDA) website.

Bologna, like many processed meats, contains sodium nitrite, which gives bologna its pink color and acts as an antimicrobial agent.

Some common antimicrobial agents that are added to food to prevent spoilage.

$$CH_3CH_2\overset{\overset{O}{\|}}{C}-O^-Na^+ \qquad \underset{\text{sodium benzoate}}{\text{C}_6H_5-\overset{\overset{O}{\|}}{C}-O^-Na^+} \qquad NaNO_2$$

sodium propionate   sodium benzoate   sodium nitrite

$$CH_3CH=CHCH=CH\overset{\overset{O}{\|}}{C}-OH \qquad CH_3CH=CHCH=CH\overset{\overset{O}{\|}}{C}-O^-K^+$$

sorbic acid   potassium sorbate

food additives has increased. Current food additives are used to preserve food, enhance its flavor or color, and maintain its appearance or physical characteristics.

All food additives are monitored and regulated by the Food and Drug Administration. The FDA regulates 1 trillion dollars' worth of products—25 cents of every consumer dollar spent. The Federal Food, Drug, and Cosmetic (FD&C) Act of 1938 gives the FDA authority over food and its ingredients. The FDA maintains a list of several hundred food additives that are generally recognized as safe (GRAS) based on their history of use in food or on published scientific evidence. The GRAS list includes familiar additives such as salt, sugar, and spices as well as less familiar additives such as sodium benzoate or ethylenediaminetetraacetic acid (EDTA). Manufacturers intending to add a new additive to the GRAS list must demonstrate that the additive is effective and safe. Recent additions to the GRAS list include fat substitutes such as Simplesse and Olestra.

We can divide food additives into five categories: antimicrobial agents, antioxidants, artificial colors, artificial flavors and flavor enhancers, and stabilizers.

ANTIMICROBIAL AGENTS

Antimicrobial agents are added to food to prevent the growth of bacteria, yeasts, and molds, some of which cause severe illness or even death. For example, the bacteria responsible for botulism produce a toxin so deadly that a few nanograms (about 1/1000 the size of a sand grain) would kill an adult. Yet its growth in food can be avoided by simple food additives. The simplest antimicrobial agents include salt, often used in drying meat and fish, and sugar, used in jams and jellies. These additives prevent microbial growth by producing dehydrating conditions; without water, microorganisms cannot survive. Less familiar antimicrobial agents include sodium propionate, sodium benzoate, sorbic acid, and potassium sorbate; these find common use in preventing microbial growth in canned, bottled, or packaged foods. The main antimicrobial agent for processed meats such as ham, hot dogs, and bologna is sodium nitrite, a substance that inhibits the growth of bacteria—especially those responsible for botulism—and imparts a pink color to the meat. Although sodium nitrite is approved by the FDA, it has come under recent scrutiny as a possible carcinogen. The evidence, however, is not currently strong enough to warrant its abandonment as a preventer of botulism.

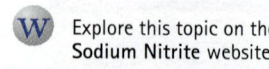

Explore this topic on the **Sodium Nitrite** website.

ANTIOXIDANTS

Antioxidants are those substances added to food to prevent oxidation. Unsaturated oils, for instance, oxidize in air, resulting in the cleavage and addition of oxygen to their hydrocarbon side chains. The products are volatile aldehydes, ketones, and acids, many with foul odors and flavors that make the oil rancid. Fruits or fruit juices also oxidize, turning brown as an apple does when its inte-

When the interior of an apple is exposed to air it oxidizes, turning brown.

Vitamin C, BHA, and BHT are all antioxidants.

rior is exposed to air. The addition of antioxidants to oils, fruits, fruit juices, and other packaged foods prevents rancidity and browning by inhibiting oxidation. Some common antioxidants include ascorbic acid (vitamin C), butylated hydroxyanisole (BHA), and butylated hydroxytoluene (BHT). Unlike other food additives that sometimes show undesirable side effects, the antioxidants have produced some surprisingly positive ones. Laboratory animals fed a diet fortified with BHA and BHT actually show slightly lower cancer rates and longer lifespans than those whose diet did not include these additives. Consequently, health food stores have rushed to sell these additives as dietary supplements, ironically on shelves right next to foods labeled "additive free." Care should be taken, however, in taking these supplements because high amounts of BHA and BHT can be toxic.

Other antioxidants include sulfites, often added to wine and fruits as a preservative and antibrowning agent, and EDTA, which immobilizes metal ions that often catalyze oxidation reactions. The presence of sulfites in foods is usually labeled because some people (especially asthmatics) are allergic to them.

ARTIFICIAL COLORS

Artificial colors are those substances added to food to improve its appearance. Some of these additives are natural plant pigments, while others are synthetically prepared. The artificial colors have perhaps come under the most scrutiny because their use is purely aesthetic. Consequently, even a minuscule amount of risk is often intolerable. For example, red dye no. 3 was banned by the FDA in 1990 for certain uses because one study in male rats showed an association with thyroid tumors. The risk, at levels found in foods, was extremely small; however, the Delaney Clause—an amendment to the FD&C Act that forbids the addition of any substance shown to cause cancer in laboratory animals regardless of dose—legally mandated its ban.

ARTIFICIAL FLAVORS AND FLAVOR ENHANCERS

Artificial flavors and flavor enhancers are added to foods to improve their taste. Artificial flavors include sweeteners like sugar, corn syrup, or aspartame (NutraSweet) as well as numerous plant flavors like oil of wintergreen, peppermint, ginger, vanilla, or almond extract. Artificial flavors can be natural or synthetic versions of the same molecule. Flavor enhancers include substances such as monosodium glutamate (MSG), which has no flavor of its own but enhances many of the flavors present in food. MSG is used extensively in Chinese cooking

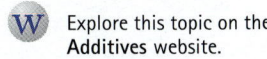
Explore this topic on the **Additives** website.

benzaldehyde (oil of almond)

methyl salicylate (oil of wintergreen)

vanillin (vanilla)

and is considered safe by the FDA. However, a number of people have shown allergic reactions to MSG that include burning or tingling in the arms, neck, back, or chest; chest pain; headaches; nausea; and rapid heartbeat.

STABILIZERS

Stabilizers improve and preserve the physical characteristics of food. **Humectants**, such as glycerine, are added to foods such as shredded coconut to keep them moist. **Anticaking agents**—such as silicon dioxide, calcium silicate, and magnesium silicate—are added to powders, such as table salt and baking powder, to keep them dry and granulized. **Emulsifiers**, such as glycerol monostearate, are added to keep mixtures of polar and nonpolar substances together. For example, emulsifiers are added to peanut butter to keep it smooth and creamy.

Emulsifiers keep substances like peanut butter mixed and smooth. The peanut butter on the left contains an emulsifier, while that on the right does not.

18.9 The Molecules Used to Grow Crops: Fertilizers and Nutrients

Plants use sunlight, water, and carbon dioxide to synthesize carbohydrates through photosynthesis:

$$\text{light} + 6CO_2 + 6H_2O \longrightarrow C_6H_{12}O_6 + 6O_2$$

However, plants also need to synthesize other compounds—such as proteins, nucleic acids, and lipids—that contain elements besides carbon, hydrogen, and oxygen; the plant gets these elements from the soil. The most important elements that a plant gets from soil are the *primary nutrients*: potassium, phosphorus, and nitrogen. When a field is used repeatedly to grow crops, the soil becomes depleted of these elements, and they must be replaced by fertilizers. The earliest fertilizers were natural products such as manure, composts, or guano. However, most commercial farmers today use chemical fertilizers containing potassium, phosphorus, and nitrogen.

Potassium is used by plants in its ionic form, K^+. Chemical potassium fertilizers are usually potassium chloride (KCl) or potassium sulfate (K_2SO_4). There are vast mineral deposits of potassium in Russia, Canada, and Germany, enough to last for the forseeable future.

Nitrogen, used by plants for the synthesis of amino acids, is normally absorbed as ammonium ions (NH_4^+) or nitrate ions (NO_3^-). Although the atmosphere is primarily nitrogen (N_2), plants cannot use nitrogen in its inert atmospheric form. Nitrogen must be *fixed*, or converted to other forms that plants can absorb. Some crops, namely, legumes such as soybeans and peas, grow nitrogen-fixing bacteria in their roots. Other crops, such as corn, have no mechanism to fix nitrogen and must depend on nitrates present in soil. Some farmers rotate crops, alternating between nitrogen-consuming crops and nitrogen-fixing ones, to replenish nitrogen in the soil. Alternatively, nitrogen can be added to soils with nitrogen fertilizers, made from nitrogen that has been fixed via the Haber process (Chapter 14):

$$3H_2 + N_2 \longrightarrow 2NH_3$$

Crops deplete the soil of nutrients such as potassium, phosphorus, and nitrogen. Fertilizers replace these nutrients in the soil.

Ammonia is applied to the soil directly as a gas, or it is mixed with water to form an ammonia solution. It can also be converted to other compounds such as urea ($CO(NH_2)_2$), ammonium nitrate (NH_4NO_3), or ammonium sulfate (($NH_4)_2SO_4$), all of which make good fertilizers.

Phosphorus is usually absorbed by plants in ionic forms *(phosphates)* such as $H_2PO_4^-$ or HPO_4^-. Phosphate fertilizers are normally made by treating phosphate rock—primarily $Ca_3(PO_4)_2$, which is insoluble in water—with sulfuric acid (H_2SO_4) to form superphosphate, a more soluble compound that is added directly to soil:

$$Ca_3(PO_4)_2 + 2H_2SO_4 \longrightarrow Ca(H_2PO_4)_2 + 2CaSO_4$$

Ammonia can be applied directly to the soil as a gas or in a water solution.

There are significant, but finite, deposits of phosphate rock in North Africa, the United States, and Russia.

Molecular Focus

Ammonium Nitrate

Formula: NH_4NO_3
Molar mass: 80.1 g/mol
Melting point: 169.6°C
Decomposes at 210°C

Ammonium nitrate forms colorless, odorless crystals at ordinary temperatures and pressures. It is usually made by the reaction of ammonia with nitric acid:

$$NH_3 + HNO_3 \longrightarrow NH_4NO_3$$

Ammonium nitrate is an excellent fertilizer because it is highly soluble in water and contains 35% nitrogen by weight. However, it is also an excellent explosive that is insensitive to shock and heat. When mixed with fuel oil, ammonium nitrate detonates on initiation with a primer explosion. The availability of ammonium nitrate for fertilization at any farm supply store makes it a likely target for abuse by terrorist groups or others in search of easy access to explosive materials. The danger of ammonium nitrate burst into public consciousness with the 1995 bombing of the Federal Building in Oklahoma City. Experts estimate that 4800 lb of ammonium nitrate and fuel oil (ANFO bomb)—contained in a Ryder rental truck—caused the blast that killed 168 people. Receipts found in the suspects' homes show that they bought 80 bags (50 lb each) of ammonium nitrate fertilizer at a co-op in Kansas—the cost: $457.48. As a result of Oklahoma City, some lawmakers have called for regulation of ammonium nitrate sales. However, no federal regulation has yet been implemented.

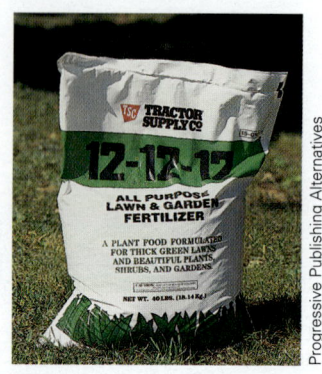

Mixed fertilizers contain nitrogen, phosphorus, and potassium in percentages indicated by three numbers on the bag.

Mixed fertilizers, containing all three of the primary nutrients, are widely used. These fertilizers are usually characterized by a set of numbers such as 5–15–5, which indicates the percentage of nitrogen, phosphorus (calculated as P_2O_5), and potassium (calculated as K_2O).

Plants need other nutrients as well—such as the *secondary nutrients* (calcium, magnesium, and sulfur) and the *micronutrients* (boron, chlorine, copper, iron, manganese, molybdenum, sodium, vanadium, and zinc)—but these are needed in smaller quantities and rarely need replenishment in soils.

18.10 The Molecules Used to Protect Crops: Insecticides and Herbicides

Insecticides and herbicides are used to control pests and weeds that carry disease, damage crops, or provide a nuisance. It is estimated that one-third of the world's total crops are destroyed by pests or weeds; therefore, the production of insecticides and herbicides is a large industry.

INSECTICIDES

Early insecticides were highly toxic substances such as hydrogen cyanide gas or arsenic compounds. These substances were effective in killing insects, but they were also highly toxic to humans. In the 1940s and 1950s, a revolution in insecticide technology occurred. Chemists synthesized new compounds such as dichlorodiphenyltrichlorethane (DDT) and hexachlorobenzene (HCB) (Figure 18). These highly stable, chlorinated hydrocarbons seemingly worked miracles. They were extremely toxic toward insects, especially many that carried diseases, but were relatively nontoxic toward humans. Farmers used DDT to increase crop yield, and public health officials used it to kill disease-carrying mosquitoes with stunning success. However, the chemical stability of chlorinated hydrocarbons allowed them to accumulate in soil and water supplies. Aquatic plants started to show traces of DDT in their cells. The fish that ate the plants began harboring DDT, as did the birds that ate the fish. Even worse, the levels of DDT concentrated as it moved up the food chain, a process called **bioamplification**. Fish and birds began to die, and our national bird, the American bald eagle, was nearly driven to extinction. Today, most chlorinated hydrocarbon insecticides have been banned in the United States.

Modern insecticides include the **organophosphate** insecticides such as *malathion* and *parathion* (Figure 19), and **carbamate** insecticides such as *carbaryl* and *aldicarb* (Figure 20). The organophosphate insecticides are primarily *broad-spectrum* insecticides, which means they attack a wide range of pests. They are, in general, more toxic than the chlorinated hydrocarbons, both to insects and to humans. They have the advantage, however, of quickly degrading in the environment into

FIGURE 18 DDT and HCB are both chlorinated organic insecticides.

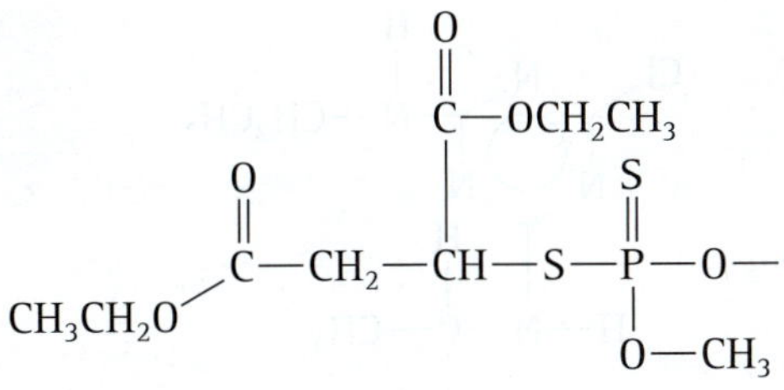

FIGURE 19 Modern organophosphate insecticides are more toxic than their chlorinated predecessors, but they quickly degrade in the environment.

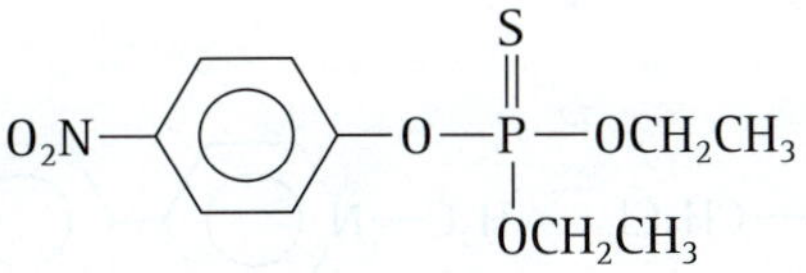

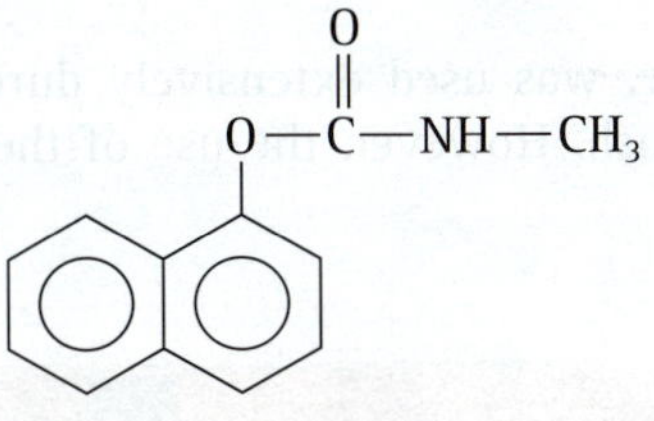

malathion

parathion

FIGURE 20 Carbaryl and aldicarb are examples of narrow-spectrum insecticides that target specific species of pests.

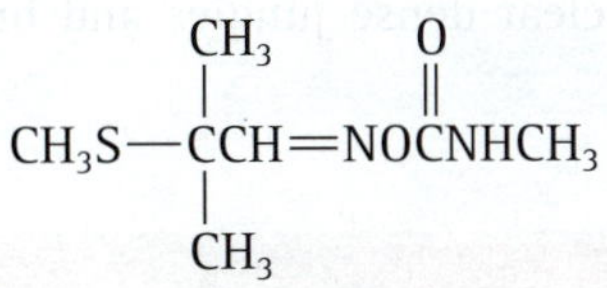

carbaryl

aldicarb

nontoxic products. Consequently, they do not accumulate in the environment and do not bioamplify in the food chain. The carbamate insecticides are similar to the organophosphates in their tendency to decompose in the environment but have the advantage of targeting specific pests. Such insecticides are termed *narrow-spectrum* insecticides. Although modern insecticides have eliminated the problem of bioamplification, they are not without risks of their own. Their high toxicity has sometimes resulted in the illness and even deaths of agricultural workers, especially in underdeveloped countries.

HERBICIDES

Herbicides are used to kill weeds and unwanted plants that can compete with and dominate crop species. In the 1960s and 1970s, popular herbicides included the **phenoxy herbicides** such as 2,4,5-trichlorophenoxyacetic acid (2,4,5-T) and 2,4-dichlorophenoxyacetic acid (2,4-D) (Figure 21). These substances are called *defoliants* because they work by causing plants to lose their leaves. A 1:1 mixture of

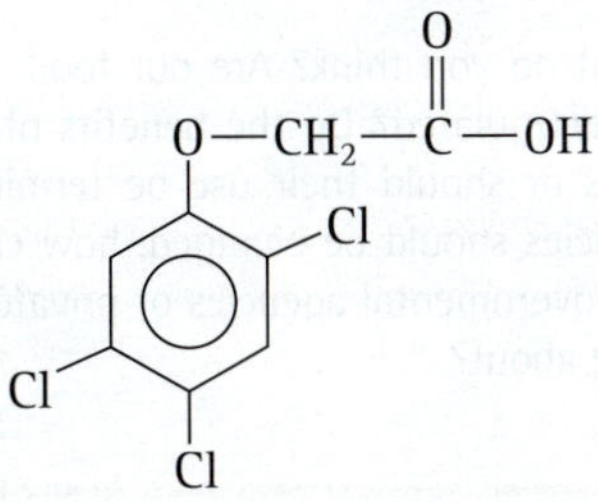

2,4,5-T

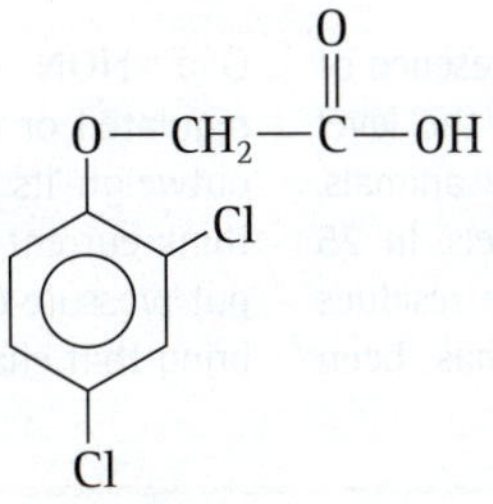

2,4-D

FIGURE 21 A mixture of 2,4,5-T and 2,4-D, called Agent Orange, was used in the Vietnam War as a defoliant to clear foliage in dense jungles.

atrazine

metolachlor

paraquat

2,4,5-T and 2,4-D, called **Agent Orange,** was used extensively during the Vietnam War to clear dense jungles and brush. However, the use of these defoliants

WHAT IF...

Pesticide Residues in Food—A Cause for Concern?

The use of insecticides and pesticides, and the danger of their residues in food and in the environment, are areas of considerable debate. On the one hand, some believe that the dangers associated with the tiny levels of these compounds in food or the environment have been grossly overstated. They point to the extreme sensitivity of modern chemical instrumentation that can detect contaminant levels so low that the same levels would have been considered to be zero just two decades ago. They also point to the lack of human fatalities due to insecticide or herbicide residues in food, as well as the huge expense associated with their discontinued use or increased regulation. On the other hand, there are those who argue that the mere presence, no matter how low, of these synthetic chemicals in the environment or in food is cause for concern. The tendency of these chemicals to accumulate in the environment, as well as their difficult-to-measure chronic effects such as cancer and birth defects, should prompt government to more strict regulation.

The FDA, for its part, has set guidelines for the presence of pesticides in foods, usually 1/100th or 1/1000th of the level found to cause any measurable effect in laboratory animals. The FDA also routinely tests foods for pesticide levels. In 25 years of testing, the FDA has seldom found pesticide residues above tolerance levels; when it has, the food has been destroyed.

The risk associated with eating unregulated food, however, may be higher. Certain waters have become increasingly polluted with chlorinated hydrocarbons, and food that originates from those waters has been found to be contaminated. Still controversial, however, are the chronic effects of eating those foods. Because the levels are not high enough to cause immediate damage, the effects are difficult to measure. Some studies have shown possible correlation between the chronic presence of these substances in the diet and diseases like cancer or developmental problems in children. For example, a multiyear study published in 1990 showed that women who routinely ate popular sport fish from Lake Michigan—which has significant levels of chlorinated hydrocarbons—had children with lower birth weights than the children of women who did not routinely eat the fish. In addition, those same children tended to score lower on developmental tests that measured verbal and memory abilities.

QUESTION: What do you think? Are our food supplies overregulated or underregulated? Do the benefits of pesticide use outweigh its risks or should their use be terminated? If you think current policies should be changed, how can consumers put pressure on governmental agencies or private industries to bring that change about?

has been banned because they were found to be contaminated with **dioxins**. The dioxins are a family of compounds with varying levels of toxicity. Of most concern is tetrachlorodibenzo-*p*-dioxin (TCDD) (Figure 22), a dioxin that is both toxic and carcinogenic. The contamination of 2,4,5-T by TCDD led to its ban in the 1980s.

Modern herbicides include the *triazine herbicides* such as *atrazine*, used to destroy weeds in corn fields. Atrazine accumulates in the environment to some degree, but its low toxicity to mammals has allowed its continued use. Other modern herbicides include *metolachlor*, which is used on soybeans and corn, and *paraquat*, which has garnered fame for its use in destroying marijuana crops. Neither metolachlor nor paraquat accumulates in the environment, nor does either bioamplify in the food chain.

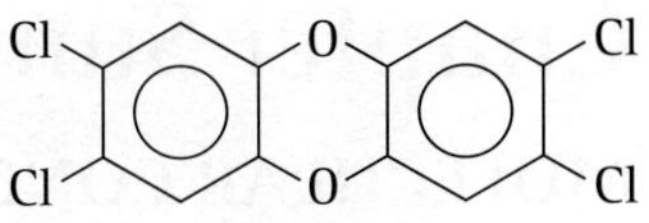

FIGURE 22 TCDD, often called dioxin, is a toxic carcinogen that contaminated Agent Orange.

Chapter Summary

MOLECULAR CONCEPT

Foods are categorized as carbohydrates, proteins, and fats/oils (18.1). The carbohydrates include sugar, starch, and fiber and contain about four nutritional calories per gram (except fiber, which contains none) (18.2). The proteins supply the necessary amino acids and contain four nutritional calories per gram; complete proteins supply all the essential amino acids—those the body cannot synthesize—in the right proportion (18.3). Fats and oils are primarily triglycerides and contain nine nutritional calories per gram. Cholesterol is a fatty substance that, along with saturated fats, increases risk of stroke and heart disease (18.4, 18.5).

The body also needs vitamins, organic substances, in small amounts. They can be divided into two groups: the fat-soluble vitamins (A, D, E, and K) and the water-soluble vitamins (C and B complex) (18.6). The body also needs minerals, nonorganic substances, in small amounts. These can be divided into the major minerals (Ca, P, Mg, Na, K, Cl, and S) and the minor minerals (Fe, Cu, Zn, I, Se, Mn, F, Cr, and Mo) (18.7).

Modern food contains many additives to preserve it and enhance its flavor and appearance. Antimicrobial agents are added to food to inhibit the growth of bacteria, yeasts, and molds. Antioxidants keep food from oxidizing when exposed to air. Artificial colors enhance the appearance of food, and artificial flavors enhance its taste. Stabilizers keep food's physical characteristics stable (18.8). Modern foods are also grown with fertilizers to replenish nutrients in soils and pesticides to protect crops from insects and weeds (18.9, 18.10).

SOCIETAL IMPACT

Our bodies are composed of molecules and atoms obtained primarily from foods. The saying "you are what you eat" is literally true—we are composed of the foods we eat, although they are usually rearranged and chemically modified. The kinds of foods we eat, and therefore the proportions of carbohydrates, proteins, fats, and oils, often vary from one culture to another. Scientists have shown, however, that certain proportions are better than others, with the ideal being about 60% carbohydrate, 10% protein, and less than 30% fat. The average North American diet is higher in protein and fat than the ideal diet.

Much of food goes to supplying the body's constant energy need, and stable weight is maintained if caloric intake matches caloric expenditure. The North American diet is often high in caloric content, so many North Americans have a tendency to be overweight (18.5).

Vitamin and mineral supplements are popular in our society. Many nutritionists, however, recommend that you get the necessary vitamins and minerals from foods. If a variety of foods are consumed, vitamin and mineral supplements are normally not necessary; however, deficiencies result in a number of adverse conditions and diseases (18.6, 18.7).

Food additives must be approved by the Food and Drug Administration. The FDA maintains that all additives on its generally recognized as safe list can be consumed over a lifetime with no adverse health effects (18.8).

Chemistry on the Web

For up-to-date URLs, visit the text website at www.chemistry.brookscole.com/tro3.

- The Food Pyramid
  http://www.mypyramid.gov/

- Counting Calories
  http://www.caloriecontrol.org/
- U.S. Food and Drug Administration (FDA)
  http://www.fda.gov/
- Sodium Nitrite
  http://www.medem.com/MedLB/article_detaillb.cfm?article_ID=ZZZ80XEN0IC&sub_cat=380
- Additives
  http://www.cfsan.fda.gov/~lrd/foodaddi.html

Key Terms

Agent Orange	emulsifiers	lipoproteins	sodium
anticaking agents	essential amino acids	magnesium	starch
B complex vitamins	fats	major minerals	sucrose
bioamplification	fiber	minor minerals	vitamin A
calcium	fructose	oils	vitamin C
carbamate	glucose	organophosphate	vitamin D
cholesterol	humectants	phenoxy herbicides	vitamin E
chromium	iodine	phosphorus	vitamin K
collagen	iron	polysaccharides	zinc
complete proteins	lactase	potassium	
copper	lactose	scurvy	
dioxins	lactose intolerance	selenium	

Exercises

 Assess your understanding of this chapter's topics with an online chapter quiz at **http://chemistry.brookscole.com/tro3**.

QUESTIONS

1. Human bodies contain approximately 10^{27} atoms. Where did all these atoms come from?
2. Briefly explain the digestion process.
3. What are carbohydrates?
4. Explain the differences among sugars, starches, and fibers.
5. Define monosaccharide, disaccharide, and polysaccharide.
6. Draw the structures of glucose and sucrose.
7. What happens to carbohydrates during digestion?
8. Describe the condition known as lactose intolerance.
9. Both starch and fiber are polysaccharides with glucose as individual units. What is the difference between them? What is the result of that difference?
10. Although fiber has no caloric content, it still plays an important role in the diet. Explain that role.
11. In a healthy diet, what percentage of total calories should come from carbohydrates?
12. What are proteins, and what functions do they serve in the body?
13. Explain how proteins are digested.
14. What kinds of foods are rich in proteins?
15. What are complete proteins? How can vegetarians obtain complete proteins?
16. What are fats and oils? What kinds of foods are high in fat?
17. What are the chemical and physical differences between saturated fats and unsaturated fats?
18. What are lipoproteins?
19. Why are high blood cholesterol levels dangerous?
20. Explain the difference between blood cholesterol level and blood LDL level.
21. In a healthy diet, what percentage of total calories should come from fat?

22. How does the first law of thermodynamics relate to food intake?
23. Explain the differences in caloric content between fats, carbohydrates, proteins, and fiber.
24. What is basal metabolism? How much energy does it consume?
25. Why does the body use fat as long-term energy storage?
26. What are vitamins?
27. What are the fat-soluble and water-soluble vitamins? For which class of vitamins is overconsumption a bigger problem? Why?
28. For each of the fat-soluble vitamins, explain its function in the body, the consequences of deficiency, and the consequences of overconsumption. Also, list the foods that are rich in the vitamin.
29. What is the role of vitamin C in the body? What are the results of vitamin C deficiency? What kinds of foods are rich in vitamin C?
30. List the B complex vitamins.
31. What are the roles of the B complex vitamins? What are the results of deficiency? What kinds of foods are rich in B complex vitamins?
32. What are nutritional minerals?
33. For Ca, P, I, Fe, and Zn, list the mineral's role in the body, the consequences of its deficiency, and the foods that are rich in the mineral.
34. What is the role of sodium in the body? What are the dangers associated with its overconsumption?
35. What are the roles of potassium and magnesium in the body? In what foods and liquids are these minerals found?
36. What are the roles of selenium, chromium, and copper in the body?
37. Why are additives often added to food? What government agency monitors these additives?
38. What is the GRAS list?
39. List several antimicrobial agents and explain their function as food additives.
40. List three antioxidants and explain their function as food additives. What are some antioxidants sold as dietary supplements?
41. What are the roles of sulfites and EDTA in food?
42. What are the roles of artificial colors in foods? Why are they so controversial?
43. List some artificial flavors and flavor enhancers that are present in foods. Why are they added to food?
44. What are the roles of humectants, anticaking agents, and emulsifiers in food?
45. What are the primary nutrients for plants?
46. Why do plants need nitrogen? Why are plants unable to use atmospheric nitrogen?
47. What industrial process is used to fix nitrogen?
48. In what form do plants usually absorb phosphorus?
49. What are the secondary nutrients and the micronutrients for plants?
50. Why were the chlorinated hydrocarbons hailed as the "dream" insecticides? What problems developed with their use?
51. Describe bioamplification.
52. How do modern insecticides differ from their chlorinated hydrocarbon predecessors?
53. What is the difference between broad-spectrum and narrow-spectrum insecticides? Give some examples of each.
54. What is Agent Orange? How was it used? Why was its use banned?
55. List some modern herbicides and their uses.

PROBLEMS

56. How many calories per day are used in basal metabolism by a 140-lb person?
57. Calculate the number of calories you burn per day due to basal metabolism alone. Use your own weight in this calculation.
58. If a person who expends 2200 Cal/day wants to lose 2 lb in 1 week, what should his or her caloric intake be?
59. If a person who expends 2800 Cal/day wants to lose 6.0 lbs in one month (30 days), what should his or her daily caloric intake be?
60. Determine the percentage of calories from fat for each of the following foods:

Food	Protein (g)	Carbohydrate (g)	Fat (g)
a. Vons tortilla chips	2	16	7
b. Nutri-Grain cereal	2	16	7
c. spaghetti	7	42	1

61. Determine the percentage of calories from fat for each of the following foods:

Food	Protein (g)	Carbohydrate (g)	Fat (g)
a. Little Debbie choco-cakes	2	37	13
b. Snickers bar	4	35	14
c. Jerseymaid peach yogurt (low fat)	9	47	2.5

62. Calculate the number of calories burned per day for a 145-lb person. Assume that he or she sleeps 8 hours per day, strenuously exercises (500 Cal/h) for 30 minutes per day, and burns approximately 100 Cal/h for the rest of the time.
63. Calculate the total number of calories you burn per day. Use basal metabolism for sleeping hours, and between 90 and 150 Cal/h, depending on your weight and activity level, for your waking hours. If you exercise regularly, include the number of calories burned during your exercise. (Moderate exercise uses about 300–500 Cal/h while strenuous exercise uses 500–700 Cal/h.)
64. Use your answer to problem 63 to calculate your caloric intake if you want to *lose* 3 lb in 1 month.
65. Use your answer to problem 63 to calculate your caloric intake if you want to *gain* 3 lb in 1 month.

66. A person who burns 2450 Cal/day eats 2625 Cal/day. How will the person's weight change over a year?
67. A person eats 2000 Cal/day and burns 2200 Cal/day. How long will it take for the person to lose 5.0 lbs?
68. As a person gains weight, the amount of energy required to maintain basal metabolism increases. How many extra calories does it take per day to maintain basal metabolism for a 200-lb person versus a 150-lb person?
69. How many extra calories does it take per day to maintain the basal metabolism of a 250-lb person versus a 170-lb person?
70. The RDA for vitamin D in young adults is 10 μg/day. One cup of vitamin D fortified milk contains 2.5 μg of vitamin D. How many cups of vitamin D fortified milk contain the RDA for vitamin D?
71. The RDA for vitamin D in young adults (19–24 yr) is 10 μg/day. Eggs contain about 0.65 μg of vitamin D. Approximately how many eggs contain the RDA for vitamin D?

POINTS TO PONDER

72. Vegetarians avoid meats for a number of reasons, including health and respect for other living things. Construct an argument for a meat-eating diet based on health. Would such a diet place limits on the amount of meat that should be consumed?
73. The selling of vitamins has become a major industry. Do you think vitamin supplements are necessary for good health? Why or why not?
74. What is ironic about health food stores that often sell antioxidants such as BHA and BHT on the shelves next to foods that stress they are additive free?
75. DDT was used for years and saved many lives by killing disease-carrying pests. How did the benefits of using DDT become overshadowed by its risks? What are the benefits and risks of some of the modern insecticides? Do you think the benefits are worth the risks?

Feature Problems and Projects

76. Space-filling diagrams for the structures of vitamin A, vitamin D, and vitamin C are shown below. Which are fat-soluble? Which are water-soluble? What is the difference between the structures of water-soluble and fat-soluble vitamins?

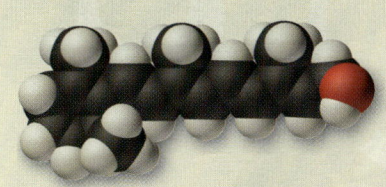

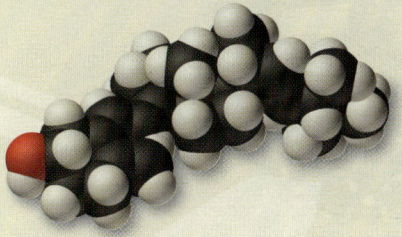

77. Make a list of the foods you ate yesterday. Use the calorie calculator at http://www.caloriecontrol.org to find the calories in each of the foods you ate and calculate the total number of calories you consumed.
78. Use your height and weight to calculate your body mass index (BMI) at http://www.caloriecontrol.org/bmi.html. A BMI over 25 is considered overweight.

19 Nanotechnology

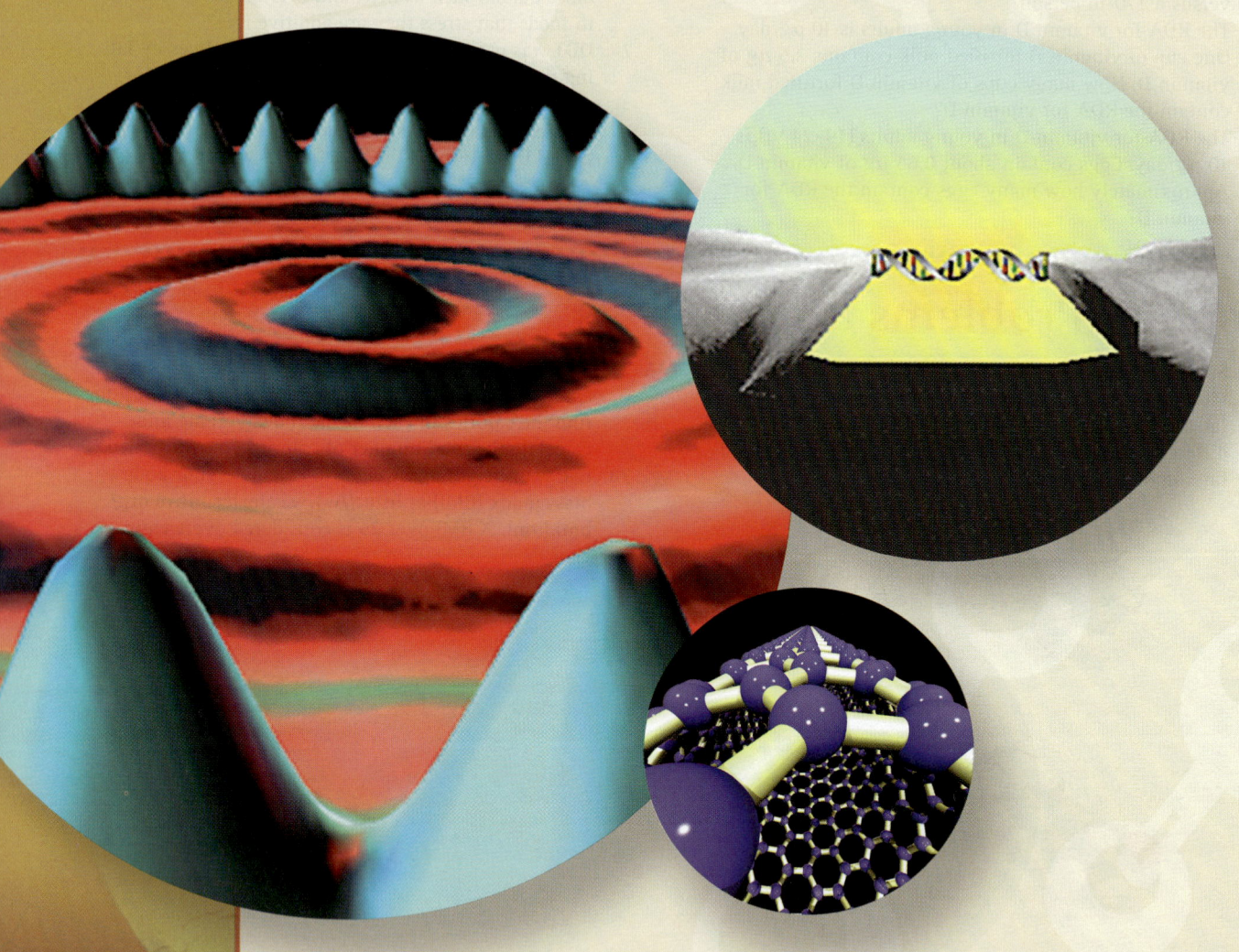

The principles of physics, as far as I can see, do not speak against the possibility of maneuvering things atom by atom. It is not an attempt to violate any laws; it is something, in principle, that can be done; but in practice, it has not been done because we are too big.

—Richard Feynman

In this chapter, we explore nanotechnology, the idea of building things on a small—very small—scale. The idea was first suggested by Richard Feynman in a 1959 speech entitled, "There's Plenty of Room at the Bottom." In that speech, Feynman outlined a new field of technology where dimensions are measured in nanometers (10^{-9} m) and where devices are constructed not from the top down, but from the bottom up, placing each atom in the exact desired location. Such technology, if realized, could revolutionize electronics, computing, and medicine. As you read this chapter, think about the future—how small can computers become? How much information will storage devices, such as hard drives and CD-ROMs, contain? Will medicine continue on its course toward being less invasive and more precise? The field of nanotechnology holds promise to affect each of these areas. How might it affect you?

QUESTIONS FOR THOUGHT

- How much smaller can things—machines, information, devices—get?
- How can making things smaller improve current technologies?
- How do scientists image individual atoms and molecules?
- What are buckyballs? What are nanotubes? What can they be used for?
- How can nanotechnology improve medicine?
- What are the problems with nanotechnology?

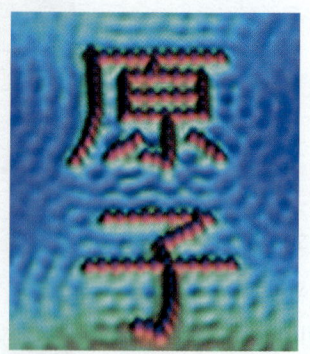

The Kanji characters for the word atom, written with iron atoms on copper and imaged with a scanning tunneling microscope at IBM laboratories. Each red bump is an iron atom that has been moved into the proper place. The blue bumps in the background are copper atoms.

 Explore this topic on the **There's Plenty of Room at the Bottom** website.

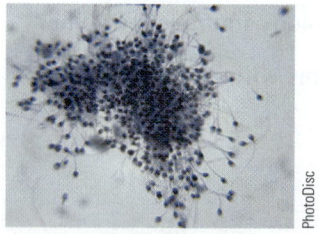

Sperm cells are similar to tiny submarines.

19.1 Extreme Miniaturization

We have learned that atoms are small. Five atoms laid end to end measure about a nanometer (10^{-9} m), an unimaginably small unit of measurement. A human hair, for example, measures about 20,000 nanometers in diameter. Nanotechnology is based on the idea that if you build things one atom at a time, things can be made smaller—much smaller—than they are now.

How small can things get? How small can we make a functional machine? To get an idea of how small things can get, consider a simple question: How small can this book be? What is the minimum space into which we can fit the information of this textbook? Figure 1 shows the initials IBM written with 35 atoms. The company's initials were written and imaged using techniques described later in this chapter. Notice that each letter requires about 12 atoms to write. If each letter in this book were written with 12 atoms, the entire book (about 3 million letters) would easily fit into a space 1000th the size of the period at the end of this sentence. When Feynman, in his 1959 speech, said that there was plenty of room at the bottom, he meant plenty. We live in a big world, and theoretically at least, things can be made much smaller.

The advantages of extreme miniaturization, however, are more than merely compactness. As nice as it would be to carry a thousand copies of *Chemistry in Focus* in a single dot, it would not be very useful. But what if we could make machines so tiny that they could navigate the bloodstream, for example? Such a machine, or army of machines, could be injected into an artery and could be programmed to search for and destroy fatty deposits on arterial walls, or they could be programmed to find cancer cells and kill them. What if surgery involved swallowing the surgeon? Only this surgeon is not human, but a tiny robot that performs the surgery from within your body. The ability to make tiny machines could revolutionize not only medicine but a variety of fields.

Is it really possible to make such tiny machines? We know that it is at least theoretically possible because it has already been done in living organisms, which are filled with tiny structures that in many respects function like micromachines. Sperm cells, for example, are similar to tiny submarines, moving through fluids with their flagellum motor. Almost all cells are a complex microfactory that reproduce themselves and fabricate everything that living things need to survive. Consequently, some see nanotechnology as bioimitation—can we construct structures that mimic living cells?

Recent advances have moved the idea of nanotechnology out of science fiction and into the laboratory. In the 1980s, scientists discovered the scanning tunneling microscope (STM), a microscope that can image and move individual atoms. Other discoveries of the eighties and nineties, such as buckyballs and buckytubes, showed that the kinds of structures from which we build large machines can also be realized on the atomic level. The new millennium has brought advances in which single molecules have been made to act as electrical conductors, electrical switches, and storage devices. Although many technical hurdles remain, these kinds of molecular devices may one day replace semiconductor-based solid-state microelectronics.

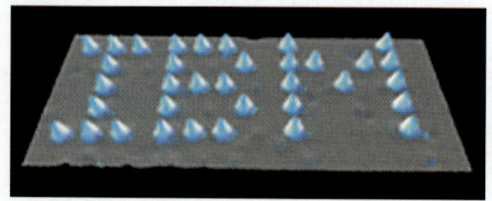

FIGURE 1 The letters IBM written with 35 individual xenon atoms and imaged with a scanning tunneling microscope at IBM laboratories.

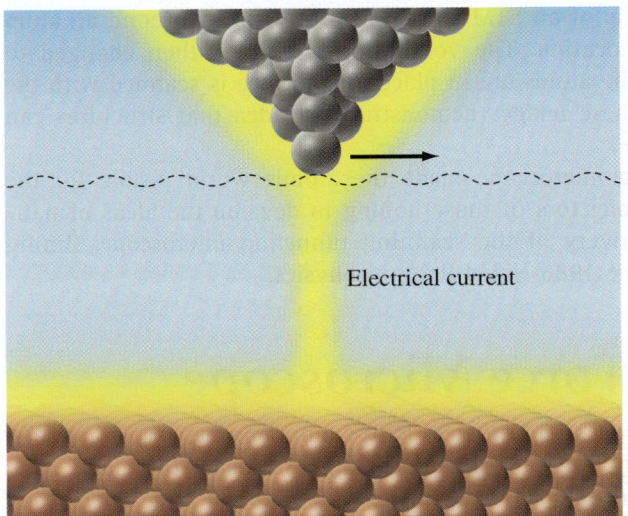

FIGURE 2 STM works by measuring the electrical current between an atomically fine tip and a metal surface.

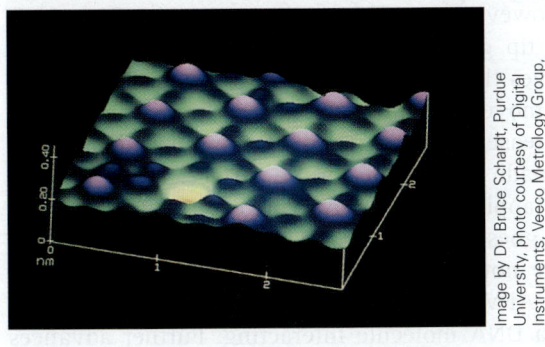

FIGURE 3 Iodine atoms adsorbed on a platinum surface and imaged with a scanning tunneling microscope. The large pink atoms are iodine and the small blue ones are platinum. The yellow spot is a vacancy, a spot on the surface that could hold another iodine atom but is not currently occupied.

19.2 Scanning Tunneling Microscope

In 1981, Gerd Binnig and Heinrich Rohrer, working at the IBM Zurich Research Laboratory, discovered a new kind of microscope with the ability to image individual atoms. The scientists were making measurements of how the electrical conductivity of metal surfaces varied from one region of the surface to another. To do this, they brought a thin wire close to a metal surface (Figure 2) and measured the electrical current between the wire tip and the surface. As they moved the tip across the surface, they noticed "bumps" in their measurements—those bumps turned out to be individual atoms. By scanning the tip back and forth over a small area, they found they could get an image of that area with atomic resolution. Modern scanning tunneling microscopes (STMs) work by moving an atomically fine metallic tip back and forth across a surface of interest. As the tip moves up and down over individual atoms, a computer records the movements and displays the result as an image of the surface—an image with individual atoms resolved (Figure 3).

Since then, scientists have used STM not only to image atoms but to move them. For example, Figures 1 and 4 show images of atomic patterns on surfaces.

 Explore this topic on the **Nanotechnology and Electronics** website.

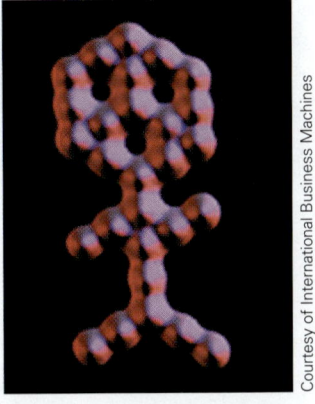

FIGURE 4 A cartoon of a person drawn with individual CO molecules on a platinum surface and imaged with a scanning tunneling microscope. Each bump is the end of a single carbon monoxide molecule.

Explore this topic on the STM Nobel Prize website.

 Explore this topic on the Scanning Probe Microscopy website.

To make these images, the tip of an STM is used as a tiny spear to bond an atom and move it to a desired location. The voltage on the tip is then changed to release the atom. Once all the atoms are in place, the surface is scanned with the STM to create an image. These images demonstrate the idea that structures can be made one atom at a time.

The scanning tunneling microscope made the atomic world visible for the first time, and it is the premier tool of those hoping to develop the ideas of nanotechnology. For their discovery of the scanning tunneling microscope, Binnig and Rohrer were awarded the 1986 Nobel Prize in physics.

19.3 Atomic Force Microscope

Scanning tunneling microscopes rely on electricity conduction between a metallic tip and the surface, so they can only image metallic surfaces. To image nonmetallic surfaces, scientists developed a technique called *atomic force microscopy* or *AFM*. In AFM, as in STM, an atomically fine tip attached to a cantilever is dragged across the surface (Figure 5). Instead of measuring the electrical current between the tip and the surface, however, an AFM monitors a laser reflected off the back of the cantilever. As the tip goes over an atom or molecule, the cantilever is deflected upward and moves the position of the laser beam. By tracking the laser movement as the tip is moved back and forth across the surface, an image is obtained. To further increase resolution and to minimize damage to samples, the tip can be tapped or made to oscillate on the surface. The amplitude of oscillation is measured as the tapping tip is moved across the surface. The amplitude gets smaller as the tip goes over atoms and bigger as it goes over vacancies. Tapping AFM can image not only nonmetallic surfaces but also biological samples. Figure 6 shows an AFM image of a DNA molecule, and Figure 7 shows two protein molecules and a DNA molecule interacting. Further advances have even allowed images to be taken quickly and sequentially, resulting in movielike video clips of atoms and molecules.

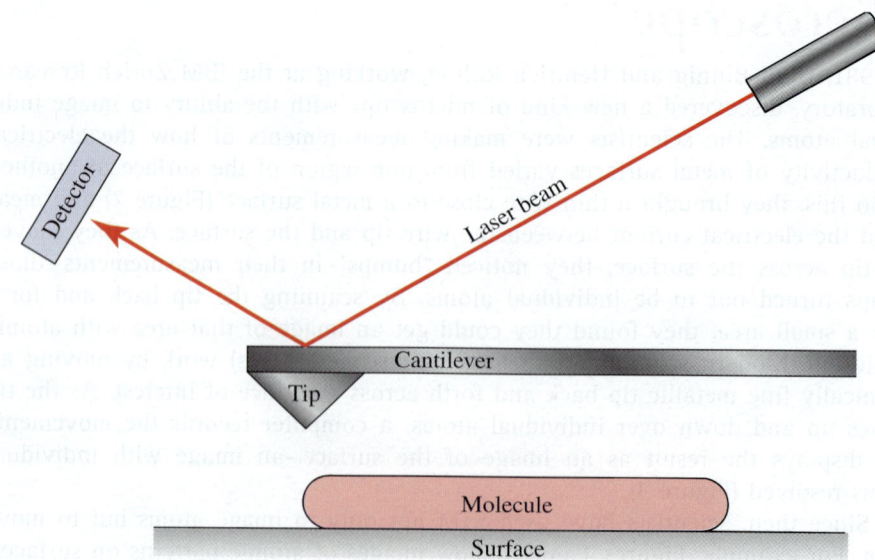

FIGURE 5 In AFM, a tip is dragged across a surface. As the tip is deflected by atoms or molecules, it causes a deflection in a laser beam, which is then detected and recorded to form an image.

Molecular Focus

Buckminsterfullerene

Formula: C_{60}
Molecular Mass: 720.66 g/mol
Melting point: 3527°C
Structure:

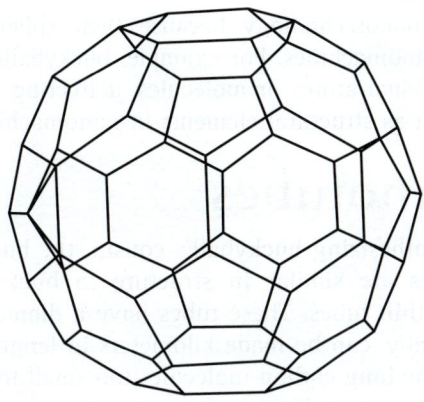

The structure of C_{60}. Double bonds have been omitted for clarity.

Three-dimensional structure:

Space-filling model of C_{60}.

Buckminsterfullerene was discovered in 1985 by Richard E. Smalley and Robert F. Curl, Jr., of the United States and Sir Harold W. Kroto of the United Kingdom. Its properties suggest that, if it can be made cheaply, it might find use as a lubricant, superconductor, or hard coating. Scientists have found evidence for the existence of C_{60} in interstellar space, in soot, and in some rock sediments. Buckminsterfullerene is currently being studied as a possible component of nanodevices.

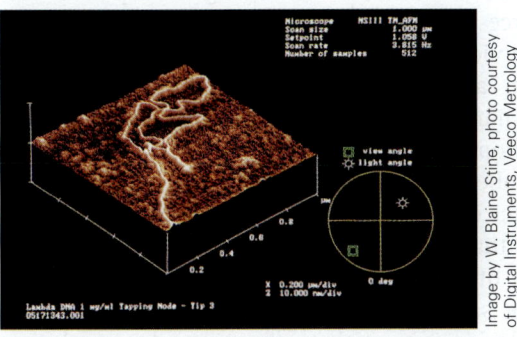

FIGURE 6 Lambda phage DNA on mica obtained by tapping-mode AFM.

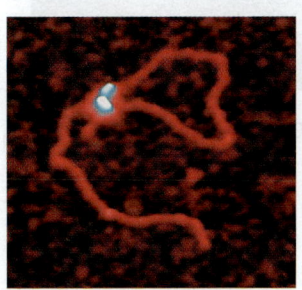

FIGURE 7 A tapping-mode AFM image of two transcription factor proteins (blue) interacting with a DNA chain (red).

19.4 Buckyballs—A New Form of Carbon

Before 1985, carbon was known to exist in only two forms, graphite and diamond. The two forms exist because carbon attaches to itself in two different ways. In graphite, carbon atoms form layered sheets, while in diamond they form a three-dimensional honeycomb-like pattern. In 1985, Richard E. Smalley and

 Explore this topic on the **Buckminsterfullerene** website.

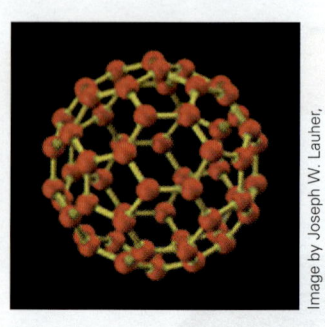

FIGURE 8 A model of C_{60}, buckminsterfullerene, or buckyball for short.

Robert F. Curl, Jr., of the United States and Sir Harold W. Kroto of the United Kingdom discovered a new form of carbon in which 60 carbon atoms bond together to form hollow spheres (Figure 8). This new form of carbon was named buckminsterfullerene, after the American architect R. Buckminster Fuller whose geodesic designs resembled the structure of C_{60}. Since the structure of buckminsterfullerene resembles a soccer ball, C_{60} molecules are sometimes called *buckyballs*. Buckyballs are very stable and have a number of interesting properties such as the ability to trap other atoms or molecules within their cagelike structure. Smalley, Curl, and Kroto were awarded the 1996 Nobel Prize in chemistry for their discovery of buckminsterfullerene.

Buckyballs are important to nanotechnology because their spherical shape might be a useful component of nanomachines. For example, buckyballs could act as nanocarriers (Figure 9) for individual atoms or molecules, delivering their cargo to specific targets, or they could act as structural elements in nanomachines.

19.5 Carbon Nanotubes

In 1991, chemists succeeded in synthesizing buckyball's cousin, the buckytube or nanotube (Figure 10). These tubes are similar in structure to buckyballs, but instead of spheres, they are long, thin tubes. These tubes have a diameter of just a few atoms, but at least theoretically, can be made kilometers in length. Imagine a single molecule—a nanotube is one long carbon molecule—too small to see, but a kilometer in length. Furthermore, these molecules are very strong—as strong as steel—and can be made to conduct electricity. So nanotubes have become a focus of nanotechnology, either as tiny rods in microscopic structures or as tiny wires that may one day carry electricity from one part of a nanomachine to another.

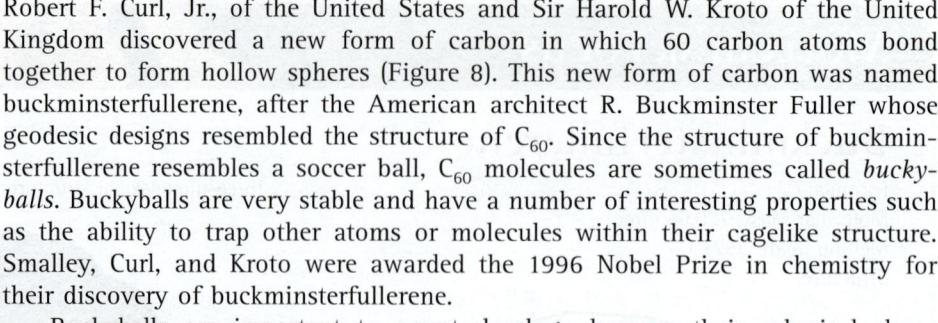

FIGURE 9 A computer representation of a buckyball with an atom trapped inside.

Figure 11 shows a carbon nanotube acting as a tiny scale to weigh objects as small as viruses. In this application, the nanotube is shaken with and without the object of interest attached; the difference in the deflection of the nanotube gives the weight of the attached object. Using this nanobalance, scientists have weighed particles as light as 20 femtograms (one femtogram is one quadrillionth of a gram).

Figure 12 shows the results of a carbon nanotube acting as a pencil. The image was taken with an atomic force microscope after using a nanotube tip to write on a silicon surface. Each individual line is about 10 nanometers across. The ability to write information in a very small space could, for example, increase the capacity of computer hard drives and other information storage technologies such as CD-ROMs and floppy disks.

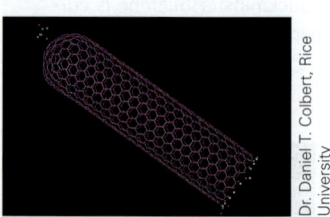

FIGURE 10 A nanotube.

Figure 13 shows a kinked nanotube acting as a wire connecting two parallel electrodes. Because nanotubes conduct electricity, they can be used to make tiny circuits, hundreds of times smaller than current technologies. Other nanotube applications include the desalination of water and flat-panel displays. Desalination of water is achieved by flowing salt water through an electrically charged nanotube forest; the sodium and chloride ions stick to the nanotubes, and pure water flows out. Flat-panel displays are made by using nanotubes as electron emitters to form glowing pixels on a display. Because the nanotubes are flexible,

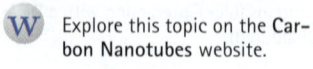

Explore this topic on the **Carbon Nanotubes** website.

FIGURE 11 A carbon nanotube acting as a nanobalance to weigh a carbon sphere with a mass of 22 femtograms (1 femtogram is 10^{-15} g).

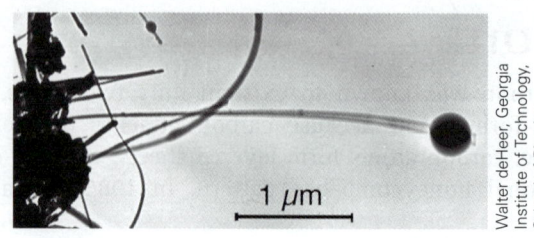

the display could also be made flexible. Imagine a computer monitor that you can roll up and carry away.

19.6 Nanomedicine

Nanotechnology has several applications in medicine. Among the simplest is a way to supply living tissues with beneficial foreign material that would normally be rejected by the immune system. For example, in a diabetic, pancreatic cells do not make sufficient insulin, a molecule that regulates blood-sugar levels. One remedy is to supply the patient with new living cells that do make insulin. These cells normally come from a compatible nonhuman source such as a pig. When such cells are injected into the body, however, they are recognized by the immune system as foreign and destroyed. Using the ideas of nanotechnology, scientists have constructed membranes with nanometer-sized pores (Figure 14). The pores are large enough for molecules such as sugar and insulin to pass through but small enough to prevent large antibodies from passing. By encapsulating the foreign, but beneficial, cells in these membranes, scientists can smuggle them into the body, beyond the reach of antibodies. Once in the body, these cells could synthesize insulin in response to blood-sugar levels, much as healthy pancreatic cells do. The technology has succeeded in small animals, and trials in larger animals are planned.

Another, more distant, possibility is the construction of artificial cells—nanomachines that mimic living cells. Perhaps the simplest artificial cells would be ones that mimic red blood cells, the body's oxygen transporters. Nanovesicles, constructed perhaps out of carbon, could be pumped full of oxygen and injected into the bloodstream. If, for some reason, the cardiovascular system failed—say the patient had a heart attack—the vesicle could slowly release the stored oxygen, allowing tissues that would ordinarily die to remain alive until cardiovascular function is restored. Because the artificial cell could theoretically carry more oxygen than a red blood cell, the person might be kept alive without brain damage for hours. Other artificial cells—including, for example, insulin-producing nanomachines for diabetics—may also be possible some day.

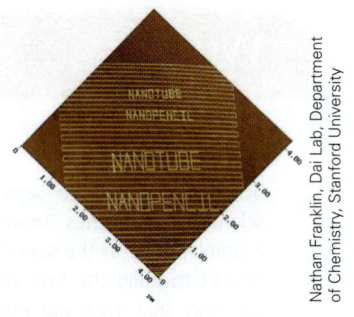

FIGURE 12 A nanotube acting as a pencil. This AFM image was taken after using a nanotube to write on a silicon surface.

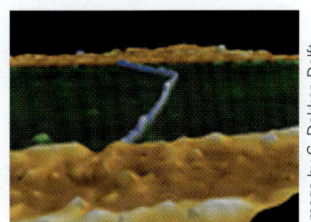

FIGURE 13 A carbon nanotube (blue) connecting two electrodes (gold). The image was taken with an AFM.

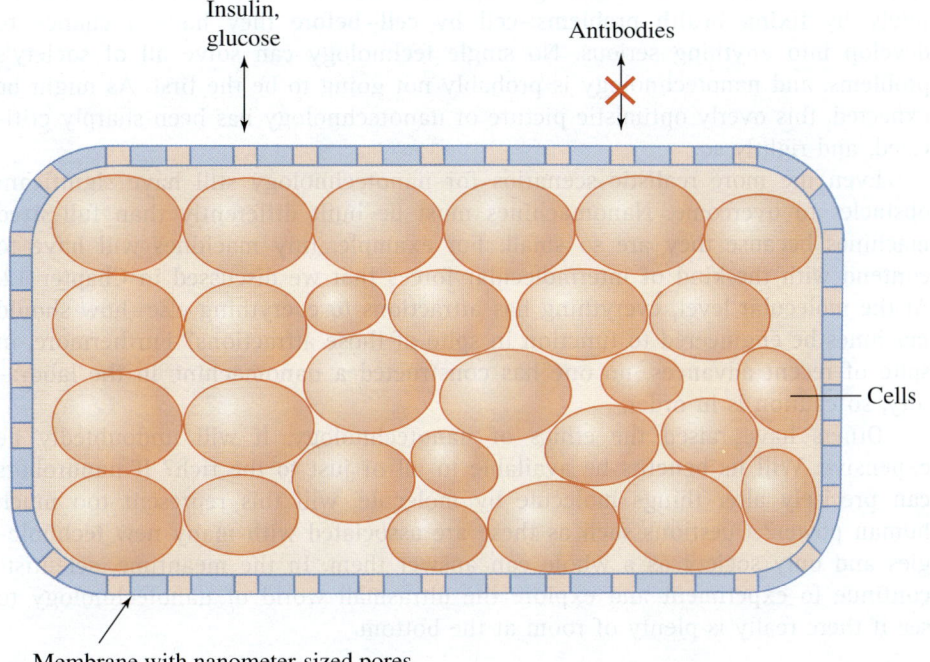

FIGURE 14 Membranes with nanometer-sized pores can protect foreign cells from antibody attack while allowing other molecules such as insulin and glucose to pass.

What if...

Value-Free Science

Scientists have often separated the ethical implications of scientific discoveries from the discoveries themselves. For example, some of the scientists who worked on the Manhattan Project to build the first atomic bomb justified their work by claiming that *they* did not make the decision to drop the bomb. They were simply advancing knowledge—not deciding how it would be used. The idea that scientific knowledge is separate from the values or ethics associated with that knowledge is called *value-free science*. It gives the scientist tremendous freedom but has been criticized as giving the scientist too much freedom.

QUESTION: What do you think of value-free science? Does the scientist have an ethical responsibility to the knowledge he or she creates? Why or why not?

Even further out into the future, scientists envision the construction of nanorobots, tiny machines that could navigate the bloodstream destroying tumors, removing fatty deposits on arterial walls, and even repairing damaged cells. Scientists are currently working on using nanotechnology for targeted drug delivery. For example, a nano-sized device could infiltrate cells and detect cancerous changes. Upon detecting such changes, the device would release toxins that kill the cancerous cell. Such selective chemotherapy could protect healthy cells from the adverse effects of chemotherapy agents and allow a more concentrated delivery of toxins to cancerous cells.

19.7 Nanoproblems

Nanotechnology is not without its problems and critics. Some nanotechnology visionaries have perhaps gone too far in their speculations about what nanotechnology may provide. They describe a society where all problems—everything from aging to food supply—are solved by arrays of nanorobots rearranging atoms. In this scenario, nanorobots recycle garbage into food and keep people alive indefinitely by fixing health problems—cell by cell—before they have a chance to develop into anything serious. No single technology can solve all of society's problems, and nanotechnology is probably not going to be the first. As might be expected, this overly optimistic picture of nanotechnology has been sharply criticized, and rightly so.

Even the more realistic scenarios for nanotechnology still have significant obstacles to overcome. Nanomachines must be built differently than full-sized machines because they are so small. For example, tiny machines will have to contend with the kind of intermolecular forces that we discussed in Chapter 12. At the molecular level, everything has attractions to everything else—how should machines be engineered to function in spite of those attractions? Furthermore, in spite of recent advances, no one has constructed a nanomachine in the laboratory, so caution is in order.

Others have raised the ethics of nanotechnology. It will undoubtedly be expensive. Will its benefits be available to all or just to the rich? If nanorobots can precisely alter things molecule by molecule, will this represent too much human power? Questions such as these are associated with many new technologies and only society as a whole can answer them. In the meantime, scientists continue to experiment and explore the ultrasmall world of nanotechnology to see if there really is plenty of room at the bottom.

Chapter Summary

MOLECULAR CONCEPT

The idea that things can be made smaller by positioning atoms one at a time was first suggested by Richard Feynman in 1959 **(19.1)**. In the 1980s and 1990s, several discoveries made that idea more plausible. Chief among those was the discovery of the scanning tunneling microscope **(19.2)** and the atomic force microscope **(19.3)**. These microscopes had the ability to image and move individual atoms, an unprecedented achievement.

The discovery of buckminsterfullerene in 1985 and carbon nanotubes in 1991 provided additional evidence that things can be made small **(19.4)**. These tiny spheres and tubes could become the building blocks of tiny machines and electronic circuits. The nanotube, with its high strength and ability to conduct electricity, has already found numerous applications such as the nanobalance, the nanopencil, and nanowires **(19.5)**.

Even though there are still many obstacles to overcome, nanotechnology is progressing at a rapid rate and should yield new and interesting devices in the years to come **(19.6, 19.7)**.

SOCIETAL IMPACT

The ability to make things smaller has the potential to impact society in several ways. The advantages are beyond mere compactness. Making machines smaller carries with it the potential to put machines into places we could not before.

The ability to image things on the nanometer scale, combined with the discovery of nanometer-sized elements (buckyballs and nanotubes), has led to a frenzy of research to make nano-sized instruments and devices that will improve human life. Chief among these are medical devices that allow, for example, the introduction of foreign cells into the human body, the construction of artificial cells, and the construction of nanorobots that may one day be able to navigate within the human body.

As with all new technology, nanotechnology carries with it ethical and societal questions about its appropriate development and use.

Chemistry on the Web

For up-to-date URLs, visit the text website at **www.chemistry.brookscole.com/tro3**.

- There's Plenty of Room at the Bottom
  http://www.zyvex.com/nanotech/feynman.html

- Nanotechnology and Electronics
  http://www.ieee-virtual-museum.org/exhibit/exhibit.php?taid=&id=159270&lid=1&seq=7

- STM Nobel Prize
  http://nobelprize.org/physics/laureates/1986/

- Scanning Probe Microscopy
  http://www.eng.yale.edu/reedlab/research/spm/spm.html

- Buckminsterfullerene
  http://www.godunov.com/Bucky/fullerene.html
  http://en.wikipedia.org/wiki/Fullerene

- Carbon Nanotubes
  http://www.pa.msu.edu/cmp/csc/nanotube.html
  http://www.research.ibm.com/nanoscience/nanotubes.html
  http://www.nanotech-now.com/nanotube-buckyball-sites.htm

Exercises

 Assess your understanding of this chapter's topics with an online chapter quiz at **http://chemistry.brookscole.com/tro3**.

QUESTIONS

1. How big is a nanometer?
2. What is the main concept behind nanotechnology?
3. What are the advantages of extreme miniaturization?
4. What did the title of Feynman's 1959 talk, "There's Plenty of Room at the Bottom," mean?
5. Why might the ideas of nanotechnology be possible?
6. What does a scanning tunneling microscope do? How does it work?
7. When and by whom was the scanning tunneling microscope (STM) discovered?
8. How does an atomic force microscope (AFM) work?
9. Why is the tip of an AFM tapped or oscillated while taking an image?
10. When and by whom was buckminsterfullerene discovered?
11. What are the formula and structure of buckminsterfullerene?
12. What are nanotubes?
13. What are some of the things for which nanotubes can be used?
14. How can nanotechnology be used to smuggle foreign cells into the human body without causing attack by the immune system?
15. How would an artificial red blood cell work?
16. How could nanorobots be used in medicine?
17. What are some problems associated with nanotechnology?

PROBLEMS

18. A dust particle measures 14×10^{-6} m in diameter. What is its diameter in nanometers?
19. A red blood cell measures about 7.8×10^{-6} m in diameter. What is its diameter in nanometers?
20. If a nanotube measures 10 nm in diameter, how many, laid side by side, would fit within the width of a human hair? A human hair is 20×10^{-6} m wide.
21. How many 10 nm diameter nanotubes, laid side by side, would it take to make a line 0.10 mm in width?
22. Examine the structure of C_{60} shown in Figure 8. What two geometric shapes are formed by the carbon atoms?
23. Examine the nanotube structure shown in Figure 10. What single geometric shape is formed by the carbon atoms?

POINTS TO PONDER

24. One of the criticisms often used against new technologies is that only the rich can afford them. Should this discourage the pursuit of nanotechnology? Why or why not?
25. Read the What If box entitled "Value-Free Science." Do you think scientists should have the freedom to pursue scientific knowledge regardless of its potential applications? Why or why not?
26. In addition to those mentioned in this chapter, what uses can you think of for nano-sized machines?

Feature Problems and Projects

 27. A U.S. manufacturer of STMs and AFMs is Veeco Instruments in Santa Barbara, CA. Visit their website at **http://www.veeco.com**. Under "products," select AFM/SPM and then click on "nanotheater." View some of the STM and AFM images shown and print out two of your favorite ones.

28. Write a short science fiction story involving a visit to the doctor in the year 2050. Include nanotechnology in your story.

APPENDIX 1
Significant Figures

A1.1 Writing Numbers to Reflect Uncertainty

If you tell someone you had three eggs for breakfast, there is little doubt about what you mean. Eggs come in integral numbers, and three eggs means three eggs; it is unlikely that you actually had 3.4 eggs. On the other hand, if you tell someone that you had one and one-half cups of milk, the meaning is less obvious. The amount of milk you actually had depends on how precisely you measured it, which in turn depends on what you used to make the measurement. While measurements of milk or eggs are probably not very critical, scientific measurements, and the uncertainties associated with them, are. Scientists report measured quantities in a way that identifies their uncertainty. For example, suppose we wanted to report our milk intake in a way that reflected the precision of the measurement. If the measurement was rough, we might write 1.5 cup of milk. If, however, we had been careful and used a precise measuring cup marked to the nearest one-tenth of a cup, we would write 1.50 cup of milk. If the measuring cup was unusually accurate and marked to the nearest one-hundredth of a cup, we would write 1.500 cup. **In general, scientific numbers are reported so that every digit is certain except the last, which is estimated.** So in our measuring cup example, if we write 1.5 cup, we mean that we are certain it is not 2 cup, but it may be 1.4 cup or 1.6 cup. Likewise, if we write 1.50 cup, we are certain it is not 1.1 cup, but it may be 1.49 cup or 1.51 cup. The number of digits reported in a measurement are called **significant figures** (or *significant digits*) and represent how precisely the measurement was made. The greater the number of significant figures, the greater the certainty in the measurement.

A1.2 Determining the Number of Significant Figures in a Reported Number

When we see a written number, such as 0.0340, how do we know how many significant figures are represented? You may mistakenly say that the number has five significant figures, but it does not. To determine the number of significant figures in a number, follow these simple steps:

1. All nonzero digits are significant.

    2.006   0.0340

2. Interior zeros are significant.

    5.0304   290.87

3. Trailing zeros after a decimal point are significant.

    4.90   89.00

4. Zeros to the left of the first nonzero number are **not significant.** They serve only to locate the decimal point.

   The number 0.0003 has only one significant figure.

5. Zeros at the end of a number, but before a decimal point, may or may not be significant. This ambiguity can be avoided by using scientific notation.

   For example, does 350 have two or three significant figures? Write the number as 3.5×10^2 to indicate two significant figures, or as 3.50×10^2 to indicate three.

6. Exact numbers, expressed either by themselves or in an equivalence statement, have an unlimited number of significant digits.

   For example, 2 atoms means 2.000000 . . . atoms; 100 cm = 1 m means 100.00000 . . . cm = 1.0000000 . . . m.

> **EXAMPLE A1.1**
>
> ## Determining the Number of Significant Figures
>
> How many significant figures are in each of the following numbers?
>
> 0.0340
>   The trailing zero is significant, as are the 3 and the 4. The leading zeros mark only the decimal and are not significant. Therefore, the number has three significant figures.
>
> 1.05
>   All digits are significant for a total of three significant figures.
>
> 8.890
>   All digits are significant for a total of four significant figures.
>
> 0.00003
>   The leading zeros mark only the decimal point and are not significant. The 3 is the only significant digit, for a total of one significant figure.
>
> 3.00×10^2
>   All digits are significant for a total of three significant figures.
>
> **YOUR TURN**
>
> ## Determining the Number of Significant Figures
>
> How many significant figures are in each of the following numbers?
>
> 0.0087
> 4.50×10^5
> 45.5
> 164.09
> 0.98850

A1.3 Significant Digits in Calculations

Calculations involving measured quantities must be done with some care so as to not gain or lose precision during the calculation.

ROUNDING

When you are rounding numbers to the correct number of significant figures, round down if the last digit being dropped is 4 or less, and round up if the last digit being dropped is 5 or more. For example:

$3.864 \rightarrow 3.86 \quad 3.867 \rightarrow 3.87 \quad 3.865 \rightarrow 3.87$

MULTIPLICATION AND DIVISION

The result of a multiplication or division should carry the same number of significant digits as the factor with the least number of significant figures (s.f.). For example:

$3.100 \times 8.01 \times 8.977 = (222.91) = 223$
(4 s.f.) (3 s.f.) (4 s.f.)   (3 s.f.)

The intermediate answer, in parentheses, is rounded to three significant figures to reflect the three significant figures in the factor 8.01.

For division, we follow the same rule:

$3.344/5.6 = (0.59714) = 0.60$
(4 s.f.)(2 s.f.)   (2 s.f.)

EXAMPLE A1.2

Significant Figures in Multiplication and Division

Carry out the following calculation to the correct number of significant figures:

$0.98 \times 0.686 \times 1.2/1.397 = (0.57747) = 0.58$

We round the intermediate answer (in parentheses) to two significant figures to reflect the two significant figures in the least accurately known quantities (1.2 and 0.98).

YOUR TURN

Significant Figures in Multiplication and Division

Carry out the following calculation to the correct number of significant figures:

$3.45 \times 0.2007 \times 0.867 \times 0.008/2.8$

ADDITION AND SUBTRACTION

The result of an addition or subtraction should have the same number of decimal places as the quantity carrying the least number of decimal places. For example:

2.0175
4.98
$\underline{0.482}$
$(7.4795) = 7.48$

The intermediate answer, in parentheses, is rounded to two decimal places because 4.98, the quantity with the least number of decimal places, has two decimal places. For subtraction, we follow the same rule. For example:

4.7
$\underline{-2.324}$
$(2.376) = 2.4$

> **EXAMPLE A1.3**
>
> ### Significant Figures in Addition and Subtraction
>
> Carry out the following calculation to the correct number of significant digits:
>
> $$0.987 + 0.1 - 1.22$$
>
> The intermediate answer is -0.1333, which is rounded to -0.1 to reflect that the quantity with the least number of decimal places (0.1) has only one decimal place.
>
> **YOUR TURN**
>
> ### Significant Figures in Addition and Subtraction
>
> Carry out the following calculation to the correct number to significant digits:
>
> $$4.342 + 2.8703 + 7.88 - 2.5$$

Exercises

1. How many significant figures are in each of the following numbers?
   a. 9.87
   b. 100.34
   c. 10.907
   d. 0.02
   e. 4.5×10^3
2. How many significant figures are in each of the following numbers?
   a. 23.555
   b. 138.0005
   c. 0.0055
   d. 1.452×10^{27}
   e. 7.0
3. Perform the following calculations to the correct number of significant figures:
   a. $3.45 \times 8.9 \times 0.0034$
   b. $4.5 \times 0.03060 \times 0.391$
   c. $5.55/8.97$
   d. $(7.890 \times 10^{12})/(6.7 \times 10^4)$
   e. $(67.8 \times 9.8)/(100.04)$
4. Perform the following calculations to the correct number of significant figures:
   a. $89.3 \times 77.0 \times 0.08$
   b. $5.01 \times 10^5/7.8 \times 10^2$
   c. $4.005 \times 74 \times 0.007$
   d. $453/2.031$
   e. $9.5 \times 10^3/34.56$
5. Perform the following calculations to the correct number of significant figures:
   a. $87.6 + 9.888 + 2.3 + 100.77$
   b. $43.7 - 2.341$
   c. $89.6 + 98.33 - 4.674$
   d. $6.99 - 5.772$
   e. $3.8 \times 10^5 - 8.45 \times 10^4$
6. Perform the following calculations to the correct number of significant figures:
   a. $1459.3 + 9.77 \div 4.32$
   b. $0.004 + 0.09879$
   c. $432 \div 7.3 - 28.523$
   d. $2.4 - 1.777$
   e. $0.00456 \div 1.0936$

APPENDIX 2
Answers to Selected Exercises

Chapter 1
QUESTIONS

1. All natural phenomena in the world we can see are the result of molecular interactions we cannot see. Examples are numerous. Some suggestions are as follows:
   a. Ice melting to water.
   b. A match being struck producing a flame.
   c. A shirt fading when exposed to light or too many washings.

3. a. The bright color of the carpet is the result of particular molecules in the rug. When particles of sunlight—photons—hit the carpet, the bright color molecules are destroyed or altered in some way.
   b. Because water molecules are attracted to salt molecules, the water breaks up the salt crystals and surrounds the individual salt molecules, a process observed as dissolving.

5. The scientific method involves first making **observations** of nature, from which **patterns** are identified. From the patterns, broadly applicable generalizations called **scientific laws** are established. A **theory** or model is then constructed to provide an interpretation of the behavior of nature. The theory is then tested by further experiments and modified if necessary.

7. Science and art are similar mostly in their creativity and observations of the world. The difference lies in what they do with their observations and how they are judged. Scientists take their observations and create a model of reality that is judged by experimentation for its validity. Artists observe the world and create a painting or sculpture that is judged by its creativity and workmanship.

9.
Galileo	Inquisition
Democritus	*atomos*
John Dalton	the atomic theory
Andreas Vesalius	human anatomy
Empedocles	four basic elements
Joseph Proust	constant composition
Copernicus	Sun-centered universe
Ernest Rutherford	the nuclear atom
Thales	all things are water
Antoine Lavoisier	conservation of mass
Robert Boyle	criticized idea of four Greek elements

11. The scientific revolution began in 1543, signaled by the publication of two books. The first book, by Copernicus, announced his Sun-centered universe theory. The second book, by Vesalius, gave an accurate description of human anatomy. The reason these books mark the beginning of the scientific revolution is the methods Copernicus and Vesalius used to learn about the natural world—they both used observation instead of pure reason.

13. A pure substance cannot be separated into simpler substances by physical means, such as filtration, chromatography, crystallization, or distillation. A mixture contains two or more pure substances, and can be separated into those individual components by selection of the appropriate physical methods.

15. The three states of matter are distinguished by the strength of the interactions between the molecules relative to their thermal energy. (For the sake of brevity we will refer here only to molecules, but it should be remembered that substances may be composed of atoms, molecules, or ions.) The relative strength of the molecular interactions decreases in the sequence solid–liquid–gas. In a solid, the interactions are strong enough to prevent free movement of the molecules, and they are locked into rigid arrangements (lattices). In a liquid, the molecular energies are strong enough to loosen the grip of the intermolecular forces sufficiently to allow free movement of molecules, but the intermolecular forces still maintain enough control to keep the molecules together. In a gas, molecular energies have increased to the extent that now all intermolecular shackles are broken and gas molecules behave completely independently. This molecular picture explains the obvious properties of solids, liquids, and gases. Solids are rigid and dense; liquids flow freely yet have a density similar to the solid; gases have a very low density and are confined only by their container.

17. John Dalton, using the laws of Lavoisier and Proust and the data from his own experiments, combined a number of ideas to formulate the **atomic theory**. Dalton's atomic theory was based on three parts:
- First, each element is composed of particles called atoms, which cannot be created nor destroyed.
- Second, all atoms of the same element have the same weight and other properties. These properties are unique characteristics of each element, and thus differ from other elements.
- Third, atoms of different types can combine to form compounds in simple whole number ratios. For example, the compound carbon dioxide is formed from one carbon atom and two oxygen atoms. The numbers 1 and 2 are simple whole numbers.

19. The only way to explain the results of the gold foil experiment was to propose a new model of the atom in which most of the atom must be empty space. This structure would allow most of the alpha particles to pass through the gold foil with little or no deflection. However, the atom must also contain a nucleus, a dense positively charged central core containing most of the mass. In the experiment, whenever an alpha particle came close to a nucleus, or hit it head on, it experienced a large repulsive force causing it to be scattered. Furthermore, since the atom is electrically neutral, it must contain an equal number of negative charges (electrons) and positive charges. The exact identity of the positive charge was later established to be the proton. Rutherford purposed the electrons were outside the nucleus.

PROBLEMS

21. a. "Two grams of hydrogen combine with sixteen grams of oxygen to form eighteen grams of water" represents an observation. It would be the result of one experiment.
    b. "Chlorine and sodium readily combine in a chemical reaction that emits much heat and light" is also an observation.
    c. "The properties of elements vary periodically with the mass of their atoms" is a law. The relationship between element properties and size statement was derived from examination of many observations. It is often referred to as the periodic law.

23. a. A silver coin is composed of the element silver.
    b. Air is a homogeneous mixture of different gases.
    c. Coffee is a homogeneous mixture of many different substances.
    d. Soil is a heterogeneous mixture since it may be composed of dirt, sand, and rocks that can be separated into distinct regions with different compositions.

25. a. Physical property. The bending of copper does not involve a composition change.
    b. Chemical property. Rusting of the iron indicates combination of iron with atmospheric oxygen and formation of iron oxide.
    c. Chemical property. Combustion is a chemical process: the reaction of the substance with oxygen. Thus flammability is a chemical property.
    d. Physical property. The smell of a substance is due to its vapor. Conversion of a liquid into vapor is a physical process since there is no change in its composition. Of course, it could be argued that the complex process of detection of the smell involves numerous chemical processes in the body and brain.

27. Physical and chemical changes are distinguished by whether or not a change in composition occurs.
    a. The crushing of salt is a physical change. There is no change in composition.
    b. The rusting of iron represents chemical change. Rust results from combination of the iron with the element oxygen to form iron oxide.
    c. The burning of natural gas in a stove is a chemical change. Burning involves combination of oxygen in the air with the carbon and hydrogen atoms of the natural gas. The energy released in the reaction provides the all-important heat we use.
    d. The vaporization of gasoline is a physical change. All changes of state are physical changes: Intermolecular forces may be gained or lost, but there are no changes in the chemical bonds between the atoms of the substances. Gasoline in the vapor or in the liquid is still gasoline, and if it was condensed it would be indistinguishable from the original liquid.

29. When gasoline is burned in an automobile engine, it does so in the presence of oxygen gas. This chemical reaction of gasoline plus oxygen produces energy, water, carbon dioxide, and carbon monoxide gases. The mass of the products is equal to the mass of the gasoline and oxygen gas that you started with.

31. a. 6 grams of hydrogen react with 48 grams of oxygen to form 54 grams of water. This is consistent with the law of conservation of mass: 6 g + 48 g = 54 g.
    b. 10 grams of gasoline react with 4.0 grams of oxygen to form 9.0 grams of carbon dioxide and 5.0 grams of water. This is consistent with the law of conservation of mass: 10 g + 4.0 g = 9.0 g + 5.0 g.

33. The reaction must follow the law of conservation of mass.
    11 g sodium + 14 g chlorine → _ g sodium chloride + 2.0 g excess sodium
    25 g reactants = 25 g products
    Therefore the amount of sodium chloride formed must be 23 grams.

35. The law of constant composition says the ratio of carbon to oxygen must always be the same.
    a. $\dfrac{12 \text{ g carbon}}{32 \text{ g oxygen}} = 0.38$
    b. $\dfrac{4.0 \text{ g carbon}}{16.0 \text{ g oxygen}} = 0.25$ inconsistent, therefore incorrect
    c. $\dfrac{1.5 \text{ g carbon}}{4.0 \text{ g oxygen}} = 0.38$
    d. $\dfrac{22.3 \text{ g carbon}}{59.4 \text{ g oxygen}} = 0.38$

37. To preserve electrical neutrality in the Rutherford atom the number of electrons will be equal to the number of protons in the nucleus (the atomic number).
    a. An atom of lithium contains 3 electrons (atomic number of lithium = 3).
    b. An atom of magnesium contains 12 electrons (atomic number of magnesium = 12).

Chapter 2

QUESTIONS

1. Curiosity is an important part of the scientific enterprise because scientists need a strong desire to investigate and learn about the behavior of nature. Science must start with the question why. The scientific method is then utilized to accumulate systematized knowledge about the physical world. A scientist's curiosity is incapable of being satisfied. Without this curious nature of the scientist, the advancement of science would not have occurred as we presently know it.
3. Measured quantities are written so that the uncertainty is contained within the last digit of the number. A volume of 30.0 mL means the volume lies in the range 29.9 – 30.1 mL.
5. Any measurement consists of a numerical value and the chosen unit. Units inform what is being measured and the scale used. In science the International System of units, or SI, is generally used.
7. Answers may vary. Three possible units for mass are grams, milligrams, and kilograms. Examples:
   Mass of a penny–grams (g)
   Mass of a straight pin–milligrams (mg)
   Mass of a bucket of water–kilograms (kg)
9. Answers may vary. Three possible units for volume are milliliters, kiloliters (gallons), and liters. Examples:
   Volume of a child's juice box–milliliters (mL)
   Volume of water in a swimming pool–gallons or kiloliters (kL)
   Volume of a bottle of soda–liters (L)
11. Graphs are a very convenient and powerful way to illustrate relationships between different quantities. Graphs can be modified to emphasize particular features. It is important to examine the range on the y axis to understand the significance of the changes plotted.
13. Density is defined as mass per unit volume. Typical units for density are g/cm^3 (commonly used for solids) and g/mL (used for liquids).

PROBLEMS

15. a. 4.5×10^{-3} g
    b. 2.3×10^4 L
    c. 2.9979×10^8 m/s
    d. 3.5×10^{-7} m
17. a. 6,400,000 m
    b. 0.000000000079 m
    c. 270,000,000 people
    d. .25
19. a. 4.0075×10^7 meters
    b. 24,901 miles
    c. 1.31480×10^8 ft
21. 12 oz is 355 mL.
23. a. 6254 mm = 6.254 m
    b. 3.28 kg = 3280 g
    c. 2566 mg = 2.566×10^{-3} kg
    d. 0.0256 L = 25.6 mL
25. a. 25.2 in = 64.0 cm
    b. 106 ft = 32.3 m
    c. 5077 yd = 4.642 km
    d. 3.1 in = 79 mm
27. a. 1.339×10^4 ft^2
    b. 1.244×10^{-3} km^2
    c. 1.244×10^5 dm^2
29. a. There are 10^6 square meters (m^2) in 1 square kilometer (km^2).
    b. There are 2.83×10^4 cm^3 in 1 ft^3.
    c. There are 9 ft^2 in 1 yd^2.
31. 3.17 miles/L
    5.10 km/L
33. a. 2.5 ppm
    b. 0.25 ppm/yr
    c. 45%
    d. 4.5%/yr
35. Density of titanium = 4.50 g/cm^3
37. Density of glycerol = 126 g/cm^3
39. a. $417 \text{ mL} \times \dfrac{1 \text{ cm}^3}{1 \text{ mL}} \times \dfrac{1.11 \text{ g}}{1 \text{ cm}^3} = 463$ g
    b. $4.1 \text{ kg} \times \dfrac{1000 \text{ g}}{1 \text{ kg}} \times \dfrac{1 \text{ cm}^3}{1.11 \text{ g}} \times \dfrac{1 \text{ L}}{1000 \text{ cm}^3} = 3.7$ L
41. a. $V = \pi r^2 h = 3.14 \times (0.55 \text{ cm})^2 \times 2.85 \text{ cm}$
    $= 2.7 \text{ cm}^3$
    $d = \dfrac{m}{V} = \dfrac{24.3 \text{ g}}{2.7 \text{ cm}^3} = 9.0 \text{ g/cm}^3$
    b. The metal is copper.

FEATURE PROBLEMS AND PROJECTS

49. penny: 1.8 cm
    nickel: 2.0 cm
    dime: 1.6 cm
    quarter: 2.3 cm
    half-dollar: 2.9 cm
    dollar: 3.8 cm
    The general trend is that as the value of the coin increases, the diameter of the coin also increases. The dime is the only coin that does not fit the general trend; it is too small.

Chapter 3

QUESTIONS

1. Atoms are the building blocks of matter. If we are to understand the macroscopic properties of an element, we must first understand the microscopic properties of its atoms. Properties studied on an atomic scale correlate directly with properties of the element on a macroscopic scale.
3. An element is a substance that cannot be separated into simpler substances by any chemical methods. Although the heaviest naturally occurring element–92, uranium–is a fairly abundant element, suggesting that there are 92 naturally occurring elements, there are in fact only 90 that occur naturally. Number 43, technetium, and number 61, promethium, do not occur naturally on earth.
5. | Chemical Symbol | Name | Atomic Number |
    |---|---|---|
    | H | Hydrogen | 1 |
    | He | Helium | 2 |
    | Li | Lithium | 3 |
    | Be | Beryllium | 4 |
    | B | Boron | 5 |
    | C | Carbon | 6 |

N	Nitrogen	7
O	Oxygen	8
F	Fluorine	9
Ne	Neon	10
Na	Sodium	11
Mg	Magnesium	12
Al	Aluminum	13
Si	Silicon	14
P	Phosphorus	15
S	Sulfur	16
Cl	Chlorine	17
Ar	Argon	18
Fe	Iron	26
Cu	Copper	29
Br	Bromine	35
Kr	Krypton	36
Ag	Silver	47
I	Iodine	53
Xe	Xenon	54
W	Tungsten	74
Au	Gold	79
Hg	Mercury	80
Pb	Lead	82
Rn	Radon	86
U	Uranium	92

7.

	Mass (g)	Mass (amu)	Charge
Proton	1.67×10^{-24}	1	+1
Neutron	1.67×10^{-24}	1	0
Electron	9.11×10^{-28}	0	−1

9. Mendeleev's biggest contribution to modern chemistry was his placement of the elements on the periodic table and thus the periodic law. He realized that in arranging the elements in order by their atomic number "... certain sets of properties reoccur periodically." From this he predicted the chemical and physical properties of undiscovered elements and unknown compounds.

11. A complete answer would require a treatise, but we can describe the distinguishing features of the quantum model as follows. Energy is not continuous but can only be transferred in discrete amounts called quanta. Light, for example, is bundled (quantized) into packages called photons. The quantum model also recognizes the unique consequences of very small size: Small particles have wave-like properties that cannot be ignored (de Broglie wave-particle duality); their location and velocity cannot be specified with absolute certainty (the Heisenberg uncertainty principle). Atomic orbitals—regions of space where electrons are most likely, but not absolutely certainly, found—follow naturally from the quantum model.

The main difference between the two models is that, while Bohr considered the electrons to be traditional particles whose motion could be described by the classical mechanics of Newton, the quantum mechanical model treats the electrons as waves. The wave properties of electrons provide a logical explanation for the existence of allowed orbits in Bohr's atomic model.

13. There are seven elements that exist as diatomic molecules: Hydrogen (H_2), Nitrogen (N_2), Oxygen (O_2), Fluorine (F_2), Chlorine (Cl_2), Bromine (Br_2), and Iodine (I_2).

Problems

15. a. 19 protons, 18 electrons
    b. 16 protons, 18 electrons
    c. 20 protons, 18 electrons
    d. 35 protons, 36 electrons
    e. 13 protons, 10 elctrons

17. a. carbon: $Z = 6$ $A = 14$
    b. aluminum: $Z = 13$ $A = 27$
    c. argon: $Z = 18$ $A = 38$
    d. copper: $Z = 29$ $A = 65$

19. a. $^{60}_{27}Co$
    b. $^{32}_{15}P$
    c. $^{131}_{53}I$
    d. $^{35}_{16}S$

21. a. protons = 92
       electrons = 92
       neutrons = 146
    b. protons = 6
       electrons = 6
       neutrons = 8
    c. protons = 11
       electrons = 10
       neutrons = 12
    d. protons = 35
       electrons = 36
       neutrons = 46
    e. protons = 8
       electrons = 10
       neutrons = 8

23. a. B: $n = 1$, $2e^-$; $n = 2$, $3e^-$
    b. Si: $n = 1$, $2e^-$; $n = 2$, $8e^-$; $n = 3$, $4e^-$
    c. Ca: $n = 1$, $2e^-$; $n = 2$, $8e^-$; $n = 3$, $8e^-$; $n = 4$, $2e^-$
    d. F: $n = 1$, $2e^-$; $n = 2$, $7e^-$
    e. Ar: $n = 1$, $2e^-$; $n = 2$, $8e^-$; $n = 3$, $8e^-$

We would predict fluorine (F) to be the most reactive element because it requires only one electron to fill the second shell, achieving a noble gas configuration. By the same token, argon will be the least reactive because its outermost shell is already filled, and therefore it will be reluctant to add or subtract electrons.

25. Number of valence electrons in elements in problem 23 are as follows:
    a. B: 3 valence electrons
    b. Si: 4 valence electrons
    c. Ca: 2 valence electrons
    d. F: 7 valence electrons
    e. Ar: 8 valence electrons

27. a. Li
    b. C

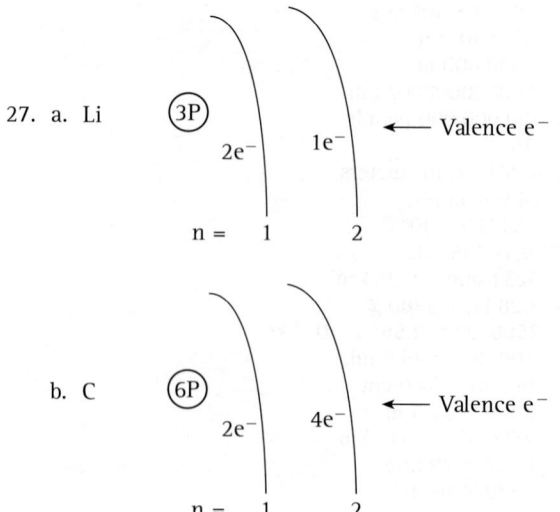

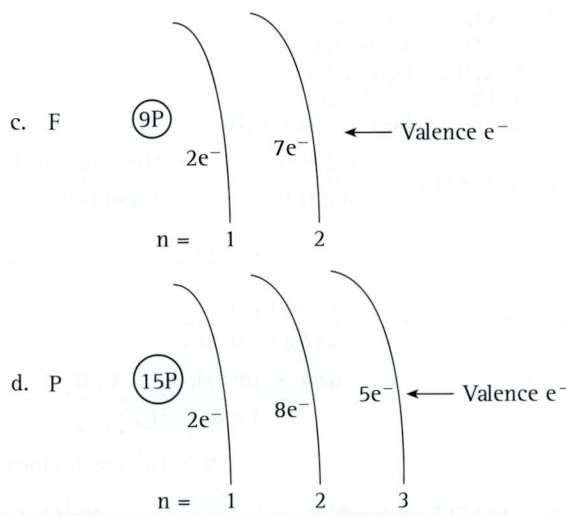

29. Mg and Ca are the most similar because they are both in the same group on the periodic table; they are both alkali earth metals.
31. Cl^- $n = 1$ $2e^-$
    $n = 2$ $8e^-$
    $n = 3$ $8e^-$
    The chloride ion (Cl^-) is more stable (less reactive) than the neutral chlorine atom (Cl) since the ion has an octet of electrons in its outer shell. The same criteria can be used to compare the reactivities of sodium (Na) and sodium ion (Na^+). The sodium ion is less reactive since it has an octet of electrons in its outer shell.
33. a. Cr—metal
    b. N—nonmetal
    c. Ca—metal
    d. Ge—metalloid
    e. Si—metalloid
35. Atomic weight = (fraction of isotope 1 × mass of isotope 1) + (fraction of isotope 2 × mass of isotope 2)
    Atomic weight = (0.9051 × 19.992 amu)
                  + (0.0027 × 20.993 amu)
                  + (0.0922 × 21.991 amu)
    Atomic weight of Ne = 20.18 amu
37. a. The percents of all the isotopes must add up to a total of 100%. Therefore, if the abundance of isotope 1 is 33.3% then the natural abundance of isotope 2 = 100% − 33.7% = 66.3%.
    b. Let x = mass of isotope 1.
    Atomic mass = (fraction of isotope 1 × mass of isotope 1) + (fraction of isotope 2 × mass of isotope 2)
    Substitute in all the numbers including x for the mass of isotope 1 and solve for x.
    29.5 amu = (0.337x) + (0.663)(30.0 amu)
    29.5 amu − 19.9 amu = 0.337 x
    $\frac{9.6 \text{ amu}}{0.337} = x$
    x = 28.5 amu = mass of isotope 1
39. 138 g Cu × $\frac{1 \text{ mol Cu}}{63.55 \text{ g Cu}}$ = 2.17 moles of Cu
41. a. 4.8 kg of Au × $\frac{1000 \text{ g Au}}{1 \text{ kg}}$ × $\frac{1 \text{ mol Au}}{196.97 \text{ g Au}}$
    = 24 moles of Au

b. 56.8 g of Na × $\frac{1 \text{ mol Na}}{22.99 \text{ g Na}}$ = 2.47 moles of Na

c. 7.9 × 10^{24} C atoms × $\frac{1 \text{ mol C}}{6.022 \times 10^{23} \text{ C atoms}}$
   = 13 moles of C

d. 4.5 × 10^{22} He atoms × $\frac{1 \text{ mol He}}{6.022 \times 10^{23} \text{ He atoms}}$
   = 0.075 moles of He

43. 38.7 g of Ag × $\frac{1 \text{ mol Ag}}{107.9 \text{ g Ag}}$ × $\frac{6.022 \times 10^{23} \text{ atoms Ag}}{1 \text{ mol Ag}}$
    = 2.16 × 10^{23} Ag atoms

45. 1.8 cm³ × $\frac{19.3 \text{ g Au}}{1 \text{ cm}^3}$ × $\frac{1 \text{ mol Au}}{196.97 \text{ g Au}}$
    × $\frac{6.022 \times 10^{23} \text{ Au atoms}}{1 \text{ mol Au}}$
    = 1.06 × 10^{23} atoms of Au

47. First, find the volume of the sphere.
    V = (4/3)π r³
    V = (4/3)π (3.4 cm)³ = 164.5 cm³
    Second, using the density find the grams of iron (Fe) and finally the number of atoms.

    164.5 cm³ × $\frac{7.86 \text{ g Fe}}{1 \text{ cm}^3}$ × $\frac{1 \text{ mol Fe}}{55.85 \text{ g Fe}}$
    × $\frac{6.022 \times 10^{23} \text{ atoms Fe}}{1 \text{ mol Fe}}$
    = 1.39 × 10^{25} atoms of Fe

Feature Problems and Projects

57. The atomic weight of element (a) is less than (b), and the atomic weight of element (c) is greater than (b). There are 84 atoms in 175 grams of element (b).

Chapter 4

Questions

1. Atoms are rarely found in nature in an uncombined state. Because most atoms do not have complete octets in their outer shell, they are very reactive. The result of their reactivity is the formation of compounds. All matter is made of atoms; the atoms are found within compounds and compounds combine to form mixtures. Therefore, it is consistent to say that all matter is made of atoms and that most common substances are either compounds or mixtures.
3. A chemical formula represents a compound or a molecule. The symbols for the elements are used to indicate the types of atoms present and subscripts are used to indicate the relative number of atoms.
5. Ionic bonds are formed by the transfer of valence electrons between a metal and a nonmetal. The bonding is a result of electrostatic attraction between a positively charged ion (metal) and a negatively charged ion (nonmetal). A crystalline structure of positive and negative ions results. Covalent bonds are formed by the sharing of valence electrons between two nonmetals. Covalent compounds are composed of clusters of two or more atoms bonded together to form molecules.

7. The properties of compounds are determined by the shape and structure of the molecule, the kinds of atoms present, and the types of bonds present.
9. Ionic compounds contain positive ions (usually a metal) and negative ions (nonmetals or polyatomic ions). The positive ion is named first and is the element name presented unchanged. The negative ion is named second. If it is a monatomic element, the element name ending is changed to $-ide$. Sodium chloride is a good example. The numbers of ions in the formula are *not* stated in the formula name.
11. Considering binary molecular compounds, the less electronegative element is named first, without change; the more electronegative element is named second, and the ending is changed to $-ide$. It is also necessary to include the number of atoms in the molecule in the formula name. The only exception here is that if the first element has only one atom, the prefix *mono* is not used.
13. The relative numbers of reactants and products in a reaction are given by the coefficients in the balanced chemical equation. These coefficients represent conversion factors similar to the subscripts in chemical formulas. For example, in the equation $CH_4 + 2O_2 \rightarrow CO_2 + 2H_2O$, the coefficients tell us that for every 1 mol of CH_4 and 2 mol of O_2 that combine, 1 mol of CO_2 and 2 mol of H_2O will be produced. These conversion factors can be used to predict the amount of reactant(s) needed in a particular reaction, or the amount of a particular product(s) that will form.
15. In a chemical reaction atoms are neither created nor destroyed. All atoms present in the reactants must be accounted for among the products. In other words, there must be the same number of each type of atom on the product side and on the reactant side of the arrow. Making sure that this rule is obeyed is called balancing a chemical equation for a reaction.

Problems

17. a. CaF_2 contains one Ca atom and two F atoms.
    b. CH_2Cl_2 contains one C atom, two H atoms, and two Cl atoms.
    c. $MgSO_4$ contains one Mg atom, one S atom, and four O atoms.
    d. $Sr(NO_3)_2$ contains one Sr atom, two N atoms, and six O atoms.
19. a. KCl is ionic.
    b. CO_2 is molecular.
    c. N_2O is molecular.
    d. $NaNO_3$ is ionic.
21. a. sodium fluoride
    b. magnesium chloride
    c. lithium oxide
    d. aluminum oxide
    e. calcium carbonate
23. a. boron trichloride
    b. carbon dioxide
    c. dinitrogen monoxide
25. a. NO_2
    b. $CaCl_2$
    c. CO
    d. $CaSO_4$
    e. $NaHCO_3$

27. a. CO 28.01 amu
    b. CO_2 44.01 amu
    c. C_6H_{14} 86.17 amu
    d. HCl 36.46 amu
29. The molecular formula is C_3H_8.
31. $10.0 \text{ g } H_2O \times \dfrac{1 \text{ mol } H_2O}{18.0 \text{ g } H_2O} \times \dfrac{6.02 \times 10^{23} \text{ molecule } H_2O}{1 \text{ mol } H_2O}$

$= 3.34 \times 10^{23} \; H_2O \text{ molecule}$

33. $5.0 \text{ g } C_{12}H_{22}O_{11} \times \dfrac{1 \text{ mol } C_{12}H_{22}O_{11}}{342 \text{ g } C_{12}H_{22}O_{11}}$

$\times \dfrac{6.02 \times 10^{23} \text{ molecule } C_{12}H_{22}O_{11}}{1 \text{ mol } C_{12}H_{22}O_{11}}$

$= 8.8 \times 10^{21} \text{ sugar molecule}$

35. a. $124 \; CCl_4 \text{ molecule} \times \dfrac{4 \text{ Cl atoms}}{1 \; CCl_4 \text{ molecule}} = 496 \text{ Cl atoms}$

    b. $38 \; HCl \text{ molecule} \times \dfrac{1 \text{ Cl atom}}{1 \; HCl \text{ molecule}} = 38 \text{ Cl atoms}$

    c. $89 \; CF_2Cl_2 \text{ molecule} \times \dfrac{2 \text{ Cl atoms}}{1 \; CF_2Cl_2 \text{ molecule}}$

    $= 178 \text{ Cl atoms}$

    d. $1368 \; CHCl_3 \text{ molecule} \times \dfrac{3 \text{ Cl atoms}}{1 \; CHCl_3 \text{ molecule}}$

    $= 4{,}104 \text{ Cl atoms}$

37. a. $0.20 \text{ mol } H_2O \times \dfrac{1 \text{ mol O}}{1 \text{ mol } H_2O} = 0.20 \text{ mol O}$

    b. $12 \text{ mol CO} \times \dfrac{1 \text{ mol O}}{1 \text{ mol CO}} = 12 \text{ mol O}$

    c. $15 \text{ mol } CO_2 \times \dfrac{2 \text{ mol O}}{1 \text{ mol } CO_2} = 30 \text{ mol O}$

    d. $25 \text{ g } NO_2 \times \dfrac{1 \text{ mol } NO_2}{46 \text{ g } NO_2} \times \dfrac{2 \text{ mol O}}{1 \text{ mol } NO_2} = 1.1 \text{ mol O}$

39. The mass of sodium in 5.8 grams of NaCl is 2.3g.
41. The mass of iron in 15.8 kg of Fe_2O_3 is 11.2 kg.
43. a. $4 HCl + O_2 \rightarrow 2 H_2O + 2 Cl_2$
    b. $3 NO_2 + H_2O \rightarrow 2 HNO_3 + NO$
    c. $CH_4 + 2O_2 \rightarrow CO_2 + 2H_2O$
45. a. $4 Al + 3 O_2 \rightarrow 2 Al_2O_3$
    b. $2 NO + O_2 \rightarrow 2 NO_2$
    c. $3 H_2 + Fe_2O_3 \rightarrow 2 Fe + 3 H_2O$
47. a. $2 H_2 + O_2 \rightarrow 2 H_2O$

    b. $2.72 \text{ mol } O_2 \times \dfrac{2 \text{ mol } H_2O}{1 \text{ mol } O_2} = 5.44 \; H_2O$

    c. $10.0 \text{ g } H_2O \times \dfrac{1 \text{ mol } H_2O}{18.0 \text{ g } H_2O} = 0.556 \text{ mol } H_2O$

49. a. $2.3 \text{ mol } CH_4 \times \dfrac{1 \text{ mol } CO_2}{1 \text{ mol } CH_4} \times \dfrac{44 \text{ g } CO_2}{1 \text{ mol } CO_2}$

    $= 1.0 \times 10^2 \text{ g } CO_2$

    b. $.52 \text{ mol } CH_4 \times \dfrac{1 \text{ mol } CO_2}{1 \text{ mol } CH_4} \times \dfrac{44 \text{ g } CO_2}{1 \text{ mol } CO_2} = 23 \text{ g } CO_2$

c. $11 \text{ g CH}_4 \times \dfrac{1 \text{ mol CH}_4}{16 \text{ g CH}_4} \times \dfrac{1 \text{ mol CO}_2}{1 \text{ mol CH}_4}$
$\times \dfrac{44 \text{ g CO}_2}{1 \text{ mol CO}_2} = 3 \text{ g CO}_2$

d. $1.3 \text{ kg CH}_4 \times \dfrac{1000 \text{ g}}{1 \text{ kg}} \times \dfrac{1 \text{ mol CH}_4}{16 \text{ g CH}_4} \times \dfrac{1 \text{ mol CO}_2}{1 \text{ mol CH}_4}$
$\times \dfrac{44 \text{ g CO}_2}{1 \text{ mol CO}_2} = 3.6 \times 10^3 \text{ g CO}_2$

Chapter 5

Questions

1. As elements, both sodium and chlorine are very reactive due to their electron configurations. Their reactivity makes them harmful to biological systems. Sodium has one valence electron, while chlorine has seven valence electrons. Thus, both elements have incomplete octets. Sodium will lose one electron and chlorine will gain one electron to form a chemical bond. Both atoms are now stable due to octets in their outer shell. Therefore, sodium and chloride when bonded together are relatively harmless.

3. Ionic bonding is represented by moving dots from the Lewis structure of the metal to the Lewis structure of the nonmetal to give both elements an octet. The dots surrounding the elements indicate the valence electrons. This corresponds to a stable configuration because the outermost occupied Bohr orbit contains an octet for each element. The metal and nonmetal each acquire a charge. Since opposite charges attract each other, there is an attractive force between the ions.

5. The Lewis theory is useful because it explains why elements combine in the observed ratios and allows one to predict the molecules that would form from certain elements. For example, fluorine, chlorine, bromine, and iodine all exist as diatomic molecules in nature, just as predicted by the Lewis theory. Also, magnesium fluoride is composed of two fluoride ions and one magnesium ion, just as predicted by the Lewis theory.

7. Na· Ȧl: ·P̈: :C̈l: :Är:

   Na and Cl would be the most chemically reactive. Ar would be the least reactive since it contains an octet of valence electrons.

9. The shape of a molecule is of primary importance in determining the properties of that substance. Polarity is a property of far-reaching importance that is largely determined by molecular shape. The distribution of charge in chemical bonds determines bond polarity, but the polarity of a molecule will depend on how the bonds are arranged: If the individual bond polarities cancel out, the molecule is nonpolar; but if they don't cancel out, the molecule is polar. Shape will determine whether or not the polarities will cancel.

11. Water is a unique substance for many reasons, and many of its properties are essential for life on earth. Water is a liquid at room temperature, whereas hydrides of the other nonmetals are gases; ice floats on water, whereas most solids are denser than their liquid phase; water has a very high heat of vaporization, which makes it a very effective cooling agent; the specific heat of water is unusually high, which means that large amounts of heat are absorbed in heating it or emitted in cooling it.

    All these properties—each of which alone is unusual, but collectively they make for a substance of very special qualities—can be explained by hydrogen bonding. Hydrogen bonding represents a form of intermolecular force resulting from the polar nature of the O—H bond. It is also found to a lesser extent in molecules containing N—H and F—H bonds. In water it is particularly strong because of the combination of the two lone electron pairs on each O atom being involved in a hydrogen bond with an H atom on each of two neighboring H_2O molecules. A network of hydrogen bonds, which makes optimum use of the two lone electron pairs and the two H atoms of every H_2O molecule, extends through the water binding the molecules together.

13. In order for a change of state to occur, intermolecular forces must be overcome by the thermal energy of the molecules. As the strength of intermolecular forces rises, so will the melting and boiling points. Polar molecules tend to have higher melting and boiling points than nonpolar molecules because the intermolecular forces are greater, owing to the electrostatic interactions between the atoms that have small positive and negative electrical charges.

Problems

15. a. ·Ċ·   b. :N̈e: c. :Ca   d. :F̈·
    Ne is chemically stable.

17. a. $[K]^+ [:\ddot{I}:]^-$ b. $[:\ddot{Br}:]^- [Ca]^{2+} [:\ddot{Br}:]^-$
    c. $[K]^+ [:\ddot{S}:]^{2-} [K]^+$ d. $[Mg]^{2+} [:\ddot{S}:]^{2-}$

19. a. $[Na]^+ [:\ddot{F}:]^-$ NaF
    b. $[:\ddot{Cl}:]^- [Ca]^{2+} [:\ddot{Cl}:]^-$ $CaCl_2$
    c. $[Ca]^{2+} [:\ddot{O}:]^{2-}$ CaO
    d. $[:\ddot{Cl}:]^- [Al]^{3+} [:\ddot{Cl}:]^-$ $AlCl_3$
       $[:\ddot{Cl}:]^-$

21. a. :Ï—Ï:

    b. 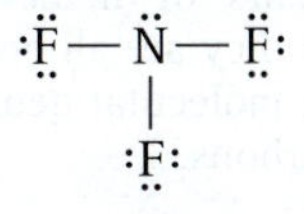 :F̈—N̈—F̈:
              |
             :F̈:

    c. 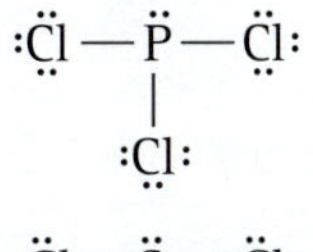 :C̈l—P̈—C̈l:
              |
             :C̈l:

    d. :C̈l—S̈—C̈l:

23. a. The Lewis structure is incorrect because Ca and O together form an ionic bond, not a covalent bond. The correct Lewis structure is as follows:

    $[Ca]^{2+} [:\ddot{O}:]^{2-}$

b. The Lewis structure is incorrect because there should be only single bonds between the oxygen and both chlorines; there should not be a double bond between the first chlorine and the oxygen. The correct Lewis structure is as follows:

:Cl̈—Ö—C̈l:

c. The Lewis structure is incorrect because the P does not have a complete octet. The correct Lewis structure is as follows:

:F̈—P̈—F̈:
   |
   :F̈:

d. The Lewis structure is incorrect because there are too many total electrons. Nitrogen has only 5 valence electrons, so between the two there should be only 10 total electrons. The correct Lewis structure is as follows:

:N≡N:

25. a. For I_2, the total number of electron pairs is one. Therefore, the electron geometry is linear. There are no lone pairs, so the resulting molecular geometry is also linear.
b. For NF_3, the total number of electron pairs is four, three bonding pairs and one nonbonding pair. The electron geometry is tetrahedral, but since one of these electron pairs is a lone pair, the resulting molecular geometry is pyramidal.
c. For PCl_3, the total number of electron pairs is four, three bonding pairs and one nonbonding pair. The electron geometry is tetrahedral, but since one of these electron pairs is a lone pair, the resulting molecular geometry is pyramidal.
d. For SCl_2, the total number of electron pairs is four, two bonding pairs and two nonbonding pairs. The electron geometry is tetrahedral, but since two of these electron pairs are lone pairs, the resulting molecular geometry is bent.

27. a. The total number of electron pairs for the Lewis structure of ClNO is three. Two pairs are bonding, and one pair is nonbonding. The electron geometry is trigonal planar, and the molecular geometry is bent.

:C̈l—N̈=O:

b. The total number of electron pairs for the Lewis structure of C_2H_6 is seven, and they are all bonding. The electron geometry and the molecular geometry are tetrahedral around both carbons.

     H H
     | |
  H—C—C—H
     | |
     H H

c. For the N_2F_2 molecule, both nitrogen atoms have three electron pairs, two bonding pairs (one double bond exists between the nitrogen atoms) and one nonbonding pair. The electron geometry is trigonal planar; however, one of the electron pairs is a lone pair. The correct molecular geometry is a bent structure on both nitrogen atoms.

:F̈—N̈=N̈—F̈:

d. For N_2H_4, both nitrogen atoms have four electron pairs, three bonding pairs and one nonbonding pair. The electron geometry is tetrahedral, but since one of these electron pairs is a lone pair, the resulting molecular geometry is pyramidal on both nitrogen atoms. The indicated three-dimensional structure is shown below.

    H⋯N̈—N̈⋯H
       /   \
      H   H

29. a. HCl: polar
    b. N_2: nonpolar
    c. O_2: nonpolar
    d. CO: polar

Chapter 6

Questions

1. Organic chemistry is the study of carbon-containing compounds.
3. Four bonds adopt a tetrahedral geometry around the carbon atom.
5. Vitalism became a popular belief because scientists were unsuccessful in preparing organic compounds. Thus it was thought that living systems had some "vital" force that enabled them to synthesize organic compounds.
7. Functionalized hydrocarbons are hydrocarbon molecules that contain an atom or group of atoms of a different element or elements. These functional groups confer very different properties on the molecule; compounds containing the same functional groups tend to have similar characteristic properties.
9. The structural formula of pentane is as follows:

     H H H H H
     | | | | |
  H—C—C—C—C—C—H
     | | | | |
     H H H H H

The condensed formula for pentane is $CH_3CH_2CH_2CH_2CH_3$.
The methyl groups have the formula CH_3, and methylene groups are CH_2.

11. As the chain length increases, the boiling point of the alkane increases.
13. Two important properties of alkanes are their flammability and nonpolar nature.
15. $2\ C_4H_{10} + 13\ O_2 = 8\ CO_2 + 10\ H_2O$
17. Ethylene is a natural ripening agent for fruit. Acetylene is a fuel used in welding torches.
19. The short answer is no, isomers do not have the same properties. Depending on the type of isomer and the degree of difference in the atomic arrangements, differences in properties can vary from slight to dramatic.

Isomers of hydrocarbons will be characterized by slightly different boiling points; whereas an ethyl alcohol and dimethyl ether (which are isomers) have very different physical and chemical properties.

21. a. Aldehyde

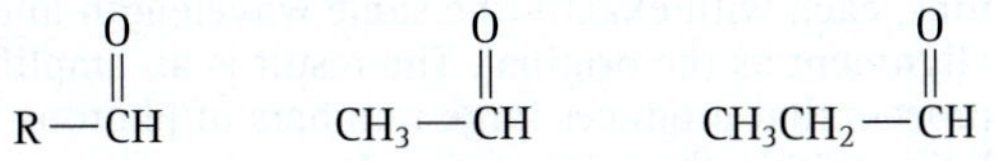

$$R-\overset{O}{\underset{\|}{C}}H \quad CH_3-\overset{O}{\underset{\|}{C}}H \quad CH_3CH_2-\overset{O}{\underset{\|}{C}}H$$

b. Ketone

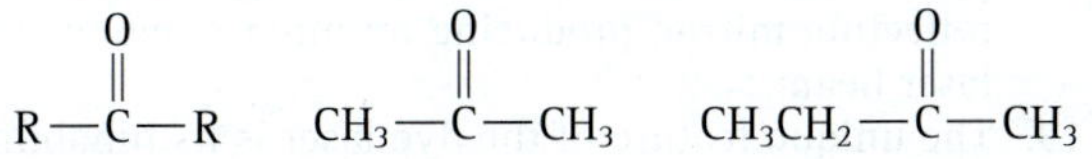

$$R-\overset{O}{\underset{\|}{C}}-R \quad CH_3-\overset{O}{\underset{\|}{C}}-CH_3 \quad CH_3CH_2-\overset{O}{\underset{\|}{C}}-CH_3$$

c. Carboxylic acid

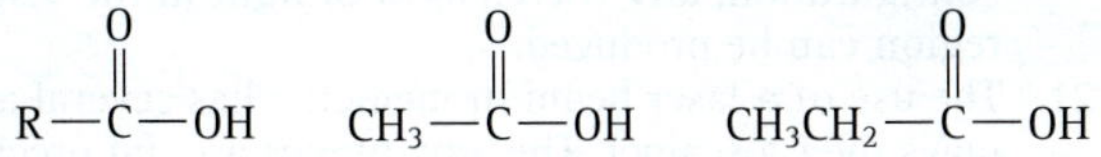

$$R-\overset{O}{\underset{\|}{C}}-OH \quad CH_3-\overset{O}{\underset{\|}{C}}-OH \quad CH_3CH_2-\overset{O}{\underset{\|}{C}}-OH$$

d. Ester

$$R-\overset{O}{\underset{\|}{C}}-OR \quad CH_3-\overset{O}{\underset{\|}{C}}-OCH_3 \quad CH_3CH_2-\overset{O}{\underset{\|}{C}}-OCH_3$$

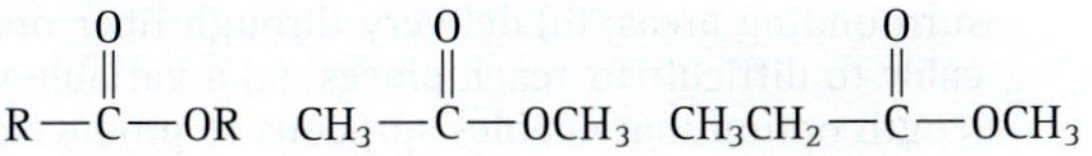

e. Ether

$$R-O-R \quad CH_3-O-CH_3 \quad CH_3CH_2-O-CH_3$$

f. Amine

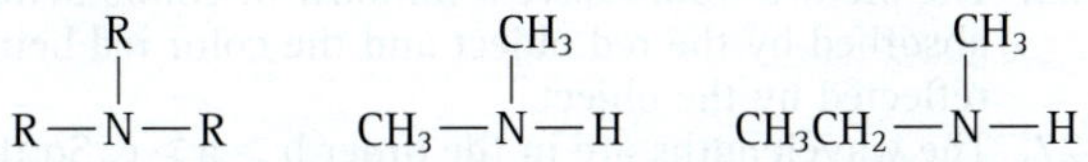

$$R-\underset{\underset{R}{|}}{N}-R \quad CH_3-\underset{\underset{H}{|}}{N}-H \quad CH_3CH_2-\underset{\underset{H}{|}}{N}-H$$

23. DDT, a chlorinated hydrocarbon, is an effective insecticide and is relatively nontoxic toward humans. However, many insects became resistant to DDT, rendering it ineffective. DDT's excellent chemical stability became a liability. DDT became concentrated in the soil and eventually moved up the food chain, killing fish and birds, including the bald eagle.

25. Ethanol functions as a depressant on the central nervous system. Excessive alcohol consumption can lead to loss of coordination, unconsciousness, and even death.

27. Due to its toxicity to bacteria, formaldehyde is used as a preservative of biological specimens.

29. Benzaldehyde is found in oil of almond, and cinnamaldehyde is found in cinnamon.

31. 2-heptanone is found in oil of cloves. Ionone is found in raspberries. Butanedione is found in butter and body odor. Butanedione is released as a waste product when bacteria attack sweat fluid.

33. Ethyl butyrate is found in pineapples. Methyl butyrate is found in apples. Ethyl formate is found in artificial rum flavor. Benzyl acetate is a component of oil of jasmine.

PROBLEMS

35.

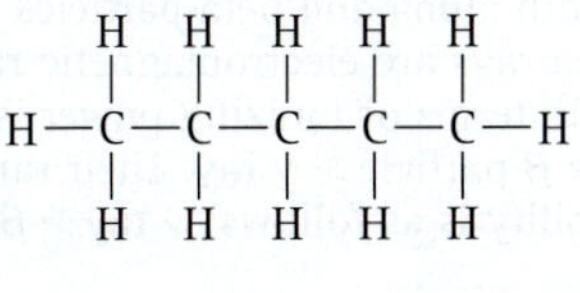

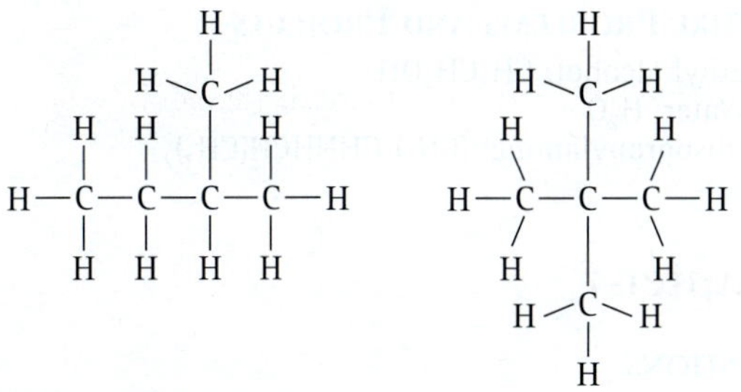

```
    H  H  H  H  H
    |  |  |  |  |
H - C- C- C- C- C - H
    |  |  |  |  |
    H  H  H  H  H
```

(Two structures shown: n-pentane and 2,2-dimethylpropane-like branched isomers)

37.

```
    H  H  H  H  H  H
    |  |  |  |  |  |
H - C= C- C- C- C- C - H
          |  |  |  |
          H  H  H  H
```

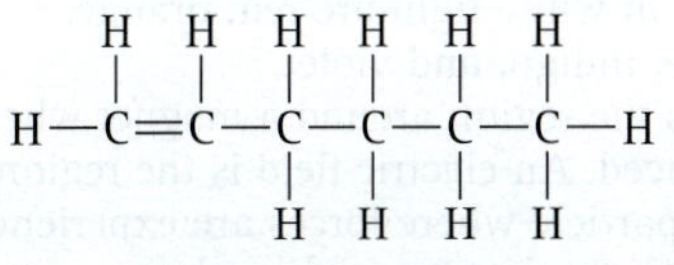

Two additional isomers are possible.

39. a. 2-methylbutane
    b. 4-ethyl-2-methylhexane
    c. 2,4-dimethylhexane
    d. 2,2-dimethylpentane

41. a. 2-butene
    b. 4-methyl-2-pentene
    c. 2-methyl-3-hexene

43. a. Propyne
    b. 3-hexyne
    c. 4-ethyl-2-hexyne

45. a. 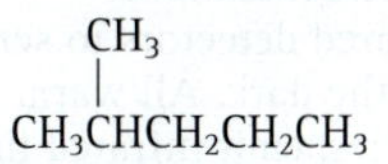 $CH_3CHCH_2CH_2CH_3$ with CH_3 branch
    b. 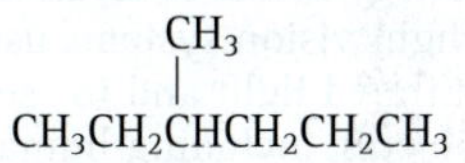 $CH_3CH_2CHCH_2CH_3$ with CH_3 branch
    c. 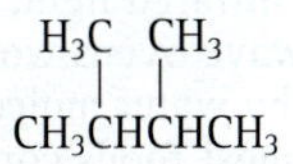 $CH_3CHCHCH_3$ with H_3C and CH_3 branches
    d. 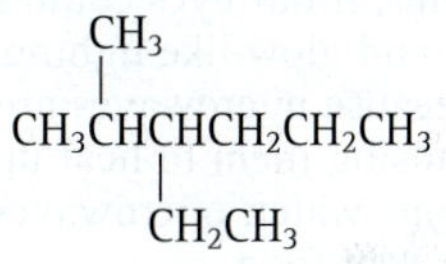 $CH_3CHCHCH_2CH_2CH_3$ with CH_3 and CH_2CH_3 branches
    e. 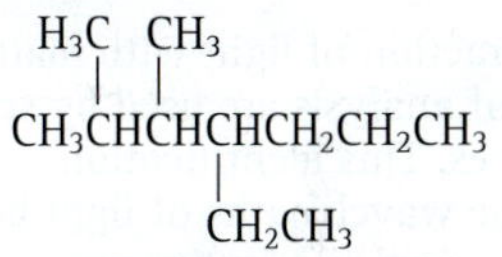 $CH_3CHCHCHCH_2CH_2CH_3$ with H_3C, CH_3, and CH_2CH_3 branches

47. a. $CH_3CH=CHCH_3$
    b. $CH_3CH_2C\equiv CCH_2CH_3$
    c. 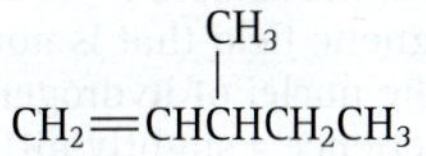 $CH_2=CHCHCH_2CH_3$ with CH_3 branch
    d. 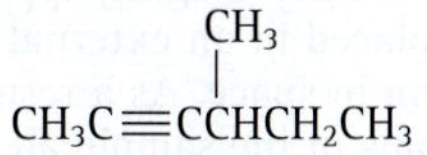 $CH_3C\equiv CCHCH_2CH_3$ with CH_3 branch

49. a. Ether
    b. Chlorinated hydrocarbon
    c. Amine
    d. Aldehyde

51. a. Carboxylic acid
    b. Aromatic hydrocarbon
    c. Alcohol
    d. Amine

53. Propanol is a polar molecule due to the presence of the OH functional group. Propane is a hydrocarbon, a nonpolar molecule. Polar molecules exhibit attractive intermolecular forces that tend to prevent the molecules from separating into the gas phase as easily as nonpolar molecules.

61. The formula $CH_3CH_2CH_3$ cannot mean the structure shown because that would mean that most of the H atoms are bonded to more than one atom. Since H forms only one bond, this would be impossible.

Feature Problems and Projects

63. Ethyl alcohol: CH_3CH_2OH
    Water: H_2O
    Diisopropylamine: $(CH_3)_2CHNHCH(CH_3)_2$

Chapter 7

Questions

1. The colors present in white light are red, orange, yellow, green, blue, indigo, and violet.
3. A magnetic field is the region around a magnet where forces are experienced. An electric field is the region around a charged particle where forces are experienced.
5. The wavelength of light determines its color. Wavelength also determines the amount of energy one of its photons carries. Wavelength and energy have an inverse relationship, indicating that as the wavelength increases, the energy decreases.
7. Sunscreens and UV-rated sunglasses contain compounds such as PABA that efficiently absorb ultraviolet light.
9. X-rays are a form of light that penetrates substances that normally block out light (opaque materials). X-rays are high-frequency, short-wavelength, high-energy forms of electromagnetic radiation.
11. Night vision systems use infrared detectors to sense infrared light and to "see" in the dark. All warm objects, including human bodies, emit infrared light. Thus, if our eyes could respond to infrared light, people would glow like lightbulbs. Microwave ovens work because microwaves are absorbed by water molecules, causing them to heat up. Because most foods contain some water, microwaves are a quick and efficient way to heat food.
13. Spectroscopy is the interaction of light with matter. Spectroscopic methods of analysis are used by scientists to identify substances. This identification procedure is based on the wavelengths of light being absorbed or emitted by a particular substance.
15. In an MRI imaging apparatus, the sample to be studied is placed in an external magnetic field that is nonuniform in space. As a result, the nuclei of hydrogen atoms in the sample all experience a slightly different magnetic field. A sequence of radio frequencies is then applied to the sample, which causes the hydrogen nuclei to come into resonance (to flip). Each nucleus in the sample flips at a slightly different radio frequency, depending on the strength of the external magnetic field it is in. At every radio frequency that caused a nucleus to flip, a line is plotted. The result is a spectrum showing an image of the object containing the sample (as shown in Figure 15). This technique is incredibly useful for obtaining medical images since our bodies have an abundance of hydrogen atoms.
17. A laser consists of three main components: a lasing medium sandwiched between two mirrors (one of which only partially reflects), and a laser cavity that encapsulates both the lasing medium and the mirrors. When a laser is turned on, molecules or atoms within the lasing medium are excited with light or electrical energy. The electrons in the molecules or atoms jump to higher energy states, and when they relax back to their lower energy state they emit light (in the form of a photon). The photon released travels through the cavity, hits the mirror, and bounces back through the lasing medium, stimulating the emission of other photons, each with exactly the same wavelength and alignment as the original. The result is an amplification process that produces large numbers of photons circulating within the laser cavity. A small fraction of the photons leave the laser cavity through the partially reflecting mirror, producing an intense, monochromatic laser beam.
19. The unique feature of the dye laser is its tunability. By choosing the proper dye and the correct laser cavity configuration, any wavelength of light in the visible region can be produced.
21. The use of a laser beam in medicine has several advantages over a scalpel. The advantages are: (a) precise cuts through skin and tissue with minimum damage to surrounding areas; (b) delivery through fiber-optic cable to difficult-to-reach places; (c) a variable wavelength option that enables surgeons to produce a variety of desirable effects.

Problems

23. 8.3 minutes
25. The picture would have a rainbow of colors being absorbed by the red object and the color red being reflected by the object.
27. The wavelengths are in the order $b > a > c$. So the photon energies are in the order $b < a < c$.
29. 3.21 m
31. 0.353 m
33. Sodium (Na)
35. Helium and hydrogen
37. Blue light would be *reflected* and not absorbed by the tattoo. Any color other than blue would be appropriate.
39. The sketch would look like Figure 6 on page 191 of the text. If that were a sketch of visible light, the UV light would have a shorter wavelength, and infrared light would have a longer wavelength.
41. X-rays are the most energetic, while microwaves are the least energetic.

Chapter 8

Questions

1. Radioactivity is the result of nuclear instability. An unstable nucleus will "decay" to achieve stability, releasing parts of its nucleus in the process. The parts released can collide with matter to produce large numbers of ions. Ions can damage biological molecules, which are contained in living organisms.
3. An alpha particle consists of two protons and two neutrons; it is a helium nucleus. A beta particle is an energetic, fast-moving electron. A gamma ray is an energetic photon. Both alpha and beta particles are matter, while gamma rays are electromagnetic radiation. Their ranking in terms of ionizing power is as follows: α particle $> \beta$ particle $> \gamma$ ray. Their ranking in terms of penetrability is as follows: γ ray $> \beta$ particle $> \alpha$ particle.

5. Radon gas is found in the soil near uranium deposits and also in the surrounding air. It represents a potential hazard because it is an unstable intermediate in the uranium radioactive decay series.
7. Albert Einstein wrote a letter to President Roosevelt describing the use of fission in the construction of very powerful bombs. American scientists were concerned that Nazi Germany would build a fission bomb first since that was where fission was discovered. President Roosevelt was convinced and decided that the United States must beat Germany to the bomb.
9. a. Los Alamos, New Mexico, is the location where America's best scientists gathered to design the nuclear bomb.
   b. Hanford, Washington, is the location where plutonium was manufactured.
   c. Oak Ridge, Tennessee, is the place where uranium was processed.
   d. Alamogordo, New Mexico, is the location where the first atomic bomb was tested.
11. During the fission reaction carried out by Fermi, a critical mass was achieved, resulting in a self-sustaining reaction. Fission was occurring in a controlled manner. The fission reaction used to produce the atomic bomb is designed to escalate in an exponential manner to produce an explosion.
13. The scientists working on the Manhattan Project rationalized that the fission bomb would better serve humanity if developed by the United States. Its use and control would be scrutinized better by the United States government than Nazi Germany.
15. Mass defect is the difference between the experimentally measured mass of a nucleus and the sum of the masses of the protons and neutrons in the nucleus. Nuclear binding energy is the energy related to the mass defect and represents the energy that holds a nucleus together.
17. "The China syndrome" is a phrase used to describe the overheated core melting through the reactor floor and into the ground (or all the way to China).
19. All nuclear waste is now being stored on the location where it was produced. The United States government now has a program to begin building permanent disposal sites for the storage of nuclear waste. These underground storage facilities would keep the nuclear waste materials isolated from the public and the environment.
21. Fusion is a nuclear reaction in which nuclei of light atoms unite to form heavier nuclei. Since the heavier ones are more stable, energy is released in the process. The following nuclear equation describes a fusion reaction: $^{2}_{1}H + ^{3}_{1}H \longrightarrow ^{4}_{2}He + ^{1}_{0}n$
23. External radioactivity can often be stopped by clothes on our bodies or by the skin. However, internal radioactivity is hazardous because the particles have direct access to vital organs.
25. Genetic defects have occurred in the offspring of laboratory animals upon exposure to high levels of radiation. However, the results of scientific studies of humans have not revealed an increase in genetic defects by radiation exposure. Few studies of humans exist because humans cannot purposely be exposed to radiation.
27. According to U-238 dating techniques, the earth is about 4.5 billion years old. To make the calculation, it is assumed no lead was originally present (rock was only uranium), that all of the $^{206}_{82}Pb$ formed over the years has remained in the rock, and the number of isotopes in the intermediate stages of decay between $^{238}_{92}U$ and $^{206}_{82}Pb$ is negligible. The last assumption is valid because once an $^{238}_{92}U$ isotope starts to decay, it reaches $^{206}_{82}Pb$ relatively fast. The age according to U-238 dating is also considered valid since we can see stars today that are billions of light years away; because we can see these stars today, we know the earth must be billions of years old.
29. One way to obtain an image of a specific internal organ or tissue is to introduce a radioactive element into the patient. This radioactive element (as a part of a compound) will concentrate in the area of interest. Detectors outside the body can scan over a region and record the activity so that a complete activity image may be reconstructed.

PROBLEMS

31. a. $^{220}_{86}Rn \rightarrow ^{216}_{84}Po + ^{4}_{2}He$
    b. $^{212}_{83}Bi \rightarrow ^{212}_{84}Po + ^{0}_{-1}e$
    c. $^{224}_{88}Ra \rightarrow ^{220}_{86}Rn + ^{4}_{2}He$
    d. $^{208}_{81}Tl \rightarrow ^{208}_{82}Pb + ^{0}_{-1}e$
33. $^{228}_{88}Ra \xrightarrow{\beta} ^{228}_{89}Ac \xrightarrow{\beta} ^{228}_{90}Th \xrightarrow{\alpha} ^{224}_{88}Ra \xrightarrow{\alpha} ^{220}_{86}Rn$
35. $^{215}_{83}Bi \rightarrow ^{211}_{81}Tl + ^{4}_{2}He \rightarrow ^{211}_{82}Pb + ^{0}_{-1}e \rightarrow ^{211}_{83}Bi + ^{0}_{-1}e$
37. 1.25 μg
39. 1.1×10^{17} gamma rays emitted
41. $^{2}_{1}H + ^{2}_{1}H \longrightarrow ^{3}_{2}He + ^{1}_{0}n$
43. 22%
45. The C-14 content of the skull is 25% of the amount in a living organism and represents two half-lives of decay from the original sample. The age of the skull can then be estimated to be about 11,500 years, since the half-life is 5,730 years.
47. Approximately 22,920 yr
49. 4.5×10^9 yr

FEATURE PROBLEMS AND PROJECTS

61. The missing particles would consist of 5 red and 7 gray; $^{16}_{7}N \longrightarrow ^{12}_{5}B + ^{4}_{2}He$

Chapter 9

QUESTIONS

1. The hotter an object is, the faster its molecules move. The cooler an object is, the slower its molecules move.
3. Heat is the random motion of molecules or atoms. Work is the use of energy to move atoms and molecules in a nonrandom or orderly fashion.
5. The United States consumes approximately 94 quads of energy per year.
7. Entropy is a measure of disorder. Entropy is important because it is the arbiter of spontaneous processes. All spontaneous processes, those that happen without continuous input of energy, occur with an increase in entropy of the universe. The world moves spontaneously toward greater disorder (entropy).

9. A perpetual motion machine is a device that continuously produces energy without the need for energy input. According to the first law of thermodynamics, this device could not exist, since energy can be neither created nor destroyed.
11. a. Heat is random atomic or molecular motion.
    b. Energy is the capacity to do work
    c. Work is the application of a force through a distance.
    d. System is any part of the universe in which we are particularly interested.
    e. Surroundings are all the rest of the universe, usually to be taken to mean just the immediate surroundings of a reaction vessel.
    f. An exothermic reaction is a chemical reaction that releases energy to the surroundings.
    g. An endothermic reaction is a chemical reaction that absorbs energy from the surroundings.
    h. Enthalpy of reaction represents the amount of heat absorbed or emitted by a chemical reaction. The units are usually in kcal/g, kJ/g, kcal/mol, or kJ/mol.
    i. Kinetic energy is energy due to motion.
    j. Potential energy is energy due to position or stored energy.
13. The temperature of the surroundings increases during an exothermic process while the temperature of the surroundings decreases during an endothermic process.
15. The heat capacity of a substance is the amount of heat energy required to change the temperature of a given amount of the substance by 1°C. A substance with a large heat capacity absorbs a lot of heat without a large increase in temperature. A low-heat-capacity substance cannot absorb a lot of heat without a large increase in temperature.
17. Petroleum—60 yr, natural gas—120 yr, coal—1500 yr, nuclear—100 yr, hydroelectric—indefinite
19. Fossil-fuel-burning power plants use heat released in combustion reactions to boil water, generating steam that turns the turbine of an electric generator. The generator is then used to create electricity that is transmitted to buildings via power lines. A typical fossil fuel plant produces 1 gigawatt (1 giga = 10^9) of power in the form of electricity, which can light 1 million homes.
21. Nitrogen dioxide (NO_2)—eye and lung irritant; ozone (O_3) and PAN ($CH_3CO_2NO_2$)—lung irritant, difficulty in breathing, eye irritant, damages rubber products, and damages crops; carbon monoxide (CO)—toxic because it diminishes the blood's capacity to carry oxygen
23. Acid rain is caused by the emission of sulfur dioxide (SO_2) from fossil fuel combustion, especially coal-burning power plants. The emission of nitrogen monoxide (NO) and nitrogen dioxide (NO_2) also contributes to acid rain.
25. The earth's atmosphere is transparent to visible light from the sun. Visible light when it strikes the earth is changed partly to infrared radiation. This infrared radiation from the earth's surface is absorbed by CO_2 and H_2O molecules in the atmosphere. In effect, the atmosphere traps some of the energy, acting like the glass in a greenhouse and keeping the earth warmer than it would be otherwise.
27. The burning of fossil fuels by human beings has increased the amount of carbon dioxide in the earth's atmosphere. Because carbon dioxide traps energy in the atmosphere, the earth is warmed. An increase in carbon dioxide increases the earth's temperature, hence the term *global warming*. Based on scientific data, the global mean temperature has increased over the last century. Therefore, global warming has been occurring.
29. Carbon dioxide levels have increased about 20% over the last century. The temperature has increased approximately 1°C corresponding to this increase in CO_2 concentration.

Problems
31. a. 1.456 Cal
    b. 1.88×10^3 J
    c. 1.7×10^7 cal
    d. 8.9×10^{19} J
33. a. 100°C
    b. −321°F
    c. 298 K
    d. 311 K
35. −62°C, 211 K
37. 0.23 kWh, 138 min
39. Electricity costs 15 cents per kilowatt-hour.
    a. $2.25
    b. $64.80
    c. $54.00
    d. $0.75
41. a. -2.6×10^5 kcal
    b. -1.4×10^7 kcal
    c. -3.7×10^5 kcal
43. a. 60 kcal
    b. 990 kcal
    c. 5000 kcal
    d. 3000 kcal
45. a. 1.1×10^2 g natural gas
    b. 6.1×10^2 g coal
47. a. 2.9×10^2 g CO_2
    b. 3.3×10^3 g CO_2
49. a. $2C_8H_{18} \times 2\, 5O_2 \rightarrow 16\, CO_2 \times 18\, H_2O$
    b. 393 moles C_8H_{18}
    c. 138 kg CO_2

Feature Problems and Projects
57. a. 58 quadrillions at an average of 1.4 quadrillion per year
    b. 34 quadrillions at an average of 0.83 quadrillion per year
    c. −25 quadrillions
    d. Production will be 91 quadrillions in 2025. Consumption will be 130 quadrillions in 2025. The gap will be 39 quadrillions

Chapter 10

Questions
1. The main obstacle to using solar energy is its low concentration by the time it hits the earth.
3. As water flows through a dam, it turns a turbine on a generator that creates electricity.
5. The heating of air masses by the sun causes them to expand and rise. As the hot air rises, cooler air rushes in to fill the void, creating wind that turns a turbine to produce electricity.

7. A solar power tower works by having hundreds of sun-tracking mirrors that focus sunlight on a central receiver located on top of a tower. The central receiver uses the focused sunlight to heat a molten salt liquid, which is circulated into a storage tank. The molten salt is pumped out of the storage tank, on demand, to generate steam, which then turns a turbine on an electrical generator.
9. The main disadvantage of solar thermal technology is the cost. Not only does it cost more per kilowatt hour than energy from fossil fuels, but there are high maintenance costs as well. The other disadvantages are the low concentration and uncontrollable weather conditions. The advantages of solar thermal technology are its efficiency, renewal capabilities, and lack of any air pollution.
11. A semiconductor is a material whose electrical conductivity can be controlled. The n-type silicon semiconductor is one doped with electron-rich material such as arsenic. The n stands for negative. The p-type silicon semiconductor is one doped with electron-poor material like boron. The p stands for positive.
13. Photovoltaic cells have two main disadvantages. The first is the cost. Semiconductors are expensive to manufacture, and photovoltaic cells are costly to produce. Second, photovoltaic cells are inefficient with respect to converting incident energy into electricity. PV cells are advantageous because they contain no moving parts, produce no noise, and are environmentally safe.
15. Biomass energy is easily obtained from plants such as corn and sugar cane. When the plants are fermented, they produce ethanol, an energy source that is easily transported and burns cleanly. Another advantage is that they do not contribute to global warming since the amount of CO_2 released during fermentation is the exact amount of CO_2 absorbed during the plants' lifetime. The major problem with biomass energy is the amount of cropland required for it to become a significant energy source.
17. A nuclear power plant is essentially a steam engine using uranium as a fuel. The advantage of using uranium as fuel instead of coal or other fossil fuels is that it does not produce the air pollutants that are associated with fossil fuels. Other advantages are that the energy produced per kilogram of fuel is large and the extraction from the land is less damaging than extraction of fossil fuels. However, the disadvantages of nuclear power are the difficulty of disposing of radioactive waste, the limited nuclear fuel supply, and the possibility of releasing damaging radioactivity into the air if an accident were to take place. The public's fear of accidents and concerns over waste disposal have suppressed the use of nuclear power in the United States.
19. Instead of using an automobile for transportation, riding a bike, ride sharing, walking, or using public transportation could be substituted; all of these options conserve energy. Conserving energy in households involves improved methods of insulation, the lowering of thermostats, and using extra sweaters or blankets.
21. It is clear that conserving our limited fuel supplies, by avoiding wasteful use of energy, should be of great concern to our society. The cost for energy will increase as energy sources are depleted because of supply and demand. It will be important to be energy efficient to minimize waste. The cost will stay down because the consumer will only pay for energy used, not energy wasted.

Problems

23. The cost will be increased by the ratio of 20/15, which makes the monthly bill $260.00.
25. The total number of kilowatt-hours used is 1633 kWh. The total number of joules is 5.88×10^9 J.
27. 12.5% efficiency
29. 238 W
31. 28 m^2
33. The number of quads of solar energy falling on earth = 3×10^6 quads per year. World energy consumption is a comparatively puny 411 quads per year, which represents 0.014% of the solar energy deposition.
35. Yes, the hot tub would receive enough energy (620 kWh/month).
37. 0.67 m^2

Feature Problems and Projects

43. a. 36 quads; 0.72 quads/yr
    b. 280% increase; 5–6% per year

Chapter 11

Questions

1. Pressure is the direct result of constant collisions between gas molecules and the surfaces around them.
3. An increase in altitude causes a decrease in pressure, which is responsible for the pain in our ears. Gas molecules are constantly being trapped in our ears, usually at the same pressure as the gas external to our ears. Equal pressures normally result because the molecules collide on either side of our eardrum equally. However, as the air pressure changes, there is an imbalance that results in more collisions on one side of the eardrum than the other. Thus, the eardrum is stressed, resulting in pain.
5. The height of the mercury column supported by the gases in the atmosphere varies with weather conditions and altitude. High-pressure areas direct storms away and are a sign of good weather, while low pressure areas draw storms in, resulting in rain. Also, winds are a direct result of changes in pressure from one geographical region to another. Air molecules migrate from regions of high pressure to regions of low pressure.
7. The volume of gas expands proportionate to its temperature. More specifically, the volume of a gas is directly proportional to the absolute temperature (in Kelvin) at constant pressure.
9. Hot air balloons achieve buoyancy due to Charles's Law. When the air in the balloon is heated, the balloon's volume increases since the pressures inside and outside must be equal. The air density inside decreases, and the balloon achieves increased buoyancy. Afternoon winds in coastal regions are a direct result of Charles's Law. As inland air masses are warmed by the sun, they expand and rise; the air from the coast, which remains cool due to the high heat capacity of the ocean, rushes in to fill the void, resulting in wind.

11. The primary gases in the atmosphere and their approximate percentages are as follows:

Gas	Percent by volume
Nitrogen (N_2)	78%
Oxygen (O_2)	21%
Argon (Ar)	0.9%
Carbon Dioxide (CO_2)	0.03%
Neon (Ne)	0.0018%
Helium (He)	0.00052%

13. Oxygen is needed for respiration by which cells extract energy from glucose.

$$C_6H_{12}O_6 + 6O_2 \rightarrow 6CO_2 + 6H_2O + \text{energy}$$

15. Plants use carbon dioxide and water to form glucose in a reaction exactly opposite to that above.

$$6CO_2 + 6H_2O + \text{energy} \rightarrow C_6H_{12}O_6 + 6O_2$$

In addition, large amounts of CO_2 are absorbed by the earth's oceans and eventually form carbonates in rocks and soil. The carbon atoms do not stay in these places indefinitely, however. In a natural cycle, CO_2 is channeled back into the atmosphere through the decay of plants, the decomposition of rocks, the eruption of volcanoes, and the respiration process in animals. The earth's carbon atoms thus travel a constant cycle, from air to plants and animals or rocks, and then back into the air. The average carbon atom in our own bodies has made this cycle approximately twenty times.

17. If the earth had no atmosphere, the sky would be black and life could not exist.

19. The six major pollutants and their primary sources are listed below:

Pollutant	Primary Sources
SO_2	Electric utilities, industrial smelters
PM-10	Agricultural tilling, construction, unpaved roads, fires
CO	Motor vehicles (indirect)
O_3	Motor vehicles, electric utilities
NO_2	Motor vehicles, smelters, battery plants
Pb	Motor vehicles

21. The Clean Air Act is a comprehensive federal law that regulates emissions and authorizes the EPA to establish the National Ambient Air Quality Standards (NAAQS). The Clean Air Act was amended in 1977 and 1990.

23. The Clean Air Act amendments of 1990 target ground-level ozone and chlorofluorocarbon emissions.

25. Ozone absorbs UV light according to the following reactions:

$$O_3 + \text{UV light} \rightarrow O_2 + O$$
$$O_2 + O \rightarrow O_3 + \text{heat}$$

27. Before CFCs were banned, they were used as refrigerants, as solvents for the electronics industry, and in foam-blowing.

29. Ozone in the mid-northern latitudes has decreased about 6% since 1979. The closer to the equator, the smaller the observed decrease in ozone.

31. HFCs are hydrofluorocarbons. Since they do not contain chlorine, they do not threaten the ozone layer.

33. The average person will be most directly impacted at the time to replace the Freon in the air conditioner of an older model automobile.

35. It is indeed true that CFCs, with molecular weights over 100 g/mol, are heavier than air, but if that prevented them from mixing with the rest of the atmosphere, then similar reasoning would lead us to believe that we should all be breathing pure oxygen. Our atmosphere is a turbulent system that quickly mixes all the gases that compose it. Consequently, the distribution of gases in our atmosphere does not correlate with molecular weight. Furthermore, CFCs have been measured in thousands of stratosphere air samples since the 1970s, and since they are purely man-made chemicals, they could not have gotten there from natural sources.

PROBLEMS

37. $0.31 \text{ atm} \times \dfrac{760 \text{ mm Hg}}{1 \text{ atm}} = 2.4 \times 10^2 \text{ mm Hg}$

39. a. $7.0 \times 10^2 \text{ mm Hg} \times \dfrac{760 \text{ torr}}{760 \text{ mm Hg}} = 7.0 \times 10^2 \text{ torr}$

b. $7.0 \times 10^2 \text{ mm Hg} \times \dfrac{1 \text{ atm}}{760 \text{ mm Hg}} = 0.92 \text{ atm}$

c. $7.0 \times 10^2 \text{ mm Hg} \times \dfrac{1 \text{ atm}}{760 \text{ mm Hg}} \times \dfrac{29.92 \text{ in Hg}}{1 \text{ atm}}$
$= 28 \text{ in Hg}$

d. $7.0 \times 10^2 \text{ mm Hg} \times \dfrac{1 \text{ atm}}{760 \text{ mm Hg}}$
$\times \dfrac{14.8 \text{ psi}}{1 \text{ atm}} = 14 \text{ psi}$

e. $7.0 \times 10^2 \text{ mm Hg} \times \dfrac{1 \text{ atm}}{760 \text{ mm Hg}} \times \dfrac{101{,}325 \text{ N/m}^2}{1 \text{ atm}}$
$= 9.3 \times 10^4 \text{ N/m}^2$

41. Apply Boyle's law:

$$V_2 = \dfrac{P_1 V_1}{P_2}$$

$$V_2 = \dfrac{0.60 \text{ atm} \times 1.5 \text{ L}}{1.0 \text{ atm}} = 0.90 \text{ L}$$

43. Apply Boyle's law:

$$V_2 = \dfrac{P_1 V_1}{P_2}$$

$$V_2 = \dfrac{760 \text{ mm Hg} \times 0.35 \text{ L}}{245 \text{ mm Hg}} = 1.1 \text{ L}$$

45. Apply Charles's law, and remember that all temperatures must be in Kelvin (°C + 273 = K).

$$V_2 = \dfrac{V_1 T_2}{T_1} = \dfrac{1.2 \text{ L} \times 77 \text{ K}}{298 \text{ K}} = 0.31 \text{ L}$$

47. Apply Charles's law, and remember that all temperatures must be in Kelvin (°C + 273 = K).

$20 + 273 = 293 \text{ K} = T_1$
$30 + 273 = 303 \text{ K} = T_2$

$$V_2 = \dfrac{V_1 T_2}{T_1} = \dfrac{1.3 \text{ L} \times 303 \text{ K}}{293 \text{ K}} = 1.34 \text{ L}$$

49. Apply the combined gas law:

$T_1 = 80 + 273 = 353 \text{ K}$

$$T_2 = 40 + 273 = 313 \text{ K}$$
$$V_2 = \frac{P_1 V_1 T_2}{T_1 P_2} = \frac{700 \text{ torr} \times 0.40 \text{ L} \times 353 \text{ K}}{313 \text{ K} \times 200 \text{ torr}} = 1.6 \text{ L}$$

51. The reading of 33.0 psi is the gauge pressure. The gauge pressure increases to 41.6 psi when the volume expands to 10.1 L at a temperature of 70.0°C. The maximum rating is exceeded.

FEATURE PROBLEMS AND PROJECTS

59. When the volume is decreased, the molecules are forced into a smaller space. They consequently have more collisions with the walls of their container and therefore exert a greater pressure.

Chapter 12

QUESTIONS

1. Liquids are spherical in shape due to cohesive forces. Molecules squeeze together into spheres to maximize their contact. A sphere has the lowest surface-area-to-volume ratio and, therefore, allows the maximum number of molecules to be completely surrounded by other molecules.
3. The differences among solids, liquids, and gases, from a molecular viewpoint, lie in the interaction of the molecules. The interaction between molecules is dependent not only on the kind of molecule present but also on their distance of separation. For gases, the attractions between molecules are weak since the molecules are separated by large distances. This molecular attraction becomes progressively stronger going to a liquid and then to a solid, as the molecules become closer and closer together.
5. a. The boiling point is the temperature at which a liquid boils. Boiling occurs when molecules are able to completely overcome their cohesive forces and leave the liquid state.
   b. The melting point is the temperature at which a solid melts. It depends on the strength of the cohesive forces of the atoms or molecules that compose it.
   c. Cohesive force is a force of attraction that exists between atoms or molecules. The strength of this force is due to the kind of atoms or molecules present and their distance apart.
7. a. Dispersion forces are the result of small fluctuations in the electron clouds of atoms and molecules. These fluctuations result in uneven distribution of electrons in an atom or molecule, which results in an instantaneous dipole.
   b. Dipole forces are a result of atoms of different electronegativities. In this case, polar bonds are formed in which electrons are unevenly shared between the bonding atoms. This results in partial negative and partial positive charges on the two bonding atoms. However, the presence of polar bonds within a molecule may or may not make the entire molecule polar, depending on molecular structure.
   c. Hydrogen bonding is a cohesive attraction between molecules just like dipole forces. Because of electronegativity differences, the molecules with hydrogen atoms bonded to F, O, or N are especially polar and result in stronger dipole forces, thus the special name—hydrogen bond.

   dispersion forces < dipole forces < hydrogen bonding

9. Soap works by having a polar end that dissolves in water and a nonpolar end that dissolves dirt and grease. The grease anchors to the nonpolar end while the polar end stays immersed in the water, which is washed down the drain. The result is clean clothes, faces, and hands.
11. The sweat glands secrete water, which evaporates from the skin. Evaporation takes heat that your body supplies, cooling you down in the process.
13. Water is an unusual molecule, for it has the properties of a large-molecular-weight molecule yet it is only 18.0 g/mol. Properties such as being a liquid at room temperature and a high boiling point can be explained by the fact that water contains polar bonds and can hydrogen bond to other water molecules. In addition, while almost every other liquid contracts upon freezing, water expands. Consequently, ice has a lower density than liquid water, which results in ice floating, keeping large bodies of water from freezing over entirely. This is important for marine life survival in winter months. Another unique property of water is its ability to dissolve many organic and inorganic compounds. Because of this ability, it is the main solvent for most of life's chemistry; without it we die.
15. Our water supply is a reusable resource that is constantly being redistributed over the earth. The hydrologic cycle is the movement of moisture from large reservoirs of water, such as oceans and seas, to the atmosphere as clouds. Clouds condense into rain and snow, which fall mostly back into the ocean. The rain and snow that collect on land in time make their way back to the oceans and again evaporate, completing the hydrologic cycle.

17. | Classification | Hardness (ppm $CaCO_3$) |
    | --- | --- |
    | Very soft | <15 |
    | Soft | 15–50 |
    | Medium | 50–100 |
    | Hard | 100–200 |
    | Very hard | >200 |

19. | Contaminants | Examples |
    | --- | --- |
    | Biological | Giardia, Legionella |
    | Inorganic | Mercury, lead |
    | Organic | Volatile—benzene |
    | | Nonvolatile—chlorobenzene |
    | Radioactive | Uranium, radon |

21. Most microorganisms (biological contaminants) are killed by boiling water.
23. The Safe Drinking Water Act (SDWA) was enacted in 1974 by Congress to ensure high water quality. The SDWA authorized the EPA to establish maximum contaminant levels (MCL) for specific contaminants likely to be found in public drinking water systems. Under the SDWA, public water suppliers must routinely sample and test the water supplied to your tap. If any contaminant is above the MCL, the supplier must act to correct the problem as soon as possible.

25. If you receive your water from a public water provider, the EPA believes home water treatment for health reasons is not necessary. However, it may be important to perform home water treatment to reduce hardness, improve clarity, or to eliminate an undesirable taste in the water.
27. Environmental groups such as the EWA and NRDC believe that water quality is not as good as it could be and consequently push for more regulation of businesses and industry. These groups issued reports stating that tap water contains toxic contaminants and bacteria that cause death and many cases of illness in the U.S. population.

PROBLEMS

29. Bromine (Br_2) is a liquid at room temperature. It has the highest molecular weight among O_2, F_2, and Cl_2 and will have the strongest dispersion forces. This results in stronger cohesive forces, keeping bromine molecules in the liquid state.
31. Water (H_2O) would have the highest boiling point. Its unusually high boiling point is due to intramolecular hydrogen bonding.
33. Volatility is a function of the strength of the cohesive forces between molecules. These liquids are all hydrocarbons with no functional groups so one can use molecular weight to decide upon volatility. Decane has the highest molecular weight and thus the strongest dispersion forces, resulting in its being the least volatile of the liquids.
35. a. Nonpolar, b. Polar, c. Nonpolar, d. Polar
37. When the perfume is first opened, the lower molecular weight molecule is smelled. After several hours, the higher molecular weight molecule is smelled.
39. 0.14 M
41. 1.8×10^3 g
43. 42 g
45. 73 ppm; medium
47. 368 mg
49. 0.053 mg/L; no
51. No, because nitrates are nonvolatile and are not removed by boiling.

FEATURE PROBLEMS AND PROJECTS

59. The water would be half as much therefore increasing the lead concentration.

Chapter 13

QUESTIONS

1. The more acidic the food, the more sour the taste.
3. Bases have a slippery feel and a bitter taste. They have the ability to react with acids and the ability to turn litmus paper blue.

5. | Acid | Use |
   |---|---|
   | Hydrochloric acid | Cleaning metals; preparation of foods; refining of ores |
   | Sulfuric acid | Manufacture of fertilizers, explosives, dyes, and glue |
   | Nitric acid | Manufacture of fertilizers, explosives, and dyes |
   | Phosphoric acid | Manufacture of fertilizer and detergents; flavor additive for food and drinks |
   | Acetic acid | Present in vinegar, manufacture of plastics and rubber, preservative in foods, solvent for resins and oils |

7. | Base | Use |
   |---|---|
   | Sodium hydroxide | Neutralization of acids; petroleum processing; manufacture of soap and plastics |
   | Potassium hydroxide | Manufacture of soap; cotton processing; electroplating; paint remover |
   | Sodium bicarbonate | Antacid; source of CO_2 in fire extinguishers and cleaning products |
   | Ammonia | Detergent; removing of stains; extracting plant color; manufacture of fertilizers, explosives, and synthetic fibers |

9. A Brønsted-Lowry acid is a proton donor; a Brønsted-Lowry base is a proton acceptor.
11. For every change of one unit of the pH scale, $[H_3O^+]$ changes by a factor of 10. Thus, a solution with a pH of 4 has $[H_3O^+] = 1 \times 10^{-4}$ M, and a solution with a pH of 3 is 10 times more acidic with a $[H_3O^+] = 1 \times 10^{-3}$ M.
13. Citric acid
15. Vinegar and salad dressing
17. Hydrochloric acid (HCl) is found in high concentrations in the stomach. Phosphoric acid is often added to soft drinks and beer to impart tartness. Carbonic acid is present in all carbonated beverages.
19. Alkaloids are nitrogen-containing organic compounds, which are basic and often poisonous.
21. Antacids are substances that neutralize excess stomach acids. They typically contain basic substances such as aluminum hydroxide, calcium carbonate, magnesium hydroxide, and sodium bicarbonate, alone or in combination, which remove acid by neturalization: acid + base = salt + water.
23. Gaseous SO_2 and NO_2 are produced during the combustion of fossil fuels. The gases represent the pollutants responsible for acid rain.
25. The pH of rain in the United States fluctuates from 4.1 to 6.1. The acidity is most likely attributed to pollutants from midwestern coal-burning power plants.
27. If the acid levels are high enough in lakes and streams, the eradication of many fish species can occur. Acid rain can also damage forests and building materials.

PROBLEMS

29. $HCl + KOH \rightarrow H_2O + KCl$
    $H^+ + OH^- \rightarrow H_2O$
31. a. $HClO_2 + H_2O \rightarrow H_3O^+ + ClO_2^-$
          acid   base
    b. $CH_3NH_2 + H_2O \rightarrow OH^- + CH_3NH_3^+$
          base   acid
    c. $HF + NH_3 \rightarrow NH_4^+ + F^-$
          acid base
33. $H-\ddot{O}-H + H-\ddot{N}(H)-H \longrightarrow {}^-{:}\ddot{O}-H + H-\overset{+}{N}(H)(H)-H$
       acid   base

35. The pH of 0.01 M CH_3COOH would be greater than 2, because a large fraction of CH_3COOH molecules do not dissociate. Consequently, significant amounts of CH_3COOH, H_3O^+, and CH_3OO^- are present in a solution of CH_3COOH. A lower concentration of H_3O^+ is present in solution and thus a higher pH (> 2, lesser acidic) would result.
37. a. pH = 4 acidic
    b. pH = 7 neutral
    c. pH = 2 acidic
    d. pH = 9 basic
39. $[H_3O^+] = 10^{-5}$ M
41. a. $NaHCO_3 + HCl \rightarrow NaCl + H_2O + CO_2$
    b. $Mg(OH)_2 + 2\ HCl \rightarrow MgCl_2 + 2\ H_2O$
43. $2\ SO_2 + O_2 + 2\ H_2O \rightarrow 2\ H_2SO_4$

FEATURE PROBLEMS AND PROJECTS

49. HF is a weak acid. Very little of the dissolved solute has ionized.

Chapter 14

QUESTIONS

1. In a piece of iron, rusting occurs as electrons transfer from iron atoms to oxygen atoms. The oxidized iron atoms (iron atoms that lost electrons) bond with the reduced oxygen atoms (oxygen atoms that gained electrons) to form iron oxide or rust. Thus, rusting requires the transfer or conduction of electrons.
3. Some common processes that involve redox reactions are the burning of coal, the function of batteries, metabolism of foods, and the corrosion of metals.
5. Reduction is defined as the loss of oxygen, the gain of hydrogen, or the gain of electrons (the most fundamental definition).
7. Carbon shares its four valence electrons with two sulfur atoms and in this sense, the electrons are not completely a part of the carbon atom anymore. The carbon has partially given away its electrons; therefore, the carbon atom takes on a partial positive charge. Because the carbon atom partially loses some of its electrons, it is oxidized.
9. The oxidizing agent is the substance that gains electrons and is reduced. It oxidizes the other compound. The reducing agent is the substance that loses electrons and is oxidized. It reduces the other compound.
11. Respiration: $C_6H_{12}O_6 + 6\ O_2 \rightarrow 6\ CO_2 + 6\ H_2O$
                    oxidized   reduced

    Photosynthesis: $6\ CO_2 + 6\ H_2O \rightarrow C_6H_{12}O_6 + 6\ O_2$
                     reduced   oxidized

    These redox reactions illustrate the fact that animals and plants depend on each other for life. Animals use oxygen to oxidize carbon, and plants use the water to reduce the carbon back again, completing the cycle.
13. An automobile battery consists of lead metal at the anode, lead (II) oxide at the cathode, and dilute sulfuric acid. Because the dissociation of H_2SO_4 does not go to completion, both HSO_4^- (aq) and SO_4^{2-}(aq) are present, but typically the two half-reactions and net reaction are expressed in terms of SO_4^{2-} (aq) as shown below:

Oxidation (anode):
$Pb(s) + SO_4^{2-}(aq) \rightarrow PbSO_4(s) + 2\ e^-$

Reduction (cathode):
$PbO_2(s) + 4\ H^+(aq) + SO_4^{2-}(aq) + 2e^- \rightarrow PbSO_4(aq) + 2\ H_2O$

Net:
$Pb(s) + PbO_2(s) + 4\ H^+(aq) + 2\ SO_4^{2-}(aq) \rightarrow 2\ PbSO_4(s) + 2\ H_2O$

Electrons flow from the lead plates to the lead oxide plates. Lead (Pb) is oxidized and forms lead (II) ions (Pb^{2+}), which react with sulfate (SO_4^{2-}) ions in solution to form solid lead (II) sulfate ($PbSO_4$). As the lead (II) oxide (PbO_2) is reduced, it also reacts with SO_4^{2-} ions to form solid $PbSO_4$.

15. Fuel cells are batteries (actually a better term is *energy converter*) with reactants that are continually renewed. In the normal chemical process within a battery, the reactants are stored within and become depleted over time. However, in a fuel cell, the reactants are constantly supplied as needed to produce electricity while the products constantly flow out of the cell.
17. In a H_2-O_2 fuel cell, the reaction of H_2 (g) and O_2 (g) in a basic medium produces $H_2O(\ell)$ according to the following redox reactions:

Oxidation: $2\ H_2(g) + 4\ OH^-(aq) \rightarrow 4\ H_2O + 4\ e^-$
Reduction: $O_2(g) + 2\ H_2O + 4e^- \rightarrow 4\ OH^-(aq)$
Net: $2\ H_2(g) + O_2(g) \rightarrow 2\ H_2O$

19. The overall reaction for the oxidation or rusting of iron is

$$2\ Fe(s) + O_2(g) + 2\ H_2O \rightarrow 2\ Fe(OH)_2(s)$$

Eventually, $Fe(OH)_2$ forms Fe_2O_3, iron (III) oxide, which we call rust. It is clear from the overall reaction that the formation of rust requires water; thus keeping iron dry helps prevent rusting.

21. The theory concerning the relationship of aging and oxidation proposes that free radicals extract electrons from large molecules within cell membranes, oxidizing the molecules and making them reactive. When the large molecules react with each other, the cell wall properties change and the result is a weakened and vulnerable body. Free radicals are atoms or molecules with unpaired electrons, which easily oxidize other molecules by extracting electrons from them. The free radicals are produced from the combustion of oxygen with pollutants or toxins present in food, water, and air.

PROBLEMS

23. The combustion of gasoline, the metabolism of foods, and the rusting of iron are all chemical processes in which electrons are being transferred from one substance to another. Thus, these processes are all classified as redox reactions.

25. a. $2K\cdot + :\ddot{C}l - \ddot{C}l: \longrightarrow 2[K^+]\ [:\ddot{C}l:^-]$
         oxidized reduced

    b. $Ca: + :\ddot{B}r - \ddot{B}r: \longrightarrow [:\ddot{B}r:^-]\ [Ca^{2+}]\ [:\ddot{B}r:^-]$
         oxidized reduced

c.

$$\cdot \dot{\underset{\text{oxidized}}{C}} \cdot + \cdot \ddot{\underset{\text{reduced}}{O}} : \longrightarrow :O \equiv C:$$

27. a. Fe – oxidized, O_2 – reduced
    b. C – oxidized, Ni – reduced
    c. H_2 – oxidized, O_2 – reduced
29. a. Zn – oxidized, Fe^{3+} – reduced
    b. Br^- – oxidized, Cl_2 – reduced
    c. Zn – oxidized, I_2 – reduced
31. a. H_2 – reducing agent; C_2H_4 – oxidizing agent
    b. H_2CO – reducing agent; H_2O_2 – oxidizing agent
    c. CO – reducing agent; Fe_2O_3 – oxidizing agent
33. $6 CO_2 + 6 H_2O \rightarrow C_6H_{12}O_6 + 6 O_2$
    CO_2 is the oxidizing agent because the carbon is reduced. H_2O is the reducing agent because the oxygen is oxidized.
35. a. $Zn + Ca^{2+} = Zn^{2+} + Ca$
    b. $2 Al + 6 H^+ = 2 Al^{3+} + 3 H_2$
37. The piece of iron covered with sulfur would rust because sulfur is not a metal; it is a nonmetal. Consequently, sulfur has a tendency to be reduced rather than oxidized, and the piece of iron will have no protection from oxygen in the atmosphere. The iron covered with strontium would not rust because the strontium is a more reactive metal, therefore it would be oxidized and the iron would be protected. When protecting iron from corrosion, one must use an active metal that will be preferentially oxidized.
39. a. The concentration of sulfuric acid would affect the lifetime of an automobile battery. The higher the concentration of sulfuric acid present in the battery, the greater the concentration of sulfate ions available for the oxidation–reduction reaction to occur ($H_2SO_4 \rightarrow 2 H^+ + SO_4^{2-}$)
    b. The size of the porous lead plates will also affect the longevity of the battery. A large lead plate could indicate more lead present and available for the oxidation process. (Pb loses electrons at the anode.) The more lead present, the more lead (II) ions formed in solution.
    c. The size of the battery casing would not necessarily affect the longevity of the battery. Various size batteries are available depending on the make of the automobile. The most important aspects of the battery are the concentrations of sulfuric acid present and the sizes of lead and lead oxide plates available inside the battery.

Feature Problems and Projects

47. The picture would have in the left cell more Zn^{2+} ions and fewer Zn atoms on the electrode. The right cell would have fewer Cu^+ ions and more Cu atoms on the electrode.

Chapter 15

Questions

1. An orange contains polar molecules, such as citric acid, which mix well with the other polar molecules. Since water is polar, it dissolves the acid molecules, removing them easily when you rinse. On the other hand, french fries are coated with grease. The molecules composing grease are nonpolar and do not mix with polar water; therefore, the grease does not dissolve and is not washed away with water alone. However, soap molecules can be used to remove grease from the hands because they have both polar and nonpolar ends. As soapy water is applied to the greasy hands, the soap molecules are attracted to both water and grease, thus removing the grease from the hands.

3. Soaps and detergent molecules, when added to water, aggregate near the surface where they can best accommodate their dual nature. At the surface, the ionic head submerges into the water while the nonpolar tail protrudes up out of the water, thus the name surfactants.

5. A colloidal suspension is a mixture of one substance dispersed in a finely divided state throughout another. In a normal solution, molecules mix on a molecular scale, but in a colloidal suspension the molecules clump together to form small particles within the other substance. Shining a light through a colloidal suspension will show a dispersion of the light beam by the particles. If this same beam of light shone into clear water, it could not be seen.

7. Certain ions in water, such as magnesium and calcium, reduce the effectiveness of soap. These ions react with soap molecules to form a slimy, gray scum called curd that deposits on skin during bathing or on the sides of the bathtub, producing that undesirable ring around the tub.

9. ABS detergents accumulate in the environment, causing suds to form in natural waters. LAS detergents are biodegradable and decompose over time, forming CO_2, H_2O, and SO_4^{2-}, all common substances in the environment.

11. Recent trends in laundry cleaners have favored a combination of anionic LAS surfactants with nonionic surfactants.

13. Sodium tripolyphosphate is no longer used in detergent additives because the presence of phosphates in wastewater causes eutrophication, which threatens marine life.

15. A combination of builders including sodium carbonate (Na_2CO_3) and zeolites have replaced tripolyphosphate in U.S. laundry formulations.

17. Solid NaOH is the main ingredient in drain cleaners. When added to water, the solid sodium hydroxide dissolves, releasing large amounts of heat and producing a hot NaOH solution. This solution melts the grease contained in clogs and converts some of the grease molecules to soap molecules. These in turn can help dissolve more grease. The NaOH solution also dissolves protein strands in hair, a component of clogged pipes. Some drain cleaners also contain small bits of aluminum. These aluminum bits dissolve in the NaOH solution and produce hydrogen gas. The bubbling action helps to physically dislodge clogs.

19. Drain cleaners contain strong bases such as sodium hydroxide, which tend to help disperse grease and/or hair that blocks drains. Toilet cleaners, on the other hand, contain acids, which remove hard water deposits from the toilet bowl. Bases and acids combine with

much liberation of heat, which in this case could lead to unpleasant results.
21. If curlers are placed into hair when it is wet, hydrogen bonds will re-form between protein chains in the position maintained by the curlers.
23. To produce a permanent curl or "perm," a reducing agent such as thioglycolic acid is applied to hair. The acid attacks the disulfide bridges and breaks the —S—S— chemical bonds. Then, the hair is set in curlers in a desired shape. A mild oxidizing agent such as dilute hydrogen peroxide is then applied to the hair, re-forming the disulfide linkages in the new shape. The result is permanently curled hair.
25. Hair rinses or conditioners help rinse out residual shampoo while softening hair and making it more manageable. Conditioners contain cationic surfactants that are quaternary ammonium compounds.
27. The lightening of hair requires the removal of the pigments, usually accomplished with a bleach such as hydrogen peroxide (H_2O_2). The H_2O_2 oxidizes the pigments to colorless products. Darkening or adding color to hair involves applying a dye of the desired color to the hair. To achieve the most permanent coloring of hair, two or more components, usually small molecules, are added to hair. These molecules diffuse into the hair strands and undergo a reaction that produces a larger, colored molecule. Since the molecule is formed within the hair strand, it tends to stay there, producing a permanent color.
29. Creams and lotions contain emulsions of water and oil. Lotions typically contain more water than oil, and the oils are usually liquids at room temperature. On the other hand, creams tend to have more oil than water, and the oils tend to be solids at room temperature. Oils used in creams include olive oil, mineral oil, and almond oil, and lanolin creams usually include waxes such as beeswax or carnauba. Also, most creams and lotions contain perfumes to add pleasant fragrances.
31. Lipstick is composed of pigments and dyes suspended in a mixture of waxes and oils.
33. The molecules from the cooking food diffuse into the air, and we inhale them through our nose. In the nose, millions of nerve endings are triggered by molecules that fit certain chemical receptors. The receptors send signals to the brain that we interpret as a particular smell.
35. The top note is the first smell one encounters when opening a bottle of perfume. The smells in the top note are caused by the more volatile molecules that quickly work their way out of the bottle and into the air, where they are detected by sensors in the nose. The middle note is detected only after a short time, and these molecules are usually larger and less volatile than those in the top note. The middle-note molecules tend to persist on the skin. Finally, the end note usually contains more animal or earthy fragrances. It is composed of large molecules that evaporate slowly (less volatility). These compounds in the end note also help hold the more volatile notes in place, preventing them from vaporizing too quickly.
37. Underarm deodorants contain antibacterial agents such as triclosan that kill bacteria and thus eliminate the odor bacteria would otherwise produce.
39. a. Monomers are the individual units that make up polymers.
b. Polymers are chainlike molecules composed of monomers.
c. Copolymers are composed of two or more dissimilar monomers.
41. Polypropylene can be drawn into fibers and is often used in athletic clothing, fabrics, and carpets. Polyvinyl chloride (PVC) is the strong, tough plastic used in pipes, cable coverings, and gaskets. Polystyrene can be compounded to make tough, clear plastic called lucite, or it can be filled with tiny bubbles during its formation to make styrofoam.
43. a. Thermoplastic is a polymer that softens when heated and hardens when cooled.
b. Addition polymers are polymers formed by the addition of monomer units, one to the other without the elimination of any atoms.
c. Condensation polymers are polymers formed by the elimination of atoms – usually water.
d. Copolymers are polymers composed of two different monomer units, which result in chains composed of alternating units, rather than a single repeating unit.
e. Elastomer is a polymer that stretches easily and returns to its original shape.
f. Vulcanization is a process in which sulfur atoms form cross-links between polyisoprene chains. These cross-links still allow the polymer chains to uncoil under stress, but they force them to snap back to the original position when the stress is released, even if the rubber is warmed.

PROBLEMS
45. Sodium stearate

$$CH_3-(CH_2)_{16}-\overset{\overset{O}{\|}}{C}-O^-\ Na^+$$

47. $-NH_3^+ + -OH^- = -NH_2 + H_2O$
49. $2\ HCl + CaCO_3 \rightarrow CaCl_2 + H_2O + CO_2$
51. As the temperature of a sample of air is lowered, it will reach a temperature where it is saturated with water. This is called the dew point. At the dew point the relative humidity is 100%. Cooling the air to a temperature below this point results in oversaturation, and the water condenses out of the air as dew.
53. You could add highly volatile molecules to the perfume to improve its top note.
55.

$$*{-}{\left[\begin{array}{c} CH-CH_2 \\ | \\ CH_3 \end{array} \right]}_n{-}*$$

polypropylene (methyl substituted for hydrogen)

$$*{-}{\left[\begin{array}{c} CH-CH_2 \\ | \\ C_6H_5 \end{array} \right]}_n{-}*$$

polystyrene (aromatic ring substituted for hydrogen)

57.
$$H_2C=\underset{\underset{CN}{|}}{\overset{\overset{H}{|}}{C}}$$

59.
$$*{-}\!\left[\overset{\overset{O}{\|}}{C}{-}CH_2{-}CH_2{-}\overset{\overset{O}{\|}}{C}{-}\overset{\overset{H}{|}}{N}{-}CH_2{-}CH_2{-}CH_2{-}CH_2{-}\overset{\overset{H}{|}}{N}\right]_n\!\!{-}*$$

Feature Problems and Projects

65. The nonpolar hydrocarbon tail is sticking into the polar water molecules. The other end of the soap molecule is the end that would be attracted to water; it should be sticking into the water with the nonpolar hydrocarbon tail sticking up.

Chapter 16

Questions

1. **Lipids** function as long-term energy storage and are the primary means for structural components of biomembranes.
   Carbohydrates function as short-term energy storage.
   Proteins function as structural components for muscle, hair, and skin and control much of the body's chemistry.
   Nucleic acids function as information storehouses used to create protein.

3. Lipids are those cellular components that are insoluble in water but extractable with nonpolar solvents. A typical triglyceride has the following general structure:

$$\begin{array}{l}CH_2{-}O{-}\overset{\overset{O}{\|}}{C}{-}R\\[2pt] CH{-}O{-}\overset{\overset{O}{\|}}{C}{-}R\\[2pt] CH_2{-}O{-}\overset{\overset{O}{\|}}{C}{-}R\end{array}$$

R groups = saturated or unsaturated hydrocarbons.

5. Carbohydrates are a class of compounds derived from **carbo**, meaning carbon, and **hydrate**, meaning water. Their formulas are approximated by multiples of the formula CH_2O, or simply $(C \cdot H_2O)_n$.

7. Carbohydrates are polyhydroxy aldehydes or ketones or their derivatives.

9. The large number of hydroxy ($-OH$) groups on carbohydrates allow for hydrogen bonding with each other and with water, making carbohydrates soluble in water and in bodily fluids.

11. A monosaccharide (glucose) contains only one carbohydrate unit. A disaccharide (sucrose) contains two monosaccharide units. Polysaccharides are complex carbohydrates (starch or cellulose), which contain many monosaccharide units.

13. Proteins are molecules composed of long-chain repeating units called amino acids. Proteins function as structural materials in much of our muscle, hair, and skin and act as enzymes that regulate countless chemical reactions. They also serve as hormones to regulate metabolic processes, transport oxygen from the lungs to cells, and act as antibodies to fight invading organisms and viruses.

15. Twenty different amino acids exist in human proteins. They differ only in their R groups on the central carbon. A general structure for an amino acid is shown below:

$$H_2N-\underset{\underset{R}{|}}{\overset{\overset{H}{|}}{C}}-\overset{\overset{O}{\|}}{C}-OH$$

These different R groups could be polar, nonpolar, acidic, basic, or neutral in character. Thus, the water solubility or water insolubility and pH in aqueous solutions of the various amino acids are controlled by these R groups.

17. The secondary structure of a protein refers to the way the chain of amino acids orients itself along its axis. This secondary structure is held together by hydrogen bonding between the oxygen (see Fig. A) and hydrogen (see Fig. B) groups in the polypeptide chains.

Figure A   Figure B

$$\begin{array}{cc}\diagdown\\C=O\\\diagup\end{array}\qquad\begin{array}{c}\diagup\\H-N\\\diagdown\end{array}$$

19. Hemoglobin is a protein composed of four polypeptide chains arranged in the proper configuration to hold four flat molecules called heme groups. These heme groups bind readily with oxygen in the lungs and carry it to cells where it is needed for glucose oxidation.

21. Lysozyme is a protein that functions as an enzyme. An enzyme is a substance that catalyzes, or promotes a specific chemical reaction. It was discovered by Alexander Fleming, who placed nasal mucus on a dish containing bacteria. The mucus was able to dissolve away the bacteria.

23.

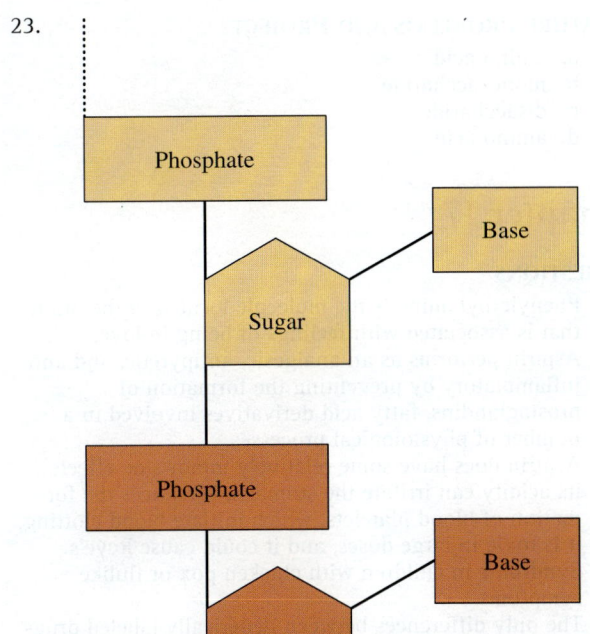

25. DNA contains four bases: adenine, guanine, cytosine, and thymine. The complementary bases are adenine-thymine and guanine-cytosine.
27. Chromosomes are biological structures that contain DNA material. Human beings have 46 chromosomes, found in the nucleus of a cell.
29. The order of codons along the DNA molecule determines the order of amino acids in a protein. A section of DNA that codes for one protein is called a gene. However, not all of the DNA is expressed in each cell. The expression of DNA requires the formation of m-RNA followed by protein synthesis at the ribosome. Cells only express the proteins specific to their function.

31.

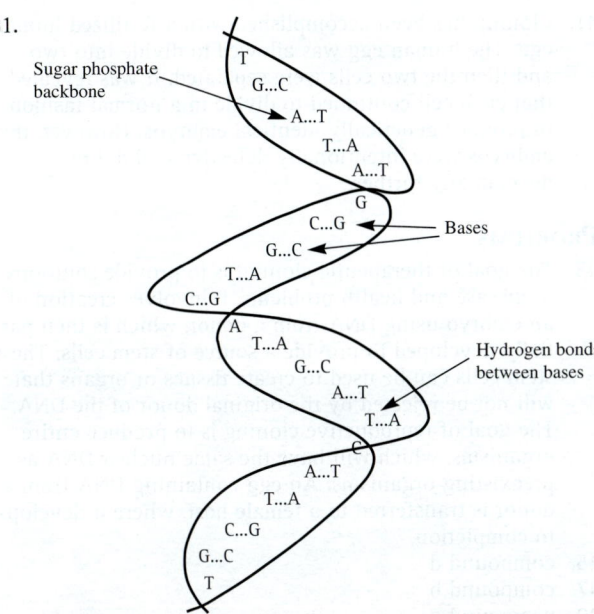

33. Recombinant DNA is the result of a transfer of DNA from one organism to another. The combination of the organism's original DNA and the transferred DNA is called recombinant DNA.
35. Scientists can now use genetic engineering to isolate a gene that codes for the production of a specific protein. The gene is transferred into a bacteria, which is then grown in a culture. As the bacteria multiplies so does the foreign DNA, resulting in millions of copies of the desired protein. The proteins now being genetically engineered include human insulin and human growth hormone.
37. DNA can be isolated from blood, semen, or almost any bodily fluid or tissue. The DNA can then be screened for genes that may increase a person's chances of developing certain diseases, such as cancer, heart disease, and Alzheimer's. By knowing early on that they are predisposed to certain diseases, a person can take preventative steps to avoid it.
39. In DNA fingerprinting, a small sample of a person's blood, hair, skin, semen, or other bodily tissue or fluid is collected at a crime scene. The DNA within the cells from the sample is isolated and cut into fragments using restriction enzymes. These DNA fragments are then separated by gel electrophoresis, and a radioactive probe, which attaches to a specific base sequence, is then applied to the mixture. The radioactive sample is allowed to expose a photographic film, which produces a pattern of light and dark. Since every person's DNA is different, the light and dark bands are unique and are called a DNA fingerprint. This procedure is useful for a trial because a positive match between the DNA fingerprints of the suspect and a sample from the crime scene links the suspect to the crime.

41. Cloning has been accomplished with a fertilized human egg. The human egg was allowed to divide into two and then the two cells were separated. It was revealed that each cell continued to divide in a normal fashion to produce genetically identical embryos. However, the embryos were intentionally defective and did not develop any further.

PROBLEMS

43. The goal of therapeutic cloning is to provide solutions to disease and health problems. It involves creation of an embryo using DNA from a donor, which is then partially developed to provide a source of stem cells. These stem cells can be used to create tissues or organs that will not be rejected by the original donor of the DNA. The goal of reproductive cloning is to produce entire organisms, which will have the same nuclear DNA as preexisting organisms. An egg containing DNA from a donor is transferred to a female host, where it develops to completion.
45. compound d
47. compound b
49. compound a
51.

53. a. monosaccharide
    b. polysaccharide
    c. disaccharide
    d. disaccharide
55. compound b
57. a. carbohydrate
    b. lipid
    c. lipid
    d. amino acid
59.

61.

You would not expect this tripeptide to mix readily with water because of its nonpolar hydrocarbon side chains. These groups give the peptide a hydrophobic nature.

63. a. thymine
    b. guanine
    c. cytosine
    d. adenine

FEATURE PROBLEMS AND PROJECTS

71. a. amino acid
    b. monosaccharide
    c. disaccharide
    d. amino acid

Chapter 17

QUESTIONS

1. Phenylethylamine is the molecule located in the brain that is associated with feelings of being in love.
3. Aspirin performs as an analgesic, antipyretic, and anti-inflammatory by preventing the formation of prostaglandins, fatty acid derivatives involved in a number of physiological processes.
5. Aspirin does have some relatively minor side effects; its acidity can irritate the stomach, it reduces the formation of blood platelets, which initiate blood clotting, it is toxic in large doses, and it could cause Reye's Syndrome in children with chicken pox or flulike symptoms.
7. The only differences between generically labeled drugs and their brand name counterparts are the price and the names.
9. Antibiotics are drugs that fight bacterial diseases by targeting the unique physiology of bacteria and selectively killing them.
11. Penicillin and cephalosporin kill bacteria by destroying their cell walls. Once the cell wall is weakened, the cell wall ruptures and the bacteria die. However, human cell walls are different and are not affected by these compounds.
13. Antibiotics can lose their effectiveness because some bacteria have evolved to be resistant to drugs. Bacterial strains have emerged that are completely drug-resistant.
15. A retrovirus consists of RNA and protein, whereas most viruses contain DNA and protein. Viral RNA enters the host cell, then is reverse transcribed into DNA. This DNA then goes on to direct the synthesis of viral protein.
17. AZT is a nucleoside analog, an antiviral drug. AZT is incorporated into DNA by viral enzymes that are fooled by the phosphorylated AZT, which resembles the nucleotide thymine. When the viral enzymes construct viral DNA, the analog is incorporated and because it has only one point of attachment, the synthesis of the DNA is terminated. Cells infected with the AIDS virus prefer the fake or phony nucleotide over the real one, which slows the progression of the disease by stopping the production of the viral DNA. Healthy cells, on the other hand, do not show this preference. Unfortunately, mutant forms of the virus developed that showed no preference for the phony nucleotide. The result for patients is that taking AZT only slows the progression of AIDS for a short time.
19. A powerful combination of drugs—a protease inhibitor, AZT, and another nucleoside analog—can kill the HIV at two different points in its replication cycle. In order for the virus to survive, it would have to develop resistance to both drugs simultaneously, an unlikely event.

21. Testosterone is a male sexual hormone that acts before the birth of a male infant to form the male sex organs. Testosterone functions at puberty and beyond to promote the growth and maturation of the male reproduction system, to develop the sexual drive, and to determine secondary sexual characteristics such as facial hair and voice deepening.
23. The birth control pill contains synthetic analogs of the two female sex hormones. The estrogen analog regulates the menstrual cycle, while the progesterone analog establishes a state of false pregnancy. Taking these hormones on a daily basis allows for the maintenance of a false pregnancy state, therefore no eggs are released, preventing conception. Nausea, weight gain, fluid retention, breast tenderness, and acne are all possible side effects of the birth control pill.
25. Steroids are chemical substances characterized by their seventeen-carbon-atom, four-fused-ring skeletal structure, as shown here.

    Three examples of steroids are progestin, cortisone, and prednisone.
27. Anabolic steroids act to increase muscle mass. However, these steroids have serious side effects; in males, lower sperm production and testicular atrophy are observed. In females, male secondary sexual characteristics, irregular menstruation, and ovulation failure are possible side effects.
29. The side effects of chemotherapy include anemia, the inability to fight infection, nausea, vomiting, infertility, hair loss, fatigue, and internal bleeding.
31. a. Alkylating agents are highly reactive compounds that add alkyl groups (e.g., —CH_2CH_3) to other organic or biological molecules. The alkylation of DNA within cells produces defects in the DNA that result in cell death.
    b. Antimetabolites are chemical substances that function by impersonating compounds normally found in the cell but possess subtle differences that usually stop DNA synthesis.
    c. Topoisomerase inhibitors function by altering the action of a topoisomerase, an enzyme that pulls apart the individual strands of DNA in preparation for DNA synthesis. The topoisomerase inhibitor results in DNA damage, and thus cell death eventually occurs. These compounds can cause cells to die whenever they initiate replication.
    d. Hormone treatment involves those cancers involving the breasts or sexual organs. Cancer cells within the tissues need a large dose of hormones to grow. Denying the tissue the needed hormones stops cancerous tumor growth.
33. The body metabolizes ethanol in the liver, where the ethanol equivalent of one drink per hour is expelled.
35. Alcoholics often drink in the morning, become depressed due to lack of drinking, drink alone or in secret, and may suffer blackouts or memory loss after episodes of heavy drinking.
37. Some common depressant inhalants used for anesthetic purposes are nitrous oxide (laughing gas), isoflurane, halothane, and enflurane.
39. The immediate effects of inhaling vapors of household chemicals can include alcohol-like intoxication, loss of consciousness, and even death. Long-term problems of vapor inhalation are brain damage, kidney damage, and liver damage.
41. Narcotics function by binding to opioid receptors, specific sites on nerve cells in the brain and spinal cord. The narcotic molecule, when bound to an opioid receptor, inhibits certain neurotransmitters involved in the transmission of pain impulses.
43. Codeine, less potent than morphine, is only effective in the relief of moderate pain. Also, its tolerance develops more slowly and it is less likely to produce addiction compared to morphine. Heroin is three times more potent than morphine and also extremely addictive. Its effect, compared to morphine, is much more intense, since it travels to the brain faster.
45. Naltrexone binds to opioid receptor sites, displacing opioids, without producing the narcotic effects. When naltrexone is taken daily, any ingestion of an opioid narcotic will be ineffective. Addicts can use this drug as part of their recovery program. The craving for the narcotic is partly overcome by the knowledge that the drug will not produce the desired high.

    Methadone works in the same way as naltrexone, binding to opioid receptor sites. Unlike other narcotics, it does not produce stupor. However, methadone does prevent the physical symptoms associated with withdrawal.
47. Stimulants are drugs that increase the activity of the central nervous system. Examples of stimulants are caffeine, amphetamines, nicotine, and cocaine.
49. Caffeine, once in the bloodstream, interacts with nerve cells to block adenosine receptors. The result is an increase of dopamine and norepinephrine neurotransmitters. The effects of increased levels of these neurotransmitters are a feeling of alertness, competence, and wakefulness.
51. Nicotine interacts with acetylcholine receptors in the central nervous system. This interaction results in a faster heart rate, elevated blood pressure, and the release of adrenaline, promoting a fight-or-flight response.
53. Hallucinogenic drugs are chemical substances that distort and disturb cognition and perception. Mescaline and LSD are examples of hallucinogenic drugs.
55. Lysergic acid diethylamine (LSD) was first synthesized by Dr. Albert Hoffman, a Swiss chemist, in 1938. Dr. Hoffman took several intentional doses of LSD after unintentionally experiencing its hallucinogenic effects while working with the drug. He documented his results, and his records indeed revealed the hallucinogenic powers of LSD.
57. LSD can produce rapid changes in mood and emotion as well as severe perceptual distortions and hallucinations. Users experience "seeing sounds" or "hearing colors" along with a significant slowing in the passage of time. Chronic use of LSD can lead to a lower capacity for abstract thinking as well as permanent brain damage.

59. Low doses of marijuana produce relaxation, relief from anxiety, an increased sense of well-being, an alteration in sense of time, and mild euphoria. As the dosage increases, the effects increase from mild hallucinations to depression, panic attacks, and paranoia. The side effects of taking marijuana are controversial; short-term effects appear to be minimal with only a mild physical dependency. The long-term effects, though, are more severe and are the result of THC buildup in the body. The effects include interference with cognitive function, and learning, reasoning, and academic impairments. Other long-term effects include lung and throat irritation, increased risk of lung cancer, and partial suppression of the immune system.

61. Since a deficiency in the neurotransmitter serotonin within the brain appears to cause depression, both Prozac and Zoloft function by elevating serotonin levels back to normal. This results in reversing the adaptive changes caused by low levels of serotonin and relieving the depression. The side effects of Prozac include nervousness, dizziness, anxiety, sexual dysfunction, insomnia, nausea, and loss of appetite. The side effects of Zoloft include dizziness, sexual dysfunction, insomnia, nausea, diarrhea, and dry mouth.

Problems

63. [Structure showing an aromatic ring with CH₃CHCH₂– group (with CH₃ branch) on one side and –CHCOH carboxylic acid group (with CH₃ branch) on the other, labeled "Aromatic" and "Carboxylic acid"]

65. 3.75 mL
67. 1400 mg or 1.4 g

Feature Problems and Projects

73. a. 5 yr (81–76)
    b. 0.16 yr/yr (5 yr/32 yr)
    c. 85 yr if born in 2030

Chapter 18

Questions

1. The atoms contained in human bodies come from food.
3. Carbohydrates are the starches, sugars, and fibers contained in food. Chemically, carbohydrates are polyhydroxy aldehydes or ketones or their derivatives.
5. *Monosaccharides* are the simplest carbohydrates; they are the unit building blocks for other, more complex carbohydrates. *Disaccharides* are simple carbohydrates that contain two monosaccharide units, which can be hydrolyzed into their respective monosaccharide units. Polysaccharides are complex carbohydrates composed of many monosaccharide units (glucose, for example) linked together.
7. Monosaccharides are the basic carbohydrate unit and can be directly transported in the blood to cells, where they are oxidized in a number of reaction steps. Disaccharides and polysaccharides (except fiber) are hydrolyzed with water at the bonds between monosaccharide units into their composite units, which are then passed into the bloodstream.
9. Both starch and fiber are composed of long chains of glucose units. However, in starch, the glucose units are linked together by alpha linkages, while in fiber, the links are beta linkages. This difference in the structures results in starch being digestible, whereas fiber is not. The intestinal enzymes recognize the alpha linkages and attack them but not the beta linkages.
11. In a healthy diet, the FDA recommends that carbohydrates compose 60% of caloric intake with 10% or less of total caloric intake from simple sugar.
13. Proteins are broken down into their amino acid components in the digestive tract. The amino acid components are then absorbed through the intestinal wall and into the bloodstream.
15. Complete proteins are proteins that contain all the essential amino acids in the right proportion. By making the right combinations, vegetarians can obtain complete proteins in their diets. Some of these right combinations include beans and rice, peanut butter and bread, and rice with tofu.
17. Saturated fats are those containing only single bonds and tend to be solids at room temperature. Unsaturated fats contain double bonds and exist as liquids at room temperature.
19. Excessive amounts of cholesterol in the blood can result in the excessive deposition of cholesterol on arterial walls, leading to a blockage of the arteries, a condition called arteriosclerosis. Such blockages are dangerous because they inhibit blood flow to important organs such as the heart or brain. The risk of stroke and heart attack increases with increasing blood cholesterol levels.
21. The FDA recommends that fats compose less than 30% of total caloric intake.
23. Fats and oils contain over twice the number of calories per gram than carbohydrates or proteins. Fiber contains no caloric content.
25. Fat is the most efficient way to store energy because it has the highest caloric content.
27. The fat-soluble vitamins include vitamins A, D, E, and K. The water-soluble vitamins are classified as the eight B vitamins and vitamin C. Fat-soluble vitamins are soluble in fatty tissue and not easily excreted and can, therefore, be overconsumed.
29. In the body, vitamin C functions in the synthesis of connective tissue called collagen, in the protection against infections, and in the absorption of iron through the intestinal wall. Vitamin C deficiency results in a condition called scurvy, in which bodily tissues and blood vessels become weakened. This condition can result in massive bleeding and, if untreated, can lead to death. Vitamin C is present in citrus fruits, green leafy vegetables, cantaloupe, strawberries, peppers, tomatoes, potatoes, papayas, and mangoes.
31. The B complex vitamins play important roles in metabolism (the extraction of energy from food), protein synthesis, and cell multiplication. Deficiencies in the B complex vitamins result in the inability of cells to extract energy from foods. The symptoms of the defi-

ciency include exhaustion, irritability, depression, forgetfulness, partial paralysis, abnormal heart action, and severe skin problems. Green leafy vegetables and the hulls of grain are rich in the B complex vitamins.

33. Calcium serves as the main structural material for teeth and bones. It also plays an important role in the transmission of nerve signals and in blood clotting. Low calcium levels in the blood will result in osteoporosis, a condition in which bones become weakened by calcium loss. Milk, cheese, sardines, oysters, and broccoli are foods rich in calcium.

    Phosphorus is involved in the structure of teeth and bones. It also assists in energy metabolism and is a component of DNA. The effects of deficiencies of phosphorus are unknown since phosphorus is contained in most diets.

    Iodine is used in the synthesis of a hormone that regulates basal metabolic rate. Its deficiency results in goiter, a condition in which the thyroid gland swells as it tries to absorb as many iodine particles as possible. Iodine deficiency, if it occurs during pregnancy, can result in severe retardation of the developing baby. Iodine is readily available in seafood, milk, and iodized salt.

    Iron is a critical component of hemoglobin, the protein that carries oxygen in the blood. Hence, loss of iron occurs through bleeding and its deficiency results in the body's inability to make enough hemoglobin. This problem results in a condition called anemia in which a person feels tired, listless, and susceptible to cold temperatures. Meat, fish, poultry, clams, legumes, and green leafy vegetables are all foods that are rich in iron.

    Zinc is essential to the function of more than 100 enzymes. It is also important in growth and development, immune function, learning, wound healing, and sperm production. Vegetarians are most at risk for zinc deficiency since plant sources contain only small amounts of zinc. Overconsumption of zinc is toxic. Animal products are excellent sources of zinc; these include meat, shellfish, and poultry.

35. Potassium and magnesium are both involved in maintaining electrolyte balance in and around cells. Magnesium also plays an important role in bone formation and in the operations of many enzymes, including those involved in the release of energy. Potassium occurs in significant amounts in fresh fruits and vegetables, including bananas, cantaloupe, and lima beans. Magnesium is found in oysters, sunflower seeds, spinach, and in "hard" drinking water.

37. Additives are often added to foods to preserve it, enhance its flavor or color, and maintain its appearance or physical characteristics. The Food and Drug Administration (FDA) monitors and regulates all food additives.

39. | Antimicrobial Agent | Function |
    |---|---|
    | Salt | Drying meat and fish |
    | Sodium nitrite | Inhibits the growth of bacteria |
    | Sodium benzoate | Prevents microbial growth in packaged foods |

41. Sulfites and EDTA are both antioxidants. Sulfites function as preservatives and antibrowning agents. EDTA acts to immobilize metal ions that catalyze oxidation reactions.

43. Artificial flavors include sugar, corn syrup, aspartame, oil of wintergreen, peppermint, ginger, vanilla, and almond extract. A typical flavor enhancer used in foods is monosodium glutamate (MSG). Artificial flavors and flavor enhancers are added to food to improve the taste.

45. The primary nutrients for plants are potassium, nitrogen, and phosphorus.

47. The Haber process is an industrial reaction used to fix nitrogen (i.e., ammonia synthesis).

49. The secondary nutrients for plants are calcium, magnesium, and sulfur. The micronutrients required for plants include boron, chlorine, copper, iron, manganese, molybdenum, sodium, vanadium, and zinc; however, these nutrients are only needed in small quantities and rarely need to be replenished in soils.

51. Bioamplification is a process in which a chemical substance becomes concentrated as it moves up the food chain (aquatic plants → fish → birds).

53. Broad-spectrum insecticides, those of the organophosphate family, are able to attack a wide range of pests. Examples of broad-spectrum insecticides are malathion and parathion. Narrow-spectrum insecticides are of the carbamate type, such as carbaryl and aldicarb, and target only specific pests.

55. | Modern Herbicides | Uses |
    |---|---|
    | Atrazine | Destroys weeds in cornfields |
    | Metalachlor | Used on soybeans and corn |
    | Paraquat | Destroys marijuana crops |

Problems

57. Answers vary by person.
59. 2100 Cal/day
61. a. 43%
    b. 45%
    c. 9%
63. Answers vary by person.
65. Answers vary by person.
67. 87.5 days or approximately 3 months.
69. 960 Cal/day
71. ≈15 eggs/day

Chapter 19

Questions

1. A nanometer is 10^{-9} meters.
3. Medicine is one of many fields that would benefit from extreme miniaturization. There would be many advantages if machines could be made small enough to navigate in the blood to unclog blocked arteries, destroy cancerous cells, and even perform surgeries inside the body. The possibility of miniature electronic devices is also appealing—you could have all the luxuries of the most powerful computers in the palm of your hand.

5. Nanotechnology has looked possible since the discovery of the scanning tunneling microscope, which can image and move individual atoms. This and other recent advances, such as the discovery of buckyballs and buckytubes, have helped move nanotechnology from merely an idea into the laboratory.

7. The scanning tunneling microscope was discovered in 1981 by Gerd Bennig and Heinrich Rohrer at the IBM Zurich Research Laboratory.

9. By tapping or oscillating while taking an image, the AFM further increases its resolution and minimizes damage to samples. Tapping makes it possible for the AFM to image not only metallic surfaces but also biological samples.

11. Buckminsterfullerenes are C_{60} and have the same structure as a soccer ball.

13. Nanotubes have already been used to act like a tiny scale, weighing objects as small as viruses; they have acted as atomic pencils and as wire, connecting two parallel electrodes. Nanotubes also have the possibility of being used to desalinate water and make flat-panel displays.

15. An artificial red blood cell would be a nanovesicle pumped full of oxygen. The nanovesicle, at the first sign of the cardiovascular system failing, would begin the slow release of oxygen to the body and allow it to survive until the cardiovascular system is restored.

17. Although nanotechnology looks promising, there are obstacles to be overcome before it becomes a reality. Nanomachines cannot be built in the same way as full-sized machines and have things like intermolecular forces to contend with. Engineers of nanomachines have yet to build one, so skepticism lingers and will probably stay around long after one is constructed, partly because of ethical considerations.

Problems

19. 7.8×10^3 nm
21. 10,000 tubes
23. A hexagon

APPENDIX 3

Answers to Your Turn Questions

Chapter 1

1.2) a. compound, b. element, c. heterogeneous mixture, d. homogeneous mixture

1.3) Much of the match's mass was converted to a gas and lost into the air. If the gain in mass of the surroundings was somehow measured, it would exactly equal the missing mass.

1.4) carbon/hydrogen = 6.0; oxygen/hydrogen = 8.0. The ratios are the same for both samples and therefore consistent with the law of constant composition.

Chapter 2

2.1) 2.3×10^{-6}

2.2) 13.4 in

2.3) 51 oz

2.4) a. 160, b. 13.3/yr, c. 15.5%, d. 1.29%/yr

2.5) 132 pool lengths

2.6) 4.67×10^4 in^3

2.7) 0.84 g/mL

2.8) 2.93×10^3 mm^3

Chapter 3

3.1) 12 protons, 10 electrons

3.2) $Z = 14$, $A = 28$

3.3) 17 protons, 18 neutrons, 18 electrons

3.4) 24.31 amu

3.5) 0.24 g

3.6) 5.2×10^{22} atoms

Chapter 4

4.1) a) 1 C; 4 Cl   b) 2 Al; 3 S; 12 O

4.2) potassium bromide

4.3) calcium carbonate

4.4) dinitrogen tetroxide

4.5) 44.02 amu

4.6) 1.50×10^{22} molecules

4.7) 9.72 mol Cl
4.8) 1.8 g O
4.9) $4\ HCl + O_2 \rightarrow 2\ H_2O + 2\ Cl_2$
4.10) 12.9 mol NH_3
4.11) 1.99×10^5 g CO_2

Chapter 5

5.1) $[:\ddot{\underset{..}{Cl}}:]\ Be^{2+}\ [:\ddot{\underset{..}{Cl}}:]$

5.2) Lewis structure: $Li^+\ [:\ddot{\underset{..}{O}}:]^{2-}\ Li^+$ Formula: Li_2O

5.3) $:\ddot{\underset{..}{Cl}}-\ddot{P}-\ddot{\underset{..}{Cl}}:$
     $\quad\quad\ \ |$
     $\quad\quad\ :\ddot{\underset{..}{Cl}}:$

5.4) $:\ddot{O}=C=\ddot{O}:$

5.5) $H-C\equiv C-H$

5.6) Electron geometry is tetrahedral; molecular geometry is pyramidal.

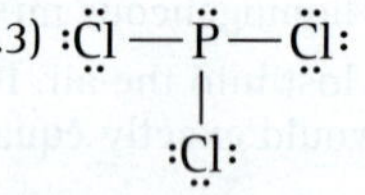

5.7) tetrahedral

5.8) trigonal planar

5.9) ●━━━●━━━● linear

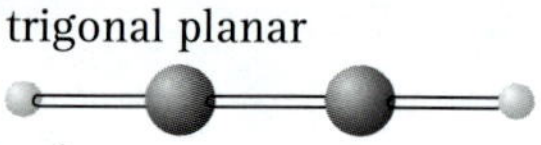

5.10) polar

Chapter 6

6.1) Molecular formula: $C_{10}H_{22}$

Structural formula:
```
     H H H H H H H H H H
     | | | | | | | | | |
  H—C—C—C—C—C—C—C—C—C—C—H
     | | | | | | | | | |
     H H H H H H H H H H
```

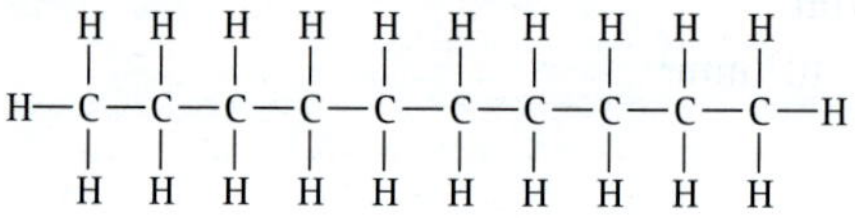

Condensed structural formula: $CH_3CH_2CH_2CH_2CH_2CH_2CH_2CH_2CH_2CH_3$ or $CH_3(CH_2)_8CH_3$

6.2)
```
        H H                      H    H
        | |                      |    |
   H—C≡C—C—C—H    or    H—C—C≡C—C—H
        | |                      |    |
        H H                      H    H
```

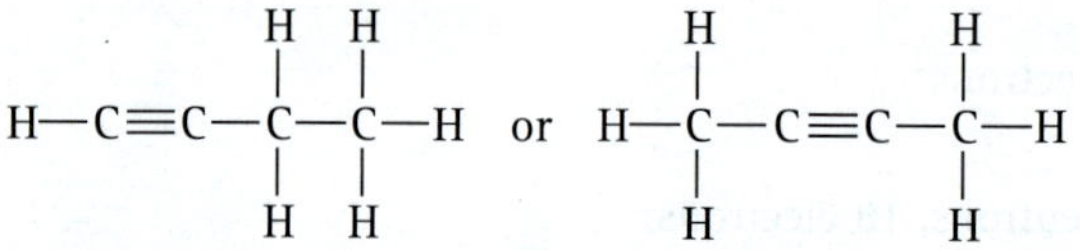

6.3)
```
         CH_3
          |
   CH_3—CH—CH_2—CH_2—CH_2—CH_2—CH_3

   CH_3—CH_2—CH_2—CH_2—CH_2—CH_2—CH_2—CH_3
```

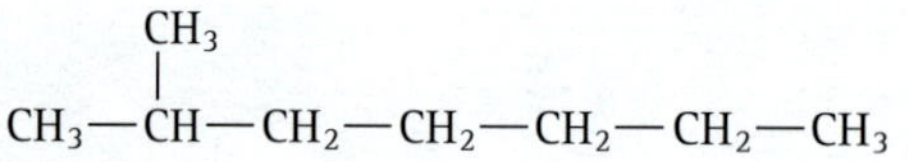

6.4) $CH_2=CH-CH_2-CH_2-CH_2-CH_3$
$CH_3-CH=CH-CH_2-CH_2-CH_3$
$CH_3-CH_2-CH=CH-CH_2-CH_3$ (any 2)

Chapter 7

7.1) 357 m

Chapter 8

8.1) $^{226}_{88}Ra \rightarrow {}^{222}_{86}Rn + {}^{4}_{2}He$

8.2) $^{214}_{82}Pb \rightarrow {}^{214}_{83}Bi + {}^{0}_{-1}e$

8.3) 16.0 mg; 1.39×10^{21} emissions

8.4) 5730 yr

8.5) Because the percentage composition falls between 0 and 1 half-life, we can safely say that the rock is younger than 4.5×10^9 yr old. Exact calculation of the rock's age is possible, but beyond the scope of this text.

Chapter 9

9.1) 2.14 kJ

9.2) $10.95

9.3) 261°F

9.4) 3.4×10^3 kcal

9.5) $2C_4H_{10} + (13) O_2 \rightarrow 8CO_2 + 10H_2O$

Chapter 10

10.1) 22%

10.2) 24W

Chapter 11

11.1) 0.50 L

11.2) 1.2 L

11.3) 0.34 L

Chapter 12

12.1) $C_{10}H_{22}$

12.2) HF, CH_3Cl, and CH_3CH_2OH are polar; the others are nonpolar.

12.3) NH_3

12.4) M = 0.58 M

12.5) 42 g

12.6) 1.1 ppm

12.7) 0.55 mg/L

12.8) 5.0×10^{-2} g NaF

12.9) 2.8×10^3 mg or 2.8 g

Chapter 13

13.1) $H_2SO_4 + 2NaOH \rightarrow 2H_2O + Na_2SO_4$
13.2) base—C_5H_5N (pyridine), acid—H_2O

Chapter 14

14.1) H_2 is oxidized and in CuO is reduced.
14.2) Reducing agent is H_2; oxidizing agent is V_2O_5.

Chapter 15

15.1) —$CH_2CHFCH_2CHFCH_2CHFCH_2CHF$—

Chapter 16

16.1) Both b. and c. are triglycerides.
16.2) Both b. and c. are unsaturated fats and will therefore be liquids at room temperature.
16.3) Structure c. is a carbohydrate.
16.4) a. monosaccharide, b. polysaccharide, c. disaccharide, d. polysaccharide
16.5) Molecule c. is an amino acid.
16.6)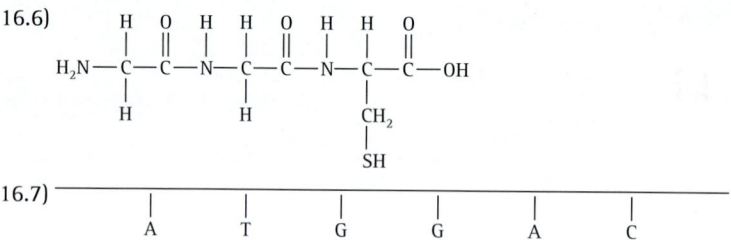
16.7)

A	T	G	G	A	C

Chapter 18

18.1) 15%
18.2) 3.17 lb

Appendix 1

A.1) Two significant figures, three signifiant figures, three significant figures, five significant figures, five significant figures
A.2) 0.002 or 2×10^{-3}
A.3) 12.6

Glossary

absorption spectrum A graph that shows electromagnetic radiation absorbed as a function of frequency.

abused inhalants Those drugs that are inhaled in their vapor state and abused for the temporary high they produce. Abused inhalants are usually depressants of the central nervous system and include glue solvents, marking pen solvents, paint thinners, and aerosol propellants.

acid A substance that produces hydrogen ions (H^+) in solution.

acid rain Rain that has been contaminated with sulfur oxides and nitrogen oxides (from fossil-fuel combustion) and consequently has turned acidic.

addition polymer A polymer formed by the addition of monomer units, one to the other without the elimination of any atoms.

adrenocortical steroids Drugs such as cortisone, hydrocortisone, prednisone, and prednisolone used primarily for their potent anti-inflammatory effects.

Agent Orange A mixture of 2,4,5-T and 2,4-D; used extensively during the Vietnam War to clear dense jungles and brush.

alchemy A partly empirical, partly magical, and entirely secretive pursuit with two main goals: the transmutation of ordinary materials into gold; and the discovery of the "elixir of life," a substance that would grant immortality to any who consumed it.

alcohols Organic compounds containing a —OH functional group bonded to a carbon atom and having the general formula R—OH.

aldehydes Organic compounds with the general formula RCHO.

alkali metals The elements in the column to the far left on the periodic table. Alkali metals are group 1A elements and are highly reactive metals.

alkaline earth metals The elements in the second column of the periodic table. The alkaline earth metals are group 2A elements and are reactive metals.

alkaloids A family of organic, nitrogen-containing bases.

alkanes Hydrocarbons in which all carbon atoms are connected by single bonds.

alkenes Hydrocarbons that contain at least one double bond between carbon atoms.

alkybenzenesulfonate (ABS) detergents Synthetic detergents with branched hydrocarbon chains. The ABS detergents work well in hard water but do not biodegrade in the environment.

alkylating agents A class of anticancer drugs. Alkylating agents are highly reactive compounds that add alkyl groups (such as CH_3CH_2—) to organic or biological molecules.

alkynes Hydrocarbons that contain at least one triple bond between carbon atoms.

alpha particle A form of radiation consisting of two protons and two neutrons and represented by the symbol ^4_2He.

amines Organic compounds that contain nitrogen and have the general formula NR_3 where R may be an alkyl group or a hydrogen atom.

amphetamine A nitrogen-containing organic compound that acts as a stimulant to the central nervous system.

anabolic steroids Illegal drugs that are synthetic analogs of testosterone, the male sex hormone. When taken in high doses and coupled with exercise or weight lifting, they act to increase muscle mass.

androgens The male sexual hormones.

anesthetic inhalants Those drugs, including nitrous oxide, isoflurane, halothane, and enflurane, that are inhaled in their vapor state to produce an anesthetic effect.

anionic detergent Any detergent in which the polar part of the detergent molecule is a negative ion.

anions Negatively charged ions.

anode The electrode in an electrochemical cell where oxidation occurs.

antagonists Drugs that undo the effects of other drugs by binding to active sites where the other drugs would otherwise bind.

anticaking agents Added to powders to keep them dry and granulated, they include silicon dioxide, calcium silicate, and magnesium silicate.

antimetabolites A class of anticancer drugs. Antimetabolites are chemical impostors; they work by mimicking compounds normally found in the cell but have subtle differences that usually stop DNA synthesis.

antimicrobial agents (as food additives) Substances added to food to prevent the growth of bacteria, yeasts, and molds.

antioxidant A substance that is easily oxidized and therefore prevents the oxidation of other substances. Antioxidants are added to foods to prevent the browning associated with oxidation.

antiseptics Substances that are applied to the skin or to minor cuts to kill bacteria and prevent infection.

Aristotle (384–321 B.C.) Greek philosopher who suggested that all matter was composed of four basic materials: air, water, fire, and earth (of Empedocles), and a fifth element, the heavenly ether, perfect, eternal, and incorruptible.

aromatic ring *See* benzene (ring).

Arrhenius, Svante (1859–1927) Swedish chemist who characterized acids and bases.

atom The smallest identifiable unit of an element.

atomic force microscope (AFM) A microscope with atomic resolution that is used to image nonmetallic surfaces. The AFM works by scanning a fine tip across a surface and using a laser beam to monitor the deflection of the tip as it moves up or down over surface structures.

atomic number (Z) The number of protons in the nucleus of an atom.

atomic theory A theory formulated by John Dalton that states that matter is ultimately composed of small indestructible particles called atoms.

atomic weight A weighted average of the masses of each naturally occurring isotope of an element.

Avogadro, Amadeo (1776–1856) A professor of physics at the University of Turin. The number of atoms in a mole (6.022×10^{23}) is called Avogadro's number in his honor.

bacteria Microorganisms that reproduce quickly and can live within the human body.

barbiturates Drugs, such as phenobarbital, that act as depressants of the central nervous system by enhancing the action of GABA, a neurotransmitter that restricts the flow of nerve signals.

barometer A device that measures pressure.

base A substance that produces hydroxide ions (OH^-) in solution.

B-complex vitamins A family of water-soluble vitamins that include thiamin, riboflavin, niacin, vitamin B6, folate, and vitamin B12. B-complex vitamins play central roles in metabolism, protein synthesis, and cell multiplication.

Becquerel, Antoine-Henri (1852–1908) French physicist who accidentally discovered radioactivity while studying X-rays.

Bennig, Gerd (1947–) Physicist who shared the 1996 Nobel prize for his joint discovery of the scanning tunneling microscope (STM).

benzene (C_6H_6) A particularly stable organic compound with six carbon atoms and alternating single and double bonds in a ring structure.

benzodiazepenes Drugs, such as diazepam (Valium), that act as depressants of the central nervous system by enhancing the action of GABA, a neurotransmitter that restricts the flow of nerve signals.

beta particle A form of radiation consisting of an electron and represented by the symbol $_{-1}^{0}e$.

bioamplification The increase in concentration of toxic chemicals within organisms as the chemicals move up the food chain.

biochemistry The study of the chemistry of living organisms.

biotechnology The application of biochemical knowledge to living organisms, especially in the area of genetic engineering.

blue baby A condition in which the ability of an infant's blood to carry oxygen is diminished by the ingestion of nitrates.

Bohr diagrams A diagram that specifies the arrangement of electrons within Bohr orbits for a particular element.

Bohr model A model for the atom in which electrons orbit around the nucleus much like planets orbit around the sun. Bohr orbits exist only at specific fixed energies and specific fixed radii.

Bohr, Niels (1885–1962) Danish physicist who formulated the Bohr model for the atom. In this model, electrons orbit the nucleus, much like planets orbit the sun.

boiling point The temperature at which the liquid and gaseous forms of a substance coexist.

bonding pair electrons The electrons between two atoms in a Lewis structure.

Boyle, Robert (1627–1691) British chemist who published *The Skeptical Chymist*, a book that criticized Greek ideas concerning a four-element explanation of matter. Boyle proposed that a substance must be tested to determine if it was an element. If a substance could be broken down into simpler substances, it was not an element.

Boyle's law States that if the pressure of a gas is increased, its volume will decrease in direct proportion.

branched structures Organic compounds whose carbon chains form branches.

breeder reactor A nuclear reactor that converts non-fissile uranium 238 into fissile plutonium 239.

brighteners Dye molecules that bind to clothes fibers and help make clothes appear "brighter than before."

Brønsted-Lowry acid A proton donor.

Brønsted-Lowry base A proton acceptor.

buckminsterfullerene A form of carbon with the formula C_{60} and a hollow spherical structure resembling a soccer ball.

buckyball A nickname for buckminsterfullerene.

builder Any substance in a cleaning formulation that increases the effectiveness of the surfactant.

caffeine A stimulant of the central nervous system found in coffee, tea, and soft drinks. Caffeine acts by increasing the concentration of neurotransmitters in the brain.

calcium As a nutritional mineral, calcium is the most abundant mineral in the body and is the main structural material of bones and teeth.

Calorie The big "C" or nutritional Calorie is a unit of energy equivalent to 1000 "little c" calories.

calorie (cal) A unit of energy equivalent to the amount of heat required to raise the temperature of 1 g of water by 1°C.

carbamate A type of insecticide that targets a narrow band of insects.

carbohydrates Polyhydroxyl aldehydes or ketones or their derivatives.

carboxylic acids Organic compounds with the general formula R—COOH.

carotenes A class of compounds responsible for the color of carrots and fall leaves.

catalyst A substance that promotes a chemical reaction without being consumed by it.

cathode The electrode in an electrochemical cell where reduction occurs.

cationic detergent Any detergent in which the polar part of the detergent molecule is a positive ion.

cations Positively charged ions.

Celsius scale A temperature scale in which the freezing point of water is defined as 0°C and the boiling point of water is defined as 100°C. Temperatures between these two points are graduated into 100 equally spaced degrees.

chalcogens Also called the oxygen family, the Group VIA elements that are moderately reactive nonmetals.

Charles's law States that if the temperature of a gas is increased, its volume will increase in direct proportion.

chemical formula A way to represent a compound. At a minimum, the chemical formula indicates the elements present in the compound and the relative number of atoms of each element.

chemical symbol A one- or two-letter abbreviation for an element.

chemistry The science that investigates the molecular reasons for those processes that are constantly occurring in our macroscopic world.

chlorofluorocarbons (CFCs) A family of compounds containing carbon, chlorine, and fluorine. CFCs are implicated in the destruction of ozone and were banned from production in the United States in 1996.

chlorophylls A class of compounds found in plant leaves that produce their green color.

cholesterol A nonpolar compound with the four-ring structure (Chapter 18, Figure 11) characteristic of steroids. Cholesterol is found in animal foods like beef, eggs, fish, poultry, and milk products and is suspected of contributing to cardiovascular disease.

chromium Important in the control of blood glucose levels.

coal Primarily long chains and rings of carbon with 200 or more atoms.

cocaine A nitrogen-containing organic compound that acts as a stimulant to the central nervous system by blocking the reuptake of neurotransmitters, especially dopamine and norepinephrine.

codeine Another opioid drug found in opium.

codon A group of three bases within a nucleic acid that codes for one amino acid.

cohesive forces The attractions between molecules that are responsible for the existence of liquids and solids.

collagen The synthesizer of connective tissue.

colloidal suspension A mixture of one substance dispersed in a finely divided state throughout another.

combustion A chemical reaction in which certain compounds, especially hydrocarbons, combine with oxygen to produce certain products, usually carbon dioxide and water.

complete proteins Proteins that contain all the essential amino acids in the right proportion.

compound A substance composed of two or more elements in fixed definite proportions.

concentration For a solution, the amount of solute relative to the amount of solvent.

condensation polymer A class of polymers that expel atoms, usually water, during their formation.

conservation of mass, law of States that in a chemical reaction matter is neither created nor destroyed.

constant composition, law of States that all samples of a given compound have the same proportions of their constituent elements.

conversion factor A quantity that can usually be written as a ratio of two quantities, each with different units. Conversion factors can be used to convert from one set of units to another.

Copernicus, Nicholas (1473–1543) Polish astronomer who claimed the sun to be the center of the universe.

copolymers Polymers that are composed of two monomers and result in chains composed of alternating units rather than a single repeating unit.

copper As a nutritional mineral, copper is needed to form Hb and collagen.

covalent bond The bond that results when two nonmetals combine in a chemical reaction. In a covalent bond, the atoms share their electrons.

critical mass The mass of uranium or plutonium required for a nuclear reaction to be self-sustaining.

crystalline solids Solids in which the constituent atoms or molecules form well-ordered, three-dimensional arrays.

crystalline structure The repeating pattern of molecules or atoms in crystalline solids.

Curie, Marie Sklodowska (1867–1934) French (Polish-born) chemist who helped understand radioactivity and also discovered two new elements together with her husband Pierre Curie.

Curie, Pierre (1859–1906) French chemist who helped understand radioactivity and also discovered two new elements together with his wife Marie Curie.

Curl, Robert, Jr. (1933–) American chemist who shared the 1996 Nobel prize in chemistry for his joint discovery of buckminsterfullerene.

Dalton, John (1766–1844) British scientist who formulated the atomic theory.

defoliants Herbicides that kill plants and trees by causing them to lose their leaves.

Democritus (460–370 B.C.) Greek philosopher who theorized that matter was ultimately composed of small, indivisible particles he called *a tomos*, or atoms, meaning *not to cut*.

density (d) For any object or substance, defined as the ratio of its mass (m) to its volume (V) ($d = m/V$). Density is often expressed in units of grams per cubic centimeter (g/cm^3).

depressants Those drugs that depress or dull the central nervous system.

dioxins A family of compounds with varying levels of toxicity.

disaccharides Carbohydrates, such as sucrose, with double-ringed structures.

dish/engine A form of solar electricity generation in which a dish focuses sunlight on an engine that uses the heat to produce electricity.

disinfectants Substances that kill bacteria and are used to sterilize and sanitize.

dispersion force A cohesive force resulting from small fluctuations in the electron clouds of atoms and molecules.

DNA (deoxyribonucleic acid) Long chainlike molecules that occur in the nucleus of cells and act as blueprints for the construction of proteins.

dopamine A neurotransmitter whose brain levels are increased by stimulant drugs such as cocaine. Increased dopamine brain levels produce feelings of alertness and power.

doping The addition of small amounts of other elements to semiconductors to modify and control their conducting properties.

double bond A bond between two atoms in which two electron pairs are shared.

Einstein, Albert (1879–1955) Physicist who discovered relativity, contributed to the discovery of quantum mechanics, and persuaded President Roosevelt to start the Manhattan Project.

elastomer A polymer that stretches easily and returns to its original shape.

electric field The area around a charged particle where forces are experienced.

electrochemical cell A device that uses the natural tendency of certain compounds to exchange electrons to produce electrical current or that uses electrical current to achieve specific chemical reactions.

electrolysis The process of using an electrical current to split water into hydrogen and oxygen.

electrolyte solutions Solutions having dissolved ions. Electrolyte solutions are good conductors of electricity due to the mobility of the charged ions.

electromagnetic radiation Radiation that is propagated by the combination of electric and magnetic fields including gamma rays, X-rays, ultraviolet (UV) light, visible light, infrared (IR) light, microwaves, and radio waves.

electron configurations A diagram that specifies the arrangement of the atom's electrons within Bohr orbits for a particular element.

electronegativity The ability to attract electrons in a covalent bond.

electronic relaxation An electronic transition in which an electron goes from a higher-energy orbit to a lower-energy one.

electronic transition When the electrons in an atom or molecule move from one energy state to another.

electrons Negatively charged particles that compose most of the atom's volume but almost none of its mass.

element A substance that cannot be broken down into simpler substances.

Empedocles (490–430 B.C.) Greek philosopher who suggested that all matter was composed of four basic materials or elements: air, water, fire, and earth.

emulsifiers Substances that keep mixtures of polar and nonpolar substances from separating into distinct layers.

emulsifying agent A substance that causes one substance to disperse in a second substance.

emulsions A mixture, aided by an emulsifying agent, of two substances that would not normally mix.

endorphins Opioid compounds produced by the body in response to pain. Endorphins bind to opioid receptors and suppress pain.

endothermic reaction A chemical reaction that absorbs energy from the surroundings.

energy The capacity to do work.

energy state The electron configuration of a molecule or atom with electrons in particular orbits and therefore at particular energies.

enthalpy of combustion (ΔH_{com}) The enthalpy of reaction for combustion.

enthalpy of reaction (ΔH_{rxn}) The amount of heat absorbed or emitted by a chemical reaction.

entropy Disorder or randomness.

enzyme A biological substance that catalyzes or promotes a specific biochemical reaction.

essential amino acids Those amino acids that are necessary for the production of human proteins but are not synthesized by the body. Essential amino acids must be obtained from food.

esters Organic compounds with the general formula R—COO—R.

estrogen Female sex hormone.

ethanol The alcohol found in alcoholic drinks.

ethers Organic compounds with the general formula R—O—R.

eutrophication The process by which overfertilization of natural waters leads to abnormally large algae blooms, which in turn consume large amounts of oxygen on their decomposition. The water becomes oxygen-poor, threatening many types of marine life.

evaporation Molecules in a liquid, undergoing constant random motion, may acquire enough energy to overcome attractions with neighbors and fly into the gas phase.

excited state An unstable state for an atom or molecule in which energy has been absorbed but not reemitted.

exothermic reaction A chemical reaction that gives off energy to the surroundings.

experiment A measurement or observation of nature that aims to validate or invalidate a scientific theory.

Fahrenheit scale A temperature scale in which the freezing point of water is defined as 32°F and the boiling point of water is defined as 212°F.

family of elements Elements such as He, Ne, and Ar that have similar outer electron configurations and therefore similar properties. Families occur in vertical columns in the periodic table.

fats Triglycerides.

fatty acid A type of lipid consisting of an organic acid with a long hydrocarbon tail.

Fermi, Enrico (1901–1954) Italian physicist who played an important role in the development of nuclear fission. In collaboration with Szilard, he constructed the first nuclear reactor.

fiber Nondigestible carbohydrates composed of long chains of glucose units linked together via beta linkages.

fillers Ingredient found in laundry-cleaning formulations that serve no laundering purpose except to increase the bulk of the detergent and control consistency and density.

first law of thermodynamics A scientific law stating that energy can neither be created nor destroyed, only transferred between a system and its surroundings.

fission, nuclear The splitting of a heavy atomic nucleus to form lighter ones.

flash freezing A process used by the frozen food industry to minimize the damage incurred on freezing. In this process, food is frozen very quickly, preventing water molecules from ordering in their ideal crystalline structure and resulting in less expansion and therefore less cell rupture.

flavor enhancers Those substances that are added to food to enhance its flavors.

flue gas scrubber A device often installed in power plant smokestacks to reduce sulfur dioxide emissions.

fluorescence The fast emission of light following electronic excitation.

fluoxetine (Prozac) The first SSRI to be available in the United States in 1988.

food additives Substances added to food to preserve it, enhance its flavor or color, and maintain its appearance or physical characteristics.

fossil fuels Those fuels—including coal, petroleum, and natural gas—that have formed from prehistoric plant and animal life.

free radicals Atoms or molecules with unpaired electrons.

freezing point depression The lowering of a freezing point of a liquid by the addition of a solute.

frequency For a wave, the number of cycles or crests that pass through a point in one second, usually reported in units of cycles per second or hertz.

fructose Consists of a single saccharide group and is found in fruits, plants, honey, and products sweetened by high fructose corn sweetener (HFCS).

functional group A set of atoms that characterize a family of organic compounds.

fusion, nuclear The combination of light atomic nuclei to form heavier ones.

Galileo (1564–1642) Italian astronomer who confirmed and expanded on Copernicus's ideas about a sun-centered universe. Galileo was chastised by the Roman Catholic Church for his views.

galvanization The coating of iron with zinc to prevent corrosion.

gamma ray The shortest wavelength and most energetic form of electromagnetic radiation. Each ray consists of an energetic photon emitted by an atomic nucleus and is represented by the symbol $^0_0\gamma$.

gel electrophoresis A method used to separate DNA fragments in which a mixture of DNA fragments is placed on a thin gel and an electric current is applied, which causes the mixture to separate into distinct bands, each representing a different DNA fragment.

geothermal energy The use of heat or steam from the earth's interior to generate electricity.

glucose A simple carbohydrate that acts as the body's primary fuel source.

Goodyear, Charles (1800–1860) Inventor of vulcanization, a process that makes rubber harder and more elastic.

greenhouse gases Those gases that allow visible light into the atmosphere but prevent heat in the form of infrared (IR) light from escaping. Increases in these gases from anthropogenic sources are believed to be causing a slight increase in the earth's average temperature.

group of elements Same as **family of elements**.

Haber process The reactions used in the reduction of atmospheric nitrogen into ammonia (nitrogen fixation).

Hahn, Otto (1879–1968) German chemist credited as one of the discoverers of nuclear fission.

half-life For a radioactive element, the time required for half the nuclei in a sample to decay.

hallucinogenic drugs Those drugs, such as mescaline and LSD, that distort and disturb cognition and perception.

halogens Group 7A elements, which are highly reactive nonmetals.

hard water Water rich in calcium and magnesium ions.

heat Random molecular or atomic motion.

heat capacity The quantity of heat energy required to change the temperature of a given amount of substance by 1°C.

hemoglobin A human protein responsible for carrying oxygen in the blood.

herbicides Chemical compounds that kill plants, usually targeted at weeds.

heroin A narcotic produced through a chemical modification of morphine. Heroin is extremely potent and addictive.

Hertz, Heinrich (1857–1894) German physicist who discovered radio waves in 1888.

histamines Biochemical substances that mediate inflammation in the body.

hormone A biochemical substance that regulates specific processes within an organism's body.

humectants Those substances added to foods to keep them moist.

hydrocarbons Organic compounds containing only carbon and hydrogen.

hydroelectric power The generation of electricity through dams that tap the energy in moving water to produce electricity.

hydrogen bond A strong cohesive force that occurs between polar molecules containing H atoms bonded to F, O, or N atoms.

hydrologic cycle The path water takes from the oceans to the clouds to rain and back to the oceans.

hypoxia Oxygen deprivation.

infrared (IR) light That portion of the electromagnetic spectrum that is immediately adjacent to red light. Infrared light is invisible to the human eye.

infrared radiation The fraction of the electromagnetic spectrum that is between the visible region and the microwave region. IR radiation is invisible to the human eye but can be felt as heat.

inhalant Any drug that is inhaled in its vapor state including those used for anesthesia, such as nitrous oxide, or those that are abused, such as paint and glue solvents. Most inhalants are depressants of the central nervous system.

insecticides Chemical compounds that kill insects.

instantaneous dipole A type of intermolecular force resulting from transient shifts in electron density within an atom or molecule.

interior carbon atoms In an organic molecule, interior carbon atoms are those located between two other carbon atoms.

iodine A trace mineral needed by the body in only very small quantities.

ion An atom or molecule that has an unequal number of electrons and protons. Ions with more electrons than protons are negatively charged while those with more protons than electrons are positively charged.

ionic bond The bond that results when a *metal* and a *nonmetal* combine in a chemical reaction. In an ionic bond, the metal transfers one or more electrons to the nonmetal.

ionosphere The highest section of the atmosphere ranging from altitudes of 80–160 km.

iron Composes a critical part of hemoglobin (Hb), the protein that carries oxygen in the blood.

isomers Molecules having the same molecular formula but different structures.

isotopes Atoms having the same number of protons but different numbers of neutrons.

joule (J) A basic unit of energy equivalent to 0.239 calories.

Joule, James (1818–1889) English scientist who demonstrated that energy could be converted from one type to another as long as the total energy was conserved.

Kekule, Friedrich August (1829–1896) German chemist who played a central role in elucidating many early organic chemical structures, especially benzene, and carbon's tendency to form four bonds.

Kelvin scale A temperature scale that avoids negative temperatures by assigning 0K to the coldest temperature possible. In the Kelvin scale, the freezing point of water is 273K and the boiling point of water is 373K.

ketones Organic compounds with the general formula RCOR.

kilogram (kg) The SI standard of mass, defined as the mass of a particular block of metal kept at the International Bureau of Weights and Measures, at Sèvres, France.

Kroto, Sir Harold (1937–) English chemist who shared the 1996 Nobel prize in chemistry for his joint discovery of buckminsterfullerene.

lactase The enzyme that catalyzes the hydrolysis of lactose.

lactose A disaccharide, consisting of the monosaccharides glucose and galactose, found primarily in milk and milk products.

lactose intolerance A common condition in which adults have low lactase levels and therefore have trouble digesting milk or milk products.

laser An acronym for light amplification by stimulated emission of radiation. Lasers produce intense, single-wavelength light.

laser cavity In a laser, the lasing medium is placed inside a laser cavity consisting of two mirrors, one of which is only partially reflecting.

lasing medium Laser light is formed by putting electrical or light energy into an element or a compound called the lasing medium.

Lavoisier, Antoine (1743–1794) French chemist who published a chemical textbook titled *Elementary Treatise on Chemistry*. Lavoisier is known as the father of modern chemistry because he was among the first to carefully study chemical reactions.

Lewis, G. N. (1875–1946) American chemist at the University of California at Berkeley who developed a simple theory for chemical bonding, now called Lewis theory.

linear alkylsulfonate (LAS) detergents Synthetic detergents with unbranched hydrocarbon chains. The LAS detergents work well in hard water and also biodegrade in the environment.

lipids Those cellular components that are insoluble in water but extractable with nonpolar solvents.

lipoproteins Specialized proteins that act as carriers for fat molecules and their derivatives in the bloodstream.

litmus paper Treated paper that turns red in acidic solution and blue in basic solution.

lone pair electrons The electrons on a single atom in a Lewis structure.

LSD Short for lysergic acid diethylamide. LSD is a hallucinogenic drug that causes distortions in perception.

magnesium As a nutritional mineral, magnesium is involved in maintaining electrolyte balance in and around cells.

magnetic field The area around a magnet where forces are experienced.

magnetic resonance imaging (MRI) A type of spectroscopy involving magnetic fields and radio waves that allows medical doctors to image organs and structures within the human body.

major minerals Compose about 4% of the body's weight and include Ca, P, Mg, Na, K, Cl, and S.

mass A measure of the quantity of matter present within an object. Mass is subtly different from weight in that weight depends on gravity and mass does not.

mass defect The difference between the experimentally measured mass of a nucleus and the sum of the masses of the protons and neutrons contained in the nucleus.

mass number (A) The sum of the number of neutrons and protons in the nucleus of a given atom.

Meitner, Lise (1878–1968) German chemist credited as one of the discoverers of nuclear fission.

melting point The temperature at which the solid and liquid forms of a substance coexist.

Mendeleev, Dmitri (1834–1907) Russian chemist who is credited with arranging the first periodic table of the elements.

mescaline A drug extracted from the peyote cactus that acts as both a stimulant and a hallucinogen.

mesosphere The region of the atmosphere ranging from altitudes of 50–80 km.

metalloids Those elements that fall in the middle-right of the periodic table and have properties between metals and nonmetals.

metals Those elements that tend toward the left side of the periodic table and tend to lose electrons in their chemical reactions.

meter (m) The SI standard of length, originally defined as 1/110,000,000 of the distance from the North Pole to the equator along a meridian passing through Paris. Today, it is defined as the distance light travels in a certain period of time, 1/299,792,458 s.

methadone A synthetic opioid drug that also binds to opioid receptor sites.

methamphetamine Also known as speed, causes the release of stored neurotransmitters to produce the stimulant effect.

methyl group A unit of organic compounds that contains one carbon atom and three hydrogen atoms (—CH$_3$).

methylene group A unit of organic compounds that contains one carbon atom and two hydrogen atoms (—CH$_2$—).

micelle A structure formed by surfactants in water where the nonpolar tails of surfactant molecules crowd into a spherical ball to maximize their interactions with one another and minimize their interaction with water molecules.

microwave radiation The fraction of the electromagnetic spectrum that is between the infrared (IR) region and the radio wave region. Microwave radiation is efficiently absorbed by water molecules and can therefore be used to heat water-containing substances.

minerals (nutritional) Those elements—other than carbon, hydrogen, nitrogen, and oxygen—needed for good health.

minor minerals Are present in the body in trace amounts and include Fe, Cu, Zn, I, Se, Mn, F, Cr, and Mo.

mixture A combination of two or more substances in variable proportions.

mixture, heterogeneous A mixture containing two or more regions with different compositions.

mixture, homogeneous A mixture with the same composition throughout.

molarity (M) A common unit of concentration defined as moles of solute per liter of solution.

mole A large number, 6.022×10^{23}, usually used when dealing with small particles such as atoms or molecules.

mole concept The idea that the molecular weight of any element or compound, in grams, is equivalent to one mole (6.022×10^{23}) of atoms (for an element) or molecules (for a compound).

molecular formula A chemical formula that specifies the actual number of each kind of atom in a molecule.

molecular weight The sum of the atomic weights of all the atoms in a molecular formula.

molecule The smallest identifiable unit of a covalent compound.

monosaccharides Carbohydrates, such as glucose or fructose, with a single-ring structure.

morphine The primary component of opium. Morphine acts on the central nervous system to produce an analgesic and sedative effect.

Muller, Paul (1899–1965) Swiss chemist who demonstrated that a chlorinated hydrocarbon, dichlorodiphenyltrichloroethane (DDT), was unusually effective in killing insects but relatively nontoxic toward humans.

nanotechnology A relatively new field of science in which microscopic structures are built one atom at a time.

nanotube A class of carbon molecules, similar in structure to buckyballs, but forming long thin tubes with diameters of just a few atoms.

narcotics Those drugs that act on the central nervous system to produce an analgesic and sedative effect.

natural gas A mixture of methane and ethane.

neurotransmitter A substance that helps transmit messages among nerve cells.

neutralization A reaction in which an acid and base react and neutralize each other.

neutrons Nuclear particles with no electrical charge and nearly the same mass as protons (1 amu).

nicotine The active drug found in cigarette smoke, chewing tobacco, nicotine gum, and nicotine skin patches.

nitrogen fixation The conversion of nitrogen molecules (N$_2$) into nitrogen compounds such as ammonia or nitrates that can be absorbed by plants.

noble gases The elements in the column on the far right of the periodic table. Noble gases are group 8A elements and are all inert gases.

nonionic detergent Any detergent in which the polar part of the detergent molecule is not ionic but maintains polarity because of OH or other polar groups.

nonmetals Those elements that tend toward the right side of the periodic table and tend to gain electrons in their chemical reactions.

nonpolar Compounds are termed nonpolar if they have no permanent dipole moment.

nonvolatile A compound is said to be nonvolatile if it does not vaporize easily.

norepinephrine A neurotransmitter whose brain levels are increased by stimulant drugs such as cocaine. Increased norepinephrine brain levels produces feelings of alertness and power.

n-type semiconductor A semiconductor doped with an element with more valence electrons than the semiconductor itself. *n*-type semiconductors have negative charge carriers.

nuclear atom, theory of Ernest Rutherford's theory of atomic structure whose main tenet was that most of the mass composing an atom was contained in a small space called the nucleus.

nuclear binding energy The energy that holds a nucleus together.

nuclear equation An equation that represents a nuclear reaction. Nuclear equations must be balanced with respect to mass number and atomic number.

nuclear magnetic resonance (NMR) A type of spectroscopy involving magnetic fields and radio waves and often used by chemists and biochemists to deduce molecular structures.

nucleic acids Biological molecules such as deoxyribonucleic acid (DNA) and ribonucleic acid (RNA) that act as the templates from which proteins are made.

nucleoside analogs A class of antiviral drugs that when in living organisms mimic nucleotides but inhibit DNA synthesis.

nucleotides The individual links—including adenine, thymine, cytosine, and guanine (abbreviated A, T, C, and G)—that compose nucleic acids.

observation A measurement of some aspect of nature. This may only involve one person making visual observations, or it may require a large team of scientists working together with complex and expensive instrumentation.

octet The number of electrons, eight, around atoms with stable Lewis structures.

oils Triglycerides.

opioid The class of narcotics derived from, or synthesized to act like, opium.

opioid antagonists Drugs that bind to opioid receptor sites, displacing opioids but not producing the narcotic effects.

opioid receptors Specific sites on nerve cells in the spinal cord and brain where narcotics bind and inhibit nerve signal transmission.

opium A narcotic, extracted from the opium poppy, that acts on the central nervous system to produce an analgesic and sedative effect.

Oppenheimer, J. R. (1904–1967) Director of the Manhattan Project, the U.S. endeavor to build the world's first nuclear weapon.

organic Compounds that contain carbon.

organic chemistry The chemistry of carbon and a few other elements such as nitrogen, oxygen, and sulfur.

organophosphate Broad-band insecticide.

oxidation The gain of oxygen, the loss of hydrogen, or the loss of electrons (the most fundamental definition).

oxidizing agents Substances that tend to gain electrons easily. They bring about the oxidation of other substances while they are themselves reduced.

oxygen family The column of elements directly below oxygen on the periodic table.

parabolic troughs A form of solar electrical generation in which large troughs track and focus sunlight onto a receiver pipe that runs along the trough's length. Oil circulating within the receiver pipe is heated and used to create steam and electricity.

penicillin An antibiotic, first discovered by Alexander Fleming, that kills bacteria by interfering with the normal development of their cell walls.

percent by mass A way of reporting the concentration of one substance in another. Percent by mass is defined as the (mass of the component substance/total mass of all the component substances) $\times 100\%$.

percent by volume A way of reporting the concentration of one substance in another. Percent by volume is defined as the (volume of the component substance/total volume of all the component substances) $\times 100\%$.

periodic table An arrangement of the elements in which atomic weight increases from left to right and elements with similar properties appear in columns called families or groups.

permanent dipole A separation of charge resulting from the unequal sharing of electrons between atoms.

perpetual motion machine A device that continuously produces energy without the need for energy input.

petroleum Contains a wide variety of hydrocarbons from pentane to hydrocarbons with 18 carbon atoms or more.

pH scale A scale used to quantify acidity or basicity. A pH of 7 is neutral; a pH lower than 7 is acidic; and a pH greater than 7 is basic.

phenoxy herbicides Herbicides, such as 2,4,5-T and 2-4D, that contain an oxygen bonded to a phenyl group.

phenyl ring A benzene ring with other substituents attached.

phosphorescence The slow emission of light following electronic excitation.

phosphorus As a nutritional mineral, phosphorus is the second most abundant mineral in the body and is largely bound with calcium in bones and teeth.

photochemical smog Air pollution consisting primarily of ozone and peroxyacetylnitrate (PAN), formed by the action of sunlight on unburned hydrocarbon fuels and nitrogen oxides.

photodecomposition The breaking of bonds within molecules due to the absorption of light.

photons Particles of light.

photosynthesis The process by which plants convert carbon dioxide, water, and sunlight into glucose: sunlight + $6CO_2$ + $6H_2O$ + $C_2H_2O_6$.

photovoltaic (PV) cells A form of solar electricity generation in which a semiconductor converts light directly into electrical current.

Plato (428–348 B.C.) Greek philosopher who asserted that reason alone was the superior way to understand the natural world.

PM-10 An abbreviation for particulate matter pollutants with a diameter less than or equal to 10 micrometers (μm).

p-n junction The junction between an n-type semiconductor and a p-type semiconductor. p-n junctions are used in photovoltaic (PV) cells for electricity generation.

polar bond A bond between elements with different electronegativities. Polar bonds have an uneven electron distribution and therefore a small charge separation.

polar molecule A molecule with polar bonds that do not cancel. One end of a polar molecule has a slight positive charge while the other end has a slight negative charge.

polarity The degree to which a molecule displays a dipole moment due to separation of electronic charge.

polyatomic ion An ion containing more than one atom.

polyethylene One of the most common polymers.

polymer A molecule with many similar units called monomers bonded together in a long chain.

polypropylene A polymer consisting of repeating propylene units and often drawn into fibers to make athletic clothing, fabrics, and carpets.

polysaccharides Carbohydrates composed of a long chain of monosaccharides linked together.

polystyrene A substituted polyethylene.

polyvinyl chloride (PVC) A substituted polyethylene.

potassium As a nutritional mineral, potassium is involved in maintaining electrolyte balance in and around cells.

power Energy per unit time, often reported in watts or joules per second.

pressure The net force created by the constant pelting of gas molecules against surfaces.

primary protein structure The amino acid sequence in a protein.

products The substances on the right-hand side of a chemical equation.

progesterone Female sex hormone.

proof For alcoholic beverages, the proof is twice the percent alcohol content by volume. For example, an eighty proof liquor contains 40% alcohol.

prostaglandins Fatty acid derivatives involved in a number of physiological processes including the propagation of pain impulses along nerves and the formation of fevers.

protease An enzyme that acts like a molecular scissors to cut freshly made proteins to the correct size.

protease inhibitors A class of antiviral drugs that work by blocking viral protein synthesis.

proteins The substances that make up much of our structure.

protons Nuclear particles with positive charge and a mass of 1 atomic mass unit (amu).

Proust, Joseph (1754–1826) French chemist who established the law of constant composition.

p-type semiconductor A semiconductor doped with an element with fewer valence electrons than the semiconductor itself. *p*-type semiconductors have positive charge carriers.

quad A unit of energy equivalent to 1.06×10^{18} joules (J).

quantum number (n) An integer that specifies the energy and the radius of a Bohr orbit. The higher the quantum number n is, the greater the distance between the electron and the nucleus and the higher its energy.

quaternary protein structure The arrangement of protein subunits in space.

radio waves The longest wavelength and least energetic form of electromagnetic radiation. Radio waves are used extensively as a medium for communication and broadcasting.

radon A radioactive gas emitted by the decay of uranium. Radon is by far the greatest single source of radiation exposure for humans.

reactants The substances on the left-hand side of a chemical equation.

recombinant DNA DNA that has been transferred from one organism to another and allowed to recombine with the new organism's DNA.

redox reaction A chemical reaction involving oxidation and reduction.

reducing agents Substances that tend to lose electrons easily. They bring about the reduction of other substances while they are themselves oxidized.

reduction The loss of oxygen, the gain of hydrogen, or the gain of electrons (the most fundamental definition).

relative humidity A measure of the actual water content of air relative to its maximum water content.

relaxation time The time required for nuclei to return to their original orientation after being pushed by electromagnetic radiation.

rem The most common unit for measuring human exposure to radiation.

resonance frequency The exact electromagnetic frequency that causes an energy transition in a molecule or atom.

respiration The process by which animals, including humans, oxidize food to obtain energy.

restriction enzymes Enzymes that cut DNA at specific places along the DNA strand.

retrovirus The human immunodeficiency virus (HIV) that causes AIDS is a retrovirus. Unlike most viruses, which consist of DNA and protein, retroviruses consist of RNA and protein.

ribosome A cellular structure where protein synthesis occurs.

RNA (ribonucleic acid) Long chainlike molecules that occur throughout cells and act as blueprints for the construction of proteins.

Roentgen, Wilhelm (1845–1923) A German physicist who discovered X-rays.

Rohrer, Heinrich (1933–) Swiss physicist who shared the 1996 Nobel prize for his joint discovery of the scanning tunneling microscope (STM).

Rutherford, Ernest (1871–1937) British physicist who studied the internal structure of the atom with his now-famous gold foil experiment. Based on this experiment, Rutherford formulated the theory of the nuclear atom.

salt An ionic compound formed by the reaction of an acid and base. The term is often used to refer specifically to sodium chloride.

saturated Containing the maximum number of hydrogen atoms and therefore no double bonds.

saturated hydrocarbons Compounds containing only carbon and hydrogen and having no double bonds between the carbon atoms.

scanning tunneling microscope (STM) A microscope with atomic resolution that is used to image metallic surfaces. The STM works by scanning a fine tip across a metallic surface and monitoring the electrical current between the tip and surface.

scientific law A broadly applicable generalization that summarizes some aspect of the behavior of the natural world. Scientific laws are usually formulated from a series of related observations or measurements.

scientific method The way that scientists learn about the natural world. The scientific method consists of observation, law, theory, and experiment.

scurvy A condition in which bodily tissues and blood vessels become weakened.

sebum Oily substance excreted by glands in the skin and scalp.

second (s) The standard SI unit of time defined by an atomic standard using a cesium clock.

second law of thermodynamics A scientific law stating that for any spontaneous process, the universe—the system and the surroundings—must become more disorderly.

secondary protein structure The way a chain of amino acids in a protein orients itself along its axis.

selenium An antioxidant.

sertraline (Zoloft) Antidepressant introduced in the U.S. shortly after fluoxetine.

Smalley, Richard (1943–) American chemist who shared the 1996 Nobel prize in chemistry for his joint discovery of buckminsterfullerene.

sodium As a nutritional mineral, sodium is involved in bodily fluid level regulation.

solar power towers A form of solar electricity generation in which the sun is focused onto a tower to generate heat, which is then used to produce steam and electricity.

solute The minority component of a solution.

solution A mixture in which a solid, liquid, or gas dissolves in a liquid.

solvent The majority component of a solution.

spectroscopy The interaction of light with matter.

starches Digestible carbohydrates composed of long chains of glucose units linked together via alpha linkages.

steroids Biochemical compounds characterized by their 17-carbon-atom, four-ring, skeletal structure.

stimulants Those drugs such as cocaine or amphetamine that stimulate the central nervous system.

Strassmann, Fritz (1902–1980) Chemist credited as one of the discoverers of nuclear fission.

stratosphere The region of the atmosphere ranging from altitudes of 10–50 km and containing the ozone (O_3) layer.

strong acid An acid that completely dissociates in solution.

strong base A base that completely dissociates in solution.

structural formula A two-dimensional representation of molecules that shows which atoms are bonded together and which are not.

sucrose A disaccharide, two monosaccharide units (glucose and fructose) linked together, that occurs naturally in sugar cane, sugar beets, and many plant nectars.

surfactant A class of compounds that have both polar and nonpolar parts. Because of their dual nature, surfactants tend to aggregate at surfaces.

surroundings In thermodynamics, the environment the subject of study is exchanging energy with.

Szilard, Leo (1898–1964) American (Hungarian-born) physicist who played an important role in the development of nuclear fission. In collaboration with Fermi, he constructed the first nuclear reactor.

terminal carbon atoms In an organic molecule, terminal carbon atoms are those located at the end of a carbon chain.

tertiary protein structure The way a protein bends and folds due to interactions between amino acids separated by significant distances on the amino acid chain.

testosterone A male sex hormone that acts before the birth of a male infant to form the male sex organs and again at puberty and beyond to promote the development of the male sexual characteristics.

Thales (624–546 B.C.) Greek philosopher who reasoned that any substance could be converted into any other substance, so all substances are in reality one basic material. Thales believed the one basic material to be water.

theory A tentative model that describes the underlying cause of physical behavior. It gives the fundamental reasons for scientific laws and predicts new behavior.

thermodynamics The study of energy and its transformation from one form to another.

thermodynamic system The subject of study in a thermodynamic experiment.

thermoplastic A polymeric material that softens when heated and hardens when cooled.

topoisomerase inhibitors A class of anticancer drugs that work by modifying the action of a topoisomerase, an enzyme that pulls the individual strands of DNA apart in preparation for DNA synthesis. Effectively, these compounds cause cells to die whenever they initiate DNA replication.

transition metals Groups 3B through 2B in the center of the periodic table. Transition metals lose electrons in their chemical reactions but do not necessarily acquire noble gas configurations.

triglyceride The combination of glycerol (a trihydroxy alcohol) with three fatty acids.

triple bond A bond between two atoms in which three electron pairs are shared.

troposphere The lowest part of the atmosphere ranging from ground level to altitudes of 10 km (6 mi).

ultraviolet (UV) light The fraction of the electromagnetic spectrum that is between the visible region and the X-ray region. UV light is invisible to the human eye.

unit Previously agreed upon quantities used to report experimental measurements. For example, centimeter is a unit used to report measurements of length.

unsaturated hydrocarbons Hydrocarbons that contain at least one double or triple bond between carbon atoms.

urea The first organic compound ever synthesized in the laboratory.

valence electrons The electrons in an element's outermost Bohr orbit.

valence shell electron pair repulsion (VSEPR) A theory that allows the prediction of the shapes of molecules based on the idea that electrons—either as lone pairs or as bonding pairs—repel one another.

vapor pressure The pressure above a liquid in equilibrium with its vapor.

Vesalius, Andreas (1514–1564) Flemish anatomist who portrayed human anatomy with unprecedented accuracy.

visible light The fraction of the electromagnetic spectrum that is visible to the human eye. The visible region is bracketed by wavelengths of 400 nanometers (nm, violet) and 780 nm (red) and contains all wavelengths in between.

vital force A mystical force possessed by living organisms that was believed to allow them to overcome physical law and synthesize organic compounds.

vitalism The nineteenth-century belief that living organisms contain a vital force allowing them to overcome ordinary physical laws and produce organic compounds.

Vitamin A This vitamin is important in vision, immune defenses, and maintenance of body lining and skin.

Vitamin C A water-soluble vitamin that serves several roles in the body, including the synthesis of collagen, protection against infection, and absorption of iron through the intestinal wall.

Vitamin D This vitamin promotes the absorption of Ca through the intestinal wall and into the blood.

Vitamin E This vitamin serves as an antioxidant in the body, preventing oxidative damage to cellular components and especially cell membranes.

Vitamin K This vitamin is necessary for the synthesis of four proteins involved in blood clotting.

vitamins Those organic compounds essential in the diet in small amounts but with little or no caloric value.

volatile Tending to vaporize easily.

volume The volume of an object is a measure of the amount of space it occupies. Volume has units such as m^3 or cm^3.

vulcanization A process in which sulfur atoms form cross-links between the polyisoprene chains in rubber. These cross-links make the rubber stronger and more durable, even when warm.

wavelength The distance between wave crests, usually reported in units of length such as meters or nanometers.

weak acid An acid that does not completely dissociate in solution.

weak base A base that does not completely dissociate in solution.

Wohler, Friedrich (1800–1882) German chemist who contributed greatly to the development of organic chemistry and was the first to synthesize an organic compound in the laboratory.

work A force acting through a distance.

X-rays The fraction of the electromagnetic spectrum that is between the ultraviolet (UV) region and the gamma ray region. X-rays can penetrate substances that normally block light and are often used in medical diagnosis.

zeolites Small porous particles with cagelike structures that can hold ions.

zinc A nutritional mineral essential to the function of more than 100 enzymes.

Index

Abortion, 487
Absorption spectrum, 200
Abuse
   alcohol, 171
   inhalants, 492–493
   prescription drugs, 502
Accidents, nuclear power, 228
Acetaminophen, 479
Acetic acid (CH_3COOH), 176, 363, 370
Acetone (CH_3COCH_3), 174–175, 340–341
Acetylene, 159
Acetylsalicylic acid ($C_9H_8O_4$), 95
   for pain, 371, 477–479
   substitutes for, 479–480
Achromycin, 481
Acid/base neutralization equations, 364
Acid rain, 312
   acidity of, 374–375
   cleaning up, 376–377
   effects of, 375
   environmental problems from, 263–264
   neutralizing effects of, 377
Acids, 361–362
   baking with, 373–374
   common, 369–372
   concentration of, 368–369
   molecular definitions, 365–366
   properties of, 362–363
   strong and weak, 367–368
   in wines, 371–372
Acquired immunodeficiency syndrome (AIDS)
   drugs for, 136, 482–486
   and viruses, 461
Acrolein, 173–174
Acyclovir, 483–484
Adams, Douglas, 35
Addictions
   narcotics, 494–495, 497
   prescription drugs, 502
   smoking, 499
Addition polymers, 420
Additives, food, 529–532
Adenine, 458
Adirondack Mountains, 264
Adrenocortical steroids, 487–488
Aerosols, 264, 375
Agent Orange, 536
Aging process, 397

Agriculture
   fertilizers in, 409–410, 532–534
   genetic engineering in, 464–465
   herbicides in, 535–537
   insecticides in, 169, 534–535
Air, 299
   air bags, 300
   air pressure, 301–304
   atmosphere. See Atmosphere
   particles in, 301
   properties of, 304–308
Air pollution
   cleaning up, 313–315
   composition of, 312–313
   smog, 261–262
Alamogordo, New Mexico, 226
Alanine, 452
Alchemy, 9–10
Alcoholism, 491
Alcohols, 440
   and barbiturates, 492
   common, 171–172
   as depressants, 490
   ethanol, 170–171
Aldehydes, 172–174, 440
Aldicarb, 534
Alizarin, 167
Alkali metals, 74
Alkaline earth metals, 74
Alkaloids, 365
Alkanes
   boiling point, 335
   naming, 163–165
   structure of, 152–156
Alkenes
   naming, 165
   structure of, 156–159
Alkylating agents, 489
Alkylbenzenesulfonates (ABS) detergents, 407–408
Alkynes
   naming, 165
   structure of, 156–159
Alloys, 10
Alpha-helix structure, 452
Alpha-Keratin protein, 452, 455–456
Alpha linkages, 443, 515
Alpha particles, 19
Alpha radiation, 216–217
Aluminum hydroxide ($Al[OH]_3$), 372–373

Aluminum oxide (Al_2O_3), 396
Alzheimer's disease, 434
Americium, 59
Amines, 178–179, 447
Amino acids, 150
   common, 448–449
   essential, 517
   identifying, 450
   structure of, 446–447
p-Aminobenzoic acid (PABA), 193–194, 414
Ammonia (NH_3), 91, 95
   as base, 366
   characteristics of, 129
   in cleaning products, 373
   Lewis structure of, 132
   for plants, 533
Ammonium nitrate (NH_4NO_3), 533
Ammonium sulfate ($[NH_4]_2SO_4$), 533
Amphetamine, 178–179, 497–498
Amplitude modulation (AM), 195
Amus, 58
Anabolic steroids, 488
Androgens, 486
Anemia
   iron deficiencies, 529
   sickle cell, 94, 434, 451
Anesthetic inhalants, 492–493
Animal engineering, 107
Anionic detergents, 408
Anions, 62, 97
Anodes, 391
Antacids, 372–373
Antagonists, 492
Antibiotics, 477, 480–482
Anticaking agents, 532
Anticancer drugs, 477
Antidepressants, 502–504
Antimetabolites, 489
Antimicrobial agents, 530
Antioxidants, 397, 525, 530–531
Antiperspirants, 418–419
Antiseptics, 388
Antiviral drugs, 477, 482–486
Argon
   in atmosphere, 310
   as inert gas, 66, 69
   valence electrons of, 71
Argon ion lasers, 204
Aristotle, 9
Aromatic hydrocarbons, 165–167

I-1

Aromatic rings, 167
Arrhenius, Svante, 365
Arrhenius definition, 365–366
Arteriosclerosis, 520
Artificial cells, 549
Artificial colors, 531
Artificial flavors, 531–532
Artists, 6–9
Ascorbic acid (vitamin C), 531
Aspartic acid, 450, 452
Aspirin, 95
    for pain, 371, 477–479
    substitutes for, 479–480
Atmosphere, 300. *See also* Air
    composition of, 308–310
    defined, 308–310
    layers in, 310–311
    ozone depletion in
        CFCs in, 316–317
        chlorine in, 75, 170
        global, 319
        and Montreal protocol, 320–321
        myths concerning, 321–322
        in polar regions, 317–319
    pollution in, 261–262
        cleaning up, 313–315
        composition of, 312–313
Atomic bomb, 213
    building of, 224–226
    ethics of, 226
Atomic force microscopes (AFMs), 546–548
Atomic mass, 64–66
Atomic number (Z), 58, 62–63
Atomic theory, 16–18, 20–21
Atoms, 9, 12, 55–58
    atomic mass, 64–66
    bent, 132
    composition of, 18–19
    electrons in, 60–62
    identifying with light, 198–199
    mole concept, 77–79
    neutrons in, 62–63
    protons in, 58–60
    quantum mechanical model of, 71–73
    specifying, 63–64
Atrazine, 537
Auden, W. H., 2
Automobiles
    air bags in, 300
    batteries for, 391–392
    electric and hybrid, 288, 291
    fuel cell, 395
    paint dulling on, 387
    rust on, 384
Avogadro, Amadeo, 77
Avogadro's number, 77–78
Azidothymidine (AZT), 483–485

B complex vitamins, 527
Bacteria
    antibiotics for, 480–482
    in water, 348

Baking powder, 373
Baking soda and bee stings, 176, 372
Balanced equations, 104–105
Barbiturates, 490–492
Barometers, 303
Bases, 361–362
    baking with, 373–374
    common, 372–374
    concentration of, 368–369
    molecular definitions, 365–366
    properties of, 363–365
    strong and weak, 367–368
Batteries, 389–391
    automobile, 391–392
    flashlight, 392–393
Becquerel, Antoine-Henri, 215
Bee stings, 176, 372
Beeswax, 413
Bent atoms, 132
Benzaldehyde, 174
Benzene (C_6H_6)
    in Kekule's dream, 165–166
    molecular formula, 93
    ring structure of, 167
    in water, 348
Benzodiazepines, 490–492
Benzophenone, 414
Benzoyl peroxide [$(C_6H_5CO)_2O_2$], 388
Benzyl acetate, 177, 416
Berkelium, 59
Beta carotene, 524
Beta links, 443, 516
Beta radiation, 218
Binnig, Gerd, 20, 545
Bioamplification, 169, 534–535
Biochemistry, 151, 433
    carbohydrates, 440–446
    cloning, 465–467
    genes, 434
    lipids and fats, 434–440
    nucleic acids, 456–462
    proteins, 446–451
        common, 454–456
        structure of, 451–454
    recombinant DNA technology, 462–465
Biological contaminants
    killing, 353–354
    in water, 348
Biomass, 286–287
Biotechnology, 107, 433
Birth control pills, 486–487
Birth defects, 490
Bismuth, 219–220
Black holes, 21
Black lights, 197
Bleaching agents, 410
Bloch, Felix, 199
Blue baby, 348
Bohr, Niels, 67, 298
Bohr diagrams, 69
Bohr model
    for periodic law, 67–71

    vs. quantum mechanical model, 71–73
Bohr orbits, 93
Boiling, 331–334
Boiling point, 333
Bonding groups, 132
Bonding pair electrons, 123
Bonds
    covalent, 93, 135–136, 454
    double, 125, 158, 161
    ionic, 92
    and molecular shapes, 134–135
    multiple, 125–126
    polar, 135–140, 336–337
    triple, 126
Boscovich, Roger Joseph, 328
Botulism, 530
Boyle, Robert, 11
Boyle's law, 304–306
Branched structures, 149, 160
Brand name drugs, 482
Breeder reactors, 287–288
Brighteners, 410
Brillat-Savarin, Anthelme, 510
Broad-spectrum insecticides, 534–535
Bronowski, Jacob, 116
Bronsted-Lowry definition, 365–366
Buckminsterfullerene (C_{60}), 547
Buckyballs, 547–548
Builders, 409–410
Butane (C_4H_{10})
    boiling point, 333
    structure of, 153, 160
n-Butane, 160
Butanedione, 175
Butene, 161
Butylated hydroxyanisole (BHA), 531
Butylated hydroxytoluene (BHT), 531

Caffeine, 498
Calcium
    in food, 527
    in hard water, 346–347
Calcium acid phosphate ($CaHPO_4$), 373
Calcium carbonate ($CaCO_3$), 97, 372
Californium, 59
Caloric intake, 521–524
Calories (Cal), 251–252
Campfires, 110, 259
Cancer
    chemicals for, 488–490
    lung, 233
    and radiation, 237
    skin, 413–414
Carbamate insecticides, 534–535
Carbaryl, 534
Carbohydrates, 434, 440–446
    complex, 442, 514–516
    identifying, 443–444
    simple, 513–514
Carbon
    buckyballs, 547–548
    characteristics of, 148–149
    in compounds, 12, 147

neutrons in, 62
protons in, 58–59
Carbon dating, 234–236
Carbon dioxide (CO$_2$)
  in atmosphere, 309–310
  chemical formula for, 91
  in global warming, 106–107
  graphic data for, 41–42
  as greenhouse gas, 265–269
  naming, 99
  polar bonds in, 137–138
  sequestering, 268
Carbon dioxide (CO$_2$) lasers, 204
Carbon filtration, 354
Carbon monoxide (CO)
  in air pollution, 313
  chemical formula for, 91
  danger from, 95
  in metal processing, 389
  in smog, 261
Carbon nanotubes, 548–549
Carbon tetrachloride (CCl$_4$)
  Lewis structure of, 124
  in water, 348
Carbonic acid, 371, 374
Carbonyl groups, 172
Carboxylic acids, 175–176, 447
Carcinogens, 167
Carezyme, 410
Carnauba, 413
Carotenes, 188–190, 524
Carvone (C$_{10}$H$_{14}$O), 173
Catalysts, 262
Catalytic converters, 262
Cathodes, 391
Cationic detergents, 408
Cations, 61–62
Cells, artificial, 549
Cellulose, 443–444
Celsius scale, 254
Centimeters (cm), 36
Cephalosporins, 481
Cesium clocks, 37
Chalcogens, 74
Challenger disaster, 286
Changes
  chemical, 16
  physical, 15
Charge (C), 63
Charles's law, 306–307
Chemical bonding, 117–119
    Lewis structures. *See* Lewis
      structures
    Lewis theory, 119–121
    molecule shapes in, 131–135
Chemical changes, 16
Chemical contaminants, 348–351
Chemical energy, 248
Chemical equations
  balanced, 104–106
  as conversion factors, 106–110
Chemical formulas
  for compounds, 91–92
  as conversion factors, 101–103

Chemical properties, 15
Chemical reactions
  for combustion, 4, 7
  energy in, 256–257
  equations for, 104–106
  importance of, 89
Chemical symbols, 58–59
Chemistry, defined, 5
Chemotherapy, 490
Chernobyl disaster, 214, 228
Children, hyperactivity in, 516
"China syndrome," 228
"China White," 495
Chlordiazepoxide (Librium), 492
Chlorinated hydrocarbons, 168–170
Chlorine (Cl$_2$)
  atomic mass of, 65–66
  color of, 59
  as disinfectant, 388
  neutrons in, 62
  in ozone layer depletion, 75, 170
  reactivity of, 58
  in salt, 118–119
Chlorobenzene in water, 348
Chlorofluorocarbons (CFCs)
  and Montreal protocol, 320–321
  in ozone layer depletion, 75,
    169–170, 316–317
Chlorohydrocarbons in
    water, 348
Chlorophylls, 189
Cholesterol (C$_{27}$H$_{46}$O), 95, 520
Chromium, 529
Chromosomes, 458
Cinnamaldehyde, 174, 416
Cisplatin, 489
Citric acid, 369–371
Civetone, 417
Classification of matter, 11–15
Clean Air Act, 43, 314–315, 320,
    376–377
Clean Air Conservancy, 376
Cleaning clothes, 404–405
Climate change
  carbon dioxide in, 106–107
  fossil-fuel use in, 264–269
  greenhouse gas data for, 41–42
  measuring, 33
Climate Change Convention, 269
Cloning, 5, 465–467
Coal, 257–259
Cocaine, 365, 372, 496–498
Codeine, 494
Codon, 458
Cohesive forces, 330, 333–334
Coke, 389
Collagen, 526
Colloidal suspension, 406
Coloring hair, 411–412
Colors
  artificial, 531
  fall, 188–190
  and temperature, 190
Combined gas law, 308

Combustion
  enthalpy of, 256
  reactions in, 4, 7, 156, 259–260,
    384–385
Commercial sector energy use,
    245–246
Common cold, 482
Common names, 96
Complete proteins, 518
Complex carbohydrates, 442, 514–516
Compounds, 12, 90–91
  chemical formulas for, 91–92
  composition of, 16, 18, 101–103
  forming and transforming, 104–106
  formula mass of, 99–100
  hydrocarbons, 168–170
  inorganic, 149–151
  ionic, 92–93, 96–98
  molar mass of, 100–101
  molecular, 93–95, 98–99
  naming, 96–99
  organic, 147, 149–151
Concentration
  of acids and bases, 368–369
  of solutions, 341
Condensation polymers, 424
Condensed structural formulas,
    152–153
Conducting polymers, 422
Consciousness, 503
Conservation of energy, 288–290
Conservation of mass, 7–8, 17
Constant composition, law of, 16, 18
Consumer chemistry, 418
Contaminants in water
  biological, 348, 353–354
  chemical, 348–351
Conversion factors
  chemical equations as, 106–110
  chemical formulas as, 101–103
  energy, 251–252
Converting measurement units,
    38–40, 47–48
Copernicus, Nicholas, 10
Copolymers, 423–424
Copper, 12
  in batteries, 389–391
  in food, 529
Corporate handouts, 396
Corrosion, 395–397
Corrosive cleaners, 410
Cortisone, 487
Cosmetics, 414–415
Covalent bonds, 93, 135–136, 454
Covalent compounds, 123–129
Creams, 412–413
Crick, Francis, 458
Critical mass, 224
Crystalline solids, 332
Crystalline structure, 332
Cubic centimeters (cm^3), 38
Cubic meters (m^3), 38
Curie, Marie Sklodowska, 215–216
Curie, Pierre, 215–216

Curies, 233
Curium, 59
Curl, Robert F., Jr., 547–548
Cyclophosphamide, 489
Cystic fibrosis (CF), 465
Cytosine, 458

Dalton, John, 8, 18
Decane, 436
Decay, nuclear, 217–218, 236–237
Defoliants, 535
Deionized water, 179
Delaney Clause, 531
Delta-9-tetrahydrocannabinol (THC), 501–502
Democritus, 9, 16, 54
Density, 46–48
Deodorants, 175, 418–419
Deoxyribonucleic acid (DNA), 150–151, 456–458
 and free radicals, 397
 recombinant techniques, 434, 462–465
 replication of, 458–460
 synthesis of, 489
 in viruses, 482–483
Deoxyribonucleic acid (DNA) fingerprint, 464
Depressants
 alcohol, 490
 barbiturates and benzodiazepines, 490–492
 endorphins, 496
 inhalants, 492–493
Depression, 478
Derived units, 37
Desalination plants, 355
Descartes, Rene, 503
Desflurane, 492
Designer drugs, 500
Detergents, 407–409
Determinism, 72
Deuterium, 62
Diabetes, 456
Diatomic molecules, 76
Diazepam (Valium), 492
Dichlorodiphenyltrichloroethane (DDT), 169, 534
Dichloromethane (CH_2Cl_2), 168–169
Diethyl ether, 334
Diglycerides, 519
Dihydroxyacetone, 414
Dimethyl ether, 178
Dinitrogen monoxide (N_2O), 99
Diode lasers, 204
Dioxins, 537
Dipeptides, 451
Dipole forces, 336–337
Dipoles, 136
Disaccharides, 442, 513–514
Disease therapy, 465
Diseases, water-borne, 345
Dish/engines, 282

Disinfectants, 388
Dispersion force, 334–335
Disulfide bridges, 454
Disulfide linkages, 411, 454
Dobson units, 318
Dopamine, 497
Doping, 283
Double bonds, 125, 158, 161
Down quarks, 65
Drain cleaners, 410
Drugs, 463–464, 477
 abuse of, 502
 for AIDS, 136, 482–486
 antibiotics, 480–482
 anticancer, 488–490
 antidepressants, 502–504
 antiviral, 477, 482–486
 depressants, 490–493
 generic vs. name brand, 482
 hallucinogens, 499–501
 marijuana, 501–502
 narcotics, 493–496
 for pain, fever, and inflammation, 478–480
 Prozac and Zoloft, 502–504
 sex hormones as, 486–487
 steroids, 487–488
 stimulants, 496–499
 street, 496
Dry cells, 392–393
Duets, 121
Dye lasers, 204

Earth, age of, 236–237
Ecstasy, 500
Efficiency of energy
 calculating, 282–283
 in conservation, 288–290
 heat tax in, 250–251
Einstein, Albert, 224
Einsteinium, 59
Elastomers, 424–425
Electric fields, 192–193
Electric vehicles, 288
Electrical charge, 58
Electricity
 from batteries, 389–393
 from fossil fuels, 260–264
Electrochemical cells, 390–391
Electrolysis, 286
Electrolyte solutions, 92
Electromagnetic radiation, 193
Electromagnetic spectrum, 193–195
Electron geometry, 131–132
Electronegativity, 136–137
Electronic relaxation, 197
Electronic transitions, 196
Electrons, 20, 60–62
 bonding pair, 123
 configurations, 69
 excited, 195–197
 lone pair, 123
 valence, 71, 119–120

Elements, 12, 56
 chemical symbols for, 58–59
 families of, 73–76
 names of, 59
 protons in, 58–60
Embryo cloning, 465–466
Embryonic stem cells, 466–467
Empedocles, 9
Emulsifiers, 406, 532
Emulsifying agents, 406
Emulsions, 406
Endorphins, 496
Endothermic reactions, 256
Energy, 242–243
 biomass, 286–287
 in chemical reactions, 256–257
 defined, 247
 efficiency of
  calculating, 282–283
  in conservation, 288–290
  heat tax in, 250–251
 electricity from fossil fuels, 260–264
 environmental problems
  acid rain, 263–264
  global warming, 264–269
  smog, 261–262
 first law of thermodynamics, 248
 future of, 290–292
 geothermal, 287
 hydroelectric, 278–280
 molecular motion in, 244–245
 nuclear, 287–288
 nuclear binding, 229
 photovoltaic, 283–284
 reliance on, 245–246
 second law of thermodynamics, 249–251
 for society, 257–260
 solar, 277, 280–283
 storage of, 285–286
 technology development, 396
 and temperature, 254–255
 transformations, 247–248
 units of, 251–253
 wind, 280
Energy expended, 522–523
Energy intake, 521–522
Energy state, 196
Energy stored, 523
Enflurane, 492
Engineering, genetic, 107
 in agriculture, 464–465
 ethics of, 467
Enthalpy, 256–257
Entropy, 249–250
Environmental problems
 acid rain, 263–264
 global warming, 264–269
 smog, 261–262
Environmental Protection Agency (EPA)
 and air pollution, 312–315

criticizing, 353
and water pollution, 348, 352–354
Enzymes
  in cleaning products, 410
  proteins as, 456
  restriction, 462
  for starches, 443
Essential amino acids, 517
Esters, 176–178
Estrogen, 486
Ethane (CH_3CH_3)
  cohesive forces in, 336
  structure of, 153
Ethanol (CH_3CH_2OH)
  as biomass fuel, 286–287
  characteristics of, 170–171
  as depressant, 490
  for methanol poisoning, 172
Ethene, 158
Ethers, 177–178
Ethics
  of atomic bomb, 226
  of genetic engineering, 467
Ethyl butyrate, 177
Ethyl formate, 177
Ethylbenzene, 348
Ethylene, 158, 419
Ethylenediaminetetraacetic acid (EDTA), 530–531
Eugenol, 416
Europium, 59
Eutrophication, 409
Evaporation, 331, 340
Excited electrons, 195–197
Excited state, 196
Exothermic reactions, 256
Experiments, 8–9
Extreme miniaturization, 544–545

Facial cosmetics, 414–415
Fahrenheit scale, 254
Fall colors, 188–190
Families of elements, 73–76
Fat-soluble vitamins, 524–526
Fats, 434–440, 518–521
Fatty acid esters, 179
Fatty acids, 435, 519
Federal Food, Drug, and Cosmetic Act, 530
Femtochemistry, 205
Fentanyl (Sublimaze), 495
Fermentation
  malolactic, 371
  pickling, 370
Fermi, Enrico, 222, 224
Fertilizers, 409–410, 532–534
Fever, reducing, 478–480
Feynman, Richard, 31, 242, 542–544
Fibers, 444, 516
Fillers, 410
Fingerprint, DNA, 464

Fire
  campfires, 110, 259
  reactions in, 4, 7
Firesticks, 4
First law of thermodynamics, 248
Fission, nuclear, 222–223
Flames, 4, 7
Flash freezing, 345
Flashlight batteries, 392–393
Flat gasoline, 342
Flavor enhancers, 531–532
Fleming, Alexander, 456, 480–481
Florey, Howard, 481
Flue gas scrubbers, 376
Flumanzil, 492
Fluorescence, 197
Fluoride, 121
Fluorine
  in Bohr model, 68
  Lewis structures for, 122
  neutrons in, 62
  reactivity of, 58, 121
Fluoxetine (Prozac), 502–504
Food, 511
  additives in, 529–532
  bodies made up of, 512–513
  caloric intake, 521–524
  carbohydrates, 513–516
  fats in, 518–521
  fertilizers for, 532–534
  herbicides for, 535–537
  insecticides for, 534–535
  minerals in, 527–529
  proteins in, 517–518
  vitamins in, 524–527
Food and Drug Administration (FDA)
  carbohydrate recommendations, 516
  drug approvals, 487
  food additives, 531
  GRAS foods, 530
  pesticide guidelines, 536
Forensics, 464
Forests, acid rain damage to, 264
Formaldehyde (H_2CO), 172, 336
Formic acid (HCOOH), 171–172, 175–176
Formula mass of compounds, 99–100
Fossil fuels
  electricity from, 260–264
  as energy source, 257–259
  environmental problems from
    acid rain, 263–264
    global warming, 265–269
    smog, 261–262
  evaluation of, 267
  subsidizing, 396
Free radicals, 397
Freezing point depression, 332
Freon, 170, 316, 320
Frequency, wavelength, 192
Frequency modulation (FM), 192, 195
Freud, Sigmund, 496

Fructose ($C_6H_{12}O_6$)
  converting, 171
  structure of, 441
  sweetness of, 514
Fuel cells, 292, 393–395
Fuels, alkanes, 152–156
Fuller, R. Buckminster, 548
Functional groups, 151, 168
Functionalized hydrocarbons, 151, 168
Fusion, 230–231

GABA, 491
Galilei, Galileo, 10–11
Gallium, 59, 66
Galvanization, 396
Gamma radiation, 219
Gamma rays, 193–194
Gas lasers, 203–204
Gases, 14–15
  particles in, 301
  properties of, 304–308
Gasoline
  alkanes, 152–156
  flat, 342
Gel electrophoresis, 462
Generally recognized as safe (GRAS) foods, 530
Generic drugs, 482
Genes, 434
Genetic defects and radiation exposure, 232
Genetic engineering, 107
  in agriculture, 464–465
  ethics of, 467
Genetic screening, 465
Geometry in molecular shape, 131–132
Geothermal power, 287
Geraniol, 180
Germanium, 59, 66
The Geysers, 287
Gigawatts (GW), 261
Global ozone depletion, 319
Global warming
  carbon dioxide in, 106–107
  fossil-fuel use in, 264–269
  greenhouse gas data for, 41–42
  measuring, 33
Glucose ($C_6H_{12}O_6$)
  converting, 171
  in foods, 513–514
  fossil fuels from, 258–259
  structure of, 440–441
Gly-val-ala-asp, 452
Glycerol, 518–519
Glycerol monostearate, 532
Glycine, 447, 452
Goodyear, Charles, 425
Grams, 37
Graphs
  extracting information from, 41–44
  reading, 40–41

Gravity and water, 330
Greek philosophers, 9–10
Greenhouse gases, 265
    graphic data for, 41–42
    in Kyoto protocol, 269
Groups
    bonding, 132
    carbonyl, 172
    element, 73–76
    functional, 151, 168
Guanine, 458

Haber process, 389
Hahn, Otto, 222
Hair products, 411–412
Half-life, 219–222
Hallucinogenic drugs, 499–501
Halogens, 74
Halothane, 492
Hard water
    characteristics of, 346–347
    surfactants for, 407–409
Harrison Narcotic Act, 494, 497
Hashish, 501
Health care costs, 201
Heat
    defined, 254–255
    in molecular motion, 244–245
    and organic compounds, 150
Heat capacity, 254–255
Heat tax, 251
Heliostats, 281, 283
Helium, 12
    in atmosphere, 310
    in Bohr model, 68
    breathing, 74
    as inert gas, 56, 66, 69–70
    protons in, 58–59
Helium-neon lasers, 203–204
Hemoglobin (Hb), 94–95, 451, 454–455
Heptane, 161
Herbicides, 535–537
Heroin, 494–495
Hertz, Heinrich, 195
Hertz (Hz), 192
Heterogeneous mixtures, 14
Hexachlorobenzene (HCB), 534
Hexane, 333
High-density lipoproteins (HDLs), 521
High-density polyethylene (HDPE), 420
Hiroshima, 226
Histamines, 479
Hitler, Adolf, 5
Hoffman, Albert, 500
Holmes, Oliver Wendell, 276
Home water treatment, 354–355
Homogeneous mixtures, 14
Honey vs. sugar, 514
Hormones
    in cancer treatment, 489–490
    insulin, 456
    sex, 486–487

Household chemistry, 403
    clothes cleaning products, 404–405
    copolymers, 423–424
    cosmetics, 414–415
    deodorants, 418–419
    hair products, 411–412
    laundry-cleaning formulations, 409–410
    perfumes, 415–418
    polymers, 419–423
    rubber, 424–425
    skin products, 412–414
    soap, 405–407
    synthetic detergents, 407–409
Human immunodeficiency virus (HIV), 136, 461, 482–483
Human life, radiation effect on, 231–234
Humectants, 532
Humidity, relative, 413
Hybrid electric vehicles, 288, 291
Hydrocarbons, 151–152
    alkanes, 152–156
    alkenes and alkynes, 156–159
    aromatic, 165–167
    incomplete combustion of, 262
    naming, 162–165
    saturated, 436
Hydrochloric acid (HCl), 362–363, 371
Hydrochlorofluorocarbons (HCFCs), 320
Hydrocortisone, 487
Hydroelectric energy, 258–259, 278–280
Hydrogen
    atomic mass of, 64
    in Bohr model, 67–68
    characteristics of, 285
    Lewis structures for, 124
    neutrons in, 62
    protons in, 58
Hydrogen bombs, 230
Hydrogen bonds, 338–339, 452, 454–455
Hydrogen ions, 365–366
Hydrogen-oxygen fuel cell, 393–394
Hydrogen peroxide (H_2O_2), 91, 388, 412
Hydrogen sulfide, 339
Hydrologic cycle, 345–346
Hydrophobic interactions, 454
Hydroxide ions, 365
Hyperactivity in children, 516
Hypotheses, 8–9
Hypoxia, 493

Ibuprofen, 479
Ice
    density of, 344–345
    structure of, 138–139
    thermal energy in, 332
Ice cream, 332
Immortality, 9–10
Industrial sector, energy use in, 245

Inflammation
    lowering, 478–480
    steroids for, 487–488
Infrared (IR) light, 194–195
Inhalants, 492–493
Inorganic compounds, 149–151
Inorganic contaminants, 348–350
Insecticides, 169, 534–535
Instantaneous dipoles, 334–335
Insulin, 456, 463
Interferon, 464
Intermittence of energy sources, 285
International Bureau of Weights and Measures, 37
International System of Units (SI units), 36–37
International Union of Pure and Applied Chemistry (IUPAC), 162
Iodine (I_2)
    as antiseptic, 388
    in food, 529
Ionic bonds, 92
Ionic compounds, 92–93, 96–98
Ionic Lewis structures, 121–123
Ionizing power, 216
Ionone, 174
Ionosphere, 311
Ions, 61–62
Iron
    in food, 529
    oxidized, 384
Iron chloride ($FeCl_3$), 96
Iron oxide (Fe_2O_3), 384, 395
Isobutane, 179
Isoflurane, 492
Isomers, 160–162
Isopropyl alcohol, 171
Isotopes, 62

Janssen, Johannes, 476
Joint European Torus (J.E.T.) project, 230
Joliot, Frederick, 216
Joliot, Irene, 216
Joule, James, 251
Joules (J), 251–252
Judson, Horace Freeland, 432

Kekule, Friedrich August, 165–166
Kekule's dream, 165–166
Kelvin (K) unit, 36
Kelvin scale, 254
Keratin, 411
Ketones, 172–175, 440
Kilograms (kg), 36–37
Kilometers (km), 36
Kilowatt-hours (KWh), 252–253
Kinetic energy, 247–248
Kroto, Harold W., 547–548
Kyoto protocol, 269

Laboratory bases, 364
Lactase, 514
Lactic acid, 370

Lactose, 514
Lactose intolerance, 514
Laser cavities, 203
Lasers, 202–203
   in chemistry, 205
   in medicine, 204–206
   types of, 203–204
LASIK process, 206
Lasing media, 203
Laughing gas, 99
Laundry-cleaning formulations, 409–410
Lavoisier, Antoine, 7, 16, 18
Law of conservation of mass, 7–8, 17
Law of constant composition, 16, 18
Laws of thermodynamics, 249–251, 285, 518
Lead
   in air pollution, 313
   in water, 349
Leclanche dry cells, 392–393
Legal stimulants, 498–499
Length
   conversion factors for, 38
   measurement units for, 36
Lewis, G. N., 119
Lewis structures, 120
   covalent, 123–129
   ionic, 121–123
   of ozone, 129–130
   writing, 126–129
Lewis theory, 119–121
Lexan, 424
Life, origin of, 150
Light, 187
   characteristics of, 190–192
   electromagnetic spectrum in, 193–195
   excited electrons in, 195–197
   fall colors, 188–190
   identifying molecules and atoms with, 198–199
   lasers, 202–206
   magnetic resonance imaging, 199–201
   ultraviolet, 193–194, 196–197, 215, 413–414
Liming, 376
Linear alkylsulfonate (LAS) detergents, 408–409
Linkages, 443, 515–516
Linseed oil, 437
Lipids, 434–440
Lipoproteins, 519
Liquids, 14–15
   nonvolatile and volatile, 339–341
   water. *See* Water
Liters (L), 38
Lithium, 68, 71
Lithium sulfide (Li_2S), 96
Litmus paper, 362, 364
Lone pair electrons, 123
Los Alamos, New Mexico, 224–226
Lotions, 412–413

Love, 478
Low-density lipoproteins (LDLs), 521
Low-density polyethylene (LDPE), 420
LP gas, 153
Lucite, 422
Lung cancer, 233
Lysergic acid diethylamide (LSD), 499–501
Lysine, 452
Lysozyme, 456

Magnesium
   in compounds, 96
   in food, 529
   in hard water, 346–347
   Lewis structures for, 122
   in rust prevention, 396
Magnesium carbonate ($MgCO_3$), 372
Magnesium fluoride (MgF_2), 122
Magnesium hydroxide ($Mg[OH]_2$), 373
Magnesium oxide (MgO), 96
Magnetic fields, 192–193
Magnetic resonance imaging (MRI), 199–201, 503
Major minerals, 527–529
Malaria control, 169
Malathion, 534
Malic acid, 371
Malolactic fermentation, 371
Manhattan Project, 224–226
Marijuana, 501–502
Mass
   atomic, 64–66
   conservation of, 7–8, 17
   conversion factors for, 38
   in density measurements, 46
   formula, 99–100
   measurement units for, 37
   molar, 100–101
Mass defect in nuclear chemistry, 229
Mass number (A), 62–63
Mathematics
   graph reading, 40–44
   scientific notation in, 33–35
   word problems in, 44–46
Matter, 9
   behavior of, 90
   classification of, 11–15
   properties of, 15–16
Maximum contaminant levels (MCLs), 351–352
Measurements
   of nuclear exposure, 231–232
   ozone, 318
   uncertainty in, 32–33
   units in, 30, 35–38
      converting, 38–40
      for density, 46–48
      for energy, 251–253
Medicine
   costs of care, 201
   drugs. *See* Drugs
   lasers in, 204–206

   magnetic resonance imaging in, 199–201
   nanotechnology in, 549–550
   nuclear, 237
   X-rays in, 193–194, 196
Meitner, Lise, 222
Meltdown, 228
Melting, 332–334
Mendeleev, Dmitri, 66
Meperidine (Demerol), 495
Mercury, 348–349
Mescaline, 499–500
Mesosphere, 311
Messenger ribonucleic acid (mRNA), 460–461
Metalloids, 75
Metals
   in periodic table, 74–75
   transition, 75, 96
Meters (m), 36
Methadone, 495
Methamphetamine, 178–179, 498
Methanal (H_2CO), 172
Methane (CH_4), 104–105, 152
Methanol
   melting point, 339
   toxicity of, 171–172
Methotrexate, 489
Methoxymethylenedioxyamphetamine (MMDA), 500
Methyl butyrate, 177
Methyl groups, 152
Methylene groups, 153–154
Methylenedioxymethamphetamine (MDMA), 500
Methylphenylpiperidylpropionate (MPPP), 496
N-methyl-phenyltetrahydropyridine (MPTP), 496
Metolachlor, 537
Micelles, 406–407
Micronutrients, 534
Microscopes
   atomic force, 546–548
   scanning tunneling, 20, 166, 545–546
Microwave radiation, 194–195
Miller, Stanley, 150
Milligrams per liter (mg/L), 342
Milliliters (mL), 38
Minerals
   major, 527–529
   minor, 529
Miniaturization, extreme, 544–545
Minor minerals, 529
Mixtures, 13–14
mm Hg measurement, 302
Molar mass, 78, 100–101
Molarity (M) of solutions, 341
Mole concept, 77–79
Molecular cause, 3
Molecular compounds, 93–95, 98–99
Molecular formulas, 93

Molecular motion, 244–245
Molecular reasons, 5–6
Molecules, 4, 12, 89
  clothes cleaning with, 404–405
  genes, 434
  identifying, 198–199
  and matter behavior, 90
  polar, 135–140
  separating, 332–334
  shapes of, 131–135
  smelling, 339–341
Molten carbonate fuel cells (MCFCs), 394–395
Monoglycerides, 519
Monomers, 419
Monosaccharides, 442–443, 513–514
Monosodium glutamate (MSG), 531–532
Monounsaturated triglycerides, 519
Montreal protocol, 320–321
Morphine, 372, 477, 494
Motion, molecular, 244–245
Muller, Paul, 169
Multiple bonds
  Lewis structure of, 125–126
  and molecular shapes, 134–135
Muriatic acid, 363
Muscone, 417
Musil, Robert, 88
Mylar, 424

n-type silicon, 283
Nagasaki, 226
Naloxone (Narcan), 495
Naltrexone (Trexan), 495
Names
  compounds, 96–99
  elements, 59
  hydrocarbons, 162–165
Nanomedicine, 549–550
Nanotechnology, 20, 542–543
  atomic force microscopes, 546–548
  buckyballs, 547–548
  carbon nanotubes, 548–549
  extreme miniaturization, 544–545
  nanomedicine, 549–550
  problems, 550
  scanning tunneling microscopes, 545–546
Nanotubes, 548–549
Nanovesicles, 549
Naphthalene ($C_{10}H_8$), 167
Naproxen, 479
Narcotics
  addictions to, 495
  for pain, 493–495
Narrow-spectrum insecticides, 535
National Ambient Air Quality Standards (NAAQS), 312, 314–315
National Surface Water Survey (NSWS), 375

Natural gas, 91
  as energy source, 257–259
  flames, 104–105
Natural phenomena, molecular causes of, 5–6
Nemerov, Howard, 301
Neodymium-yttrium-aluminum-garnet (Nd:YAG), 203
Neon
  in atmosphere, 310
  as inert gas, 56, 66, 69
  valence electrons of, 71
Neurotransmitters, 478, 503
Neutralization reactions, 362
Neutron stars, 21
Neutrons, 21, 62–63
Newton, Isaac, 186, 188
Nicotine, 372, 498–499
Night blindness, 524
Nitrates
  in soil, 533
  in water, 348–349
Nitric acid (HNO_3), 363, 367
Nitrogen
  in atmosphere, 300, 309
  Lewis structures for, 125
  for plants, 532–534
Nitrogen dioxide (NO_2)
  in acid rain, 263
  in air pollution, 313
Nitrogen fixation, 309
Nitrogen monoxide (NO), 263
Nitrogen oxides, 261–262
Nitrous oxide, 492
Nobelium, 59
Noble gases
  boiling point, 335
  in periodic table, 74
Nonionic detergents, 408–409
Nonmetals, 74–75
Nonpolar molecules, 138
Nonvolatile liquids, 339–341
Norepinephrine, 497
Norplant, 487
Nuclear binding energy, 229
Nuclear chemistry, 212–213
  carbon dating, 234–236
  Chernobyl disaster, 214, 228
  discovery of, 215–216
  fission, 222–223
  fusion, 230–231
  half-life, 219–222
  Manhattan Project, 224–226
  mass defect in, 229
  medicine, 237
  power. See Nuclear power
  radiation effects on human life, 231–234
  radioactivity, 216–219
  Shroud of Turin, 234–236
Nuclear equation, 217
Nuclear fission, 222–223

Nuclear magnetic resonance (NMR), 166, 199–200
Nuclear medicine, 237
Nuclear power, 226–227, 287–288
  accidents, 228
  Chernobyl disaster, 214, 228
  as energy source, 258–259
  waste disposal, 227–228
Nuclear theory, 18–21
Nuclear transfer technique, 465–466
Nuclear Waste Policy Act, 227–228
Nuclei, 20–21
Nucleic acids, 434
  as blueprint for proteins, 456–462
  summary, 462
Nucleic chains, 456–458
Nucleoside analogs, 483–484
Nucleotides, 456–458
Nutrition
  carbohydrates in, 513–516
  fats in, 518–521
  minerals in, 527–529
  proteins in, 517–518
  vitamins in, 524–527
Nylon, 423–424

Observation
  vs. reason, 10
  in scientific method, 7
Octane (C_8H_{18}), 94, 436
Octet rule, 120–121
Oil of almond, 174
Oil of jasmine, 177
Oils, 518–521
Olestra, 530
Olfactory receptors, 180
Opioid agonists, 495
Opioid antagonists, 495
Opioid receptors, 493–494
Opium, 494
Oppenheimer, J. R., 212, 224, 226
Orbitals, 71–72
Organic chemistry
  carbon in, 148–149
  defined, 147
  smell, 180
  structures in, 166
  vital force in, 149–151
Organic compounds, 149–151
  alcohols, 170–172
  aldehydes, 172–174
  amines, 178–179
  carboxylic acids, 175–176
  esters and ethers, 176–178
  hydrocarbons, 151–159
    aromatic, 165–167
    chlorinated, 168–170
    functionalized, 168
    naming, 162–165
  isomers, 160–162
  ketones, 172–175
Organic contaminants, 348
Organophosphates, 534–535

Origin of life, 150
Osmosis, reverse, 355
Oxidation, 385–386
Oxidation-reduction reactions, 383
  agents in, 388–389
  aging, 397
  batteries, 389–393
  combustion, 384–385
  definitions, 384–387
  fuel cells, 393–395
  respiration and photosynthesis, 389
  rust, 384, 395–397
Oxidized iron, 384
Oxidizing agents
  common, 388–389
  defined, 386
Oxybenzone, 414
Oxygen, 12
  in atmosphere, 309
  Lewis structures for, 124
Oxygen family, 74
Ozone
  in air pollution, 313
  in atmosphere, 311
  characteristics of, 318
  depletion of
    CFCs in, 316–317
    chlorine in, 75, 170
    global, 319
    and Montreal protocol, 320–321
    myths concerning, 321–322
    in polar regions, 317–319
  Lewis structure of, 129–130
  measuring, 318
  in photochemical smog, 262

p-n junctions, 283
p-type silicon, 283
Pain
  endorphins for, 496
  narcotics for, 493–496
  relieving, 478–480
Palmitic acid, 179
Parabolic troughs, 281–282
Paraquat, 537
Parathion, 534
Parts per million (ppm), 342
Pascals (Pa), 302
Pauling, Linus, 402
Paxil, 502
Penicillins, 480–481
Pentane, 179
Peptides, 450–451
Percent by mass of solutions, 341
Percent by volume of solutions, 341
Perfumes, 339–341, 415–418
Periodic law, 67–71
Periodic table
  atomic mass in, 64
  elements in, 60, 74–75
Permanent dipoles, 336
Peroxyacetylnitrate (PAN), 262
Perpetual motion machines, 251

Pesticide residues, 536
Petroleum, 257–259
pH scale, 368–369
Pharmaceuticals. *See* Drugs
Pharmacology, 10
Phase in matter classification, 14–15
Phenoxy herbicides, 535
Phenyl rings, 167
Phenylethylamine, 478
Philosophy, 72
Phosphates
  in laundry formulations, 409–410
  for plants, 533
Phosphorescence, 197, 215
Phosphoric acid (H_3PO_4), 371
Phosphorus
  in food, 527
  for plants, 532–534
Phosphorylated acyclovir, 484
Photochemical smog, 262
Photochemistry, 205
Photodecomposition, 197
Photons, 191–192
Photosynthesis, 258, 286, 389
Photovoltaic energy, 283–284
Photovoltaic (PV) cells, 283–284
Physical changes, 15
Physical properties, 15
Pickling, 370
Pizza, 107–110
Plastics, 419–423
Plato, 9
Pleated sheets, 452–453
Plutonium, 224–225
PM-10, 312–313
Polar bonds, 135–140, 336–337
Polar molecules, 135–140, 336
Polar regions, ozone depletion over, 317–319
Polar stratospheric clouds (PSCs), 317–319
Pollution
  air, 312–315
  water, 345–346, 348–351
Polonium, 215
Polyatomic ions, 97
Polycarbonates, 423–424
Polycyclic aromatic hydrocarbons, 167
Polyesters, 177, 424
Polyethylene, 419–422
Polyethylene terephthalate (PET), 424
Polyisoprene, 424–425
Polymers
  conducting, 422
  structure of, 177, 419–422
Polyoxyethylene ($C_{14}H_{30}O_2$), 408
Polypeptides, 451
Polypropylene, 420
Polysaccharides, 442, 514–515
Polystyrene, 420–422
Polytetrafluoroethylene (PTFE), 422
Polyunsaturated triglycerides, 519
Polyvinyl chloride (PVC), 420–422

Potassium
  in Bohr model, 69
  in compounds, 97
  in food, 529
  in periodic law, 66
  for plants, 532–534
  valence electrons of, 71
Potassium chloride (KCl), 533
Potassium hydroxide (KOH), 363, 393
Potassium nitrate, 97
Potassium sorbate, 530
Potassium sulfate (K_2SO_4), 363, 533
Potential energy, 247
Power, 252–253
Precision in measurements, 32–33
Prednisolone, 487
Prednisone, 487
Prefixes
  for compound names, 99
  for multipliers, 36
Pregnancy, 161, 487
Prescription drug abuse, 502
Pressure
  air, 301–304
  in air bags, 300
  and volume, 304–306
Primary nutrients, 532
Primary structure of proteins, 452
Problem solving, 44–46
Products of equations, 104
Progesterone ($C_{21}H_{30}O_2$), 161–162, 486–487
Progestin, 487
Proof of liquor, 171
Propane (C_3H_8), 94, 153
Properties of matter, 15–16
Prostaglandins, 371, 479
Protease inhibitors, 136, 484–486
Proteases, 484
Protein molecules, 94
Proteins, 434, 446–451
  animal engineering for, 107
  common, 454–456
  nucleic acids as blueprint for, 456–462
  and nutrition, 517–518
  structure of, 451–454
  summary, 462
  synthesis of, 460–461
Proton acceptors, 366
Proton donors, 366
Protons, 21
  in element determination, 58–60
  in ions, 61
Proust, Joseph, 16, 18
Prozac, 502–504
Psi (pounds per square inch), 302
Psychedelics, 499–501
Pu-239, 225
Public water treatment, 352–354
Purcell, Edward, 199
Pure substances, 12
Putrescine, 178

Pyramidal geometry, 132
Pyrene, 167
Pyrogens, 479

Quads, 245, 251–252
Quantum mechanical model of atoms, 71–73
Quantum numbers, 67
Quantum shells, 67
Quarks, 65
Quaternary structures of proteins, 453–454

Radiation. *See also* Nuclear chemistry
   alpha, 216–217
   beta, 218
   detecting, 231–232
   effects on human life, 231–234, 237
   electromagnetic, 193
   gamma, 219
   microwave, 194–195
Radio waves, 192, 195
Radioactive contaminants, 350–351
Radioactivity, 216–219
Radium
   discovery of, 216
   in water, 350
Radon
   in homes, 232–234
   in water, 350
Raffinose ($C_{18}H_{32}O_{16}$), 446
Rain, acid, 312
   acidity of, 374–375
   cleaning up, 376–377
   effects of, 375
   environmental problems from, 263–264
   neutralizing effects of, 377
Reactants, 104
Reaction stoichiometry, 106–110
Reason vs. observation, 10
Recombinant DNA technology, 434, 462–465
Redox reactions. *See* Oxidation-reduction reactions
Reducing agents, 386, 388–389
Reduction, 385–386
Reforming process, 394
Relative humidity, 413
Relaxation, electronic, 197
Relaxation time, 201
Rems, 231
Reproductive cloning, 466–467
Reserpine, 478
Residential sector, energy use in, 245–246
Resonance frequency, 200
Resonance structures, 130
Respiration, 384, 389
Restriction enzymes, 462
Retinal, 206
Retroviruses, 482–483

Reverse osmosis, 355
Reverse transcription, 483
Reye's syndrome, 479
Ribonucleic acid (RNA), 456
   for protein synthesis, 460–461
   in viruses, 461, 482–483
Ribosomes, 460–461
Roentgen, Wilhelm, 193–194
Rohrer, Heinrich, 20, 545
Roosevelt, Franklin D., 224
RU-486 pill, 487
Rubber, 424–425
Rubbing alcohol, 171
Rust, 384, 395–397
Rutherford, Ernest, 18–21, 216

Saccharides, 444–445
Safe Drinking Water Act (SDWA), 329, 351–352
Salicylic acid, 371
Salt, 91–93, 96
   as food additive, 530
   as inorganic compound, 150
   in neutralization reactions, 362
   reactions forming, 118–119
Salt bridges, 454
Saturated fats, 439, 518–519
Saturated hydrocarbons, 157, 436
Savory, Theodore H., 382
Scandium, 66
Scanning tunneling microscopes (STMs), 20, 166, 545–546
Schawlow, A. L., 204
Schrodinger, Erwin, 72
Science
   reasons for studying, 5
   value-free, 550
Scientific law, 7–8
Scientific method, 7–9, 11
Scientific notation, 33–35
Scientific revolution, 10–11
Scientists, 6–9
Scuba diving, 305
Scurvy, 526
Sebum, 411–412
Second law of thermodynamics, 249–251, 285, 518
Secondary nutrients, 534
Secondary structure of proteins, 452–453
Seconds (s), 37
Selenium, 529
Semiconductor lasers, 204
Separating molecules, 332–334
Sequestering
   carbon dioxide, 268
   in laundry formulations, 409
Serine, 447, 450
Serotonin, 478, 503
Serotonin-specific re-uptake inhibitors (SSRIs), 504
Sertraline (Zoloft), 502–504
Sex hormones, 486–487
Shampoos, 411

Shapes of molecules, 131–135
Shroud of Turin, 234–236
SI units, 36–37
Sickle cell anemia, 94, 434, 451
Significant figures, 32–33
Silicon, 283
Simple carbohydrates, 513–514
Simplesse, 530
Skeletal structures, 126
Skin cancer, 413–414
Skin products, 412–414
Skin protection factor (SPF), 414
Smalley, Richard E., 547–548
Smells
   perfumes, 339–341, 415–418
   sensing, 180
Smog, 261–262
Smoke detectors, 233
Smoking, 499
Snow, C. P., 226
Soap, 338, 405–407
Society and alcohol, 172
Sodium
   in Bohr model, 69–71
   in compounds, 96, 118–119
   electrons in, 61
   in food, 527
   neutrons in, 62
   in periodic law, 66
   valence electrons of, 71
Sodium aluminum sulfate ($NaAl[SO_4]_2$), 373
Sodium benzoate, 530
Sodium bicarbonate ($NaHCO_3$), 372
Sodium carbonate ($NaCO_3$), 410
Sodium chloride (NaCl), 91–93, 96
   as food additive, 530
   as inorganic compound, 150
   in neutralization reactions, 362
   reactions forming, 118–119
Sodium fluoride, 121
Sodium hydroxide (NaOH), 362
Sodium hypochlorite (NaOCl), 388
Sodium nitrite, 530
Sodium propionate, 530
Sodium tripolyphosphate ($Na_5P_3O_{10}$), 409
Solar energy, 277, 280–283
Solar power towers
   Solar One, 280
   Solar Two, 280–281
Solid-state lasers, 203
Solids
   in classifications, 14–15
   and water, 331–332
Solutes in solutions, 341
Solutions, 341–343
Solvents in solutions, 341
Sorbic acid, 530
Sorbitol, 179
Spectroscopy, 199–201
Speed, 498
Stabilizers, 532
Stain removal, 404

Starches, 442–443, 514–516
Stearic acid, 435, 518
Stem cells, 466–467
Steroids, 487–488
Stimulants
    cocaine, 496–498
    legal, 498–499
Stings, 176, 372
Storage of energy, 285–286
Straight chain alkanes, 155
Strassmann, Fritz, 222
Stratosphere
    ozone depletion in, 316–319
    range of, 311
Street drugs, 496
Strong acids and bases, 367–368
Structural formulas
    for alkanes, 152–153
    for isomers, 160–161
Styrene-butadiene rubber (SBR), 425
Sucrose ($C_{12}H_{22}O_{11}$), 442, 513–514
Sugar, 91, 95
    carbohydrates, 440–446, 513–514
    decomposing, 12
    as food additive, 530
    vs. honey, 514
    hyperactivity from, 516
    as organic compound, 149–150
Sulfate aerosols, 375
Sulfur aerosols, 264
Sulfur bridges, 455
Sulfur dioxide (SO_2)
    in acid rain, 263–264, 375
    in air pollution, 312
    characteristics of, 263
    concentrations of, 43
Sulfuric acid (H_2SO_4), 363, 367
Sun, 277
Sunlight, 188
Sunscreens, 413–414
Surfactants
    for hard water, 407–409
    soap, 405–407
Surroundings, 248
Synthetic detergents, 407–409
Systematic names, 96
Szent-Gyorgyi, Albert, 360
Szilar, Leo, 224

Tartaric acid, 371
Technetium-99m, 237
Technology, cost of, 201
Temperature, 254–255
    and color, 190
    global, 33
    measurement units for, 36, 254–255
    and volume, 306–307
Terpenes, 173
Terramycin, 481
Tertiary structure of proteins, 453–454
Testosterone, 486
Tetra-chlorodibenzo-p-dioxin (TCDD), 537
Tetracyclines, 481

Tetrahedral geometry, 131
Tetrahydrocannabinol (THC), 162
Thales, 9
Theories, 8–9
Therapeutic cloning, 5, 466–467
Thermal energy, 247
Thermodynamic systems, 248
Thermodynamics, 247, 393
    first law of, 248
    second law of, 249–251, 285, 518
Thermoplastics, 420
Three-Mile Island, 228
Thymine, 458
Time measurement units, 37
Topoisomerase inhibitors, 489
Total Ozone Measuring Spectrometers (TOMS), 318
Traits, 434
Transfer ribonucleic acid (tRNA), 460
Transformations, energy, 247–248
Transition metals, 75, 96
Transitions
    electromagnetic, 200
    electronic, 196
Transportation sector, energy use in, 245
Triazine herbicides, 537
Trichloroethylene (TCE), 348, 350
Tricyclic antidepressants, 502
Triethanolamine, 179
Triglycerides, 436–438, 518–519
Trilinolenin, 437
Trimethylamine, 178
Triolein, 437
Tripeptides, 451
Triple bonds, 126
Tristearin, 436–437
Tritium, 62
Troposphere
    air pollution in, 312–315
    range of, 310–311
Truman, Harry S, 226
Tuberculosis (TB), 481–482

U-235, 224–225
U-238, 219, 236–237
Ultrashort pulsed lasers, 205
Ultraviolet (UV) light, 215
    energy in, 193
    in photodecomposition, 196–197
    sunscreens for, 413–414
Units, measurement, 30, 35–38
    converting, 38–40
    for density, 46–48
    for energy, 251–253
Unsaturated fats, 439–440
Unsaturated hydrocarbons, 156–157
Up quarks, 65
Uranium, 64, 215
    in Earth age determinations, 236–237
    in fission, 222–223
    in water, 350
Urea, 150–151, 533

UV-A light, 316, 413–414
UV-B light, 316, 413–414

Vaccinations, 483
Valence electrons, 71, 119–120
Valence shell electron pair repulsion (VSEPR) theory, 119, 131, 133
Valine, 447
Valium, 501
Value-free science, 550
Vapor pressure, 340
Vegetarians, 518
Vesalius, Andreas, 10
Vinegar, 176
Viruses, 461, 482–486
Visible light, 191, 193
Visible region, 193
Vision improvement, lasers for, 205
Vital force, 149–151
Vitalism, 150
Vitamins
    fat-soluble, 524–526
    Vitamin A, 524–525
    Vitamin C, 526–527, 531
    Vitamin D, 525
    Vitamin E, 525
    Vitamin K, 526
    water-soluble, 526–527
Volatile liquids, 340–341
Volcanoes, pollution from, 322
Volume
    conversion factors for, 38
    in density measurements, 46
    measurement units for, 37–38
    and pressure, 304–306
    and temperature, 306–307
Vulcanization, 425

Waste disposal, nuclear, 227–228
Water, 329
    characteristics of, 343–345
    as compound, 12, 90–91
    contaminants in
        biological, 348, 353–354
        chemical, 348–351
    drinking, 351–352
    forces on, 334–339
    formula mass of, 100
    and gravity, 330
    hard, 346–347
    heat capacities of, 255
    home treatment of, 354–355
    Lewis structures for, 124, 126, 135–136
    melting and boiling, 332–334
    molar mass of, 100
    molecular properties of, 135–140
    from neutralization reactions, 362
    public treatment of, 352–354
    pure and polluted, 345–346, 348–351
    and solids and liquids, 331–332
    in solutions, 341–343
Water-borne diseases, 345
Water softening, 354–355

Water-soluble vitamins, 526–527
Watson, James, 458
Watts, 252–253
Wavelength, 191–195, 198–199
Weak acids and bases, 367–368
Weather forecasting, 303–304
White light, 188
Whyte, Lancelot L., 146
Wigner, Eugene Paul, 28
Wind power, 280
Wines, acids in, 371–372

Wohler, Friedrich, 150, 433
Wool, 455
Word problems, solving, 44–46
Work, 245, 247

X-ray crystallography, 166, 454
X-rays
 dangers with, 196
 discovery of, 193–194, 215–216
 energy in, 194

Yeasts, 374
Yucca Mountain, Nevada, 227–228

Zeolites, 355, 410
Zewail, Ahmed, 205
Zinc
 in batteries, 389–391
 in food, 529
 in rust prevention, 396
Zoloft, 502–504